Variance Components

Variance Components

SHAYLE R. SEARLE
GEORGE CASELLA
CHARLES E. McCULLOCH

WILEY-INTERSCIENCE

A JOHN WILEY & SONS, INC., PUBLICATION

Published by John Wiley & Sons, Inc., Hoboken, New Jersey.
Published simultaneously in Canada.

Library of Congress Cataloging-in-Publication is available.

ISBN-13 978-0-470-00959-8
ISBN-10 0-470-00959-4

10 9 8 7 6 5 4 3 2 1

To Helen
— a wonderful wife
— and a caring friend

PREFACE

This book presents a broad coverage of its topic: variance components estimation and mixed models analysis. Although the use of variance components has a long history dating back to the 1860s, it is only in the last forty years or so that variance components have attracted much attention in the statistical research literature. Numerous books have maybe a chapter or two on the subject but few are devoted solely to variance components. This book is designed to make amends for that situation.

The introductory Chapter 1 describes fixed, random and mixed models and uses nine examples to illustrate them. This is followed by a chapter that surveys the history of variance components estimation. Chapter 3 describes the 1-way classification in considerable detail, both for balanced data (equal numbers of observations in the classes) and for unbalanced data (unequal numbers of observations). That chapter, for the 1-way classification, details four main methods of estimation: analysis of variance (ANOVA), maximum likelihood (ML), restricted (or residual) maximum likelihood (REML) and Bayes.

Chapters 4 and 5 deal with ANOVA estimation in general, Chapter 4 for balanced data and 5 for unbalanced. Chapter 6 covers ML and REML estimation and Chapter 7 describes the prediction of random effects using best prediction (BP), best linear prediction (BLP) and best linear unbiased prediction (BLUP). Chapters 8–12 are more specialized than 1–7. They cover topics that are of current research interest: computation of ML and REML estimates in 8; Bayes estimation and hierarchical models in 9; binary and discrete data in 10; estimation of covariance components and criteria-based estimation in 11; and the dispersion-mean model and fourth moments in 12.

This broad array of topics has been planned to appeal to research workers, to students and to the wide variety of people who have interests in the use of mixed models and variance components for statistically analyzing data. This includes data from such widely disparate disciplines as animal breeding, biology in general, clinical trials, finance, genetics, manufacturing processes, psychology, sociology and so on. For students the book is suitable for linear models courses that include something on mixed models, variance components and prediction;

and, of course, it provides ample material for a graduate course on variance components with the pre-requisite of a linear models course. Finally, the book will also serve as a reference for a broad spectrum of topics for practicing statisticians who, from time to time, need to use variance components and prediction.

More specifically, for graduate teaching there are at least four levels at which the book can be used. (1) When variance components are to be part of a solid linear models course, use Chapters 1, 3 and 4 with Chapter 2 (history) being supplementary reading. This would introduce students to random effects and mixed models in Chapter 1, and in Chapter 3, for the 1-way classification, they would cover all the major topics of ANOVA and ML estimation, and prediction. (As time and interests allowed, additional aspects of these topics could also be selected from Chapters 5, 6 and 7.) And Chapter 4 provides results and methodology for a variety of commonly occurring balanced data situations. (2) This same material, presented slowly and in detail, could also be the basis for an easy-going course on variance components. (3) For an advanced course we would recommend using Chapters 1 and 2 for an easy introduction, followed by a quick overview of Chapters 3–5 (1-way classification, and ANOVA estimation from balanced and unbalanced data) and then Chapters 6 and 7 in detail (ML and REML, and prediction). We suggest following this with sections 8.1–8.3, (introduction to computing ML and REML) and all of Chapters 10 (binary and discrete data) and 11 (covariance components and criteria-based estimation). Then, for a general overview of Bayes, ML and REML, use Sections 9.1–9.4, and for a mathematical synthesis of ML and REML from a pseudo least squares viewpoint, Chapter 12 is appropriate. (4) Finally, of course, Chapters 1–7, and then 8–12, could constitute a detailed 2-semester (or 2-quarter) course on variance components.

Considering the paucity of books devoted solely to variance components, we have attempted a broad coverage of the subject. But we have not, of course, succeeded in a complete coverage—undoubtedly that is impossible. Some readers will therefore be irked by some of our omissions or slim treatment of certain topics. For example, much emphasis is placed on point estimation, with only some attention to interval estimation. The latter, for ANOVA estimation, is very difficult, with only a modicum known about exact intervals (e.g., Table 3.4); although, for ML and REML asymptotic properties of the estimators provide straightforward derivation. Also, even for estimation we chose to concentrate on methodology with sparse attention to interpreting analyses of specific data sets—and thus few numerical examples or illustrations will be found. And topics that receive slim treatment are criteria-based estimation and non-negative estimation (in Sections 11.3 and 12.7 respectively). The former (e.g., minimum norm estimation) is not, in our opinion, a procedure to be recommended in practice; and it already has its own book-length presentation.

Sections within chapters are numbered in the form 1.1, 1.2, 1.3, . . . ; e.g., Section 1.3 is Section 3 of Chapter 1. These numbers are also shown in the running head of each page: e.g., [1.3] is found on page 7. Equations are numbered

(1), (2),... throughout each chapter. Equation references across chapters are few, but include explicit mention of the chapter concerned; otherwise "equation (4)" or just "(4)" means the equation numbered (4) in the chapter concerned. Exercises are in the final section of each chapter (except Chapters 1 and 2), with running heads such as [E 5] meaning exercises of Chapter 5. Reference to exercise 2 of Chapter 5, for example, is then in the form E 5.2.

Grateful thanks go to Harold V. Henderson and Friedrich Pukelsheim for comments on early drafts of some of the chapters; and to students in Cornell courses and in a variety of short courses both on and off campus who have also contributed many useful ideas. Special thanks go to Norma Phalen for converting handwritten scrawl to the word processor with supreme care and accuracy; and to Pamela Archin, Colleen Bushnell, April Denman and Jane Huling for patiently and efficiently dealing with occasional irascibility and with almost endless revisions for finalizing the manuscript: such helpful support is greatly appreciated.

SHAYLE R. SEARLE
GEORGE CASELLA
CHARLES E. MCCULLOCH

Ithaca, New York
April 1991

LIST OF CHAPTERS

CONTENTS

Variance Components

CHAPTER 1

INTRODUCTION

Statistics is concerned with the variability that is evident in any body of data. A traditional (and exceedingly useful) method of summarizing that variability is known as the analysis of variance table. This not only presents a partitioning of observed variability, but it also summarizes calculations that enable us to test, under certain (normality) assumptions, for significant differences among means of certain subsets of the data. Sir Ronald Fisher, the originator of analysis of variance, had this to say about it in a letter to George Snedecor dated 6/Jan/'34 that was on display at the 50th Anniversary Conference of the Statistics Department at Iowa State University, June, 1983:

> The analysis of variance is (not a mathematical theorem but) a simple method of arranging arithmetical facts so as to isolate and display the essential features of a body of data with the utmost simplicity.

Initially this analysis of variance technique was developed for considering differences between means, but later came to be adapted to estimating variance components—as indicated in Chapter 2 and presented in detail in Chapters 3–5.

Although Fisher's descriptions of analysis of variance methodology were in terms of sums of squares of differences among observed averages, the trend in recent decades has been to present many of the ideas behind the analysis of variance in terms of what are called linear models, particularly that class of linear models known as fixed effects models (or just fixed models). These are described in Section 1.3. Numerous books are available on this topic at varying levels of theory and application. Eight examples of those that are at least somewhat theoretic are: Searle (1971), which emphasizes unbalanced data; Rao (1973), with its broad-based mathematical generality; Graybill (1976), which emphasizes balanced data; Seber (1977), with its concentration on the full-rank model; Arnold (1981), which uses a co-ordinate-free approach and emphasizes similarities between univariate and multivariate analyses; Guttman (1982), which is mainly an introduction; Hocking (1985), which is very wide-ranging;

and Searle (1987), which is confined to unbalanced data, needs no matrix algebra for its first six chapters, and does offer some brief comments on statistical computing packages.

Variation among data can also be studied through a different class of linear models, those known as random effects models (or just random models)—see Section 1.3; and also those called mixed models, which are models that have a mixture of the salient features of fixed and random models. For some situations, data analysis using these models is closely allied to traditional analysis of variance, but in many instances it is not. The various analysis techniques that are available for random and for mixed models have been developed over many years in the research literature, with certain facets of those methods being available in a chapter or two of a number of books, e.g., Anderson and Bancroft (1952), Scheffé (1956), Searle (1971), Rao (1973), Neter and Wasserman (1974), Graybill (1976), Hocking (1985) and Searle (1987), to name a few. In contrast, this book is devoted entirely to random and mixed models, with particular concentration on estimating the variances (the components of variance, as they are called), which is the feature of these models that makes them very different from fixed effects models. We begin with some useful terminology and then, through a series of examples, illustrate and explain fixed effects and random effects. Chapter 2 is a brief history of the development of methods for estimating variance components, and as such it serves as an introductory survey of the array of methods available. Chapter 3 begins the description of those methods in detail.

1.1. FACTORS, LEVELS, CELLS AND EFFECTS

In studying the variability that is evident in data, we are interested in attributing that variability to the various categorizations of the data. For example, consider a clinical trial where three different tranquilizer drugs are used on both men and women, some of whom are married and some not. The resulting data could be arrayed in the tabular form indicated by Table 1.1.

TABLE 1.1. A FORMAT FOR SUMMARIZING DATA

	Marital Status					
	Married			Not Married		
	Drug			Drug		
Sex	A	B	C	A	B	C
Male						
Female						

The three classifications, sex, drug and marital status, that identify the source of each datum are called *factors*. The individual classes of a classification are the *levels* of the factor; e.g., the three different drugs are the three levels of the factor "drug"; and male and female are the two levels of the factor "sex". The subset of data occurring at the "intersection" of one level of every factor being considered is said to be in a *cell* of the data. Thus with the three factors, sex (2 levels), drug (3 levels) and marital status (2 levels), there are $2 \times 3 \times 2 = 12$ cells.

In classifying data in terms of factors and their levels the feature of interest is the extent to which different levels of a factor affect the variable of interest. We refer to this as the *effect* of a level of a factor on that variable.

The effects of a factor are always one or other of the two kinds, as has already been indicated. First are *fixed effects*, which are the effects attributable to a finite set of levels of a factor that occur in the data and which are there because we are interested in them. In Table 1.1 the effects for the factor sex are fixed effects, as are those for the factors drug and marital status. Further quality discussion of fixed effects is in Kempthorne (1975). In a different context the effect on crop yield of three levels of a factor called fertilizer could correspond to the three different fertilizer regimes used in an agricultural experiment. They would be three regimes of particular interest, the effects of which we would want to quantify from the data to be collected from the experiment.

The second kind of effects are *random effects*. These are attributable to a (usually) infinite set of levels of a factor, of which only a random sample are deemed to occur in the data. For example, four loaves of bread are taken from each of six batches of bread baked at three different temperatures. Whereas the effects due to temperature would be considered fixed effects (presumably we are interested in the particular temperatures used), the effects due to batches would be considered random effects because the batches chosen would be considered a random sample of batches from some hypothetical, infinite population of batches. Since there is definite interest in the particular baking temperatures used, the statistical concern is to estimate those temperature effects; they are fixed effects. No assumption is made that the temperatures are selected at random from a distribution of temperature values. Since, in contrast, this kind of assumption has then been made about the batch effects, interest in them lies in estimating the variance of those effects. Thus such data are considered as having two sources of random variation: batch variance and, as usual, error variance. These two variances are known as *variance components*: their sum is the variance of the variable being observed.

Models in which the only effects are fixed effects are called *fixed effects models*, or sometimes just *fixed models*. Models that contain both fixed and random effects are called *mixed models*. And those having (apart from a single, general mean common to all observations) only random effects are called *random effects models* or, more simply, *random models*. Further examples and properties of fixed effects and of random effects are given in Sections 1.3 and 1.4.

1.2. BALANCED AND UNBALANCED DATA

a. Balanced data

Data can be usefully characterized in several ways that depend on whether or not each cell contains the same number of observations. When these numbers are the same, the data shall be described as *balanced data*; they typically come from designed factorial experiments that have been executed as planned.

A formal, rigorous, mathematical definition of balanced data is elusive. Although the word "balanced" is used in a variety of contexts in statistics (see Speed, 1983), its use as a descriptor of equal-subclass-numbers data is now more widely accepted and has been formalized by a number of authors, e.g., Smith and Hocking (1978), Seifert (1979), Searle and Henderson (1979) and Anderson *et al.* (1984); and an explicit definition of a very broad class of balanced data is given in Searle (1987). These details are not pursued here.

b. Special cases of unbalanced data

In a general sense all data that are not balanced are, quite clearly, unbalanced. Nevertheless, there are at least two special cases of that broad class of unbalanced data that need to be identified. So far as analysis of variance is concerned, they can be dispensed with because their analyses come within the purview of the standard (so-called) analyses of balanced data. These analyses can be used for variance components estimation either by adapting the techniques for balanced data (Chapter 4), or by using the methods available for unbalanced data in general (Chapters 5–12). However, in most instances of these special cases of unbalanced data estimating variance components would not be judicious because there are often impractically too few levels of the factors. Nevertheless, we briefly illustrate both cases, to ensure that the reader realizes we deem them to be outside the ken of what we generally refer to as unbalanced data.

-i. Planned unbalancedness. Certain experimental designs are planned so that they yield unbalanced data. There are no observations on certain, carefully planned combinations of levels of the factors involved, e.g., latin squares, balanced incomplete blocks and their many extensions. We call this *planned unbalancedness*. An example shown in Table 1.2 is a particular one-third of a 3-factor experiment (of rows, columns and treatments with 3 levels of each), that is a latin square of order 3, as shown in Table 1.3. In each of the 9 cells defined by the 3 rows and 3 columns, only one treatment occurs, and not all three treatments. This is displayed in Table 1.2 as unbalanced data (planned unbalancedness) with zero or one observation per cell of the 27 cells of a $3 \times 3 \times 3$ (3-factor) experiment. The customary display of this latin square is shown in Table 1.3.

Another example of planned unbalancedness is an experiment involving 3 fertilizer treatments A, B and C, say, used on 3 blocks of land in which one of the 3 treatment pairs A and C, A and B, and B and C is used in each block. This is a simple example of a balanced incomplete blocks experiment that can

TABLE 1.2. AN EXAMPLE OF PLANNED UNBALANCEDNESS: THE LATIN SQUARE

	Number of Observations								
	Treatment								
	A			B			C		
	Column			Column			Column		
Row	1	2	3	1	2	3	1	2	3
1	1	0	0	0	1	0	0	0	1
2	0	1	0	0	0	1	1	0	0
3	0	0	1	1	0	0	0	1	0

TABLE 1.3. A LATIN SQUARE OF ORDER 3 (TREATMENTS A, B, C)

	Column		
Row	1	2	3
1	A	B	C
2	C	A	B
3	B	C	A

TABLE 1.4. NUMBER OF OBSERVATIONS IN A BALANCED INCOMPLETE BLOCKS EXPERIMENT

	Block		
Treatment	1	2	3
A	1	1	0
B	0	1	1
C	1	0	1

be represented as a 2-factor experiment, each factor having 3 levels, with certain cells empty, as shown in Table 1.4.

Analyses of variance of data exhibiting planned unbalancedness of the nature just illustrated are well known and are often found in the same places as those describing the analysis of variance of balanced data. In a manner more general than either of the two preceding examples, planned unbalancedness need not require that a planned subset of cells be empty; it could be that subsets of cells are just used unequally; e.g., Table 1.4 with every 0 and 1 being a 1 and 2, respectively, would still represent planned unbalancedness.

-ii. Estimating missing observations. The second special case of unbalanced data is when the number of observations in every cell is the same, except that in a very few cells the number of observations is just one or two less than all the other cells. This usually occurs when some intended observations have inadvertently been lost or gone missing somehow, possibly due to misadventure during the course of an experiment. Maybe in a laboratory experiment, equipment got broken or animals died; or in an agricultural experiment, farm animals broke fences and ate some experimental plots. Under these circumstances there are many well-known classical techniques for estimating such *missing observations* [e.g., Steel and Torrie (1980), pp. 209, 227 and 388], as well as some newer, computer-intensive techniques (see Little and Rubin, 1987). After estimating the missing observations, one uses standard analyses of variance for balanced data. We therefore give no further consideration to estimating missing observations.

c. Unbalanced data

After defining balanced data and excluding from all other data those that can be described as exhibiting planned unbalancedness or involving just a few missing observations, we are left with what shall be called unbalanced data. This is data where the numbers of observations in the cells (defined by one level of each factor) are not all equal, and may in fact be quite unequal. This can include some cells having no data but, in contrast to planned unbalancedness, with those cells occurring in an unplanned manner. Survey data are often like this, where data are sometimes collected simply because they exist and so the numbers of observations in the cells are just those that are available. Records of many human activities are of this nature; e.g., yearly income for people classified by age, sex, education, education of each parent, and so on. This is the kind of data that shall be called *unbalanced data.*

In describing unbalanced data this way we give no consideration to whatever mechanism it was that led to the inequality of the numbers of observations in the cells. For example, with milk yields of dairy cows sired by artificial insemination, bulls that are genetically superior have more daughters than other bulls, simply because of that superiority. This effect should, of course, be taken into account in estimating between-bull variance, as was considered by Harville (1967, 1968). We do not deal with such difficulties and so, in the sense that unbalanced data are balanced data with some observations missing, we are in effect assuming that those are what Little and Rubin (1987) call missing-at-random observations.

Within the class of unbalanced data we make two divisions. One is for data in which all cells contain data; none are empty. We call these *all-cells-filled* data. Complementary to this are *some-cells-empty* data, wherein there are some cells that have no data. This division is vitally useful in the analysis of fixed effects models (Searle, 1987) where Yates' (1934) weighted-squares-of-means analysis is very useful for all-cells-filled data but is not applicable to some-cells-empty data. This is of less importance for random and mixed models than for fixed effects models, although the weighted-squares-of-means analysis is indeed one possible basis for variance components estimation from all-cells-filled data. (See Chapter 5.)

1.3. FIXED EFFECTS AND RANDOM EFFECTS

The two classes of effects, fixed and random, have been specified and described in general terms. We now illustrate the nature of both classes, using some illustrative examples to do so, and emphasizing properties of random effects in the process.

a. Fixed effects models

Example 1 (Tomato varieties). Consider a home gardener carrying out a small experiment with 24 tomato plants, 6 plants of each of 4 varieties that the gardener is particularly interested in, through having tried them occasionally in recent summers. Comparison of the four varieties is now to be made in the $12' \times 8'$ garden space available. Each plant is allocated randomly to one of the $2' \times 2'$ squares. If y_{ij} is the yield of fruit from plant j of variety i (for $i = 1, \ldots, 4$ and $j = 1, \ldots, 6$), a possible model for y_{ij} would be

$$E(y_{ij}) = \mu_i, \tag{1}$$

where E represents expectation and μ_i is the expected yield from a plant of variety i. If we wanted to write $\mu_i = \mu + \alpha_i$ we would then have

$$E(y_{ij}) = \mu + \alpha_i, \tag{2}$$

where μ is a general mean and α_i is the effect on yield of tomatoes due to the plant being variety i.

In this modelling of the expected value of y_{ij} each μ_i (or μ and each α_i) is considered as a fixed unknown constant, the magnitudes of which we wish, in some general sense, to estimate; e.g., we might want to estimate μ_1 and μ_4 or $\mu_1 - \mu_4$. In doing this the μ_is, (or the α_is) correspond to the four different varieties that the gardener is interested in. They are four very specific varieties of interest, and in using them the gardener has no thought for any other varieties. This is the concept of fixed effects. Attention is fixed upon just the varieties in the experiment, upon these and no others, and so the effects are called *fixed effects*. And because all the effects in (2) are fixed effects, the model is called a *fixed effects model*. It is also called *Model I*, so named by Eisenhart (1947).

Armed with (2), we now define the deviation of y_{ij} from its expected value $E(y_{ij})$ as residual error:

$$e_{ij} = y_{ij} - E(y_{ij}) = y_{ij} - (\mu + \alpha_i)\,.$$

This gives

$$y_{ij} = \mu + \alpha_i + e_{ij}, \tag{3}$$

or equivalently $y_{ij} = \mu_i + e_{ij}$.

A consequence of the definition of the residual e_{ij} is that it is a random variable with mean zero:

$$E(e_{ij}) = E[y_{ij} - E(y_{ij})] = 0 \; . \tag{4}$$

But that definition carries no consequences so far as second moments are concerned. Therefore we attribute a variance and covariance structure to the e_{ij}s: first, that every e_{ij} has the same variance, σ_e^2, and second, that the e_{ij}s are independently and identically distributed and that pairs of different e_{ij}s have zero covariance. Thus, using var$(\cdot)$ to denote variance and cov$(\cdot, \cdot)$ for covariance,

$$\text{var}(e_{ij}) = \sigma_e^2 \quad \forall \; i \quad \text{and} \quad j \tag{5}$$

(with $\forall$ meaning "for all") and

$$\text{cov}(e_{ij}, e_{i'j'}) = 0 \quad \text{except for } i = i' \text{ and } j = j' \; . \tag{6}$$

In light of (4), this means that

$$\text{var}(e_{ij}) = \sigma_e^2 = E(e_{ij}^2)$$

and

$$E(e_{ij}e_{i'j'}) = 0 \quad \text{except for } i = i' \text{ and } j = j' \; . \tag{7}$$

The manner in which data are obtained always affects inferences that can be drawn from them. We therefore describe a sampling process pertinent to this fixed effects model. The data are envisaged as being one possible set of data involving these same tomato varieties that could be derived from repetitions of the experiment, repetitions for each of which a different sample of 6 plants of each variety would be used. This would lead on each occasion to a set of *e*s that would be a random sample from a population of error terms having zero mean, variance σ_e^2 and zero covariances. It is the probability distribution associated with the *e*s that provides the means for making inferences about functions of the μ_is (or of μ and the α_is) and about σ_e^2 .

The all-important feature of fixed effects is that they are deemed to be constants representing the effects on the response variable y of the different levels of the factor concerned, in this case the varieties of tomatoes. These varieties are the levels of the factor of particular interest, chosen because of interest in those varieties in the experiment. But they could just as well be different fertilizers applied to a corn crop, different forage crops grown in the same region, different machines used in a manufacturing process, different drugs given for the same illness, and so on. The possibilities are legion, as are the varieties of models and their complexities, reaching far beyond those of (1)–(7). We briefly offer two more examples.

Example 2 (Medications). Consider a clinical trial designed for testing the efficacy of a placebo and 3 different medications intended for reducing blood pressure. The placebo and drug are administered to 24 executives of the same N.Y. City corporation, all aged 40–45 and earning salaries in the range

\$100,000–\$250,000 per annum. Six executives are chosen at random to receive the placebo and six others for each of the three medications. After 30 days of treatment, blood pressure is measured again, and for each of the 24 executives the change in blood pressure from before treatment to after treatment is recorded. The difference for the jth patient on treatment i, for $i = 1, 2, 3, 4$ and $j = 1, 2, \ldots, 6$, is denoted y_{ij}. Then for studying the effect of treatment on change in blood pressure, the same model could be used as that suggested in Example 1 for studying effects of the four varieties on yield of tomatoes. Just as with the four different varieties of tomatoes, so with the four different treatments (placebo and 3 medications) on the executives: the μ_is (or μ and α_is) are considered as fixed, unknown constants. This is because the four treatments being used are the four treatments that have been decided upon as being of interest. They are the treatments on which our attention is fixed. The effects in the model corresponding to those treatments are therefore fixed effects.

Although medications other than those we have used could be envisaged, the ones chosen for the experiment are, insofar as the experiment is concerned, the treatments of interest. In no way are the four chosen treatments deemed to be a sample from a larger array of possible treatments.

Example 3 (Soils and fertilizers). The growth of a potted plant depends on the potting soil and the fertilizer it is grown in. Suppose 30 chrysanthemum plants, all of the same variety and age, are randomly allocated to 30 pots, one per pot, where each pot contains one combination of each of 6 soil mixtures with each of 5 fertilizers. A suitable linear model for y_{ij}, the growth of the plant in soil i used with fertilizer j would be

$$E(y_{ij}) = \mu + \alpha_i + \beta_j \tag{8}$$

where μ is a general mean, α_i is the effect on growth due to soil i and β_j is the effect on growth of fertilizer j. Since the 6 soils and the 5 fertilizers have been specifically chosen as being the soils and fertilizers of interest, the α_is and β_js are fixed effects corresponding to those soils and fertilizers—with $i = 1, 2, \ldots, 6$ and $j = 1, 2, \ldots, 5$. As with the drug treatments, the soils and fertilizers in the experiment are the specific items of interest and under no circumstances can they be deemed as having been chosen randomly from a larger array of soils and fertilizers. Thus the α_is and β_js are fixed effects. This is just a simple extension of Examples 1 and 2, which each embody only one factor: variety effects on yield of tomatoes, and treatment effects on blood pressure. With the chrysanthemums there are two factors: soil effects and fertilizer effects on growth.

b. Random effects models

Example 4 (Clinics). Suppose a new form of injectable insulin is being tested using 15 different clinics in New York State. It is not unreasonable to think of those clinics (as do Chakravorti and Grizzle, 1975) as a randomly chosen sample of clinics from a population of clinics (i.e., doctors who administer the injections). If clinic i has n_i patients in the trial and the measured response of patient j in clinic i is y_{ij} then a possible model would be

$$E(y_{ij}) = \mu + \alpha_i \quad \text{for } i = 1, \ldots, n_i \ . \tag{9}$$

Although (9) is algebraically the same as (2) for Examples 1 and 2, some assumptions underlying it are different. In Example 2 each α_i is a fixed effect, the effect on blood pressure of the patient having received treatment i, a treatment that is a pre-decided treatment of interest. But in (9) α_i is the effect on blood-sugar level of the observed patient having been injected in clinic i; and clinic i is just one clinic, that one from among the randomly chosen clinics that happened to be numbered i in the clinical trial. Since the clinics have been chosen randomly with the object of treating them as a representation of the population of all clinics in New York State, and from which inferences can and will be made about that population, the one labelled i is of no particular interest of itself to the trial; it is of interest solely as being one of the 15 clinics randomly chosen from a larger population of clinics. This is a characteristic of random effects: they can be used as the basis for making inferences about populations from which they have come. Thus α_i is a *random effect*. As such it is, indeed, a random variable.

More precisely, α_i corresponding to the clinic that has been assigned label i is the (unknowable) realization of a random variable "clinic effect" appropriate to that clinic labeled i. However, for notational convenience we judiciously ignore the distinction between a random variable and a realized value of it and let α_i do double duty for both.

With the α_is being treated as random variables, we must attribute probability properties to them. There are two that are customarily employed: first, that all α_is are independently and identically distributed (i.i.d.); second, that they have zero mean, and then, that they all have the same variance, σ_α^2. We summarize this as

$$\alpha_i \sim \text{i.i.d.}(0, \sigma_\alpha^2) \quad \forall\ i \ .$$

Consequences of this are

$$E(\alpha_i) = 0 \quad \forall\ i, \tag{10}$$

$$\text{var}(\alpha_i) = E[\alpha_i - E(\alpha_i)]^2 = E(\alpha_i^2) = \sigma_\alpha^2, \tag{11}$$

and

$$\text{cov}(\alpha_i, \alpha_k) = 0 \quad \forall\ i \neq k \ . \tag{12}$$

There are, of course, properties other than these that could be used, e.g., non-zero values for $\text{cov}(\alpha_i, \alpha_k)$. In point of fact, these elementary properties lead to enough difficulties insofar as estimation is concerned that alternatives seldom get used. Nevertheless, some of these alternatives are mentioned briefly in Chapter 3.

A second outcome of treating the α_is as random variables is that we must consider $E(y_{ij}) = \mu + \alpha_i$ of (9) with more forethought, because it is really a conditional mean. Suppose for the moment that α^* represents the random variable "clinic effect", and that for the clinic labeled i, α_i is the realized (but unobservable) value of α^*. Then in reality (9) is the expected value of y_{ij} given that α^* is α_i, i.e., that $\alpha^* = \alpha_i$. Hence (9) is the conditional mean

$$E(y_{ij}|\alpha^* = \alpha_i) = \mu + \alpha_i \ . \tag{13}$$

As already indicated, for notational simplicity we drop the use of α^* and write (13) as

$$E(y_{ij}|\alpha_i) = \mu + \alpha_i \ . \tag{14}$$

Taking expectation over α_i, as in (10), then gives

$$E(y_{ij}) = \mu \ . \tag{15}$$

Note that $E(\alpha_i) = 0$ of (10) involves no loss of generality in (14) and (15) because if $E(\alpha_i) \neq 0$ but $E(\alpha_i) = \tau$, say, then $E(y_{ij}|\alpha_i) = \mu + \tau + \alpha_i - \tau$ gives $E(y_{ij}) = \mu + \tau$. Therefore, on defining μ' as $\mu' = \mu + \tau$ and $\alpha_i' = \alpha_i - \tau$, we have $E(y_{ij}|\alpha_i) = \mu' + \alpha_i'$, which is (14) with μ' in place of μ and α_i' in place of α_i, and the form of (14) and (15) is retained.

Finally, we introduce the residual, similar to $e_{ij} = y_{ij} - E(y_{ij})$ defined earlier for fixed effects models. The definition here is

$$e_{ij} = y_{ij} - E(y_{ij}|\alpha_i) = y_{ij} - (\mu + \alpha_i), \tag{16}$$

so that e_{ij} is a random variable; and (16) gives an equation

$$y_{ij} = \mu + \alpha_i + e_{ij}, \tag{17}$$

similar to (3). Then e_{ij} has properties similar to those of Example 1. Thus $E(e_{ij}) = 0$ and attributing uniform variance σ_e^2 to each e_{ij} gives

$$\text{var}(e_{ij}) = \sigma_e^2 = E(e_{ij}^2), \tag{18}$$

similar to (5). Furthermore, we also treat the e_{ij}s as being independent of each other and of every α_i so that

$$\text{cov}(e_{ij}, e_{i'j'}) = 0 \quad \forall\ i, i' \text{ and } j, j' \text{ except } i = i' \text{ and } j = j'$$

and (19)

$$\text{cov}(e_{ij}, \alpha_k) = 0 \quad \forall\ i, j \text{ and } k \ .$$

Hence, for the same values of i, i', j, j' and k

$$E(e_{ij}e_{i'j'}) = 0 \quad \text{and} \quad E(e_{ij}\alpha_k) = 0 \ . \tag{20}$$

In view of (17), the variance of y_{ij} is

$$\text{var}(y_{ij}) = \text{var}(\mu + \alpha_i + e_{ij}),$$

which is

$$\sigma_y^2 = \sigma_\alpha^2 + \sigma_e^2, \qquad [\text{Use (18)–(20)}] \ . \tag{21}$$

In this way we see that σ_α^2 and σ_e^2 are components of σ_y^2, the variance of y; thus they have attracted the name *components of variance* or *variance components*.

Nevertheless, we also note that σ_α^2 is the intra-class covariance, i.e., the covariance between every pair of observations in the same clinic:

$$\text{cov}(y_{ij}, y_{i,j'}) = \text{cov}(\mu + \alpha_i + e_{ij}, \mu + \alpha_i + e_{ij'}) = \sigma_\alpha^2 \quad \text{for } j \neq j' .$$

Parenthetical statements. The phrase enclosed in square brackets in (21) is the first use of a practice that re-occurs in subsequent chapters: an isolated phrase that indicates to the reader the reasoning behind the derivation of the equality on that line. In (21) that reasoning is quite straightforward, but such is not always so.

It is important to recognize that a model is not just its equation such as (17), but also everything that prescribes properties of the elements in that equation. Thus the model is not just equation (17) but it is that and all the equations and other properties described between equations (10) and (20). It is these properties that distinguish this model from that used in Examples 1 and 2. In Example 1, for instance, the model equation is (3), but the model is (3) and everything from (2) down to (7).

Model equation (17) has μ as a fixed effect, and α_i and e_{ij} as random. Thus everything except μ is random. This is the characteristic of what is called a *random effects* model, or just a *random model*. It was named *Model II* by Eisenhart (1947), a name that is somewhat disappearing from use.

Example 5 (Dairy bulls). It is common practice for dairy farmers today, rather than mating their own bulls to their own cows, to have their cows inseminated by a technician who is supplied with bull semen from an artificial breeding corporation. That corporation's business is to own bulls that generally sire daughter cows that are high-yielding producers of milk. It can achieve that by each year buying some 80–150 young bulls that are considered to be a random sample from the population of bulls (of some particular breed—mainly Holstein, in the U.S.A.). Then semen from those bulls is used enough so that three years later there will be approximately 60 daughter cows per bull that have milk production records. Letting y_{ij} be the record of the jth daughter of the ith sire, an appropriate model for y_{ij} is then precisely the same as in Example 4, based on the model equation

$$y_{ij} = \mu + \alpha_i + e_{ij}$$

but for $i = 1, \ldots, 150$ and $j = 1, \ldots, 60$. Because the bulls are considered random, each α_i is a random effect, with $\text{var}(\alpha_i) = \sigma_\alpha^2$ and $\text{cov}(\alpha_i, \alpha_k) = 0$ for $i \neq k$, and with all the other specifications given in Example 4.

To the animal breeder and farmer, who are both interested in using breeding to help increase the production of economically important products from farm animals (e.g., eggs, milk, butter, wool, tallow and bacon), the variance components σ_e^2 and σ_α^2 are of much interest. They are needed, for example, for the ratio $h = 4\sigma_\alpha^2/(\sigma_\alpha^2 + \sigma_e^2)$, which is a parameter called heritability that is of great importance in genetics, not only in the breeding of animals but of plants too. A similar ratio is $\sigma_\alpha^2/(\sigma_\alpha^2 + \sigma_e^2)$, the intra-class correlation; it also occurs in psychological and educational testing, where it has the connotation of reliability.

Another aspect of this model that is often of particular interest is that of predicting the value of α_i corresponding to the bull labeled i. This is important for ranking the bulls on the basis of the predicted values of the α_is because those predictions act as estimates of the bulls' genetic merits as sires of daughter cows that are valued for high milk production.

Note how this example differs from the preceding ones. On one hand, it is quite reasonable to assume that the bull effects are a random sample of possible values, and we accordingly treat bull as a random factor. On the other hand, we are interested in predicting the value of α_i *for a specific bull* that occurs in the data. Thus the distinction between fixed and random effects centers on whether we are willing to assume that the levels of a factor are sampled randomly from a distribution, not whether we are specifically interested in the levels of that factor.

Given that the exact genetic contribution (a random half of his genetic make-up) of a bull to his daughter cows is, in fact, different for each daughter, we cannot estimate α_i in the sense that we estimate fixed effects. With α_i being a random variable, the best we can do is to consider the expected value of α_i, given the records that we have from all the daughters of all 150 of the bulls. Thus we seek to estimate the conditional mean $E(\alpha_i|\mathbf{y})$, where $\mathbf{y}$ is the vector of all records. This estimator (which is nowadays called a predictor) turns out to be, as derived in (40) of Chapter 3,

$$\tilde{\alpha}_i = \frac{n_i\sigma_\alpha^2}{\sigma_e^2 + n_i\sigma_\alpha^2}(\bar{y}_{i\cdot} - \tilde{\mu}), \tag{21}$$

where bull i has n_i daughters with mean record $\bar{y}_{i\cdot} = \Sigma_{j=1}^{n_i} y_{ij}/n_i$ and $\tilde{\mu}$ is the estimator of μ:

$$\tilde{\mu} = \frac{\Sigma_i w_i \bar{y}_{i\cdot}}{\Sigma_i w_i} \quad \text{for} \quad w_i = \frac{1}{\text{var}(\bar{y}_{i\cdot})} = \frac{n_i}{n_i\sigma_\alpha^2 + \sigma_e^2}, \tag{22}$$

as derived in (34) of Chapter 3. The estimator (21), which is known as the *b*est *l*inear *u*nbiased *p*redictor (BLUP), can also be written as

$$\tilde{\alpha}_i = \frac{n_i h}{4 + (n_i - 1)h}(\bar{y}_{i\cdot} - \tilde{\mu}),$$

where $h = 4\sigma_\alpha^2/(\sigma_\alpha^2 + \sigma_e^2)$, as before. Details of these results and generalization of them are given in Chapter 7. Clearly, in order to use $\tilde{\alpha}_i$ in practice, estimates of σ_α^2 and σ_e^2 are needed. That is what this book is all about, estimating variance components.

Example 6 (Ball bearings and calipers). Consider the problem of manufacturing ball bearings to a specified diameter that must be achieved with a high degree of accuracy. Suppose each of 100 different ball bearings is measured with each of 20 different micrometer calipers, all of the same brand. Then a suitable model equation for y_{ij}, the diameter of the ith ball bearing measured with the jth caliper, could be

$$E(y_{ij}) = \mu + \alpha_i + \beta_j \,. \tag{23}$$

This is the same model equation as (8) in Example 3; but it is the equation of a different model because the α_i and the β_j are here random effects corresponding, respectively, to the 100 ball bearings being considered as a random sample from the production line, and to the 20 calipers that were being considered as a random sample of calipers from some population of available calipers. Hence in (23) each α_i and β_j is treated just as is α_i in Examples 4 and 5, with the additional property of taking the α_is and β_js as being independent of one another. Similar independence is taken for the α_is and β_js and e_{ijk}s. Thus

$$\text{cov}(\alpha_i, \beta_j) = \text{cov}(\alpha_i, e_{i'j'k}) = \text{cov}(\beta_j, e_{i'j'k}) = 0$$

and similar results for expected values of corresponding products, as in (7) and (12).

c. Mixed models

Example 7 (Medications and clinics). Suppose in Example 2 that blood pressure studies were made at 15 different, randomly chosen clinics throughout New York City, with 5 patients on each of 4 treatments (placebo and 3 medications) at each clinic. In this case a suitable model equation for the kth patient on treatment i at clinic j would be

$$E(y_{ijk}) = \mu + \alpha_i + \beta_j + \gamma_{ij}, \tag{24}$$

where α_i, β_j and γ_{ij} are the effects due to treatment i, clinic j and treatment-by-clinic interaction, respectively. The range of values for i, j and k are $i = 1, \ldots, 4$, $j = 1, \ldots, 15$ and $k = 1, \ldots, 5$. Since, as before, the treatments are the treatments of interest, α_i is a fixed effect. But the clinics that have been used were chosen randomly, and so β_j is a random effect. Then, because γ_{ij} is an interaction between a fixed effect and a random effect, it is a random effect, too. Thus the model equation (24) has a mixture of both fixed effects, the α_is, and random effects, the β_js and γ_{ij}s. It is thus called a *mixed model*. It incorporates problems relating to the estimation of both fixed effects and variance components.

In application to real-life situations, mixed models have broader use than random models, because so often it is appropriate (by the manner in which data have been collected) to have both fixed effects and random effects in the same model. Indeed, every model that contains a μ is a mixed model, because it also contains a residual error term, and so automatically has a mixture of fixed and random elements. In practice, however, the name mixed model is usually reserved for any model having both fixed effects (other than μ) and random effects, as well as the customary random residuals.

Example 8 (Varieties and gardens). Example 1 deals with 4 different varieties of tomatoes. Suppose they are to be compared in 15 different gardens in Tompkins County of New York State. Then (24) of Example 7 would be an appropriate model, with α_i being the fixed effect for variety of tomato, and β_j the random effect for garden; and γ_{ij} the interaction effect—a random effect. This and Example 7 would then be essentially the same—4 different treatments (fixed effects) being used in each of 15 randomly chosen places (random effects).

1.4. FIXED OR RANDOM?

Equation (8) involving soils and fertilizers is indistinguishable from (23) for the ball bearings and calipers. But the complete models in these two cases are different because of the interpretation attributed to the effects: in the one case fixed, and in the other, random. In these and the other examples most of the effects are readily seen to be categorically fixed or random: thus tomato varieties and medications are fixed effects, whereas clinics and dairy bulls are random effects. But such clear answers to the question "fixed or random"? are not necessarily the norm. Consider the following example.

Example 9 (Mice and technicians). A laboratory experiment designed to study the maternal ability of mice uses litter weights of ten-day-old litters as a measure of maternal ability. Suppose there are four female mice, each of which has six litters. The experiment is supervised by a laboratory technician, a different technician for each successive pair of litters that the mice had. One possible model for y_{ijk}, the weight of the kth litter from mouse i with the experiment being supervised by technician j, would be

$$E(y_{ijk}) = \mu + m_i + \tau_j + \phi_{ij}, \tag{25}$$

where μ is a general mean, m_i is the effect due to mouse i, τ_j is the effect due to technician j and ϕ_{ij} is an interaction effect.

Consider the m_is and the mice they represent. The data relate to maternal ability, a variable that is assuredly subject to variation from animal to animal. The prime concern of the experiment is therefore unlikely to center specifically on those four animals used in the experiment. After all, they are only a sample from a population of mice: and so the m_is are random effects. But what of the τ_js, the technician effects? If the technicians each came and went, as a random sample of employees, so to speak, with many more such people also being available, then the τ_js could reasonably be treated as random effects. But suppose three particular people were the only candidates available for the position of technician, and each wanted it as long-term employment. Then we are specifically interested in just those three technicians and want to assess differences between them, and pick for the job the one deemed best. In that case we would be unwilling to assume that the technician effects were sampled from a population of values, and they would be fixed effects, not random effects. Thus it is that the situation to which a model applies is the deciding factor in determining whether effects are to be considered as fixed or random. Extensive discussion of this is to be found in the landmark paper of Eisenhart (1947), with further comment available in Kempthorne (1975) and Searle (1971).

In some situations the decision as to whether certain effects are fixed or random is not immediately obvious. Take the case of year effects, for example, in studying wheat yields: are the effects of years on yield to be considered fixed or random? The years themselves are unlikely to be random, for they will probably be a group of consecutive years over which the data have been gathered or the experiments run. But the effects on yield may reasonably be considered random, subject, perhaps, to correlation between yields in successive years. Of course, if one was interested in comparing specific years for some purposes, then treating years as random would not be appropriate.

In endeavoring to decide whether a set of effects is fixed or random, the context of the data, the manner in which they were gathered and the environment from which they came are the determining factors. In considering these points the important question is that of inference: are the levels of the factor going to be considered a random sample from a population of values? "Yes"—then the effects are to be considered as random effects. "No"—then, presumably, inferences will be made just about the levels occurring in the data and the effects are considered as fixed effects. Thus when inferences will be made about a population of effects from which those in the data are considered to be a random sample, the effects are considered as random; and when inferences are going to be confined to the effects in the model, the effects are considered fixed.

Another way of putting it is to ask the questions "Do the levels of a factor come from a probability distribution"? and "Is there enough information about a factor to decide that the levels of it in the data are like a random sample"? Negative answers to these questions mean that one treats the factor as a fixed effects factor and estimates the effects of the levels. Affirmative answers mean treating the factor as a random effects factor and estimating the variance component due to that factor. In that case, if one is also interested in the realized values of those random effects that occur in the data, then one also uses a prediction procedure for those values (see Section 3.4).

It is to be emphasized that the assumption of randomness does not carry with it the assumption of normality. Often this assumption *is* made for random effects, but it is a separate assumption made subsequent to that of assuming effects are random. Although most estimation procedures for variance components do not require normality, if distributional properties of the resulting estimators are to be investigated then normality of the random effects is often assumed.

1.5. FINITE POPULATIONS

Random effects occurring in data are assumed to be from a population of effects. The populations are usually considered to have infinite size, as is, for example, the population of all possible crosses between two varieties of tomato. They could be crossed an infinite number of times. However, the definition of random effects does not demand infinite populations of such effects. They can be finite. In addition, finite populations may be very large, indeed so large as to be considered infinite for most purposes; an example would be all the mice in New York State on July 4, 1990! Hence random effects factors can be conceptual populations of three kinds insofar as their size is concerned: infinite, finite but so large as to be deemed infinite, and finite.

We shall be concerned with random effects coming solely from populations assumed to be of infinite size, either because this is the case or because, although finite, the population is large enough to be taken as infinite. These are the most oft-occurring situations found in practical problems. Finite populations, *a propos* variance components, are discussed in several places, e.g., Bennett and Franklin (1954, p. 404) and Gaylor and Hartwell (1969). Rules for converting the estimation procedure of any infinite-population situation into one of finite populations are given in Searle and Fawcett (1970).

1.6. SUMMARY

a. Characteristics of the fixed effects model and the random effects model for the 1-way classification

Characteristic	Fixed Effects Model	Random Effects Model
Model equation	$y_{ij} = \mu + \alpha_i + e_{ij}$	$y_{ij} = \mu + \alpha_i + e_{ij}$
Mean of y_{ij}	$E(y_{ij}) = \mu + \alpha_i$	$E(y_{ij}\|\alpha_i) = \mu + \alpha_i$ $E(y_{ij}) = \mu$
α_i	Fixed, unknowable constant	$\alpha_i \sim \text{i.i.d.}(0, \sigma_\alpha^2)$
e_{ij}	$e_{ij} = y_{ij} - E(y_{ij})$ $= y_{ij} - (\mu + \alpha_i)$ $e_{ij} \sim \text{i.i.d.}(0, \sigma_e^2)$	$e_{ij} = y_{ij} - E(y_{ij}\|\alpha_i)$ $= y_{ij} - (\mu + \alpha_i)$ $e_{ij} \sim \text{i.i.d.}(0, \sigma_e^2)$
$E(e_{ij}\alpha_i)$	$E(e_{ij}\alpha_i) = \alpha_i E(e_{ij}) = 0$	$E(e_{ij}\alpha_i) = 0$
$\text{var}(y_{ij})$	$\text{var}(y_{ij}) = \sigma_e^2$	$\text{var}(y_{ij}) = \sigma_\alpha^2 + \sigma_e^2$
$\text{cov}(y_{ij}, y_{i'j'})$	$\text{cov}(y_{ij}, y_{i'j'}) = \begin{cases} \sigma_e^2 & \text{for } i = i' \text{ and } j = j' \\ 0 & \text{otherwise} \end{cases}$	$\text{cov}(y_{ij}, y_{i'j'}) = \begin{cases} \sigma_\alpha^2 + \sigma_e^2 & \text{for } i = i' \text{ and } j = j' \\ \sigma_\alpha^2 & \text{for } i = i' \text{ and } j \neq j' \\ 0 & \text{otherwise} \end{cases}$

b. Examples

No.	Page	Content	Classification	Model
1	7	Tomato varieties	1-way	Fixed
2	8	Medications	1-way	Fixed
3	9	Soils and fertilizers	2-way	Fixed
4	9	Clinics	1-way	Random
5	12	Dairy bulls	1-way	Random
6	13	Ball bearings and calipers	2-way	Random
7	14	Medications and clinics	2-way	Mixed
8	14	Varieties and gardens	2-way	Mixed
9	15	Mice and technicians	2-way	Mixed or random

c. Fixed or random?

For any factor, the following decision tree has to be followed in order to decide whether the factor is to be considered as fixed or random.

IS IT REASONABLE TO ASSUME THAT LEVELS OF THE FACTOR COME FROM A PROBABILITY DISTRIBUTION?

- **No**
  ↓
  Treat factor as fixed
- **Yes**
  ↓
  Treat factor as random
  ↓
  WHERE DOES INTEREST LIE?
  - ONLY IN THE DISTRIBUTION OF THE RANDOM EFFECTS
    ↓
    Estimate the variance of the random effects
  - IN BOTH THE DISTRIBUTION AND THE REALIZED VALUES OF THE RANDOM EFFECTS
    ↓
    Estimate the variance of the random effects and calculate predictors (BLUP) of the realized values of the random effects

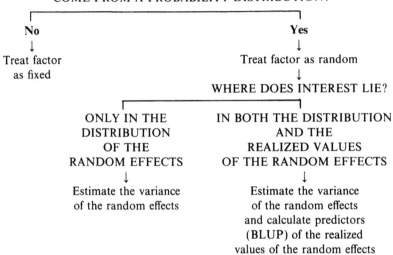

CHAPTER 2

HISTORY AND COMMENT

This chapter gives a brief history of the different methods now available for estimating variance components. In so doing, it provides a skeleton survey of many of the topics that are detailed in ensuing chapters. We begin with an introductory section on analysis of variance because, historically, that is the starting point of methods of estimating variance components.

2.1. ANALYSIS OF VARIANCE

The starting point of Fisher's analysis of variance table was the array of different means or averages available from a body of data. Thus in Example 1 of Chapter 1, where y_{ij} is the yield from plant j of variety i, there are variety means $\bar{y}_{i\cdot}$ and the overall mean $\bar{y}_{..}$. From that example of the 1-way classification we generalize notation to have

$$\bar{y}_{i\cdot} = \frac{\sum_{j=1}^{n} y_{ij}}{n} \quad \text{and} \quad \bar{y}_{..} = \frac{\sum_{i=1}^{a}\sum_{j=1}^{n} y_{ij}}{an} = \frac{\sum_{i=1}^{a} \bar{y}_{i\cdot}}{a} \tag{1}$$

for $i = 1, \ldots, a$ and $j = 1, \ldots, n$. In that example, $a = 4$ and $n = 6$. Thus, in general, a is the number of groups or classes (tomato varieties in the example) with the number of observations (plants) in each being n.

We begin with the identity

$$y_{ij} - \bar{y}_{..} = (y_{ij} - \bar{y}_{i\cdot}) + (\bar{y}_{i\cdot} - \bar{y}_{..})\,. \tag{2}$$

Squaring each side of (2) and summing over i and j gives, in contrast to the linear identity of (2), what can be called a quadratic identity:

$$\sum_{i=1}^{a}\sum_{j=1}^{n} (y_{ij} - \bar{y}_{..})^2 = \sum_{i=1}^{a}\sum_{j=1}^{n} (y_{ij} - \bar{y}_{i\cdot})^2 + \sum_{i=1}^{a}\sum_{j=1}^{n} (\bar{y}_{i\cdot} - \bar{y}_{..})^2\,. \tag{3}$$

This arises because in squaring the right-hand side of (2), the cross-product term is zero:

$$\sum_{i=1}^{a}\sum_{j=1}^{n}(y_{ij}-\bar{y}_{i.})(\bar{y}_{i.}-\bar{y}_{..})=\sum_{i=1}^{a}\left[\sum_{j=1}^{n}(y_{ij}-\bar{y}_{i.})\right](\bar{y}_{i.}-\bar{y}_{..})=\sum_{i=1}^{a}0(\bar{y}_{i.}-\bar{y}_{..})=0\,.$$

Note that each term in (3) is analogous to the sum of squares used in the customary estimation of variance from a simple sample of k observations, $x_1, x_2, \ldots, x_k$, namely

$$s^2=\frac{\sum_{r=1}^{k}(x_r-\bar{x})^2}{r-1}\quad\text{for}\quad \bar{x}=\frac{\sum_{r=1}^{k}x_r}{r}\,. \tag{4}$$

Thus the term on the left-hand side of (3) is the total sum of squares of deviations of all the observations from their mean, and the two terms on the right-hand side of (3) are sums of squares of deviations of observations from their group means, $\bar{y}_{i.}$, and of those group means from their mean, $\bar{y}_{..}$. Thus (3) is a partitioning of the *total sum of squares* (or total sum of squares corrected for the mean) into two other sums of squares, all three of them being available for calculating estimated variances after the manner of s^2 in (4). This partitioning, namely the identity (3), is easily summarized in tabular form as in Table 2.1, wherein the labels SSA, SSE and SST_m have been given to the sums of squares. SST_m, the total sum of squares adjusted for the mean (a.f.m), is used for distinction from $\text{SST} = \Sigma_{i=1}^{a}\Sigma_{j=1}^{n} y_{ij}^2$, with $\text{SST}_\text{m} = \text{SST} - an\bar{y}_{..}^2 = \Sigma_i\Sigma_j(y_{ij}-\bar{y}_{..})^2$.

All of this is just straightforward algebra. Now we introduce certain statistical properties that originate from the customary assumptions of independence and normality: that the y_{ij} are realized values of independent random variables that are normally distributed with $E(y_{ij}) = \mu_i$ and $\text{var}(y_{ij}) = \sigma_e^2$. Under these circumstances it was Fisher's work that showed that SSA and SSE are each distributed as a multiple of a χ^2-distribution, that they are stochastically

TABLE 2.1. PARTITIONING THE SUM OF SQUARES IN THE 1-WAY CLASSIFICATION, BALANCED DATA

Source of Variation	Sum of Squares
Groups	$\text{SSA} = \sum_{i=1}^{a}\sum_{j=1}^{n}(\bar{y}_{i.}-\bar{y}_{..})^2$
Within groups	$\text{SSE} = \sum_{i=1}^{a}\sum_{j=1}^{n}(y_{ij}-\bar{y}_{i.})^2$
Total (a.f.m.)	$\text{SST}_\text{m} = \sum_{i=1}^{a}\sum_{j=1}^{n}(y_{ij}-\bar{y}_{..})^2$

TABLE 2.2. ANALYSIS OF VARIANCE FOR A 1-WAY CLASSIFICATION WITH BALANCED DATA

Source of Variation	d.f.[1]	Sum of Squares	Mean Square	F-Statistic
Groups	$a-1$	$\text{SSA} = n\sum_{i=1}^{a}(\bar{y}_{i.}-\bar{y}_{..})^2$	$\text{MSA} = \text{SSA}/(a-1)$	$F = \dfrac{\text{MSA}}{\text{MSE}}$
Within groups	$a(n-1)$	$\text{SSE} = \sum_{i=1}^{a}\sum_{j=1}^{n}(y_{ij}-\bar{y}_{i.})^2$	$\text{MSE} = \text{SSE}/a(n-1)$	
Total a.f.m.	$an-1$	$\text{SST}_{\text{m}} = \sum_{i=1}^{a}\sum_{j=1}^{b}(y_{ij}-\bar{y}_{..})^2$		

[1] d.f. = degrees of freedom

independent and that

$$F = \frac{\text{SSA}/(a-1)}{\text{SSE}/a(n-1)} \sim \mathscr{F}^{a-1}_{a(n-1)}, \tag{5}$$

meaning that F is distributed according to Fisher's F-distribution (so named by Snedecor) with $a-1$ and $a(n-1)$ degrees of freedom for the numerator and denominator, respectively. This calculation and its intermediate steps are summarized in the familiar analysis of variance table of Table 2.2, which is simply an expansion of Table 2.1.

The simplest use for which Fisher designed the analysis of variance table is that in Table 2.2, on assuming normality and the model equation

$$E(y_{ij}) = \mu + \alpha_i, \tag{6}$$

the F-statistic of (5) and Table 2.2 is a test statistic for testing the hypothesis

$$\text{H:} \quad \alpha_1 = \alpha_2 = \cdots = \alpha_a . \tag{7}$$

As has been said, this is for the fixed effects model. But for the random effects model, which is more pertinent to this book, the important question is "How does Table 2.2 get used in the random model?" This is answered by considering two questions that are more specific.

The first is "In the random model, what use is F?" The difference between the fixed model and the random model is what is essential here. In the fixed model, the αs correspond to specific, carefully chosen, levels (e.g. tomato varieties) of specific interest; and in the random model the αs correspond to a random sample of levels from some larger population (e.g., a sample of bulls). In the fixed effects case we are most interested in just the particular αs that occur in the data—and in only those effects. In the random effects case we are interested in the effects that occur in the data only inasmuch as they are a sample from a population and can therefore be used to make inferences about that population—in particular about its variance. Hence fixed effects models focus concern upon means: random effects models focus concern upon variances.

Thus for the fixed effects model the assumption of normality leads to $F = \text{MSA}/\text{MSE}$ [of (5) and Table 2.2] being a test statistic for the hypothesis H: $\alpha_1 = \alpha_2 = \cdots = \alpha_a$ shown in (7). But in the random model that same F-statistic tests H: $\sigma_\alpha^2 = 0$.

The preceding $F = \text{MSA}/\text{MSE}$ is a test statistic for H: $\sigma_\alpha^2 = 0$ only for the 1-way classification (see Sections 3.5d-v and 3.6d-v). But the reader is cautioned that this obviously useful result, of an F-statistic being available for testing a hypothesis that a variance component is zero, does not extend to every F-statistic that arises in analysis of variance tables of data of all mixed or random models—not even for balanced data. And this caution leads to another. Users of computer packages that have F-values among their output must be totally certain that they know precisely what the hypothesis is that can be tested by each such F-value. This is so both for fixed effects models (e.g., Searle, 1987), and for random and mixed models, too. Thus for the 1-way classification of Table 2.2, with the random model the statistic F does not have an F-distribution when using unbalanced data, unless $\sigma_\alpha^2 = 0$.

The second question concerning Table 2.2, which is particularly pertinent to this book, is "What part does the analysis of variance table play in estimating components?" The answer to this question occupies Chapters 3–5 that follow, dealing not just with the analysis of variance of Table 2.2, but with many extensions for both balanced and unbalanced data.

2.2. EARLY YEARS, 1861–1949

a. Sources

The following brief history emphasizes the development of methods of estimating variance components, much of it being akin to Searle (1988a, 1989). The early years of 1861–1939 are dealt with in more detail than is 1940 onwards, because publications are sparser and, for many readers, harder to locate than those since 1940. For this early history we draw heavily, plagiaristically in some cases, on Anderson (1978, pp. 11–25), with his kind permission; and he, in turn, utilized Scheffé (1956). For the more recent period, heavy reliance is placed on (and free use made of) the excellent survey of Khuri and Sahai (1985)—again, with their kind permission. The proliferation of papers in the last fifteen years or so is extremely well summarized by those authors and the interested reader is encouraged to read their article and use their comprehensive bibliography as an entré to almost all aspects of variance components. In relying on their survey, the account given here of the recent years does, for some topics, refer to just an early paper and a recent one, so providing the reader with both a starting point and something up-to-date. To encompass all the literature would be to repeat Khuri and Sahai's (1985) paper, the reference list of which is extensive; and even more so are the bibliographies of Sahai (1979) and Sahai, Khuri and Kapadia (1985).

Much of the early story of variance components revolves around the 1-way classification that has already been set out in Chapter 1, summarized as follows:

$$y_{ij} = \mu + \alpha_i + e_{ij}, \quad \text{with } i = 1, 2, \ldots, a; \tag{8}$$

$$\text{var}(\alpha_i) = \sigma_\alpha^2; \quad \text{var}(e_{ij}) = \sigma_e^2; \quad \text{all covariances zero;} \tag{9}$$

$$j = 1, 2, \ldots, n \quad \forall\ i, \quad \text{for balanced data;} \tag{10}$$

$$j = 1, 2, \ldots, n_i, \quad \text{for unbalanced data .} \tag{11}$$

For consistent notation, changes have been made to what some authors have used (even within direct quotations), and a unified set of equation numbers has been employed, with authors' numbers shown in square brackets.

b. Pre-1900

An excellent telling of the early history of variance components is given in Scheffé (1956) and is enlarged upon in Anderson (1978, 1979a). Both of these accounts are drawn on extensively in what follows.

Legendre (1806) and Gauss (1809) are well known as the independent fathers of the method of least squares. Plackett (1972) has an intriguing discussion of their relative rights to priority. An interesting aspect of those two early papers is, as pointed out by Scheffé (1956), that they were both published in books concerned with problems arising from astronomy: the orbits of the comets were Legendre's concern and Gauss dealt with conic sections. But what is even more interesting is that whereas Legendre and Gauss were implicitly dealing with fixed effects aspects of linear models (although they wrote no model equations as would be recognized today), the subject of random effects models also seems to have originated from problems in astronomy.

The first known formulation of a random effects model (although not called such) seems to be that of Airy (1861, especially Part IV). Scheffé (1956) refers to this work as being "very explicit use of a variance-components model for the one-way layout ... with all the subscript notation necessary for clarity." He (Scheffé) describes the work (Airy, 1861, Sec. 118; Sec. 113 in the 3rd edition) as being concerned with making telescopic observations on the same phenomenon for a nights, n_i observations on the ith night. It is noteworthy (as remarked upon by Anderson, 1978) that in this earliest known use of a variance components model there is provision for unequal numbers of observations on the different nights. Then, with a footnote that he has changed Airy's capital letters to lower case, and that he has "added the general mean μ since he [Airy] writes the equations for the observations minus μ instead of for the observations," Scheffé describes Airy's model as follows [but now using the notation we have set out in (8)–(11)]:

Airy assumes the following structure for the jth observation on the ith night:

$$[2.1] \qquad y_{ij} = \mu + \alpha_i + e_{ij}, \tag{12}$$

where μ is the general mean or "true" value, and the $\{\alpha_i\}$ and $\{e_{ij}\}$ are random effects with the following meanings: He calls α_i the "constant error", meaning it is constant on the ith night; we would call it the ith night effect; it is caused by the "atmospheric and personal circumstances" peculiar to the ith night. The $\{e_{ij}\}$ for fixed i we would call the errors about the (conditional) mean $\mu + \alpha_i$ on the ith night. It is implied by Airy's discussion that he assumes all the e_{ij} independently and identically distributed, similarly for the α_i, that the $\{e_{ij}\}$ are independent of the $\{\alpha_i\}$, and that all have zero means. Let us denote the variances of the $\{e_{ij}\}$ and the $\{\alpha_i\}$ by σ_e^2 and σ_α^2.

Nowadays this seems to be accepted as the first occurrence of a random effects model in the literature. Yet Airy himself must not have thought of it as being the first, for in the preface of his book, quoted by Anderson (1978), he writes "No novelty, I believe, of fundamental character, will be found in these pages."; and "... the work has been written without reference to or distinct recollection of any other treatise (excepting only Laplace's *Théorie des Probabilités*)...." As Anderson (1978) says, this, insofar as attempts at establishing the exact origin of the components of variance concept are concerned, is an unfortunate style of writing.

Quoting from Scheffé (1956, p. 256) again, it is interesting to note that Airy estimates what we would call σ_e^2 by first calculating

$$\hat{\sigma}_{e,i}^2 = \frac{\Sigma_j(y_{ij} - \bar{y}_{i\cdot})^2}{n_i - 1} \qquad (13)$$

for the ith night and then averages the square roots of the values given by (13) to estimate σ_e^2 by

$$\hat{\sigma}^2 = \left[\sum_{i=1}^{a} (\hat{\sigma}_{e,i}^2)^{\frac{1}{2}}/a\right]^2 . \qquad (14)$$

It is noteworthy to see such an early use of $n_i - 1$ as denominator of (13), although Anderson (1978) states that this is not an original use. "In establishing a criterion for the rejection of discordant observations," he writes, "Pierce (1852) specified "the sum of squares of all errors' as being $(N - m)\varepsilon^2$, where N is the total number of observations, m is the number of unknown quantities contained in the observations and ε^2 is the mean error (sample variance). Clearly, astronomers understood the concept of degrees of freedom (but without using the term) as early as the year 1852."

The second user of a random effects model appears, according to Scheffé, to be Chauvenet (1863, Vol. II, Articles 163 and 164), who, although he did not write model equations, certainly implied such models and derived the variance of $\bar{y}_{\cdot\cdot} = \Sigma_{i=1}^a \Sigma_{j=1}^n y_{ij}/an$ of (1) as

$$\text{var}(\bar{y}_{\cdot\cdot}) = \frac{\sigma_\alpha^2 + \sigma_e^2/n}{a} .$$

Chauvenet suggests that there is little practical advantage in having n greater than 5, and refers to Bessel (1820) for this idea; but Scheffé says that the reference is wrong, although it "does contain a formula for the probable error of a sum of independent random variables which could be the basis for such a conclusion. Probably Bessel made the remark elsewhere." If so, the question is "Where?"; and might that other reference be the first germ of an idea about optimal design? Preitschopf (1987) has searched the 1820–1826 and 1828 yearbooks containing Bessel (1920) and finds not even a hint about not having "n greater than 5"; the only pertinent remark is on page 166 of the 1823 yearbook which has, with x_i being the "random error of part i, $i = 1, \ldots, n$, total error is $y = \sqrt{x_1^2 + \ldots + x_n^2}$."

Apart from some inconsequential comments by Yule (1911, Chap. XI) that indicate his unawareness of Airy (1861) and Chauvenet (1863), the next and major foundational ideas on estimating variance components are seen in the work of R. A. Fisher.

c. 1900–1939

-i. R. A. Fisher. In an essay on the status of quantitative genetic theory, Kempthorne (1977) remarks: "Without doubt, the basic and seminal paper in the theory of quantitative genetics is that of Fisher (1918)." However, considering that the motivation for Fisher's paper was his having foreseen that the basis for "...a more exact analysis of the causes of human variability" lay in reconciling the continuous variation of a metric trait with the discrete nature of Mendelian inheritance processes, Kempthorne's remark can also be applied to Fisher's contribution to variance component theory. In this connection, some notable aspects of Fisher's paper are [adapting freely from Anderson (1978)]:

(i) Inceptive use of the terms "variance" and "analysis of the variance".

(ii) Implicit, but unmistakable, use of variance components models.

(iii) Definitive ascription of percentages of a total variance to constituent causes; e.g., that dominance deviations accounted for 21% of the total variance in human stature.

Following that genetics paper, Fisher's book (1925, Sec. 40) made a major contribution to variance component models through initiating what has come to be known as the analysis of variance method of estimation: equate sums of squares from an analysis of variance to their expected values (taking expectations under the appropriate random or mixed model) and thereby obtain a set of equations that are linear in the variance components to be estimated. This idea arose from using an analysis of variance for deriving an estimate of an intra-class correlation from data from a completely randomized design. The pertinent passage in Fisher (1925, p. 190) is as follows.

> Let a quantity be made up of two parts, each normally and independently distributed; let the variance of the first part be A, and that of the second part, B; then it is easy to see that the variance of the total quantity is $A + B$. Consider a sample of n' values of the first part, and to each of these add a sample of k in

each case. We then have n' classes of values with k in each class. In the infinite population from which these are drawn the correlation between pairs of numbers in the same class will be

$$\rho = \frac{A}{A+B} . \tag{15}$$

From such a set of kn' values we may make estimates of the values of A and B, or in other words we may analyze the variance into the portions contributed by the two causes; the intraclass correlation will be merely the fraction of the total variance due to the cause which observations in the same class have in common. The value of B may be estimated directly, for variation within each class is due to this cause alone, consequently

$$\mathop{\mathrm{S}}_{1}^{kn'} (x - \bar{x}_p)^2 = n'(k-1)B . \tag{16}$$

The mean of the observations in any class is made up of two parts, the first part with variance A, and a second part, which is the mean of k values of the second parts of the individual values, and has therefore a variance B/k; consequently from the observed variation of the means of the classes, we have

$$k \mathop{\mathrm{S}}_{1}^{n'} (\bar{x}_p - \bar{x})^2 = (n'-1)(kA + B) . \tag{17}$$

S in (16) and (17) represents summation; the notation of Table 2.1 has these equations as

$$E \sum_{i=1}^{a} \sum_{j=1}^{n} (y_{ij} - \bar{y}_{i\cdot})^2 = a(n-1)\sigma_e^2,$$

i.e., (18)

$$E(\mathrm{SSE}) = a(n-1)\sigma_e^2,$$

and

$$En \sum_{i=1}^{a} (\bar{y}_{i\cdot} - \bar{y}_{\cdot\cdot})^2 = (a-1)(n\sigma_\alpha^2 + \sigma_e^2),$$

i.e., (19)

$$E(\mathrm{SSA}) = (a-1)(n\sigma_\alpha^2 + \sigma_e^2) .$$

Fisher did not write the expectation operator E, nor did he even use the phrase "expected value", but he clearly had that idea in mind when, preceding (15), he wrote "In the infinite population from which these are drawn..."—even though it applies there to the correlation of (15) and not to the sums of squares of (16) and (17). But it is definitely implicit in (16) and (17), and therein hangs Fisher's germinal contribution to the analysis of variance (ANOVA) method of estimating variance components. For that is precisely what (16) and (17) represent, as we see from rewriting their equivalent forms (18) and (19) as

$$\text{SSE} = a(n-1)\hat{\sigma}_e^2$$

and (20)

$$\text{SSA} = (a-1)(n\hat{\sigma}_\alpha^2 + \hat{\sigma}_e^2)\,.$$

Definitive priority for this *idea* undoubtedly goes to Fisher, based on his sentence "The value of B may be estimated directly..." immediately prior to (16), although the actual doing of it must go to some reader of Fisher (1925) who did what (at least nowadays seems) was obviously intended, namely

$$\hat{\sigma}_e^2 = \frac{\text{SSE}}{a(n-1)} = \text{MSE},$$

$$\hat{\sigma}_\alpha^2 = \frac{\text{SSA}/(a-1) - \hat{\sigma}_e^2}{n} = \frac{\text{MSA} - \text{MSE}}{n}\,. \tag{21}$$

These, for balanced data (of a classes with n observations in each), in a 1-way classification random model, are what are known as the ANOVA (analysis of variance) estimators of the variance components. They are akin to method-of-moments estimators.

Had Fisher foreseen even a small part of the methodology for estimating variance components that was heralded by (16) and (17) he might have given more attention to this topic. But he did not. Section 40 of Fisher (1925) remains quite unchanged in subsequent editions (e.g. 8th ed., 1941, p. 215 and 12th ed., 1954, p. 221), even after variance component principles were well established. Furthermore, even when Fisher extended the analysis of variance to a 1-way classification model with unbalanced data, to a 2-factor model with interaction and to more complex settings, he did not address the estimation of variance components in those settings.

-ii. L. C. Tippett. As noted by Urquhart, Weeks and Henderson (1973), Fisher "did not use linear models to explain the analysis of variance of designed experiments even though his writings on regression and correlation (both simple and multiple) lean toward linear models." In contrast, Tippett (1931; Secs. 6.1, 6.2 and 10.3) not only clarified the analysis of variance method of estimating variance components from balanced data but also extended it (apparently for the first time) to the 2-way crossed classification, without interaction, random model. The following quote from Tippett (1931, p. 89) illustrates this point.

> Let it be assumed, for example, that a quantity x is subject to random variations, and to others associated with two factors A and B; then the value of any one observation of x is
>
> $$x = \bar{\xi} + \alpha + \beta + \xi',$$

where $\bar{\xi}$ is the mean, α and β are deviations arising from A and B, and ξ' is the random deviation. The square of its [i.e., x's] deviation from the mean is

$$(x - \bar{\xi})^2 = \alpha^2 + \beta^2 + \xi'^2 + 2\alpha\beta + 2\alpha\xi' + 2\beta\xi'$$

and this may be summed for a sample of N individuals, and divided by the degrees of freedom (N in this case, since we have not found the mean $\bar{\xi}$ from the sample, but have assumed it). Thus we obtain [with $\mathrm{S} \equiv \Sigma$]

$$\frac{\mathrm{S}(x - \bar{\xi})^2}{N} = \frac{\mathrm{S}\alpha^2}{N} + \frac{\mathrm{S}\beta^2}{N} + \frac{\mathrm{S}\xi'^2}{N} + \frac{2\,\mathrm{S}\alpha\beta}{N} + \frac{2\,\mathrm{S}\alpha\xi'}{N} + \frac{2\,\mathrm{S}\beta\xi'}{N}$$

and as N becomes indefinitely large, the last three terms of this equation tend to zero if α, β and ξ' are independent; the other terms are the squares of the standard deviations or variances, so that finally

$$\sigma_x^2 = \sigma_\alpha^2 + \sigma_\beta^2 + \sigma_{\xi'}^2 \quad \ldots \quad [16] \qquad (22)$$

Hence the variance of x is the sum of the random variance and of those due to A and B.

It is interesting that by relying on the notation of uncorrelated random "deviations", Tippett (especially in Sec. 10.3) overlooked the possibility of having interaction effects in linear models whereas Fisher (1925, Sec. 42), despite his non-usage of a linear model, not only used the term "interaction" (p. 200), but also described an interaction effect between two factors A and B.

Tippett (1931, Sec. 6.2, pp. 92–93) describes the analysis of variance method for estimating variance components as follows. It yields the estimators a little more explicitly than do (16) and (17) from Fisher (1925).

If v_s^2 is the mean variance between shrubs [amongst classes mean square], and v_r^2 the mean variance within a shrub [error mean square], as found from the sample

$$\left.\begin{array}{l} v_s^2 \to n\sigma_s^2 + \sigma_r^2 \\ v_r^2 \to n\sigma_r^2 \end{array}\right\} \quad \ldots \quad [18] \qquad (23)$$

where n is the number of readings per shrub, σ_r^2 is the variance "within a shrub", σ_s^2 is the mean variance "between shrubs" and $\to$ denotes "that the quantity on the left is an estimate of that on the right, and that the former approaches the latter as the size of the sample" (number of degrees of freedom in both parts) increases indefinitely. Having for a set of data obtained values of v_s^2 and v_r^2 of 261.492 and 3.057, respectively, and with $n = 100$, Tippett continues (p. 93)

Using the relations of equations [18]

$$261.492 \to 100\sigma_s^2 + \sigma_r^2$$

$$3.057 \to \sigma_r^2 \quad \ldots$$

whence

$$258.435 \to 100\sigma_s^2$$

$$2.584 \to \sigma_s^2\,.$$

The preceding methodology is extended in Tippett (1931, Sec. 10.3, p. 180) to the 2-way crossed classification model, using deviations from the grand mean, an analysis of variance table, and expected mean squares; and the preceding expressions for calculating variance components estimates of Tippett's first edition of 1931 were extended in the second edition of 1937 (p. 182), in terms of an example.

With a method of estimating variance components established (notwithstanding its restriction to balanced data), the problem of selecting an optimal sample design for any particular experiment could be studied definitively. Thus the "best way of distributing the observations between and within groups" for a 1-way model was addressed by Tippett (1931, Sec. 10.1, p. 182), as it had been by Chauvenet (1863) and perhaps Bessel (1820).

-iii. The late 1930s. Despite Tippett's consideration of optimal design just mentioned, the comprehensive study on sampling for yield in cereal experiments by Yates and Zacopanay (1935), which dealt with designs corresponding to higher-order models, would appear to be an early beginning to optimal sampling design. In the same year Neyman, Iwaszkiewicz and Kolodziejczyk (1935) examined the comparative efficiency of randomized blocks and Latin squares designs and, in contradistinction to all previous studies, they made extensive use of linear models (including mixed models) and associated mathematical concepts.

Neyman *et al.* (1935) also have some claim to originating the term "variance component". In an acrimonious review of that paper, Fisher (1935) used the term "components of variation", which, coupled with the paper's use of the term "error components", undoubtedly influenced ultimate adoption of "components of variance" (or "variance component"). However, this cannot be asserted unequivocally because Daniels (1939), who appears to have been the first to use the phrase "components of variance", did not mention either Neyman *et al.* (1935) or Fisher (1935) in this regard when he wrote that variability

> ...is the result of factors..., each factor being responsible for its quota of the dispersion, and it is natural to use the analysis of variance techniques not only to detect possible sources of variation but to arrive at estimates of the components of total variance assignable to each factor. The components of variance can then be used to establish an efficient sampling scheme....

Both Daniels (1939) and, a few months later (across the Atlantic), Winsor and Clarke (1940) derive the equivalent of (16) and (17) that Fisher (1925) has. In doing so, both papers use the "expected value" concept; Daniels mentions Tippett (1931) but not Fisher. whereas Winsor and Clark describe their derivation as being "a straightforward extension of the suggestions of R. A. Fisher in his *Statistical Methods for Research Workers* [Sec. 40]." Presumably this is the seventh edition, published in 1938, in which Sec. 40 is the section dealing with the intraclass correlation, exactly as does the same section, unchanged, in both the first edition of 1925 and the twelfth edition of 1954. Yet, as we have seen, although Fisher (1925) has the idea of taking expected values, he has not there specifically formulated it using the E operator as do Daniels, and Winsor and Clarke.

At about the same time as both the Daniels and the Winsor and Clarke papers were published (the latter in what, even at that time, must have been somewhat of an obscure journal for statisticians), Snedecor's third edition (1940) became available with, as far as can be seen, no reference to variance components at all. Page 205 contains discussion of estimating the intra-class correlation as $A/(A + B)$, just as does the 1938 seventh edition of Fisher (1925). The nearest thing to characterizing A as a variance component is the description that "A is the same for all ... samples—it is the common element, analogous to covariance." And that is, of course, the case: the covariance between y_{ij} and $y_{ij'}$ for $j \neq j'$ is σ_α^2.

The work of Daniels (1939) was significant in two other respects:

(i) Sampling variances of variance component estimates were derived, for balanced data, up to the complexity of a 3-way crossed classification random effects model, complete with all interactions.

(ii) In deriving expected mean squares, account was taken of the possibility that the population of effects for a random factor could be of finite size. This was motivated by Tippett's (1937, Sec. 10.13) treatment of the 1-way random model for which he derived the estimator of the variance component due to classes (σ_α^2) as $(1 - 1/n)^{-1}(\text{MSA} - \text{MSE})$. This estimator differs from the corresponding infinite population estimator of (21) through multiplication by $(1 - 1/n)^{-1}$; i.e., it is $(1 - 1/n)^{-1}$ times the estimator given by [15] of Fisher (1925).

Since linear models have nowadays become an integral part of describing variance components, it is interesting to note that this had become widely accepted by 1939; e.g., Neyman *et al.* (1935), Welch (1936), Daniels (1939) and Jackson (1939). Moreover, the models specified here were surprisingly up-to-date in some cases. Consider the following sentence from the appendix of Welch (1936):

> x_{ti} are a set of $N = kn$ observations consisting of k groups ($t = 1, 2, \ldots, k$) of n individuals in each group ($i = 1, 2, \ldots, n$) such that $x_{ti} = \alpha + y_t + z_{ti}$ where y and z are normally and independently distributed about zero with S.D.s [standard deviations] σ_1 and σ_2, respectively.

Welch then utilizes properties of χ^2-variables to derive essentially the same results as Fisher (1925), shown earlier as (16) and (17). Stemming from his reliance on normality for deriving expectation of sums of squares, Welch's sentence about the unbiasedness of the resulting variance component estimators suggests that he may not have realized that the assumption of normality is not necessary for establishing that unbiasedness.

Jackson (1939) also assumed normality for random effects and error terms in his description of a mixed model for a no-interaction 2-factor situation with one factor random and the other non-random. He writes the model as $y_{st} = A + B_s + C_t + z_{st}$, with A being "a measure of the effect common to all individuals...", B_s as being "a measure of the trial effect", C_t as "a measure of the individual effect" and z_{st} as "the error of measuring...". This seems to be the first occurrence in the literature of the word "effect" in what is now its customary usage in the context of linear models; and this description of a mixed model, although not so called at that time, may well be its first occurrence in the literature also.

Considering the detail of the descriptions that Welch (1936) and Jackson (1939) give to their models, it is surprising that it was not until Eisenhart (1947) that the first precise distinction was made between "fixed" and "random" models (Eisenhart's Models I and II, respectively), and that the name "mixed model" or "mixed analysis of variance" had not been suggested before 1947. Clearly, it was recognized before then that there is a need to specify which of the effects in a linear model are fixed and which are random. Albeit, it is a distinction that Yates (1967) later took great exception to.

-iv. Unbalanced data. Almost all of the work described so far concerns balanced data; e.g., k observations in each of the n' classes of Fisher's description of the 1-way classification. The case of unbalanced data was given but a passing comment by Tippett (1931, Sec. 6.5, p. 96): "In such cases, the relations [18] do not hold, for in summing the squares of the deviations of the group means from the grand mean, each group has been given a different weight, n_s [the number of observations in group s]." Nevertheless, Section 9.6 (p. 166) subsequently provides an approximation to allow the calculation of an intraclass correlation coefficient from such data. In contrast, Snedecor (1934, Sec. 31, p. 20) simply stated "The direct relation between analysis of variance and intraclass correlation disappears if there are unequal frequencies in the classes."

It is nowadays well known that estimating variance components from balanced data is generally much easier than from unbalanced data. A comment on the history of this state of affairs is that although Airy (1861) made provision for unbalanced data—see (12) and (13)—and estimation from balanced data first appeared (implicitly) in Fisher (1925), it was to be fourteen years before something appeared for unbalanced data—in Cochran (1939). And this was for only the simplest case, the 1-way classification random model. With data consisting of a groups having n_i observations in group i, Cochran states that "the mean square variance between groups is an estimate of $\sigma_w^2 + (\Sigma_i n_i - \Sigma_i n_i^2/\Sigma_i n_i)\sigma_g^2/(a-1)$, where σ_w^2 is the variance within groups and σ_g^2 the true variance between groups." This expression is, of course, the expected value of the between-group mean square. Although Cochran goes on to use his result in a manner that we might not use today, he certainly seems to be the first in print with a procedure for handling unbalanced data—albeit for the simplest possible case, the 1-way classification random model.

Cochran follows the result with the comment that "if $n_i = n$ in all groups, the coefficient of σ_g^2 reduces to n; otherwise the coefficient is somewhat smaller than the average number of sampling units per group." With $\bar{n}_.$ denoting this average, this latter observation is valid because

$$\bar{n}_. - \frac{\Sigma_i n_i - \Sigma_i n_i^2/\Sigma_i n_i}{a-1} = \frac{\Sigma_i n_i}{a} - \left[\frac{\Sigma_i n_i}{a-1} - \frac{\Sigma_i n_i^2}{(a-1)\Sigma_i n_i}\right]$$

$$= \left[\Sigma_i n_i^2 - \frac{(\Sigma_i n_i)^2}{a}\right]\frac{1}{(a-1)\Sigma_i n_i}$$

$$= \frac{\Sigma_i (n_i - \bar{n}_.)^2}{(a-1)\Sigma n_i} > 0\,.$$

Whereas Cochran (1939) was not specifically concerned with estimating variance components from unbalanced data in the 1-way random model, Winsor and Clarke (1940) certainly were. The essence of their results is the pair of expectations

$$E\sum_{i=1}^{a} n_i(\bar{y}_{i\cdot} - \bar{y}_{\cdot\cdot})^2 = (a-1)\left[\frac{\Sigma_i n_i - \Sigma_i n_i^2/\Sigma_i n_i}{a-1}\sigma_\alpha^2 + \sigma_e^2\right]$$

and (24)

$$E\sum_{i=1}^{a}\sum_{j=1}^{n_i}(y_{ij} - \bar{y}_{i\cdot})^2 = (n_. - a)\sigma_e^2,$$

for unbalanced data, something that Daniels (1939) does not address himself to. Interestingly enough, Snedecor (1st edn, 1937) touches obliquely on this subject in Example 10.21 (p. 195), where, in referring to unbalanced data of Table 10.8, he asks the question "Why can't you calculate intraclass correlation accurately?" for such data. Winsor and Clarke's results (24) would show that you could. Needless to say, that example does not appear in the completely rewritten fourth edition (1947) nor, of course, in Snedecor and Cochran (1989).

Notation In (24) E represents the expectation operator. It is often written in the form $E(\cdot)$ or $E[\cdot]$ but for clarity, as in (24), we also use E followed by a space, to mean the expectation of the expression that follows that space.

d. The 1940s

The general method of estimating variance components by equating analysis of variance mean squares (or, quite equivalently, sums of squares) to their expected values, under either mixed models or random models, is now known as the ANOVA method of estimation. It was firmly in place by 1934. The 1940s saw a number of extensions to that method; they were but a prelude to the flood of developments that came later. For example, Ganguli (1941) applied it to the k-way nested classification, and Crump (1946) to the 2-way crossed classification, random model, with interaction.

Both Ganguli (1941) and Crump (1946) drew attention to a deficiency of this method of estimating variance components, namely that it can, depending on the data, produce negative estimates. And, as a method of estimation, it does, of course, have no provision for preventing this embarrassment (of having a negative estimate of a parameter that, by definition, is positive). Whenever this does occur, both authors suggested truncating negative values to zero; but this sacrifices the property of unbiasedness that is implicit in the ANOVA method.

Under normality assumptions, Crump (1947) also derived sampling variances of this class of estimators for the 1-way and the 2-way crossed classification random models. Sampling variances for the 1-way model were also derived by Hammersley (1949), but for arbitrary distributional form. However, to obtain "usable" results, fourth cumulants of the random effects distributions had to be set to zero (their correct value under normality). Crump (1947) also invoked normality for considering maximum likelihood estimation, as summarized in Crump (1951), a procedure later used by Hartley and J. N. K. Rao (1967) in developing quite general results (see Section 2.4a which follows).

Three other papers in the 1940 decade are of particular note: Satterthwaite (1946), who dealt with approximate sampling distributions of variance component estimates (and in doing so also gave us the procedure still known by his name for calculating approximate degrees of freedom for approximate F-statistics in random models), and Wald (1940, 1941), who considered confidence intervals for ratios of variance components in 1-way and 2-way classifications with unbalanced data.

2.3. GREAT STRIDES: 1950–1969

The years from 1950 to 1969 brought major developments in methods of estimating variance components, starting with important extensions of the methodology already in place and ending with establishment of new methods based on maximum likelihood and minimum norm criteria.

Early on came the Anderson and Bancroft (1952) book, the first to contain substantial discussion (four chapters) of variance components. This really set the subject on a firm footing, and solidly established the procedure of equating analysis of variance sums of squares to their expectations as a method of estimating variance components. The book deals very thoroughly with estimation from balanced data for both mixed and random models; it also deals with unbalanced data for nested classifications and, after considering incomplete blocks designs, it poses a number of pertinent research problems, many of which have still not been answered satisfactorily. In all, the book is a milestone in the history of variance components estimation. It was followed two years later by Bennett and Franklin (1954) who, in their long (160-page) chapter on analysis of variance, show numerous expected mean squares in terms of variance components, including details pertaining to finite-sized populations, a subject later taken up by Searle and Fawcett (1971).

a. The Henderson methods

A landmark paper dealing with the difficult problem of how to use unbalanced data for estimating variance components is Henderson (1953). The paper was motivated by what was to be its author's lifetime work with the statistical analysis of dairy cow records [e.g., Example 5 of Chapter 1, and see Henderson (1984)]. The paper in 1953 is important because it presents three different ways of using unbalanced data, from random or mixed models, with as many crossed and/or nested classifications as one wishes. All three are adaptations of the ANOVA method of equating (for balanced data) analysis of variance sums of squares to their expected values. Those three adaptations have come to be known as the three Henderson methods. Method I uses sums of squares that are unbalanced-data analogues of those used with balanced data; Method II adjusts the data for whatever fixed effects are in the model, and then uses Method I on those adjusted data; and Method III is based on sums of squares that result from fitting a linear model and its submodels (i.e., from the method of fitting constants). Details of these three methods, based largely on Searle's (1968) matrix reformulation of them, are given in Chapter 5. All three have been used extensively, in a wide variety of applications.

With the hope of providing a criterion for assessing relative optimality, several papers between 1956 and 1968 developed formulae for (or that could lead to) sampling variances of ANOVA estimators and of Henderson methods estimators in particular. The unbiased property of ANOVA estimators demands no distributional assumptions of the random effects and the residual error terms in a model, but all sampling variance results [save those of Hammersley (1949) mentioned earlier] have been developed on the basis of assuming normality. With this, and for unbalanced data, the following cases have been dealt with: extending the 1-way classification results of Crump (1951) to include covariance components, Searle (1956); Method I estimation for the random model, for the 2-way crossed classification in Searle (1958), for the 2-way nested classification in Searle (1961), for the 3-way nested in Mahamunulu (1963) and the 3-way crossed in Blischke (1966). Method III estimation for the 2-way crossed classification without interaction was dealt with by Low (1964). And very general results for Method III are given in Rohde and Tallis (1969). Except for the latter, all of these results are set out in Searle (1971, Chap. 11), and all of them lead to the sampling variance of almost every estimator except $\hat{\sigma}_e^2$ being a quadratic function of the population σ^2s having very complicated functions of the numbers of observations as their coefficients. Despite this, Ahrens (1965) provides a mechanism (described in Searle, 1971, Section 10.2) for estimating such a variance unbiasedly, provided unbiased estimates of the σ^2s are available. This is always the case with ANOVA estimation methodology. Nevertheless, the only currently available expressions for the sampling variances, to which we can apply Ahrens' method, are those derived under normality assumptions, and even then, closed form expressions for the distributions of estimated variance components are unknown (save, in many cases, that of $\hat{\sigma}_e^2$).

b. ANOVA estimation, in general

The ANOVA name given to the method of estimating variance components by the procedure of equating sums of squares to their expected values initially applied to balanced data for which it is particularly apt, because with such data the sums of squares that are used are indeed those of the analysis of variance of those data. But for unbalanced data there is no unique set of sums of squares that can be used. Nevertheless, the method is still called the ANOVA method; and the Henderson methods are just three of the many possible variations of the ANOVA method. Other possibilities are, for example, to use the sums of squares from the weighted squares of means analysis or from the analysis of unweighted means—when the data have all cells filled (see Yates, 1934). Indeed, almost any set of quadratic functions of the observations can be used—as is discussed in detail subsequently.

The 1950–1969 era includes many published results on properties of estimators obtained by the ANOVA method. We comment briefly on some of them.

A first description of ANOVA estimation in its general form is as follows. Let $\boldsymbol{\sigma}^2$ be the vector of variance components to be estimated in some model, and let $\mathbf{s}$ be a vector of sums of squares. Then, when each sum of squares has an expected value that is a linear function of the variance components, $E(\mathbf{s})$ is a vector of such linear functions, which we will represent as $\mathbf{C}\boldsymbol{\sigma}^2$, so that

$$E(\mathbf{s}) = \mathbf{C}\boldsymbol{\sigma}^2 . \tag{25}$$

Hence, for non-singular $\mathbf{C}$ the ANOVA estimator of $\boldsymbol{\sigma}^2$ is based on (25) and is the solution for $\hat{\boldsymbol{\sigma}}^2$ to

$$\mathbf{s} = \mathbf{C}\hat{\boldsymbol{\sigma}}^2,$$

namely

$$\hat{\boldsymbol{\sigma}}^2 = \mathbf{C}^{-1}\mathbf{s} . \tag{26}$$

*-i. **Negative estimates.*** It is clear from (26) that each element of $\hat{\boldsymbol{\sigma}}^2$, i.e., each estimated variance component, is a linear combination of the sums of squares in $\mathbf{s}$. Moreover, there is nothing inherent in (26) to ensure that every element of $\hat{\boldsymbol{\sigma}}^2$ is always non-zero. Thus it is that ANOVA estimates can be negative. For example, $\hat{\sigma}_\alpha^2$ of (21) will be negative whenever MSA < MSE. And whether this inequality occurs or not is simply a function of whatever the data are that are used in calculating MSA and MSE. And when it does occur it produces the embarrassment of having a negative estimate of a parameter that, by definition, is positive. Nevertheless, this is a characteristic of ANOVA estimators: they can yield negative estimates. What to do about them is discussed in Chapters 3 and 4, as in Searle (1971).

-ii. Unbiasedness. The estimator in (26) is always unbiased:

$$E(\hat{\boldsymbol{\sigma}}^2) = \mathbf{C}^{-1}E(\mathbf{s}) = \mathbf{C}^{-1}\mathbf{C}\boldsymbol{\sigma}^2 = \boldsymbol{\sigma}^2 .$$

This is the case for all ANOVA estimators. They are unbiased.

Although unbiasedness is a property of estimators that is deemed to have merit in the case of estimating means (e.g., in designed experiments), there are at least two reasons for questioning its merit when estimating variance components. The first is that if the unbaisedness of ANOVA estimators is attractive, using such estimators can nevertheless yield negative estimators of positive parameters, which can be rather awkward, to say the least. Explaining to someone in a subject-matter discipline that we will use a negative estimate of an essentially positive parameter is not easy. Estimators that avoid this embarrassment, even if not unbiased, may therefore be appealing.

A second reason for questioning the merit of unbiasedness stems from the concept underlying it. In the situation of a designed experiment, for example, the concept of unbiasedness is that over many repetitions of exactly the same experiment the average value of the (unbiased) estimator of a parameter would be the parameter itself. The trouble with this is that, when estimating variance components, the data available often do not come from carefully designed and executed experiments, for which many repetitions can be idealized; instead, data for estimating variance components are often voluminous and come from situations where repetition of exactly the same data-gathering process is a totally unrealistic idea; e.g., gathering milk yield from exactly the same sample of, say, 400,000 Holstein cows in New York and Pennsylvania as were available in 1989. Repeated data-gathering can be envisaged but, especially in the case of unbalanced data, not necessarily with the same pattern of unbalancedness nor with the same set of (random) effects in subsequent data sets. Replications of data are not, therefore, just replications of data from the same structure as in an initial data set. Indeed, not only might the whole idea of re-sampling inherent in the idea of unbiasedness be impractical but the data may be so voluminous, 1,500,000 records, say, that one might want to think of a variance component estimate more as a descriptor of those data than as a sample of one from the sampling distribution of the estimator being used. Mean unbiasedness may therefore no longer be pertinent, and replacing it with some other criterion might be considered. Modal unbiasedness is one possibility, suggested by Searle (1968, discussion), although Harville (1969b) doubts if modally unbiased estimators exist and questions the justification of such a criterion on decision-theoretic grounds. Nevertheless, as Kempthorne (1968) points out, mean unbiasedness in estimating fixed effects "... leads to residuals which do not contain systematic effects and is therefore valuable ... and is fertile mathematically in that it reduces the class of candidate statistics (or estimates)". However, "... in the variance component problem it does not lead to a fertile smaller class of statistics". Unbiasedness is therefore, in our opinion, not necessarily a property of variance components estimators that should be slavishly accepted as meritorious. We say this at this juncture, in the midst of this brief history, because unbiasedness appears so often in the development of methods of estimating variance components that we feel that comments about its lack of merit in this context deserve to be mentioned early on.

-iii. ***Best unbiasedness.*** The general property of an estimator being best unbiased is that among all unbiased estimators of a parameter that which has minimum variance is called best unbiased. For balanced data Graybill (1954) investigated sampling variances of ANOVA estimators of variance components, and for the general k-fold nested, random-effects model showed that in the class of quadratic functions of the observations that are unbiased estimators of variance components, ANOVA estimators have minimum variance; i.e., ANOVA estimators are *b*est *q*uadratic *u*nbiased *e*stimators (BQUE). With the added assumption of normality, Graybill and Wortham (1956) showed for any random model (with balanced data) that ANOVA estimators are unbiased functions of jointly complete sufficient statistics, and therefore by the Lehmann–Scheffé Theorem (Casella and Berger, 1990, p. 344) they are uniformly *b*est *u*nbiased *e*stimators (BUE); that is, in the class of all unbiased estimators (as distinct from just the quadratic unbiased subclass), ANOVA estimators under normality have minimum variance; i.e., they are BUE. As well as reiterating the latter result, Graybill and Hultquist (1961) extended Graybill (1954) to apply to all models; namely, without any distributional assumptions at all (save a fully random model and balanced data), ANOVA estimators are BQUE. The same results for mixed models were established by Albert (1976). Thus ANOVA estimators from balanced data are BQUE, and they are BUE under normality, whether the underlying model is a mixed model or a random model. Anderson (1978) rightly notes that such or kindred optimality properties have yet to be demonstrated for mixed models that include a covariate, which is not surprising because the presence of covariates effectively converts balanced data into unbalanced data.

In contrast to balanced data, variance component estimators that are uniformly best do not exist in the case of unbalanced data. The essential problem is well summarized by Scheffé (1959; Sec. 7.2): Although the ANOVA

> procedure is commonly used also in the unbalanced cases, it loses there the intuitive justification it has for this writer. At the present writing, the "best" tests and estimates in the unbalanced cases of random-effects models and mixed models are not known, even in a rough intuitive sense. The basic trouble is that the distribution theory gets so much more complicated. We have nothing to offer the reader on the unbalanced cases outside the fixed-effects models except for some results for the completely nested cases in Sec. 7.6.

And as a footnote to the penultimate sentence of the preceding quotation, Scheffé adds

> In the one-way layout, for example, there are three unknown parameters, μ, σ_A^2, and σ_e^2. In the case of balance the (minimal) number of (real) sufficient statistics

is three; in the case of unbalanced it is greater. The sum of squares between groups, $\Sigma_i w_i(y_i - \bar{y}.)^2$, where $\bar{y}. = \Sigma_i w_i y_i/\Sigma_i w_i$, is not distributed as a constant times a non-central chi-square, no matter what (known) weights $w_i > 0$ are used. There is no unbiased quadratic estimate of σ_A^2 of uniformly minimum variance, etc.

An interesting omission of Scheffé's is that of not citing Henderson (1953), especially since in his Sec. 7.6 (referred to above) he in fact uses Henderson's Method I [the same procedure as that of Ganguli (1941), also uncited] in estimating variance components for the random effects nested model with unbalanced data. Moreover, Scheffé did not discuss in detail estimating variance components from unbalanced data with mixed models. A remark from the preface of his book is revealing:

> What I feel most apologetic about is the little I have to offer the reader on the unbalanced cases for the random-effects models and mixed models. They cannot be generally avoided in planning biological experiments, especially in genetics, the situation being unlike that in physical science.

This promotes the question as to what prompted his reference to genetics and thus why was there no reference to Henderson (1953) of six years prior to Scheffé (1959). The earlier book, Anderson and Bancroft (1952), had dealt with the random effects nested model with unbalanced data; in that, not only was the work of Ganguli (1941) clearly outlined (Sec. 22.4), but so too was that of Cochran (1939).

-iv. ***Minimal sufficient statistics.*** For balanced data, minimal sufficient statistics for a random model are, on the basis of normality assumptions, the arithmetic mean of the data and the sums of squares of the analysis of variance. The ANOVA estimators of variance components, being linear functions of those sums of squares, are (with $\bar{y}$) therefore minimal sufficient statistics. They are also complete. These properties of ANOVA estimators were first derived by Graybill and Wortham (1956). Details for the 1-way and 2-way crossed classifications, and for several nested classifications (all with balanced data) are available in Graybill (1976, Chapter 15); see also Hultquist and Graybill (1965).

For unbalanced data, the situation is much more difficult because, even under the usual normality assumptions, for the sums of squares "the distribution theory gets so much more complicated", as Scheffé (1959) says, and there are more minimal sufficient statistics than there are variance components. This is commensurate with the general lack of uniqueness of the ANOVA method for unbalanced data.

-v. ***Lack of uniqueness.*** We have already mentioned that the three Henderson methods are simply three sets of possible sums of squares that can be used as elements of **s** in (25) and (26). Indeed, there is even greater generality in being able to use not just sums of squares as elements of **s** but also a limitless range of quadratic forms of the observations (which includes sums of squares,

of course). This is so because if $\mathbf{q}$ is a vector of quadratic forms such that

$$E(\mathbf{q}) = \mathbf{B}\boldsymbol{\sigma}^2, \tag{27}$$

then, if $\mathbf{B}$ is non-singular, (27) yields $\hat{\boldsymbol{\sigma}}^2 = \mathbf{B}^{-1}\mathbf{q}$ as an unbiased estimator of $\boldsymbol{\sigma}^2$, just like (26).

Even more generality can be introduced. Suppose we use more elements in $\mathbf{q}$ than there are elements in $\boldsymbol{\sigma}^2$. Then, provided the form of (27) still applies, but with $\mathbf{B}$ having full column rank,

$$\boldsymbol{\sigma}^2 = (\mathbf{B}'\mathbf{B})^{-1}\mathbf{B}'\mathbf{q} \tag{28}$$

is an ANOVA estimator of $\boldsymbol{\sigma}^2$. It is unbiased, too. And it is $\mathbf{B}^{-1}\mathbf{q}$ when $\mathbf{B}$ is non-singular.

So there is a broad array of specific uses of the ANOVA method of estimating variance components. If the resulting estimates were invariant to what one used as elements of $\mathbf{q}$, there would be no problem of a lack of uniqueness about ANOVA methodology. But this is not so for unbalanced data. In broad terms this situation does not arise with balanced data because analysis of variance sums of squares used in $\mathbf{q}$ have been shown (see Sec. 2.3b-ii) to have attractive properties. But with unbalanced data, the lack of uniqueness is a real problem. It is avoided in Henderson's Methods I and II, but only by definition, since Method II uses Method I and Henderson (1948) specifically defined his Method I to be that procedure which utilizes "analogous sums of squares" (analogous to those used with balanced data). But it does arise in Method III, and this has brought criticism of Henderson's methods, as has the complete absence of any criteria for deciding which of the three methods is optimal in any sense. An example of this criticism is that of Rao (1971b):

> Essentially, analysis of variance techniques are used but the theoretical basis is not clear. The procedures suggested are *ad hoc* in nature and much seems to depend on intuition. No general method is put forward to cover all experimental situations and, where alternative methods are suggested, no principle is laid down for choosing one among them as appropriate in a given problem.

Blischke's (1968) phrase "methods of a basically *ad hoc* nature" refers to methods more general than Henderson's but certainly includes them. And the label is appropriate, for *any* use of the ANOVA method, because the method can be applied to almost any quadratic function of the observations. Thus in Example 4 of Chapter 1, one naïve application of (27) is

$$E\ (y_{1,3} - y_{1,7})^2 = 2\sigma_e^2 \quad \text{and} \quad E\ (y_{1,3} - y_{2,5})^2 = 2(\sigma_\alpha^2 + \sigma_e^2)\,.$$

That, like each of the Henderson methods and like any other application of the ANOVA method, yields estimators that are unbiased, but having no general analytic properties that can be used to determine relative optimality of any one application of the general ANOVA method over another. There are some features of the Henderson methods that condition their applicability to certain models (as shall be discussed later), but none of them really contribute to the difficulty of being unable to judge relative optimality of the different applications.

2.4. INTO THE 1970S AND BEYOND

The realization that ANOVA estimation had serious weaknesses was slow to dawn. Not only did Henderson's Method I provide a procedure for estimating variance components from unbalanced data, where none had been previously available, but it was also reasonably computable for those pre-computer days—at least when judged by the standards of computing feasibility of those days. Method II was a little more difficult (see, e.g., Henderson, Searle and Schaeffer, 1974), and Method III was almost totally impractical from the computing point of view. Nevertheless, whatever computability considerations there were, the weaknesses of ANOVA estimators remained: negativity, lack of distributional properties and no useful way to compare different applications of ANOVA methodology. In light of these weaknesses it was natural that an alternative would be sought, and so maximum likelihood estimation duly came to be considered.

a. Maximum likelihood (ML)

Estimation by ML demands attributing a distribution to the data, which, in the case of random and mixed models, suggests doing just that for the random effects. This is, of course, not a requirement of ANOVA estimation, other than requiring finite variance components and, as in (25), that $E(\mathbf{q})$ contain no terms in the fixed effects.

To date, nearly all closed-form results for ML estimation of variance components are on the basis of normality assumptions: e.g., for the 1-way classification of (8)–(11), that the random effects have the first- and second-moment properties well defined, and are additionally taken as being normally distributed. It is under these conditions, and their direct extension to multi-way classification, that the development of ML methodology has proceeded.

The beginning appears to lie with Crump (1947, 1951), who dealt with the 1-way classification for both balanced and unbalanced data, in the latter case deriving equations that have to be solved iteratively. Herbach (1959) derived explicit maximum likelihood (ML) estimators for certain balanced data models and took account of the necessity that such estimators must be non-negative (because the method of maximum likelihood prescribes maximization over the parameter space—and variance components are non-negative). Corbeil and Searle (1976b) summarize a number of these balanced data cases, showing their biases and sampling variances.

The landmark paper for ML estimation in general is Hartley and J. N. K. Rao (1967), wherein a methodology is developed for a very wide class of models: all mixed and random models, with or without covariates, balanced or unbalanced data. One may wonder why there was a delay of some forty or so years between Fisher's (1922, 1925) derivation of the method of maximum likelihood and its general application to the estimation of variance components: undoubtedly it was the matrix specification of a mixed model that Hartley and Rao (1967) used that was instrumental to their deriving ML equations for the general case. Solving those equations for a data set, and calculating ML estimates, involves iterative calculations on the ML equations—and for some years this was an impediment to any widespread use of ML estimation of variance components. Computing methods have to be able to deal with sparse matrices of very large dimension, with equations that are very non-linear, with iterative procedures that lead to a global rather than a local maximum, and with adapting those procedures to take account of the ultimate non-negativity of the estimates. Fortunately, with the advent of supercomputers and the development of new computing packages (e.g., Thompson, 1980; Giesbrecht, 1983, 1985), these problems are getting to be circumvented.

Miller (1973, 1977) also worked on ML estimation, dealing with both balanced and unbalanced data. For the 2-way classification, random model, with or without interaction, he showed very explicitly that the maximum likelihood equations can be written with (relatively) disarmingly looking simplicity, but that they cannot be solved analytically. Miller also looked at asymptotic properties of the estimators; and Searle (1970) derived an expression for the large-sample dispersion matrix of ML estimators in the general unbalanced data case.

b. Restricted maximum likelihood (REML)

W. A. Thompson (1962) also considered ML estimation, and it was he who introduced the idea of maximizing that part of the likelihood which is invariant to the location parameters of the model; i.e., to the fixed effects. This has now come to be known as restricted maximum likelihood (REML), and is sometimes called marginal (or, in Europe, residual) maximum likelihood. It was put on a broad basis for unbalanced data by Patterson and R. Thompson (1971). The computational difficulties of ML are also equally as pertinent to REML as to ML, since REML methodology is effectively (see Harville, 1977) no more than ML on certain linear combinations of the data rather than on the data themselves. One of the interesting features of REML is that for balanced data, solutions to REML equations are identical to ANOVA estimators. Also, the REML methodology takes account of the implicit degrees of freedom associated with the fixed effects, whereas ML does not. ML and REML are coming to be the preferred method of estimation, especially from unbalanced data.

c. Minimum norm estimation

Attempts at finding minimum variance quadratic unbiased estimators of variance components (an analogue of best linear unbiased estimation of the mean in linear models) began with Townsend (1968), Harville (1969a) and Townsend and Searle (1971). This was quickly followed by LaMotte's (1970, 1971, 1973a,b, 1976) work on minimum variance estimation and C. R. Rao's (1970, 1971a,b, 1972) papers on minimum-norm quadratic unbiased equation

(MINQUE). The resulting estimators have, in some broad sense, a minimized generalized variance, stemming from the minimizing of a Euclidean norm, which, under normality, equates to a minimum variance property.

MINQUE estimation demands no distributional properties of the random effects or error terms in the model. Nor does it involve iteration, just the solution of linear equations. However, estimators obtained by MINQUE are functions of *a priori* values used in place of the variance components in the estimation procedure itself. Thus the MINQUE procedure has what we deem to be a serious deficiency: the minimality property applies only at those *a priori* values. It also has the feature that from the same data set and the same model, N different people, each with their own set of *a priori* values, could yield N different sets of estimators. Nevertheless, no matter what the *a priori* values are, MINQUE estimators are unbiased.

For a given set of *a priori* values, the MINQUE equations are linear in the variance component estimators and can thus be solved without iteration. But the presence of the *a priori* values suggests iterating on those equations using successive solutions as *a priori* values. The resulting solutions, once convergence is reached, are called I-MINQUE estimates. They are the same as REML estimates (Hocking and Kutner, 1975), and under large sample theory are normally distributed (Brown, 1976). Similarly, any MINQUE estimate is the same as a first-round iterate from REML, using *a priori* values needed for MINQUE as the starting values for REML iteration. These connections of MINQUE to REML add weight, we feel, to our opinion that MINQUE is not a practical method of estimating variance components. Readers who disagree with us are referred to Rao and Kleffe (1988), a book that is devoted almost entirely to MINQUE. And we do briefly describe the method in Section 11.3d.

d. The dispersion-mean model

Consider a vector having elements that are all the squares of, and products two-at-a-time of, the observations. A particular variant of that vector was shown by Pukelsheim (1976) to have expected value that can be expressed as a set of linear combinations of the variance components. In this way one has a linear model with the vector of variance components being the parameters to be estimated. It is called the dispersion-mean model and is described in Chapter 12. Generalized least squares applied to this dispersion-mean model yields MINQUE, and applied to a mild variation of the model it yields ML (Anderson, 1978). Brown (1978), using a vector of residuals, also developed MINQUE in a similar way.

e. Bayes estimation

Estimation of variance components using Bayesian principles is found in Hill (1965, 1967), who dealt with balanced data from the 1-way classification model. So did numerous other workers, followed thereafter by similar work on the 2-way classification, both nested and crossed; see Khuri and Sahai (1985, pp. 283–284). As those authors write (p. 290), "There have been only a few published papers on ... unbalanced models", i.e., unbalanced data. Gnot and

Kleffe (1983) is another good paper on this topic. We offer Chapter 9 on this topic.

f. The recent decade

In contrast to 1940–1980 there seems to have been only one major development of a new methodology for estimating variance components over the last ten years or so. This is the work of Smith and Murray (1986) and of Hocking *et al.* (1989) for balanced data [and Green (1988) for unbalanced data], who formulate variance components as covariances and then use the ANOVA procedure of equating quadratic forms of the data to their expected values. This formulation is described in Section 11.2.

But, as opposed to new estimation procedures, there has been work in a variety of other topics. With the plethora of methods already available for unbalanced data, one emphasis has been the attention given to comparing different methods, mostly by the use of relatively small sets of simulated data. The papers range from Townsend and Searle (1971), for the 1-way classification without an overall mean, to Swallow and Monahan (1984). Their results "indicate that unless data are severely unbalanced and $\sigma_\alpha^2/\sigma_e^2 > 1$, ANOVA estimators are adequate" (Khuri and Sahai, 1985, p. 291). Comparisons have also been made for the 2-way crossed classification models (Corbeil and Searle, 1976b) and the split-plot design (Li and Klotz, 1978). Generally speaking, maximum likelihood is the favoured methodology in these studies: or perhaps REML is even more favored.

A second topic that has attracted research is that of designing experiments so that variance components can be estimated with some optimal properties. This has long been an interest of R. L. Anderson who, along with co-workers, has published a series of papers on the subject dating from Bush and Anderson (1963) and Anderson (1975) to Muse, Anderson and Thitakamol (1982). Khuri and Sahai (1985) provide an extensive collection of references (many of them by Anderson's students) and a delightfully clear survey of them.

Another matter of current interest is estimating variance components from discrete data, of which binary data are an important case. Chapter 10 describes methods for doing this.

Developing confidence intervals for variance components and for functions of them has attracted considerable interest in recent years, especially for F. A. Graybill and colleagues. Some of the earliest work is that of Satterthwaite (1941). For the 1-way classification, random model, with balanced data, a summary of confidence intervals for the variance components and some ratios of them is given in Searle (1971, Table 9.14), and a comprehensive survey of numerous papers on the subject is given in Khuri and Sahai (1985). They have a similar account for unbalanced data, ranging from Wald (1940) to the comprehensive review of Burdick and Graybill (1984); and a more recent survey is Burdick and Graybill (1988).

Finally, a current topic of great importance is that of successful computing procedures for calculating ML and REML estimates. Some of the difficulties involved are listed in Section 6.4, and further details are given in Chapter 8.

CHAPTER 3

THE 1-WAY CLASSIFICATION

The collecting of patient data from 15 clinics discussed as Example 4 in Chapter 1 is an example of a 1-way classification: clinics are the only way of classifying the data. This chapter deals with the 1-way classification more generally, introducing *inter alia* many topics concerning variance components that re-occur in subsequent chapters in more complicated situations and with more detail than is needed here. So, as well as dealing with the 1-way classification in its own right, this chapter also introduces a variety of topics dealt with in depth in subsequent chapters.

3.1. THE MODEL

Describing the random model for the 1-way classification is somewhat repetitious of some of Section 1.3b, but it is done for the sake of completeness. The situation envisaged is that of having data that are grouped by classes, those classes being considered a random sample from some population of classes. The model equation that shall be used is

$$y_{ij} = \mu + \alpha_i + e_{ij}, \tag{1}$$

where y_{ij} is the jth observation in the ith class, μ is a general mean, α_i is the effect on the y-variable of its being observed on an observational unit that is in the ith class, and e_{ij} is a residual error. The number of classes in the data shall be denoted by a, and the number of observations in the ith class by n_i. Thus $i = 1, 2, \ldots, a$ and $j = 1, 2, \ldots, n_i$, for $n_i \geqslant 1$. For balanced data there is the same number of observations in every class, n say, so that $n_i = n$ for every class, i.e., $n_i = n \ \forall \ i$.

a. The model equation

In the fixed effects model of Section 1.3a, both μ and α_i are taken as fixed constants and the starting point is to assume $E(y_{ij}) = \mu + \alpha_i$. Then e_{ij} is defined as $e_{ij} = y_{ij} - E(y_{ij})$, from which $y_{ij} = \mu + \alpha_i + e_{ij}$.

For the random model we must take account of α_i being a random variable. To do so, we first assume that

$$E_1(\alpha_i) = 0 . \tag{2}$$

where E_1 represents expectation over the population of αs. Lest it be thought that (2) implies some loss of generality, the reader is referred to the paragraph following (15) in Chapter 1.

Now consider some particular class, and label it the ith class. Its n_i observations, y_{ij} for $j = 1,\ldots, n_i$, are considered to be a random sample from that class. Then, for E_2 representing expectation over repeated sampling from class i, the expected value of y_{ij} for that class is $\mu + \alpha_i$. We denote this by the conditional expected value.

$$E_2(y_{ij}|\alpha_i) = \mu + \alpha_i . \tag{3}$$

Then, analogously to defining $e_{ij} = y_{ij} - E(y_{ij})$ in the fixed effects model, we define e_{ij} for the random model as

$$e_{ij} = y_{ij} - E_2(y_{ij}|\alpha_i) = y_{ij} - (\mu + \alpha_i) . \tag{4}$$

This gives the model equation

$$y_{ij} = \mu + \alpha_i + e_{ij} . \tag{5}$$

b. First moments

From the definition of e_{ij} in (4)

$$E_2(e_{ij}|\alpha_i) = E_2(y_{ij}|\alpha_i) - E_2(y_{ij}|\alpha_i) = 0, \tag{6}$$

and on using E to represent expectation over repeated sampling from class i and E_1 for expectation over all classes,

$$E(e_{ij}) = E_1 E_2(e_{ij}|\alpha_i) = 0 . \tag{7}$$

Similarly, using (5) and (6),

$$E_2(y_{ij}|\alpha_i) = E_2(\mu + \alpha_i + e_{ij}|\alpha_i) = \mu + \alpha_i,$$

which is our starting point (3). And on using (2)–(7),

$$E(y_{ij}) = E_1 E_2(y_{ij}|\alpha_i) = E_1(\mu + \alpha_i) = \mu . \tag{8}$$

c. Second moments

The first moments of (2), (7) and (8) are either definitions or direct consequences of definitions. But those definitions produce no comparable results for second moments. In contrast we have to attribute second-moment properties to the α_is and the e_{ij}s. Insofar as covariances are concerned, it is usual in random models to define all covariances as zero:

$$\text{cov}(e_{ij}, e_{i'j'}) = 0 \quad \text{except for } i = i' \text{ and } j = j' . \tag{9}$$

This means that the covariance between every pair of different e_{ij} terms is zero; similarly, for the α_i terms,

$$\text{cov}(\alpha_i, \alpha_{i'}) = 0 \quad \forall\ i \neq i'; \tag{10}$$

and likewise for the covariance of each α_i with every e_{ij}:

$$\text{cov}(\alpha_i, e_{i'j'}) = 0 \quad \forall\ i, i' \text{ and } j' . \tag{11}$$

Whenever stochastic independence of the e_{ij}s, of the α_is, and of the α_is and e_{ij}s is assumed, these zero covariances are, of course, a direct consequence of those independencies. Conversely, on assuming normality of the α_is and the e_{ij}s (usually just called the "normality assumptions"), these zero covariances imply independence.

Now consider (9) and (10) for $i = i'$ and $j = j'$. These lead to variances, defined as follows:

$$\text{var}(e_{ij}) = \sigma_e^2 \quad \forall\ i \text{ and } j, \quad \text{and} \quad \text{var}(\alpha_i) = \sigma_\alpha^2 \quad \forall\ i . \tag{12}$$

These variances, σ_e^2 and σ_α^2, are called *variance components* because they are the components of the variance of an observation:

$$\sigma_y^2 = \text{var}(y_{ij}) = \text{var}(\mu + \alpha_i + e_{ij}) = \sigma_\alpha^2 + \sigma_e^2 . \tag{13}$$

Note also, starting from the definition of variance and covariance, and using $E(e_{ij}) = 0$ and $E(\alpha_i) = 0$, that

$$\begin{aligned} \sigma_e^2 &= \text{var}(e_{ij}) = E[e_{ij} - E(e_{ij})]^2 = E(e_{ij}^2), \\ \sigma_\alpha^2 &= \text{var}(\alpha_i) = E(\alpha_i^2), \end{aligned} \tag{14}$$

$$\begin{aligned} \text{cov}(\alpha_i, \alpha_{i'}) &= E(\alpha_i \alpha_{i'}) = 0 \quad \forall\ i \neq i', \\ \text{cov}(\alpha_i, e_{i'j}) &= E(\alpha_i e_{i'j}) = 0 \quad \forall\ i \text{ and } i', \\ \text{cov}(e_{ij}, e_{i'j'}) &= 0 \quad \text{except for } i = i' \text{ and } j = j' . \end{aligned} \tag{15}$$

Moreover, although α_i and e_{ij} are uncorrelated, the y_{ij}s are not. For those in the same class

$$\text{cov}(y_{ij}, y_{ij'}) = \text{cov}(\mu + \alpha_i + e_{ij}, \mu + \alpha_i + e_{ij'}) = \sigma_\alpha^2 \quad \text{for } j \neq j',$$

whereas for those in different classes

$$\text{cov}(y_{ij}, y_{i'j'}) = \text{cov}(\mu + \alpha_i + e_{ij}, \mu + \alpha_{i'} + e_{i'j'}) = 0 \quad \text{for } i \neq i' .$$

Equations (2), (3), (4) and (9)–(12) specify the usual random model. Although these details have been given as applying to the 1-way classification, they are, in fact, the definitions and assumptions used in most variance components models. That is, any random effect in most such models usually has attributed to it the same properties as have been given for the α_i in (2), (10), (11) and (12), namely zero mean, zero covariances with each other and with residual terms, and homoscedastic variances. Also, when there is more than one random factor, covariances of effects of one factor with those of another

TABLE 3.1. AN EXAMPLE OF THE 1-WAY CLASSIFICATION
(3 CLASSES WITH 4 OBSERVATIONS EACH)

	Data y_{ij}				Total	Mean
Class	y_{i1}	y_{i2}	y_{i3}	y_{i4}	$y_{i\cdot}$	$\bar{y}_{i\cdot}$
$i = 1$	3	3	12	2	20	5
$i = 2$	11	13	17	7	48	12
$i = 3$	4	2	1	33	40	10
			Grand total, $y_{\cdot\cdot} = 108$			$9 = \bar{y}_{\cdot\cdot}$ = grand mean

are also usually taken as zero. These properties are used extensively in all that follows, with little further mention of the details shown here.

3.2. MATRIX FORMULATION OF THE MODEL

A matrix formulation of the model is introduced by means of an example.

a. Example 1

Suppose we have 4 observations on each of 3 classes, as in Table 3.1. The model equations (1) for the observations in Table 3.1 are

$$\begin{bmatrix} y_{11} \\ y_{12} \\ y_{13} \\ y_{14} \\ \hdashline y_{21} \\ y_{22} \\ y_{23} \\ y_{24} \\ \hdashline y_{31} \\ y_{32} \\ y_{33} \\ y_{34} \end{bmatrix} = \begin{bmatrix} 3 \\ 3 \\ 12 \\ 2 \\ \hdashline 11 \\ 13 \\ 17 \\ 7 \\ \hdashline 4 \\ 2 \\ 1 \\ 33 \end{bmatrix} = \begin{bmatrix} 1 & 1 & \cdot & \cdot \\ 1 & 1 & \cdot & \cdot \\ 1 & 1 & \cdot & \cdot \\ 1 & 1 & \cdot & \cdot \\ \hdashline 1 & \cdot & 1 & \cdot \\ 1 & \cdot & 1 & \cdot \\ 1 & \cdot & 1 & \cdot \\ 1 & \cdot & 1 & \cdot \\ \hdashline 1 & \cdot & \cdot & 1 \\ 1 & \cdot & \cdot & 1 \\ 1 & \cdot & \cdot & 1 \\ 1 & \cdot & \cdot & 1 \end{bmatrix} \begin{bmatrix} \mu \\ \alpha_1 \\ \alpha_2 \\ \alpha_3 \end{bmatrix} + \begin{bmatrix} e_{11} \\ e_{12} \\ e_{13} \\ e_{14} \\ \hdashline e_{21} \\ e_{22} \\ e_{23} \\ e_{24} \\ \hdashline e_{31} \\ e_{32} \\ e_{33} \\ e_{34} \end{bmatrix} \qquad (16)$$

where the vectors and matrix have been partitioned corresponding to the three classes in the data, and a dot as an element of a matrix represents zero. Denote by **y** and **e** the vectors of observations and residual errors, respectively, in (16).

Also define

$$\boldsymbol{\alpha} = [\alpha_1 \quad \alpha_2 \quad \alpha_3]' \quad \text{and} \quad \boldsymbol{\beta} = [\mu \quad \alpha_1 \quad \alpha_2 \quad \alpha_3]' = [\mu \quad \boldsymbol{\alpha}']' . \tag{17}$$

Then (16) is

$$\mathbf{y} = \mathbf{X}\boldsymbol{\beta} + \mathbf{e}$$

for

$$\mathbf{X}\boldsymbol{\beta} = \begin{bmatrix} 1 & 1 & \cdot & \cdot \\ 1 & 1 & \cdot & \cdot \\ 1 & 1 & \cdot & \cdot \\ 1 & 1 & \cdot & \cdot \\ 1 & \cdot & 1 & \cdot \\ 1 & \cdot & 1 & \cdot \\ 1 & \cdot & 1 & \cdot \\ 1 & \cdot & 1 & \cdot \\ 1 & \cdot & \cdot & 1 \\ 1 & \cdot & \cdot & 1 \\ 1 & \cdot & \cdot & 1 \\ 1 & \cdot & \cdot & 1 \end{bmatrix} \begin{bmatrix} \mu \\ \alpha_1 \\ \alpha_2 \\ \alpha_3 \end{bmatrix} = \begin{bmatrix} 1 \\ 1 \\ 1 \\ 1 \\ 1 \\ 1 \\ 1 \\ 1 \\ 1 \\ 1 \\ 1 \\ 1 \end{bmatrix} \mu + \begin{bmatrix} 1 & \cdot & \cdot \\ 1 & \cdot & \cdot \\ 1 & \cdot & \cdot \\ 1 & \cdot & \cdot \\ \cdot & 1 & \cdot \\ \cdot & 1 & \cdot \\ \cdot & 1 & \cdot \\ \cdot & 1 & \cdot \\ \cdot & \cdot & 1 \\ \cdot & \cdot & 1 \\ \cdot & \cdot & 1 \\ \cdot & \cdot & 1 \end{bmatrix} \begin{bmatrix} \alpha_1 \\ \alpha_2 \\ \alpha_3 \end{bmatrix} . \tag{18}$$

We now utilize the summing vector $\mathbf{1}_k = [1 \quad 1 \quad \ldots \quad 1]'$ of k elements 1 and in doing so introduce the reader to Appendix M following Chapter 12. It contains a variety of definitions and reminders about matrix algebra. Equation (18) can then be rewritten as

$$\mathbf{X}\boldsymbol{\beta} = \mathbf{1}_{12}\mu + \begin{bmatrix} \mathbf{1}_4 & \cdot & \cdot \\ \cdot & \mathbf{1}_4 & \cdot \\ \cdot & \cdot & \mathbf{1}_4 \end{bmatrix} \boldsymbol{\alpha} . \tag{19}$$

By expressing $\mathbf{1}_{12}$ and the 12×3 matrix as direct products (Appendix M.2), the model equation becomes

$$\mathbf{y} = (\mathbf{1}_3 \otimes \mathbf{1}_4)\mu + (\mathbf{I}_3 \otimes \mathbf{1}_4)\boldsymbol{\alpha} + \mathbf{e} . \tag{20}$$

b. The general case

Appendix M.3 introduces new notation for writing $\mathbf{A}$ of order $r \times c$, namely $\{a_{ij}\}$ for $i = 1, \ldots, r$ and $j = 1, \ldots, c$, where a_{ij} is the element in row i and column j of $\mathbf{A}$. It is

$$\mathbf{A} = \{_{\text{m}}\, a_{ij}\}_{i=1,\, j=1}^{r \quad c} = \{_{\text{m}}\, a_{ij}\}_{ij} = \{_{\text{m}}\, a_{ij}\} ,$$

with the i and j and/or their ranges being omitted for brevity provided context permits. The m indicates that the elements a_{ij} are arrayed as a matrix, and by the use of r, c and d one similarly represents rows, columns and diagonal matrices; e.g., $\{_\mathrm{r}\, b_j\}_{j=1}^{k}$ is a $1 \times k$ row vector. This notation is useful because it can be used operationally without having to give a matrix symbol to every matrix involved. For example,

$$\{_\mathrm{m}\, a_{ij}\}\{_\mathrm{c}\, t_j\} = \{_\mathrm{c}\, \Sigma_j a_{ij} t_j\}_i$$

avoids having to write "$\mathbf{A}$ is a matrix $\{a_{ij}\}$ and $\mathbf{t}$ is a column of elements t_j and therefore $\mathbf{At}$ is a column of elements $\Sigma_j a_{ij} t_j$."

We now use this notation to define vectors of observations and error terms, respectively, as

$$\mathbf{y} = \{_\mathrm{c}\, \{_\mathrm{c}\, y_{ij}\}_{j=1}^{n_i}\}_{i=1}^{a} = \{_\mathrm{c}\, y_{ij}\}_{j=1,\, i=1}^{n_i,\quad a} \tag{21}$$

and

$$\mathbf{e} = \{_\mathrm{c}\, \{_\mathrm{c}\, e_{ij}\}_{j=1}^{n_i}\}_{i=1}^{a} = \{_\mathrm{c}\, e_{ij}\}_{j=1,\, i=1}^{n_i,\quad a}, \tag{22}$$

in each of which the elements are arranged in lexicon order, ordered by j within i. Then with $\boldsymbol{\alpha}$ defined as $\boldsymbol{\alpha} = [\alpha_1 \quad \alpha_2 \quad \ldots \quad \alpha_a]'$ for the general case of a classes, the model equations for n observations in each class are, like (20),

$$\mathbf{y} = (\mathbf{1}_a \otimes \mathbf{1}_n)\mu + (\mathbf{I}_a \otimes \mathbf{1}_n)\boldsymbol{\alpha} + \mathbf{e}\,. \tag{23}$$

Searle and Henderson (1979) and Anderson *et al.* (1984) use extensions of this formulation for multi-way classifications to develop a variety of properties of random models. It has also been used by many other writers: e.g., Seifert (1981) and Smith and Murray (1984).

A distinction between balanced and unbalanced data (see Section 1.2) must be noted. Although the example has balanced data (4 observations in each class), the definitions in (21) and (22) provide for unbalanced data (n_i observations in class i). But with unbalanced data, the direct product formulation of (23) does not exist because, for example, the diagonal terms $\mathbf{1}_4$ of (19) will no longer be all the same. Thus, if $n_1 = 3$, $n_2 = 4$ and $n_3 = 2$, those terms would be $\mathbf{1}_3$, $\mathbf{1}_4$ and $\mathbf{1}_2$, and this would not permit of a direct product multiplying $\boldsymbol{\alpha}$ in (20). (See Section 3.2d, and Exercise E 3.1.)

c. Dispersion matrices

-i. The traditional random model. The dispersion (variance–covariance) matrices of $\mathbf{y}$, $\boldsymbol{\alpha}$ and $\mathbf{e}$ are from (9) and (12)

$$\mathrm{var}(\mathbf{e}) = \sigma_e^2 \mathbf{I}_{an}; \tag{24}$$

and similarly from (10) and (12)

$$\mathrm{var}(\boldsymbol{\alpha}) = \sigma_\alpha^2 \mathbf{I}_a\,. \tag{25}$$

Then from (9)–(12) and (23)

$$\begin{aligned}\mathbf{V} = \operatorname{var}(\mathbf{y}) &= (\mathbf{I}_a \otimes \mathbf{1}_n)\sigma_\alpha^2 \mathbf{I}_a (\mathbf{I}_a \otimes \mathbf{1}_n)' + \sigma_e^2 \mathbf{I}_{an} \\ &= \sigma_\alpha^2(\mathbf{I}_a \otimes \mathbf{J}_n) + \sigma_e^2(\mathbf{I}_a \otimes \mathbf{I}_n) = \mathbf{I}_a \otimes (\sigma_\alpha^2 \mathbf{J}_n + \sigma_e^2 \mathbf{I}_n)\,. \end{aligned} \tag{26}$$

These forms of dispersion matrices arise directly from the variance–covariance structures attributed in (9)–(12) to the random α_is and e_{ij}s in the traditional form of the random and mixed models. But, although they are the structures most frequently employed and to which most of this book is therefore directed, they are by no means the only structures that could be envisaged. The possibilities are almost endless. We show but three in the following paragraph.

-ii. Other alternatives. First, although forms of var($\boldsymbol{\alpha}$) and var($\mathbf{e}$) other than (24) and (25) are sometimes employed, one property of $\boldsymbol{\alpha}$ and $\mathbf{e}$ that is almost universally adopted is to take $\operatorname{cov}(\alpha_i, e_{kj}) = 0$ for all i, j and k, as in (11). This gives

$$\operatorname{cov}(\boldsymbol{\alpha}, \mathbf{e}') = \mathbf{0}, \quad \text{of order } a \times an, \tag{27}$$

the orders of $\boldsymbol{\alpha}$ and $\mathbf{e}'$, namely a and an, respectively, determining the order of $\operatorname{cov}(\boldsymbol{\alpha}, \mathbf{e}')$. But for var($\boldsymbol{\alpha}$) there may be situations when adopting

$$\operatorname{cov}(\alpha_i, \alpha_{i'}) = \rho\sigma_\alpha^2 \quad \text{for } i \neq i'$$

is reasonable. This gives, for $a = 5$ (for ease of illustration),

$$\operatorname{var}(\boldsymbol{\alpha}) = \sigma_\alpha^2 \begin{bmatrix} 1 & \rho & \rho & \rho & \rho \\ \rho & 1 & \rho & \rho & \rho \\ \rho & \rho & 1 & \rho & \rho \\ \rho & \rho & \rho & 1 & \rho \\ \rho & \rho & \rho & \rho & 1 \end{bmatrix},$$

with its general form being

$$\operatorname{var}(\boldsymbol{\alpha}) = \sigma_\alpha^2[(1-\rho)\mathbf{I}_a + \rho\mathbf{J}_a]\,. \tag{28}$$

An example of this in animal genetics could be where the classes were sires that were all paternal half-sibs.

In a different context, if time series data are under consideration, with $i = 1, \ldots, a$ representing a series of time intervals, it may be appropriate to adopt either the structure

$$\operatorname{cov}(\alpha_i, \alpha_{i'}) = \begin{cases} \rho\sigma_\alpha^2 & \forall\, i - i' = \pm 1, \\ 0 & \text{otherwise} \end{cases}$$

or

$$\operatorname{cov}(\alpha_i, \alpha_{i'}) = \sigma_\alpha^2 \rho^{|i-i'|} \quad \forall\, i \neq i',$$

in which case (for $a = 5$ again)

$$\text{var}(\boldsymbol{\alpha}) = \sigma_\alpha^2 \begin{bmatrix} 1 & \rho & \cdot & \cdot & \cdot \\ \rho & 1 & \rho & \cdot & \cdot \\ \cdot & \rho & 1 & \rho & \cdot \\ \cdot & \cdot & \rho & 1 & \rho \\ \cdot & \cdot & \cdot & \rho & 1 \end{bmatrix} \quad \text{and} \quad \text{var}(\boldsymbol{\alpha}) = \sigma_\alpha^2 \begin{bmatrix} 1 & \rho & \rho^2 & \rho^3 & \rho^4 \\ \rho & 1 & \rho & \rho^2 & \rho^3 \\ \rho^2 & \rho & 1 & \rho & \rho^2 \\ \rho^3 & \rho^2 & \rho & 1 & \rho \\ \rho^4 & \rho^3 & \rho^2 & \rho & 1 \end{bmatrix},$$

respectively.

Along with any of the above, a variety of possibilities exists for var($\mathbf{e}$). For example, if a model is to provide for different variances within each class, e.g., $\text{var}(e_{ij}) = \sigma_i^2$, then

$$\text{var}(\mathbf{e}) = \begin{bmatrix} \sigma_1^2 \mathbf{I}_n & & & & \\ & \ddots & & & \\ & & \sigma_i^2 \mathbf{I}_n & & \\ & & & \ddots & \\ & & & & \sigma_a^2 \mathbf{I}_n \end{bmatrix},$$

which, with the notation of Appendix M.3, can also be written as

$$\text{var}(\mathbf{e}) = \{_d\, \sigma_i^2 \mathbf{I}_n\}_{i=1}^{a} = \bigoplus_{i=1}^{a} \sigma_i^2 \mathbf{I}_n = \{_d\, \sigma_i^2\}_{i=1}^{a} \otimes \mathbf{I}_n\,.$$

Covariances among e_{ij}s could also be incorporated; for example, the adoption of $\text{cov}(e_{ij}, e_{ij'}) = \rho_i \sigma_i^2 \ \forall\ j \neq j'$ leads to $\sigma_i^2 \mathbf{I}_n$ in the preceding expressions being replaced by $\sigma_i^2[(1 - \rho_i)\mathbf{I}_n + \rho_i \mathbf{J}_n]$, similar to (28). Clearly, the variations are manifold.

d. Unbalanced data

*-i. **Example 2.*** A small example of unbalanced data is shown in Table 3.2. The model equations for these data are, like (16),

$$\begin{bmatrix} y_{11} \\ y_{12} \\ y_{13} \\ \hdashline y_{21} \\ y_{22} \\ y_{23} \\ y_{24} \\ \hdashline y_{31} \\ y_{32} \end{bmatrix} = \begin{bmatrix} 3 \\ 3 \\ 12 \\ \hdashline 11 \\ 13 \\ 17 \\ 7 \\ \hdashline 4 \\ 2 \end{bmatrix} = \begin{bmatrix} 1 & 1 & \cdot & \cdot \\ 1 & 1 & \cdot & \cdot \\ 1 & 1 & \cdot & \cdot \\ \hdashline 1 & \cdot & 1 & \cdot \\ 1 & \cdot & 1 & \cdot \\ 1 & \cdot & 1 & \cdot \\ 1 & \cdot & 1 & \cdot \\ \hdashline 1 & \cdot & \cdot & 1 \\ 1 & \cdot & \cdot & 1 \end{bmatrix} \begin{bmatrix} \mu \\ \alpha_2 \\ \alpha_2 \\ \alpha_3 \end{bmatrix} + \begin{bmatrix} e_{11} \\ e_{12} \\ e_{13} \\ \hdashline e_{21} \\ e_{22} \\ e_{23} \\ e_{24} \\ \hdashline e_{31} \\ e_{32} \end{bmatrix}.$$

TABLE 3.2. UNBALANCED DATA FROM A 1-WAY CLASSIFICATION

Class	Data y_{ij}: y_{i1}	y_{i2}	y_{i3}	y_{i4}	Total $y_{i.}$	n_i	Mean $\bar{y}_{i.}$
$i = 1$	3	3	12		18	3	6
$i = 2$	11	13	17	7	48	4	12
$i = 3$	4	2			6	2	3
					Grand total, $y_{..} = 72$	$n_. = 9$	$8 = \bar{y}_{..}$ = grand mean

These still have the form $\mathbf{y} = \mathbf{X\beta} + \mathbf{e}$, but now $\mathbf{X\beta}$ is

$$\mathbf{X\beta} = \mathbf{1}_9\mu + \begin{bmatrix} \mathbf{1}_3 & \cdot & \cdot \\ \cdot & \mathbf{1}_4 & \cdot \\ \cdot & \cdot & \mathbf{1}_2 \end{bmatrix} \boldsymbol{\alpha} . \tag{29}$$

The matrix multiplying $\boldsymbol{\alpha}$ is still a diagonal matrix of summing vectors, but in contrast to (19) of the balanced data case where those summing vectors all have the same order they now have, for unbalanced data, different orders, namely the numbers of observations in the classes.

-ii. ***The general case.*** Generalizing from (29), the model equation for unbalanced data is, for $N = n_. = \Sigma_i n_i$,

$$\mathbf{y} = \mathbf{1}_N\mu + \{_{\mathrm{d}}\, \mathbf{1}_{n_i}\}_{i=1}^{a}\, \boldsymbol{\alpha} + \mathbf{e} . \tag{30}$$

-iii. ***Dispersion matrix.*** Correspondingly, the dispersion matrix of $\mathbf{y}$ is

$$\begin{aligned} \mathbf{V} = \mathrm{var}(\mathbf{y}) &= \{_{\mathrm{d}}\, \mathbf{1}_{n_i}\}_{i=1}^{a}\, \sigma_\alpha^2 \mathbf{I}_a \{_{\mathrm{d}}\, \mathbf{1}'_{n_i}\}_{i=1}^{a} + \sigma_e^2 \mathbf{I}_N \\ &= \sigma_\alpha^2 \{_{\mathrm{d}}\, \mathbf{J}_{n_i}\}_i + \sigma_e^2 \mathbf{I}_N \\ &= \{_{\mathrm{d}}\, \sigma_\alpha^2 \mathbf{J}_{n_i} + \sigma_e^2 \mathbf{I}_{n_i}\}_i, \end{aligned} \tag{31}$$

where i takes values $i = 1, \ldots, a$. Whenever $n_i = n\ \forall\ i$, the form in (31) does, of course, reduce to (26) for balanced data.

3.3. ESTIMATING THE MEAN

In a fixed effects model represented as $\mathbf{y} = \mathbf{X\beta} + \mathbf{e}$ with $\mathrm{var}(\mathbf{y}) = \mathrm{var}(\mathbf{e}) = \mathbf{V}$, the ordinary least squares estimator (OLSE) of the estimable functions $\mathbf{X\beta}$ is

$$\mathrm{OLSE}(\mathbf{X\beta}) = \mathbf{X}(\mathbf{X}'\mathbf{X})^{-}\mathbf{X}'\mathbf{y} = \mathbf{XX}^{+}\mathbf{y}, \tag{32}$$

where $(\mathbf{X}'\mathbf{X})^-$ is a generalized inverse of $\mathbf{X}'\mathbf{X}$ and $\mathbf{X}^+$ is the Moore–Penrose inverse of $\mathbf{X}$ (see Appendix M.4). This estimator makes no use of any information about covariances that is contained in $\mathbf{V}$. In contrast, this information *is* utilized in the generalized least squares estimator

$$\text{GLSE}(\mathbf{X}\boldsymbol{\beta}) = \mathbf{X}(\mathbf{X}'\mathbf{V}^{-1}\mathbf{X})^-\mathbf{X}'\mathbf{V}^{-1}\mathbf{y}\,. \tag{33}$$

Derivation of (32) and (33) is briefly described in Section S.1. (Appendix S contains short reminders of some results in mathematical statistics.)

We utilize (33) to estimate μ in the unbalanced data, one-way classification, random effects model of (30), wherein $E(\mathbf{y}) = \mu\mathbf{1}_N$, having $\mathbf{X}$ of (33) as $\mathbf{X} = \mathbf{1}_N$ and $\text{var}(\mathbf{y}) = \mathbf{V}$ of (31). For this (33) is

$$\text{GLSE}(\mathbf{1}_N\mu) = \mathbf{1}_N(\mathbf{1}'_N\{_d\,\sigma_\alpha^2\mathbf{J}_{n_i} + \sigma_e^2\mathbf{I}_{n_i}\}^{-1}\mathbf{1}_N)^{-1}\mathbf{1}'_N\{_d\,\sigma_\alpha^2\mathbf{J}_{n_i} + \sigma_e^2\mathbf{I}_{n_i}\}^{-1}\mathbf{y},$$

where, for clarity, the limits of the indicator variable from $i = 1$ to $i = a$ have been omitted. (This omittance is continued whenever context permits.) Then, since $\mathbf{1}_N$ is a vector, $\text{GLSE}(\mathbf{1}_N\mu)$ yields

$$\text{GLSE}(\mu) = \frac{\mathbf{1}'_N\{_d\,(\sigma_\alpha^2\mathbf{J}_{n_i} + \sigma_e^2\mathbf{I}_{n_i})^{-1}\}\mathbf{y}}{\mathbf{1}'_N\{_d\,(\sigma_\alpha^2\mathbf{J}_{n_i} + \sigma_e^2\mathbf{I}_{n_i})^{-1}\}\mathbf{1}_N}$$

Because, using (ii) of Section M.1,

$$(\sigma_\alpha^2\mathbf{J}_{n_i} + \sigma_e^2\mathbf{I}_{n_i})^{-1} = \frac{1}{\sigma_e^2}\left(\mathbf{I}_{n_i} - \frac{\sigma_\alpha^2}{\sigma_e^2 + n_i\sigma_\alpha^2}\mathbf{J}_{n_i}\right),$$

$$\text{GLSE}(\mu) = \frac{\dfrac{1}{\sigma_e^2}\sum_{i=1}^{a}\left(y_{i\cdot} - \dfrac{n_i\sigma_\alpha^2 y_{i\cdot}}{\sigma_e^2 + n_i\sigma_\alpha^2}\right)}{\dfrac{1}{\sigma_e^2}\sum_{i=1}^{a}\left(n_i - \dfrac{n_i^2\sigma_\alpha^2}{\sigma_e^2 + n_i\sigma_\alpha^2}\right)} = \frac{\sum_{i=1}^{a}\dfrac{n_i\bar{y}_{i\cdot}}{\sigma_e^2 + n_i\sigma_\alpha^2}}{\sum_{i=1}^{a}\dfrac{n_i}{\sigma_e^2 + n_i\sigma_\alpha^2}} = \frac{\sum_{i=1}^{a}\bar{y}_{i\cdot}/\text{var}(\bar{y}_{i\cdot})}{\sum_{i=1}^{a}1/\text{var}(\bar{y}_{i\cdot})} \tag{34}$$

for $\text{var}(\bar{y}_{i\cdot}) = \sigma_\alpha^2 + \sigma_e^2/n_i$ being the variance of $\bar{y}_{i\cdot}$, since the model equation for $\bar{y}_{i\cdot}$ from (1) is $\bar{y}_{i\cdot} = \mu + \alpha_i + \bar{e}_{i\cdot}$.

In the final form of (34) we see that $\text{GLSE}(\mu)$ is the weighted mean of the cell means, weighted by the inverse of their variances. And the second form then shows very easily that when $n_i = n$, we have

$$\text{for balanced data}\quad \text{GLSE}(\mu) = \bar{y}_{\cdot\cdot}\,. \tag{35}$$

It can also be noticed from (32) that

$$\text{in all cases}\quad \text{OLSE}(\mu) = \bar{y}_{\cdot\cdot}\,. \tag{36}$$

The juxtaposition of (35) and (36) may prompt the question "When does a GLSE of a parameter (or function of parameters) equal the OLSE from the same model?" This is discussed in Sections 4.9, 5.10 and 12.4b.

Both $\text{GLSE}(\mu)$ of (34) and $\text{OLSE}(\mu) = \bar{y}_{\cdot\cdot}$ of (36) are unbiased estimators of μ; so also is $\Sigma_{i=1}^{a}\,\bar{y}_{i\cdot}/a$. All three are special cases of a weighted average of the observed class means, the $\bar{y}_{i\cdot}$:

$$\hat{\mu}_w = \Sigma_i w_i \bar{y}_{i\cdot} / \Sigma_i w_i,$$

where

$$\text{GLSE}(\mu) \text{ of (34) is } \hat{\mu}_w \text{ with } w_i = \frac{1}{\sigma_\alpha^2 + \sigma_e^2/n_i},$$

$$\text{OLSE}(\mu) = \bar{y}_{\cdot\cdot} \text{ is } \hat{\mu}_w \text{ with } w_i = n_i$$

and

$$\sum_{i=1}^{a} \bar{y}_{i\cdot}/a \text{ is } \hat{\mu}_w \text{ with } w_i = 1 .$$

In the random model all of these are unbiased for μ. But they have different variances, with

$$\text{var}[\text{GLSE}(\mu)] = \left(\Sigma_i \frac{n_i}{n_i\sigma_\alpha^2 + \sigma_e^2}\right)^{-1} . \tag{37}$$

This, as shown by Searle and Pukelsheim (1986), never exceeds var($\hat{\mu}_w$), no matter what values are used for the w_i in $\hat{\mu}_w$; i.e., in the random model, GLSE(μ) is that weighted average of the class means which has smallest variance among all weighted averages.

In the fixed effects model, OLSE(μ) = $\bar{y}_{\cdot\cdot}$ would be the estimator used for μ. In that model it has variance that never exceeds that of $\hat{\mu}_w$; moreover, the variance of OLSE(μ) in the fixed effects model never exceeds the variance in the random model of GLSE(μ), which in turn never exceeds the variance of OLSE(μ) in the random model. Exercise 3.4 is concerned with these results.

3.4. PREDICTING RANDOM EFFECTS

The model equation $y_{ij} = \mu + \alpha_i + e_{ij}$ for an observation leads to $\bar{y}_{i\cdot} = \mu + \alpha_i + \bar{e}_{i\cdot}$ for the mean of n_i observations in class i. Suppose $\bar{y}_{i\cdot}$ is the average of n_i IQ test scores of college freshman Ronnie Fysher, with α_i being his true, unobservable, IQ value. If Ronnie Fysher is considered as randomly chosen from some population of college freshman then, insofar as true IQ values of that population are concerned, α_i is just a random sample of one from the population of α-values corresponding to the population of freshmen. Although Ronnie Fysher is a specific person, his α_i-value is just one of the population of α-values, one that happens to have some name attached to it. The value α_i can thus be considered as the realized (albeit unobservable) value of a random variable representing true IQ values.

Although α_i is unobservable (just as it is when it is a fixed effect) we do have some information about it, namely $\bar{y}_{i\cdot}$, the average of Ronnie Fysher's n_i scores. A natural question to ask is, therefore, "How can we put some numerical value to α_i based on $\bar{y}_{i\cdot}$?" Let us denote that value by $\tilde{\alpha}_i$. Whatever we do to derive $\tilde{\alpha}_i$, we do not call it an estimator [as Henderson (1950) first did] because estimation applies to parameters and α_i in the random model is not a parameter, it is a random variable. Instead, $\tilde{\alpha}_i$ is called a *predictor* or *prediction* of α_i.

In predicting the (unobservable) realized value of a random variable, which is what we want to do, it might seem sensible to take as the predictor the mean of the random variable; i.e., take $\tilde{\alpha}_i$ as $E(\alpha_i)$. But $E(\alpha_i) = 0$. And $\tilde{\alpha}_i = E(\alpha_i) = 0$ makes no use of data. Yet, if the average of Ronnie Fysher's test scores, namely $\bar{y}_{i\cdot}$, were considerably above the overall freshman average then we would expect α_i to be positive. With this thought in mind we are motivated to use the conditional mean $E(\alpha_i \mid \bar{y}_{i\cdot})$ rather than $E(\alpha_i) = 0$ as our assessment of the true IQ of a freshman having an average IQ test score of $\bar{y}_{i\cdot}$. This means that for each freshman having the same number of IQ tests as Ronnie Fysher and whose average test score is the same $\bar{y}_{i\cdot}$ as Ronnie Fysher's, we assess his or her α-value as the mean of the α-values of all freshmen that have or might have the same n_i and same test score $\bar{y}_{i\cdot}$. Thus our predictor is

$$\tilde{\alpha}_i = E(\alpha_i \mid \bar{y}_{i\cdot}), \tag{38}$$

meaning that $\tilde{\alpha}_i$ is the expected value of α-values of the sub-population of freshmen for each of whom average test score, on the same number of tests, $n_{i\cdot}$, is (or would be if it were to be available) the observed value that has, for some paticular i, been labeled $\bar{y}_{i\cdot}$.

In Chapter 7 we show from several viewpoints that (38) is a reasonable predictor of α_i. In the meantime, we give an easy derivation of an expression for $\tilde{\alpha}_i$ that is more practical than simply $E(\alpha_i \mid \bar{y}_{i\cdot})$. To do so we invoke normality assumptions for the α_i and the e_{ij} in $y_{ij} = \mu + \alpha_i + e_{ij}$. Doing this with the usual first- and second-moment properties detailed in (2)–(12) leads to α_i and $\bar{y}_{i\cdot}$ being jointly distributed with a bivariate normal density having mean and variance

$$E\begin{bmatrix} \alpha_i \\ \bar{y}_{i\cdot} \end{bmatrix} = \begin{bmatrix} 0 \\ \mu \end{bmatrix} \quad \text{and} \quad \text{var}\begin{bmatrix} \alpha_i \\ \bar{y}_{i\cdot} \end{bmatrix} = \begin{bmatrix} \sigma_\alpha^2 & \sigma_\alpha^2 \\ \sigma_\alpha^2 & \sigma_\alpha^2 + \sigma_e^2/n_i \end{bmatrix}. \tag{39}$$

Moreover, from a well-known property of the bivariate normal distribution (see Appendix S.2), we have

$$E(\alpha_i \mid \bar{y}_{i\cdot}) = E(\alpha_i) + \text{cov}(\alpha_i, \bar{y}_{i\cdot})\,[\text{var}(\bar{y}_{i\cdot})]^{-1}[\bar{y}_{i\cdot} - E(\bar{y}_{i\cdot})],$$

which, from (39), is

$$E(\alpha_i \mid \bar{y}_{i\cdot}) = \frac{\sigma_\alpha^2}{\sigma_\alpha^2 + \sigma_e^2/n_i}(\bar{y}_{i\cdot} - \mu)\,.$$

Thus our predictor of α_i is

$$\tilde{\alpha}_i = \frac{n_i\sigma_\alpha^2}{n_i\sigma_\alpha^2 + \sigma_e^2}(\bar{y}_{i\cdot} - \mu)\,. \tag{40}$$

Notice that (40) has here been derived from (38), which was introduced solely on the grounds of its seeming to be reasonable. In Chapter 7 we show that (38) and (40) are special cases of general prediction procedures derived from several different starting points, such as regression, best prediction, best linear prediction and Bayes estimation. In the meantime we observe that $\tilde{\alpha}_i$ explicitly involves n_i, not just in its occurrence in $\bar{y}_{i\cdot}$ but also in the coefficient multiplying $\bar{y}_{i\cdot} - \mu$. For large n_i-values, $\tilde{\alpha}_i$ is closer to $\bar{y}_{i\cdot} - \mu$ than it is for small n_i-values.

Moreover, it is easy to rewrite $\tilde{\alpha}_i$ as

$$\tilde{\alpha}_i = (\bar{y}_{i\cdot} - \mu) - \frac{\sigma_e^2/n_i}{\sigma_\alpha^2 + \sigma_e^2/n_i}(\bar{y}_{i\cdot} - \mu),$$

which shows that $\tilde{\alpha}_i$ regresses towards $\bar{y}_{i\cdot} - \mu$ as n_i increases. This means that when $\bar{y}_{i\cdot}$ exceeds μ then $\tilde{\alpha}_i$ is less than $\bar{y}_{i\cdot} - \mu$; whereas for $\bar{y}_i$ less than μ then $\tilde{\alpha}_i$ exceeds $\bar{y}_{i\cdot} - \mu$. Hence

$$\tilde{\alpha}_i \gtrless \bar{y}_{i\cdot} - \mu \quad \text{according as} \quad \bar{y}_{i\cdot} \lessgtr \mu \ .$$

Thus when $\bar{y}_{i\cdot}$ exceeds μ, which suggests that α_i is better than average (i.e. better than 0), we predict it to be better but only by a fraction of $\bar{y}_{i\cdot} - \mu$, and not by $\bar{y}_{i\cdot} - \mu$ itself. Conversely, when $\bar{y}_{i\cdot}$ is less than μ, we predict α_i as being poorer than average but only by a fraction of $\bar{y}_{i\cdot} - \mu$ and not by $\bar{y}_{i\cdot} - \mu$ itself. For example, $\sigma_\alpha^2 = 90$, $\sigma_e^2 = 60$, $n_i = 6$ and $\mu = 100$ give $\tilde{\alpha}_i = 0.9(\bar{y}_{i\cdot} - 100)$; and when $\bar{y}_{i\cdot} = 110 > 100$, the predictor is $\tilde{\alpha}_i = 9 < 10 = 110 - 100$, but when $\bar{y}_{i\cdot} = 80$ then $\tilde{\alpha}_i = -18 > -20 = 80 - 100$.

The example used for introducing the idea of predicting a random variable has been that of predicting IQ from test scores. It is an idea that applies in many other situations, some of the most notable being in agriculture, where the production of economically important animal products can be increased through well-planned breeding programs. In increasing milk production, for example, it is very useful to be able to predict a bull's genetic value (α_i) from the milk production of his daughter cows. Animal breeders do this using $\tilde{\alpha}_i$ of (40). With h defined as $h = 4\sigma_\alpha^2/(\sigma_\alpha^2 + \sigma_e^2)$, a parameter well known to geneticists as heritability (which leads to $\sigma_e^2/\sigma_\alpha^2 = 4/h - 1$), the predictor (40) becomes

$$\tilde{\alpha}_i = \frac{n_i}{n_i + (4/h - 1)}(\bar{y}_{i\cdot} - \mu) = \frac{n_i h}{(n_i - 1)h + 4}(\bar{y}_{i\cdot} - \mu), \tag{41}$$

a familiar expression to animal breeders.

Note that (40) and its equivalent (41) are in terms of parameters σ_α^2, σ_e^2 and μ. To have a numerical value of $\tilde{\alpha}_i$ therefore demands having estimates of these parameters. Estimating μ by its GLSE(μ) of (34) and using this in $\tilde{\alpha}_i$ gives what is known as BLUP of α_i, the *b*est *l*inear *u*nbiased *p*redictor:

$$\text{BLUP}(\alpha_i) = \frac{n_i\sigma_\alpha^2}{\sigma_e^2 + n_i\sigma_\alpha^2}[\bar{y}_{i\cdot} - \text{GLSE}(\mu)] \ .$$

Its general derivation is given in Chapter 7. It gets its name from the fact that it is a linear function of the observations, it is unbiased in the sense that its expected value equals the expected value of α_i, i.e.,

$$E[\text{BLUP}(\alpha_i)] = E(\alpha_i),$$

which in this case is zero; and among all linear functions of the observations that have expected value $E(\alpha_i)$ it is the one with minimum variance. It also has other optimal properties, which are discussed in Chapter 7. They are also dealt with by Peixoto and Harville (1986), who consider bias and mean-squared error properties of a variety of different predictors of α_i, of which $\text{BLUP}(\alpha_i)$ is one special case.

A natural extension of $\text{BLUP}(\alpha_i)$ is

$$\begin{aligned}\text{BLUP}(\mu + \alpha_i) &= \text{GLSE}(\mu) + \text{BLUP}(\alpha_i)\\ &= \text{GLSE}(\mu) + \frac{n_i\sigma_\alpha^2}{n_i\sigma_\alpha^2 + \sigma_e^2}[\bar{y}_{i\cdot} - \text{GLSE}(\mu)]\ . \qquad (42)\end{aligned}$$

Both are special cases of BLUP in general, the direct derivation of which is given in Chapter 7. Practical usage of them requires, of course, estimates of the variance components, in this case σ_α^2 and σ_e^2. This is an example of what motivates the subject of this book.

3.5. ANOVA ESTIMATION—BALANCED DATA

a. Expected sums of squares

As indicated in Table 2.1, the two sums of squares that are the basis of the analysis of variance of balanced data from a 1-way classification are

$$\text{SSA} = \sum_{i=1}^{a} n(\bar{y}_{i\cdot} - \bar{y}_{\cdot\cdot})^2$$

and (43)

$$\text{SSE} = \sum_{i=1}^{a}\sum_{j=1}^{n} (y_{ij} - \bar{y}_{i\cdot})^2,$$

totaling to

$$\text{SST}_{\text{m}} = \sum_{i=1}^{a}\sum_{j=1}^{n} (y_{ij} - \bar{y}_{\cdot\cdot})^2\ . \qquad (44)$$

The ANOVA method of estimation is based on deriving the expected values of SSA and SSE from the definitions and their consequences in (1)–(15). One then equates observed and expected values and solves for estimators. We show some details of one method of deriving the expected values for balanced data, and a slightly different but equivalent method in the next section for unbalanced data.

-i. A direct derivation. From the model equation (1),

$$y_{ij} = \mu + \alpha_i + e_{ij}, \quad \text{for } i = 1, \ldots, a \text{ and } j = 1, \ldots, n,$$

we get

$$\bar{y}_{i\cdot} = \mu + \alpha_i + \bar{e}_{i\cdot}, \quad \text{for } \bar{e}_{i\cdot} = \sum_{j=1}^{n} e_{ij}/n,$$

and

$$\bar{y}_{\cdot\cdot} = \mu + \bar{\alpha}_{\cdot} + \bar{e}_{\cdot\cdot}, \quad \text{for } \bar{\alpha}_{\cdot} = \sum_{i=1}^{a} \alpha_i/a \text{ and } \bar{e}_{\cdot\cdot} = \sum_{i=1}^{a} \bar{e}_{i\cdot}/a\ .$$

Therefore

$$\begin{aligned} E(\text{SSA}) &= E\left[n \sum_{i=1}^{a} (\bar{y}_{i\cdot} - \bar{y}_{\cdot\cdot})^2\right] = n \sum_{i=1}^{a} E\ [(\alpha_i - \bar{\alpha}_{\cdot}) + (\bar{e}_{i\cdot} - \bar{e}_{\cdot\cdot})]^2 \\ &= n \sum_{i=1}^{a} [E(\alpha_i - \bar{\alpha}_{\cdot})^2 + E(\bar{e}_{i\cdot} - \bar{e}_{\cdot\cdot})^2], \end{aligned}$$

using $E(\alpha_i e_{i'j'}) = 0$ of (15). Then, on using $E(\alpha_i) = 0 = E(e_{ij})$ of (4) and (5),

$$\begin{aligned} E(\text{SSA}) &= n \sum_{i=1}^{a} [\text{var}(\alpha_i - \bar{\alpha}_{\cdot}) + \text{var}(\bar{e}_{i\cdot} - \bar{e}_{\cdot\cdot})] \\ &= n \sum_{i=1}^{a} \left(\sigma_\alpha^2 + \frac{\sigma_\alpha^2}{a} - \frac{2\sigma_\alpha^2}{a}\right) + n \sum_{i=1}^{a} \left(\frac{\sigma_e^2}{n} + \frac{\sigma_e^2}{an} - \frac{2n\sigma_e^2}{nan}\right) \\ &= n(a-1)\sigma_\alpha^2 + (a-1)\sigma_e^2 = (a-1)(n\sigma_\alpha^2 + \sigma_e^2), \end{aligned}$$

as shown in (19) of Section 2.2c-i. And with MSA = SSA/$(a - 1)$ this gives

$$E(\text{MSA}) = \frac{E(\text{SSA})}{a-1} = n\sigma_\alpha^2 + \sigma_e^2\ .$$

It is left to the reader (as Exercise 3.7) to use the same methods to derive

$$E(\text{MSE}) = \frac{E(\text{SSE})}{a(n-1)} = \frac{a(n-1)\sigma_e^2}{a(n-1)} = \sigma_e^2\ .$$

-ii. Using the matrix formulation. When $E(\mathbf{y}) = \boldsymbol{\theta}$ and $\text{var}(\mathbf{y}) = \mathbf{V}$, we write $\mathbf{y} \sim (\boldsymbol{\theta}, \mathbf{V})$ and then, as in Theorem S1 of Appendix S.5,

$$E(\mathbf{y}'\mathbf{A}\mathbf{y}) = \text{tr}(\mathbf{A}\mathbf{V}) + \boldsymbol{\theta}'\mathbf{A}\boldsymbol{\theta}\ . \tag{45}$$

Now for $\mathbf{y}$ of (23) we have $\mathbf{y} \sim (\boldsymbol{\theta}, \mathbf{V})$ with

$$\boldsymbol{\theta} = \mathbf{1}_{an}\mu = (\mathbf{1}_a \otimes \mathbf{1}_n)\mu \quad \text{and} \quad \mathbf{V} = \mathbf{I}_a \otimes (\sigma_\alpha^2 \mathbf{J}_n + \sigma_e^2 \mathbf{I}_n), \tag{46}$$

where $\otimes$ is the Kronecker product operator (see Appendix M.2). Then, since

$$\text{SSA} = n \sum_{i=1}^{a} (\bar{y}_{i\cdot} - \bar{y}_{\cdot\cdot})^2 = n \sum_{i=1}^{a} \bar{y}_{i\cdot}^2 - an\bar{y}_{\cdot\cdot}^2\ ,$$

we can also express SSA, using the $\bar{\mathbf{J}}_n$ definition of Appendix M.1, as

$$\mathrm{SSA} = \mathbf{y}'(\{_{\mathrm{d}}\,\bar{\mathbf{J}}_n\}_{i=1}^{a} - \bar{\mathbf{J}}_{an})\mathbf{y} = \mathbf{y}'(\mathbf{I}_a \otimes \bar{\mathbf{J}}_n - \bar{\mathbf{J}}_{an})\mathbf{y} = \mathbf{y}'\mathbf{A}_1\mathbf{y}, \tag{47}$$

where $\mathbf{A}_1$ is defined as

$$\mathbf{A}_1 = \mathbf{I}_a \otimes \bar{\mathbf{J}}_n - \bar{\mathbf{J}}_{an} = \mathbf{I}_a \otimes \bar{\mathbf{J}}_n - \bar{\mathbf{J}}_a \otimes \bar{\mathbf{J}}_n = (\mathbf{I}_a - \bar{\mathbf{J}}_a) \otimes \bar{\mathbf{J}}_n\,. \tag{48}$$

Hence, using (47) and (46) in (45) gives

$$\begin{aligned} E(\mathrm{SSA}) &= \mathrm{tr}\{[(\mathbf{I}_a - \bar{\mathbf{J}}_a) \otimes \bar{\mathbf{J}}_n][\mathbf{I}_a \otimes (\sigma_\alpha^2 \mathbf{J}_n + \sigma_e^2 \mathbf{I}_n)]\} \\ &\quad + \mu(\mathbf{1}'_a \otimes \mathbf{1}'_n)[(\mathbf{I}_a - \bar{\mathbf{J}}_a) \otimes \bar{\mathbf{J}}_n](\mathbf{1}_a \otimes \mathbf{1}_n)\mu \\ &= \mathrm{tr}[(\mathbf{I}_a - \bar{\mathbf{J}}_a) \otimes (\sigma_\alpha^2 \mathbf{J}_n + \sigma_e^2 \bar{\mathbf{J}}_n)] + 0, \end{aligned} \tag{49}$$

the zero because $\mathbf{1}'_a(\mathbf{I}_a - \bar{\mathbf{J}}_a) = \mathbf{1}'_a - \mathbf{1}'_a = \mathbf{0}$. Thus

$$E(\mathrm{SSA}) = [\mathrm{tr}(\mathbf{I}_a - \bar{\mathbf{J}}_a)][\mathrm{tr}(\sigma_\alpha^2 \bar{\mathbf{J}}_n + \sigma_e^2 \bar{\mathbf{I}}_n)] = (a-1)(n\sigma_\alpha^2 + \sigma_e^2),$$

as before. It is left to the reader to use similar methods to derive $E(\mathrm{SSE})$.

Naturally one gets the same results as when using the direct derivations; and although the matrix methodology is cumbersome in this instance, it is extremely useful for later, more complicated (usually unbalanced data) situations. The preceding details are foundation for those cases.

b. ANOVA estimators

Having derived

$$E(\mathrm{SSA}) = (a-1)(n\sigma_\alpha^2 + \sigma_e^2) \tag{50}$$

and

$$E(\mathrm{SSE}) = a(n-1)\sigma_e^2, \tag{51}$$

we use the "equate sums of squares (or, equivalently, mean squares) to their expected values" principle, which is called the ANOVA method of estimation. The resulting equations, which are linear in the variance components, are now written using the estimators $\hat{\sigma}_e^2$ and $\hat{\sigma}_\alpha^2$, so that the equations are, from (50) and (51),

$$\mathrm{SSA} = (a-1)(n\hat{\sigma}_\alpha^2 + \hat{\sigma}_e^2) \tag{52}$$

and

$$\mathrm{SSE} = a(n-1)\hat{\sigma}_e^2\,. \tag{53}$$

These yield the estimators

$$\hat{\sigma}_e^2 = \frac{\mathrm{SSE}}{a(n-1)} = \mathrm{MSE} \tag{54}$$

and

$$\hat{\sigma}_\alpha^2 = \left(\frac{\mathrm{SSA}}{a-1} - \hat{\sigma}_e^2\right)\bigg/ n = \frac{\mathrm{MSA} - \mathrm{MSE}}{n}\,. \tag{55}$$

TABLE 3.3. 1-WAY CLASSIFICATION, BALANCED DATA

Source of Variation	d.f.	Sum of Squares	Expected Value of Sum of Squares Under Random Model
		General Case	
Classes	$a-1$	$\text{SSA} = n\Sigma_i(\bar{y}_{i.} - \bar{y}_{..})^2$	$(a-1)(n\sigma_\alpha^2 + \sigma_e^2)$
Within classes	$a(n-1)$	$\text{SSE} = \Sigma_i\Sigma_j(y_{ij} - \bar{y}_{i.})^2$	$a(n-1)\sigma_e^2$
Total (a.f.m.)	$an-1$	$\text{SST}_m = \Sigma_i\Sigma_j(y_{ij} - \bar{y}_{..})^2$	
		Example 1 (Table 3.1.)	
Classes	2	SSA = 104	$2(4\sigma_\alpha^2 + \sigma_e^2)$
Within classes	9	SSE = 828	$9\sigma_e^2$
Total (a.f.m.)	11	$\text{SST}_m = 932$	

They are unbiased estimators: $E(\hat{\sigma}_e^2) = \sigma_e^2$ and $E(\hat{\sigma}_\alpha^2) = \sigma_\alpha^2$, as the reader may easily verify (Exercise 3.8).

The expected values of (50) and (51) are summarized in the format of an analysis of variance table in Table 3.3, the lower part of which shows the calculated values for the data of Table 3.1. Using these, the estimation equations (54) and (55) therefore give for the example

$$\hat{\sigma}_e^2 = \text{MSE} = 828/9 = 92$$

and

$$\hat{\sigma}_\alpha^2 = (\text{MSA} - \text{MSE})/n = (104/2 - 92)/4 = -10\,, \tag{56}$$

where $\text{MSA} = \text{SSA}/(a-1) = 104/2$. These are the ANOVA estimates of σ_e^2 and σ_α^2 from the data of Table 3.1.

c. Negative estimates

The estimate of σ_α^2 in (56), the ANOVA estimate, is negative, -10. This negativity is not universal. Indeed, one always hopes that $\hat{\sigma}_\alpha^2$ will not be negative; but it will be whenever MSA < MSE. And such an occurrence is a characteristic of data: with some data it will happen that MSA < MSE, and with some data it will not happen. There is nothing in the ANOVA method of estimation that will prevent a negative estimate occurring should MSA < MSE. This leads to some embarrassment: a negative estimate of a parameter which by definition is non-negative. Variances are never negative.

Two questions immediately arise: (1) What can be done with a negative estimate? (2) How can negative estimates be avoided? The broad answer to (1) is that a negative estimate either may be indicative of using a wrong model, in which case we could try changing the model, or it may be an indication that the true value of the variance component is zero, i.e., that $\sigma_\alpha^2 = 0$. If $\hat{\sigma}_\alpha^2$ is especially large and negative, that and its unbiasedness might well be suggestive that $\sigma_\alpha^2 = 0$. If this is taken to be so then it effectively reduces the model to be $y_{ij} = \mu + e_{ij}$, for which the ANOVA estimator of σ_e^2 is $\hat{\sigma}_e^2 = \text{SST}_m/(an - 1)$.

To avoid negative estimates, two trite answers to question (2) would be first to check the data for erroneous values and to check one's arithmetic, and second to collect more data in the hope that the total set of data would then yield positive estimates. A more serious alternative would be to use a method of estimation that explicitly excludes the possibility of negative estimates—maximum likelihood (ML), restricted maximum likelihood (REML) and Bayes estimation are three such methods. Alternatively, if one has strong prior information on the true value of the components, one might try a minimum norm method (MINQUE) method of estimation. All these methods are described in their general forms, in Chapters 6, 9 and 11.

This problem of negative estimates is discussed in a more general setting than here in LaMotte (1973a) and Styan and Pukelsheim (1981). Changing the model is considered by Hocking (1973, 1985) and Smith and Murray (1984), who, instead of modeling y_{ij} in the manner done here, simply define a variance–covariance structure for the y_{ij}s as

$$\operatorname{cov}(y_{ij}, y_{i'j'}) = \begin{cases} 0 & \text{for } i \neq i', \\ \rho\sigma^2 & \text{for } i = i' \text{ and } j \neq j', \\ \sigma^2 & \text{for } i = i' \text{ and } j = j' . \end{cases}$$

Thus σ^2 is the variance of each y_{ij}, and ρ is the correlation between y_{ij}s in the same class. Then for the n observations in class i the dispersion matrix is

$$\operatorname{var}\{_{\mathrm{c}}\, y_{ij}\}_{j=1}^{n} = \sigma^2[(1-\rho)\mathbf{I}_n + \rho\mathbf{J}_n] = \mathbf{V}_{\mathrm{c}}$$

say, so that

$$\mathbf{V} = \{_{\mathrm{d}}\, \mathbf{V}_{\mathrm{c}}\}_{i=1}^{a} = \mathbf{I}_a \otimes \mathbf{V}_{\mathrm{c}} .$$

Then, although Hocking *et al.* (1989) in extending Smith and Murray (1984) use a variation of the ANOVA method that appears to be different from usual, it is in fact precisely the same as (54) and (55) except that those equations are now

$$\text{MSE} = (1-\hat{\rho})\hat{\sigma}^2 \quad \text{and} \quad \text{MSA} = (n\hat{\rho} + 1 - \hat{\rho})\hat{\sigma}^2 .$$

These lead to

$$\hat{\sigma}^2 = \frac{\text{MSA}}{n} + \left(1 - \frac{1}{n}\right)\text{MSE} = \hat{\sigma}_\alpha^2 + \hat{\sigma}_e^2$$

and

$$\hat{\rho} = \frac{\text{MSA} - \text{MSE}}{(n-1)\text{MSE} + \text{MSA}} \quad \text{with } \hat{\rho}\hat{\sigma}^2 = \hat{\sigma}_\alpha^2 .$$

In the context of ρ being a correlation parameter, the occurrence of negative $\hat{\rho}$ when MSA < MSE is now of no concern. But on wanting to convert these parameters ρ and σ^2 into σ_α^2 and σ_e^2, one comes right back to (54) and (55), with $\hat{\sigma}_\alpha^2$ of (55) being negative when MSA < MSE. Moreover, in using this correlation model, which involves equal correlation between all pairs of observations in the same class (for every class), ρ is not entirely free to be any value in the $(-1, 1)$ range as is usually the case for a correlation. This is so because $\mathbf{V}_c = \sigma^2[(1-\rho)\mathbf{I}_n + \rho\mathbf{J}_n]$ is, through being a dispersion matrix, non-negative definite, and so has a non-negative determinant. This leads to ρ having to satisfy $\rho > -1/(n-1)$. Hence, if n is 11 or larger, ρ cannot be more negative than -0.1, which is somewhat of a limitation to its being a correlation that can be negative. Non-negative, minimum biased estimators are given by Hartung (1981) as $n\text{MSA}/(1+n^2)$ and MSE. Estimators of this nature are discussed more fully by Kleffe and Rao (1986), Rao and Kleffe (1988) and Mathew *et al.* (1991a).

Further comment on negative estimates is given in Sections 4.4 and 12.7.

d. Normality assumptions

Except for a brief mention of normality in Section 3.4 (predicting α_i), it is to be noted that up to this point no assumptions have been made about the form of the probability density functions of the random α_is and e_{ij}s, other than that they have zero means and finite variances. The ANOVA method of estimation, although it uses sums of squares traditionally encountered in an analysis of variance table, does not invoke normality. Neither is normality needed, of course, in the analysis of variance table itself until F-statistics calculated from those sums of squares are used in a confidence interval or hypothesis-testing context.

The assumptions that are called the usual normality assumptions in the random model are that the α_is and e_{ij}s are taken as being normally distributed, with the first- and second-moment properties of (6)–(12). Stated succinctly in matrix notation, this means that

$$\begin{bmatrix} \boldsymbol{\alpha} \\ \mathbf{e} \end{bmatrix} \sim \mathcal{N}\left(\begin{bmatrix} \mathbf{0} \\ \mathbf{0} \end{bmatrix}, \begin{bmatrix} \sigma_\alpha^2 \mathbf{I}_a & \mathbf{0} \\ \mathbf{0} & \sigma_e^2 \mathbf{I}_{an} \end{bmatrix} \right) \tag{57}$$

and, using (26)

$$\mathbf{y} \sim \mathcal{N}(\mu \mathbf{1}_{an}, \mathbf{V}) \quad \text{for } \mathbf{V} = \mathbf{I}_a \otimes (\sigma_\alpha^2 \mathbf{J}_n + \sigma_e^2 \mathbf{I}_n)] . \tag{58}$$

-i. χ^2-distributions of sums of squares. From Theorem S2 of Section S.4 we have for

$$\mathbf{y} \sim \mathcal{N}(\boldsymbol{\theta}, \mathbf{V}) \quad \text{that} \quad \mathbf{y}'\mathbf{A}\mathbf{y} \sim \chi^2(r_\mathbf{A}, \tfrac{1}{2}\boldsymbol{\theta}'\mathbf{A}\boldsymbol{\theta}) \tag{59}$$

if $\mathbf{AV}$ is idempotent, where $r_\mathbf{A}$ is the rank of $\mathbf{A}$.

Notation. $\chi^2(r, \lambda)$ represents the non-central χ^2-distribution with r degrees of freedom and non-centrality parameter λ. Details and references are given in Appendix S.4. χ_r^2 represents the central χ^2-distribution with r degrees of freedom, just as $\mathscr{F}_d^n$ is for the F-distribution in (5) of Chapter 2.

This notation shall be used in two ways. One is when we want shorthand for statements like "u has a χ^2-distribution with k degrees of freedom", which shall be written as $u \sim \chi_k^2$. An example is (59). A second usage will be when "the probability that $u \sim \chi_k^2$ is less than some value c" is abbreviated to $\Pr(\chi_k^2 < c)$.

To apply (59) to SSA we write $\text{SSA} = \mathbf{y}'\mathbf{A}_1\mathbf{y}$ for $\mathbf{A}_1 = (\mathbf{I}_a - \bar{\mathbf{J}}_a) \otimes \bar{\mathbf{J}}_n$ of (48) and with $\mathbf{V}$ of (58) we find, using the algebra of $\mathbf{J}$-matrices and of direct (or Kronecker) products set out in Appendix M, that $[\mathbf{A}_1/(n\sigma_\alpha^2 + \sigma_e^2)]\mathbf{V}$ is an idempotent matrix with $\mathbf{A}_1$ having rank $a - 1$. Also, $\frac{1}{2}\boldsymbol{\theta}'\mathbf{A}\boldsymbol{\theta}$ of (59) here has $\boldsymbol{\theta}'\mathbf{A} = \mu\mathbf{1}'_{an}\mathbf{A}_1 = \mathbf{0}$, because $\mathbf{1}'\mathbf{A}_1 = \mathbf{0}$. Hence (59) gives

$$\text{SSA}/(n\sigma_\alpha^2 + \sigma_e^2) \sim \chi_{a-1}^2, \qquad (60)$$

a χ^2-distribution with $a - 1$ degrees of freedom. A similar use of (59) shows that for

$$\text{SSE} = \mathbf{y}'\mathbf{A}_2\mathbf{y} \quad \text{with} \quad \mathbf{A}_2 = \mathbf{I}_{an} - \{_{\text{d}}\,\bar{\mathbf{J}}_n\}$$

$$\text{SSE}/\sigma_e^2 \sim \chi_{a(n-1)}^2 \,. \qquad (61)$$

-ii. ***Independence of sums of squares.*** Theorem S3 of Appendix S.5 shows that for $\mathbf{y} \sim \mathcal{N}(\mu, \mathbf{V})$,

$$\mathbf{y}'\mathbf{A}\mathbf{y} \text{ and } \mathbf{y}'\mathbf{B}\mathbf{y} \text{ are independent if } \mathbf{AVB} = \mathbf{0}\,. \qquad (62)$$

Applying this to $\mathbf{A}_1\mathbf{V}\mathbf{A}_2$ of the preceding paragraph, we find that $\mathbf{A}_1\mathbf{V}\mathbf{A}_2$ reduces to $\mathbf{0}$ and so therefore

$$\text{SSA and SSE are independent.} \qquad (63)$$

This is a simple example of a well-known property of sums of squares of balanced data.

-iii. ***Sampling variances of estimators.*** The independence (under normality) of SSA and SSE has been established and each has a distribution that is proportional to a χ^2; and the variance of the (central) χ_f^2-distribution is $2f$. From this we derive sampling variances of the estimators. The important results (Appendix S.3b) are that for a sum of squares SS being distributed proportionate to a χ^2 with f degrees of freedom and mean square $\text{MS} = \text{SS}/f$,

$$\text{var}(\text{MS}) = \frac{2[E(\text{MS})]^2}{f} \qquad (64)$$

and

$$\frac{(\text{MS})^2}{f+2} \text{ is an unbiased estimator of } \frac{[E(\text{MS})]^2}{f}\,. \qquad (65)$$

Hence

$$\text{var}(\hat{\sigma}_e^2) = \text{var}(\text{MSE}) = \frac{2[E(\text{MSE})]^2}{a(n-1)} = \frac{2\sigma_e^4}{a(n-1)} \tag{66}$$

is unbiasedly estimated by

$$\hat{\text{var}}(\hat{\sigma}_e^2) = \frac{2\hat{\sigma}_e^4}{a(n-1)+2}. \tag{67}$$

Similarly

$$\text{var}(\hat{\sigma}_\alpha^2) = \text{var}\left(\frac{\text{MSA} - \text{MSE}}{n}\right) = \frac{2}{n^2}\left\{\frac{[E(\text{MSA})]^2}{a-1} + \frac{[E(\text{MSE})]^2}{a(n-1)}\right\} \tag{68}$$

$$= \frac{2}{n^2}\left[\frac{(n\sigma_\alpha^2 + \sigma_e^2)^2}{a-1} + \frac{\sigma_e^4}{a(n-1)}\right], \tag{69}$$

which is unbiasedly estimated by

$$\hat{\text{var}}(\hat{\sigma}_\alpha^2) = \frac{2}{n^2}\left[\frac{(n\hat{\sigma}_\alpha^2 + \hat{\sigma}_e^2)^2}{a+1} + \frac{\hat{\sigma}_e^4}{a(n-1)+2}\right]. \tag{70}$$

And the covariance of $\hat{\sigma}_\alpha^2$ with $\hat{\sigma}_e^2$ is

$$\text{cov}(\hat{\sigma}_\alpha^2, \hat{\sigma}_e^2) = \frac{\text{cov}[(\text{MSA} - \text{MSE}), \text{MSE}]}{n} = -\frac{\text{var}(\text{MSE})}{n} = \frac{-2\sigma_e^4}{an(n-1)}, \tag{71}$$

for which an unbiased estimator is

$$\hat{\text{cov}}(\hat{\sigma}_\alpha^2, \hat{\sigma}_e^2) = \frac{-2\hat{\sigma}_e^4}{n[a(n-1)+2]}. \tag{72}$$

Thus, although the sampling (co)variances in (66), (69) and (71) are quadratic functions of σ_α^2 and σ_e^2, we can estimate those sampling (co)variances by replacing σ_α^2 and σ_e^2 therein by $\hat{\sigma}_\alpha^2$ and $\hat{\sigma}_e^2$, and adding 2 to denominator degrees of freedom. This gives unbiased estimators of those sampling (co)variances, as shown in (67), (70) and (72). And best invariant unbiased estimators of mean square errors of $\hat{\sigma}_\alpha^2$ and $\hat{\sigma}_e^2$ are derived by Hartung and Voet (1986).

Using Edgeworth series and third and fourth cumulants, Singha (1984) develops, in the absence of normality, approximate expressions for the variance of $\hat{\sigma}_\alpha^2$ and $\hat{\sigma}_e^2$ and for the means and variances of various ratio functions of $\hat{\sigma}_\alpha^2$ and $\hat{\sigma}_e^2$.

-iv. An F-statistic to test H: $\sigma_\alpha^2 = 0$. In the fixed effects model (where the α_is are fixed effects), $F = \text{MSA}/\text{MSE}$ tests the hypothesis H: α_is *all equal*. In that model, we have, under that hypothesis, that $F \sim \mathscr{F}_{a(n-1)}^{a-1}$, the F-distribution (Section S.4) on $a - 1$ and $a(n-1)$ degrees of freedom.

In the random effects model, provided the data are balanced, the χ^2 and independence properties of SSA and SSE of (60), (61) and (63) lead to

$$\frac{\text{MSA}}{n\sigma_\alpha^2 + \sigma_e^2}\Bigg/\frac{\text{MSE}}{\sigma_e^2} \sim \mathscr{F}_{a(n-1)}^{a-1};$$

i.e., to

$$\frac{\sigma_e^2 F}{n\sigma_\alpha^2 + \sigma_e^2} \sim \mathscr{F}^{a-1}_{a(n-1)} \,. \tag{73}$$

Therefore, under the hypothesis H: $\sigma_\alpha^2 = 0$, the left-hand side of (73) reduces to F, which then has an $\mathscr{F}$-distribution and so $F = \text{MSA}/\text{MSE}$ provides a test of H: $\sigma_\alpha^2 = 0$. Note that in the random effects model $F = \text{MSA}/\text{MSE}$ has a distribution that is a multiple of a central $\mathscr{F}$-distribution, whereas in the fixed effects model F has a non-central $\mathscr{F}$-distribution when H: *α_is all equal* is not true.

-v. ***Confidence intervals.*** Exact confidence intervals are available for σ_e^2, $\sigma_\alpha^2/(\sigma_\alpha^2 + \sigma_e^2)$, $\sigma_e^2/(\sigma_\alpha^2 + \sigma_e^2)$ and $\sigma_\alpha^2/\sigma_e^2$. Define $\chi^2_{k,\,\text{L}}$ and $\chi^2_{k,\,\text{U}}$ by the probability statement

$$\Pr\{\chi^2_{k,\,\text{L}} \leqslant \chi^2_k \leqslant \chi^2_{k,\,\text{U}}\} = 1 - \alpha$$

for some probability value $\alpha(= .05$, say). Then (61) gives

$$\Pr\left\{\chi^2_{a(n-1),\,\text{L}} \leqslant \frac{\text{SSE}}{\sigma_e^2} \leqslant \chi^2_{a(n-1),\,\text{U}}\right\} = 1 - \alpha,$$

which is equivalent to

$$\Pr\left\{\frac{\text{SSE}}{\chi^2_{a(n-1),\,\text{U}}} \leqslant \sigma_e^2 \leqslant \frac{\text{SSE}}{\chi^2_{a(n-1),\,\text{L}}}\right\} = 1 - \alpha,$$

so leading to the $100(1-\alpha)\%$ confidence level $\text{SSE}/\chi^2_{a(n-1),\,\text{U}}$ to $\text{SSE}/\chi^2_{a(n-1),\,\text{L}}$ shown on the first line of Table 3.4.

Similarly, on defining upper and lower points of the $\mathscr{F}$-distribution as F_U and F_L by

$$\Pr\{F_\text{L} \leqslant \mathscr{F}^{a-1}_{a(n-1)} \leqslant F_\text{U}\} = 1 - \alpha,$$

we have, for $F = \text{MSA}/\text{MSE}$,

$$\Pr\left\{F_\text{L} \leqslant \frac{\sigma_e^2 F}{n\sigma_\alpha^2 + \sigma_e^2} \leqslant F_\text{U}\right\} = 1 - \alpha\,.$$

But the two-sided inequality within this probability statement is equivalent to

$$\frac{F_\text{L}}{F} \leqslant \frac{\sigma_e^2}{n\sigma_\alpha^2 + \sigma_e^2} \leqslant \frac{F_\text{U}}{F}$$

$$\Leftrightarrow \frac{F}{F_\text{U}} - 1 \leqslant \frac{n\sigma_\alpha^2 + \sigma_e^2}{\sigma_e^2} - 1 \leqslant \frac{F}{F_\text{L}} - 1$$

$$\Leftrightarrow \frac{F/F_\text{U} - 1}{n} \leqslant \frac{\sigma_\alpha^2}{\sigma_e^2} \leqslant \frac{F/F_\text{L} - 1}{n}\,.$$

The confidence interval shown in the last line of Table 3.4 comes from the preceding statement. Further manipulations of that statement, similar to those

TABLE 3.4. CONFIDENCE INTERVALS ON VARIANCE COMPONENTS AND FUNCTIONS THEREOF, IN THE 1-WAY CLASSIFICATION, RANDOM MODEL, BALANCED DATA

		Confidence Interval[1]		
Line	Parameter	Lower Limit	Upper Limit	Confidence Coefficient
1	σ_e^2	$\dfrac{\text{SSE}}{\chi^2_{a(n-1),\text{U}}}$	$\dfrac{\text{SSE}}{\chi^2_{a(n-1),\text{L}}}$	$1-\alpha$
2	σ_α^2	$\dfrac{\text{SSA}(1-F_\text{U}/F)}{n\chi^2_{a-1,\text{U}}}$	$\dfrac{\text{SSA}(1-F_\text{L}/F)}{n\chi^2_{a-1,\text{L}}}$	$1-2\alpha$
3	$\dfrac{\sigma_\alpha^2}{\sigma_\alpha^2+\sigma_e^2}$	$\dfrac{F/F_\text{U}-1}{n+F/F_\text{U}-1}$	$\dfrac{F/F_\text{L}-1}{n+F/F_\text{L}-1}$	$1-\alpha$
4	$\dfrac{\sigma_e^2}{\sigma_\alpha^2+\sigma_e^2}$	$\dfrac{n}{n+F/F_\text{L}-1}$	$\dfrac{n}{n+F/F_\text{U}-1}$	$1-\alpha$
5	$\dfrac{\sigma_\alpha^2}{\sigma_e^2}$	$\dfrac{F/F_\text{U}-1}{n}$	$\dfrac{F/F_\text{L}-1}{n}$	$1-\alpha$

[1] Notation:

$$F = \text{MSA}/\text{MSE},$$
$$\Pr\{\chi^2_{k,\text{L}} \leqslant \chi^2_k \leqslant \chi^2_{k,\text{U}}\} = 1-\alpha$$
$$\Pr\{F_\text{L} \leqslant \mathscr{F}^{a-1}_{a(n-1)} \leqslant F_\text{U}\} = 1-\alpha$$

that established it, yield the confidence intervals for $\sigma_\alpha^2/(\sigma_\alpha^2+\sigma_e^2)$ and for $\sigma_e^2/(\sigma_\alpha^2+\sigma_e^2)$ shown in Table 3.4. (Exercise 3.10.)

For σ_α^2 there is no exact confidence interval, but by considering the intersection of confidence intervals on $\sigma_\alpha^2+\sigma_e^2$ and on $\sigma_e^2/(n\sigma_\alpha^2+\sigma_e^2)$, based on the distributions of SSA and of F, Williams (1962) derived the confidence interval for σ_α^2 shown on the second line of Table 3.4. An excellent description of deriving this Williams interval is g' en in Graybill (1976, pp. 618–620). The intervals for $\sigma_\alpha^2/(\sigma_\alpha^2+\sigma_e^2)$ and $\sigma_e^2/(\sigma_\alpha^2+\sigma_e^2)$ are given in Graybill (1961, p. 379; 1976, pp. 617–618), and for $\sigma_\alpha^2/\sigma_e^2$ by Scheffé (1959, p. 229).

-vi. Probability of a negative estimate. Section 3.5c describes how it is possible for $\hat{\sigma}_\alpha^2$ to be negative, and gives a trite example thereof. From the distribution of F in (73) one can derive the probability of such negativity occurring:

$$\begin{aligned}\Pr\{\hat{\sigma}_\alpha^2 < 0\} &= \Pr\{\text{MSA} < \text{MSE}\}\\ &= \Pr\{F < 1\}\\ &= \Pr\left\{\left(\frac{n\sigma_\alpha^2}{\sigma_e^2}+1\right)\mathscr{F}^{a-1}_{a(n-1)} < 1\right\}, \quad \text{from (73)}\\ &= \Pr\left\{\mathscr{F}^{a-1}_{a(n-1)} < \frac{\sigma_e^2}{\sigma_e^2+n\sigma_\alpha^2}\right\},\end{aligned}$$

We now use the well-known fact that the reciprocal of $\mathscr{F}_d^n$ has the $\mathscr{F}_n^d$ distribution. We also define τ as the ratio of the variance components and ρ as the intra-class correlation:

$$\tau = \frac{\sigma_\alpha^2}{\sigma_e^2}, \quad \text{and } \rho = \frac{\sigma_\alpha^2}{\sigma_\alpha^2 + \sigma_e^2} = \frac{\tau}{1+\tau};$$

and then have the probability of $\hat{\sigma}_\alpha^2$ being negative as

$$P = \Pr\{\hat{\sigma}_\alpha^2 < 0\} = \Pr\{\mathscr{F}_{a-1}^{a(n-1)} > 1 + n\tau\}\,. \tag{74}$$

Calculated values of this probability are shown in Table 3.5 for a, the number of classes, being 2, 5, 10, 25 and 50 and for n, the number of observations per class, being 5, 25 and 100. For the resultant fifteen pairs of (a, n) values, P is shown for $\tau = .01, .05, .10$ and $.25$ (and for each τ the corresponding values of ρ and h are also shown). The most noticeable feature of these values is, as would be expected, that P decreases as either a or n increases. This characteristic of P is very evident in Figure 3.1, which shows, when $\tau = .01$, contour lines for $P = .4, .3, .2, .1, .05$ and $.01$ plotted on (a, n) co-ordinates, ranging from 2 to 360 for a and from 2 to 100 for n. When similar plots were made for other values of τ it was found, as would be indicated by the P-values in Table 3.5,

TABLE 3.5. PROBABILITY OF THE ANOVA ESTIMATOR OF σ_α^2 BEING NEGATIVE WHEN OBTAINED FROM BALANCED DATA OF a CLASSES EACH WITH n OBSERVATIONS, UNDER NORMALITY ASSUMPTIONS

$$P = \Pr\{\hat{\sigma}_\alpha^2 < 0\} = P\{\mathscr{F}_{a-1}^{a(n-1)} > 1 + n\tau\} \text{ for } \tau = \sigma_\alpha^2/\sigma_e^2;$$

$$\rho = \frac{\sigma_\alpha^2}{\sigma_\alpha^2 + \sigma_e^2} = \frac{\tau}{1+\tau} \text{ and } h = \frac{4\sigma_\alpha^2}{\sigma_\alpha^2 + \sigma_e^2} = 4\rho$$

	$\tau = .01$			$\tau = .05$		
	$\rho = .0099$ $h = .0396$			$\rho = .0476$ $h = .1904$		
a	$n = 5$	$n = 25$	$n = 100$	$n = 5$	$n = 25$	$n = 100$
2	.65	.62	.52	.60	.49	.32
5	.55	.47	.26	.46	.22	.04
10	.51	.38	.12	.38	.09	0
25	.47	.26	.02	.27	0	0
50	.43	.16	0	.18	0	0

	$\tau = .10$			$\tau = .25$		
	$\rho = .0909$ $h = .3636$			$\rho = .20$ $h = .80$		
a	$n = 5$	$n = 25$	$n = 100$	$n = 5$	$n = 25$	$n = 100$
2	.56	.40	.24	.48	.29	.16
5	.38	.11	.01	.22	.03	0
10	.27	.02	0	.10	0	0
15	.13	0	0	.01	0	0
50	.05	0	0	0	0	0

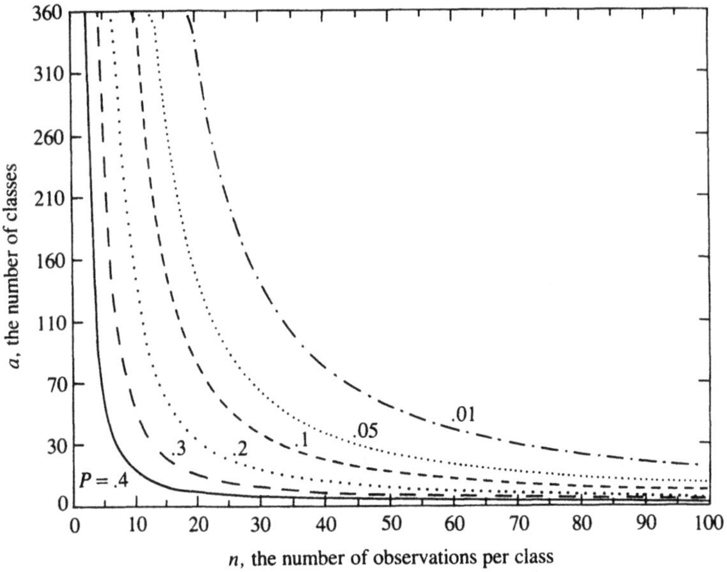

Figure 3.1. Contours of $P = \Pr\{\hat{\sigma}^2_\alpha < 0\} = \Pr\{\mathscr{F}^{a(n-1)}_{a-1} > 1 + n\tau\} = .4, .3, .2, .1, .05$ and $.01$, for $\tau = \sigma^2_\alpha/\sigma^2_e = .01$.

that in all cases the pattern of contour lines was similar to those in Figure 3.1. For $\tau < .01$ the lines were more separated than those in Figure 3.1, spreading out more and more away from (0, 0). This implies that as τ gets very small, the probability of a negative $\hat{\sigma}^2_\alpha$ is appreciable over a larger range of (a, n) values. Conversely, for $\tau > .01$, the contour lines bunched up more and more towards (0, 0), until for $\tau \geqslant 1$ there were, from a practical point of view, no lines of any consequence at all. For example, with $\tau = 1$, P is effectively zero (less than .01) for all (a, n) pairs further from (0, 0) along either axis than (3, 99), (4, 25), (5, 12), (6, 8) and (7, 6); and for $a \geqslant 4$, $P \leqslant .09$ for $n \geqslant 5$. For τ larger than unity, $\tau = 5$, say, this occurrence of non-zero P is even closer to (0, 0); at (2, 5), P is only .15, and it is zero everywhere beyond (5, 5). For $\tau = 10$, P is .11 at (2, 5) and .03 at (2, 100) and is zero everywhere beyond (3, 5), where it is only .02.

This discussion of P, the probability of $\hat{\sigma}^2_\alpha$ having a negative value, yields the following useful conclusions.

(i) For any a (or any n) P decreases as n (or a) increases, decreasing faster for large a (or n) than for small.

(ii) For $\sigma^2_\alpha > \sigma^2_e$, P is zero except for small values of a, and it exceeds .1 only for $a < 4$.

(iii) For $\sigma^2_\alpha \leqslant \frac{1}{10}\sigma^2_e$, P can be appreciably large, e.g., for $a = 10$ and $n = 5$, $P = .27$ at $\sigma^2_\alpha = \frac{1}{10}\sigma^2_e$ and $P = .51$ at $\sigma^2_\alpha = \frac{1}{100}\sigma^2_e$, as is evident in Table 3.5.

In general, there seems to be no need for the data analyst to worry about the possibility of having $\hat{\sigma}^2_\alpha$ negative provided the number of classes is not too small; having many classes is more important than having many observations per class. This is what one would expect: numerous classes are needed if one wants to estimate the class variance component with any degree of optimality. Three hundred observations on each of two classes is only giving information about two classes; it is better information on those classes than five observations on each of them would be, but it is still only two classes.

-vii. Distribution of estimators. The distributional result $\text{SSE}/\sigma_e^2 \sim \chi^2_{a(n-1)}$ in (61) gives

$$\hat{\sigma}_e^2 = \text{MSE} \sim \frac{\sigma_e^2}{a(n-1)}\chi^2_{a(n-1)},$$

by which notation is meant that $\hat{\sigma}_e^2$ is distributed as a $\chi^2_{a(n-1)}$-variable multiplied by $\sigma_e^2/[a(n-1)]$; more precisely $a(n-1)\text{MSE}/\sigma^2 \sim \chi^2_{a(n-1)}$. In contrast, for $\hat{\sigma}_\alpha^2 = (\text{MSA} - \text{MSE})/n$, although MSA and MSE are each distributed as a multiple of a χ^2, and are independent, MSA − MSE is *not* distributed as a multiple of a χ^2. Therefore neither is $\hat{\sigma}_\alpha^2$. In fact $\hat{\sigma}_\alpha^2$ has no simple, closed form distribution. A somewhat complicated and more general form can be obtained from Fleiss (1971), and an alternative form based on the confluent hypergeometric function is available in Robinson (1965).

3.6. ANOVA ESTIMATION—UNBALANCED DATA

The analysis of variance sums of squares for unbalanced data are

$$\text{SSA} = \sum_{i=1}^{a} n_i(\bar{y}_{i\cdot} - \bar{y}_{\cdot\cdot})^2 = \Sigma_i n_i \bar{y}_{i\cdot}^2 - N\bar{y}_{\cdot\cdot}^2$$

and (75)

$$\text{SSE} = \sum_{i=1}^{a}\sum_{j=1}^{n_i}(y_{ij} - \bar{y}_{i\cdot})^2 = \Sigma_i\Sigma_j y_{ij}^2 - \Sigma_i n_i \bar{y}_{i\cdot}^2$$

these being the same as in (43) for balanced data, except for having n_i in place of n, and with $N = \Sigma n_i$.

Notation. In the right-most expressions of (75) two notational conventions have been adopted that will be used throughout: Σ_i and Σ_j represent $\Sigma_{i=1}^{a}$ and $\Sigma_{j=1}^{n_i}$, respectively, and N is the total number of observations.

a. Expected sums of squares

As in Section 3.5a, we show details of two methods for deriving expected values of sums of squares.

-i. A direct derivation. The balanced data sums of squares in Section 3.5a-i were handled as sums of squares of deviations among means, similar to the first expression in (75). But unbalanced data sums of squares are sometimes more tractable when expressed as linear combinations of crude sums of squares, as in the second expression of (75).

Expected values of the three different terms in those expressions are now given. Derivation is, of course, based on exactly the same model as described in Section 3.1, making particular use of the results in (12)–(15), such as $E(\alpha_i) = 0$, $E(\alpha_i^2) = \sigma_\alpha^2$ and $E(\alpha_i\alpha_{i'}) = 0$ for $i \neq i'$; and $E(e_{ij}) = 0$, $E(e_{ij}^2) = \sigma_e^2$, $E(e_{ij}e_{i'j'}) = 0$ unless $i = i'$ and $j = j'$, and $E(\alpha_i e_{i'j'}) = 0$. Thus we have

$$\begin{aligned} E(N\bar{y}_{..}^2) &= NE(\mu + \Sigma_i n_i\alpha_i/N + \bar{e}_{..})^2 \\ &= NE\left(\mu^2 + \frac{\Sigma_i n_i^2\alpha_i^2 + \underset{i \neq i'}{\Sigma\Sigma}\, n_i n_{i'}\alpha_i\alpha_{i'}}{N^2} + \frac{\Sigma_i\Sigma_j e_{ij}^2 + \underset{i \neq i'\; j \neq j'}{\Sigma\Sigma\;\Sigma\Sigma}\, e_{ij}e_{i'j'}}{N^2}\right. \\ &\qquad \left. + \frac{2\mu\Sigma_i n_i\alpha_i}{N} + 2\mu\bar{e}_{..} + \frac{2\Sigma_i n_i\bar{e}_{..}\alpha_i}{N}\right) \\ &= N\mu^2 + \frac{N\Sigma_i n_i^2\sigma_\alpha^2}{N^2} + 0 + \frac{N\Sigma_i\Sigma_j\sigma_e^2}{N^2} + 0 + 0 + 0 + 0 \\ &= N\mu^2 + \sigma_\alpha^2\,\Sigma_i n_i^2/N + \sigma_e^2\,. \end{aligned} \tag{76}$$

Similarly

$$E(\Sigma_i n_i\bar{y}_{i.}^2) = \Sigma_i n_i E(\mu + \alpha_i + \bar{e}_{i.})^2 = N\mu^2 + N\sigma_\alpha^2 + a\sigma_e^2$$

and

$$E(\Sigma_i\Sigma_j y_{ij}^2) = \Sigma_i\Sigma_j E(\mu + \alpha_i + e_{ij})^2 = N\mu^2 + N\sigma_\alpha^2 + N\sigma_e^2\,.$$

Using these in (75) gives

$$E(\mathrm{SSA}) = (N - \Sigma_i n_i^2/N)\sigma_\alpha^2 + (a-1)\sigma_e^2 \tag{77}$$

and

$$E(\mathrm{SSE}) = (N - a)\sigma_e^2\,. \tag{78}$$

-ii. Using the matrix formulation. It is not difficult to confirm that the expressions in (75) are equivalent to

$$\mathrm{SSA} = \mathbf{y}'\mathbf{A}_1\mathbf{y} \quad \text{for } \mathbf{A}_1 = \{_{\mathrm{d}}\,\bar{\mathbf{J}}_{n_i}\} - \bar{\mathbf{J}}_N \tag{79}$$

and

$$\mathrm{SSE} = \mathbf{y}'\mathbf{A}_2\mathbf{y} \quad \text{for } \mathbf{A}_2 = \mathbf{I}_N - \{_{\mathrm{d}}\,\bar{\mathbf{J}}_{n_i}\} \tag{80}$$

where the $\{_{\mathrm{d}}\ \}$ notation for block diagonal matrices is as described in Appendix M.3. That (79) is a generalization of (47), from balanced to unbalanced data, can be noted; i.e., (47) is (79) with $n_i = n\ \forall\ i$.

With

$$\mathbf{y} \sim (\mathbf{1}_N\mu, \mathbf{V} = \{_{\mathrm{d}}\,\sigma_\alpha^2\mathbf{J}_{n_i} + \sigma_e^2\mathbf{I}_{n_i}\}) \tag{81}$$

from (30) and (31), we now use (79) and (81) in Theorem S1, just as in (45), to derive the expected value of SSA as

$$E(\text{SSA}) = E(\mathbf{y}'\mathbf{A}_1\mathbf{y}) = \text{tr}(\mathbf{A}_1\mathbf{V}) + E(\mathbf{y}')\mathbf{A}_1 E(\mathbf{y})$$
$$= \text{tr}[(\{_d\, \bar{\mathbf{J}}_{n_i}\} - \bar{\mathbf{J}}_N)\{_d\, \sigma_\alpha^2 \mathbf{J}_{n_i} + \sigma_e^2 \mathbf{I}_{n_i}\}] + \mu \mathbf{1}'_N(\{_d\, \bar{\mathbf{J}}_{n_i}\} - \bar{\mathbf{J}}_N)\mathbf{1}_N\mu\,.$$

Recalling that $\mathbf{1}'\mathbf{Q}\mathbf{1}$ for any $\mathbf{Q}$ is the sum of all elements of $\mathbf{Q}$ gives

$$E(\text{SSA}) = \sigma_\alpha^2[\text{tr}\{_d\, \bar{\mathbf{J}}_{n_i}\mathbf{J}_{n_i}\} - \text{tr}(\bar{\mathbf{J}}_N\{_d\, \mathbf{J}_{n_i}\})]$$
$$+ \sigma_e^2(\text{tr}\{_d\, \bar{\mathbf{J}}_{n_i}\} - \text{tr}\, \bar{\mathbf{J}}_N) + \mu^2\left(\Sigma_i \frac{n_i^2}{n_i} - \frac{N^2}{N}\right)$$
$$= \sigma_\alpha^2\left(\Sigma_i \frac{n_i^2}{n_i} - \frac{\Sigma_i n_i^2}{N}\right) + \sigma_e^2\left(\Sigma_i \frac{n_i}{n_i} - \frac{N}{N}\right)$$
$$= \left(N - \frac{\Sigma_i n_i^2}{N}\right)\sigma_\alpha^2 + (a-1)\sigma_e^2,$$

just as in (77). Similar manipulations yield

$$E(\text{SSE}) = \text{tr}(\mathbf{A}_2\mathbf{V}) + \mu\mathbf{1}'_N\mathbf{A}_2\mathbf{1}_N\mu = (N-a)\sigma_e^2,$$

as in (78). These derivations, although appearing tedious, involve methods that are very useful in more complicated models.

b. ANOVA estimators

Using exactly the same reasoning that led from (50) and (51) to the estimation equations (52) and (53) for balanced data, of equating sums of squares in their expected values, gives

$$\text{SSA} = \left(N - \frac{\Sigma_i n_i}{N}\right)\hat{\sigma}_\alpha^2 + (a-1)\hat{\sigma}_e^2 \quad \text{and} \quad \text{SSE} = (N-a)\hat{\sigma}_e^2$$

for unbalanced data. Therefore

$$\hat{\sigma}_e^2 = \text{MSE} \tag{82}$$

and

$$\hat{\sigma}_\alpha^2 = \frac{\text{MSA} - \text{MSE}}{(N - \Sigma_i n_i^2/N)/(a-1)}\,. \tag{83}$$

are the ANOVA estimators for unbalanced data. They do, of course, reduce to those for balanced data ($n_i = n\ \forall\ i$) in (54) and (55).

Example 1 (continued). The data of Example 2 are in Table 3.2. They and calculation of SSA and SSE are shown in Table 3.6, from which the analysis of variance table is shown in Table 3.7.

The estimation equations (82) and (83) give

$$\hat{\sigma}_e^2 = 108/6 = 18 \tag{84}$$

TABLE 3.6. DATA OF TABLE 3.2 AND ANOVA CALCULATIONS

	y_{ij}				
	$i=1$	$i=2$	$i=3$		
	3	11	4		$\text{SSA} = 3(6-8)^2 + 4(12-8)^2 + 2(3-8)^2 = 126$
	3	13	2		
	12	17			
	—	7	—		$\text{SSE} = 2(3^2) + 6^2 + 2(1^2 + 5^2) + 2(1^2) = 108$
$y_{i.}$	18	48	6	$y_{..} = 72$	
n_i	3	4	2	$N = 9$	
$\bar{y}_{i.}$	6	12	3	$\bar{y}_{..} = 8$	$N - \Sigma_i n_i^2/N = 9 - (3^2 + 4^2 + 2^2)/9 = \frac{52}{9}$

TABLE 3.7. 1-WAY CLASSIFICATION, UNBALANCED DATA (TABLE 3.2)

Source of Variation	d.f.	Sum of Squares	Expected Value of Sum of Squares Under a Random Model
Classes	$a - 1 = 2$	$\text{SSA} = 126$	$\frac{52}{9}\sigma_\alpha^2 + 2\sigma_e^2$
Residual	$N - a = 6$	$\text{SSE} = 108$	$6\sigma_e^2$
Total	$N - 1 = 8$	$\text{SST}_\text{m} = 234$	

and

$$\hat{\sigma}_\alpha^2 = \frac{63 - 18}{(52/9)/2} = 15\tfrac{15}{26}\,. \tag{85}$$

c. Negative estimates

Data being unbalanced does not eliminate the possibility of obtaining a negative ANOVA estimate for σ_α^2. As illustration, suppose in the example of Table 3.6 that the data for $i = 2$ are 2, 2, 37 and 7 instead of 11, 13, 17 and 7. The values of n_2, N, $\bar{y}_{2.}$ and $\bar{y}_{..}$ are unchanged so that SSA is also unchanged. But SSE is then $108 - 2(1^2 + 5^2) + (10^2 + 10^2 + 25^2 + 5^2) = 906$. Hence, with that replacing 108 in Table 3.7,

$$\hat{\sigma}_\alpha^2 = \frac{63 - 906/6}{52/18} = -\tfrac{198}{13}\,.$$

As with balanced data, there is nothing inherent in the ANOVA method of estimation that prevents the possibility of such negativity. Mathew *et al.* (1991b) consider non-negative estimators from unbalanced data for models that have two variance components, of which the 1-way classification is a special case.

d. Normality assumptions

Similar to (58) but now using $\mathbf{V}$ of (81), the normality assumptions are

$$\begin{bmatrix}\boldsymbol{\alpha}\\ \mathbf{e}\end{bmatrix} \sim \mathcal{N}\left(\begin{bmatrix}\mathbf{0}\\ \mathbf{0}\end{bmatrix}, \begin{bmatrix}\sigma_\alpha^2\mathbf{I}_a & \mathbf{0}\\ \mathbf{0} & \sigma_e^2\mathbf{I}_N\end{bmatrix}\right)$$

and

$$\mathbf{y} \sim \mathcal{N}[\mu\mathbf{1}_N, \mathbf{V} = \{_{\mathrm{d}}\, \sigma_\alpha^2\mathbf{J}_{n_i} + \sigma_e^2\mathbf{I}_{n_i}\}] \tag{86}$$

-i.* χ^2*-distributions of sums of squares. As is usual, the basis for considering χ^2 properties is Theorem S2 summarized in (59). For $\mathrm{SSE} = \mathbf{y}'\mathbf{A}_2\mathbf{y}$ of (80),

$$\mathbf{A}_2\mathbf{V} = (\mathbf{I}_N - \{_{\mathrm{d}}\, \bar{\mathbf{J}}_{n_i}\})\{_{\mathrm{d}}\, \sigma_\alpha^2\mathbf{J}_{n_i} + \sigma_e^2\mathbf{I}_{n_i}\} \tag{87}$$

$$= \sigma_\alpha^2\{_{\mathrm{d}}\, \mathbf{J}_{n_i} - \bar{\mathbf{J}}_{n_i}\mathbf{J}_{n_i}\} + \sigma_e^2\{_{\mathrm{d}}\, \mathbf{I}_{n_i} - \bar{\mathbf{J}}_{n_i}\}$$

$$= \sigma_e^2\{_{\mathrm{d}}\, \mathbf{I}_{n_i} - \bar{\mathbf{J}}_{n_i}\} = \sigma_e^2\mathbf{A}_2\,. \tag{88}$$

Therefore, since $\mathbf{A}_2\mathbf{V}/\sigma_e^2$ is idempotent, SSE/σ_e^2 has a χ^2-distribution with degrees of freedom $r(\mathbf{A}_2) = \mathrm{tr}(\mathbf{A}_2) = N - a$; i.e.,

$$\mathrm{SSE}/\sigma_e^2 \sim \chi^2_{N-a}\,. \tag{89}$$

We begin similarly for $\mathrm{SSA} = \mathbf{y}'\mathbf{A}_1\mathbf{y}$ of (79):

$$\mathbf{A}_1\mathbf{V} = (\{_{\mathrm{d}}\, \bar{\mathbf{J}}_{n_i}\} - \bar{\mathbf{J}}_N)\{_{\mathrm{d}}\, \sigma_\alpha^2\mathbf{J}_{n_i} + \sigma_e^2\mathbf{I}_{n_i}\} \tag{90}$$

$$= \sigma_\alpha^2(\{_{\mathrm{d}}\, \bar{\mathbf{J}}_{n_i}\mathbf{J}_{n_i}\} - \bar{\mathbf{J}}_N\{_{\mathrm{d}}\, \mathbf{J}_{n_i}\}) + \sigma_e^2(\{_{\mathrm{d}}\, \bar{\mathbf{J}}_{n_i}\} - \bar{\mathbf{J}}_N)$$

$$= \sigma_\alpha^2(\{_{\mathrm{d}}\, \mathbf{J}_{n_i}\} - \mathbf{1}_N\{_{\mathrm{r}}\, n_i\mathbf{1}'_{n_i}\}/N) + \sigma_e^2(\{_{\mathrm{d}}\, \bar{\mathbf{J}}_{n_i}\} - \bar{\mathbf{J}}_N)\,. \tag{91}$$

Inspection of $(\mathbf{A}_1\mathbf{V})^2$ using (91) reveals that in general neither $\mathbf{A}_1\mathbf{V}$ nor any multiple of it is idempotent; it is if $n_i = n\ \forall\ i$, or $\sigma_\alpha^2 = 0$. Therefore neither SSA nor a multiple of it has a χ^2-distribution. This is in *sharp* contrast to the balanced data situation where $n_i = n\ \forall\ i$ reduces (91) to

$$\mathbf{A}_1\mathbf{V} = \sigma_\alpha^2(\mathbf{I}_a \otimes \mathbf{J}_n - \mathbf{1}_{an}\mathbf{1}'_{an}/a) + \sigma_e^2(\mathbf{I}_a \otimes \bar{\mathbf{J}}_n - \bar{\mathbf{J}}_{an})$$

$$= \sigma_\alpha^2(\mathbf{I}_a \otimes \mathbf{J}_n - \bar{\mathbf{J}}_a \otimes \mathbf{J}_n) + \sigma_e^2(\mathbf{I}_a \otimes \bar{\mathbf{J}}_n - \bar{\mathbf{J}}_a \otimes \bar{\mathbf{J}}_n)$$

$$= (n\sigma_\alpha^2 + \sigma_e^2)[(\mathbf{I}_a - \bar{\mathbf{J}}_a) \otimes \bar{\mathbf{J}}_n]; \tag{92}$$

and so for balanced data $\mathbf{A}_1\mathbf{V}/(n\sigma_\alpha^2 + \sigma_e^2) = [(\mathbf{I}_a - \bar{\mathbf{J}}_a) \otimes \bar{\mathbf{J}}_n]$ which is idempotent, so yielding the χ^2 result in (60). SSA not having a χ^2-distribution in the random model with unbalanced data is also in sharp contrast to the fixed effects model with unbalanced data. In that case, under the hypothesis H: α_i*s all equal*, one effectively has σ_α^2 as zero in (91) and so then $\mathbf{A}_1\mathbf{V}/\sigma_e^2 = \{_{\mathrm{d}}\, \bar{\mathbf{J}}_{n_i}\} - \bar{\mathbf{J}}_N$, which is idempotent. Hence SSA/σ_e^2 has a χ^2 (non-central) distribution—as is well known. But with unbalanced data, for $\sigma_\alpha^2 \neq 0$, there is no χ^2-distribution associated with SSA.

-ii.* *Independence of sums of squares. Despite SSA not having a χ^2-density, SSA and SSE are independent, just as in both the random model, balanced data case, and in the fixed effects model with either balanced or unbalanced data. Using (79), (80) and (86) in (62) gives

$$\mathbf{A}_1\mathbf{V}\mathbf{A}_2 = [\sigma_\alpha^2(\{_\mathrm{d}\,\mathbf{J}_{n_i}\} - \mathbf{1}_N\{_\mathrm{r}\, n_i\mathbf{1}'_{n_i}\}/N) + \sigma_e^2(\{_\mathrm{d}\,\bar{\mathbf{J}}_{n_i}\} - \bar{\mathbf{J}}_N)][\mathbf{I}_N - \{_\mathrm{d}\,\bar{\mathbf{J}}_{n_i}\}] \tag{93}$$

$$\begin{aligned} &= \sigma_\alpha^2(\{_\mathrm{d}\,\mathbf{J}_{n_i} - \mathbf{J}_{n_i}\bar{\mathbf{J}}_{n_i}\} - \mathbf{1}_N\{_\mathrm{r}\, n_i\mathbf{1}'_{n_i} - n_i\mathbf{1}'_{n_i}\mathbf{1}_{n_i}\mathbf{1}'_{n_i}/n_i\}/N) \\ &\quad + \sigma_e^2(\{_\mathrm{d}\,\bar{\mathbf{J}}_{n_i} - \bar{\mathbf{J}}_{n_i}^2\} - \bar{\mathbf{J}}_N + \mathbf{1}_N\{_\mathrm{r}\, n_i\mathbf{1}'_{n_i}/n_i\}/N) \\ &= \sigma_\alpha^2(\mathbf{0} - \mathbf{0}) + \sigma_e^2(\mathbf{0} - \bar{\mathbf{J}}_N + \bar{\mathbf{J}}_N) \\ &= \mathbf{0}\,. \end{aligned} \tag{94}$$

Therefore, with $\mathbf{y}$ having been assumed normally distributed, SSA and SSE are independent.

-iii. Sampling variances of estimators. Two of the three results are easy. First, because $\mathrm{SSE}/\sigma_e^2 \sim \chi^2_{N-a}$,

$$\mathrm{var}(\hat{\sigma}_e^2) = \mathrm{var}(\mathrm{MSE}) = \frac{2\sigma_e^4}{N-a}\,. \tag{95}$$

Second, using (82) and (83),

$$\begin{aligned} \mathrm{cov}(\hat{\sigma}_\alpha^2, \hat{\sigma}_e^2) &= \frac{\mathrm{cov}(\mathrm{MSA} - \mathrm{MSE}, \mathrm{MSE})}{(N - \Sigma_i n_i^2/N)/(a-1)} \\ &= \frac{-2\sigma_e^4}{(N-a)(N - \Sigma_i n_i^2/N)/(a-1)}\,. \end{aligned} \tag{96}$$

On writing

$$\hat{\sigma}_\alpha^2 = \frac{\mathrm{MSA} - \mathrm{MSE}}{n_\mathrm{u}} \quad \text{for } n_\mathrm{u} = \frac{N - \Sigma n_i^2/N}{a-1}, \tag{97}$$

and

$$n_\mathrm{u}\hat{\sigma}_\alpha^2 = \mathbf{y}'\mathbf{B}\mathbf{y} \quad \text{for } \mathbf{B} = \frac{\mathbf{A}_1}{a-1} - \frac{\mathbf{A}_2}{N-a}, \tag{98}$$

the tedious derivation is obtaining $\mathrm{var}(\hat{\sigma}_\alpha^2)$ from Theorem S4, that

$$\mathbf{y} \sim \mathcal{N}(\boldsymbol{\mu}, \mathbf{V}) \Rightarrow \mathrm{var}(\mathbf{y}'\mathbf{B}\mathbf{y}) = 2\,\mathrm{tr}(\mathbf{B}\mathbf{V})^2 + 4\boldsymbol{\mu}'\mathbf{B}\mathbf{V}\mathbf{B}\boldsymbol{\mu}\,. \tag{99}$$

In using $\mathbf{B}$ of (98) in (99) we find, with $E(\mathbf{y}) = \boldsymbol{\mu} = \mu\mathbf{1}_N$ of (83), that $\boldsymbol{\mu}'\mathbf{B}$ is null, because it is a linear combination of $\mathbf{1}'\mathbf{A}_1$ and $\mathbf{1}'\mathbf{A}_2$, each of which, from (79) and (80), is null. Therefore (99) becomes

$$\begin{aligned} \mathrm{var}(n_\mathrm{u}\hat{\sigma}_\alpha^2) &= 2\,\mathrm{tr}\left(\frac{\mathbf{A}_1\mathbf{V}}{a-1} - \frac{\mathbf{A}_2\mathbf{V}}{N-a}\right)^2 \\ &= 2\left[\frac{\mathrm{tr}(\mathbf{A}_1\mathbf{V})^2}{(a-1)^2} + \frac{\mathrm{tr}(\mathbf{A}_2\mathbf{V})^2}{(N-a)^2}\right] \end{aligned} \tag{100}$$

because $\mathbf{A}_1\mathbf{V}\mathbf{A}_2 = \mathbf{0}$, as in (94). Hence from (91) and (88)

$$\text{var}(n_u\hat{\sigma}_\alpha^2) = \frac{2}{(a-1)^2}\text{tr}[\sigma_\alpha^2(\{_\text{d}\, \mathbf{J}_{n_i}\} - \mathbf{1}\{_\text{r}\, n_i\mathbf{1}'_{n_i}\}/N) + \sigma_e^2(\{_\text{d}\, \bar{\mathbf{J}}_{n_i}\} - \bar{\mathbf{J}}_N)]^2$$
$$+ \frac{2}{(N-a)^2}\sigma_e^4\,\text{tr}\left\{_\text{d}\, \mathbf{I}_{n_i} - \bar{\mathbf{J}}_{n_i}\right\}^2. \tag{101}$$

After considerable usage (left to the reader as Exercise 3.12) of the algebra of $\mathbf{1}$, $\mathbf{J}$ and $\bar{\mathbf{J}}$ as in Appendix M.1, the ultimate simplification of (101) is

$$\text{var}(\hat{\sigma}_\alpha^2) = \frac{2N}{(N^2 - \Sigma_i n_i^2)}$$
$$\times\left[\frac{N(N-1)(a-1)}{(N-a)(N^2 - \Sigma_i n_i^2)}\sigma_e^4 + 2\sigma_e^2\sigma_\alpha^2\right.$$
$$\left. + \frac{N^2\Sigma_i n_i^2 + (\Sigma_i n_i^2)^2 - 2N\Sigma_i n_i^3}{N(N^2 - \Sigma n_i^2)}\sigma_\alpha^4\right]. \tag{102}$$

Crump (1951) was the first to derive this result; it occurs again in Searle (1956) with $2\sigma_e^2\sigma_\alpha^2$ erroneously shown as $\sigma_e^2\sigma_\alpha^2$, and is correct in Searle (1971). An extension of (102) to the rth cumulant is developed by Singh (1989), and Chatterjee and Das (1983) develop best asymptotical normal (BAN) estimators.

-iv. The effect of unbalancedness on sampling variances. A question of long-standing interest is to what extent does unbalancedness of data affect the minimum variance properties (Section 2.3b-iii) of the ANOVA estimators? At first thought one might expect that a satisfactory method of answering this question would be to study the behavior of the sampling variances of, and covariance between, $\hat{\sigma}_e^2$ and $\hat{\sigma}_\alpha^2$, for different degrees of unbalancedness, i.e., for different sets of n_i-values for given N and a.

This is easy for $\text{var}(\hat{\sigma}_e^2)$ and $\text{cov}(\hat{\sigma}_\alpha^2, \hat{\sigma}_e^2)$, since each is just a multiple of σ_e^2. Clearly, $\text{var}(\hat{\sigma}_e^2)$ of (95) for given N and a is unaffected by unbalancedness; and $\text{cov}(\hat{\sigma}_\alpha^2, \hat{\sigma}_e^2)$ of (96) is affected only to the extent that $\Sigma_i n_i^2$ is. And since $\Sigma_i n_i^2$ for $\Sigma_i n_i = N$ and $n_i \geqslant 1$ is at its maximum for one n_i being $N - (a-1)$ and the others being unity, and is at its minimum for all n_i being the same (or as nearly so as every n_i being an integer will allow), we see that $\text{cov}(\hat{\sigma}_\alpha^2, \hat{\sigma}_e^2)$ increases, numerically, as the degree of unbalancedness increases; or, put the other way, the closer that data are to being balanced, the smaller (numerically) is that covariance.

Unfortunately, the behavior of $\text{var}(\hat{\sigma}_\alpha^2)$ of (102) is not monotonic for changes in the n_i-values, given N and a. The coefficients of σ_e^4 and of $\sigma_e^2\sigma_\alpha^2$ each increase as unbalancedness increases, but this is not the case for the coefficient of σ_α^2 in (102). For example, with $N = 25$ and $a = 5$, the last term of (102) is $1.1066\sigma_\alpha^4$ for the n_i being 1, 1, 1, 11 and 11, whereas it is $.7963\sigma_\alpha^4$ for the more unbalanced data case of the n_i being 1, 1, 1, 1 and 21. A consequence of this is that, after

TABLE 3.8. EXAMPLES OF $\text{var}(\hat{\sigma}_\alpha^2)$ OF (102) FOR $N = 25$, $a = 5$ AND $\sigma_e^2 = 1$

	$\tau = \sigma_\alpha^2/\sigma_e^2$			
n_i-Values	.25	.5	1	10
1, 1, 1, 11, 11	.14	.41	1.374	113
1, 1, 1, 1, 21	.20	.49	1.370	85

expressing $\text{var}(\hat{\sigma}_\alpha^2)$ as σ_e^4 multiplying a quadratic in $\tau = \sigma_\alpha^2/\sigma_e^2$, it will be found that $\text{var}(\hat{\sigma}_\alpha^2)$ for some values of τ increases as between the first of these sets of n_i-values and the second, and it decreases for other values of τ. Examples are shown in Table 3.8. Thus, for some values of τ, $\text{var}(\hat{\sigma}_\alpha^2)$ increases as unbalancedness increases, and for other values it decreases.

The value of τ at which $\text{var}(\hat{\sigma}_\alpha^2)$ changes from increasing to decreasing as unbalancedness increases is, of course, not the same for all situations, but depends in no simple way on N, a and the n_i-values.

-v. F-statistics. The ratio $F = \text{MSA}/\text{MSE}$ can, of course, be calculated. In the fixed effects model F has a non-central $\mathscr{F}$-distribution. In the random model with balanced data F is distributed as a multiple of an $\mathscr{F}$-distribution—see (73). But in the random model with unbalanced data F does *not* have even a multiple of an $\mathscr{F}$-distribution when $\sigma_\alpha^2 > 0$. This is because, even though MSA and MSE are independent with σ_α^2 being non-zero, MSA is then not distributed as a multiple of a χ^2. Nevertheless, on defining

$$w_i = \frac{\sigma_e^2}{\text{var}(\bar{y}_{i\cdot})} = \frac{\sigma_e^2}{\sigma_\alpha^2 + \sigma_e^2/n_i} = \frac{n_i}{1 + n_i\tau},$$

and

$$F^* = \frac{h(\tau)}{(a-1)\text{MSE}} \quad \text{for } h(\tau) = \sum_{i=1}^{a} w_i\left(\bar{y}_{i\cdot} - \frac{\Sigma_i w_i \bar{y}_{i\cdot}}{\Sigma_i w_i}\right)^2 .$$

it is shown in Wald (1940) that $F^* \sim \mathscr{F}_{N-a}^{a-1}$. Moreover, $\sigma_\alpha^2 = 0$ simplifies F^* to be $F = \text{MSA}/\text{MSE}$, thus providing F as a test statistic for H: $\sigma_\alpha^2 = 0$, even though F is not distributed as an $\mathscr{F}$-variable when $\sigma_\alpha^2 > 0$. [In concert with $F^* \sim \mathscr{F}_{N-a}^{a-1}$ when $\sigma_\alpha^2 = 0$, it can also be shown from (91) that $\text{SSA} \sim \sigma_e^2\chi_{a-1}^2$ when $\sigma_\alpha^2 = 0$.] Spjøtvoll (1967) suggests that this test is nearly optimal for large values of τ, and Westfall (1988, 1989) and LaMotte *et al.* (1988) have made comparisons of this test with other exact tests. A summary of these results is shown in Table 3.9. A further test, which is locally best, invariant, and unbiased is developed by Das and Sinha (1987). They also consider both other models and robustness.

-vi. Confidence intervals. Because $\text{SSE}/\sigma_e^2 \sim \chi_{N-a}^2$, a confidence interval on σ_e^2 is easily derived in the same manner as for balanced data, in Section

TABLE 3.9. F-DISTRIBUTIONS AND F-TESTS IN THE 1-WAY CLASSIFICATION

$$F = \frac{\text{MSA}}{\text{MSE}} \text{ and } F^* = \frac{\Sigma_i w_i(\bar{y}_{i\cdot} - \Sigma_i w_i \bar{y}_{i\cdot}/\Sigma w_i)^2}{(a-1)\text{MSE}} \text{ with } w_i = \left(\tau + \frac{1}{n_i}\right)^{-1}$$

Model	Balanced Data	Unbalanced Data	Balanced and Unbalanced Data
Fixed[1]	$F \sim \mathscr{F}'[a-1, a(n-1), \lambda]$	$F \sim \mathscr{F}'(a-1, N-a, \lambda)$	F tests H: α_is *all equal*[2]
Random	$\dfrac{F}{1+\tau} \sim \mathscr{F}^{a-1}_{a(n-1)}$	$F^* \sim \mathscr{F}^{a-1}_{N-a}$	F tests H: $\sigma^2_\alpha = 0$[3]

[1] $\mathscr{F}'(\cdot,\cdot,\cdot)$ represents the non-central $\mathscr{F}$ distribution: see Appendix S.4.
[2] Putting all α_is equal reduces λ to zero.
[3] $\sigma^2_\alpha = 0$ reduces both $F/(1+\tau)$ and F^* to F.

3.5d-v, namely

$$\Pr\left\{\frac{\text{SSE}}{\chi^2_{N-a,\,U}} \leqslant \sigma^2_e \leqslant \frac{\text{SSE}}{\chi^2_{N-a,\,L}}\right\} = 1 - \alpha .$$

But because there is no readily tractable density function, no algebraic confidence interval is available for σ^2_α. There is, however, a variety of approximate intervals, which are fully reviewed by Burdick and Graybill (1988). The $1-\alpha$ approximate interval that they say performs well is one developed by Burdick and Eickman (1986), based on the following parade of definitions.

$\mathscr{F}^n_d$ is a random variable having an $\mathscr{F}$-distribution with numerator and denominator degrees of freedom n and d, respectively. $\mathscr{F}^n_d(\alpha)$ is the point on the real line beyond which there is an area α in the distribution of $\mathscr{F}^n_d$; i.e.

$$\Pr\{\mathscr{F}^n_d > \mathscr{F}^n_d(\alpha)\} = \alpha .$$

For $\alpha_{11} + \alpha_{21} = \alpha = \alpha_{12} + \alpha_{22}$, which provides two opportunities for dividing the α probability level into two pieces, define

$$f_1 = \mathscr{F}^{a-1}_{\infty}(\alpha_{11}), \qquad f_2 = \mathscr{F}^{a-1}_{a(n-1)}(\alpha_{12}),$$

$$f_3 = \mathscr{F}^{a-1}_{\infty}(1-\alpha_{21}), \quad f_4 = \mathscr{F}^{a-1}_{a(n-1)}(1-\alpha_{22}) .$$

$$m = \min\{n_i\}, \quad M = \max\{n_i\} \quad k = [\Sigma_i(1/n_i)]^{-1}$$

$$\text{MSA}_t = \frac{\Sigma_i(\bar{y}_{i\cdot} - \Sigma_i \bar{y}_{i\cdot}/a)^2}{a-1},$$

$$L = \frac{\text{MSA}_t}{f_2 \text{MSE}} - \frac{1}{m}, \quad U = \frac{\text{MSA}_t}{f_4 \text{MSE}} - \frac{1}{M} .$$

Then,

$$\Pr\left\{\frac{kL(\text{MSA}_t)}{f_1(1+kL)} \leqslant \sigma^2_\alpha \leqslant \frac{kU(\text{MSA}_t)}{f_3(1+kU)}\right\} \doteqdot 1 - \alpha .$$

An exact confidence interval for $\tau = \sigma_\alpha^2/\sigma_e^2$ is proposed by Wald (1940), based on his F^* shown in the title to Table 3.9. Because $F^* \sim \mathscr{F}_{N-a}^{a-1}$, a $1-\alpha$ confidence interval for F^* is

$$F_{\mathrm{L}} \leqslant \frac{h(\tau)}{(a-1)\mathrm{MSE}} \leqslant F_{\mathrm{U}},$$

where $h(\tau)$ is the numerator of F^* and F_{L} and F_{U} are lower and upper limits of the $\mathscr{F}_{N-a}^{a-1}$ distribution, respectively, similar to their definition used in Table 3.4.

Wald (1940) shows that $h(\tau)$ decreases as τ increases. Therefore the confidence limits on τ, say τ_{L}^* and τ_{U}^*, are based on the solutions to

$$h(\tau_{\mathrm{L}}) = [(a-1)\mathrm{MSE}]F_{\mathrm{U}} \quad \text{and} \quad h(\tau_{\mathrm{U}}) = [(a-1)\mathrm{MSE}]F_{\mathrm{L}}\,.$$

Because $h(\tau)$ is decreasing in τ, with $h(0) = \mathrm{MSA}/\mathrm{MSE}$ and $h(\infty) = 0$ (as may be easily verified), there may be no solutions to either or both of these equations when $h(0)$ is less than F_{L} or F_{U}. When that occurs, the corresponding limit for τ, namely τ_{U}^* and τ_{L}^*, respectively, is taken as zero. Thus, in summary, a $1-\alpha$ confidence interval for τ is $(\tau_{\mathrm{L}}^*, \tau_{\mathrm{U}}^*)$, where

$$\tau_{\mathrm{L}}^* = \begin{cases} \tau_{\mathrm{L}} & \text{when } h(0) > F_{\mathrm{U}}, \\ 0 & \text{otherwise,} \end{cases}$$

and

$$\tau_{\mathrm{U}}^* = \begin{cases} \tau_{\mathrm{U}} & \text{when } h(0) > F_{\mathrm{L}}, \\ 0 & \text{otherwise}\,. \end{cases}$$

The corresponding confidence interval for $\rho = \sigma_\alpha^2/(\sigma_\alpha^2 + \sigma_e^2)$ is

$$\frac{\tau_{\mathrm{L}}^*}{1+\tau_{\mathrm{L}}^*} \leqslant \rho \leqslant \frac{\tau_{\mathrm{U}}^*}{1+\tau_{\mathrm{U}}^*}\,.$$

A broad review of inference procedures for ρ is given by Donner (1986), including much of what is in Shoukri and Ward (1984). Confidence intervals in the unbalanced data case are considered by Burdick, Maqsood and Graybill (1986). They begin at the last line of Table 3.4 and extend Wald's (1940) methods and in doing so compare a variety of methods for deriving confidence intervals. Extension to the 2-way nested classification is considered in Burdick, Birch and Graybill (1986).

3.7. MAXIMUM LIKELIHOOD ESTIMATION (MLE)

Maximum likelihood estimation of variance components from data on a continuous variable is often confined to situations based on the normality assumptions. For unbalanced data, with

$$\mathbf{y} \sim \mathscr{N}(\mu\mathbf{1}_N, \mathbf{V} = \{_{\mathrm{d}}\, \sigma_\alpha^2 \mathbf{J}_{n_i} + \sigma_e^2 \mathbf{I}_{n_i}\}),$$

the likelihood function is then defined as

$$L = L(\mu, \mathbf{V} \mid \mathbf{y}) = \frac{\exp[-\frac{1}{2}(\mathbf{y} - \mu\mathbf{1}_N)'\mathbf{V}^{-1}(\mathbf{y} - \mu\mathbf{1}_N)]}{(2\pi)^{\frac{1}{2}N}|\mathbf{V}|^{\frac{1}{2}}} . \tag{103}$$

This is the same as the density function for $\mathbf{y}$, usually denoted as $f(\mathbf{y} \mid \mu, \mathbf{V})$; i.e.,

$$L(\mu, \mathbf{V} \mid \mathbf{y}) \equiv f(\mathbf{y} \mid \mu, \mathbf{V}) .$$

Although these two symbols represent exactly the same function, each is used in its own particular context. The notation $f(\mathbf{y} \mid \mu, \mathbf{V})$ is used when interest lies in the density of $\mathbf{y}$, with μ and $\mathbf{V}$ being treated as fixed. In contrast, $L(\mu, \mathbf{V} \mid \mathbf{y})$ is used when we want to emphasize that the same function, namely the right-hand side of (103), can also be viewed as a function of μ and $\mathbf{V}$ for some given vector of data, $\mathbf{y}$. This is the context in which (103) is used as a basis of maximum likelihood estimation of μ and $\mathbf{V}$, the MLE of $\mathbf{V}$ being $\mathbf{V}$ with σ_α^2 and σ_e^2 replaced by their MLEs.

With $\mathbf{V}$ in (103) involving the form $a\mathbf{I} + b\mathbf{J}$ that is discussed at the end of Section M.1, we get from that section

$$\mathbf{V}^{-1} = \left\{_{\rm d}\, \frac{1}{\sigma_e^2}\left(\mathbf{I}_{n_i} - \frac{\sigma_\alpha^2}{\sigma_e^2 + n_i\sigma_\alpha^2}\mathbf{J}_{n_i}\right)\right\} \quad \text{and} \quad |\mathbf{V}| = \prod_{i=1}^{a} \sigma_e^{2(n_i - 1)}(\sigma_e^2 + n_i\sigma_\alpha^2),$$

leading to

$$L = \frac{\exp\left\{-\dfrac{1}{2\sigma_e^2}\left[\Sigma_i\Sigma_j(y_{ij} - \mu)^2 - \sum_i \dfrac{\sigma_\alpha^2}{\sigma_e^2 + n_i\sigma_\alpha^2}(y_{i\cdot} - n_i\mu)^2\right]\right\}}{(2\pi)^{\frac{1}{2}N}\sigma_e^{2[\frac{1}{2}(N-a)]}\prod_{i=1}^{a}(\sigma_e^2 + n_i\sigma_\alpha^2)^{\frac{1}{2}}} . \tag{104}$$

Since parameter values maximizing L are equal to those maximizing its natural logarithm, and because log L, which we denote by l, is often a more tractable function than L, we deal with

$$\begin{aligned} l = \log L &= \log[L(\mu, \mathbf{V} \mid \mathbf{y})] \\ &= -\tfrac{1}{2}N\log 2\pi - \tfrac{1}{2}(N - a)\log\sigma_e^2 - \tfrac{1}{2}\Sigma_i \log(\sigma_e^2 + n_i\sigma_\alpha^2) \\ &\quad - \frac{\Sigma_i\Sigma_j(y_{ij} - \mu)^2}{2\sigma_e^2} + \frac{1}{2\sigma_e^2}\Sigma_i \frac{\sigma_\alpha^2}{\sigma_e^2 + n_i\sigma_\alpha^2}(y_{i\cdot} - n_i\mu)^2 . \end{aligned}$$

a. Balanced data

-i. Likelihood. Balanced data has $n_i = n \;\forall\; i$. This greatly simplifies log L so that it becomes

$$\begin{aligned} l = \log L = &-\tfrac{1}{2}N\log 2\pi - \tfrac{1}{2}a(n-1)\log\sigma_e^2 - \tfrac{1}{2}a[\log(\sigma_e^2 + n\sigma_\alpha^2)] \\ &- \frac{\Sigma_i\Sigma_j(y_{ij} - \mu)^2}{2\sigma_e^2} + \frac{n^2\sigma_\alpha^2\Sigma_i(\bar{y}_{i\cdot} - \mu)^2}{2\sigma_e^2(\sigma_e^2 + n\sigma_\alpha^2)} . \end{aligned}$$

The last two terms can be rearranged so as to display SSA and SSE:

$$-\frac{\Sigma_i\Sigma_j(y_{ij}-\mu)^2}{2\sigma_e^2}+\frac{n^2\sigma_\alpha^2\Sigma_i(\bar{y}_{i.}-\mu)^2}{2\sigma_e^2(\sigma_e^2+n\sigma_\alpha^2)}$$

$$=-\frac{1}{2\sigma_e^2}\left[\Sigma_i\Sigma_j(y_{ij}-\bar{y}_{i.}+\bar{y}_{i.}-\mu)^2-\frac{n\sigma_\alpha^2}{\sigma_e^2+n\sigma_\alpha^2}\Sigma_i n(\bar{y}_{i.}-\mu)^2\right]$$

$$=-\frac{1}{2\sigma_e^2}\left[\text{SSE}+\left(1-\frac{n\sigma_\alpha^2}{\sigma_e^2+n\sigma_\alpha^2}\right)\Sigma_i n(\bar{y}_{i.}-\bar{y}_{..}+\bar{y}_{..}-\mu)^2\right]$$

$$=-\frac{1}{2\sigma_e^2}\left\{\text{SSE}+\frac{\sigma_e^2}{\sigma_e^2+n\sigma_\alpha^2}[\text{SSA}+an(\bar{y}_{..}-\mu)^2]\right\}.$$

Notation. Because the MLE of a function of parameters is that same function of the MLEs of the parameters, we simplify notation by writing

$$\lambda=\sigma_e^2+n\sigma_\alpha^2\,. \tag{105}$$

Then l becomes

$$\begin{aligned}l=\log L=&-\tfrac{1}{2}N\log 2\pi-\tfrac{1}{2}a(n-1)\log\sigma_e^2\\&-\tfrac{1}{2}a\log\lambda-\frac{\text{SSE}}{2\sigma_e^2}-\frac{\text{SSA}}{2\lambda}-\frac{an(\bar{y}_{..}-\mu)^2}{2\lambda}\,.\end{aligned} \tag{106}$$

-ii. ML equations and their solutions. The maximum likelihood equations are those equations obtained by equating to zero the partial derivatives of log L with respect to μ, σ_e^2 and λ:

$$\begin{aligned}\frac{\partial l}{\partial\mu}&=\frac{an(\bar{y}_{..}-\mu)}{\lambda},\\ \frac{\partial l}{\partial\sigma_e^2}&=\frac{-a(n-1)}{2\sigma_e^2}+\frac{\text{SSE}}{2\sigma_e^4}=\frac{-a(n-1)}{2\sigma_e^4}\left[\sigma_e^2-\frac{\text{SSE}}{a(n-1)}\right],\\ \frac{\partial l}{\partial\lambda}&=\frac{-a}{2\lambda}+\frac{\text{SSA}}{2\lambda^2}+\frac{an(\bar{y}_{..}-\mu)^2}{2\lambda^2}=\frac{-a}{2\lambda^2}\left(\lambda-\frac{\text{SSA}}{a}\right)+\frac{an(\bar{y}_{..}-\mu)^2}{2\lambda^2}\,.\end{aligned} \tag{107}$$

In equating these partial derivatives to zero we change the parameter symbols μ, σ_e^2 and λ to be the symbols $\dot{\mu}$, $\dot{\sigma}_e^2$ and $\dot{\lambda}$ representing solutions to those equations, and from the form of the derivatives get those solutions as

$$\dot{\mu}=\bar{y}_{..},\quad \dot{\sigma}_e^2=\text{MSE},\quad \dot{\lambda}=\frac{\text{SSA}}{a}=\left(1-\frac{1}{a}\right)\text{MSA}$$

and (108)

$$\dot{\sigma}_\alpha^2=\frac{\dot{\lambda}-\dot{\sigma}_e^2}{n}=\frac{(1-1/a)\text{MSA}-\text{MSE}}{n}\,.$$

These are the solutions to the maximum likelihood equations. But they are not necessarily the maximum likelihood estimators, even though $L(\dot{\mu}, \dot{\mathbf{V}} \mid \mathbf{y})$, which is identical to $L(\dot{\mu}, \dot{\sigma}_e^2, \dot{\sigma}_\alpha^2 \mid \mathbf{y})$, is the maximum of $L(\mu, \sigma_e^2, \sigma_\alpha^2 \mid \mathbf{y})$ for variation in μ, σ_e^2 and σ_α^2.

-iii. ML estimators. Denote the MLEs by $\tilde{\mu}$, $\tilde{\sigma}_e^2$ and $\tilde{\sigma}_\alpha^2$. The reason that not all of $\dot{\mu}$, $\dot{\sigma}_e^2$ and $\dot{\sigma}_\alpha^2$ are $\tilde{\mu}$, $\tilde{\sigma}_e^2$ and $\tilde{\sigma}_\alpha^2$, respectively, is that all of $\dot{\mu}$, $\dot{\sigma}_e^2$ and $\dot{\sigma}_\alpha^2$ do not always lie in the parameter space for μ, σ_e^2 and σ_α^2. In particular, $\dot{\sigma}_\alpha^2$ of (108) can be negative, and when it is it is not in the parameter space $(0, \infty)$ defined by σ_α^2 being a variance and thus non-negative, i.e., $0 \leqslant \sigma_\alpha^2 < \infty$. Thus, in general, the solution $\dot{\sigma}_\alpha^2$ will be the MLE $\tilde{\sigma}_\alpha^2$ only when it is non-negative. The very definition of maximum likelihood demands that the likelihood be maximized over the *parameter space.* Hence MLEs must be in the parameter space, which means $\tilde{\sigma}_e^2 > 0$ and $\tilde{\sigma}_\alpha^2 \geqslant 0$.

Fortunately, in the 1-way classification $\dot{\sigma}_\alpha^2$ is the only one of the three ML solutions $\dot{\mu}$, $\dot{\sigma}_e^2$ and $\dot{\sigma}_\alpha^2$ that is not necessarily in the parameter space, which is, from the nature of the parameters, $-\infty < \mu < \infty, 0 < \sigma_e^2 < \infty$ and $0 \leqslant \sigma_\alpha^2 < \infty$. We consider the solutions $\dot{\mu}$, $\dot{\sigma}_e^2$ and $\dot{\sigma}_\alpha^2$ in turn. First, $\dot{\mu}$ does not depend on $\dot{\sigma}_e^2$ or $\dot{\sigma}_\alpha^2$, and since $\dot{\mu} = \bar{y}_{..}$ is clearly in the space of μ it is the MLE of μ:

$$\text{MLE}(\mu) = \tilde{\mu} = \dot{\mu} = \bar{y}_{..}\,.$$

And $\dot{\sigma}_e^2 = \text{MSE}$ is in the parameter space for σ_e^2, since MSE is never negative (and we exclude the naïve case where $y_{ij} = \bar{y}_{i.}\ \forall\ i$ and j, which would give $\dot{\sigma}_e^2 = 0$). But since $\dot{\sigma}_\alpha^2$ depends on $\dot{\sigma}_e^2$, we must ensure not just that $\tilde{\sigma}_e^2$ is in the parameter space for σ_e^2 but that the pair of estimators $(\tilde{\sigma}_e^2, \tilde{\sigma}_\alpha^2)$ is in the 2-space defined by $(\sigma_e^2, \sigma_\alpha^2)$. As a result, we find that there is a condition under which $\dot{\sigma}_e^2 = \text{MSE}$ is not the MLE of σ_e^2.

We now consider $\dot{\sigma}_e^2$ and $\dot{\sigma}_\alpha^2$ and invoke an argument similar to the original one of Herbach (1959) to derive ML estimators $\tilde{\sigma}_e^2$ and $\tilde{\sigma}_\alpha^2$ from the ML solutions $\dot{\sigma}_e^2$ and $\dot{\sigma}_\alpha^2$. To do so, we consider σ_e^2 and $\lambda = \sigma_e^2 + n\sigma_\alpha^2$, the latter in place of σ_α^2. Then L is a positive function of positive parameters σ_e^2 and λ. It could be plotted in a 3-dimensional figure with L being the third dimension above the positive quadrant of the (σ_e^2, λ)-plane of Figure 3.2.

Consider the line $\sigma_e^2 = \lambda$ shown in Figure 3.2. Since $\sigma_\alpha^2 \geqslant 0$ implies $\lambda = \sigma_e^2 + n\sigma_\alpha^2 \geqslant \sigma_e^2$, and in Figure 3.2 all points for which $\lambda \geqslant \sigma_e^2$ are those on and below the $\lambda = \sigma_e^2$ line, this is the region of the figure in which the MLE point $(\tilde{\lambda}, \tilde{\sigma}^2)$ must lie. It is called the *feasible region.* Whenever $(\dot{\lambda}, \dot{\sigma}_e^2)$ is in that region, it is the MLE point. In other words,

$$\text{when } \dot{\sigma}_\alpha^2 \geqslant 0, \quad \tilde{\sigma}_\alpha^2 = \dot{\sigma}_\alpha^2 \text{ and } \tilde{\sigma}_e^2 = \dot{\sigma}_e^2\,. \tag{109}$$

This leaves us needing the MLEs when $\dot{\sigma}_\alpha^2 < 0$, which is when $\dot{\lambda} < \dot{\sigma}_e^2$, an example of which is shown in Figure 3.2. We argue by contradiction that when $\dot{\sigma}_\alpha^2 < 0$, $\tilde{\sigma}_\alpha^2$ must equal zero. Therefore assume that $\dot{\sigma}_\alpha^2 < 0$ but $\tilde{\sigma}_\alpha^2 > 0$, i.e., $\tilde{\lambda} > \tilde{\sigma}_e^2$. We consider the two cases $\tilde{\sigma}_e^2 < \dot{\sigma}_e^2$ and $\tilde{\sigma}_e^2 \geqslant \dot{\sigma}_e^2$ separately. In the first

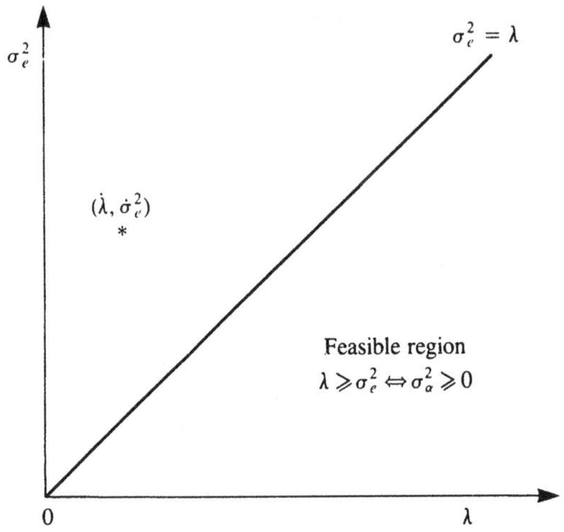

Figure 3.2. The positive quadrant, for $\lambda = \sigma_e^2 + n\sigma_\alpha^2$, of the (λ, σ_e^2)-plane, showing the $\sigma_e^2 = \lambda$ line, the feasible region (that line and all points between it and the λ-axis) and a solution point $(\dot{\lambda}, \dot{\sigma}_e^2)$ that is not in the feasible region.

case $\tilde{\sigma}_e^2 < \dot{\sigma}_e^2$, and from (107) we have, at the MLE,

$$\left.\frac{\partial \log L}{\partial \sigma_e^2}\right|_{\sigma_e^2 = \tilde{\sigma}_e^2} = \frac{-a(n-1)}{2\tilde{\sigma}_e^4}\left[\tilde{\sigma}_e^2 - \frac{\text{SSE}}{a(n-1)}\right] = \frac{-a(n-1)}{2\tilde{\sigma}_e^4}(\tilde{\sigma}_e^2 - \dot{\sigma}_e^2) > 0 \,.$$

Therefore we can increase the log likelihood which, of course, is a function of λ and σ_e^2, by increasing σ_e^2 from $\tilde{\sigma}_e^2$ and leaving λ at $\tilde{\lambda}$. This is a contradiction to $\tilde{\sigma}_e^2$ being the MLE. Now consider the second case, $\tilde{\sigma}_e^2 \geqslant \dot{\sigma}_e^2$, where

$$\tilde{\lambda} \geqslant \tilde{\sigma}_e^2 \geqslant \dot{\sigma}_e^2 > \dot{\lambda} \,. \tag{110}$$

The first inequality is the requirement that the MLEs be in the feasible region; the second inequality is our second case and the third inequality follows because $\dot{\sigma}_\alpha^2 < 0$. Now at the MLE, from (107) we have

$$\begin{aligned}\left.\frac{\partial \log L}{\partial \lambda}\right|_{\substack{\mu = \bar{y}_{..} \\ \lambda = \tilde{\lambda}}} &= \frac{-a}{2\tilde{\lambda}^2}\left(\tilde{\lambda} - \frac{\text{SSA}}{a}\right) + \frac{an(\bar{y}_{..} - \bar{y}_{..})^2}{2\tilde{\lambda}^2} \\ &= \frac{-a}{2\tilde{\lambda}^2}(\tilde{\lambda} - \dot{\lambda}) \\ &< 0, \quad \text{from (110)} \,.\end{aligned}$$

Hence we can increase the log likelihood by decreasing λ from $\tilde{\lambda}$ and leaving σ_e^2 at $\tilde{\sigma}_e^2$. This is a contradiction to $\tilde{\lambda}$ being the MLE. Therefore we have contradicted the statement $\tilde{\sigma}_\alpha^2 > 0$ when $\dot{\sigma}_\alpha^2 < 0$, and so $\tilde{\sigma}_\alpha^2$ must be zero; i.e., $\tilde{\sigma}_\alpha^2 = 0$ and, equivalently, $\tilde{\lambda} = \tilde{\sigma}_e^2$ when $\dot{\sigma}_\alpha^2 < 0$.

Thus, in order to find the ML estimator of μ and of σ_e^2 when $\tilde{\sigma}_\alpha^2 = 0$, we must obtain them by maximizing $\log L$ subject to $\lambda = \sigma_e^2$. This does not mean that we are taking $\lambda = \sigma_e^2$ (and hence $\sigma_\alpha^2 = 0$) in the model, but that we are simply going to maximize log L confined to the plane $\lambda = \sigma_e^2$ in the 3-dimensional space (L, σ_e^2, λ). Thus, on denoting the log likelihood when $\lambda = \sigma_e^2$ as $l(\lambda = \sigma_e^2)$, we find that putting $\lambda = \sigma_e^2$ in (106) gives

$$l(\lambda = \sigma_e^2) = -\tfrac{1}{2}N \log 2\pi - \tfrac{1}{2}N \log \sigma_e^2 - \frac{\text{SSA} + \text{SSE}}{2\sigma_e^2} - \frac{an(\bar{y}_{..} - \mu)^2}{2\sigma_e^2}. \quad (111)$$

Maximizing this with respect to μ and σ_e^2 leads to $\tilde{\mu} = \bar{y}_{..} = \dot{\mu}$ and

$$\tilde{\sigma}_e^2 = \frac{\text{SSA} + \text{SSE}}{\text{N}} = \frac{\text{SST}_{\text{m}}}{an} \quad (112)$$

$$= \frac{(a-1)\text{MSA} + a(n-1)\text{MSE}}{an} = \left(1 - \frac{1}{n}\right)\text{MSE} + \frac{1}{n}\left(1 - \frac{1}{a}\right)\text{MSA},$$

$$= \text{MSE} + \frac{1}{n}\left[\left(1 - \frac{1}{a}\right)\text{MSA} - \text{MSE}\right]$$

$$= \dot{\sigma}_e^2 + \dot{\sigma}_\alpha^2 . \quad (113)$$

$$< \dot{\sigma}_e^2 . \qquad [\dot{\sigma}_\alpha^2 < 0]$$

Note that $\dot{\sigma}_\alpha^2$ can never be sufficiently negative for (113) to be negative (because $\tilde{\sigma}_e^2 = \text{SST}_{\text{m}}/an$ and SST_{m} is never negative). Thus when $\dot{\sigma}_\alpha^2 < 0$, the MLEs are $\tilde{\sigma}_\alpha^2 = 0$ and $\tilde{\sigma}_e^2 = \text{SST}_{\text{m}}/an$. In summary, then, the MLEs are as follows:

$$\tilde{\sigma}_e^2 = \begin{cases} \text{MSE} & \text{if } \left(1 - \frac{1}{a}\right)\text{MSA} \geqslant \text{MSE}, \\ \dfrac{\text{SST}_{\text{m}}}{an} & \text{if } \left(1 - \frac{1}{a}\right)\text{MSA} < \text{MSE}, \end{cases} \quad (114)$$

and

$$\tilde{\sigma}_\alpha^2 = \begin{cases} \left[\left(1 - \frac{1}{a}\right)\text{MSA} - \text{MSE}\right]\Big/ n & \text{if } \left(1 - \frac{1}{a}\right)\text{MSA} \geqslant \text{MSE}, \\ 0 & \text{if } \left(1 - \frac{1}{a}\right)\text{MSA} < \text{MSE}. \end{cases} \quad (115)$$

Although this is certainly the correct way of stating the MLEs, we also state them in a manner that may well be more immediately readable for data analysts. This is because we state the data conditions first:

$$\text{if } \left(1 - \frac{1}{a}\right)\text{MSA} \geqslant \text{MSE} \quad \text{then} \quad \tilde{\sigma}_\alpha^2 = \left[\left(1 - \frac{1}{a}\right)\text{MSA} - \text{MSE}\right]\Big/ n$$

$$\text{and} \quad \tilde{\sigma}_e^2 = \text{MSE}, \quad (116)$$

$$\text{if } \left(1 - \frac{1}{a}\right)\text{MSA} < \text{MSE} \quad \text{then} \quad \tilde{\sigma}_\alpha^2 = 0 \text{ and } \tilde{\sigma}_e^2 = \frac{\text{SST}_{\text{m}}}{an}, \quad (117)$$

the latter being $\tilde{\sigma}_e^2$ of (109).

Example. The balanced data of Table 3.1, with analysis of variance in Table 3.3, have

$$a = 3, \quad \text{SSA} = 104, \quad \text{MSA} = 52,$$
$$n = 4, \quad \text{SSE} = 828, \quad \text{MSE} = 92,$$
$$\text{SST}_{\text{m}} = 932 .$$

From (56) the ANOVA estimates are

$$\hat{\sigma}_e^2 = \text{MSE} = 92 \quad \text{and} \quad \hat{\sigma}_\alpha^2 = (\text{MSA} - \text{MSE})/n = \tfrac{1}{4}(52 - 92) = -10 .$$

The ML solutions of (108) are

$$\dot{\sigma}_e^2 = \text{MSE} = 92$$

and

$$\dot{\sigma}_\alpha^2 = \frac{1}{n}\left[\left(1 - \frac{1}{a}\right)\text{MSA} - \text{MSE}\right] = \frac{1}{4}\left[\frac{2(52)}{3} - 92\right] = -14\tfrac{1}{3}$$
$$= \hat{\sigma}_\alpha^2 - \text{MSA}/an = -10 - 13/3 = -14\tfrac{1}{3} . \tag{118}$$

Since $\dot{\sigma}_\alpha^2 < 0$, the MLEs are, from (115)

$$\tilde{\sigma}_\alpha^2 = 0 \quad \text{and} \quad \tilde{\sigma}_e^2 = \dot{\sigma}_e^2 + \dot{\sigma}_\alpha^2 = 92 - 14\tfrac{1}{3} = 77\tfrac{2}{3} .$$

-iv. Expected values and bias. $E(\text{MSE}) = \sigma_e^2$ and $E(\text{MSA}) = n\sigma_\alpha^2 + \sigma_e^2$ so it is not difficult to derive

$$E(\dot{\sigma}_e^2) = \sigma_e^2 \quad \text{and} \quad E(\dot{\sigma}_\alpha^2) = \left(1 - \frac{1}{a}\right)\sigma_\alpha^2 - \frac{1}{an}\sigma_e^2 .$$

Thus the solution $\dot{\sigma}_e^2$ is an unbiased estimator of σ_e^2; and the solution $\dot{\sigma}_\alpha^2$ is a biased estimator of σ_α^2.

Finding expected values of the ML estimators is more difficult because the form of the estimators depends on whether $\dot{\sigma}_\alpha^2$ is positive or negative. Thus the expectations depend on p of (74):

$$p = \Pr\{\dot{\sigma}_\alpha^2 < 0\}$$
$$= \Pr\{\mathscr{F}_{a-1}^{a(n-1)} > (1 - 1/a)(1 + n\tau)\} \tag{119}$$

for $\tau = \sigma_\alpha^2/\sigma_e^2$.

First consider $\tilde{\sigma}_\alpha^2$. It is $\dot{\sigma}_\alpha^2$ with probability $1 - p$ and zero with probability p. Hence its expectation is $(1 - p)E(\dot{\sigma}_\alpha^2 \mid \dot{\sigma}_\alpha^2 \geqslant 0)$. The expectation involved here, over only the non-negative part of the real line, is not easily derived because the density of $\dot{\sigma}_\alpha^2$ is not a tractable function.

The expected value of $\dot{\sigma}_e^2$ is no more tractable. Since $\tilde{\sigma}_e^2 = \text{MSE}$ when $\dot{\sigma}_\alpha^2 > 0$ and $\tilde{\sigma}_e^2 = \text{SST}_{\text{m}}/an$ when $\dot{\sigma}_\alpha^2 < 0$, and because $\dot{\sigma}_\alpha^2 > 0$ with probability $1 - p$,

$$E(\tilde{\sigma}_e^2) = (1 - p)E(\text{MSE} \mid \dot{\sigma}_\alpha^2 \geqslant 0) + pE(\text{SST}_{\text{m}} \mid \dot{\sigma}_\alpha^2 < 0)/an . \tag{120}$$

Again these expectations, being conditional expectations over the non-negative part of the real line, in one case, and over the negative part in the other, are not tractable functions.

-v. ***Sampling variances.*** As a basis for comparison, we restate scalar results from (66), (69) and (71) as the sampling dispersion matrix of the ANOVA estimators:

$$\text{var}\begin{bmatrix} \hat{\sigma}_e^2 \\ \hat{\sigma}_\alpha^2 \end{bmatrix} = \text{var}\begin{bmatrix} \text{MSE} \\ \dfrac{\text{MSA} - \text{MSE}}{n} \end{bmatrix} = 2\sigma_e^4 \begin{bmatrix} \dfrac{1}{a(n-1)} & \dfrac{-1}{an(n-1)} \\ \dfrac{-1}{an(n-1)} & \dfrac{1}{n^2}\left(\dfrac{\lambda^2/\sigma_e^4}{a-1} + \dfrac{1}{a(n-1)}\right) \end{bmatrix}. \tag{121}$$

And for the solutions of the maximum likelihood equations in (108)

$$\text{var}\begin{bmatrix} \dot{\sigma}_e^2 \\ \dot{\sigma}_\alpha^2 \end{bmatrix} = \text{var}\begin{bmatrix} \text{MSE} \\ \dfrac{(1 - 1/a)\text{MSA} - \text{MSE}}{n} \end{bmatrix}$$

$$= 2\sigma_e^4 \begin{bmatrix} \dfrac{1}{a(n-1)} & \dfrac{-1}{an(n-1)} \\ \dfrac{-1}{an(n-1)} & \dfrac{1}{n^2}\left(\dfrac{\lambda^2/\sigma_e^4}{a^2/(a-1)} + \dfrac{1}{a(n-1)}\right) \end{bmatrix}. \tag{122}$$

In both cases $\lambda^2/\sigma_e^4 = (1 + n\tau)^2$ for $\tau = \sigma_\alpha^2/\sigma_e^2$.

The only difference between (121) and (122) is that in the (2, 2) element of (122) the term in λ^2/σ_e^4 has denominator $a^2/(a-1)$ whereas in (121) it has $a - 1$. And since $a^2/(a-1) > a - 1$, we have $\text{var}(\dot{\sigma}_\alpha^2) < \text{var}(\hat{\sigma}_\alpha^2)$.

To derive the asymptotic large-sample dispersion matrix for the MLEs, we need the negative of the expected value of the matrix of second derivatives of the likelihood (see Appendix S.7). Let l_θ denote $\partial \log L/\partial\theta$ for some scalar θ; and similarly let $l_{\theta,\phi}$ denote $\partial l_\theta/\partial\phi$ for scalar ϕ. Then from (107)

$$l_\mu = \frac{an(\bar{y}_{..} - \mu)}{\lambda}, \quad l_{\sigma_e^2} = \frac{-a(n-1)}{2\sigma_e^2} + \frac{\text{SSE}}{2\sigma_e^4} \tag{123}$$

and

$$l_\lambda = \frac{-a}{2\lambda} + \frac{\text{SSA}}{2\lambda^2} + \frac{an(\bar{y}_{..} - \mu)^2}{2\lambda^2}. \tag{124}$$

From this it is not difficult (Exercise 3.13) to obtain second derivatives, e.g., $l_{\mu,\mu} = -an/\lambda$, and the negative of their expected values, e.g., $-E(l_{\mu,\mu}) = an/\lambda$, and hence derive the large-sample dispersion matrix of $\tilde{\mu}, \tilde{\sigma}_e^2$ and $\tilde{\lambda} = \tilde{\sigma}_e^2 + n\tilde{\sigma}_\alpha^2$ as

$$\text{var}\begin{bmatrix} \tilde{\mu} \\ \tilde{\sigma}_e^2 \\ \tilde{\lambda} \end{bmatrix} \simeq \begin{bmatrix} -E(l_{\mu,\mu}) & -E(l_{\mu,\sigma_e^2}) & -E(l_{\mu,\lambda}) \\ -E(l_{\mu,\sigma_e^2}) & -E(l_{\sigma_e^2,\sigma_e^2}) & -E(l_{\sigma_e^2,\lambda}) \\ -E(l_{\mu,\lambda}) & -E(l_{\sigma_e^2,\lambda}) & -E(l_{\lambda,\lambda}) \end{bmatrix}^{-1} \tag{125}$$

$$= \begin{bmatrix} \dfrac{an}{\lambda} & 0 & 0 \\ 0 & \dfrac{a(n-1)}{2\sigma_e^4} & 0 \\ 0 & 0 & \dfrac{a}{2\lambda^2} \end{bmatrix}^{-1} = \begin{bmatrix} \dfrac{\lambda}{an} & 0 & 0 \\ 0 & \dfrac{2\sigma_e^4}{a(n-1)} & 0 \\ 0 & 0 & \dfrac{2\lambda^2}{a} \end{bmatrix}.$$

Thus $\tilde{\mu}$ has zero asymptotic covariance with $\tilde{\sigma}_e^2$ and $\tilde{\lambda}$. Since $\tilde{\sigma}_\alpha^2 = (\tilde{\lambda} - \tilde{\sigma}_e^2)/n$, the large-sample dispersion matrix for the MLEs of the variance components is, provided $\sigma_\alpha^2 > 0$,

$$\text{var}\begin{bmatrix} \tilde{\sigma}_e^2 \\ \tilde{\sigma}_\alpha^2 \end{bmatrix} \simeq 2\sigma_e^4 \begin{bmatrix} \dfrac{1}{a(n-1)} & \dfrac{-1}{an(n-1)} \\ \dfrac{-1}{an(n-1)} & \dfrac{1}{n^2}\left(\dfrac{\lambda^2/\sigma_e^4}{a} + \dfrac{1}{a(n-1)}\right) \end{bmatrix}. \tag{126}$$

Notice from (121), (122) and (126) that all three leading elements are the same, i.e., all three estimators of σ_e^2 have the same variance, $2\sigma_e^4/a(n-1)$, and the same covariance with $\tilde{\sigma}_\alpha^2$, namely $-2\sigma_e^4/an(n-1)$. This is true even though $\hat{\sigma}_e^2 = \text{MSE} = \dot{\sigma}_e^2$, but $\tilde{\sigma}_e^2$ can be SST_m/an if $\dot{\sigma}_\alpha^2 < 0$. The reason that $\text{var}(\tilde{\sigma}_e^2)$ of (126) is not different from $\text{var}(\hat{\sigma}_e^2)$ of (121) is that when $\dot{\sigma}_\alpha^2 > 0$, $\tilde{\sigma}_e^2 = \hat{\sigma}_e^2 = \dot{\sigma}_e^2$; and $\text{var}(\tilde{\sigma}_e^2)$ of (126) is an asymptotic, large-sample variance; and in that asymptotic situation $\tilde{\sigma}_\alpha^2$, which is consistent, cannot be negative, and so in the limit $\tilde{\sigma}_e^2 = \text{MSE}$ always, with variance equal to $\text{var(MSE)} = 2\sigma_e^4/a(n-1)$.

In contrast, the exact variance of $\tilde{\sigma}_e^2$ is, recalling using $p = \Pr\{\dot{\sigma}_\alpha^2 < 0\}$,

$$\begin{aligned}\text{var}(\tilde{\sigma}_e^2) &= E(\tilde{\sigma}_e^4) - \left[E(\tilde{\sigma}_e^2)\right]^2 \\ &= (1-p)E(\text{MSE}^2 \mid \dot{\sigma}_\alpha^2 \geqslant 0) + pE(\text{SST}_\text{m}^2 \mid \dot{\sigma}_\alpha^2 < 0)/a^2n^2 - [E(\tilde{\sigma}_e^2)]^2 .\end{aligned} \tag{127}$$

Again, intractability is apparent, and numerical evaluation would be necessary in particular cases. [See Yu, Searle and McCulloch (1991).]

b. Unbalanced data

*-i. **Likelihood.*** Under the usual normality assumptions of (103) the log likelihood is, as shown just before Section 3.7a,

$$\begin{aligned} l = &-\tfrac{1}{2}N\log 2\pi - \tfrac{1}{2}(N-a)\log\sigma_e^2 - \tfrac{1}{2}\Sigma_i \log\lambda_i \\ &- \frac{1}{2\sigma_e^2}\Sigma_i\Sigma_j(y_{ij}-\mu)^2 + \frac{1}{2\sigma_e^2}\Sigma_i \frac{\sigma_\alpha^2 n_i^2(\bar{y}_{i\cdot}-\mu)^2}{\lambda_i}, \end{aligned} \tag{128}$$

where

$$\lambda_i = \sigma_e^2 + n_i\sigma_\alpha^2 \tag{129}$$

as a generalization of (105). It is left to the reader as an exercise (see E 3.15) to simplify the last two terms of (128) so that it becomes

$$l = -\tfrac{1}{2}N\log 2\pi - \tfrac{1}{2}(N-a)\log\sigma_e^2 - \tfrac{1}{2}\Sigma_i\log\lambda_i - \frac{\text{SSE}}{2\sigma_e^2} - \Sigma_i\frac{n_i(\bar{y}_{i\cdot}-\mu)^2}{2\lambda_i}. \tag{130}$$

The simplification is easily derived using the identity (where $\tau = \sigma_\alpha^2/\sigma_e^2$)

$$1 - \frac{n_i\sigma_\alpha^2}{\lambda_i} = \frac{\sigma_e^2}{\lambda_i} = \frac{1}{1+n_i\tau}. \tag{131}$$

-ii. ML equations and their solutions. With $\partial\lambda_i/\partial\sigma_e^2 = 1$ and $\partial\lambda_i/\partial\sigma_\alpha^2 = n_i$ we differentiate $\log L$ of (130) to get (using $l_\theta \equiv \partial\log L/\partial\theta$)

$$l_\mu = \Sigma_i\frac{n_i(\bar{y}_{i\cdot}-\mu)}{\lambda_i}, \tag{132a}$$

$$l_{\sigma_e^2} = \frac{-(N-a)}{2\sigma_e^2} - \tfrac{1}{2}\Sigma_i\frac{1}{\lambda_i} + \frac{\text{SSE}}{2\sigma_e^4} + \Sigma_i\frac{n_i(\bar{y}_{i\cdot}-\mu)^2}{2\lambda_i^2}, \tag{132b}$$

and

$$l_{\sigma_\alpha^2} = -\tfrac{1}{2}\Sigma_i\frac{n_i}{\lambda_i} + \Sigma_i\frac{n_i^2(\bar{y}_{i\cdot}-\mu)^2}{2\lambda_i^2}. \tag{132c}$$

The ML equations are obtained by equating the right-hand sides of (132) to zero, after replacing the symbols μ, σ_e^2 and λ_i by $\dot{\mu}$, $\dot{\sigma}_e^2$ and $\dot{\lambda}_i = \dot{\sigma}_e^2 + n_i\dot{\sigma}_\alpha^2$, respectively. Then $\dot{\mu}$, $\dot{\sigma}_e^2$ and $\dot{\sigma}_\alpha^2$ are the solutions to the ML equations. Carrying out this procedure with l_μ of (132a) gives

$$\dot{\mu} = \sum_i\frac{n_i\bar{y}_{i\cdot}}{\dot{\lambda}_i}\Bigg/\sum_i\frac{n_i}{\dot{\lambda}_i} = \frac{\displaystyle\sum_i\frac{n_i\bar{y}_{i\cdot}}{\dot{\sigma}_e^2+n_i\dot{\sigma}_\alpha^2}}{\displaystyle\sum_i\frac{n_i}{\dot{\sigma}_e^2+n_i\dot{\sigma}_\alpha^2}} = \frac{\Sigma_i\bar{y}_{i\cdot}/\dot{\text{var}}(\bar{y}_{i\cdot})}{\Sigma_i[1/\dot{\text{var}}(\bar{y}_{i\cdot})]}. \tag{133}$$

This, it can be noted, is the same as GLSE(μ) of (34) in Section 3.3, only with σ_e^2 and σ_α^2 replaced by $\dot{\sigma}_e^2$ and $\dot{\sigma}_\alpha^2$; i.e.,

$$\dot{\text{var}}(\bar{y}_{i\cdot}) = \dot{\sigma}_\alpha^2 + \dot{\sigma}_e^2/n_i.$$

Derivation of $\dot{\sigma}_\alpha^2$ and $\dot{\sigma}_e^2$ comes from equating the right-hand sides of (132b) and (132c) to zero, so giving

$$\frac{\text{SSE}}{\dot{\sigma}_e^4} - \frac{N-a}{\dot{\sigma}_e^2} + \sum_i\frac{n_i(\bar{y}_{i\cdot}-\dot{\mu})^2}{\dot{\lambda}_i^2} - \sum_i\frac{1}{\dot{\lambda}_i} = 0 \tag{134}$$

and

$$\sum_i \frac{n_i^2(\bar{y}_{i\cdot} - \dot{\mu})^2}{\dot{\lambda}_i^2} = \sum_i \frac{n_i}{\dot{\lambda}_i}. \tag{135}$$

With $\dot{\lambda}_i = \dot{\sigma}_e^2 + n_i\dot{\sigma}_\alpha^2$ occurring in the denominators of the terms being summed (over i) in these equations, there is clearly no analytic solution for the estimators; there is when the data are balanced, i.e. $n_i = n$ and $\lambda_i = \lambda \ \forall\ i$.

-iii. ML estimators. As with balanced data, solutions $\dot{\mu}$, $\dot{\sigma}_e^2$ and $\dot{\sigma}_\alpha^2$ are ML estimators only if the triplet $(\dot{\mu}, \dot{\sigma}_e^2, \dot{\sigma}_\alpha^2)$ is in the 3-space of $(\mu, \sigma_e^2, \sigma_\alpha^2)$. And in ensuring that this is achieved, the negativity problem raises its head again. For each data set, equations (134) and (135) have to be solved numerically, using some iterative method suited to the numerical solution of non-linear equations. After doing this, we derive the ML estimators as follows:

when $\dot{\sigma}_\alpha^2 \geqslant 0$,

$$\tilde{\sigma}_e^2 = \dot{\sigma}_e^2, \quad \tilde{\sigma}_\alpha^2 = \dot{\sigma}_\alpha^2 \quad \text{and} \quad \tilde{\mu} = \dot{\mu}; \tag{136}$$

when $\dot{\sigma}_\alpha^2 < 0$,

$$\tilde{\sigma}_e^2 = \text{SST}_\text{m}/N, \quad \tilde{\sigma}_\alpha^2 = 0 \quad \text{and} \quad \tilde{\mu} = \bar{y}_{\cdot\cdot}\,. \tag{137}$$

In the latter case, when $\dot{\sigma}_\alpha^2 < 0$, the argument for having $\tilde{\sigma}_\alpha^2 = 0$ is essentially the same as with balanced data, whereupon it is left to the reader to show that $\log L$ reduces to being such that on equating its derivatives to zero one obtains $\tilde{\sigma}_e^2 = \text{SST}_\text{m}/N$, as in (115) for balanced data and $\tilde{\mu} = \bar{y}_{\cdot\cdot}$. (See E 3.16.)

Having been derived by the method of maximum likelihood, the estimators in (136) and (137) are, as is well known, asymptotically normally distributed. Their relationship to a weighted least squares approach is considered by Chatterjee and Das (1983).

The question might well be raised as to what to do if the numerical solution of (134) and (135) yields a negative value for $\dot{\sigma}_e^2$. Fortunately, it can be shown that $L \to 0$ as σ_e^2 tends either to zero or to infinity, and so L must have a maximum at a positive value of σ_e^2. (See E 3.21.)

-iv. Bias. With balanced data we were able to specify p, the probability of the solution for $\dot{\sigma}_\alpha^2$ to the ML equations being negative—in (117). But with unbalanced data $F = \text{MSA}/\text{MSE}$ does not have a distribution that is proportional to an F, so this probability cannot be easily specified. Moreover, although we know that $\tilde{\sigma}_e^2 = \text{SST}_\text{m}/N$ with probability p, and the expected value of SST_m is readily derived, the expected value of $\tilde{\sigma}_e^2$ when $\dot{\sigma}_\alpha^2 < 0$ cannot be easily derived. Thus, in general, the bias in the solutions obtained to (134) cannot be derived analytically.

-v. Sampling variances. Large-sample variances come from a matrix similar to (125), namely the inverse of the negative of the expected value of the Hessian (matrix of second derivatives) of $\log L$ with respect to μ, σ_e^2 and σ_α^2. Keeping in mind that, by definition, $\sigma_\alpha^2 > 0$ (because if $\sigma_\alpha^2 = 0$ the model and L change)

TABLE 3.10. SECOND DERIVATIVES OF $\log L$ AND THEIR EXPECTED VALUES

Second Derivative of log L	−(Expected Value)
$l_{\mu,\mu} = -\sum_i \frac{n_i}{\lambda_i}$	$\sum_i \frac{n_i}{\lambda_i}$
$l_{\mu,\sigma_e^2} = -\sum_i \frac{n_i(\bar{y}_{i\cdot} - \mu)}{\lambda_i^2}$	0
$l_{\mu,\sigma_\alpha^2} = -\sum_i \frac{n_i^2(\bar{y}_{i\cdot} - \mu)}{\lambda_i^2}$	0
$l_{\sigma_e^2,\sigma_e^2} = \frac{N-a}{2\sigma_e^4} + \frac{1}{2}\sum_i \frac{1}{\lambda_i^2} - \frac{\text{SSE}}{\sigma_e^6} - \sum_i \frac{n_i(\bar{y}_{i\cdot} - \mu)^2}{\lambda_i^3}$	$\frac{N-a}{2\sigma_e^4} + \frac{1}{2}\sum_i \frac{1}{\lambda_i^2}$
$l_{\sigma_e^2,\sigma_\alpha^2} = \frac{1}{2}\sum_i \frac{n_i}{\lambda_i^2} - \sum_i \frac{n_i^2(\bar{y}_{i\cdot} - \mu)^2}{\lambda_i^3}$	$\frac{1}{2}\sum_i \frac{n_i}{\lambda_i^2}$
$l_{\sigma_\alpha^2,\sigma_\alpha^2} = \frac{1}{2}\sum_i \frac{n_i^2}{\lambda_i^2} - \sum_i \frac{n_i^3(\bar{y}_{i\cdot} - \mu)^2}{\lambda_i^3}$	$\frac{1}{2}\sum_i \frac{n_i^2}{\lambda_i^2}$

differentiating the three terms of (132) gives the derivatives and expected values shown in Table 3.10.

The expected values are easily derived (E 3.17) utilizing

$$E(\bar{y}_{i\cdot}) = \mu \quad \text{and} \quad E(\bar{y}_{i\cdot} - \mu)^2 = \sigma_\alpha^2 + \sigma_e^2/n_i\,.$$

Hence arraying these expected values in the matrix similar to (125) gives

$$\text{var}\begin{bmatrix} \tilde{\mu} \\ \tilde{\sigma}_e^2 \\ \tilde{\sigma}_\alpha^2 \end{bmatrix} \simeq \begin{bmatrix} \sum_i \frac{n_i}{\lambda_i} & 0 & 0 \\ 0 & \frac{N-a}{2\sigma_e^4} + \frac{1}{2}\sum_i \frac{1}{\lambda_i^2} & \frac{1}{2}\sum_i \frac{n_i}{\lambda_i^2} \\ 0 & \frac{1}{2}\sum_i \frac{n_i}{\lambda_i^2} & \frac{1}{2}\sum_i \frac{n_i^2}{\lambda_i^2} \end{bmatrix}^{-1}. \tag{138}$$

Therefore

$$\text{var}(\tilde{\mu}) \simeq \left(\sum_i \frac{n_i}{\lambda_i}\right)^{-1} = \left(\sum_i \frac{n_i}{\sigma_e^2 + n_i\sigma_\alpha^2}\right)^{-1}$$

and (see overleaf for D)

$$\text{var}\begin{bmatrix} \tilde{\sigma}_e^2 \\ \tilde{\sigma}_\alpha^2 \end{bmatrix} \simeq \frac{2}{D}\begin{bmatrix} \sum_i \frac{n_i^2}{\lambda_i^2} & -\sum_i \frac{n_i}{\lambda_i^2} \\ -\sum_i \frac{n_i}{\lambda_i^2} & \frac{N-a}{\sigma_e^4} + \sum_i \frac{1}{\lambda_i^2} \end{bmatrix} \tag{139}$$

for

$$D = \frac{N-a}{\sigma_e^4}\Sigma_i \frac{n_i^2}{\lambda_i^2} + \Sigma_i \frac{1}{\lambda_i^2}\Sigma_i \frac{n_i^2}{\lambda_i^2} - \left(\Sigma_i \frac{n_i}{\lambda_i^2}\right)^2 .$$

These are the asymptotic (large-sample) variances—although var($\tilde{\mu}$) is also the exact variance of $\tilde{\mu}$ and of GLSE(μ) of (34).

Three points are worth noting.

(A) The terms in (139) reduce, for balanced data, to what one would expect, namely (126).

(B) D in (139) can also be expressed as $\sigma_e^4 D = N\Sigma_i w_i^2 - (\Sigma_i w_i)^2$ for $w_i = n_i/\lambda_i$ as used in Searle (1956).

(C) The matrix in (139) can also be derived using the result from Searle (1970) that

$$\text{var}(\tilde{\sigma}^2) \simeq 2\left[\left\{ {}_{\text{m}}\, \text{tr}\left(\mathbf{V}^{-1}\frac{\partial \mathbf{V}}{\partial \sigma_i^2}\mathbf{V}^{-1}\frac{\partial \mathbf{V}}{\partial \sigma_j^2}\right)\right\}_{i,j=0}^{r}\right]^{-1},$$

where $\text{var}(\mathbf{y}) = \mathbf{V}$ and there are $r+1$ variance components in the model, $\sigma_0^2 = \sigma_e^2$ and σ_i^2 for $i = 1, \ldots, r$.

Establishing (A) and (B) is left to the reader in E 3.18; and (C) is derived in Section 11.1e-ii.

3.8. RESTRICTED MAXIMUM LIKELIHOOD ESTIMATION (REML)

An adaptation of ML is to maximize just that part of the likelihood that is said to be location invariant. [The reader interested in invariance more generally will find a good discussion of it in Casella and Berger (1990).] In terms of the 1-way classification this means maximizing that part of the likelihood that does not involve μ. It is an idea that seems to have had its genesis in Anderson and Bancroft (1952, p. 320) and was later extended by W.A. Thompson (1962) and generalized by Patterson and R. Thompson (1971). We discuss REML in some generality in Chapter 6, but here just demonstrate its applicability to the 1-way classification, random model.

We can note in passing a characteristic of REML estimation that is often considered to be one of its merits: it is a maximum likelihood method that, even though it is not concerned with estimating fixed effects, does take into account the degrees of freedom associated with the fixed effects of the model. (REML estimation is also an example of marginal likelihood estimation discussed in Chapter 9.) An elementary example of this is the case of estimating the variance from a simple sample of n independent observations distributed $\mathcal{N}(\mu, \sigma^2)$. If $x_1, \ldots, x_n$ are those data then

$$\hat{\sigma}^2 = \frac{\Sigma_i(x_i - \bar{x})^2}{n-1} \quad \text{and} \quad \tilde{\sigma}^2 = \frac{\Sigma_i(x_i - \bar{x})^2}{n}$$

are two well-known estimators of σ^2. The first, $\hat{\sigma}^2$, is unbiased; the second, $\tilde{\sigma}^2$, is the ML estimator under normality. But $\hat{\sigma}^2$ is also the REML estimator; and we see that it has taken into account the single degree of freedom needed for estimating μ. (See E 3.19.)

a. Balanced data

-i. Likelihood. For balanced data this restricted likelihood (as it is called nowadays) is easily derived. In doing so, we utilise the $L(\cdot\,|\,\cdot)$ notation of (104) to provide clarification of different parts of the likelihood, the starting point of which is

$$L(\mu, \sigma_e^2, \sigma_\alpha^2 \,|\, \mathbf{y}) = \frac{\exp\left\{-\dfrac{1}{2}\left[\dfrac{\text{SSE}}{\sigma_e^2} + \dfrac{\text{SSA}}{\lambda} + \dfrac{(\bar{y}_{..} - \mu)^2}{\lambda/an}\right]\right\}}{(2\pi)^{\frac{1}{2}an}\sigma_e^{2[\frac{1}{2}a(n-1)]}\lambda^{\frac{1}{2}a}},$$

as reconstituted from $\log L$ of (106). Observe that since $\bar{y}_{..}$ is independent of both SSE and SSA, the likelihood of the preceding expression can be factored as

$$L(\mu, \sigma_e^2, \sigma_\alpha^2 \,|\, \mathbf{y}) = L(\mu \,|\, \bar{y}_{..})L(\sigma_e^2, \sigma_\alpha^2 \,|\, \text{SSA, SSE}),$$

where $L(\mu \,|\, \bar{y}_{..})$ is the likelihood of μ given $\bar{y}_{..}$, namely

$$L(\mu \,|\, \bar{y}_{..}) = \frac{\exp\left[\dfrac{-(\bar{y}_{..} - \mu)^2}{2\lambda/an}\right]}{(2\pi)^{\frac{1}{2}}(\lambda/an)^{\frac{1}{2}}}, \tag{140}$$

and

$$L(\sigma_e^2, \sigma_\alpha^2 \,|\, \text{SSE, SSA}) = \frac{\exp\left[-\dfrac{1}{2}\left(\dfrac{\text{SSE}}{\sigma_e^2} + \dfrac{\text{SSA}}{\lambda}\right)\right]}{(2\pi)^{\frac{1}{2}(an-1)}\sigma_e^{2[\frac{1}{2}a(n-1)]}\lambda^{\frac{1}{2}(a-1)}(an)^{\frac{1}{2}}} \tag{141}$$

is the likelihood function of σ_e^2 and σ_α^2 given SSA and SSE. Note also that (141) can be expressed as

$$L(\sigma_e^2, \sigma_\alpha^2 \,|\, \text{SSE, SSA}) = \int L(\mu, \sigma_e^2, \sigma_\alpha^2 \,|\, \mathbf{y})\, d\mu \tag{142}$$

showing the marginal likelihood relationship. This, for the 1-way classification with balanced data, is known as the restricted likelihood, or sometimes as the marginal likelihood, the latter by analogy with the concept of marginal density functions.

ii. REML equations and their solutions. REML estimation consists of obtaining estimators for σ_e^2 and σ_α^2 that maximize (141) within the parameter space $\sigma_e^2 > 0$ and $\sigma_\alpha^2 \geqslant 0$. Denote the logarithm of that function by l_{R}:

$$\begin{aligned} l_{\text{R}} &= \log L(\sigma_e^2, \sigma_\alpha^2 \,|\, \text{SSE, SSA}) \\ &= -\tfrac{1}{2}(an - 1)\log 2\pi - \tfrac{1}{2}\log an - \tfrac{1}{2}a(n-1)\log \sigma_e^2 \\ &\quad - \tfrac{1}{2}(a-1)\log\lambda - \frac{\text{SSE}}{2\sigma_e^2} - \frac{\text{SSA}}{2\lambda}. \end{aligned} \tag{143}$$

The derivatives of this are

$$l_{\mathrm{R},\sigma_e^2} = \frac{-a(n-1)}{2\sigma_e^2} + \frac{\mathrm{SSE}}{2\sigma_e^4} \tag{144}$$

and

$$l_{\mathrm{R},\lambda} = \frac{-(a-1)}{2\lambda} + \frac{\mathrm{SSA}}{2\lambda^2}\,.$$

Equating these to zero and replacing σ_e^2 and λ by the solutions $\dot{\sigma}_{e,\mathrm{R}}^2$ and $\dot{\lambda}_{\mathrm{R}}$, we get those solutions as being $\dot{\lambda}_{\mathrm{R}} = \mathrm{SSA}/(a-1) = \mathrm{MSA}$ and

$$\dot{\sigma}_{e,\mathrm{R}}^2 = \frac{\mathrm{SSE}}{a(n-1)} = \mathrm{MSE}; \quad \text{and thus } \dot{\sigma}_{\alpha,\mathrm{R}}^2 = \frac{1}{n}(\mathrm{MSA} - \mathrm{MSE})\,. \tag{145}$$

These are the REML solutions.

-iii. REML estimators. Similar to the situation with ML, the preceding REML solutions are REML estimators only when both are non-negative. $\dot{\sigma}_{e,\mathrm{R}}^2$ can never be negative, but $\dot{\sigma}_{\alpha,\mathrm{R}}^2$ can be, whereupon we have to maximize l_{R} subject to $\dot{\sigma}_{\alpha,\mathrm{R}}^2 = 0$, which leads to $\dot{\sigma}_{e,\mathrm{R}}^2$ then being $\mathrm{SST_m}/(an-1)$. Thus the REML estimators are

$$\begin{aligned} &\text{when } \dot{\sigma}_{\alpha,\mathrm{R}}^2 \geqslant 0, \quad \tilde{\sigma}_{e,\mathrm{R}}^2 = \mathrm{MSE} \quad \text{and} \quad \tilde{\sigma}_{\alpha,\mathrm{R}}^2 = \frac{1}{n}(\mathrm{MSA} - \mathrm{MSE}); \\ &\text{when } \dot{\sigma}_{\alpha,\mathrm{R}}^2 < 0, \quad \tilde{\sigma}_{e,\mathrm{R}}^2 = \frac{\mathrm{SST_m}}{an-1} \quad \text{and} \quad \tilde{\sigma}_{\alpha,\mathrm{R}}^2 = 0\,. \end{aligned} \tag{146}$$

-iv. Comparison with ANOVA and ML. Comparing $\dot{\sigma}_{e,\mathrm{R}}^2$ and $\dot{\sigma}_{\alpha,\mathrm{R}}^2$ of (145) with the ANOVA estimators in (54) and (55), we see that they are the same; i.e., that solutions of the REML equations are the ANOVA estimators. This result is, in fact, true generally for all cases of balanced data (see Section 6.7f).

Comparing the REML estimators of (146) with the ML estimators of (114) and (115), we see that the condition for a negative solution for σ_α^2 is not quite the same in the two cases: in REML it is MSA < MSE, whereas in ML it is $(1 - 1/a)\mathrm{MSA} < \mathrm{MSE}$; and the positive estimator is similarly slightly different: $(\mathrm{MSA} - \mathrm{MSE})/n$ in REML but $[(1 - 1/a)\mathrm{MSA} - \mathrm{MSE}]/n$ in ML. Also, when there is a negative solution for σ_α^2, the resulting estimator of σ_e^2 is not the same in the two cases: $\mathrm{SST_m}/(an-1)$ in REML but $\mathrm{SST_m}/an$ in ML. Each of these differences has a common feature: that with REML we see SSA being divided by $a-1$ where it is divided by a in ML; and in REML the divisor of $\mathrm{SST_m}$ is $an-1$ whereas it is an in ML. In both instances the REML divisor is one less than the ML divisor. In this way REML is taking account of the degree of freedom that gets utilized in estimating μ—even though REML does not explicitly involve the estimation of μ. Nevertheless, it is a general feature of REML estimation of variance components from balanced data that degrees of freedom for fixed effects get taken into account.

In many applications $\tau = \sigma_\alpha^2/\sigma_e^2$ is at least as useful a parameter as σ_α^2 or σ_e^2. Under ANOVA estimation its estimator would be $(F - 1)/n$ for $F = \text{MSA}/\text{MSE}$; under REML estimation it would be the positive part of this; and under ML it would be the positive part of $[(1 - 1/a)F - 1]/n$. Loh (1986) considers the admissability of these estimators and suggests that an improvement is the positive part of

$$\left[\left(1 - \frac{1}{a}\right)\left(a - \frac{4}{n-1}\right)\frac{F}{a+1} - 1\right]\Big/ n\,.$$

-v. *Bias.* What has just been said about REML might lead one to surmise that REML estimators are unbiased. They are not. The same need for non-negative estimates arises as with ML estimation. Similar to (119) we define, for balanced data

$$\begin{aligned} p_{\text{R}} &= \Pr\{\dot\sigma^2_{\alpha,\text{R}} < 0\} = \Pr\{\text{MSA} < \text{MSE}\} \\ &= \Pr\{\mathscr{F}^{a(n-1)}_{a-1} > 1 + n\tau\}, \end{aligned} \tag{147}$$

akin to (119). Then, based on (146), the expected value of $\tilde\sigma^2_{e,\text{R}}$ is

$$E(\tilde\sigma^2_{e,\text{R}}) = (1 - p_{\text{R}})E(\text{MSE}\,|\,\dot\sigma^2_{\alpha,\text{R}} \geqslant 0) + p_{\text{R}}E(\text{SST}_{\text{m}}\,|\,\dot\sigma^2_{\alpha,\text{R}} < 0)/(an - 1)\,. \tag{148}$$

-vi. *Sampling variances.* Based on the derivatives in (144), we can easily find the large-sample dispersion matrix,

$$\text{var}\begin{bmatrix}\tilde\sigma^2_{e,\text{R}} \\ \tilde\lambda_{\text{R}}\end{bmatrix} \simeq \begin{bmatrix} -E(l_{\text{R},\sigma_e^2,\sigma_e^2}) & -E(l_{\text{R},\sigma_e^2,\lambda}) \\ -E(l_{\text{R},\lambda,\sigma_e^2}) & -E(l_{\text{R},\lambda,\lambda})\end{bmatrix}^{-1},$$

which leads to exactly the same results as in (121). And something comparable but not very different from (127) could be derived if deemed worthwhile.

b. Unbalanced data

As in (104), the likelihood function for unbalanced data is

$$L(\mu, \sigma_e^2, \sigma_\alpha^2\,|\,\mathbf{y}) = \frac{\exp\left\{-\left[\dfrac{\Sigma_i\Sigma_j(y_{ij} - \mu)^2}{2\sigma_e^2} - \Sigma_i\dfrac{n_i^2\sigma_\alpha^2(\bar y_{i\cdot} - \mu)^2}{2\sigma_e^2(\sigma_e^2 + n_i\sigma_\alpha^2)}\right]\right\}}{(2\pi)^{\frac{1}{2}N}\sigma_e^{2[\frac{1}{2}(N-a)]}\prod_{i=1}^{a}(\sigma_e^2 + n_i\sigma_\alpha^2)^{\frac{1}{2}}}$$

There is no straightforward factoring of this likelihood that permits separating out a function of μ in the manner of (140) for balanced data. Nevertheless, equations for REML estimators can be established—as a special case of the equations for the general case. This is left until Chapter 6.

For the 1-way classification, random model, with unbalanced data, Westfall (1987) makes numerical and analytical comparisons of a variety of estimators: ANOVA, ML, REML and several forms of MINQUE (see Chapter 11). For a few small designs, Khatree and Gill (1988) make similar comparisons and conclude that for estimating σ_α^2 REML seems to be the favored method, whereas for σ_e^2 it is ANOVA; and for simultaneous estimation of σ_α^2 and σ_e^2 ML is favored. MINQUE(0), a variant of MINQUE (see Chapter 11), seems to be the worst of the methods compared.

3.9. BAYES ESTIMATION

A brief introduction to the basic ideas of Bayes estimation is given in Appendix S.6. The salient result is the one labeled (3) there. It states that $\pi(\theta \mid y)$, the posterior density for the parameter θ that occurs in the density function $f(y \mid \theta)$ for the random variable y representing the data, is

$$\pi(\theta \mid y) = \frac{f(y \mid \theta)\pi(\theta)}{\int_{R_\theta} f(y \mid \theta)\pi(\theta)\, d\theta} . \tag{149}$$

$\pi(\theta)$ is the prior density of θ and R_θ is the range of possible values of θ. All these terms are briefly described in Appendix S.6.

a. A simple example

From $\mathbf{x} = [x_1, \ldots, x_n]' \sim \mathcal{N}(\mu\mathbf{1}, \sigma^2\mathbf{I})$ we consider the estimation of σ^2. The usual unbiased estimator is

$$s^2 = \frac{\sum_{i=1}^{n} (x_i - \bar{x})^2}{n-1}, \quad \text{with} \quad \frac{(n-1)s^2}{\sigma^2} \sim \chi^2_{n-1} . \tag{150}$$

It is also well known that the ML estimator is

$$\tilde{\sigma}^2 = (1 - 1/n)s^2 . \tag{151}$$

Define

$$m = n - 1 . \tag{152}$$

Then from (150)

$$f(s^2 \mid \sigma^2) = \frac{(m/\sigma^2)^{\frac{1}{2}m} s^{2(\frac{1}{2}m - 1)} e^{-\frac{1}{2}ms^2/\sigma^2}}{\Gamma(\frac{1}{2}m)2^{\frac{1}{2}m}} . \tag{153}$$

For Bayes estimation we need a prior distribution (see Appendix S.6), for σ^2, for which we will use inverted gamma distribution. This is a common choice in estimating variances of normal distributions, not only because it is quite realistic for a positive random variable, but also because it leads to a tractable form for the resulting posterior density (Appendix S.6b). The general form of the inverted gamma density is

$$f(x) = \frac{x^{-(a+1)} e^{-1/bx}}{\Gamma(a)b^a}, \tag{154}$$

with

$$E(x) = \frac{1}{(a-1)b} \quad \text{and} \quad \operatorname{var}(x) = \frac{1}{(a-1)^2(a-2)b^2}\,. \tag{155}$$

We use this with $a = 2$ and $b = 1$ as the prior for σ^2: i.e., from (154) and (155)

$$\pi(\sigma^2) = (\sigma^2)^{-3}e^{-1/\sigma^2}; \quad \text{with } E(\sigma^2) = 1 \quad \text{and} \quad \operatorname{var}(\sigma^2) = \infty\,. \tag{156}$$

Because $\operatorname{var}(\sigma^2) = \infty$, the prior density in (156) is imparting rather vague information. It is chosen for its apparent lack of subjectivity and its mathematical tractability.

As described in Section S.6, we need the posterior density of σ^2, which on using (149) is

$$\pi(\sigma^2 \,|\, s^2) = \frac{f(s^2 \,|\, \sigma^2)\pi(\sigma^2)}{\int_{R_{\sigma^2}} f(s^2 \,|\, \sigma^2)\pi(\sigma^2)\, d\sigma^2}\,. \tag{157}$$

Its numerator is

$$f(s^2, \sigma^2) = f(s^2 \,|\, \sigma^2)\pi(\sigma^2) = \frac{(m/\sigma^2)^{\frac{1}{2}m} s^{2(\frac{1}{2}m - 1)} e^{-ms^2/2\sigma^2}[(\sigma^2)^{-3}e^{-1/\sigma^2}]}{\Gamma(\frac{1}{2}m)2^{\frac{1}{2}m}} \tag{158}$$

$$= \frac{m^{\frac{1}{2}m}}{\Gamma(\frac{1}{2}m)2^{\frac{1}{2}m}} \frac{s^{2(\frac{1}{2}m-1)}e^{-(\frac{1}{2}ms^2+1)/\sigma^2}}{\sigma^{2(3+\frac{1}{2}m)}}\,. \tag{159}$$

And the denominator of (157) is

$$f(s^2) = \int_0^\infty f(s^2, \sigma^2)\, d\sigma^2 = \frac{m^{\frac{1}{2}m}s^{2(\frac{1}{2}m-1)}}{\Gamma(\frac{1}{2}m)2^{\frac{1}{2}m}} \int_0^\infty \frac{e^{-(\frac{1}{2}ms^2+1)/\sigma^2}\, d\sigma^2}{\sigma^{2(3+\frac{1}{2}m)}}\,.$$

To carry out the integration, make the transformation

$$\frac{\frac{1}{2}ms^2 + 1}{\sigma^2} = u, \quad \text{with} \quad \left|\frac{\partial \sigma^2}{\partial u}\right| = \frac{\frac{1}{2}ms^2 + 1}{u^2}\,.$$

This gives

$$f(s^2) = \frac{m^{\frac{1}{2}m}s^{2(\frac{1}{2}m-1)} \int_0^\infty e^{-u}u^{1+\frac{1}{2}m}\, du}{\Gamma(\frac{1}{2}m)2^{\frac{1}{2}m}(\frac{1}{2}ms^2+1)^{2+\frac{1}{2}m}} = \frac{m^{\frac{1}{2}m}s^{2(\frac{1}{2}m-1)}\Gamma(2+\frac{1}{2}m)}{\Gamma(\frac{1}{2}m)2^{\frac{1}{2}m}(\frac{1}{2}ms^2+1)^{2+\frac{1}{2}m}}\,. \tag{160}$$

Using (159) and (160) as the numerator and denominator, respectively, in (157) gives

$$\pi(\sigma^2 \,|\, s^2) = \frac{f(s^2, \sigma^2)}{f(s^2)} = \frac{e^{-(\frac{1}{2}ms^2+1)/\sigma^2}}{\sigma^{2(3+\frac{1}{2}m)}} \frac{(\frac{1}{2}ms^2+1)^{2+\frac{1}{2}m}}{\Gamma(2+\frac{1}{2}m)}$$

$$= \frac{(\sigma^2)^{-(2+\frac{1}{2}m+1)}}{\Gamma(2+\frac{1}{2}m)} \frac{e^{-1/[1/(\frac{1}{2}ms^2+1)]\sigma^2}}{[1/(\frac{1}{2}ms^2+1)]^{2+\frac{1}{2}m}} = \frac{(\sigma^2)^{-(a+1)}e^{-1/b\sigma^2}}{\Gamma(a)b^a} \tag{161}$$

for

$$a = 2 + \tfrac{1}{2}m \quad \text{and} \quad b = \frac{1}{\tfrac{1}{2}ms^2 + 1}. \tag{162}$$

Thus (161) is (154) with a and b of (162). Hence the posterior distribution (161) is an inverted gamma distribution, the same form of density as is the prior, in this case (156) with $a = 2$ and $b = 1$. This is a defining property of what is called a *conjugate prior*: it leads to a posterior density that is in the same family of densities as is the prior. In this case both are inverted gamma densities.

Comparing (161) with (154) and thus using (162) in (155) gives the mean of the conditional variable $\sigma^2 \mid s^2$ as

$$\begin{aligned} E(\sigma^2 \mid s^2) &= \frac{1}{(a-1)b} = \frac{\tfrac{1}{2}ms^2 + 1}{\tfrac{1}{2}m + 1} = \frac{(n-1)s^2 + 2}{n+1} \\ &= \frac{n}{n+1}\left(1 - \frac{1}{n}\right)s^2 + \frac{2}{n+1}(1) = \frac{n}{n+1}\tilde{\sigma}^2 + \frac{2}{n+1}E(\sigma^2), \end{aligned} \tag{163}$$

where $\tilde{\sigma}^2$ is the ML estimator from (151), and $E(\sigma^2) = 1$ is the expected value of σ^2 from $\pi(\sigma^2)$ of (156). This weighted averaging is similar to (13) of Section S.6.

Similarly, (162) and (155) also yield

$$\operatorname{var}(\sigma^2 \mid s^2) = \frac{2(ms^2 + 2)^2}{m(m+2)^2}.$$

On choosing to use $E(\sigma^2 \mid s^2)$ of (163) as a Bayes estimator, call it $\hat{\sigma}^2_{\mathrm{B}}$, we could derive its mean and variance in the usual (classical) manner. Thus with

$$\hat{\sigma}^2_{\mathrm{B}} = E(\sigma^2 \mid s^2) = \frac{(n-1)s^2 + 2}{n+1}$$

from (163), and $E(s^2) = \sigma^2$ and $\operatorname{var}(s^2) = 2\sigma^4/(n-1)$, it is clear that

$$E(\hat{\sigma}^2_{\mathrm{B}}) = \frac{(n-1)\sigma^2 + 2}{n+1} \quad \text{and} \quad \operatorname{var}(\hat{\sigma}^2_{\mathrm{B}}) = \frac{2(n-1)\sigma^4}{(n+1)^2}.$$

In comparing $\hat{\sigma}^2_{\mathrm{B}}$ and s^2 we find that $\hat{\sigma}^2_{\mathrm{B}}$ has bias $-2(\sigma^2 - 1)/(n-1)$ whereas s^2 is unbiased; and

$$\operatorname{var}(\hat{\sigma}^2_{\mathrm{B}}) = \frac{2(n-1)\sigma^4}{(n+1)^2} < \frac{2\sigma^4}{n-1} = \operatorname{var}(s^2).$$

Since $\hat{\sigma}^2_{\mathrm{B}}$ is biased, a better comparison than that of variances is to compare mean square errors: variance plus squared bias.

$$\operatorname{MSE}(\hat{\sigma}^2_{\mathrm{B}}) = \operatorname{var}(\hat{\sigma}^2_{\mathrm{B}}) + \frac{4(\sigma^2 - 1)^2}{(n+1)^2} = \frac{2[(n+1)\sigma^4 - 4\sigma^2 + 2]}{(n+1)^2},$$

and, because s^2 is unbiased,

$$\text{MSE}(s^2) = \text{var}(s^2) = \frac{2\sigma^4}{n-1}.$$

Therefore, $\hat{\sigma}^2_{\text{B}}$ has smaller MSE than s^2 when

$$(n-1)[(n+1)\sigma^4 - 4\sigma^2 + 2] - (n+1)^2\sigma^4 < 0;$$

i.e., when

$$\frac{n+1}{n-1}\sigma^4 + 2\sigma^2 > 1,$$

which certainly occurs whenever $\sigma^2 > \frac{1}{2}$.

b. The 1-way classification, random model

From (104), the likelihood is (with $\lambda_i = \sigma_e^2 + n_i\sigma_\alpha^2$)

$$L(\mu, \sigma_e^2\, \sigma_\alpha^2 \,|\, \mathbf{y}) = \frac{\exp\left\{-\dfrac{1}{2\sigma_e^2}\left[\Sigma_i\Sigma_j(y_{ij}-\mu)^2 - \sum_i \dfrac{\sigma_\alpha^2}{\lambda_i}(y_{i\cdot} - n_i\mu)^2\right]\right\}}{(2\pi)^{\frac{1}{2}N}\sigma_e^{2[\frac{1}{2}(N-a)]}\prod_{i=1}^{a}\lambda_i^{\frac{1}{2}}}.$$

Notice that

$$\Sigma_i\Sigma_j(y_{ij}-\mu)^2 - \sum_i \frac{\sigma_\alpha^2}{\lambda_i}(y_{i\cdot} - n_i\mu)^2$$

$$= \Sigma_i\Sigma_j(y_{ij} - \bar{y}_{i\cdot} + \bar{y}_{i\cdot} - \mu)^2 - \sum_i \frac{n_i^2\sigma_\alpha^2}{\lambda_i}(\bar{y}_{i\cdot} - \mu)^2 \tag{164}$$

$$= \Sigma_i\Sigma_j(y_{ij} - \bar{y}_{i\cdot})^2 + \Sigma_i\left(n_i - \frac{n_i^2\sigma_\alpha^2}{\lambda_i}\right)(\bar{y}_{i\cdot} - \mu)^2 \tag{165}$$

$$= \text{SSE} + \sigma_e^2\sum_i \frac{n_i(\bar{y}_{i\cdot} - \mu)^2}{\lambda_i}. \tag{166}$$

The cross-product term from (164) disappears in (165) because $\Sigma_i\Sigma_j(y_{ij} - \bar{y}_{i\cdot}) = 0$; and (166) comes from (165) by the definition of SSE and through $\lambda_i = \sigma_e^2 + n_i\sigma_\alpha^2$. Thus, from (166), writing it as a density in the manner following (104),

$$f(\mathbf{y} \,|\, \mu, \sigma_e^2, \sigma_\alpha^2) = \frac{\exp\left(-\dfrac{\text{SSE}}{2\sigma_e^2}\right)}{\sigma_e^{2[\frac{1}{2}(N-a)]}} \frac{\exp\left[-\frac{1}{2}\sum_i \dfrac{n_i(\bar{y}_{i\cdot} - \mu)^2}{\sigma_e^2 + n_i\sigma_\alpha^2}\right]}{(2\pi)^{\frac{1}{2}N}\prod_{i=1}^{a}(\sigma_e^2 + n_i\sigma_\alpha^2)^{\frac{1}{2}}}. \tag{167}$$

The density in (167) has three parameters: μ, σ_e^2 and σ_α^2. Using it in Bayes estimation would require a prior, $\pi(\mu, \sigma_e^2, \sigma_\alpha^2)$, on all three parameters. Then we would, in the manner of (149), develop a posterior density

$$\pi(\mu, \sigma_e^2, \sigma_\alpha^2 \mid \mathbf{y}) = \frac{f(\mathbf{y}, \mu, \sigma_e^2, \sigma_\alpha^2)}{\iiint f(\mathbf{y}, \mu, \sigma_e^2, \sigma_\alpha^2)\, d\mu\, d\sigma_e^2\, d\sigma_\alpha^2},$$

where

$$f(\mathbf{y}, \mu, \sigma_e^2, \sigma_\alpha^2) = f(\mathbf{y} \mid \mu, \sigma_e^2, \sigma_\alpha^2)\pi(\mu, \sigma_e^2, \sigma_\alpha^2),$$

just as was done for σ^2 in (157). This expression is customarily utilized with prior densities of the form

$$\pi(\mu, \sigma_e^2, \sigma_\alpha^2) = \pi(\mu)\pi(\sigma_e^2, \sigma_\alpha^2), \tag{168}$$

thus assuming that μ is independent of σ_e^2 and σ_α^2. We then proceed in stages. First, use $\pi(\mu)$ to derive

$$f(\mathbf{y} \mid \sigma_e^2, \sigma_\alpha^2) = \int f(\mathbf{y} \mid \mu, \sigma_e^2, \sigma_\alpha^2)\pi(\mu)\, d\mu; \tag{169}$$

second, obtain the posterior distribution of σ_e^2 and σ_α^2 using

$$\pi(\sigma_e^2, \sigma_\alpha^2 \mid \mathbf{y}) = \frac{f(\mathbf{y} \mid \sigma_e^2, \sigma_\alpha^2)\pi(\sigma_e^2, \sigma_\alpha^2)}{\iint f(\mathbf{y} \mid \sigma_e^2, \sigma_\alpha^2)\pi(\sigma_e^2, \sigma_\alpha^2)\, d\sigma_e^2\, d\sigma_\alpha^2}. \tag{170}$$

In the absence of any good prior information on a location parameter such as μ, we use the non-informative prior $\pi(\mu) = 1$. Then the integration in (169) requires "completing the square" for the quadratic in μ that occurs in (167). That quadratic is, using $t_i \equiv n_i/\lambda_i$,

$$\begin{aligned}
\sum_i \frac{n_i(\bar{y}_{i\cdot} - \mu)^2}{\lambda_i} &= \mu^2 \sum_i \frac{n_i}{\lambda_i} - 2\mu \sum_i \frac{n_i \bar{y}_{i\cdot}}{\lambda_i} + \sum_i \frac{n_i \bar{y}_{i\cdot}^2}{\lambda_i} \\
&= (\Sigma_i t_i)\left(\mu - \frac{\Sigma_i t_i \bar{y}_{i\cdot}}{\Sigma_i t_i}\right)^2 + \Sigma_i t_i \bar{y}_{i\cdot}^2 - \frac{(\Sigma_i t_i \bar{y}_{i\cdot})^2}{\Sigma_i t_i} \\
&= (\Sigma_i t_i)\left(\mu - \frac{\Sigma_i t_i \bar{y}_{i\cdot}}{\Sigma_i t_i}\right)^2 + \Sigma_i t_i\left(\bar{y}_{i\cdot} - \frac{\Sigma_i t_i \bar{y}_{i\cdot}}{\Sigma_i t_i}\right)^2.
\end{aligned}$$

Therefore the integration (169) becomes

$$\begin{aligned}
&\int \exp\left[-\tfrac{1}{2}\sum_i \frac{n_i(\bar{y}_{i\cdot} - \mu)^2}{\lambda_i}\right] d\mu \\
&= \exp\left[-\tfrac{1}{2}\sum_i t_i\left(\bar{y}_{i\cdot} - \frac{\Sigma_i t_i \bar{y}_{i\cdot}}{\Sigma_i t_i}\right)^2\right] \int \exp\left[-\tfrac{1}{2}\frac{1}{(\Sigma_i t_i)^{-1}}\left(\mu - \frac{\Sigma_i t_i \bar{y}_{i\cdot}}{\Sigma_i t_i}\right)^2\right] d\mu \\
&= \exp\left[-\tfrac{1}{2}\sum_i t_i\left(\bar{y}_{i\cdot} - \frac{\Sigma_i t_i \bar{y}_{i\cdot}}{\Sigma_i t_i}\right)^2\right] (2\pi)^{\frac{1}{2}}(\Sigma_i t_i)^{-\frac{1}{2}}.
\end{aligned} \tag{171}$$

Hence using (167) in (169) and then (171) gives

$$f(\mathbf{y}\,|\,\sigma_e^2, \sigma_\alpha^2) = \frac{\exp(-\text{SSE}/2\sigma_e^2)}{\sigma_e^{2[\frac{1}{2}(N-a)]}} \frac{\exp[-\frac{1}{2}\Sigma_i t_i(\bar{y}_{i\cdot} - \Sigma_i t_i \bar{y}_{i\cdot}/\Sigma_i t_i)^2](2\pi)^{\frac{1}{2}}(\Sigma_i t_i)^{-\frac{1}{2}}}{(2\pi)^{\frac{1}{2}N} \prod_{i=1}^{a} (\sigma_e^2 + n_i\sigma_\alpha^2)^{\frac{1}{2}}}. \tag{172}$$

Clearly, this is not very tractable. Nevertheless, it is the likelihood, $L(\sigma_e^2, \sigma_\alpha^2\,|\,\mathbf{y})$, and for maximum likelihood estimation of σ_e^2 and σ_α^2 one could use numerical maximization. However, because of the implicit intractability of (172) we turn to balanced data.

c. Balanced data

Balanced data have $n_i = n\ \forall\ i$, whereupon each $\lambda_i = \lambda = \sigma_e^2 + n\sigma_\alpha^2$ and $t_i = n/\lambda$. Then (172) reduces to

$$f(\mathbf{y}\,|\,\sigma_e^2, \sigma_\alpha^2) = \frac{\exp(-\text{SSE}/2\sigma_e^2)}{\sigma_e^{2[\frac{1}{2}a(n-1)]}} \frac{\exp[-\frac{1}{2}(n/\lambda)\Sigma_i(\bar{y}_{i\cdot} - \bar{y}_{\cdot\cdot})^2]\,(2\pi)^{\frac{1}{2}}(an/\lambda)^{-\frac{1}{2}}}{(2\pi)^{\frac{1}{2}N}\lambda^{\frac{1}{2}a}}$$

$$= \frac{\exp(-\text{SSE}/2\sigma_e^2)}{\sigma_e^{2[\frac{1}{2}a(n-1)]}} \frac{\exp(-\text{SSA}/2\lambda)}{\lambda^{\frac{1}{2}(a-1)}} \frac{1}{(2\pi)^{\frac{1}{2}(an-1)}} \frac{1}{(an)^{\frac{1}{2}}}, \tag{173}$$

which is the same as the restricted likelihood function in (141). Thus (173) is, on replacing λ by $\sigma_e^2 + n\sigma_\alpha^2$,

$$f(\mathbf{y}\,|\,\sigma_e^2, \sigma_\alpha^2) = \frac{\exp(-\text{SSE}/2\sigma_e^2)\exp[-\frac{1}{2}\text{SSA}/(\sigma_e^2 + n\sigma_\alpha^2)]}{\sigma_e^{2[\frac{1}{2}a(n-1)]}(\sigma_e^2 + n\sigma_\alpha^2)^{\frac{1}{2}(a-1)}(2\pi)^{\frac{1}{2}(an-1)}(an)^{\frac{1}{2}}}. \tag{174}$$

Equation (174) must now be used in (170), which also requires specifying the prior, $\pi(\sigma_e^2, \sigma_\alpha^2)$. And (174) is both used in the numerator of (170), and has to be integrated in the denominator of (170). All this is not at all tractable. Hill (1965) tried all manner of approaches to simplify the process, for he had no computing facilities of today's power. Fortunately, on many occasions one does not really need to consider the denominator of (170). It is just a function of $\mathbf{y}$ and so, given $\mathbf{y}$, is effectively a constant. Therefore the numerator of (170) commands attention. That requires looking at (174) and at the same time ascertaining if we can have a prior density $\pi(\sigma_e^2, \sigma_\alpha^2)$ that is both realistic and tractable.

The right-hand side of (174) looks a little like a product of inverted gammas—except that the ranges of σ_e^2 and $\sigma_e^2 + n\sigma_\alpha^2$ are connected through $\sigma_e^2 + n\sigma_\alpha^2$ never being less than σ_e^2. That makes for complications. For $\pi(\sigma_e^2, \sigma_\alpha^2)$ one possibility is a product of inverted gammas defined in (154):

$$\pi(\sigma_e^2, \sigma_\alpha^2) = k\,\frac{e^{-1/q\sigma_e^2}}{\sigma_e^{2(p+1)}}\frac{e^{-1/b\sigma_\alpha^2}}{\sigma_\alpha^{2(c+1)}}, \tag{175}$$

for a constant k that is a function of c, b, p and q such that $\int \pi(\sigma_e^2, \sigma_\alpha^2)\,d\sigma_e^2\,d\sigma_\alpha^2 = 1$.

Then, on substituting (174) and (175) into (170), and ignoring terms that are not functions of σ_e^2 and σ_α^2, the log of the numerator of (170) is

$$\log f(\mathbf{y}, \sigma_e^2, \sigma_\alpha^2) = \frac{\text{SSE}}{2\sigma_e^2} - \frac{a(n-1)}{2}\log \sigma_e^2 - \frac{\text{SSA}}{2\lambda} - \frac{a-1}{2}\log \lambda$$

$$- \frac{1}{q\sigma_e^2} - (p+1)\log \sigma_e^2 - \frac{1}{b\sigma_\alpha^2} - (c+1)\log \sigma_\alpha^2$$

$$= -\frac{1}{\sigma_e^2}\left(\frac{\text{SSE}}{2} + \frac{1}{q}\right) - \left[\frac{a(n-1)}{2} + p + 1\right]\log \sigma_e^2 - \frac{\text{SSA}}{2(\sigma_e^2 + n\sigma_\alpha^2)}$$

$$- \frac{a-1}{2}\log(\sigma_e^2 + n\sigma_\alpha^2) - \frac{1}{b\sigma_\alpha^2} - (c+1)\log \lambda\,.$$

Differentiating this expression with respect to σ_e^2 and σ_α^2 and equating the derivatives to zero will yield the posterior modes, which are reasonable Bayes estimators of variances:

$$\frac{\partial \log f(\mathbf{y}, \sigma_e^2, \sigma_\alpha^2)}{\partial \sigma_e^2} = \left(\frac{\text{SSE}}{2} + \frac{1}{q}\right)\frac{1}{\sigma_e^4} - \left[\frac{a(n-1)}{2} + p + 1\right]\frac{1}{\sigma_e^2} + \frac{\text{SSA}}{2(\sigma_e^2 + n\sigma_\alpha^2)^2}$$

$$- \frac{a-1}{2(\sigma_e^2 + n\sigma_\alpha^2)},$$

$$\frac{\partial \log f(\mathbf{y}, \sigma_e^2, \sigma_\alpha^2)}{\partial \sigma_\alpha^2} = \frac{n\text{SSA}}{2(\sigma_e^2 + n\sigma_\alpha^2)^2} - \frac{n(a-1)}{2(\sigma_e^2 + n\sigma_\alpha^2)} + \frac{1}{b\sigma_\alpha^4} - \frac{c+1}{\sigma_\alpha^2}\,.$$

Equating these two expressions to zero and denoting the solutions by $\mathring{\sigma}_e^2$ and $\mathring{\sigma}_\alpha^2$ gives

$$\mathring{\sigma}_e^2 = \frac{\frac{1}{2}\text{SSE} + \dfrac{1}{q} + \frac{1}{2}\text{SSA}\left(\dfrac{\mathring{\sigma}_e^2}{\mathring{\sigma}_e^2 + n\mathring{\sigma}_\alpha^2}\right)^2}{\frac{1}{2}a(n-1) + p + 1 + \dfrac{\frac{1}{2}(a-1)\mathring{\sigma}_e^2}{\mathring{\sigma}_e^2 + n\mathring{\sigma}_\alpha^2}}$$

and (176)

$$\mathring{\sigma}_\alpha^2 = \frac{\frac{1}{2}n\text{SSA}\left(\dfrac{\mathring{\sigma}_\alpha^2}{\mathring{\sigma}_e^2 + n\mathring{\sigma}_\alpha^2}\right)^2 + \dfrac{1}{b}}{\dfrac{\frac{1}{2}(a-1)n\mathring{\sigma}_\alpha^2}{\mathring{\sigma}_e^2 + n\mathring{\sigma}_\alpha^2} + c + 1}\,.$$

Solution of these equations and plotting of the solutions then indicates the behavior of these as Bayes estimators. Figures 3.3a and 3.3b show an example of plots of ANOVA and Bayes estimates of σ_e^2, and Figures 3.4a and 3.4b show similar plots for σ_α^2. All four of these plots are for $a = 12$ and $n = 5$, and for the parameters of the prior distribution (175) being $p = 10$, $q = 1$, $b = 2$ and

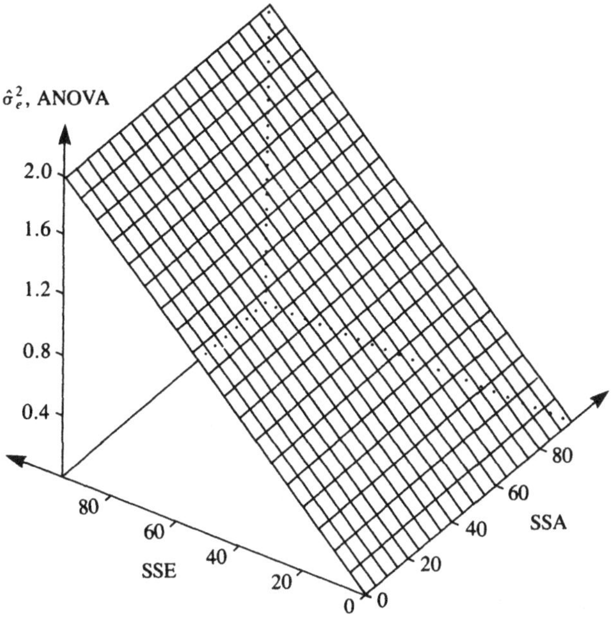

Figure 3.3a. ANOVA estimator $\hat{\sigma}_e^2 = \text{MSE} = \frac{1}{48}\text{SSE}$ of (54), for $a = 12$ and $n = 5$.

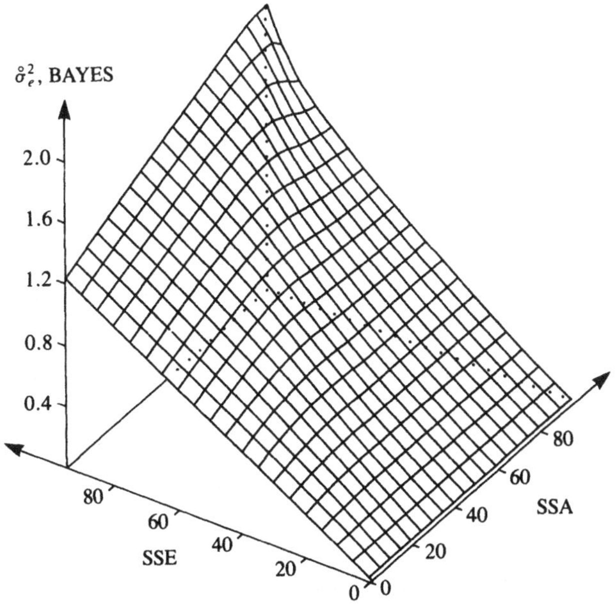

Figure 3.3b. Bayes estimator $\mathring{\sigma}_e^2$ of (176), for $a = 12$ and $n = 5$, with parameters of the prior distribution (175) being $p = 10$, $q = 1$, $b = 2$ and $c = 6$.

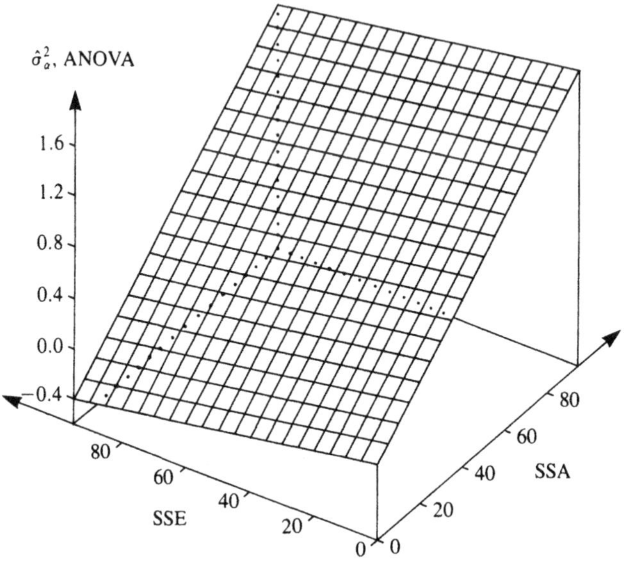

Figure 3.4a. ANOVA estimator $\hat{\sigma}^2_\alpha = (\text{MSA} - \text{MSE})/n = [\text{SSA}/(a-1) - \text{SSE}/a(n-1)]/n = \frac{1}{55}\text{SSA} - \frac{1}{240}\text{SSE}$ of (55), when $a = 12$ and $n = 5$.

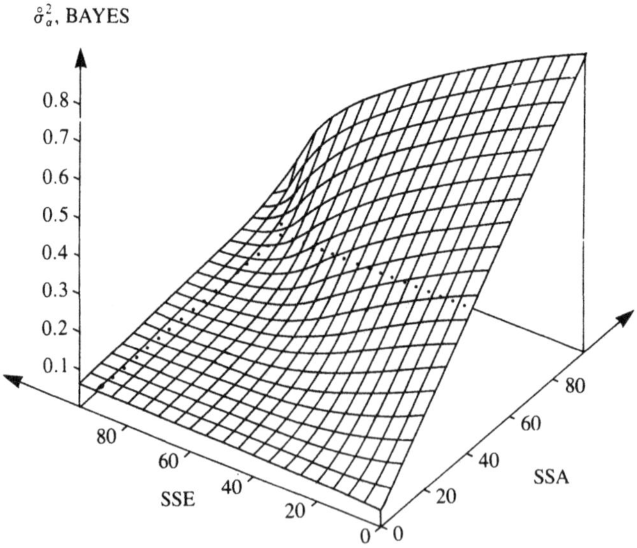

Figure 3.4b Bayes estimator $\mathring{\sigma}^2_\alpha$ of (176), for $a = 12$ and $n = 5$, with parameters of the prior distribution (175) being $p = 10$, $q = 1$, $b = 2$ and $c = 6$.

$c = 6$. Two features of the figures warrant comment. First, negative ANOVA estimates $\hat{\sigma}_\alpha^2$ are evident in Figure 3.4a, but the Bayes estimates $\mathring{\sigma}_\alpha^2$ in Figure 3.4b are never negative. (This is, of course, also true for any values of a and n and for any prior distribution.) Second, applying (155) to the two inverted gammas of (175) gives $E(\sigma_e^2) = 1/[q(p-1)] = \frac{1}{9}$ and $E(\sigma_\alpha^2) = 1/[b(c-1)] = \frac{1}{10}$. Equating these means to the ANOVA estimators, i.e.,

$$\tfrac{1}{9} = \text{MSE} = \tfrac{1}{48}\text{SSE} \quad \text{and} \quad \tfrac{1}{10} = \tfrac{1}{5}(\tfrac{1}{11}\text{SSA} - \tfrac{1}{48}\text{SSE}),$$

gives SSE $= 5.33$ and SSA $= 6.72$, and indicates approximately how these values relate to SSE and SSA. Examination of Figures 3.3 and 3.4 suggests how Bayes estimation pulls the ANOVA estimates towards prior values $\sigma_e^2 = \frac{1}{9}$ and $\sigma_\alpha^2 = \frac{1}{10}$ at SSA $= 6.72$ and SSE $= 5.33$. This effect is, perhaps, easiest to see as between Figures 3.3a and b. There, Bayes estimation of σ_e^2 in Figure 3.3b seems to represent a non-linear pull of the ANOVA estimate in Figure 3.3a toward (0, 0), which is an approximation of the prior values.

3.10. A SUMMARY

We list here a brief summary of many of the main results from this chapter. Alongside each of them is the equation number (or other reference) where it is derived.

The chapter is sectionalized (aside from Sections 3.5 and 3.6) by methods; in contrast, this summary is dichotomized into balanced and unbalanced data, so that results will be easily available for the analyst whose data will always be either one or the other.

a. Balanced data

Model

$$y_{ij} = \mu + \alpha_i + e_{ij}, \quad \text{with } i = 1, \ldots, a \text{ and } j = 1, \ldots, n; \tag{5}$$

$$\boldsymbol{\alpha} \sim (\mathbf{0}, \sigma_\alpha^2 \mathbf{I}_a), \quad \mathbf{e} \sim (\mathbf{0}, \sigma_e^2 \mathbf{I}_{an}), \quad \text{cov}(\boldsymbol{\alpha}, \mathbf{e}') = \mathbf{0}_{a \times an},$$
$$E(\mathbf{y}) = \mu\mathbf{1} \quad \text{and} \quad \text{var}(\mathbf{y}) = \mathbf{V} = \{_d\, \sigma_e^2 \mathbf{I}_n + \sigma_\alpha^2 \mathbf{J}_n\}_{i=1}^{a}. \tag{6)–(14}$$

Estimating μ

$$\text{GLSE}(\hat{\mu}) = \bar{y}_{..} = \text{OLSE}(\mu). \tag{35), (36}$$

Predicting α_i

$$\tilde{\alpha}_i = \frac{n\sigma_\alpha^2}{n\sigma_\alpha^2 + \sigma_e^2}(\bar{y}_{i.} - \mu); \tag{40}$$

$$\text{BLUP}(\mu + \alpha_i) = \bar{y}_{..} + \frac{n\sigma_\alpha^2}{n\sigma_\alpha^2 + \sigma_e^2}(\bar{y}_{i.} - \bar{y}_{..}). \tag{42}$$

Sums of (and mean) squares

$$\text{SSA} = n\Sigma_i(\bar{y}_{i.} - \bar{y}_{..})^2, \quad \text{SSE} = \Sigma_i\Sigma_j(y_{ij} - \bar{y}_{i.})^2, \tag{43}$$

$$\text{MSA} = \frac{\text{SSA}}{a-1}, \qquad \text{MSE} = \frac{\text{SSE}}{a(n-1)}.$$

Normality assumptions

$$\boldsymbol{\alpha} \sim \mathcal{N}(\mathbf{0}, \sigma_\alpha^2 \mathbf{I}_a) \quad \text{and} \quad \mathbf{e} \sim \mathcal{N}(\mathbf{0}, \sigma_e^2 \mathbf{I}_{an}). \tag{57}$$

Everything that follows, except what is marked $\mathcal{N}$nn (normality not needed), is based on normality assumptions.

ANOVA estimation

$$\hat{\sigma}_\alpha^2 = \frac{1}{n}(\text{MSA} - \text{MSE}) \quad [\mathcal{N}\text{nn}]; \tag{55}$$

$$\text{var}(\hat{\sigma}_\alpha^2) = \frac{2}{n^2}\left[\frac{(n\sigma_\alpha^2 + \sigma_e^2)^2}{a-1} + \frac{\sigma_e^2}{a(n-1)}\right]. \tag{69}$$

Unbiased estimation of (69):

$$\text{var}(\hat{\sigma}_\alpha^2) \text{ is estimated unbiasedly by } \frac{2}{n^2}\left[\frac{(n\hat{\sigma}_\alpha^2 + \hat{\sigma}_e^2)^2}{a+1} + \frac{\hat{\sigma}_e^4}{a(n-1)+2}\right]; \tag{70}$$

$$\hat{\sigma}_e^2 = \text{MSE} \quad [\mathcal{N}\text{nn}]; \tag{54}$$

$$\text{var}(\hat{\sigma}_e^2) = \frac{2\sigma_e^4}{a(n-1)}, \quad \text{estimated unbiasedly by } \frac{2\hat{\sigma}_e^4}{a(n-1)+2}; \tag{66), (67}$$

$$\text{cov}(\hat{\sigma}_\alpha^2, \hat{\sigma}_e^2) = \frac{-2\sigma_e^4}{an(n-1)}, \quad \text{estimated unbiasedly by } \frac{-2\hat{\sigma}_e^4}{n[a(n-1)+2]}. \tag{71), (72}$$

Testing H: $\sigma_\alpha^2 = 0$

$$F = \frac{\text{MSA}}{\text{MSE}} \sim \mathscr{F}^{a-1}_{a(n-1)}. \tag{73}$$

Confidence intervals (Section 3.5d-v). Based on normality assumptions (Section 3.5d),

$$\Pr\left\{\frac{\text{SSE}}{\chi^2_{a(n-1),\text{U}}} \leqslant \sigma_e^2 \leqslant \frac{\text{SSE}}{\chi^2_{a(n-1),\text{L}}}\right\} = 1 - \alpha,$$

$$\Pr\left\{\frac{F_\text{U}/F - 1}{n} \leqslant \frac{\sigma_\alpha^2}{\sigma_e^2} \leqslant \frac{F_\text{L}/F - 1}{n}\right\} = 1 - \alpha.$$

For σ_α^2 see Table 3.4.

Probability of negative $\hat{\sigma}_\alpha^2$

$$\Pr\{\hat{\sigma}_\alpha^2 < 0\} = \Pr\{\mathscr{F}^{a(n-1)}_{a-1} > 1 + n\sigma_\alpha^2/\sigma_e^2\}. \tag{74}$$

Maximum likelihood estimation

ML solutions:

$$\dot{\sigma}_e^2 = \text{MSE}, \quad \dot{\sigma}_\alpha^2 = \frac{(1 - 1/a)\text{MSA} - \text{MSE}}{n}. \tag{108}$$

ML estimators:

$$\text{when } \dot{\sigma}_\alpha^2 \geqslant 0, \quad \tilde{\sigma}_e^2 = \dot{\sigma}_e^2, \qquad \tilde{\sigma}_\alpha^2 = \dot{\sigma}_\alpha^2; \tag{115}$$

$$\text{when } \dot{\sigma}_\alpha^2 < 0, \quad \tilde{\sigma}_e^2 = \frac{\text{SST}_\text{m}}{an}, \quad \tilde{\sigma}_\alpha^2 = 0. \tag{115}$$

Large-sample variances:

$$\text{var}(\tilde{\sigma}_e^2) = \frac{2\sigma_e^4}{a(n-1)}, \quad \text{cov}(\tilde{\sigma}_e^2, \tilde{\sigma}_\alpha^2) = -\frac{\text{var}(\tilde{\sigma}_e^2)}{n},$$

$$\text{var}(\tilde{\sigma}_\alpha^2) = \frac{2\sigma_e^4}{n^2}\left[\frac{(\sigma_e^2 + n\sigma_\alpha^2)^2/\sigma_e^4}{a} + \frac{1}{a(n-1)}\right]. \tag{126}$$

Restricted maximum likelihood estimation

When ANOVA $\hat{\sigma}_\alpha^2 > 0$, REML estimators are ANOVA estimators;

$$\text{when ANOVA } \hat{\sigma}_\alpha^2 \leqslant 0, \quad \text{REML}(\tilde{\sigma}_e^2) = \frac{\text{SST}_\text{m}}{an-1}, \quad \text{REML}(\tilde{\sigma}_\alpha^2) = 0. \tag{146}$$

Large-sample variances: these are the same as (126) above.

Bayes estimation

Simple example:

$$\mathbf{x} = [x_1, \ldots, x_n]' \sim \mathcal{N}(\mu\mathbf{1}, \sigma^2\mathbf{I}_n),$$

$$s^2 = \frac{\Sigma_i(x_i - \bar{x})^2}{n-1}.$$

Inverted gamma density with $a = 2$ and $b = 1$ as prior:

$$\pi(\sigma^2) = (\sigma^2)^{-3}e^{-1/\sigma^2}, \tag{156}$$

$$\text{Bayes}(\hat{\sigma}_\text{B}^2) = \frac{(n-1)s^2 + 2}{n+1}, \tag{163}$$

$$\text{var}[\text{Bayes}(\hat{\sigma}_\text{B}^2)] = \frac{2\sigma^4(n-1)}{(n+1)^2}. \tag{164}$$

1-way classification: this demands numerical solution. See (176).

b. Unbalanced data

Model

$$y_{ij} = \mu + \alpha_i + e_{ij}, \quad \text{with } i = 1, \ldots, a \text{ and } j = 1, \ldots, n_i; \tag{5}$$

$$N = n_{.} = \sum_{i=1}^{a} n_i;$$

$$\boldsymbol{\alpha} \sim (\mathbf{0}, \sigma_\alpha^2 \mathbf{I}_a), \quad \mathbf{e} \sim (\mathbf{0}, \sigma_e^2 \mathbf{I}_N), \quad \text{cov}(\boldsymbol{\alpha}, \mathbf{e}') = \mathbf{0}_{a \times N},$$

$$E(\mathbf{y}) = \mu \mathbf{1}, \quad \mathbf{V} = \text{var}(\mathbf{y}) = \{_{\text{d}}\, \sigma_e^2 \mathbf{I}_{n_i} + \sigma_\alpha^2 \mathbf{J}_{n_i}\}_{i=1}^{a}.$$

Estimating μ

$$\text{BLUE}(\mu) = \text{GLSE}(\mu) = \frac{\sum_i \dfrac{n_i \bar{y}_{i.}}{\sigma_e^2 + n_i \sigma_\alpha^2}}{\sum_i \dfrac{n_i}{\sigma_e^2 + n_i \sigma_\alpha^2}} \quad [\mathcal{N}\text{nn}]; \tag{34}$$

$$\text{var}[\text{GLSE}(\mu)] = \frac{1}{\sum_i \dfrac{n_i}{\sigma_e^2 + n_i \sigma_\alpha^2}} \quad [\mathcal{N}\text{nn}]. \tag{37}$$

Predicting α_i

$$\tilde{\alpha}_i = \frac{n_i \sigma_\alpha^2}{n_i \sigma_\alpha^2 + \sigma_e^2} (\bar{y}_{i.} - \mu); \tag{40}$$

$$\text{BLUP}(\mu + \alpha_i) = \text{BLUE}(\mu) + \frac{n_i \sigma_\alpha^2}{n_i \sigma_\alpha^2 + \sigma_e^2} [\bar{y}_{i.} - \text{BLUE}(\mu)] \quad [\mathcal{N}\text{nn}]. \tag{42}$$

Sums of (and mean) squares

$$\text{SSA} = \Sigma_i n_i (\bar{y}_{i.} - \bar{y}_{..})^2, \quad \text{SSE} = \Sigma_i \Sigma_j (y_{ij} - \bar{y}_{i.})^2,$$

$$\text{MSA} = \frac{\text{SSA}}{a-1}, \qquad \text{MSE} = \frac{\text{SSE}}{N-a}. \tag{75}$$

ANOVA estimation

$$\hat{\sigma}_\alpha^2 = \frac{\text{MSA} - \text{MSE}}{(N - \Sigma_i n_i^2/N)/a - 1} \quad [\mathcal{N}\text{nn}]; \tag{83}$$

for $S_2 \equiv \Sigma_i n_i^2$ and $S_3 = \Sigma_i n_i^3$

$$\text{var}(\hat{\sigma}_\alpha^2) = \frac{2N}{N^2 - S_2} \left[\frac{N(N-1)(a-1)\sigma_e^4}{(N-a)(N^2 - S_2)} + 2\sigma_e^2 \sigma_\alpha^2 + \frac{N^2 S_2 + S_2^2 - 2N S_3}{N(N^2 - S_2)} \sigma_\alpha^4 \right]; \tag{102}$$

$$\hat{\sigma}_e^2 = \text{MSE} \quad [\mathcal{N}\text{nn}]; \tag{82}$$

$$\text{var}(\hat{\sigma}_e^2) = \frac{2\sigma_e^4}{N-a}; \tag{95}$$

$$\text{cov}(\hat{\sigma}_\alpha^2, \hat{\sigma}_e^2) = \frac{-2\sigma_e^4}{(N-a)(N - \Sigma_i n_i^2/N)/(a-1)}. \tag{96}$$

Unbiased estimates. The terms (102), (95) and (96) can be estimated unbiasedly: see Section 5.2e.

Testing H: $\sigma_\alpha^2 = 0$. Use $F = \text{MSA}/\text{MSE}$. See Table 3.9.

Confidence intervals. For σ_e^2: $\Pr\left\{\frac{\text{SSE}}{\chi^2_{N-a,\text{U}}} \leqslant \sigma_e^2 \leqslant \frac{\text{SSE}}{\chi^2_{N-a,\text{L}}}\right\} = 1 - \alpha$.

For σ_α^2 see Section 3.6d-vi.

Maximum likelihood estimation. First obtain solutions $\dot{\mu}$, $\dot{\sigma}_e^2$ and $\dot{\lambda}$ to

$$\dot{\mu} = \sum_i \frac{n_i \bar{y}_{i\cdot}}{\dot{\sigma}_e^2 + n_i\dot{\sigma}_\alpha^2} \Big/ \sum_i \frac{n_i}{\dot{\sigma}_e^2 + n_i\dot{\sigma}_\alpha^2}, \tag{133}$$

$$\frac{\text{SSE}}{\dot{\sigma}_e^4} - \frac{N-a}{\dot{\sigma}_e^2} + \sum_i \frac{n_i(\bar{y}_{i\cdot} - \dot{\mu})^2}{\dot{\lambda}_i^2} - \sum_i \frac{1}{\dot{\lambda}_i} = 0 \tag{134}$$

and

$$\sum_i \frac{n_i^2(\bar{y}_{i\cdot} - \dot{\mu})^2}{\dot{\lambda}_i^2} = \sum_i \frac{n_i}{\dot{\lambda}_i}. \tag{135}$$

$$\begin{aligned} &\text{When } \dot{\sigma}_\alpha^2 > 0, \quad \tilde{\mu} = \dot{\mu}, \quad \tilde{\sigma}_e^2 = \dot{\sigma}_e^2 \quad \text{and} \quad \tilde{\sigma}_\alpha^2 = \dot{\sigma}_\alpha^2; \\ &\text{when } \dot{\sigma}_\alpha^2 < 0, \quad \tilde{\mu} = \bar{y}_{\cdot\cdot}, \quad \tilde{\sigma}_e^2 = \frac{\text{SST}_\text{m}}{N} \quad \text{and} \quad \tilde{\sigma}_\alpha^2 = 0. \end{aligned} \tag{137}$$

Large-sample variances:

$$\text{var}(\tilde{\mu}) = \left(\sum_i \frac{n_i}{\sigma_e^2 + n_i\sigma_e^2}\right)^{-1}; \tag{138}$$

$$\text{var}\begin{bmatrix} \tilde{\sigma}_e^2 \\ \tilde{\sigma}_\alpha^2 \end{bmatrix} \simeq \frac{2}{D}\begin{bmatrix} \sum_i \frac{n_i^2}{\lambda_i^2} & -\sum_i \frac{n_i}{\lambda_i^2} \\ -\sum_i \frac{n_i}{\lambda_i^2} & \frac{N-a}{\sigma_e^4} + \sum_i \frac{1}{\lambda_i^2} \end{bmatrix}, \tag{139}$$

with

$$\begin{aligned} D &= \frac{N-a}{\sigma_e^4}\sum_i \frac{n_i^2}{\lambda_i^2} + \sum_i \frac{1}{\lambda_i^2}\sum_i \frac{n_i^2}{\lambda_i^2} - \left(\sum_i \frac{n_i}{\lambda_i^2}\right)^2 \\ &= [N\Sigma_i w_i^2 - (\Sigma_i w_i)^2]\sigma_e^{-4} \quad \text{for } w_i = \frac{n_i}{\lambda_i} = \frac{n_i}{\sigma_e^2 + n_i\sigma_\alpha^2} = \frac{1}{\text{var}(\bar{y}_{i\cdot})}. \end{aligned}$$

Restricted maximum likelihood estimation. The general procedure of Section 3.6 has to be used.

Bayes estimation. This is very intractable. See (172).

3.11. EXERCISES

E 3.1. For each data set A and B write the model equation (1) in matrix and vector form including the use of direct products where appropriate.

Data A

	$j=1$	$j=2$	$j=3$
$i=1$:	10	12	8
$i=2$:	10	12	14
$i=3$:	6	11	7
$i=4$:	18	17	7

Data B

	$j=1$	$j=2$	$j=3$	$j=4$
$i=1$:	12	8	6	10
$i=2$:	17	13		
$i=3$:	16	11	15	

E 3.2. Suppose Data A are to be analyzed using the model equation $y_{ij} = \mu + \alpha_i + \beta_j + e_{ij}$. Write this equation for Data A, both with and without using direct products.

E 3.3. Suppose a data set consists of n observations in each of b columns that are nested within each of a rows. Write the model equation

$$y_{ijk} = \mu + \alpha_i + \beta_{ij} + e_{ijk}$$

for such data, using direct products.

E 3.4. With the notation

$$\hat{\mu}_{\mathrm{r}} = \mathrm{GLSE}(\mu), \quad \hat{\mu}_{\mathrm{n}} = \mathrm{OLSE}(\mu),$$

and with $\mathrm{var_R}(\cdot)$ and $\mathrm{var_F}(\cdot)$ denoting variance in the random and fixed models, respectively, show that

(a) in the random model, $\mathrm{var_R}(\hat{\mu}_{\mathrm{r}}) \leqslant \mathrm{var_R}(\hat{\mu}_{\mathrm{w}})$

(b) in the fixed effects model, $\mathrm{var_F}(\hat{\mu}_{\mathrm{n}}) \leqslant \mathrm{var_F}(\hat{\mu}_{\mathrm{w}})$

(c) $\mathrm{var_F}(\hat{\mu}_{\mathrm{n}}) \leqslant \mathrm{var_R}(\hat{\mu}_{\mathrm{r}}) \leqslant \mathrm{var_R}(\hat{\mu}_{\mathrm{n}})$.

Hint: Use the Cauchy–Schwartz inequality

$$\left(\sum_{i=1}^{n} x_i y_i\right)^2 \leqslant \left(\sum_{i=1}^{n} x_i^2\right)\left(\sum_{j=1}^{n} y_i^2\right).$$

E 3.5. For the model having equation (30) and dispersion matrix $\mathbf{V}$ of (31), show that $\tilde{\alpha}_i$ of (40) comes from

$$\tilde{\boldsymbol{\alpha}} = \sigma_\alpha^2 \mathbf{Z}'\mathbf{V}^{-1}(\mathbf{y} - \mu\mathbf{1}) \quad \text{for } \mathbf{Z} = \{_{\mathrm{d}}\, \mathbf{1}_n\}_{i=1}^{a}.$$

E 3.6. Why in (28) must ρ exceed $-1/(a-1)$?

E 3.7. Use the method (a) of subsection i, and (b) of subsection ii of Section 3.5a, to derive $E(\text{MSE}) = \sigma_e^2$.

E 3.8. Show that $\hat{\sigma}_e^2$ and $\hat{\sigma}_\alpha^2$ of (54) and (55) are unbiased.

E 3.9. Using Theorem S2 of Section S.4, together with the algebra of J-matrices and Kronecker products in Appendix M, derive the results in (60), (61) and (63).

E 3.10. Derive the confidence intervals for $\sigma_\alpha^2/(\sigma_\alpha^2 + \sigma_e^2)$ and for $\sigma_e^2/(\sigma_\alpha^2 + \sigma_e^2)$ shown in Table 3.4.

E 3.11. Suppose an experimenter sets out to estimate σ_α^2 and σ_e^2 from an experiment of 4 observations from each of 5 classes, with prior knowledge that σ_α^2 is likely to be $51\frac{1}{2}\%$ of σ_e^2. What is the probability that the ANOVA estimate of σ_α^2 will be negative?

E 3.12. Derive (102) and (101).

E 3.13. Derive (126).

E 3.14. Why from (106) can the MLE of μ be derived without differentiation?

E 3.15. Show that

$$-\frac{1}{2\sigma_e^2}\left[\Sigma_i\Sigma_j(y_{ij} - \mu)^2 - \Sigma_i \frac{n_i^2\sigma_\alpha^2(\bar{y}_{i\cdot} - \mu)^2}{\lambda_i}\right]$$

$$= -\frac{\text{SSE}}{2\sigma_e^2} - \Sigma_i \frac{n_i(\bar{y}_{i\cdot} - \mu)^2}{2\lambda_i}.$$

E 3.16. (a) Using $\log L$ of (130), show that ML estimators when $\tilde{\sigma}_\alpha^2 = 0$ are $\tilde{\sigma}_e^2 = \text{SST}_{\text{m}}/N$ and $\tilde{\mu} = \bar{y}_{\cdot\cdot}$.
(b) Show that (134) and (135) reduce for balanced data to (108).

E 3.17. Derive the expected values in Table 3.10.

E 3.18. Verify (A) and (B) following (139).

For (B) recall that $\Sigma_i a_i^2 \Sigma_i b_i^2 - (\Sigma_i a_i b_i)^2 = \sum\sum_{i \neq j} (a_i b_j - a_j b_i)^2$ and that $w_i = n_i/\lambda_i$ implies $w_i/n_i = (1 - w_i\sigma_\alpha^2)/\sigma_e^2$.

E 3.19. Suppose $\mathbf{x} = [x_1, \ldots, x_n]' \sim \mathcal{N}(\mu\mathbf{1}, \sigma^2\mathbf{I})$.
(a) Using $L(\mu, \sigma^2 | \mathbf{x})$, show that the ML estimator of σ^2 is s^2/n for $s^2 = \Sigma_i(x_i - \bar{x})^2$.

(b) Derive $L(\sigma^2 \mid s^2)$ and use it to show that the REML estimator of σ^2 is $s^2/(n-1)$.

E 3.20. Show that minimizing (141) subject to $\tilde{\sigma}^2_{\alpha(r)} = 0$ yields $\tilde{\sigma}^2_{e(r)} = \text{SST}_m/(an-1)$.

E 3.21. (a) Fc- fixed σ^2_α find the limiting value of $\tilde{\mu} = \dot{\mu}$ of (133) and (136):

(i) as $\sigma^2_e \to 0$ and (ii) as $\sigma^2_e \to \infty$.

(b) For fixed σ^2_e, find the limiting value of $\tilde{\mu} = \dot{\mu}$ of (133) and (136):

(i) as $\sigma^2_\alpha \to 0$ and (ii) as $\sigma^2_\alpha \to \infty$.

(c) What is an explanation for your results in (a) and (b)?

(d) Prove that the likelihood function implicit in (128) can have a maximum (i) at a negative value of σ^2_α, but (ii) at only positive values of σ^2_e.

E 3.22. Derive (14) and (15) of Section S.6.

E 3.23. Verify (155).

E 3.24. For the following data from a 1-way classification having $a = 3$ and $N = 7$

(a) calculate ANOVA estimates of variance components, their sampling variances and unbiased estimates thereof;

(b) try to repeat (a), using ML estimation: at least write out the ML equations, and the terms of the asymptotic sampling dispersion matrix for σ^2_α and σ^2_e.

Data

$i=1$	$i=2$	$i=3$
10	3	17
14	7	
18		
22		

E 3.25. Consider the following three separate variations on the 1-way classification, random model, with $\text{var}(\boldsymbol{\alpha}) = \sigma^2_\alpha \mathbf{I}$ and $\text{cov}(\boldsymbol{\alpha}, \mathbf{e}') = \mathbf{0}$ as is usual. With $\mathbf{e}_i = [e_{i1} \quad e_{i2} \cdots e_{in_i}]'$ for $i = 1, \ldots, a$, and with every $\mathbf{e}_i$ and $\mathbf{e}_{i'}$ having zero covariance,

(i) $\text{var}(\mathbf{e}_i) = \sigma^2_i \mathbf{I}_{n_i}$

(ii) $\text{var}(\mathbf{e}_i) = \sigma^2_e \mathbf{I}_{n_i} + \rho\sigma^2_e(\mathbf{J}_{n_i} - \mathbf{I}_{n_i})$

(iii) $\text{var}(\mathbf{e}_i) = \sigma^2_e \mathbf{I}_{n_i} + \rho_i\sigma^2_e(\mathbf{J}_{n_i} - \mathbf{I}_{n_i})$

(a) For balanced data derive ANOVA estimators of σ^2_α and σ^2_e.

(b) For balanced data under normality, derive sampling variances of your estimators in (i), and unbiased estimators of those sampling variances.

(c) Try repeating (b) for (ii) and (iii).

E 3.26. Try using equations (176) for Data A of E 3.1.

CHAPTER 4

BALANCED DATA

Balanced data, as discussed in Section 1.2, are defined by the nature of the numbers of observations in the cells of the data. A "cell" is a subclass of the data defined by one level of each factor by which the data are being classified. To emphasize this we might use the descriptor "sub-most cell". For example, in Table 1.1, the data for married men receiving drug A are a sub-most cell of the data; married men are a subclass, too, but not a sub-most cell. We defined balanced data as data wherein every sub-most cell has the same number of observations. This omits what we have called planned unbalancedness such as Latin squares and variants thereof (see Section 1.2b-i).

Estimating variance components from balanced data is, generally speaking, much easier than from unbalanced data. We therefore devote a chapter to balanced data. Admittedly, balanced data are usually the outcome of a designed experiment, wherein the number of levels of each factor is usually relatively small, say 6, 10 or maybe 20. This is not an ideal situation for estimating the variance component for any such factor, because if that factor has, say, 6 levels, then no matter how many observations there are in each level there are still only 6 levels, and the situation is akin to estimating a variance from 6 observations. Indeed, a sample of only 6 effects (from a hypothesized population of effects) occurs in the data. Nevertheless, there are many circumstances where researchers do want to estimate variance components from balanced data. Such data have a number of interesting characteristics that lead in many cases to ANOVA estimators of the variance components being not only easy to calculate (e.g., Tables 4.8, 4.10, 4.12 and 4.14) but having attractive optimal features. Also, although ML methodology does not require ANOVA tables, ML estimators from balanced data are, in a number of cases, simple functions of the mean squares in an ANOVA table, e.g., Tables 4.9, 4.11, 4.13 and 4.15. On the other hand, there are also some balanced data cases for which MLEs do not exist in any explicit form, e.g., Sections 4.7f-i and ii; they can be derived only as numerical solutions of non-linear equations.

We begin with ANOVA estimation from balanced data that are classified in a factorial manner, consisting of crossed and nested classifications and combinations thereof. For such data, there are easy rules of thumb by which, no matter how complicated the classifications are, nor how numerous, the ANOVA estimators of variance components can be straightforwardly derived. These rules lay out procedures for determining (1) the lines in the analysis of variance, (2) their degrees of freedom, (3) formulae for calculating sums of squares and (4) expected values of mean squares. Most of the rules are based on Henderson (1969) except that Rule 9 comes from Millman and Glass (1967), who rely heavily on the Henderson paper for a similar set of rules.

The description of the rules is purposefully brief, with no attempt at substantiation. For this the reader is referred to Lum (1954) and Schultz (1955).

4.1. ESTABLISHING ANALYSIS OF VARIANCE TABLES

a. Factors and levels

The analysis of variance table is described in terms of factors A, B, $C, \ldots,$ with the number of levels in them being $n_a, n_b, n_c, \ldots,$ respectively. When one factor is nested within another the notation will be $C{:}B$ for factor C within factor B, and $C{:}BA$ for C within AB subclasses, and so on. A letter on the left of the colon represents the nested factor and letters on the right of the colon represent the factors within which the nested factor is found. With a nested factor, C for example, n_c is the number of levels of factor C within each of the factors in which it is nested. Factors that are not nested, namely those forming cross-classifications, will be called crossed factors.

Within every sub-most cell of the data we assume there is the same number of observations, n_w, either one or more than one. In either case these observations can, as Millman and Glass (1967) point out, be referred to as replications within all other subclasses. Following Henderson (1969), we refer to these as the "within" factor, using the notation $W{:}ABC \ldots$, the number of levels of the "within" factor (i.e., number of replicates) being n_w. The total number of observations is then the product of the ns, namely $N = n_a n_b n_c \ldots n_w$.

b. Lines in the analysis of variance table

Rule 1. There is one line for each factor (crossed or nested), for each interaction, and for "within".

c. Interactions

Interactions are obtained symbolically as products of factors, both factorial and nested. Any possible products of two, three, four, ... factors can be considered. For the sake of generality all crossed factors are assumed to have a colon to the right of the symbol; e.g., $A{:}$ and $B{:}$ and so on.

Rule 2. Every interaction is of the form $ABC \ldots : XYZ \ldots$, where $ABC \ldots$ is the product on the left of the colon of the factors being combined and $XYZ \ldots$ is the product on the right of the colon of the factors so associated with A, B and $C \ldots$.

Rule 3. Repeated letters on the right of the colon are replaced by one of their kind.

Rule 4. If any letter occurs on both sides of a colon that interaction does not exist.

Examples

Factors	*Interaction*	
A and B	AB	(Rule 2)
A and $C{:}B$	$AC{:}B$	(Rule 2)
$A{:}B$ and $C{:}B$	$AC{:}BB \equiv AC{:}B$	(Rule 3)
$A{:}B$ and $B{:}DE$	$AB{:}BDE$, nonexistent	(Rule 4)

The symbolic form $W{:}ABC\ \ldots$ for replicates does, by Rule 4, result in no interactions involving W. Furthermore, the line in the analysis of variance labeled $W{:}ABC\ \ldots$, being the "within" line, is the residual error line.

d. Degrees of freedom

Each line in an analysis of variance table refers either to a crossed factor (such as $A{:}$), to a nested factor (such as $C{:}B$) or to an interaction (e.g., $AC{:}B$). Any line can therefore be typified by the general expression given for an interaction in Rule 2, namely $ABC\ \ldots{:}\ XYZ\ \ldots$.

Rule 5. Degrees of freedom for the line denoted by

$$AB{:}XY \quad \text{are} \quad (n_a - 1)(n_b - 1)n_x n_y\,.$$

The rule is simple. Degrees of freedom are the product of terms like $(n_a - 1)$ for every letter A on the left of the colon and of terms like n_x for every letter X on the right of the colon.

Rule 6. The sum of all degrees of freedom is $N - 1$, with $N = n_w n_a n_b n_c \ldots$.

e. Sums of squares

The symbols that specify a line in the analysis of variance are used to establish the corresponding sum of squares. The basic elements are taken to be the uncorrected sums of squares (see Section 4.1f) with notation

$a \equiv$ uncorrected sum of squares for the A-factor,

$ab \equiv$ uncorrected sum of squares for the AB-interaction factor

$=$ (number of observations in each level of the AB-interaction factor)

$\times$ (sum of squares of the observed mean in each level of the AB-interaction factor),

and so on, and

$$1 \equiv N\bar{y}^2,$$

the correction factor for the mean, where $\bar{y}$ is the grand mean of the N data values.

Rule 7. The sum of squares for the line denoted by

$$AB{:}XY \quad \text{is} \quad (a-1)(b-1)xy = abxy - axy - bxy + xy\,.$$

Again the rule is simple: symbolically, a sum of squares is the product of terms like $(a-1)$ for every letter A on the left of the colon and of terms like x for every letter X on the right of the colon. This rule is identical to Rule 5 for degrees of freedom: if in the expression for degrees of freedom every n_f is replaced by f, the resulting expansion is, symbolically, the sum of squares: e.g.,

$$(n_a-1)(n_b-1)n_xn_y \quad \text{becomes} \quad (a-1)(b-1)xy = abxy - axy - bxy + xy\,.$$

After expansion, interpretation of these products of lower case letters is as uncorrected sums of squares, as given by Rule 6.

Note that all sums of squares are expressed essentially in terms of crossed factors. Even when a factor is nested, sums of squares are expressed in terms of uncorrected sums of squares calculated as if the nested factor were a crossed factor. For example, the sum of squares for $A{:}B$ (A within B) is $(a-1)b = ab - b$, where ab is the uncorrected sum of squares of the AB subclasses.

Rule 8. The total of all sums of squares is $\Sigma y^2 - N\bar{y}^2$, where Σy^2 represents the sum of squares of the individual observations, $wabc\ldots$ in the above notation, and where $N\bar{y}^2$ is the correction factor for the mean, as in Rule 6.

Example. Table 4.1 shows the analysis of variance derived from these rules for the case of two crossed classifications A and B, a classification C nested within B, namely $C{:}B$, their interactions and the within factor $W{:}ABC$. Application of these rules is indicated at appropriate points in the table.

f. Calculating sums of squares

The uncorrected sums of squares denoted by lower case letters such as a and ab in Rule 7 have so far been defined solely in words; for example, ab is the uncorrected sum of squares for AB subclasses. Henderson (1969) has no formal algebraic definition of these terms. As the uncorrected sum of squares for the AB subclasses, ab is the sum over all such subclasses of the square of each

TABLE 4.1. EXAMPLE OF RULES 1–8: ANALYSIS OF VARIANCE FOR FACTORS A, B, $C{:}B$, THEIR INTERACTIONS AND $W{:}ABC$

Line (Rules 1–4)	Degrees of Freedom (Rule 5)	Sum of Squares (Rule 7)
A	$n_a - 1$	$(a-1) = a-1$
B	$n_b - 1$	$(b-1) = b-1$
$C{:}B$	$(n_c-1)n_b$	$(c-1)b = bc - b$
AB	$(n_a-1)(n_b-1)$	$(a-1)(b-1) = ab - a - b + 1$
$AC{:}B$	$(n_a-1)(n_c-1)n_b$	$(a-1)(c-1)b = abc - ab - bc + b$
$W{:}ABC$	$(n_w-1)n_an_bn_c$	$(w-1)abc = wabc - abc$
Total	$N-1$ (Rule 6)	$\Sigma y^2 - N\bar{y}^2 \equiv wabc$ (Rule 8) $-\,1$

subclass total, the sum being divided by the number of observations in such a subclass (the same number in each). However, Millman and Glass (1967) give a neat procedure for formalizing this. It starts from an expression for the total of all the observations. We state the rule using as an example the uncorrected sum of squares bc in a situation where y_{hijk} is the observation in levels h, i, j and k of factors A, B, C and W, respectively.

Rule 9.
(i) Write down the total of all observations:

$$\sum_{h=1}^{n_a} \sum_{i=1}^{n_b} \sum_{j=1}^{n_c} \sum_{k=1}^{n_w} y_{hijk} .$$

(ii) Re-order the summation signs so that those pertaining to the letters in the symbolic form of the uncorrected sum of squares of interest (bc, in this case) come first, and enclose the remainder of the sum in parentheses:

$$\sum_{i=1}^{n_b} \sum_{j=1}^{n_c} \left(\sum_{h=1}^{n_a} \sum_{k=1}^{n_w} y_{hijk} \right) .$$

(iii) Square the parenthesis and divide by the product of the ns therein. The result is the required sum of squares: e.g.,

$$bc = \frac{\displaystyle\sum_{i=1}^{n_b} \sum_{j=1}^{n_c} \left(\sum_{h=1}^{n_a} \sum_{k=1}^{n_w} y_{hijk} \right)^2}{n_a n_w} .$$

As a workable rule this is patently simple.

4.2. EXPECTED MEAN SQUARES, E(MS)

Mean squares are sums of squares divided by degrees of freedom. Expected values of mean squares, to be denoted generally by E(MS), can be obtained by an easy set of rules. They are based on using means, variances and covariances of random effects that are applications of equations (6)–(15) of Chapter 3 to all random effects factors. This means that all the effects of each factor are assumed to have zero mean, the same variance and zero covariance with each other [as in equations (7), (14) and (10), respectively, of Chapter 3]. Furthermore, all effects of each random factor are assumed to have zero covariance with those of each other factor and with the error terms. And error terms all have zero means and the same variance and zero covariance with each other [e.g., equations (11) and (13) of Chapter 3].

Rule 10. Denote variances by σ^2 with appropriate subscripts. There will be as many σ^2s, with corresponding subscripts, as there are lines in the analysis and variance table. The variance corresponding to the W-factor is the error variance: $\sigma^2_{w:abc} = \sigma^2_e$.

Example. Where there is an $AC{:}B$ interaction, there is a variance $\sigma^2_{ac:b}$. When $n_w = 1$, there is no W-line in the analysis of variance, although it may be appropriate to envisage σ^2_e as existing.

Rule 11. Whenever a σ^2 appears in any $E(\text{MS})$, its coefficient is the product of all ns whose subscripts do not occur in the subscript of that σ^2.

Example. When the factors are A, B, $C{:}B$ and $W{:}ABC$, the coefficient of $\sigma^2_{ac:b}$ is n_w.

This rule implies that the coefficient of $\sigma^2_{w:abc\ldots}$ is always unity.

Rule 12. Each $E(\text{MS})$ contains only those σ^2s (with coefficients) whose subscripts include all letters pertaining to the MS.

Example. For the $AC{:}B$ line $E[\text{MS}(AC{:}B)] = n_w\sigma^2_{ac:b} + \sigma^2_{w:abc}$.

According to this rule, $\sigma^2_e = \sigma^2_{w:abc\ldots}$ occurs in every $E(\text{MS})$ expression.

The above examples of Rules 10–12 are part of the expected values shown in Table 4.2. These are the expected values, under the random model, of the mean squares of the analysis of variance of Table 4.1.

Rule 13. If the model is completely random, leave as is; for a fixed or mixed model, σ^2-terms corresponding to fixed effects and interactions of fixed effects get changed into quadratic functions of these fixed effects. All other σ^2-terms remain, including those pertaining to interactions of fixed and random effects.

This rule is equivalent to that given by Henderson (1969) but differs from that of the 1959 first edition of that paper, where it is stated that some σ^2-terms "disappear" from some of the expectations of mean squares. Explanation of this difference is included in the discussion of the 2-way classification that now follows.

TABLE 4.2. EXAMPLE OF RULES 10–12: EXPECTED VALUES, UNDER THE RANDOM MODEL, OF MEAN SQUARES OF TABLE 4.1

Mean Square	Variances (Rule 10) and Coefficients (Rule 11)					
	$n_bn_cn_w\sigma^2_a$	$n_an_cn_w\sigma^2_b$	$n_an_w\sigma^2_{c:b}$	$n_cn_w\sigma^2_{ab}$	$n_w\sigma^2_{ac:b}$	$\sigma^2_{w:abc} = \sigma^2_e$
	Terms included (Rule 12)					
MS(A)	*			*	*	*
MS(B)		*	*	*	*	*
MS($C{:}B$)			*		*	*
MS(AB)				*	*	*
MS($AC{:}B$)					*	*
MS($W{:}ABC$)						*

* denotes a σ^2-term that is included; e.g., $n_bn_cn_w\sigma^2_a$ is part of $E[\text{MS}(A)]$.

4.3. THE 2-WAY CROSSED CLASSIFICATION

a. Introduction

We now introduce a model for one of the most important and useful applications of the linear model. It is useful both in the practicalities of analyzing data and in illustrating many of the numerous mathematical and statistical difficulties that can arise in more general applications of the linear model. Indeed it is the simplest situation that illustrates these difficulties.

Suppose data can be classified by two factors. For example, in horticulture plants can be classified by variety of plant and fertilizer treatment used in growing them; in animal agriculture beef cattle can be classified by breed and the feed regimen they are given; in clinical trials patients can be classified by clinic and medication; and so on. In these examples, and whenever data can be classified according to the levels of two factors, those data can be conveniently arrayed in tabular form where the levels of one factor are rows of the table and those of the other factor are columns. We henceforth refer to the factors generically as rows and columns, letting y_{ijk} be the kth observation in the (i, j)-cell, namely the cell defined by the ith row and jth column. Denote by a, b and n the number of rows, columns and number of observations per cell, respectively, so that $i = 1, \ldots, a$, $j = 1, \ldots, b$ and $k = 1, \ldots, n$. Then the model we use is

$$E(y_{ijk}) = \mu + \alpha_i + \beta_j + \gamma_{ij}, \tag{1}$$

where μ represents a general mean, α_i is the effect due to y_{ijk} being an observation in the ith row, β_j is the effect due to column j and γ_{ij} is the interaction effect of row i with column j. Defining the residual error as

$$e_{ijk} = y_{ijk} - E(y_{ijk}) \tag{2}$$

gives the customary model equation for y_{ijk} as

$$y_{ijk} = \mu + \alpha_i + \beta_j + \gamma_{ij} + e_{ijk}\,. \tag{3}$$

The definition of e_{ijk} in (2) gives the expected value as $E(e_{ijk}) = 0$, and the variance–covariance properties we attribute to the e_{ijk}s are

$$\text{var}(e_{ijk}) = \sigma_e^2$$

and

$$\text{cov}(e_{ijk}, e_{i'j'k'}) = 0 \quad \text{unless} \quad i = i', j = j' \text{ and } k = k'\,. \tag{4}$$

b. Analysis of variance table

The analysis of variance table for balanced data of a 2-way crossed classification is as shown in Table 4.3, where the means of the data in row i, column j, cell (i, j) and the grand mean are, respectively,

$$\bar{y}_{i..} = \frac{1}{bn}\sum_{j=1}^{b}\sum_{k=1}^{n} y_{ijk}, \quad \bar{y}_{.j.} = \frac{1}{an}\sum_{i=1}^{a}\sum_{k=1}^{n} y_{ijk}.$$

$$\bar{y}_{ij.} = \frac{1}{n}\sum_{k=1}^{n} y_{ijk} \quad \text{and} \quad \bar{y}_{...} = \frac{1}{abn}\sum_{i=1}^{a}\sum_{j=1}^{b}\sum_{k=1}^{n} y_{ijk}\,. \tag{5}$$

And the corresponding totals are these means with denominators omitted:

$$y_{i..} = bn\bar{y}_{i..} = \sum_{j=1}^{b} \sum_{k=1}^{n} y_{ijk} \,.$$

A popular method of estimating variance components from balanced data is the analysis of variance method (ANOVA). For the 2-way classification this consists of equating expected values of the sums of squares (or equivalently mean squares) of Table 4.3 to their observed values. Depending on whether none, all or some of the effects α_i, β_j and γ_{ij} of the model equation (3) are taken as random effects, the model will be a fixed, random or mixed model. And although the prime concern of this book is variance components we shall, for the sake of completeness, look at expected mean squares here for all three models.

c. Expected mean squares

The starting point is to substitute the model equation (3) into the means of (5) and then put those into the sums of squares of Table 4.3. We then want expected values of those sums of squares. Derivation begins with taking μ and the αs, βs and γs as fixed effects so that with $E(e_{ijk}) = 0$

$$E(\mu e_{ijk}) = E(\alpha_i e_{ijk}) = E(\beta_j e_{ijk}) = E(\gamma_{ij} e_{ijk}) = 0 \quad \forall\ i, j \text{ and } k;$$

and from (4)

$$\begin{aligned} E(\bar{e}_{i..}^2) &= \sigma_e^2/bn, \\ E(\bar{e}_{i..}\bar{e}_{...}) &= E(\bar{e}_{.j.}\bar{e}_{...}) = E(\bar{e}_{ij.}\bar{e}_{...}) = \sigma_e^2/abn, \\ E(\bar{e}_{i..}\bar{e}_{.j.}) &= \sigma_e^2/abn \end{aligned} \tag{6}$$

and

$$\begin{aligned} E(\bar{e}_{i..} - \bar{e}_{...})^2 &= (a-1)\sigma_e^2/abn, \\ E(\bar{e}_{.j.} - \bar{e}_{...})^2 &= (b-1)\sigma_e^2/abn, \\ E(\bar{e}_{ij.} - \bar{e}_{i..} - \bar{e}_{.j.} + \bar{e}_{...})^2 &= (a-1)(b-1)\sigma_e^2/abn \end{aligned} \tag{7}$$

and

$$E(e_{ijk} - \bar{e}_{ij.})^2 = (n-1)\sigma_e^2/n \,.$$

Expected values of the mean squares in Table 4.3 then simplify to be

$$\begin{aligned} E(\text{MSA}) &= \frac{bn}{a-1} \sum_{i=1}^{a} E(\alpha_i - \bar{\alpha}. + \bar{\gamma}_{i.} - \bar{\gamma}_{..})^2 + \sigma_e^2, \\ E(\text{MSB}) &= \frac{an}{b-1} \sum_{j=1}^{b} E(\beta_j - \bar{\beta}. + \bar{\gamma}_{.j} - \bar{\gamma}_{..}) + \sigma_e^2, \\ E(\text{MSAB}) &= \frac{n}{(a-1)(b-1)} \sum_{i=1}^{a} \sum_{j=1}^{b} E(\gamma_{ij} - \bar{\gamma}_{i.} - \bar{\gamma}_{.j} + \bar{\gamma}_{..})^2 + \sigma_e^2 \end{aligned} \tag{8}$$

and

$$E(\text{MSE}) = \sigma_e^2 \,.$$

TABLE 4.3. ANALYSIS OF VARIANCE FOR A 2-WAY CROSSED CLASSIFICATION INTERACTION MODEL, WITH BALANCED DATA

Source of Variation	d.f.	Sum of Squares	Mean Square
A-factor	$a-1$	$\text{SSA} = bn \sum_{i=1}^{a} (\bar{y}_{i..} - \bar{y}_{...})^2$	$\text{MSA} = \dfrac{\text{SSA}}{a-1}$
B-factor	$b-1$	$\text{SSB} = an \sum_{j=1}^{b} (\bar{y}_{.j.} - \bar{y}_{...})^2$	$\text{MSB} = \dfrac{\text{SSB}}{b-1}$
AB-interaction	$(a-1)(b-1)$	$\text{SSAB} = n \sum_{i=1}^{a} \sum_{j=1}^{b} (\bar{y}_{ij.} - \bar{y}_{i..} - \bar{y}_{.j.} + \bar{y}_{...})^2$	$\text{MSAB} = \dfrac{\text{SSAB}}{(a-1)(b-1)}$
Residual error	$ab(n-1)$	$\text{SSE} = \sum_{i=1}^{a} \sum_{j=1}^{b} \sum_{k=1}^{n} (y_{ijk} - \bar{y}_{ij.})^2$	$\text{MSE} = \dfrac{\text{SSE}}{ab(n-1)}$
Total	$abn-1$	$\text{SST}_\text{m} = \sum_{i=1}^{a} \sum_{j=1}^{b} \sum_{k=1}^{n} (y_{ijk} - \bar{y}_{...})^2$	

Notice that these expected values still contain expectation operators. This is because we have not yet specified a model vis-à-vis the α_i, β_j and γ_{ij} effects; i.e., we must specify the model as fixed, random or mixed. This is now done, based on the expressions in (8) which hold whether the model is fixed, random or mixed. Each model determines the consequence of the expectation operations shown on the right-hand sides of (8).

-i. ***The fixed effects model.*** In the fixed effects model all the αs, βs and γs are fixed effects. Therefore the expectation operations on the right-hand sides of (8) just involve dropping the E-symbol. The results are shown in Table 4.4, where

$$\bar{\alpha}. = \sum_{i=1}^{a} \frac{\alpha_i}{a}, \quad \bar{\beta}. = \sum_{j=1}^{b} \frac{\beta_j}{b},$$

and $\bar{\gamma}_{.i}$, $\bar{\gamma}_{.j}$ and $\bar{\gamma}_{..}$ are defined similarly.

Readers may wonder why the expected values in Table 4.4 contain expressions such as $\Sigma_i(\alpha_i - \bar{\alpha}. + \bar{\gamma}_{i.} - \bar{\gamma}_{..})^2$ rather than the more familiar $\Sigma_i\alpha_i^2$. This is because Table 4.4 does not involve what are sometimes called the "usual restrictions" or "Σ-restrictions" such as $\Sigma_i\alpha_i = 0$ and $\Sigma_j\gamma_{ij} = 0 \ \forall \ i$. It is these restrictions that reduce $\Sigma_i(\alpha_i - \bar{\alpha}. + \bar{\gamma}_{i.} - \bar{\gamma}_{..})^2$ to $\Sigma_i\alpha_i^2$. They are equivalent to defining

$$E(y_{ijk}) = \mu_{ij}$$

and

$$\dot{\mu} = \bar{\mu}_{..}, \quad \dot{\alpha}_i = \mu_{i.} - \bar{\mu}_{..}, \quad \dot{\beta}_j = \bar{\mu}_{.j} - \bar{\mu}_{..} \quad \text{and} \quad \dot{\gamma}_{ij} = \mu_{ij} - \bar{\mu}_{i.} - \bar{\mu}_{.j} + \bar{\mu}_{..} \,. \tag{9}$$

Then the model equation is

$$y_{ijk} = \dot{\mu} + \dot{\alpha}_i + \dot{\beta}_j + \dot{\gamma}_{ij} + e_{ijk}, \tag{10}$$

TABLE 4.4. EXPECTED MEAN SQUARES OF A 2-WAY CROSSED CLASSIFICATION INTERACTION MODEL, WITH BALANCED DATA

Fixed Effects Model
$E(\mathrm{MSA}) = \dfrac{bn}{a-1}\sum_{i=1}^{a}(\alpha_i - \bar{\alpha}. + \bar{\gamma}_{i.} - \bar{\gamma}_{..})^2 + \sigma_e^2$
$E(\mathrm{MSB}) = \dfrac{an}{b-1}\sum_{j=1}^{b}(\beta_j - \bar{\beta}. + \bar{\gamma}_{.j} - \bar{\gamma}_{..})^2 + \sigma_e^2$
$E(\mathrm{MSAB}) = \dfrac{n}{(a-1)(b-1)}\sum_{i=1}^{a}\sum_{j=1}^{b}(\gamma_{ij} - \bar{\gamma}_{i.} - \bar{\gamma}_{.j} + \bar{\gamma}_{..})^2 + \sigma_e^2$
$E(\mathrm{MSE}) = \sigma_e^2$

with, for example, $\Sigma_{i=1}^{a}\dot{\alpha}_i = \Sigma_{i=1}^{a}(\bar{\mu}_{i.} - \bar{\mu}_{..}) = 0$; i.e., the dotted terms in (9) satisfy the Σ-restrictions

$$\Sigma_i\dot{\alpha} = 0,\quad \Sigma_j\dot{\beta}_j = 0,\quad \Sigma_i\dot{\gamma}_{ij} = 0\ \forall\ i \quad\text{and}\quad \Sigma_j\dot{\gamma}_{ij} = 0\ \forall\ i\,. \tag{11}$$

Nevertheless, on comparing (10) and (3), we can conclude from Table 4.4 that the expected value of MSA is

$$E(\text{MSA}) = \frac{bn}{a-1}\sum_{i=1}^{a}(\dot{\alpha}_i - \bar{\dot{\alpha}}_. + \bar{\dot{\gamma}}_{i.} - \bar{\dot{\gamma}}_{..})^2 = \frac{bn}{a-1}\Sigma_i\dot{\alpha}_i^2, \tag{12}$$

because of (11). Hence $\Sigma_i\dot{\alpha}_i^2$ has precisely the same meaning as $\Sigma_i(\alpha_i - \bar{\alpha}_. - \bar{\gamma}_{i.} + \bar{\gamma}_{..})^2$ of Table 4.4. It is to be emphasised that the Σ-restricted models (10) and (11) are equivalent in this manner to the unrestricted model (3) only for balanced data. This equivalence does not occur for unbalanced data, because the sums of squares used with such data have expected values that do not involve the means of the effects in such a simple manner as with balanced data. [See, e.g., Searle (1987, Table 4.8).] Restrictions that are in terms of weighted sums of the effects are sometimes suggested for unbalanced data, although these have no simplifying effect when there are empty cells, as is often the case with unbalanced data.

-ii. The random effects model. In the random model all the α-, β- and γ-effects are taken as being random, with zero means, variances σ_α^2, σ_β^2 and σ_γ^2, respectively, and all covariances zero:

$$E(\alpha_i) = 0,\quad E(\beta_j) = 0,\quad E(\gamma_{ij}) = 0, \tag{13}$$

$$\text{var}(\alpha_i) = E(\alpha_i^2) = \sigma_\alpha^2,\quad \text{cov}(\alpha_i\alpha_{i'}) = E(\alpha_i, \alpha_{i'}) = 0 \quad \forall\ i \neq i'; \tag{14}$$

with similar statements for the βs and γs. Also

$$\text{cov}(\alpha_i, \beta_j) = 0 = \text{cov}(\alpha_i, \gamma_{ij}) = \text{cov}(\alpha_i, e_{ijk}), \tag{15}$$

with similar statements for the βs and γs. They represent the customary formulation of random effects in random or mixed models, and as such are a direct and natural extension of equations (9)–(15) in Chapter 3. Applying this formulation to (8) leads to the expected values shown in Table 4.5. It is left to the reader (E 4.6) to derive those expected values.

-iii. The mixed model. Suppose the α-effects are fixed effects and the βs are random. Then the γs, being interactions of the α-factor with the β-factor, are random. The expectation operations on the right-hand sides of (8) therefore involve dropping the E-symbol insofar as it pertains to αs and using properties like those of (13), (14) and (15) for the βs and γs. This leads to the results shown in Table 4.6.

The difference between the random and mixed models is that the αs are random effects in the random model and are fixed effects in the mixed model. Since only the first equation in (8) involves αs, only the first entry in Table 4.6

TABLE 4.5. EXPECTED MEAN SQUARES OF A 2-WAY CROSSED CLASSIFICATION INTERACTION MODEL, WITH BALANCED DATA

Random Effects Model	
$E(\text{MSA}) = bn\sigma_\alpha^2$	$+ n\sigma_\gamma^2 + \sigma_e^2$
$E(\text{MSB}) =$	$an\sigma_\beta^2 + n\sigma_\gamma^2 + \sigma_e^2$
$E(\text{MSAB}) =$	$n\sigma_\gamma^2 + \sigma_e^2$
$E(\text{MSE}) =$	σ_e^2

TABLE 4.6. EXPECTED MEAN SQUARES OF A 2-WAY CROSSED CLASSIFICATION INTERACTION MODEL, WITH BALANCED DATA

Mixed model: αs fixed, βs and γs random	
$E(\text{MSA}) = \dfrac{bn}{a-1}\sum_{i=1}^{a}(\alpha_i - \bar{\alpha}_.)^2$	$+ n\sigma_\gamma^2 + \sigma_e^2$
$E(\text{MSB}) =$	$an\sigma_\beta^2 + n\sigma_\gamma^2 + \sigma_e^2$
$E(\text{MSAB}) =$	$n\sigma_\gamma^2 + \sigma_e^2$
$E(\text{MSE}) =$	σ_e^2

differs from the corresponding entry in Table 4.5, and then only through having a quadratic term in the αs instead of a term in σ_α^2.

-iv. A mixed model with Σ-restrictions. The expected mean squares in Table 4.5 for the random model have been arrived at without any use of Σ-restrictions of the kind shown in (11) for the fixed effects model. This is appropriate, because with the α_is that occur in the model equations for the data being taken as realized values of random variables, it is not realistic to have them summing to zero, i.e., $\Sigma_i \alpha_i = 0$ is not appropriate. Moreover, having $\Sigma_i \alpha_i = 0$ is never even considered in the case of unbalanced data, for which expected mean square derivations all reduce to those of Table 4.5 for balanced data.

Likewise, Σ-restrictions are not involved in Table 4.6 either. Neither do we think they should be. Nevertheless, some presentations of the mixed model (for the case being considered here, the 2-way crossed classification, mixed model with one factor fixed, balanced data) do incorporate, either explicitly or implicitly, Σ-restrictions. This leads to expected mean squares that differ from those in Table 4.6. We therefore proceed to give an explanation of those differences, similar to Searle (1971, Chap. 9). Other explanations are available as, e.g., in Hocking (1973, 1985), and Samuels *et al.* (1991), but they are effectively equivalent to what follows.

In the model equation

$$y_{ijk} = \mu + \alpha_i + \beta_j + \gamma_{ij} + e_{ijk} \tag{16}$$

where α_i is taken as fixed effect and β_j as random, γ_{ij} is then the interaction between a fixed and random effect. It is therefore taken as a random effect. That is how it is treated in deriving Table 4.6. Nevertheless, there is debate over whether or not, because it is the interaction involving a fixed effect, it should be defined subject to partial Σ-restrictions in the form that sums of γ_{ij} over the levels of the fixed effects factor be defined as zero, i.e., $\Sigma_{i=1}^{a} \gamma_{ij} = 0 \ \forall \ j$. To follow the effect of such a definition we write the model as

$$y_{ijk} = \mu' + \alpha'_i + \beta'_j + \gamma'_{ij} + e_{ijk} \tag{17}$$

with the restrictions

$$\sum_{i=1}^{a} \alpha'_i = 0 \quad \text{and} \quad \sum_{i=1}^{a} \gamma'_{ij} = \gamma'_{\cdot j} = 0 \quad \text{for all } j\,. \tag{18}$$

The prime notation used here distinguishes the model (17) with the restrictions (18) from the model (16) with no such restrictions; and it is also distinctive from the model (10). In (17) the α's are fixed effects and the β's and γ's are random effects with zero means and variances $\sigma^2_{\beta'}$ and $\gamma^2_{\gamma'}$, respectively, and with the β's and γ's being uncorrelated with each other and the es. All this is exactly the same as in the mixed model described earlier, except for (18).

In (18) it is the restriction on the γ'_{ij}s that is particularly noteworthy. It implicitly defines a covariance between every pair of γ'_{ij}s that are in the same column, i.e., between γ'_{ij} and $\gamma'_{i'j}$ for $i \neq i'$. Suppose this covariance is the same, for all pairs:

$$\operatorname{cov}(\gamma'_{ij}, \gamma'_{i'j}) = c \quad \forall \ i \neq i'\,. \tag{19}$$

Then, from (18)

$$\operatorname{var}\left(\sum_{i=1}^{a} \gamma'_{ij}\right) = 0$$

and so

$$a\sigma^2_{\gamma'} + a(a-1)c = 0,$$

giving

$$c = -\sigma^2_{\gamma'}/(a-1)\,. \tag{20}$$

Note that this covariance pertains only to γ's within the same level of the β-factor, arising as it does from (18). The covariance between γ's in the same level of the α-factor is zero, as usual:

$$\operatorname{cov}(\gamma'_{ij}, \gamma'_{ij'}) = 0 \quad \text{for all } i \text{ and } j \neq j'\,. \tag{21}$$

Prior to utilizing (18), the expected mean squares for the model (17) can be derived from equations (8) with primes on μ, the αs, βs and γs. Upon invoking

$\bar{\gamma}'_{.j} = 0$ from (18), and hence $\bar{\gamma}'_{..} = 0$, equations (8) become

$$E(\text{MSA}) = \frac{bn}{a-1}\left[\sum_{i=1}^{a} \alpha_i'^2 + \sum_{i=1}^{a} E(\bar{\gamma}'_{i.})^2\right] + \sigma_e^2,$$

$$E(\text{MSB}) = \frac{an}{b-1}\sum_{j=1}^{b} E(\beta'_j - \bar{\beta}'_.)^2 + \sigma_e^2, \tag{22}$$

$$E(\text{MSAB}) = \frac{n}{(a-1)(b-1)}\sum_{i=1}^{a}\sum_{j=1}^{b} E(\gamma'_{ij} - \bar{\gamma}'_{i.})^2 + \sigma_e^2$$

and

$$E(\text{MSE}) = \sigma_e^2 .$$

In carrying out the expectation operations in $E(\text{MSA})$ and $E(\text{MSAB})$, use is made of (21) to give

$$E(\bar{\gamma}'_{i.})^2 = \sigma_{\gamma'}^2\left[\frac{1}{b} + \frac{b(b-1)0}{b^2}\right] = \frac{\sigma_{\gamma'}^2}{b}$$

and

$$E(\gamma'_{ij} - \bar{\gamma}'_{i.})^2 = \sigma_{\gamma'}^2\left(1 + \frac{1}{b} - \frac{2}{b}\right) = \frac{(b-1)\sigma_{\gamma'}^2}{b} .$$

As a result, expressions (22) reduce to those shown in Table 4.7.

The results in Table 4.7 differ from those in Table 4.6 in two important ways. First, in Table 4.7, whenever σ_γ^2 occurs it does so in the form $n\sigma_{\gamma'}^2/(1-1/a)$, whereas in Table 4.6 the term appears as just $n\sigma_\gamma^2$. Second, $E(\text{MSB})$ in Table 4.7 has no term in $\sigma_{\gamma'}^2$, whereas in Table 4.6 it contains $n\sigma_\gamma^2$. This is the reason why Rule 13 at the end of Section 4.1d differs from the first edition (1959) of Henderson (1969) but is the same as in the second. The first edition specifies a general rule that leads to the absence of σ_γ^2, from $E(\text{MSB})$ on the basis of $\gamma'_{.j} = 0$, as in (18), whereas the second specifies a general rule that retains σ_γ^2 in $E(\text{MSB})$ as in Table 4.6, using a model that has no restrictions like (18).

TABLE 4.7. EXPECTED MEAN SQUARES OF A 2-WAY CROSSED CLASSIFICATION INTERACTION MODEL, WITH BALANCED DATA

Mixed Model, with Restrictions on Interaction Effects: $\gamma'_{.j} = 0$ for All j			
$E(\text{MSA}) =$	$\frac{bn}{a-1}\sum_{i=1}^{a} \alpha_i'^2$		$+\, n\sigma_{\gamma'}^2/(1-1/a) + \sigma_e^2$
$E(\text{MSB}) =$		$an\sigma_{\beta'}^2$	$+\, \sigma_e^2$
$E(\text{MSAB}) =$			$n\sigma_{\gamma'}^2/(1-1/a) + \sigma_e^2$
$E(\text{MSE}) =$			σ_e^2

This dual approach to the mixed model is evident in many places. For example, Mood (1950, p. 344) and Kirk (1968, p. 137) use the Table 4.6 expectations whereas Anderson and Bancroft (1952, p. 339), Scheffé (1959, p. 269), Graybill (1961, p. 398) and Snedecor and Cochran (1989, p. 322) use those akin to Table 4.7. Mood and Graybill (1963) do not discuss the topic. Although results like Table 4.7 predominate in the literature, those of Table 4.6 are consistent with the results for unbalanced data, and this fact, as Hartley and Searle (1969) point out, is strong argument for using Table 4.6.

The second difference between Tables 4.6 and 4.7 is the occurrence of $1/(1-1/a)$ in the terms in the interaction variance component in Table 4.7. This is a consequence of the restriction $\gamma'_{.j}=0$ of (18), as shown also, for example, in Steel and Torrie (1960, pp. 214, 246).

One criterion for deciding between the two forms of the mixed model is the following. Consider, momentarily, the possibility of redefining the βs as fixed. If that would lead to redefining the γs as fixed then one should, when the βs are random, have $\Sigma_i\gamma_{ij}=0$ as part of the mixed model. But if redefining the βs as fixed would not lead to redefining the γs as fixed but would leave them as random then the Σ-restrictions $\Sigma_i\gamma_{ij}=0$ should not be part of the mixed model. The difficulty with $\Sigma_i\gamma_{ij}=0\ \forall\ j$ is that after using it with unbalanced data the resulting analysis does not for $n_{ij}=n\ \forall\ i, j$ reduce to the well-known analysis for balanced data.

A connection between Tables 4.6 and 4.7 can be established as follows. The model for Table 4.6 is (16). Suppose it is rewritten as

$$y_{ijk}=(\mu+\bar{\alpha}.)+(\alpha_i-\bar{\alpha}.)+(\beta_j+\bar{\gamma}_{.j})+(\gamma_{ij}-\bar{\gamma}_{.j})+e_{ijk}\,.$$

Then, on defining

$$\mu'=\mu+\bar{\alpha}.,\quad \alpha'_i=\alpha_i-\bar{\alpha}.,\quad \beta'_j=\beta_j+\bar{\gamma}_{.j}\quad\text{and}\quad \gamma'_{ij}=\gamma_{ij}-\bar{\gamma}_{.j}, \tag{23}$$

we have exactly the models (17) and (18) used for Table 4.7. Not only do the definitions in (23) satisfy (18), but it is easily shown (see E 4.6) that (19)–(21) are also satisfied and that

$$\sigma^2_{\gamma'}=(1-1/a)\sigma^2_\gamma, \tag{24}$$

as is evident from comparing the values of $E(\text{MSA})$ and $E(\text{MSAB})$ in Table 4.6 with those in Table 4.7. Moreover,

$$\sigma^2_{\beta'}=\sigma^2_\beta+\sigma^2_\gamma/a\,. \tag{25}$$

The question of which form of the mixed model to use, that without the Σ-restrictions (Table 4.6) or that with them (Table 4.7) remains open; and seems likely to remain so. It is irrelevant to ask "Which model is best?", because this question really has no definitive, universally acceptable answer. The important thing is to understand that there are two models and that although they are different, they are closely related. Then in analyzing any particular set of data one is in a position of understanding the two models and their relationship, and can decide which is the more appropriate to the data at hand. Lengthy discussion of the models (17) and (18) that lead to Table 4.7 is to be found in such papers as Wilk and Kempthorne (1955, 1956) and Cornfield and Tukey (1956) as well as in Scheffé (1959). Nevertheless, the model that is used for unbalanced data (of which balanced data are a special case) is the one without the Σ-restrictions that leads to Table 4.6. And that table is what one gets when simplifying the procedures for unbalanced data to the case of balanced data. Also, when there is no within-cell replication, i.e., $n = 1$, and hence no SSE, Table 4.6 provides a test of H: $\sigma_\beta^2 = 0$ whereas Table 4.7 does not then provide a test of the analogous, but distinctly different, hypothesis H: $\sigma_{\beta'}^2 = 0$.

d. ANOVA estimators of variance components

The ANOVA method of estimating variance components from balanced data is to equate mean squares of the analysis of variance to the expected values. The latter are linear combinations of variance components. The resulting equations are solved for the variance components and the solutions are the estimated variance components. All this is just as was done in Chapter 3 for the 1-way classification. Applying it to the fixed effects model of the 2-way classification means applying it to just $E(\text{MSE}) = \sigma_e^2$ of Table 4.4 and so $\hat{\sigma}_e^2 = \text{MSE}$. Applied to the random model of Table 4.5 it yields

$$\hat{\sigma}_e^2 = \text{MSE}, \qquad \hat{\sigma}_\beta^2 = \frac{\text{MSB} - \text{MSAB}}{an},$$

$$\hat{\sigma}_\gamma^2 = \frac{\text{MSAB} - \text{MSE}}{n}, \quad \hat{\sigma}_\alpha^2 = \frac{\text{MSA} - \text{MSAB}}{bn}. \tag{26}$$

For the mixed model, without Σ-restrictions, Table 4.6 leads to

$$\hat{\sigma}_e^2 = \text{MSE}, \quad \hat{\sigma}_\gamma^2 = \frac{\text{MSAB} - \text{MSE}}{n} \quad \text{and} \quad \hat{\sigma}_\beta^2 = \frac{\text{MSB} - \text{MSAB}}{an}. \tag{27}$$

These estimates, it will be noticed, are the same as in (26) for the random model, except that there is no estimator for σ_α^2—because, of course, the αs are being taken as fixed effects. This is the situation with all mixed models using balanced data. There will always be as many mean squares having expectation that contain no fixed effects as there are variance components to be estimated. In the case of Table 4.6 this number is 3. And the estimated variance components will thus always be a subset of those obtained if all the fixed effects (except μ) were taken as random.

For the mixed model with Σ-restrictions we use Table 4.7. This gives

$$\hat{\sigma}_e^2 = \text{MSE}, \quad \hat{\sigma}_{\gamma'}^2 = \frac{(\text{MSAB} - \text{MSE})(1 - 1/a)}{n} \quad \text{and} \quad \hat{\sigma}_{\beta'}^2 = \frac{\text{MSB} - \text{MSE}}{an}. \tag{28}$$

Other than $\hat{\sigma}_e^2$, these are different from the estimators of the mixed model without Σ-restrictions. But the two sets of estimators are related:

$$\hat{\sigma}_{\gamma'}^2 = \frac{\text{MSAB} - \text{MSE}}{n}\left(1 - \frac{1}{a}\right) = \left(1 - \frac{1}{a}\right)\hat{\sigma}_\gamma^2$$

from (27), and in accord with (24). And

$$\begin{aligned}
\hat{\sigma}_{\beta'}^2 &= \frac{\text{MSB} - \text{MSE}}{an}, \quad \text{from (28)}, \\
&= \frac{\text{MSB} - \text{MSAB} + \text{MSAB} - \text{MSE}}{an} \\
&= \frac{an\hat{\sigma}_\beta^2 + n\hat{\sigma}_\gamma^2}{an}, \quad \text{from (27)}, \\
&= \hat{\sigma}_\beta^2 + \hat{\sigma}_\gamma^2/a,
\end{aligned}$$

in accord with (25). Thus the connection between the two sets of estimators is very simple and, naturally, is the same as that between the two sets of components in (24) and (25).

4.4. ANOVA ESTIMATION

Having derived estimators, the next step would be to consider their properties: e.g., sampling variances, confidence intervals, and so on. But before doing so we introduce some general results, which can then be applied not only to the 2-way classification but also to any combination of crossed and nested fixed or random factors.

The methodology of ANOVA estimation of variance components from balanced data is clearly demonstrated using the expected mean squares of Table 4.5 and the estimators in (26); and equally so for the mixed model by using the last three lines of Table 4.6 and (27). Each of these is a special case of equating mean squares (of the analysis of variance table) to the expected values and using the solutions for the variance components as the estimators thereof. The generalization of this is to let $\mathbf{m}$ be the vector of mean squares, having the same order as $\boldsymbol{\sigma}^2$, the vector of variance components in the model. Suppose $\mathbf{P}$ is such that

$$E(\mathbf{m}) = \mathbf{P}\boldsymbol{\sigma}^2 . \tag{29}$$

Then the ANOVA estimator of $\boldsymbol{\sigma}^2$ is $\hat{\boldsymbol{\sigma}}^2$, obtained from $\mathbf{m} = \mathbf{P}\hat{\boldsymbol{\sigma}}^2$ as

$$\hat{\boldsymbol{\sigma}}^2 = \mathbf{P}^{-1}\mathbf{m} \tag{30}$$

provided **P** is nonsingular, as is the case in most standard analyses. In the case of Table 4.6

$$E(\mathbf{m}) = \begin{bmatrix} bn & 0 & n & 1 \\ 0 & an & n & 1 \\ 0 & 0 & n & 1 \\ 0 & 0 & 0 & 1 \end{bmatrix} \begin{bmatrix} \sigma_\alpha^2 \\ \sigma_\beta^2 \\ \sigma_\gamma^2 \\ \sigma_e^2 \end{bmatrix},$$

with

$$\mathbf{P}^{-1} = \begin{bmatrix} \frac{1}{bn} & \frac{-1}{bn} & 0 & 0 \\ 0 & \frac{1}{an} & \frac{-1}{an} & 0 \\ 0 & 0 & \frac{1}{n} & -\frac{1}{n} \\ 0 & 0 & 0 & 1 \end{bmatrix},$$

commensurate with (26).

Clearly, the estimators in (30) are unbiased, because

$$E(\hat{\boldsymbol{\sigma}}^2) = \mathbf{P}^{-1}E(\mathbf{m}) = \mathbf{P}^{-1}\mathbf{P}\boldsymbol{\sigma}^2 = \boldsymbol{\sigma}^2 .$$

Thus, when using balanced data, for either random or mixed models, ANOVA estimators of all variance components are unbiased. But this unbiasedness is not necessarily a feature of all applications of ANOVA methodology to unbalanced data—as is discussed in Chapter 5.

The estimators in (30) have the smallest variance of all estimators that are both quadratic functions of the observations and unbiased. That is to say, they are minimum variance, quadratic unbiased (MVQU). This property was established by Graybill and Hultquist (1961) and applies to all ANOVA estimators from balanced data. Note that it does not demand any assumption of normality. When such assumptions are made, the estimators in (30) have the smallest variance from among all unbiased estimators, both those that are quadratic functions of the observations and those that are not. And this, too, is the case for all ANOVA estimators from balanced data. Thus, under normality, ANOVA estimators of variance components are minimum variance, unbiased (MVU). This result is presented in Graybill (1954) and Graybill and Wortham (1956). It is to be emphasized that it applies only to balanced data.

The possibility of ANOVA methodology yielding a negative estimate of a variance component is discussed briefly in Section 3.5c. Clearly, such an occurrence is an embarassment, because variances are positive parameters (well, at least non-negative), and so interpretation of a negative estimate is a problem. Several courses of action exist, few of them satisfactory.

(i) Accept the estimate, despite its distastefulness, and use it as evidence that the true value of the component is zero. Although this interpretation may be appealing, the unsatisfying nature of the negative estimate still remains. This is particularly so if the negative estimate is used in estimating a sum of components. The estimated sum can be less than the estimate of an individual component. For example, in (56) of Chapter 3 we got $\hat{\sigma}_e^2 = 92$ and $\hat{\sigma}_\alpha^2 = -10$, giving $\text{var}(y) = \hat{\sigma}_\alpha^2 + \hat{\sigma}_e^2 = -10 + 92 = 82 < \hat{\sigma}_e^2$, which does not make sense.

(ii) Accept the negative estimate as evidence that the true value of the corresponding component is zero and hence, as the estimate, use zero in place of the negative value. Although this seems a logical replacement such a truncation procedure disturbs the properties of the estimates as otherwise obtained. For example, they are no longer unbiased, but their mean squared error is less.

(iii) Use the negative estimate as indication of a zero component to ignore that component in the model, but retain the factor so far as the lines in the analysis of variance table are concerned. This leads to ignoring the component estimated as negative and re-estimating the others. Thompson (1961, 1962) gives rules for doing this, known as "pooling minimal mean squares with predecessors", and gives an application in Thompson and Moore (1963).

(iv) Interpret the negative estimate as indication of a wrong model and re-examine the source of one's data to look for a new model. In this connection, Searle and Fawcett (1970) suggest that finite population models may be viable alternatives because they sometimes give positive estimates when infinite population models have yielded negative estimates. Their use is likely to be of limited extent, however. In contrast, Nelder (1965a, b) suggests that at least for split plot and randomized block designs, randomization theory indicates that negative variance components can occur in some situations. Such an apparent inconsistency can arise from the intra-block correlation of plots being less than the inter-block correlation.

(v) Interpret the negative estimate as throwing question on the method that yielded it, and use some other method that yields non-negative estimators. Two possibilities exist. One is to use a maximum likelihood procedure, as discussed in Chapter 6. A second possibility is to use a Bayes estimator, for which the reader is referred to Section 3.9 for an introduction and to Chapter 9 for more general consideration.

(vi) Take the negative estimate as indication of insufficient data, and follow the statistician's last hope: collect more data and analyze them, either on their own or pooled with those that yielded the negative estimate. If the estimate from the pooled data is negative that would be additional evidence that the corresponding component is indeed zero.

Obtaining a negative estimate from the analysis of variance method is solely a consequence of the data and the method. It in no way depends on normality. However, when normality *is* assumed, it is possible in certain cases to derive the probability of obtaining a negative estimate, as illustrated in Section 3.5d-vi. Generalization of this is shown in Section 4.5, which follows.

4.5. NORMALITY ASSUMPTIONS

No particular form for the distribution of error terms or of the random effects in a model has been assumed in this chapter up to now. All the preceding results in the chapter are true for any distribution. We now make the normality assumptions that the error terms and the random effects factor are normally distributed with zero means and the variance-covariance structure discussed at the start of Section 4.2. That is, the effects of each random factor have a variance–covariance matrix that is their variance (component) multiplied by an identity matrix; and the effects of each random factor are independent of those of every other factor and of the error terms. Under these conditions we assume normality. Thus, for example, for the 2-way crossed classification of (1) and (2) we define

$$\boldsymbol{\alpha} = \{_c\, \alpha_i\}_{i=1}^{a}, \quad \boldsymbol{\beta} = \{_c\, \beta_j\}_{j=1}^{b},$$
$$\boldsymbol{\gamma} = \{_c\, \gamma_{ij}\}_{i,j} \quad \text{and} \quad \mathbf{e} = \{_c\, e_{ijk}\}_{i,j,k}, \tag{31}$$

where in $\boldsymbol{\gamma}$ and $\mathbf{e}$ the elements are arrayed in lexicon order, i.e., ordered respectively by j within i, and by k within j within i. Then the normality assumptions, based on (13)–(15), are

$$\begin{bmatrix} \boldsymbol{\alpha} \\ \boldsymbol{\beta} \\ \boldsymbol{\gamma} \\ \mathbf{e} \end{bmatrix} \sim \mathcal{N}\left(\mathbf{0}, \begin{bmatrix} \sigma_\alpha^2 \mathbf{I}_a & \mathbf{0} & \mathbf{0} & \mathbf{0} \\ \mathbf{0} & \sigma_\beta^2 \mathbf{I}_b & \mathbf{0} & \mathbf{0} \\ \mathbf{0} & \mathbf{0} & \sigma_\gamma^2 \mathbf{I}_{ab} & \mathbf{0} \\ \mathbf{0} & \mathbf{0} & \mathbf{0} & \sigma_e^2 \mathbf{I}_{abn} \end{bmatrix}\right). \tag{32}$$

a. Distribution of mean squares

Let f_i, S_i and M_i be the degrees of freedom, sum of squares and mean square

$$M_i = S_i/f_i \tag{33}$$

in a line of an analysis of variance of balanced data. Under the normality assumptions just described it can be shown that

$$\frac{S_i}{E(M_i)} \sim \chi^2_{f_i}; \quad \text{and the } S_i\text{s are independent}\,.$$

Hence (34)

$$\frac{f_i M_i}{E(M_i)} \sim \chi^2_{f_i}; \quad \text{and the } M_i\text{s are independent}\,.$$

Result (34) is derived by writing $S_i/E(M_i)$ as a quadratic form of $\mathbf{y}'\mathbf{A}\mathbf{y}$ in the observation vector $\mathbf{y}$, and applying Theorems S2 and S3 of Appendix S. In applying these theorems to random or mixed models $\mathbf{V}$ is not $\sigma_e^2\mathbf{I}$, as it is in the fixed model, but is a matrix whose elements are functions of the σ^2s of the model, as illustrated in (58) of Chapter 3. Nevertheless, for the $\mathbf{A}$-matrices involved in expressing each $S_i/E(M_i)$ as a quadratic form $\mathbf{y}'\mathbf{A}\mathbf{y}$ it will be found that $\mathbf{AV}$ is always idempotent. Furthermore, for the random model, $\boldsymbol{\mu}$ has the form $\mu\mathbf{1}$, and $\boldsymbol{\mu}'\mathbf{A}\boldsymbol{\mu} = \mu\mathbf{1}'\mathbf{A}\mathbf{1}\mu$ will, by the nature of $\mathbf{A}$, always be zero. Hence, for the random model the χ^2s are central, as indicated in (34). Pairwise independence is established from Theorem S3, whereupon the underlying normality leads to independence of all the Ss (and Ms). For the mixed model, (34) will also apply for all sums of squares whose expected values do not involve fixed effects; those that do involve fixed effects will be non-central χ^2s. This is illustrated further in Section 4.6.

b. Distribution of estimators

The ANOVA method of estimation, that of equating mean squares to their expected values, yields estimators of variance components that are linear functions of mean squares. These mean squares have the properties given in (34). The resulting variance components estimators are therefore linear functions of multiples of χ^2-variables, some of them with negative coefficients. No closed form exists for the distribution of such functions and, furthermore, the coefficients are themselves functions of the population variance components.

Example. In Table 4.5

$$\frac{(a-1)\text{MSA}}{bn\sigma_\alpha^2 + n\sigma_\gamma^2 + \sigma_e^2} \sim \chi^2_{a-1}$$

and, independently,

$$\frac{(a-1)(b-1)\text{MSAB}}{n\sigma_\gamma^2 + \sigma_e^2} \sim \chi^2_{(a-1)(b-1)} .$$

Therefore

$$\hat{\sigma}_\alpha^2 = \frac{\text{MSA} - \text{MSAB}}{bn} \tag{35}$$

$$\overset{d}{=} \left[\frac{bn\sigma_\alpha^2 + n\sigma_\gamma^2 + \sigma_e^2}{bn(a-1)}\chi^2_{a-1} - \frac{n\sigma_\gamma^2 + \sigma_e^2}{bn(a-1)(b-1)}\chi^2_{(a-1)(b-1)}\right], \tag{36}$$

where the notation $x \overset{d}{=} y$ means that x and y have the same distribution. No closed form of the distribution of (36) can be derived, because linear combinations of independent χ^2-variables (other than simple sums) do not have a χ^2-distribution. This state of affairs is true for these kinds of variance components estimators generally. Were the coefficients of the χ^2s known, the methods of

Robinson (1965), Wang (1967) or Fleiss (1971), or numerical integration, could be employed to obtain the distributions.

In contrast to other components, the distribution of $\hat{\sigma}_e^2$ is always known exactly, under normality assumptions:

$$\hat{\sigma}_e^2 = \text{MSE} \sim \frac{\sigma_e^2}{f_{\text{MSE}}}\chi^2(f_{\text{MSE}}), \tag{37}$$

where f_{MSE} is the degrees of freedom associated with MSE.

Generalization of (36) arises from (30), which is $\hat{\boldsymbol{\sigma}}^2 = \mathbf{P}^{-1}\mathbf{m}$. The elements of $\mathbf{m}$ follow (34) and so, for example, $M_i \sim E(M_i)f_i^{-1}\chi^2(f_i)$. Now write

$$\mathbf{C} = \{_{\text{d}}\, f_i^{-1}\chi_{f_i}^2\}_{i=1}^{k},$$

where there are k lines in the analysis of variance being used. Then from (30)

$$\hat{\boldsymbol{\sigma}}^2 \overset{d}{=} \mathbf{P}^{-1}\mathbf{C}E(\mathbf{m}) \overset{d}{=} \mathbf{P}^{-1}\mathbf{C}\mathbf{P}\boldsymbol{\sigma}^2\,. \tag{38}$$

In this way the vector of estimators is expressed as a vector of multiples of central χ^2-variables.

c. Tests of hypotheses

Expected values of mean squares (derived by the rules of Section 4.2) often suggest which mean squares are the appropriate denominators for testing hypotheses that certain variance components are zero. Thus in Table 4.6 MSAB/MSE is appropriate for testing the hypothesis H: $\sigma_\gamma^2 = 0$; and MSB/MSAB is the F-statistic for testing H: $\sigma_\beta^2 = 0$. In the random model all ratios of mean squares are proportional to central F-distributions, because all mean squares follow (34). In the mixed model the same is true of ratios of mean squares whose expected values contain no fixed effects.

The table of expected values will not always suggest the "obvious" denominator for testing a hypothesis. For example, suppose from Table 4.2 we wished to test the hypothesis $\sigma_b^2 = 0$. From that table we have, using M_1, M_2, M_3 and M_4, respectively, for MS(B), MS(C:B), MS(AB) and MS(AC:B),

$$\begin{aligned}
E(M_1) &= E[\text{MS}(B)] &&= k_1\sigma_b^2 + k_2\sigma_{c:b}^2 + k_3\sigma_{ab}^2 + k_4\sigma_{ac:b}^2 + \sigma_e^2,\\
E(M_2) &= E[\text{MS}(C{:}B)] &&= \qquad\quad k_2\sigma_{c:b}^2 \qquad\qquad\; + k_4\sigma_{ac:b}^2 + \sigma_e^2,\\
E(M_3) &= E[\text{MS}(AB)] &&= \qquad\qquad\qquad\qquad k_3\sigma_{ab}^2 + k_4\sigma_{ac:b}^2 + \sigma_e^2,\\
E(M_4) &= E[\text{MS}(AC{:}B)] &&= \qquad\qquad\qquad\qquad\qquad\quad k_4\sigma_{ac:b}^2 + \sigma_e^2,
\end{aligned}$$

where we have here written the coefficients of the σ^2s, the products of ns shown in the column headings of Table 4.2, as ks: e.g., $k_1 = n_a n_c n_w$. It is clear from these expected values that no mean square in the table is suitable as a denominator to M_1 for an F-statistic to test H: $\sigma_b^2 = 0$, because there is no mean square whose expected value is $E(M_1)$ with the σ_b^2 term omitted, namely

$$E(M_1) - k_1\sigma_b^2 = k_2\sigma_{c:b}^2 + k_3\sigma_{ab}^2 + k_4\sigma_{ac:b}^2 + \sigma_e^2\,. \tag{39}$$

However, there is a linear function of the other mean squares whose expected value equals $E(M_1) - k_1\sigma_b^2$, namely

$$E(M_2) + E(M_3) - E(M_4) = k_2\sigma_{c:b}^2 + k_3\sigma_{ab}^2 + k_4\sigma_{ac:b}^2 + \sigma_e^2 . \tag{40}$$

From this we show how to use the mean squares in (39) and (40) to calculate a ratio that is approximately distributed as a central F-distribution.

In (40) some of the mean squares are involved negatively. But from (39) and (40) together it is clear that

$$E(M_1) + E(M_4) = k_1\sigma_b^2 + E(M_2) + E(M_3) .$$

From this let us generalize to

$$E(M_r + \cdots + M_s) = k\sigma_\alpha^2 + E(M_m + \cdots + M_n) \tag{41}$$

and consider testing the hypothesis H: $\sigma_\alpha^2 = 0$ where σ_α^2 is some component of a model. The statistic suggested by Satterthwaite (1946) for testing this hypothesis is

$$F = \frac{M'}{M''} = \frac{M_r + \cdots + M_s}{M_m + \cdots + M_n}, \quad \text{which is approximately} \sim \mathscr{F}_q^p, \tag{42}$$

where it is implicitly assumed that no mean square occurs in both numerator and denominator of (42), and where

$$p = \frac{(M_r + \cdots + M_s)^2}{M_r^2/f_r + \cdots + M_s^2/f_s} \quad \text{and} \quad q = \frac{(M_m + \cdots + M_n)^2}{M_m^2/f_m + \cdots + M_n^2/f_n} . \tag{43}$$

In p and q the term f_i is the degrees of freedom associated with the mean square M_i. Furthermore, of course, p and q are not necessarily integers and so, in comparing F against tabulated values of the $\mathscr{F}$-distribution, interpolation will be necessary.

The basis of this test is that under H: $\sigma_\alpha^2 = 0$ both numerator and denominator of (42) are distributed approximately as multiples of central χ^2-variables (each mean square in the analysis is distributed as a multiple of a central χ^2). Furthermore, in (42) there is no mean square that occurs in both numerator and denominator, which are therefore independent, and so F of (42) is distributed approximately as $\mathscr{F}_q^p$ as shown.

Both M' and M'' in (42) are sums of mean squares. p of (43) was derived by Satterthwaite (1946) from matching the first two moments of $pM'/E(M')$ to those of a central χ^2 with p degrees of freedom. This yielded p of (43) with $pM'/E(M')$ being distributed approximately as χ_p^2. (A similar result holds for M'' with q degrees of freedom.) More generally, consider the case where some mean squares are included negatively. Suppose

$$M_0 = M_1 - M_2 > 0,$$

where M_1 and M_2 are now *sums* of mean squares having f_1 and f_2 degrees of freedom, respectively. Let

$$\rho = \frac{E(M_1)}{E(M_2)} \quad \text{and} \quad \hat{\rho} = \frac{M_1}{M_2} \geqslant 1,$$

and

$$\hat{f}_0 = \frac{(\hat{\rho} - 1)^2}{(\hat{\rho}/f_1 + 1/f_2)^2} .$$

Then, simulation studies by Gaylor and Hopper (1969) suggest that

$$\frac{\hat{f}_0 M_0}{E(M_0)} \quad \text{is approximately} \quad \sim \chi^2_{\hat{f}_0}$$

provided

$$\rho > \mathscr{F}^{f_2}_{f_1, 0\cdot975}, \quad f_1 \leqslant 100 \quad \text{and} \quad f_1 \leqslant 2f_2,$$

where $\mathscr{F}^{f_2}_{f_1, 0\cdot975}$ is defined by $\Pr\{\mathscr{F}^{f_2}_{f_1} < \mathscr{F}^{f_2}_{f_1, 0\cdot975}\} = 0.975$. They further suggest that $\rho > \mathscr{F}^{f_2}_{f_1, 0\cdot975}$ "appears to be fulfilled reasonably well" when

$$\hat{\rho} > \mathscr{F}^{f_2}_{f_1, 0\cdot975} \times \mathscr{F}^{f_1}_{f_2, 0\cdot50} .$$

Under these conditions, Satterthwaite's procedure in (42) and (43) can be adapted to functions of mean squares that involve differences as well as sums.

d. Confidence intervals

In ANOVA tables of balanced data, mean squares are, under normality assumptions, distributed independently as multiples of χ^2-distributions, as in (34). Therefore an exact $1 - \alpha$ confidence interval on any $E(M_i)$ is, similar to line 1 in Table 3.4,

$$\frac{f_i M_i}{\chi^2_{f_i, \mathrm{U}}} \leqslant E(M_i) \leqslant \frac{f_i M_i}{\chi^2_{f_i, \mathrm{L}}}, \tag{44}$$

where $\chi^2_{f, \mathrm{U}}$ and $\chi^2_{f, \mathrm{L}}$ are, for the χ^2-variable χ^2_f, defined by

$$\Pr\{\chi^2_{f, \mathrm{L}} \leqslant \chi^2_f < \chi^2_{f, \mathrm{U}}\} = 1 - \alpha . \tag{45}$$

$E(M_i)$ is, of course, a linear combination of σ^2s, e.g., $an\sigma^2_\beta + n\sigma^2_\gamma + \sigma^2_e$ of Table 4.5.

Likewise, a $1 - \alpha$ confidence interval on a ratio $E(M_i)/E(M_j)$ is

$$\frac{M_i}{M_j}\mathscr{F}_{f_j, f_i, \mathrm{L}} \leqslant \frac{E(M_i)}{E(M_j)} \leqslant \frac{M_i}{M_j}\mathscr{F}_{f_j, f_i, \mathrm{U}}, \tag{46}$$

where the $\mathscr{F}$-values for the $\mathscr{F}$-distribution with f_j and f_i degrees of freedom are defined by

$$\Pr\{\mathscr{F}_{f_j, f_i, \mathrm{L}} \leqslant \mathscr{F}_{f_j, f_i} \leqslant \mathscr{F}_{f_j, f_i, \mathrm{U}}\} = 1 - \alpha . \tag{47}$$

The intervals (44) and (46) are those of Theorem 15.3.6 of Graybill (1976, p. 625).

Suppose, as is often the case with balanced data, that two expected mean squares differ by only a (multiple of a) σ^2; e.g.,

$$E(M_i) = k\sigma^2 + E(M_j) \, . \tag{48}$$

Then a $1 - 2\alpha$ approximate confidence interval on σ^2 is

$$\frac{f_i M_i(1 - \mathscr{F}_{f_i, f_j, \text{U}}/F)}{k\chi^2_{f_i, \text{U}}} \leqslant \sigma^2 \leqslant \frac{f_i M_i(1 - \mathscr{F}_{f_i, f_j, \text{L}}/F)}{k\chi^2_{f_i, \text{L}}}, \tag{49}$$

where $F = M_i/M_j$ and the $\mathscr{F}$ are just as in (47), except that there f_j is the numerator degrees of freedom and here it is f_i. This is a special case of Theorem 15.3.5 of Graybill (1976, p. 624), which has a linear combination of variance components in place of σ^2 in (48) and hence in (49) also. That in turn is an extension of the Williams (1962) result given as line 2 of Table 3.4.

One general method for deriving approximate confidence intervals on a linear function of expected mean squares is that given by Graybill (1961, p. 361; 1976, p. 642) using upper and lower limits of the χ^2-density, as defined in (45). An approximate confidence interval on $\Sigma_i k_i E(M_i)$, provided $\Sigma_i k_i M_i > 0$, is given by

$$\Pr\left\{\frac{r\Sigma_i k_i M_i}{\chi^2_{r,\text{U}}} \leqslant \Sigma_i k_i E(M_i) \leqslant \frac{r\Sigma_i k_i M_i}{\chi^2_{r,\text{L}}}\right\} = 1 - \alpha, \tag{50}$$

where

$$r = \frac{(\Sigma_i k_i M_i)^2}{\Sigma_i k_i^2 M_i^2/f_i},$$

analogous to (43). Since r will seldom be an integer, $\chi^2_{r,\text{L}}$ and $\chi^2_{r,\text{U}}$ are obtained from tables of the central χ^2-distribution, using either interpolation or the nearest (or next largest) integer to r. An adjustment to $\chi^2_{r,\text{U}}$ and $\chi^2_{r,\text{L}}$ in (50), when $r < 30$, is given by Welch (1956) and recommended by Graybill (1961, p. 370), where details may be found.

An improved confidence interval for $\Sigma_i k_i E(M_i)$ when every k_i is positive is given by Graybill and Wang (1980). For $u \sim \mathscr{F}^n_d$ we define $\mathscr{F}^n_{d,\alpha}$, akin to F_L and F_U of Chapter 3, by $\Pr\{u > \mathscr{F}^n_{d,\alpha}\} = \alpha$ so that

$$\Pr\{u < \mathscr{F}^n_{d,\alpha}\} = 1 - \alpha, \quad \text{and hence } \Pr\{u < \mathscr{F}^n_{d,1-\alpha}\} = \alpha \, .$$

For independent mean squares M_i with f_i degrees of freedom, define

$$p_i = 1 - \frac{1}{\mathscr{F}^{f_i}_{\infty, 1-\alpha_i}} \quad \text{and} \quad q_i = \frac{1}{\mathscr{F}^{f_i}_{\infty, \alpha_i}} - 1$$

for $1 > \alpha_i > 0$ and $i = 1, 2, \ldots, s$. Then an approximate $1 - \alpha$ confidence interval on $\Sigma_i k_i E(M_i)$ is

$$\Sigma_i k_i M_i - \sqrt{\Sigma_i (p_i k_i M_i)^2} \leqslant \Sigma_i k_i E(M_i) \leqslant \Sigma_i k_i M_i + \sqrt{\Sigma_i (q_i k_i M_i)^2} \, .$$

Khuri (1984) shows, with balanced data, how to deal with the general mixed model so as to develop simultaneous confidence intervals for functions of the variance components based on Khuri (1981); and, when the ANOVA table provides no exact confidence intervals on estimable functions of fixed effects, Khuri (1984) develops a method for deriving suitable intervals. His work is inspired by results given in Scheffé (1956; 1959, Chap. 8) for the 2-way classification, extended by Imhof (1960).

Other methods for finding simultaneous confidence intervals on ratios of variance components are to be found in Broemeling (1969). And Boardman (1974) contains results of some Monte Carlo studies for comparing some of these intervals for cases like (48). These and a host of other (mostly approximate) confidence intervals are reviewed in very readable form in Burdick and Graybill (1988).

e. Probability of a negative estimate

Whenever M_i and M_j are such that (48) is true, the ANOVA estimator of σ^2 is

$$\hat{\sigma}^2 = (M_i - M_j)/k \,. \tag{51}$$

Then the probability of $\hat{\sigma}^2$ being negative is

$$\begin{aligned} \Pr\{\hat{\sigma}^2 \text{ is negative}\} &= \Pr\{M_i/M_j < 1\} \\ &= \Pr\left\{\frac{M_i/E(M_i)}{M_j/E(M_j)} < \frac{E(M_j)}{E(M_i)}\right\} \\ &= \Pr\left\{\mathscr{F}_{f_j}^{f_i} < \frac{E(M_j)}{E(M_i)}\right\}. \end{aligned} \tag{52}$$

This provides a means of calculating the probability of the estimator (51) being negative. It requires giving numerical values to the variance components being estimated because $E(M_j)$ and $E(M_i)$ are functions of the components. However, in using a series of arbitrary values for these components, calculation of (52) provides some general indication of the probability of obtaining a negative estimate. The development of this procedure is given by Leone *et al.* (1968). Clearly, it could also be extended to use the approximate F-statistic of (42) for finding the probability that the estimate of σ_α^2 of (41) would be negative. Verdooren (1982) also deals with the probability of obtaining negative estimates.

An example of (52) is $\Pr(\hat{\sigma}_\alpha^2 < 0)$ given in Section 3.5d-vi.

f. Sampling variances of estimators

Sampling variances of variance component estimators that are linear functions of χ^2-variables can be derived even though the distribution functions of the estimators, generally speaking, cannot be. The variances are, of course, functions of the unknown components.

From $\hat{\boldsymbol{\sigma}}^2 = \mathbf{P}^{-1}\mathbf{m}$ of (30), where $\mathbf{m}$ is a vector of mean squares,

$$\text{var}(\hat{\boldsymbol{\sigma}}^2) = \mathbf{P}^{-1}\,\text{var}(\mathbf{m})\,\mathbf{P}^{-1'} \,. \tag{53}$$

And with

$$\text{var}(M_i) = \frac{2[E(M_i)]^2}{f_i} \quad \text{and} \quad \text{cov}(M_i, M_j) = 0, \tag{54}$$

from (34)

$$\operatorname{var}(\hat{\boldsymbol{\sigma}}^2) = \mathbf{P}^{-1}\left\{ \frac{2[E(M_i)]^2}{f_i} \right\}_{\mathrm{d}} \mathbf{P}^{-1'} . \tag{55}$$

Then on using $M_i^2/(f_i + 2)$ as an unbiased estimator of $[E(M_i)]^2/f_i$, from Appendix S.3b, we have an unbiased estimator of $\operatorname{var}(\hat{\boldsymbol{\sigma}}^2)$ as

$$\hat{\operatorname{var}}(\hat{\boldsymbol{\sigma}}^2) = \mathbf{P}^{-1}\left\{ \frac{2M_i^2}{f_i + 2} \right\}_{\mathrm{d}} \mathbf{P}^{-1'} . \tag{56}$$

An example is given in Section 3.5d-iii.

4.6. A MATRIX FORMULATION OF MIXED MODELS

Section 3.2 shows a matrix formulation of the model for balanced data from a 1-way classification, introduced there largely by means of an example. We now extend that formulation, first to be applicable to any mixed model and then show the specifications for balanced data.

a. The general mixed model

The starting point is the traditional fixed effects linear model written as

$$\mathbf{y} = \mathbf{X}\boldsymbol{\beta} + \mathbf{e},$$

where $\mathbf{y}$ is an $N \times 1$ vector of data, $\boldsymbol{\beta}$ is a $p \times 1$ vector of fixed effects parameters occurring in the data, $\mathbf{X}$ is a known $N \times p$ coefficient matrix and $\mathbf{e}$ is an error vector defined as $\mathbf{e} = \mathbf{y} - E(\mathbf{y}) = \mathbf{y} - \mathbf{X}\boldsymbol{\beta}$ and thus has $E(\mathbf{e}) = \mathbf{0}$. To $\mathbf{e}$ is usually attributed the dispersion matrix $\operatorname{var}(\mathbf{e}) = \sigma_e^2\mathbf{I}_N$. $\mathbf{X}$ is often a matrix of zeros and ones, in which case it is known as an *incidence matrix*, because then, in the expected value of the data vector, it indicates the incidence of the parameters that are in $\boldsymbol{\beta}$. But $\mathbf{X}$ can also include columns of covariates, and in regression these may, apart from a column that is $\mathbf{1}_N$, be its only columns. To cover all three of these possibilities, $\mathbf{X}$ is nowadays called a *model matrix* (Kempthorne, 1980).

In variance components models the random effects of a model can be represented as $\mathbf{Zu}$, of a nature that parallels $\mathbf{X}\boldsymbol{\beta}$ when $\mathbf{X}$ is an incidence matrix. $\mathbf{u}$ will be the vector of the random effects that occur in the data and $\mathbf{Z}$ the corresponding matrix, usually an incidence matrix. Moreover, $\mathbf{u}$ can be partitioned into a series of r sub-vectors

$$\mathbf{u} = [\mathbf{u}_1' \quad \mathbf{u}_2' \quad \ldots \quad \mathbf{u}_r']', \tag{57}$$

where each sub-vector is a vector of effects representing all levels of a single factor occurring in the data, be it a main effects factor, an interaction factor or a nested factor. r represents the number of such random factors. For example, the 1-way classification random model, with model equation $y_{ij} = \mu + \alpha_i + e_{ij}$, has $r = 1$ and $\mathbf{u}_1 = \boldsymbol{\alpha}$. The 2-way classification random model, $y_{ijk} = \mu + \alpha_i + \beta_j + \gamma_{ij} + e_{ijk}$, has $r = 3$ with $\mathbf{u}_1 = \boldsymbol{\alpha}$, $\mathbf{u}_2 = \boldsymbol{\beta}$ and $\mathbf{u}_3 = \boldsymbol{\gamma}$.

Incorporating $\mathbf{u}$ of (57) into $\mathbf{y} = \mathbf{X}\boldsymbol{\beta} + \mathbf{e}$ gives a general form of model equation for a mixed model as

$$\mathbf{y} = \mathbf{X}\boldsymbol{\beta} + \mathbf{Z}\mathbf{u} + \mathbf{e}, \tag{58}$$

with $\boldsymbol{\beta}$ representing fixed effects and $\mathbf{u}$ being for random effects. $\mathbf{X}$ and $\mathbf{Z}$ are the corresponding model matrices, with $\mathbf{Z}$ often an incidence matrix, and $\mathbf{e}$ is a vector of residual errors. Definition of $\mathbf{e}$ is based on first defining

$$E(\mathbf{y}) = \mathbf{X}\boldsymbol{\beta} \quad \text{and} \quad E(\mathbf{y}\,|\,\mathbf{u}) = \mathbf{X}\boldsymbol{\beta} + \mathbf{Z}\mathbf{u} \tag{59}$$

and then

$$\mathbf{e} = \mathbf{y} - E(\mathbf{y}\,|\,\mathbf{u})\,. \tag{60}$$

$E(\mathbf{y}\,|\,\mathbf{u})$ is the conditional mean of $\mathbf{y}$, given that $\mathbf{u}$ represents the actual random effects as they occur in the data. Put more carefully, by $E(\mathbf{y}\,|\,\mathbf{u})$ we would mean

$$E(\mathbf{Y}\,|\,\mathbf{U} = \mathbf{u}) = \mathbf{X}\boldsymbol{\beta} + \mathbf{Z}\mathbf{u}, \tag{61}$$

where $\mathbf{Y}$ and $\mathbf{U}$ would be *vectors* of random variables for which $\mathbf{y}$ and $\mathbf{u}$ are the realizations in the data. Thus (61) would be the expected value of the random variable $\mathbf{Y}$, given that the random variable $\mathbf{U}$ has the value $\mathbf{u}$. This use of capital letters as random variables is standard in much of the writing of mathematical statistics, but when used as here in vector form it conflicts with our preferred use of capital letters as matrices. Therefore the notation of (58) is retained: $\mathbf{y}$ and $\mathbf{u}$ do double duty as random variables and as realizations thereof.

For (58) we therefore have

$$E(\mathbf{y}) = \mathbf{X}\boldsymbol{\beta} \quad \text{and} \quad E(\mathbf{e}) = \mathbf{0}\,. \tag{62}$$

To $\mathbf{e}$ we now attribute the usual variance–covariance structure for error terms: every element of $\mathbf{e}$ has variance σ_e^2 and every pair of elements has covariance zero, i.e.,

$$\text{var}(\mathbf{e}) = \sigma_e^2\mathbf{I}_N\,. \tag{63}$$

Similar properties are attributed to the elements of each $\mathbf{u}_i$:

$$\text{var}(\mathbf{u}_i) = \sigma_i^2\mathbf{I}_{q_i} \quad \forall\ i, \tag{64}$$

with q_i being the number of elements in $\mathbf{u}_i$, i.e., the number of levels of the factor corresponding to $\mathbf{u}_i$ that are represented in the data. And to elements of $\mathbf{u}_i$ and to those of $\mathbf{u}_j$ are attributed a zero covariance. Thus

$$\text{cov}(\mathbf{u}_i, \mathbf{u}_j') = \mathbf{0} \quad \forall\ i \neq j; \tag{65}$$

and similarly for all elements of $\mathbf{u}$ and $\mathbf{e}$:

$$\text{cov}(\mathbf{u}, \mathbf{e}') = \mathbf{0}\,. \tag{66}$$

Models incorporating (64)–(66) have all possible covariances zero, and so provide no opportunity for dealing with situations where components of covariance would be appropriate. Models that do include components of covariance are discussed in Section 11.1.

Utilizing (64)–(66), the variance structure of $\mathbf{u}$ is

$$\mathbf{D} = \text{var}(\mathbf{u}) = \begin{bmatrix} \sigma_1^2\mathbf{I}_{q_1} & & & \\ & \sigma_2^2\mathbf{I}_{q_2} & & \\ & & \ddots & \\ & & & \sigma_r^2\mathbf{I}_{q_r} \end{bmatrix} = \{_\text{d}\, \sigma_i^2\mathbf{I}_{q_i}\} . \tag{67}$$

Then partitioning $\mathbf{Z}$ conformably with $\mathbf{u}$ of (57) as

$$\mathbf{Z} = [\mathbf{Z}_1 \quad \mathbf{Z}_2 \quad \dots \quad \mathbf{Z}_r]$$

gives

$$\mathbf{y} = \mathbf{X}\boldsymbol{\beta} + \mathbf{Z}\mathbf{u} + \mathbf{e} = \mathbf{X}\boldsymbol{\beta} + \sum_{i=1}^{r} \mathbf{Z}_i\mathbf{u}_i + \mathbf{e} . \tag{68}$$

Hence, from (58)–(67)

$$\mathbf{V} = \text{var}(\mathbf{y}) = \mathbf{Z}\mathbf{D}\mathbf{Z}' + \sigma_e^2\mathbf{I} = \sum_{i=1}^{r} \sigma_i^2\mathbf{Z}_i\mathbf{Z}_i' + \sigma_e^2\mathbf{I}_N . \tag{69}$$

A useful extension of this is to observe that since $\mathbf{e}$ is a vector of random variables just as is each $\mathbf{u}_i$, we can define $\mathbf{e}$ as another $\mathbf{u}$-vector, $\mathbf{u}_0$ say, and incorporate it into (68); i.e., define

$$\mathbf{u}_0 \equiv \mathbf{e}, \quad \mathbf{Z}_0 \equiv \mathbf{I}_N \quad \text{and} \quad \sigma_0^2 \equiv \sigma_e^2$$

and so have

$$\mathbf{y} = \mathbf{X}\boldsymbol{\beta} + \sum_{i=0}^{r} \mathbf{Z}_i\mathbf{u}_i \tag{70}$$

and

$$\mathbf{V} = \sum_{i=0}^{r} \mathbf{Z}_i\mathbf{Z}_i'\sigma_i^2 . \tag{71}$$

The originators of this formulation were Hartley and Rao (1967), who use it to great advantage for unbalanced data. We now illustrate this formulation for balanced data from a 2-way classification, from which we generalize to any multi-factored model for balanced data.

b. The 2-way crossed classification

*-i. **Model equation.*** Section 3.2a develops the general form

$$\mathbf{y} = (\mathbf{1}_a \otimes \mathbf{1}_n)\mu + (\mathbf{I}_a \otimes \mathbf{1}_n)\boldsymbol{\alpha} + \mathbf{e}$$

and

$$\mathbf{V} = (\mathbf{I}_a \otimes \mathbf{J}_n)\sigma_\alpha^2 + (\mathbf{I}_a \otimes \mathbf{I}_n)\sigma_e^2$$

for the 1-way classification having model equation $y_{ij} = \mu + \alpha_i + e_{ij}$ for $i = 1, \ldots, a$ and $j = 1, \ldots, n$. We now do the same for the 2-way crossed classification with interaction with model equation, of (3),

$$y_{ijk} = \mu + \alpha_i + \beta_j + \gamma_{ij} + e_{ijk}, \tag{72}$$

for $i = 1, \ldots, a$, $j = 1, \ldots, b$ and $k = 1, \ldots, n$. Suppose $a = 2$, $b = 3$ and $n = 2$; then arraying the y_{ijk} in lexicon order (ordered by k within j within i) in a 12×1 vector gives

$$\mathbf{y} = \begin{bmatrix} 1\\1\\1\\1\\1\\1\\1\\1\\1\\1\\1\\1 \end{bmatrix}\mu + \begin{bmatrix} 1 & \cdot \\ 1 & \cdot \\ 1 & \cdot \\ 1 & \cdot \\ 1 & \cdot \\ 1 & \cdot \\ \cdot & 1 \\ \cdot & 1 \\ \cdot & 1 \\ \cdot & 1 \\ \cdot & 1 \\ \cdot & 1 \end{bmatrix}\begin{bmatrix} \alpha_1 \\ \alpha_2 \end{bmatrix} + \begin{bmatrix} 1 & \cdot & \cdot \\ 1 & \cdot & \cdot \\ \cdot & 1 & \cdot \\ \cdot & 1 & \cdot \\ \cdot & \cdot & 1 \\ \cdot & \cdot & 1 \\ 1 & \cdot & \cdot \\ 1 & \cdot & \cdot \\ \cdot & 1 & \cdot \\ \cdot & 1 & \cdot \\ \cdot & \cdot & 1 \\ \cdot & \cdot & 1 \end{bmatrix}\begin{bmatrix} \beta_1 \\ \beta_2 \\ \beta_3 \end{bmatrix} + \begin{bmatrix} 1 & \cdot & \cdot & \cdot & \cdot & \cdot \\ 1 & \cdot & \cdot & \cdot & \cdot & \cdot \\ \cdot & 1 & \cdot & \cdot & \cdot & \cdot \\ \cdot & 1 & \cdot & \cdot & \cdot & \cdot \\ \cdot & \cdot & 1 & \cdot & \cdot & \cdot \\ \cdot & \cdot & 1 & \cdot & \cdot & \cdot \\ \cdot & \cdot & \cdot & 1 & \cdot & \cdot \\ \cdot & \cdot & \cdot & 1 & \cdot & \cdot \\ \cdot & \cdot & \cdot & \cdot & 1 & \cdot \\ \cdot & \cdot & \cdot & \cdot & 1 & \cdot \\ \cdot & \cdot & \cdot & \cdot & \cdot & 1 \\ \cdot & \cdot & \cdot & \cdot & \cdot & 1 \end{bmatrix}\begin{bmatrix} \gamma_{11} \\ \gamma_{12} \\ \gamma_{13} \\ \gamma_{21} \\ \gamma_{22} \\ \gamma_{23} \end{bmatrix} + \mathbf{e}\,.$$

$$= \mathbf{1}_{12}\mu + (\mathbf{I}_2 \otimes \mathbf{1}_6)\boldsymbol{\alpha} + \begin{bmatrix} \mathbf{I}_3 \otimes \mathbf{1}_2 \\ \mathbf{I}_3 \otimes \mathbf{1}_2 \end{bmatrix}\boldsymbol{\beta} + (\mathbf{I}_6 \otimes \mathbf{1}_2)\boldsymbol{\gamma} + \mathbf{I}_{12}\mathbf{e}$$

$$= (\mathbf{1}_2 \otimes \mathbf{1}_3 \otimes \mathbf{1}_2)\mu + (\mathbf{I}_2 \otimes \mathbf{1}_3 \otimes \mathbf{1}_2)\boldsymbol{\alpha} + (\mathbf{1}_2 \otimes \mathbf{I}_3 \otimes \mathbf{1}_2)\boldsymbol{\beta}$$
$$+ (\mathbf{I}_2 \otimes \mathbf{I}_3 \otimes \mathbf{1}_2)\boldsymbol{\gamma} + (\mathbf{I}_2 \otimes \mathbf{I}_3 \otimes \mathbf{I}_2)\mathbf{e}\,.$$

For the general 2-way crossed classification this form is

$$\mathbf{y} = (\mathbf{1}_a \otimes \mathbf{1}_b \otimes \mathbf{1}_n)\mu + (\mathbf{I}_a \otimes \mathbf{1}_b \otimes \mathbf{1}_n)\boldsymbol{\alpha} + (\mathbf{1}_a \otimes \mathbf{I}_b \otimes \mathbf{1}_n)\boldsymbol{\beta}$$
$$+ (\mathbf{I}_a \otimes \mathbf{I}_b \otimes \mathbf{1}_n)\boldsymbol{\gamma} + (\mathbf{I}_a \otimes \mathbf{I}_b \otimes \mathbf{I}_n)\mathbf{e}\,. \tag{73}$$

Several features of (73) need to be noted. First, every coefficient matrix in (73) is a Kronecker product (KP) of three terms: three, because it is a 2-way classification and $3 = 2 + 1$ (two main effect factors plus error). Second, every term in every KP is a $\mathbf{1}$ or an $\mathbf{I}$. Third, the orders of the three terms in every KP are a, b and n, respectively, the number of levels of the two factors and the number of observations in each cell; and the sequence of these orders, a, b and then n matches, and is determined by, the nature of the lexicon ordering of the data in $\mathbf{y}$, i within which j is ordered, within which k is ordered.

Fourth is a characteristic that determines the form of each KP that is a coefficient matrix in (73): every term in every KP is a $\mathbf{1}$ except that it is $\mathbf{I}$ for the term corresponding to the parameter vector which that KP is multiplying. For example, in $(\mathbf{I}_a \otimes \mathbf{1}_b \otimes \mathbf{1}_n)\boldsymbol{\alpha}$ the vector being multiplied is $\boldsymbol{\alpha}$, and so the first term in the KP is $\mathbf{I}_a$ and not $\mathbf{1}_a$. Similarly, in $(\mathbf{1}_a \otimes \mathbf{I}_b \otimes \mathbf{1}_n)\boldsymbol{\beta}$ the second term of the KP is $\mathbf{I}_b$, corresponding to the $\boldsymbol{\beta}$ being multiplied. This principle easily adapts itself to the other KP matrices in (73). Thus by thinking of $\boldsymbol{\gamma}$ as requiring $\boldsymbol{\alpha}$ and $\boldsymbol{\beta}$ for its definition (since γ_{ij} is an interaction effect of the $\boldsymbol{\alpha}$-factor with the $\boldsymbol{\beta}$-factor), then in the term $(\mathbf{I}_a \otimes \mathbf{I}_b \otimes \mathbf{1}_n)\boldsymbol{\gamma}$ in (73) the $\mathbf{I}_a$ and $\mathbf{I}_b$ occur rather than $\mathbf{1}_a$ and $\mathbf{1}_b$. Likewise for the last term, $(\mathbf{I}_a \otimes \mathbf{I}_b \otimes \mathbf{I}_n)\mathbf{e}$, thinking of e_{ijk} as being nested within the (i, j) cell then requires $\boldsymbol{\alpha}$, $\boldsymbol{\beta}$ and $\mathbf{e}$ itself for defining e_{ijk}, and so $\mathbf{I}_a$, $\mathbf{I}_b$ and $\mathbf{I}_n$ is appropriate. And finally in the first term, $(\mathbf{1}_a \otimes \mathbf{1}_b \otimes \mathbf{1}_n)\mu$, defining μ requires neither $\boldsymbol{\alpha}$, $\boldsymbol{\beta}$ nor $\mathbf{e}$, and so the KP has $\mathbf{1}_a$, $\mathbf{1}_b$ and $\mathbf{1}_n$.

-ii. ***Random or mixed?*** In comparing (73) with $\mathbf{y} = \mathbf{X}\boldsymbol{\beta} + \mathbf{Z}\mathbf{u} + \mathbf{e}$, the determination of which parts of (73) constitute $\mathbf{X}\boldsymbol{\beta}$ and which are $\mathbf{Z}\mathbf{u}$ depends entirely on the decision as to which parts of (73) are fixed and which are random; and this decision is quite external to the algebraic form of (73). μ is always fixed: and so in random models $\mathbf{X}\boldsymbol{\beta}$ is $(\mathbf{1}_a \otimes \mathbf{1}_b \otimes \mathbf{1}_n)\mu$. If $\boldsymbol{\alpha}$ is also to be considered fixed then in the resulting mixed model $\mathbf{X}\boldsymbol{\beta}$ is the first two terms of (73).

-iii. ***Dispersion matrix.*** The terms of (73) that determine $\mathbf{V}$ do, of course, depend upon which terms of (73) are taken as random. Whichever are so defined, they are assumed to have variance and covariance properties in accord with (64)–(66). Hence, for example, on taking $\boldsymbol{\alpha}$ as random

$$
\begin{aligned}
\operatorname{var}[(\mathbf{I}_a \otimes \mathbf{1}_b \otimes \mathbf{1}_n)\boldsymbol{\alpha}] &= (\mathbf{I}_a \otimes \mathbf{1}_b \otimes \mathbf{1}_n)\sigma_\alpha^2 \mathbf{I}_a (\mathbf{I}_a \otimes \mathbf{1}_b \otimes \mathbf{1}_n)' \\
&= (\mathbf{I}_a \otimes \mathbf{1}_b \otimes \mathbf{1}_n)\sigma_\alpha^2 (\mathbf{I}'_a \otimes \mathbf{1}'_b \otimes \mathbf{1}'_n) \\
&= \sigma_\alpha^2 (\mathbf{I}_a \otimes \mathbf{J}_b \otimes \mathbf{J}_n),
\end{aligned}
$$

after using the properties of KPs in Appendix M.2. Using this kind of result, and results like $\operatorname{cov}(\boldsymbol{\alpha}, \boldsymbol{\beta}') = \mathbf{0}$ and $\operatorname{cov}(\boldsymbol{\alpha}, \mathbf{e}') = \mathbf{0}$ from (65) and (66), it is easily seen that for the αs, βs and γs all being taken as random in (73)

$$
\begin{aligned}
\mathbf{V} = {} & (\mathbf{I}_a \otimes \mathbf{J}_b \otimes \mathbf{J}_n)\sigma_\alpha^2 + (\mathbf{J}_a \otimes \mathbf{I}_b \otimes \mathbf{J}_n)\sigma_\beta^2 \\
& + (\mathbf{I}_a \otimes \mathbf{I}_b \otimes \mathbf{J}_n)\sigma_\gamma^2 + (\mathbf{I}_a \otimes \mathbf{I}_b \otimes \mathbf{I}_n)\sigma_e^2 \,. \qquad (74)
\end{aligned}
$$

And for a mixed model with $\boldsymbol{\alpha}$ representing fixed effects the term in σ_α^2 in (74) would be dropped.

c. The 2-way nested classification

The model equation for a 2-way nested classification, with β_{ij} nested within α_i, is

$$
y_{ijk} = \mu + \alpha_i + \beta_{ij} + e_{ijk}, \qquad (75)
$$

with, for balanced data, $i = 1, \ldots, a$, $j = 1, \ldots, b$ for each i, and $k = 1, \ldots, n$. Algebraically, this is the same as (72) for the 2-way crossed classification except that β_j in (72) is deleted and γ_{ij} is replaced by β_{ij} in (75). Making these changes in (73) leads to the vector form of the model equation (75) being

$$\begin{aligned} \mathbf{y} = (\mathbf{1}_a \otimes \mathbf{1}_b \otimes \mathbf{1}_n)\mu + (\mathbf{I}_a \otimes \mathbf{1}_b \otimes \mathbf{1}_n)\boldsymbol{\alpha} \\ + (\mathbf{I}_a \otimes \mathbf{I}_b \otimes \mathbf{1}_n)\boldsymbol{\beta} + (\mathbf{I}_a \otimes \mathbf{I}_b \otimes \mathbf{I}_n)\mathbf{e}, \end{aligned} \tag{76}$$

with, for the random model,

$$\mathbf{V} = (\mathbf{I}_a \otimes \mathbf{J}_b \otimes \mathbf{J}_n)\sigma_\alpha^2 + (\mathbf{I}_a \otimes \mathbf{I}_b \otimes \mathbf{J}_n)\sigma_\beta^2 + (\mathbf{I}_a \otimes \mathbf{I}_b \otimes \mathbf{I}_n)\sigma_e^2 . \tag{77}$$

d. Interaction or nested factor?

In the with-interaction model equation of (73), note that the coefficient of $\boldsymbol{\gamma}$, namely $(\mathbf{I}_a \otimes \mathbf{I}_b \otimes \mathbf{1}_n)$, is the same as the coefficient of $\boldsymbol{\beta}$ in the nested model of (76). How then, one well might ask, does one identify $\boldsymbol{\gamma}$ in (73) as representing interactions and $\boldsymbol{\beta}$ in (76) as representing a nested factor? The answer relies on characteristics of this way of formulating models. Consider $\boldsymbol{\gamma}$ in (73). It represents interaction between factors A and B, and so they are both needed for defining $\boldsymbol{\gamma}$. Therefore its KP has $\mathbf{I}$ for both A and B and so is $(\mathbf{I}_a \otimes \mathbf{I}_b \otimes \mathbf{1}_n)$. Next consider $\boldsymbol{\beta}$ in (76). It represents a B-factor that is nested within A. Therefore it needs both A and B for defining it, and so it too has $(\mathbf{I}_a \otimes \mathbf{I}_b \otimes \mathbf{1}_n)$ as its KP. Therefore, since both $\boldsymbol{\gamma}$ in (73) and $\boldsymbol{\beta}$ in (76) are pre-multiplied by the same KP, the question is how, without knowing that $\boldsymbol{\gamma}$ in (73) and $\boldsymbol{\beta}$ in (76) represent different kinds of factors, can we ascertain this distinction just from the model equations (73) and (76)? Easily. In (73) there is a term that represents the A-factor, and there is also one that represents just the B-factor: therefore, in answering our question, the $\boldsymbol{\gamma}$-term that has both A- and B-factors represented by $\mathbf{I}$ in its KP is an interaction term. And complementary to this, in (76), where there is no term representing a main effect B-factor, that tells us that $\boldsymbol{\beta}$ (which also has both A- and B-factors represented by $\mathbf{I}$ in its KP) must represent a factor nested within A. It is that absence of a term in just B in (76) that triggers the conclusion that $\boldsymbol{\beta}$ represents nesting within A.

e. The general case

The preceding examples and as many more as one cares to consider will provide convincing evidence that (even in the absence of rigorous proof) this style of formulation applies quite generally. As Cornfield and Tukey (1956) so rightly say in a similar context, in carrying out detailed steps such as those of the examples, the "systematic algebra can take us deep into the forest of notation. But the detailed manipulation will, sooner or later, blot out any understanding we may have started with." Furthermore, having accomplished this, we would then, so far as developing general results is concerned, be no more than "ready for another step of induction and so on" as Cornfield and Tukey aptly put it. We therefore give the general result towards which this type of induction apparently leads.

-i. ***Model equation.*** In the model equation for a linear model of $p-1$ main effect factors (crossed and/or nested) the coefficient of the vector of effects corresponding to each factor, and to each interaction of factors, can be represented by a KP of p matrices each being an **I**-matrix or a **1**-vector. Thus each term in the model equation can be represented in the form

$$(\mathbf{1}_{n_p}^{i_p} \otimes \mathbf{1}_{n_{p-1}}^{i_{p-1}} \otimes \mathbf{1}_{n_{p-2}}^{i_{p-2}} \otimes \cdots \otimes \mathbf{1}_{n_1}^{i_1})\boldsymbol{\alpha}, \tag{78}$$

where $\mathbf{1}_{n_p}$ is a summing vector of order n_p, with $i_p = 0$ or 1, with $\mathbf{1}_{n_p}^0$ being the zero'th power of the $\mathbf{1}_{n_p}$-vector and hence (in accord with scalar algebra where, for example, $7^0 = 1$) is given the value **I**. Since in (78) it is the values of $i_p, \ldots, i_1$ that determine the nature of the KP, and which are determined by what the factor (represented by $\boldsymbol{\alpha}$) is, we can in fact use $\mathbf{i} = [i_p \quad \ldots \quad i_1]$ and write any linear model equation as

$$\mathbf{y} = \sum_{\mathbf{i}=\mathbf{0}}^{\mathbf{1}_p} (\mathbf{1}_{n_p}^{i_p} \otimes \mathbf{1}_{n_{p-1}}^{i_{p-1}} \otimes \cdots \otimes \mathbf{1}_{n_1}^{i_1})\boldsymbol{\alpha}_{\mathbf{i}}\,. \tag{79}$$

In this general formulation, every element of **i** is 0 or 1. When every element is unity, $\mathbf{i} = \mathbf{1}_p$, every term in the KP is a **1**-vector, and the corresponding $\boldsymbol{\alpha}_{\mathbf{1}}$ is μ; hence the term in $\boldsymbol{\alpha}_{\mathbf{1}}$ is $\mu\mathbf{1}$; and when every element in **i** is zero, $\mathbf{i} = \mathbf{0}$, every term in the KP is an **I**-matrix and the corresponding $\boldsymbol{\alpha}_{\mathbf{0}}$ is **e**. And since every value of **i** in (79) is a binary number, there are 2^p possible terms in (79). In practice, of course, the $\boldsymbol{\alpha}_{\mathbf{i}}$ corresponding to many of those binary numbers will not exist in the model and so (79) will have fewer than 2^p terms. An example is (73) where $p - 1 = 2$, with $2^p = 8$, but only 5 terms occur in (73).

This formulation of a model for balanced data has been used by a variety of authors; e.g., Nelder (1965a, b), Nerlove (1971), Balestra (1973), Smith and Hocking (1978), Searle and Henderson (1979), Seifert (1979) and Anderson *et al.* (1984).

We can also note that the order of $\boldsymbol{\alpha}_{\mathbf{i}}$ is the product of the n_{i_t}s that correspond to non-zero i_t in **i**. This can be written as $\Pi_{t=1}^{p}[1 + (n_{i_t} - 1)i_t]$ since i_t is either 0 or 1.

-ii. ***Dispersion matrix.*** When (79) represents a random model, with every $\boldsymbol{\alpha}_{\mathbf{i}}$ therein being a random effect (except $\boldsymbol{\alpha}_{\mathbf{1}} = \mu$) the variance–covariance matrix of **y** is

$$\mathbf{V} = \text{var}(\mathbf{y}) = \sum_{\mathbf{i}=\mathbf{0}}^{\mathbf{1}_p} \theta_{\mathbf{i}}(\mathbf{J}_{n_p}^{i_p} \otimes \mathbf{J}_{n_{p-1}}^{i_{p-1}} \otimes \cdots \otimes \mathbf{J}_{n_1}^{i_1}), \tag{80}$$

where

$$\theta_{\mathbf{i}} = \text{var}(\boldsymbol{\alpha}_{\mathbf{i}})\,.$$

Searle and Henderson (1979) have a useful result for the inverse of **V**. First define

$$\mathbf{T}_p = \begin{bmatrix} 1 & 1 \\ 0 & n_p \end{bmatrix} \otimes \begin{bmatrix} 1 & 1 \\ 0 & n_{p-1} \end{bmatrix} \otimes \cdots \otimes \begin{bmatrix} 1 & 1 \\ 0 & n_1 \end{bmatrix}$$

and

$$\boldsymbol{\theta}_p = [\theta_{0\ldots00} \quad \theta_{0\ldots01} \quad \theta_{0\ldots10} \quad \cdots \quad \theta_{1\ldots11}]' .$$

Then the vector of 2^p (possibly) distinct eigenvalues of $\mathbf{V}$ is given by

$$\boldsymbol{\lambda}_p = \mathbf{T}_p\boldsymbol{\theta}_p . \tag{81}$$

Now define

$$\mathbf{v}_p = \left[\frac{1}{\lambda_{0\ldots00}} \quad \frac{1}{\lambda_{0\ldots01}} \quad \frac{1}{\lambda_{0\ldots10}} \quad \frac{1}{\lambda_{1\ldots11}}\right], \tag{82}$$

the vector of reciprocals of the eigenvalues of $\mathbf{V}$ (and thus the eigenvalues of $\mathbf{V}^{-1}$). Then write

$$\mathbf{T}_p^{-1} = \frac{1}{n_p}\begin{bmatrix} n_p & -1 \\ 0 & 1 \end{bmatrix} \otimes \frac{1}{n_{p-1}}\begin{bmatrix} n_{p-1} & -1 \\ 0 & 1 \end{bmatrix} \otimes \cdots \otimes \frac{1}{n_1}\begin{bmatrix} n_1 & -1 \\ 0 & 1 \end{bmatrix} . \tag{83}$$

Calculate

$$\boldsymbol{\tau}_p = \mathbf{T}_p^{-1}\mathbf{v}_p,$$

and the inverse of $\mathbf{V}$ is

$$\mathbf{V}_p^{-1} = \sum_{\mathbf{i}=\mathbf{0}}^{\mathbf{1}_p} \tau_{\mathbf{i}}(\mathbf{J}_{n_p}^{i_p} \otimes \mathbf{J}_{n_{p-1}}^{i_{p-1}} \otimes \cdots \otimes \mathbf{J}_{n_1}^{i_1}) . \tag{84}$$

It is to be noticed that $\mathbf{V}^{-1}$ in (84) is a linear combination of the same matrices (KPs of **J**s to the power of 0 or 1) as is $\mathbf{V}$ of (80). The only difference is in the coefficients: θs in (80), which are σ^2s, and τs in (84); and through (81)–(83) the τs are derived from the θs. An interesting feature of this relationship is that the pattern of non-zero τs is not necessarily the same as the pattern of non-zero θs. Thus it is possible to have a zero $\theta_{\mathbf{i}}$ for some $\mathbf{i}$ and $\tau_{\mathbf{i}}$ for the same $\mathbf{i}$ can be non-zero. An example of this is shown in Searle and Henderson (1979), wherein the only non-zero θs for the random model of the 2-way classification with interaction (Table 4.5) are $\theta_{000} = \sigma_e^2$, $\theta_{001} = \sigma_\gamma^2$, $\theta_{011} = \sigma_\alpha^2$ and $\theta_{101} = \sigma_\beta^2$. These are the four non-zero coefficients in $\mathbf{V}$. In $\mathbf{V}^{-1}$, the four τ-values with the same subscripts as these θs are non-zero, but so also is τ_{111}, whereas θ_{111} is zero.

The determinant of $\mathbf{V}$ is also available from Searle and Henderson. With the eigenvalues of $\mathbf{V}$ being given by (81), and labeled $\lambda_{\mathbf{0}}$ through $\lambda_{\mathbf{1}}$, as indicated in (82),

$$\text{the multiplicity of } \lambda_{\mathbf{i}} = \lambda_{i_p i_{p-1}\ldots i_1} \text{ is } m_{\mathbf{i}} = \prod_{r=1}^{p} (n_r - 1)^{1-i_r},$$

where the number of levels of the rth factor is n_r. Therefore

$$|\mathbf{V}| = \prod_{\mathbf{i}=\mathbf{0}}^{\mathbf{1}} (\lambda_{\mathbf{i}})^{m_{\mathbf{i}}} .$$

4.7. MAXIMUM LIKELIHOOD ESTIMATION (ML)

Many of the general principles of estimating variance components by maximum likelihood are discussed in Section 3.7: the use of normality, the derivation of ML equations, the solutions of which [e.g., (108) of Chapter 3] are ML estimators only if they are in what is called a feasible region (Section 3.7a-iii) wherein σ_e^2 is positive and all other σ^2s are zero or positive. And if one or more solution is negative then the whole set of solutions has to be adapted [e.g., as in (114) and (115) of Chapter 3] in order to have estimators that are in the feasible region.

a. Estimating the mean in random models.

As noted following (79), the term in $\boldsymbol{\alpha}_1$ in (79) is $\mu\mathbf{1}$, or, more exactly, $\mu\mathbf{1}_N$ when $N = \Pi_{t=1}^p n_t$ is the number of observations. Hence the likelihood under normality is

$$L(\mu, \mathbf{V} \mid \mathbf{y}) = L = \frac{\exp[-\frac{1}{2}(\mathbf{y} - \mu\mathbf{1})'\mathbf{V}^{-1}(\mathbf{y} - \mu\mathbf{1})]}{(2\pi)^{\frac{1}{2}N} |\mathbf{V}|^{\frac{1}{2}}}, \tag{85}$$

as following (103) of Chapter 3. Differentiating $\log L$ with respect to μ, and equating to zero, gives as the equation for the ML solution of μ, namely $\tilde{\mu}$, as

$$\mathbf{1}'\tilde{\mathbf{V}}^{-1}\mathbf{1}\tilde{\mu} = \mathbf{1}'\tilde{\mathbf{V}}^{-1}\mathbf{y}, \tag{86}$$

where $\tilde{\mathbf{V}}^{-1}$ is $\mathbf{V}_p^{-1}$ of (84) but with ML estimators of the σ^2s replacing the σ^2s themselves. $\mathbf{V}^{-1}$ is given by (83); and for (86) we want

$$\mathbf{1}'\mathbf{V}^{-1} = (\mathbf{1}'_{n_p} \otimes \mathbf{1}'_{n_{p-1}} \otimes \cdots \otimes \mathbf{1}'_{n_1}) \sum_{\mathbf{i}=\mathbf{0}}^{\mathbf{1}} \tau_{\mathbf{i}}(\mathbf{J}_{n_p}^{i_p} \otimes \mathbf{J}_{n_{p-1}}^{i_{p-1}} \otimes \cdots \otimes \mathbf{J}_{n_1}^{i_1})$$

$$= \sum_{\mathbf{i}=\mathbf{0}}^{\mathbf{1}} \tau_{\mathbf{i}} \bigotimes_{t=p}^{1} (\mathbf{1}'_{n_t}\mathbf{J}_{n_t}^{i_t}) .$$

But, because $i_t = 0$ or 1,

$$\mathbf{1}'_{n_t}\mathbf{J}_{n_t}^{i_t} = n_t^{i_t}\mathbf{1}'_{n_t} .$$

Hence

$$\mathbf{1}'\mathbf{V}^{-1} = \sum_{\mathbf{i}=\mathbf{0}}^{\mathbf{1}} \tau_{\mathbf{i}} \left(\prod_{t=1}^{p} n_t^{i_t} \right) \mathbf{1}'_N = q\mathbf{1}'_N$$

and so (86) is $q\mathbf{1}'_N\mathbf{1}_N\tilde{\mu} = q\mathbf{1}'_N\mathbf{y}$ for $q = \sum_{\mathbf{i}=\mathbf{0}}^{\mathbf{1}} \tau_{\mathbf{i}} \left(\prod_{t=1}^{p} n_t^{i_t} \right)$, thus giving

$$\tilde{\mu} = \bar{y}, \tag{87}$$

the grand mean of all N observations. Thus for all random models the ML estimator of μ is the grand mean of the data.

b. Four models with closed form estimators.

We here summarize the ML estimators of variance components obtainable from balanced data of four different models. The first, for the sake of completeness is the 1-way classification of Section 3.7. For none of them do we show the derivation, preferring to simply quote the results so that they are readily available for use by an interested reader who can, if so inclined, carry out the derivation based as it is, on assuming that **y** follows a multi-normal distribution. There are at least two ways of doing this. One starts with partitioning the sum of squares in the exponent of the likelihood of y. That partitioning is done so as to be in terms of the sums of squares that occur in the analysis of variance for the model at hand. Then differentiating the likelihood will lead, via what is often tedious algebra, to ML equations for the variance components. An example of this is the 1-way classification, random model, balanced data, in Section 3.7a-i.

A second method is to use the result for the general model

$$\mathbf{y} = \mathbf{X}\boldsymbol{\beta} + \sum_{i=0}^{r} \mathbf{Z}_i\mathbf{u}_i$$

introduced in (70). That result, derived in Section 6.2b, is that the ML equations are

$$\{_{\mathrm{m}}\ \mathrm{sesq}(\mathbf{Z}_i'\dot{\mathbf{V}}^{-1}\mathbf{Z}_j)\}_{i,j=0}^{r}\dot{\boldsymbol{\sigma}}^2 = \{_{\mathrm{c}}\ \mathrm{sesq}(\mathbf{Z}_i\dot{\mathbf{P}}\mathbf{y})\}_{i=0}^{r},$$

where sesq (**A**) represents the sum of squares of every element of **A**; and

$$\mathbf{P} = \mathbf{V}^{-1} - \mathbf{V}^{-1}\mathbf{X}(\mathbf{X}'\mathbf{V}^{-1}\mathbf{X})\mathbf{X}'\mathbf{V}^{-1}\,.$$

Whichever of these methods is used the algebra can, as just mentioned, get to be quite tedious.

*-i. **The 1-way random model.*** The model equation is

$$y_{ij} = \mu + \alpha_i + e_{ij};$$
$$i = 1, \ldots, a, \quad \text{and} \quad j = 1, \ldots, n\,.$$

TABLE 4.8. ANALYSIS OF VARIANCE OF A 1-WAY CLASSIFICATION

Source	d.f.	Sum of Squares	Mean Square
A	$a-1$	$\mathrm{SSA} = \Sigma_i n(\bar{y}_{i.} - \bar{y}_{..})^2$	$\mathrm{MSA} = \mathrm{SSA}/(a-1)$
Residual	$a(n-1)$	$\mathrm{SSE} = \Sigma_i\Sigma_j(y_{ij} - \bar{y}_{i.})^2$	$\mathrm{MSE} = \mathrm{SSE}/a(n-1)$
Total	$an-1$	$\mathrm{SST_m} = \Sigma_i\Sigma_j(y_{ij} - \bar{y}_{..})^2$	
ANOVA estimators:			
	$\hat{\sigma}_e^2 = \mathrm{MSE}$	$\hat{\sigma}_\alpha^2 = (\mathrm{MSA} - \mathrm{MSE})/n$	
ML solutions:			
	$\dot{\sigma}_e^2 = \mathrm{MSE}$	$\dot{\sigma}_\alpha^2 = [(1 - 1/a)\mathrm{MSA} - \mathrm{MSE}]/n$	

TABLE 4.9. ML ESTIMATORS OF σ_α^2 AND σ_e^2 IN A 1-WAY RANDOM MODEL

Conditions satisfied by the ML solutions	ML Estimators	
	$\tilde{\sigma}_e^2$	$\tilde{\sigma}_\alpha^2$
$\dot{\sigma}_\alpha^2 \geqslant 0$	$\dot{\sigma}_e^2 = \text{MSE}$	$\dot{\sigma}_\alpha^2 = [(1 - 1/a)\text{MSA} - \text{MSE}]/n$
$\dot{\sigma}_\alpha^2 < 0$	SST_m/an	0

-ii. ***The 2-way nested random model.*** The model equation is

$$y_{ijk} = \mu + \alpha_i + \beta_{ij} + e_{ijk};$$

$$i = 1, \ldots, a, \quad j = 1, \ldots, b \quad \text{and} \quad k = 1, \ldots, n\,.$$

TABLE 4.10. ANALYSIS OF VARIANCE OF A 2-WAY NESTED, RANDOM MODEL

Source	d.f.	Sum of Squares	Mean Square
A	$a - 1$	$\text{SSA} = \Sigma_i bn(\bar{y}_{i..} - \bar{y}_{...})^2$	$\text{MSA} = \text{SSA}/(a - 1)$
B within A	$a(b - 1)$	$\text{SSB:A} = \Sigma_i \Sigma_j n(\bar{y}_{ij.} - \bar{y}_{i..})^2$	$\text{MSB:A} = \text{SSB:A}/a(b - 1)$
Residual	$ab(n - 1)$	$\text{SSE} = \Sigma_i \Sigma_j \Sigma_k (y_{ijk} - \bar{y}_{ij.})^2$	$\text{MSE} = \text{SSE}/ab(n - 1)$
Total	$abn - 1$	$\text{SST}_\text{m} = \Sigma_i \Sigma_j \Sigma_k (y_{ijk} - \bar{y}_{...})^2$	

ANOVA estimators:

$$\hat{\sigma}_\alpha^2 = \frac{\text{MSA} - \text{MSB:A}}{bn}, \quad \hat{\sigma}_\beta^2 = \frac{\text{MSB:A} - \text{MSE}}{n} \quad \text{and} \quad \hat{\sigma}_e^2 = \text{MSE}\,.$$

ML solutions:

$$\dot{\sigma}_\alpha^2 = \frac{(1 - 1/a)\text{MSA} - \text{MSB:A}}{bn}, \quad \dot{\sigma}_\beta^2 = \hat{\sigma}_\beta^2 \quad \text{and} \quad \dot{\sigma}_e^2 = \hat{\sigma}_e^2 = \text{MSE}\,.$$

The ML estimators are as shown in Table 4.11.

The formal statement of the ML estimator of each variance component is obtained by reading down the columns of Table 4.11, e.g.,

$$\tilde{\sigma}_\alpha^2 = \begin{cases} \dfrac{\text{SSA}/a - \text{MSB:A}}{bn} & \text{when } \dot{\sigma}_\alpha^2 \geqslant 0 \text{ and } \dot{\sigma}_\beta^2 \geqslant 0, \\[2ex] \dfrac{\text{SSA}/a - \tilde{\sigma}_e^2}{bn} & \text{when } \dot{\sigma}_\alpha^2 \geqslant 0 \text{ and } \dot{\sigma}_\beta^2 < 0, \\[2ex] 0 & \text{otherwise}\,. \end{cases}$$

In contrast, each row of Table 4.11 indicates what the ML estimators are for a particular set of circumstances that the data can produce, vis-à-vis positive and negative values of $\dot{\sigma}_\alpha^2$ and $\dot{\sigma}_\beta^2$.

TABLE 4.11. ML ESTIMATORS OF σ_α^2, σ_β^2 AND σ_e^2 IN A 2-WAY NESTED CLASSIFICATION, RANDOM MODEL

Conditions satisfied by the ML solutions	MLE		
	$\tilde{\sigma}_\alpha^2$	$\tilde{\sigma}_\beta^2$	$\tilde{\sigma}_e^2$
$\dot{\sigma}_\alpha^2 \geqslant 0, \dot{\sigma}_\beta^2 \geqslant 0$	$\dfrac{\text{SSA}/a - \text{MSB:A}}{bn}$	$\dfrac{\text{MSB:A} - \text{MSE}}{n}$	MSE
$\dot{\sigma}_\alpha^2 \geqslant 0, \dot{\sigma}_\beta^2 < 0$	$\dfrac{\text{SSA}/a - \tilde{\sigma}_e^2}{bn}$	0	$\dfrac{\text{SSE} + \text{SSB:A}}{a(bn-1)}$
$\dot{\sigma}_\alpha^2 < 0, \dot{\sigma}_\beta^2 \geqslant 0$	0	$\dfrac{1}{n}\left(\dfrac{\text{SSA} + \text{SSB:A}}{ab} - \text{MSE}\right)$	MSE
$\dot{\sigma}_\alpha^2 < 0, \dot{\sigma}_\beta^2 < 0$	0	0	$\dfrac{\text{SST}_\text{m}}{abn}$

-*iii*. ***The 2-way crossed, with interaction, mixed model.*** The model equation is

$$y_{ijk} = \mu + \alpha_i + \beta_j + \gamma_{ij} + e_{ijk};$$
$$i = 1, \ldots, a, \quad j = 1, \ldots, b \quad \text{and} \quad k = 1, \ldots, n; \quad \alpha_i\text{s fixed}.$$

TABLE 4.12. ANALYSIS OF VARIANCE FOR THE 2-WAY CROSSED CLASSIFICATION WITH INTERACTION

Source	d.f.	Sum of Squares	Mean Square
A	$a-1$	$\text{SSA} = \Sigma_i bn(\bar{y}_{i..} - \bar{y}_{...})^2$	$\text{MSA} = \dfrac{\text{SSA}}{a-1}$
B	$b-1$	$\text{SSB} = \Sigma_j an(\bar{y}_{.j.} - \bar{y}_{...})^2$	$\text{MSB} = \dfrac{\text{SSB}}{b-1}$
AB	$(a-1)(b-1)$	$\text{SSAB} = \Sigma_i\Sigma_j n(\bar{y}_{ij.} - \bar{y}_{i..} - \bar{y}_{.j.} + \bar{y}_{...})^2$	$\text{MSAB} = \dfrac{\text{SSAB}}{(a-1)(b-1)}$
Error	$ab(n-1)$	$\text{SSE} = \Sigma_i\Sigma_j\Sigma_k(y_{ijk} - \bar{y}_{ij.})^2$	$\text{MSE} = \dfrac{\text{SSE}}{ab(n-1)}$
Total	$abn-1$	$\text{SST}_\text{m} = \Sigma_i\Sigma_j\Sigma_k(y_{ijk} - \bar{y}_{...})^2$	

ANOVA estimators:

$$\hat{\sigma}_\beta^2 = \frac{\text{MSB} - \text{MSAB}}{an}, \quad \hat{\sigma}_\gamma^2 = \frac{\text{MSAB} - \text{MSE}}{n}, \quad \hat{\sigma}_e^2 = \text{MSE}.$$

ML solutions:

$$\dot{\sigma}_\beta^2 = \frac{(1 - 1/b)(\text{MSB} - \text{MSAB})}{an}, \quad \dot{\sigma}_\gamma^2 = \frac{(1 - 1/b)\text{MSAB} - \text{MSE}}{n}, \quad \dot{\sigma}_e^2 = \text{MSE}.$$

TABLE 4.13. ML ESTIMATORS OF σ_β^2, σ_γ^2 AND σ_e^2 IN THE 2-WAY CROSSED CLASSIFICATION, MIXED MODEL

Conditions satisfied by the ML solutions	MLE		
	$\tilde{\sigma}_\beta^2$	$\tilde{\sigma}_\gamma^2$	$\tilde{\sigma}_e^2$
$\dot{\sigma}_\beta^2 \geqslant 0, \dot{\sigma}_\gamma^2 \geqslant 0$	$\dot{\sigma}_\beta^2$	$\dot{\sigma}_\gamma^2$	$\dot{\sigma}_e^2$
$\dot{\sigma}_\beta^2 \geqslant 0, \dot{\sigma}_\gamma^2 < 0$	$\dfrac{1}{an}\left(\dfrac{\text{SSB}}{b} - \dfrac{\text{SSE} + \text{SSAB}}{abn - a - b + 1}\right)$	0	$\dfrac{\text{SSE} + \text{SSAB}}{abn - a - b + 1}$
$\dot{\sigma}_\beta^2 < 0, \dot{\sigma}_\gamma^2 \geqslant 0$	0	$\dfrac{\Sigma_i\Sigma_j n(\bar{y}_{ij.} - \bar{y}_{...})^2}{ab - 1}$	$\dot{\sigma}_e^2$
$\dot{\sigma}_\beta^2 < 0, \dot{\sigma}_\gamma^2 < 0$	0	0	$\dfrac{\text{SST}_m}{abn}$

-*iv*. ***The 2-way crossed, no interaction, mixed model.*** The model equation is

$$y_{ijk} = \mu + \alpha_i + \beta_j + e_{ijk};$$

$$i = 1,\ldots, a, \quad j = 1,\ldots, b \quad \text{and} \quad k = 1,\ldots, n; \quad \alpha_i\text{s fixed} .$$

TABLE 4.14. ANALYSIS OF VARIANCE FOR A 2-WAY CROSSED CLASSIFICATION, NO INTERACTION MODEL

Source	d.f.	Sum of Squares	Mean Square
A	$a - 1$	$\text{SSA} = \Sigma_i bn(\bar{y}_{i..} - \bar{y}_{...})^2$	$\text{MSA} = \text{SSA}/(a - 1)$
B	$b - 1$	$\text{SSB} = \Sigma_j an(\bar{y}_{.j.} - \bar{y}_{...})^2$	$\text{MSB} = \text{SSB}/(b - 1)$
Error	$abn - a - b + 1$	$\text{SSE} = \Sigma_i\Sigma_j\Sigma_k(y_{ijk} - \bar{y}_{i..} - \bar{y}_{.j.} + \bar{y}_{...})^2$	$\text{MSE} = \text{SSE}/(abn - a - b + 1)$
Total	$abn - 1$	$\text{SST}_m = \Sigma_i\Sigma_j\Sigma_k(y_{ijk} - \bar{y}_{...})^2$	

ANOVA estimators:

$$\hat{\sigma}_\beta^2 = \frac{\text{MSB} - \text{MSE}}{an} \quad \text{and} \quad \hat{\sigma}_e^2 = \text{MSE} .$$

ML solutions:

$$\dot{\sigma}_\beta^2 = \frac{\text{SSB}/b - \dot{\sigma}_e^2}{an} \quad \text{and} \quad \dot{\sigma}_e^2 = \left[1 - \frac{a - 1}{b(an - 1)}\right]\text{MSE} .$$

Notice that the form of the ML estimators in Table 4.15 is the same as that in Table 4.9 for the 1-way classification.

TABLE 4.15. ML ESTIMATORS OF σ_β^2 AND σ_e^2 IN A 2-WAY CROSSED CLASSIFICATION, MIXED MODEL, α-EFFECTS FIXED

Conditions satisfied by the ML solutions	MLE	
	$\tilde{\sigma}_\beta^2$	$\tilde{\sigma}_e^2$
$\dot{\sigma}_\beta^2 \geqslant 0$	$\dot{\sigma}_\beta^2$	$\dot{\sigma}_e^2$
$\dot{\sigma}_\beta^2 < 0$	0	$\dfrac{\text{SST}_{\text{m}}}{abn}$

c. Unbiasedness

Most ML estimators are biased, and so are many ML solutions if they are used as estimators. For instance, the example of Table 4.12 has $E(\dot{\sigma}_e^2) = E(\text{MSE}) = \sigma_e^2$, and so $\dot{\sigma}_e^2$ is unbiased. But

$$E(\dot{\sigma}_\gamma^2) = \frac{(1 - 1/b)E(\text{MSAB}) - E(\text{MSE})}{n}$$

$$= \frac{(1 - 1/b)(n\sigma_\gamma^2 + \sigma_e^2) - \sigma_e^2}{n} = \sigma_\gamma^2 - \frac{n\sigma_\gamma^2 + \sigma_e^2}{bn},$$

showing tha $\dot{\sigma}_\gamma^2$ is not unbiased for σ_γ^2. Second, the solutions of the ML equations are not, as has been emphasized, the ML estimators. The estimators are truncated versions of the solutions, as for example in Table 4.13; and this truncation further negates unbiasedness. For example, suppose $\sigma_\alpha^2 = 0$ in the 1-way classification. The ANOVA estimator $\hat{\sigma}_\alpha^2$ is unbiased yet it can be negative. And it will be negative often enough to balance out the occurrence of positive values to average zero. Deleting those negative values and substituting zero, to get the ML $\tilde{\sigma}_\alpha^2$ therefore gives $\tilde{\sigma}_\alpha^2$ as biased upwards. Moreover, in Table 4.13, the ML estimator of σ_e^2 is represented by the last column of the table, and the expected value of that estimator will involve the probabilities of solutions $\dot{\sigma}_\beta^2$ and $\dot{\sigma}_\gamma^2$ being negative either singly or together—just as such a probability is illustrated in Section 3.7a-iv.

Readers interested in the bias of the solutions of the ML equations for the four models of sub-section b will find details in Corbeil and Searle (1976b)—wherein the solutions are wrongly referred to as ML estimators!

d. The 2-way crossed classification, random model

Lest the reader be led astray by the preceding examples into thinking that ML estimators of variance components from balanced data are always in closed form (as in those examples), we now consider the 2-way crossed classification random model, for which the ML estimators from balanced data are not in closed form.

Replacing μ in L of (85) by $\tilde{\mu} = \bar{y}$ of (87) leads to what is called the profile likelihood, which can be used for deriving ML estimators of variance components of the random model. Using balanced data, it is

$$L = \frac{\exp[-\frac{1}{2}(\mathbf{y} - \bar{y}\mathbf{1})'\mathbf{V}^{-1}(\mathbf{y} - \bar{y}\mathbf{1})]}{(2\pi)^{N/2}|\mathbf{V}|^{\frac{1}{2}}}. \tag{88}$$

where $\mathbf{V}$ is the appropriate form of (80) and $\mathbf{V}^{-1}$ is available from (84). ML equations are obtained by equating to zero $\partial L/\partial\sigma^2$ for each σ^2 in the model. Applying this methodology to any particular random model involves nothing that is particularly difficult except that the whole process can be somewhat tedious (see E 4.11 for the 2-way nested classification). An alternative method of derivation, applicable to unbalanced data, but for which balanced data are a simplifying case, is given in Chapter 6, and even that simplification can be tedious. We therefore omit derivation of the ML equations that follow for the 2-way crossed classification, random model.

-i. ***With interaction.*** Details of this model are given in Section 4.3. For writing the ML equations we define the following linear combinations of the variance components:

$$\theta_0 = \sigma_e^2, \qquad \theta_{11} = \sigma_e^2 + n\sigma_\gamma^2 + bn\sigma_\alpha^2,$$
$$\theta_1 = \sigma_e^2 + n\sigma_\gamma^2, \quad \theta_{12} = \sigma_e^2 + n\sigma_\gamma^2 + an\sigma_\beta^2$$

and (89)

$$\theta_4 = \sigma_e^2 + n\sigma_\gamma^2 + bn\sigma_\alpha^2 + an\sigma_\beta^2 = \theta_{11} + \theta_{12} - \theta_1 .$$

Then, using the sums of squares defined in Table 4.3, the ML equations are

$$\begin{aligned}
&\frac{1}{\dot\theta_4} + \frac{a-1}{\dot\theta_{11}} + \frac{b-1}{\dot\theta_{12}} + \frac{(a-1)(b-1)}{\dot\theta_1} + \frac{ab(n-1)}{\dot\theta_0} = \frac{\text{SSA}}{\dot\theta_{11}^2} + \frac{\text{SSB}}{\dot\theta_{12}^2} + \frac{\text{SSAB}}{\dot\theta_1^2} + \frac{\text{SSE}}{\dot\theta_0^2}\\
&\frac{1}{\dot\theta_4} + \frac{a-1}{\dot\theta_{11}} = \frac{\text{SSA}}{\dot\theta_{11}^2}\\
&\frac{1}{\dot\theta_4} + \frac{b-1}{\dot\theta_{12}} = \frac{\text{SSB}}{\dot\theta_{12}^2}\\
&\frac{1}{\dot\theta_4} + \frac{a-1}{\dot\theta_{11}} + \frac{b-1}{\dot\theta_{12}} + \frac{(a-1)(b-1)}{\dot\theta_1} = \frac{\text{SSA}}{\dot\theta_{11}^2} + \frac{\text{SSB}}{\dot\theta_{12}^2} + \frac{\text{SSAB}}{\dot\theta_1^2}.
\end{aligned} \tag{90}$$

Despite the tantalizingly apparent simplicity of these equations [which were first derived by Miller (1977)], they are in fact nonlinear in the $\dot\theta$s (even after using $\theta_4 \equiv \theta_{11} + \theta_{12} - \theta_1$) except, through subtracting the last equation from the first, for $\dot\sigma_e^2 = \text{MSE}$. Otherwise, there is no closed form of solution for the $\dot\sigma^2$s. It has to be found, for each particular set of data, by using numerical methods. Since this is also the manner in which ML estimates have to be calculated from unbalanced data, discussion of such techniques is left until Chapter 8.

A re-writing of (90) that might be useful for iterative purposes is the following. First, requiring no iteration, subtracting the last equation of (90) from the first gives

$$\dot{\theta}_0 = \frac{\text{SSE}}{ab(n-1)} = \text{MSE} .$$

Then the second and third equations can be written as

$$\dot{\theta}_{11} = \text{MSA} - \frac{\dot{\theta}_{11}^2}{(a-1)(\dot{\theta}_{11} + \dot{\theta}_{12} - \dot{\theta}_1)},$$

$$\dot{\theta}_{12} = \text{MSB} - \frac{\dot{\theta}_{12}^2}{(b-1)(\dot{\theta}_{11} + \dot{\theta}_{12} - \dot{\theta}_1)} .$$

And subtracting the second and third equation from the fourth gives

$$\dot{\theta}_1 = \text{MSAB} + \frac{\dot{\theta}_1^2}{(a-1)(b-1)(\dot{\theta}_{11} + \dot{\theta}_{12} - \dot{\theta}_1)} .$$

These last three equations are clearly amenable to iteration.

-ii. No interaction. If the no-interaction form of the model is used, its ML equations are derived from (89) and (90) by putting $\sigma_\gamma^2 = 0$ and combining SSAB and SSE, and omitting the last equation of (90). Thus with

$$\theta_1 = \theta_0 = \sigma_e^2, \quad \theta_{11} = \sigma_e^2 + bn\sigma_\alpha^2, \quad \theta_{12} = \sigma_e^2 + an\sigma_\beta^2 \tag{91}$$

and

$$\theta_4 = \sigma_e^2 + an\sigma_\beta^2 + bn\sigma_\alpha^2 \equiv \theta_{11} + \theta_{12} - \theta_0$$

the ML equations are

$$\begin{aligned} \frac{1}{\dot{\theta}_4} + \frac{a-1}{\dot{\theta}_{11}} + \frac{b-1}{\dot{\theta}_{12}} + \frac{abn-a-b+1}{\dot{\theta}_0} &= \frac{\text{SSA}}{\dot{\theta}_{11}^2} + \frac{\text{SSB}}{\dot{\theta}_{12}^2} + \frac{\text{SSAB} + \text{SSE}}{\dot{\theta}_0^2}, \\ \frac{1}{\dot{\theta}_4} + \frac{a-1}{\dot{\theta}_{11}} &= \frac{\text{SSA}}{\dot{\theta}_{11}^2}, \\ \frac{1}{\dot{\theta}_4} + \frac{b-1}{\dot{\theta}_{12}} &= \frac{\text{SSB}}{\dot{\theta}_{12}^2} . \end{aligned} \tag{92}$$

These two are nonlinear in the $\dot{\theta}$s and have no closed form solution for them or the $\dot{\sigma}^2$s.

e. Existence of explicit solutions

The absence of explicit, closed form solutions to the ML equations in the preceding example may come as a surprise, considering that it is for balanced data, for which one so often has a strong intuitive feeling along the lines that "everything is straightforward". But this is not so. Indeed Szatrowski and Miller

(1980) have a theorem that tells us when explicit ML solutions exist for balanced data from a mixed model.

Suppose a mixed model has r random factors (excluding the error terms); and denote the number of different symbols used as subscripts in the model equation by s. Define $\mathbf{t}'_p$ as a row vector of order s that is null except for unity as its qth element when the pth random factor of the model has the qth subscript; and define $\mathbf{t}_0 = \mathbf{1}_s$. Let $\mathbf{T}_{(r+1)\times s} = [\mathbf{t}_0 \quad \cdots \quad \mathbf{t}_r]'$. Let $\mathbf{w}_1, \ldots, \mathbf{w}_s$ be the columns of $\mathbf{T}$ and $\mathbf{w}_0 = \mathbf{1}$.

Recall $\mathbf{a} * \mathbf{b}$ as the Hadamard product of two vectors $\mathbf{a}$ and $\mathbf{b}$, it being a vector having elements $a_i b_i$. Define $\mathscr{W}$ as the smallest set containing $\mathbf{w}_0, \mathbf{w}_1, \ldots, \mathbf{w}_s$ closed under Hadamard multiplication of vectors. Let $n(\mathscr{W})$ be the number of distinct columns in $\mathscr{W}$; distinct, not necessarily linearly independent.

Theorem. The model has explicit ML solutions if and only if $n(\mathscr{W}) = r + 1$.

Example 1. The 2-way crossed classification, with interaction, random model has equation $y_{ijk} = \mu + \alpha_i + \beta_j + \gamma_{ij} + e_{ijk}$ with $r = 3$ and $s = 3$.

$$\mathbf{T} = \begin{bmatrix} 1 & 1 & 1 \\ 1 & 0 & 0 \\ 0 & 1 & 0 \\ 1 & 1 & 0 \end{bmatrix}, \quad \mathscr{W} = \begin{bmatrix} 1 & 1 & 1 & 1 & 1 \\ 1 & 1 & 0 & 0 & 0 \\ 1 & 0 & 1 & 0 & 0 \\ 1 & 1 & 1 & 0 & 1 \end{bmatrix} \Rightarrow n(\mathscr{W}) = 5 > 4 = 3 + 1 = r + 1\,.$$

Therefore this model has, as we have seen in (90), no explicit ML solutions.

Example 2. The 2-way crossed classification, with interaction, mixed model, with αs fixed has $r = 2$ and $s = 3$.

$$\mathbf{T} = \begin{bmatrix} 1 & 1 & 1 \\ 0 & 1 & 0 \\ 1 & 1 & 0 \end{bmatrix}, \quad \mathscr{W} = \begin{bmatrix} 1 & 1 & 1 \\ 1 & 0 & 0 \\ 1 & 1 & 0 \end{bmatrix} \Rightarrow n(\mathscr{W}) = 3 = 2 + 1 = r + 1\,.$$

Therefore this model has explicit ML solutions, as in Table 4.15.

f. Asymptotic sampling variances for the 2-way crossed classification

The general theory of maximum likelihood estimation has it that the large-sample dispersion matrix of a vector of ML estimators is the inverse of the matrix whose elements are minus the expected value of the second derivatives of the likelihood function. Details of this are shown in Chapter 6 in terms of the matrix formulation of a linear model

$$\mathbf{y} = \mathbf{X}\boldsymbol{\beta} + \sum_{i=1}^{r} \mathbf{Z}_i \mathbf{u}_i + \mathbf{e}$$

of (68) with $\mathbf{V}$ of (69). The result is that

$$\text{var}(\tilde{\sigma}^2) \simeq 2[\{_m\, \text{tr}(\mathbf{V}^{-1}\mathbf{Z}_i\mathbf{Z}'_i\mathbf{V}^{-1}\mathbf{Z}_j\mathbf{Z}'_j)\}_{i,j=0}^{r}]^{-1}, \tag{93}$$

consequences and implications of which are dealt with in Chapter 6. It suffices to say here that (93) can be used for the 2-way crossed classification, with interaction, random model of the preceding subsection, c-i. That in turn can be used for the no-interaction case of d-ii, and also for all of the four cases of the preceding subsection b, each of which can be treated as a special case of c-i. We deal with these six cases.

-i. The 2-way crossed classification, with interaction, random model. The result of applying (93) to the 2-way crossed classification, with interaction, random model is that

$$\operatorname{var}\begin{bmatrix}\tilde{\sigma}_e^2\\ \tilde{\sigma}_\alpha^2\\ \tilde{\sigma}_\beta^2\\ \tilde{\sigma}_\gamma^2\end{bmatrix} \simeq 2\begin{bmatrix} t_{ee} & t_{\alpha\alpha}/bn & t_{\beta\beta}/an & t_{\gamma\gamma}/n\\ & t_{\alpha\alpha} & abn^2/\theta_4^2 & t_{\alpha\alpha}/b\\ & \text{symmetric} & t_{\beta\beta} & t_{\beta\beta}/a\\ & & & t_{\gamma\gamma}\end{bmatrix}^{-1} \tag{94}$$

for

$$t_{ee} = \frac{ab(n-1)}{\theta_0^2} + \frac{t_{\gamma\gamma}}{n^2}, \quad t_{\alpha\alpha} = b^2n^2\left(\frac{a-1}{\theta_{11}^2} + \frac{1}{\theta_4^2}\right),$$

$$t_{\beta\beta} = a^2n^2\left(\frac{b-1}{\theta_{12}^2} + \frac{1}{\theta_4^2}\right) \quad \text{and} \quad t_{\gamma\gamma} = n^2\left[\frac{(a-1)(b-1)}{\theta_1^2} + \frac{a-1}{\theta_{11}^2} + \frac{b-1}{\theta_{12}^2} + \frac{1}{\theta_4^2}\right], \tag{95}$$

with the θs of (89). Note that (94) is an asymptotic equality, because it is a large-sample result.

Note: The occurrence of "symmetric" (or "sym") in a matrix as in (94) indicates that the matrix is symmetric.

-ii. The 2-way crossed classification, no interaction, random model. For this model put $\sigma_\gamma^2 = 0$ in (95) and delete the last row and column of (94). This gives for the θs of (91), with

$$\phi = \frac{abn - a - b + 1}{\theta_0^2} + \frac{a-1}{\theta_{11}^2} + \frac{b-1}{\theta_{12}^2} + \frac{1}{\theta_4^2}$$

$$\operatorname{var}\begin{bmatrix}\tilde{\sigma}_e^2\\ \tilde{\sigma}_\alpha^2\\ \tilde{\sigma}_\beta^2\end{bmatrix} \simeq 2\begin{bmatrix} \phi & bn\left(\frac{a-1}{\theta_{11}^2} + \frac{1}{\theta_4^2}\right) & an\left(\frac{b-1}{\theta_{12}^2} + \frac{1}{\theta_4^2}\right)\\ & b^2n^2\left(\frac{a-1}{\theta_{11}^2} + \frac{1}{\theta_4^2}\right) & \frac{abn^2}{\theta_4^2}\\ \text{symmetric} & & a^2n^2\left(\frac{b-1}{\theta_{12}^2} + \frac{1}{\theta_4^2}\right)\end{bmatrix}^{-1} \tag{96}$$

We can now use (94) and (95) to derive large-sample dispersion matrices for the four cases in sub-section b—the cases where the ML solutions are in closed form. In each of those cases the solutions are linear functions of independently χ^2-distributed mean squares and so sampling variances and covariances of those solutions could easily be found, as was done in Section 4.5f for ANOVA estimators. This was also done for the 1-way classification in (123) of Chapter 3. And at (127) of that chapter an expression for $\text{var}(\tilde{\sigma}_e^2)$ was obtained taking into account the probability of $\dot{\sigma}_\alpha^2$ being negative. Neither of these derivations are being made here. What is being derived is large-sample variances and covariances, akin to (126) of Chapter 3. Indeed, we show in detail how that result is derived from (96).

-iii. The 2-way crossed, with interaction, mixed model. With αs taken as fixed effects all we need to do is use $\sigma_\alpha^2 = 0$ in the θs of (89) and delete the second row and column of (94). This gives, with

$$\theta_0 = \sigma_e^2, \quad \theta_1 = \sigma_e^2 + n\sigma_\gamma^2 = \theta_{11},$$
$$\theta_{12} = \sigma_e^2 + n\sigma_\gamma^2 + an\sigma_\beta^2 = \theta_4, \tag{97}$$

$$t_{ee} = \frac{ab(n-1)}{\theta_0^2} + \frac{t_{\gamma\gamma}}{n^2}, \quad t_{\beta\beta} = \frac{a^2bn^2}{\theta_{12}^2} \quad \text{and} \quad t_{\gamma\gamma} = bn^2\left(\frac{a-1}{\theta_1^2} + \frac{1}{\theta_{12}^2}\right) \tag{98}$$

$$\text{var}\begin{bmatrix} \tilde{\sigma}_e^2 \\ \tilde{\sigma}_\beta^2 \\ \tilde{\sigma}_\gamma^2 \end{bmatrix} \simeq 2\begin{bmatrix} t_{ee} & t_{\beta\beta}/an & t_{\gamma\gamma}/n \\ & t_{\beta\beta} & t_{\beta\beta}/a \\ \text{symmetric} & & t_{\gamma\gamma} \end{bmatrix}^{-1}$$

$$= \frac{2}{b}\begin{bmatrix} \dfrac{a(n-1)}{\theta_0^2} + w & \dfrac{an}{\theta_{12}^2} & nw \\ & \dfrac{a^2n^2}{\theta_{12}^2} & \dfrac{an^2}{\theta_{12}^2} \\ \text{symmetric} & & n^2w \end{bmatrix}^{-1}, \quad \text{for } w = \frac{a-1}{\theta_1^2} + \frac{1}{\theta_{12}^2} \tag{99}$$

$$= \frac{2}{b}\begin{bmatrix} \dfrac{\sigma_e^4}{a(n-1)} & 0 & \dfrac{-\sigma_e^4}{an(n-1)} \\ & \dfrac{(\sigma_e^2 + n\sigma_\gamma^2)^2/(a-1) + \theta_{12}^2}{a^2n^2} & \dfrac{-(\sigma_e^2 + n\sigma_\gamma^2)^2}{an^2(a-1)} \\ \text{symmetric} & & \dfrac{1}{n^2}\left[\dfrac{(\sigma_e^2 + n\sigma_\gamma^2)^2}{a-1} + \dfrac{\sigma_e^4}{a(n-1)}\right] \end{bmatrix}. \tag{100}$$

-iv. The 2-way crossed, no interaction, mixed model. In taking the αs as fixed effects we adapt the preceding case to have no interactions by putting

$\sigma_\gamma^2 = 0$. This gives (97) and (98) as

$$\theta_0 = \sigma_e^2 = \theta_1 = \theta_{11} \quad \text{and} \quad \theta_{12} = \sigma_e^2 + an\sigma_\beta^2 = \theta_4$$

and

$$t_{\gamma\gamma} = bn^2\left[\frac{a-1}{\sigma_e^4} + \frac{1}{(\sigma_e^2 + an\sigma_\beta^2)^2}\right], \tag{101}$$

so that

$$t_{ee} = \frac{b(an-1)}{\sigma_e^4} + \frac{b}{(\sigma_e^2 + an\sigma_\beta^2)^2} \quad \text{and} \quad t_{\beta\beta} = \frac{a^2bn^2}{(\sigma_e^2 + an\sigma_\beta^2)^2}.$$

This leads to (99), after deleting its last row and column, being

$$\text{var}\begin{bmatrix} \tilde{\sigma}_e^2 \\ \tilde{\sigma}_\beta^2 \end{bmatrix} \simeq \frac{2}{b}\begin{bmatrix} \dfrac{an-1}{\sigma_e^4} + \dfrac{1}{(\sigma_e^2 + an\sigma_\beta^2)^2} & \dfrac{an}{(\sigma_e^2 + an\sigma_\beta^2)^2} \\ \dfrac{an}{(\sigma_e^2 + an\sigma_\beta^2)^2} & \dfrac{a^2n^2}{(\sigma_e^2 + an\sigma_\beta^2)^2} \end{bmatrix}^{-1} \tag{102}$$

$$= \frac{2\sigma_e^4}{b(an-1)}\begin{bmatrix} 1 & \dfrac{-1}{an} \\ \dfrac{-1}{an} & \dfrac{1 + (an-1)(1 + an\sigma_\beta^2/\sigma_e^2)^2}{a^2n^2} \end{bmatrix}. \tag{103}$$

Note that although putting $\sigma_\gamma^2 = 0$ in (99) and deleting its last row and column yields (102), the same operations on (100) do not yield (103)—as neither they should. (The inverse of a submatrix is not necessarily part of the inverse of the matrix of which it is a submatrix.)

g. Asymptotic sampling variances for two other models

-i. The 2-way nested classification, random model. Taking the model equation as $y_{ijk} = \mu + \alpha_i + \beta_{ij} + e_{ijk}$ of sub-section 4b-ii, we can derive its information matrix from (94) by putting $\sigma_\beta^2 = 0$ in (89), changing σ_γ^2 to σ_β^2 and deleting the third row and column of (94). The changes in (89) and (95) lead to

$$\theta_0 = \sigma_e^2, \quad \theta_1 = \sigma_e^2 + n\sigma_\beta^2 = \theta_{12} \quad \text{and} \quad \theta_4 = \sigma_e^2 + n\sigma_\beta^2 + bn\sigma_\alpha^2 = \theta_{11}$$

with (104)

$$t_{ee} = \frac{ab(n-1)}{\theta_0^2} + \frac{t_{\beta\beta}}{n^2}, \quad t_{\alpha\alpha} = \frac{ab^2n^2}{\theta_{11}^2} \quad \text{and} \quad t_{\beta\beta} = an^2\left(\frac{b-1}{\theta_1^2} + \frac{1}{\theta_{22}^2}\right).$$

These ts are exactly the same as in (98) except for notation changes: $t_{\alpha\alpha}$ instead of $t_{\beta\beta}$, $t_{\beta\beta}$ in place of $t_{\alpha\alpha}$, σ_β^2 in place of σ_γ^2, σ_α^2 in place of σ_β^2, a and b interchanged, and θ_{11} in place of θ_{12}. Therefore the dispersion matrix will be the same as

(100), only with these same changes. Thus, with

$$\delta = \frac{(\sigma_e^2 + n\sigma_\beta^2)^2}{b-1},$$

$$\operatorname{var}\begin{bmatrix} \tilde{\sigma}_e^2 \\ \tilde{\sigma}_\alpha^2 \\ \tilde{\sigma}_\beta^2 \end{bmatrix} \simeq \frac{2}{a}\begin{bmatrix} \dfrac{\sigma_e^4}{b(n-1)} & 0 & \dfrac{-\sigma_e^4}{bn(n-1)} \\ & \dfrac{\delta + \theta_{11}^2}{b^2n^2} & \dfrac{-\delta}{bn^2} \\ \text{symmetric} & & \dfrac{1}{n^2}\left[\delta + \dfrac{\sigma_e^4}{b(n-1)}\right] \end{bmatrix}. \tag{105}$$

-ii. ***The 1-way classification, random model.*** The 2-way crossed classification, no interaction, random model can be converted to the 1-way model equation by dropping β_{ij} and effectively putting $\sigma_\beta^2 = 0$ and $b = 1$. Doing this in (91) gives

$$\theta_0 = \sigma_e^2 = \theta_1 = \theta_{12} \quad \text{and} \quad \theta_{11} = \sigma_e^2 + n\sigma_\alpha^2 = \theta_4 \,.$$

Making these changes in (96), along with dropping its last row and column, gives

$$\operatorname{var}\begin{bmatrix} \tilde{\sigma}_e^2 \\ \tilde{\sigma}_\alpha^2 \end{bmatrix} \simeq 2\begin{bmatrix} \dfrac{a(n-1)}{\sigma_e^4} + \dfrac{a}{(\sigma_e^2 + n\sigma_\alpha^2)^2} & \dfrac{an}{(\sigma_e^2 + n\sigma_\alpha^2)^2} \\ \dfrac{an}{(\sigma_e^2 + n\sigma_\alpha^2)^2} & \dfrac{an^2}{(\sigma_e^2 + n\sigma_\alpha^2)^2} \end{bmatrix}^{-1}$$

$$= \frac{2\sigma_e^4}{a(n-1)}\begin{bmatrix} 1 & \dfrac{-1}{n} \\ \dfrac{-1}{n} & \dfrac{1 + (n-1)(1 + n\sigma_\alpha^2/\sigma_e^2)^2}{n^2} \end{bmatrix}, \tag{106}$$

which is, of course, the same as (126) of Chapter 3.

h. Locating results

The sequence chosen for presenting the preceding results was governed by ease of derivation. As a consequence, locating results for each of the six cases presented might be found a little confusing. Table 4.16 should ease this confusion.

4.8. RESTRICTED MAXIMUM LIKELIHOOD (REML)

The general concept of REML estimation is introduced in Section 3.8. Details of REML applicable to unbalanced data are given in Chapter 6. Moreover, in Chapter 11 it is shown that when we seek minimum variance quadratic unbiased

TABLE 4.16. SUBSECTIONS CONTAINING RESULTS FOR ML ESTIMATION

Classification and Model	Subsection: Estimators	Asymptotic Dispersion Matrix
1-way, random	b-i	g-ii
2-way nested, random	b-ii	g-i
2-way crossed		
with interaction, mixed	b-iii	f-iii
no interaction, mixed	b-iv	f-iv
with interaction, random	d-i	f-i
no interaction, random	d-ii	f-ii

estimators of variance components for unbalanced data generally, under the usual normality assumptions, we arrive at the same equations as are used for REML. But since, for balanced data, we already know that ANOVA estimators under normality assumptions are minimum variance unbiased, and are quadratic, we therefore have the result for balanced data that solutions of the REML equations are the same as ANOVA estimators, i.e.,

for balanced data: REML solutions = ANOVA estimators .

Other derivations of this result are available in Anderson (1979b) and Pukelsheim and Styan (1979), who add the telling phrase that this result "need not be checked explicitly" (as they do for unbalanced data—see Section 6.7). We therefore say no more about REML solutions for balanced data.

REML estimators are obtained from REML solutions by applying the same procedures to ensure non-negativity requirements as is done with deriving ML estimators from ML solutions. This has been illustrated in Tables 4.9, 4.11, 4.13 and 4.15.

4.9. ESTIMATING FIXED EFFECTS IN MIXED MODELS

In random models the only fixed effect is μ, and its MLE has been shown in (87) to be $\tilde{\mu} = \bar{y}$. Its sampling variance is $\text{var}(\tilde{\mu}) = \mathbf{1}'\mathbf{V}\mathbf{1}/N^2$, using $\text{var}(\mathbf{y}) = \mathbf{V}$ as in Section 4.6a. In that same section the fixed effects of the mixed model equation $\mathbf{y} = \mathbf{X}\boldsymbol{\beta} + \mathbf{Z}\mathbf{u} + \mathbf{e}$ are $\boldsymbol{\beta}$ with the elements of $\mathbf{X}\boldsymbol{\beta}$ always being estimable functions of elements of $\boldsymbol{\beta}$. The ordinary least squares estimator of $\boldsymbol{\beta}$ (see Appendix S.1) is

$$\text{OLSE}(\mathbf{X}\boldsymbol{\beta}) = \mathbf{X}(\mathbf{X}'\mathbf{X})^{-}\mathbf{X}'\mathbf{y} . \tag{107}$$

This, it will be noticed, does not involve $\mathbf{V}$, and so $\text{OLSE}(\mathbf{X}\boldsymbol{\beta})$ can be derived without having to estimate variance components. Moreover, for balanced data, $\mathbf{X}$ can always be written as a partitioned matrix of submatrices that are

Kronecker products of **I**-matrices and **1**-vectors. For example, in the 2-way crossed classification of (73), were $\boldsymbol{\alpha}$ to be taken as fixed (along with μ) then **X** would be

$$\mathbf{X} = [\mathbf{1}_a \otimes \mathbf{1}_b \otimes \mathbf{1}_n \qquad \mathbf{I}_a \otimes \mathbf{1}_b \otimes \mathbf{1}_n] \,. \tag{108}$$

A consequence of this is (e.g., Searle, 1988b) that OLSE($\mathbf{X}\boldsymbol{\beta}$) of (107) always has a simple form. For example, **X** of (108) reduces (107) to be the familiar

$$\text{OLSE}(\mu + \alpha_i) = \bar{y}_{i\cdot\cdot} \,. \tag{109}$$

This style of result is true for all cases of balanced data from mixed models: OLS estimators of estimable functions of fixed effects are based on cell means and factor level means, the kind of result that one sees in standard analyses of designed experiments.

The variance components are taken into account when estimating fixed effects by utilizing **V** as in

$$\text{GLSE}(\mathbf{X}\boldsymbol{\beta}) = \mathbf{X}(\mathbf{X}'\mathbf{V}^{-1}\mathbf{X})^{-}\mathbf{X}'\mathbf{V}^{-1}\mathbf{y}, \tag{110}$$

as outlined in Appendix S.1. But for balanced data **V** is a linear combination of Kronecker products of **I**- and **J**-matrices, as exemplified in (74). Indeed, for the 2-way crossed classification with αs fixed, for which (108) is the **X**-matrix, **V** is (74) without the σ_α^2-term:

$$\mathbf{V} = (\mathbf{J}_a \otimes \mathbf{I}_b \otimes \mathbf{J}_n)\sigma_\beta^2 + (\mathbf{I}_a \otimes \mathbf{I}_b \otimes \mathbf{J}_n)\sigma_\gamma^2 + (\mathbf{I}_a \otimes \mathbf{I}_b \otimes \mathbf{I}_n)\sigma_e^2 \,. \tag{111}$$

Using this and **X** of (108) in (110), it will be found (E 4.23) that (110) reduces to

$$\text{GLSE}(\mu + \alpha_i) = \bar{y}_{i\cdot\cdot}, \tag{112}$$

the same as OLSE of (109). Thus in this case

$$\text{GLSE}(\mathbf{X}\boldsymbol{\beta}) = \text{OLSE}(\mathbf{X}\boldsymbol{\beta}) \,. \tag{113}$$

However, this is a result that is true for all cases of balanced data from any customary mixed model (e.g., Searle, 1988b) excluding the use of covariates, which effectively causes data to be unbalanced. Result (113) is true, as shown by Zyskind (1969), if and only if there is some matrix **Q** for which $\mathbf{VX} = \mathbf{XQ}$. For unbalanced data this is established using (70) developed in Chapter 12, namely that there does exist a $\mathbf{Q}_i$ such that $\mathbf{Z}_i\mathbf{Z}_i'\mathbf{X} = \mathbf{X}\mathbf{Q}_i$. Then it follows for $\mathbf{V} = \Sigma_{i=0}^r \sigma_i^2 \mathbf{Z}_i\mathbf{Z}_i'$ of (71) in this chapter that

$$\mathbf{VX} = \sum_{i=0}^{r} \sigma_i^2 \mathbf{Z}_i\mathbf{Z}_i'\mathbf{X} = \mathbf{X} \sum_{i=0}^{r} \sigma_i^2 \mathbf{Q}_i = \mathbf{XQ}$$

for $\mathbf{Q} = \Sigma_{i=0}^r \sigma_i^2 \mathbf{Q}_i$.

Maximum likelihood estimation (under normality) of fixed effects in a mixed model is simple for balanced data. This is so because the ML equation for $\mathbf{X}\tilde{\boldsymbol{\beta}}$ is

$$\mathbf{X}'\tilde{\mathbf{V}}^{-1}\mathbf{X}\tilde{\boldsymbol{\beta}} = \mathbf{X}'\tilde{\mathbf{V}}^{-1}\mathbf{y},$$

which leads to

$$\mathbf{X}\tilde{\boldsymbol{\beta}} = \mathbf{X}(\mathbf{X}'\tilde{\mathbf{V}}^{-1}\mathbf{X})^{-}\mathbf{X}'\tilde{\mathbf{V}}^{-1}\mathbf{y}, \tag{114}$$

$\tilde{\mathbf{V}}^{-1}$ being the ML estimator of $\mathbf{V}$, namely $\mathbf{V}$ with each σ^2 replaced by its corresponding $\tilde{\sigma}^2$. But since, for balanced data, $\mathbf{X}\tilde{\boldsymbol{\beta}}$ of (114) is the same as GLSE($\mathbf{X}\boldsymbol{\beta}$) of (110), only with $\tilde{\mathbf{V}}$ in place of $\mathbf{V}$, and from (113) we have GLSE($\mathbf{X}\boldsymbol{\beta}$) = OLSE($\mathbf{X}\boldsymbol{\beta}$), which involves no variance components, so likewise for $\mathbf{X}\tilde{\boldsymbol{\beta}}$. Thus, for balanced data

$$\text{OLSE}(\mathbf{X}\boldsymbol{\beta}) = \text{GLSE}(\mathbf{X}\boldsymbol{\beta}) = \text{MLE}(\mathbf{X}\boldsymbol{\beta}) = \text{BLUE}(\mathbf{X}\boldsymbol{\beta}), \tag{115}$$

where the last of these four is the best linear unbiased estimator.

Insofar as REML estimation is concerned it is an estimation method applicable only to variance components and it gives no direction whatever on how to estimate $\mathbf{X}\boldsymbol{\beta}$. But in view of (115) this is of no concern.

4.10. SUMMARY

Establishing analysis of variance tables: Section 4.1

Lines in the table:	Rule 1
Interactions:	Rules 2, 3 and 4
Degrees of freedom:	Rules 5 and 6
Sums of squares:	Rules 7 and 8
Calculating sums of squares:	Rule 9
Expected mean squares: Section 4.2	Rules 10, 11, 12 and 13

The 2-way crossed classification: Section 4.3

Sums of squares and mean squares:	Table 4.3
Expected mean squares:	
Fixed effects model:	Table 4.4
Random effects model:	Table 4.5
Mixed model:	Table 4.6
Mixed model with Σ-restrictions:	Table 4.7

ANOVA estimators of variance components: Section 4.3d

ANOVA estimation and negative estimates: Section 4.4

$$E(\mathbf{m}) = \mathbf{P}\boldsymbol{\sigma}^2; \tag{29}$$

$$\hat{\boldsymbol{\sigma}}^2 = \mathbf{P}^{-1}\mathbf{m}\,. \tag{30}$$

Normality assumptions: Section 4.5

$$\frac{f_i M_i}{E(M_i)} \sim \chi^2_{f_i}; \tag{34}$$

$$\hat{\boldsymbol{\sigma}}^2 \sim \mathbf{P}^{-1}\{_{\text{d}}\, \chi^2_{f_i}/f_i\}\mathbf{P}\boldsymbol{\sigma}^2. \tag{38}$$

Hypothesis testing: Section 4.5c
Satterthwaite procedures (42), (43)
Confidence intervals: Section 4.5d
Probability of a negative estimate: Section 4.5e
Sampling variances: Section 4.5f

$$\text{var}(\hat{\boldsymbol{\sigma}}^2) = \mathbf{P}^{-1}\left\{_{\text{d}}\ \frac{2[E(M_i)]^2}{f_i}\right\}\mathbf{P}^{-1'}, \tag{55}$$

$$\hat{\text{var}}(\hat{\boldsymbol{\sigma}}^2) = \mathbf{P}^{-1}\left\{_{\text{d}}\ \frac{2M_i^2}{f_i + 2}\right\}\mathbf{P}^{-1'}. \tag{56}$$

Matrix formulation: Section 4.6

$$\mathbf{y} = \mathbf{X}\boldsymbol{\beta} + \mathbf{Z}\mathbf{u} + \mathbf{e} = \mathbf{X}\boldsymbol{\beta} + \sum_{i=1}^{r} \mathbf{Z}_i\mathbf{u}_i + \mathbf{e}; \qquad (58), (68)$$

$$E(\mathbf{y}) = \mathbf{X}\boldsymbol{\beta}, \quad E(\mathbf{e}) = \mathbf{0}, \quad \text{var}(\mathbf{e}) = \sigma_e^2\mathbf{I}_N; \qquad (62), (63)$$

$$\mathbf{D} = \text{var}(\mathbf{u}) = \{_{\text{d}}\ \text{var}(\mathbf{u}_i)\} = \{_{\text{d}}\ \sigma_i^2\mathbf{I}_{q_i}\}; \qquad (64), (67)$$

$$\mathbf{V} = \text{var}(\mathbf{y}) = \mathbf{Z}\mathbf{D}\mathbf{Z}' + \sigma_e^2\mathbf{I} = \sum_{i=1}^{r} \mathbf{Z}_i\mathbf{Z}_i'\sigma_i^2 + \sigma_e^2\mathbf{I}; \tag{69}$$

$$\mathbf{u}_0 \equiv \mathbf{e}, \quad \mathbf{Z}_0 \equiv \mathbf{I}_N, \quad \sigma_0^2 \equiv \sigma_e^2;$$

$$\mathbf{y} = \mathbf{X}\boldsymbol{\beta} + \sum_{i=0}^{r} \mathbf{Z}_i\mathbf{u}_i, \quad \mathbf{V} = \sum_{i=0}^{r} \mathbf{Z}_i\mathbf{Z}_i'\sigma_i^2. \qquad (70), (71)$$

Kronecker product notation: Sections 4.6b, c, d, and e

$$\text{e.g.,} \quad (\mathbf{I}_a \otimes \mathbf{1}_b \otimes \mathbf{1}_n)\boldsymbol{\alpha}. \tag{73}$$

Maximum likelihood: Section 4.7
Estimating the mean in random models: Section 4.7a

$$\hat{\mu} = \bar{y}. \tag{87}$$

Closed-form estimators: Section 4.7b
Classification and model

(b-i)	1-way random	Tables 4.8 and 4.9
(b-ii)	2-way nested random	Tables 4.10 and 4.11
(b-iii)	2-way crossed classification mixed	Tables 4.12 and 4.13
(b-iv)	2-way crossed, no interaction, mixed	Tables 4.14 and 4.15

No closed-form estimators: Section 4.7d
The 2-way crossed classification, random model:
(d-i) With interaction (90)
(d-ii) No interaction (92)

Existence of explicit solutions: Section 4.7e
Asymptotic samp!' ıg variances
The 2-way crossed classification: Section 4.7f
Model
(f-i) With interaction, random (94)
(f-ii) No interaction, random (96)
(f-iii) With interaction, mixed (100)
(f-iv) No interaction, mixed (103)

Two other models
(g-i) The 2-way nested random (105)
(g-ii) The 1-way random (106)

Note: The preceding sub-sections pair up as follows: Table 4.16

b-i and g-ii; b-ii and g-i;
b-iii and f-iii; b-iv and f-iv;
d-i and f-i; d-ii and f-ii .

Restricted maximum likelihood: Section 4.8

Estimating fixed effects: Section 4.9

$$\text{OLSE}(\mathbf{X}\boldsymbol{\beta}) = \mathbf{X}(\mathbf{X}'\mathbf{X})^{-}\mathbf{X}'\mathbf{y}, \tag{107}$$

$$\text{GLSE}(\mathbf{X}\boldsymbol{\beta}) = \mathbf{X}(\mathbf{X}'\mathbf{V}^{-1}\mathbf{X})^{-}\mathbf{X}'\mathbf{V}^{-1}\mathbf{y}. \tag{110}$$

4.11. EXERCISES

E 4.1. Suppose you have balanced data from a model having factors *A*, *B*, *C* within *AB*-subclasses, and *D* within *C*. Set up the analysis of variance table, and give expected values of mean squares for (i) the random model, (ii) the mixed model when *A* is a fixed effects factor and (iii) the mixed model when both *A* and *B* are fixed effects factors.

E 4.2. Repeat E 4.1 for a model having factors *A*, *B*, *D*, and *C* within *AB*.

E 4.3. A split plot experiment, whose main plots form a randomized complete blocks design, can be analyzed with the model equation

$$y_{ijk} = \mu + \alpha_i + \rho_j + \delta_{ij} + \beta_k + \theta_{jk} + e_{ijk},$$

where α_i represents a treatment effect, ρ_j is a block effect and β_k is the effect due to the kth sub-plot treatment. Set up the analysis of variance table, and give expected values of mean squares for the following cases:

(a) random model;

(b) mixed model, ρs and δs random;

(c) mixed model, only the βs fixed;

(d) mixed model, only the αs fixed.

E 4.4. (a) From (4) derive (6) and (7).

(b) Derive (8).

(c) Using (11), show that $E(\text{MSB})$ of Table 4.4 is $an\Sigma_j\dot{\beta}_j^2/(b-1)$.

E 4.5. Use (13), (14) and (15) to derive Table 4.5 from (8).

E 4.6. In the context in which (23) is used, show that it leads to (18)–(21) being satisfied; and to

$$\sigma_\gamma^2 = \frac{\sigma_{\gamma'}^2}{1-1/a}, \quad \text{cov}(\beta'_j, \gamma'_{ik}) = 0 \quad \text{and} \quad \text{cov}(\gamma'_{ij}, \gamma'_{ij'}) = 0 \quad \text{for } j \neq j' .$$

E 4.7. Calculate ANOVA estimates of variance components from the following data of 3 rows, 4 columns and 2 observations per cell. Use

(a) the random model;

(b) the mixed model with rows fixed;

(c) the mixed model of (b), with Σ-restrictions.

Data

10	16	12	9
14	22	18	19
23	17	24	18
25	21	32	24
13	8	16	7
17	12	12	19

E 4.8. (a) Use (49) and (50) to derive confidence intervals on σ_α^2 and σ_β^2 of Table 4.5.

(b) Use the data of E 4.7 to calculate values of the intervals derived in (a).

E 4.9. (a) Use (56) to derive unbiased estimators of sampling variances of and covariances between estimated components of variance derived from Table 4.5.

(b) Use the data of E 4.7 to calculate estimates from the unbiased estimators in (a).

E 4.10. For the model of Table 4.5 find $\mathbf{V} = \text{var}(\mathbf{y})$ and $\mathbf{V}^{-1}$.

E 4.11. Use (76) of Section 4.6c, together with (80) and (81), to derive the ML solution of Section 4.7b-ii.

E 4.12. Confirm $\mathbf{V}^{-1}$ and $|\mathbf{V}|$ obtained in E 4.11 by writing $\mathbf{V}$ as

$$\mathbf{V} = \mathbf{I}_a \otimes (\mathbf{D} - \mathbf{C}\mathbf{A}^{-1}\mathbf{B})$$

for $\mathbf{D} = \mathbf{I}_b \otimes (\sigma_\beta^2 \mathbf{J}_n + \sigma_e^2 \mathbf{I}_n)$ with $\mathbf{A}^{-1} = -\sigma_\alpha^2$ and $\mathbf{C} = \mathbf{B}' = \mathbf{1}_b \otimes \mathbf{1}_n$ and use Appendix M.5.

E 4.13. Derive the dispersion matrix (100) from equation (99).

E 4.14. Consider data sets A and B:

A

8	8	11
4	10	19

B

1, 7	2, 6	4, 7
1, 3	4, 6	8, 11

Treating Data Set A as a 1-way classification of 3 classes and 2 observations, calculate for a random model

(a) ANOVA estimates of variance components;

(b) unbiased estimates of the sampling variances and covariances of estimators, assuming normality, used in (a).

E 4.15. Repeat E 4.14 with A but treat it as having 2 rows and 3 columns.

E 4.16. Repeat E 4.14 with B but treat it as having 2 rows and 3 columns with 2 observations per cell.

E 4.17. Data sets P–U are examples of a 2-way nested classification with $a = 3$, $b = 2$ and $n = 4$.

P

a_1		a_2		a_3	
6	4	4	10	5	4
4	2	5	6	6	8
5	3	6	8	3	1
5	3	6	8	2	3

Q

a_1		a_2		a_3	
3	2	2	3	3	2
4	3	1	2	2	1
3	1	3	3	1	1
2	2	1	1	1	1

R

a_1		a_2		a_3	
2	3	1	3	6	2
2	2	1	3	7	1
1	6	2	1	4	4
3	5	0	5	7	1

S

a_1		a_2		a_3	
2	14	8	4	8	6
4	10	5	0	6	4
6	10	5	4	12	6
4	14	6	8	14	8

T

a_1		a_2		a_3	
3	2	8	1	10	7
4	1	11	4	14	10
7	4	10	6	12	5
6	5	11	5	16	6

U

a_1		a_2		a_3	
6	2	0	1	11	10
8	8	1	2	7	5
7	1	2	6	10	7
3	5	1	3	12	2

(a) Calculate ANOVA estimates of the variance components for the random model.
(b) Under normality assumptions, derive sampling variances and covariances of the estimators in (a).
(c) Derive unbiased estimators of the sampling variances in (b).
(d) Re-do your calculations using $b = 4$ and $n = 2$.

E 4.18. Suppose a clothing manufacturer has collected data on the number of defective socks it makes. There are 6 subsidiary companies (factor C) that make knitted socks. At each company there are 5 brands (B) of knitting machine, with 20 machines of each brand in each company. All machines, of all brands, are used on the different types of yarn (Y) from which socks are made: nylon, cotton and wool. At each company data have been collected from just two machines (M) of each brand, when operated by each of 4 locally resident women (F), using each of the yarns; and on each occasion the number of defective socks in two replicate samples of 100 socks was recorded.

(a) Key-out the analysis of variance for these data: for each line in the analysis give a label (both symbolic and verbal); and give the degrees of freedom.
(b) Decide which factors are to be considered random–and give brief reasons for your decisions.
(c) For each of just the random main effects factors, and their interaction with each other,

(i) derive the expected mean square in the analysis of variance of part (a) using (b);
(ii) using x for an observation, with appropriate subscripts (that include those for the factors in the sequence C, B, Y, M, F) show what the sum of squares is corresponding to each expected mean square of (i).
(iii) Write down the terms of $\mathbf{V} = \text{var}(\mathbf{y})$ that involve the variance components $\sigma^2_{\text{M:CB}}$, $\sigma^2_{\text{F:C}}$ and $\sigma^2_{\text{FM:CB}}$. Use Kronecker products of **I**-matrices and **J**-matrices.

E 4.19. In the 2-way crossed classification show that

$$\frac{\sum_{i=1}^{a}\sum_{j\neq j'=1}^{b}(\bar{y}_{ij\cdot}-\bar{y}_{\cdot j\cdot})(\bar{y}_{ij'\cdot}-\bar{y}_{\cdot j'\cdot})}{b(a-1)(b-1)}=\frac{\text{MSA}-\text{MSAB}}{bn}$$

and hence that its expected value is σ^2_α. This is the estimator suggested by Hocking *et al.* (1989), as providing an explanation as to why $\hat{\sigma}^2_\alpha$ can be negative. It is a pooled product–moment correlation, of the $\bar{y}_{ij\cdot}$-means in each pair of columns, pooled over all pairs.

E 4.20. Using either the likelihood function or the general results given in Chapter 6, derive ML results given in Section 4.7 (sub-sections containing the results are shown in parentheses):

(a) 1-way random model (b-i and g-ii);
(b) 2-way nested random model (b-ii and g-i);
(c) 2-way crossed, with interaction, mixed model (b-iii and f-iii);
(d) 2-way crossed, no interaction, mixed model (b-iv and f-iv);
(e) 2-way crossed, with interaction, random model (d-i and f-i);
(f) 2-way crossed, no interaction, random model (d-ii and f-ii).

E 4.21. Show for balanced data from a 3-way crossed classification with all interactions and one fixed effects factor, that explicit ML solutions for the variance components do not exist.

E 4.22. Derive (109) from (107) using (108) for the model equation (72).

E 4.23. Derive (112) from (110) using (108) and (111).

CHAPTER 5

ANALYSIS OF VARIANCE ESTIMATION FOR UNBALANCED DATA

The previous chapter describes the ANOVA method of estimating variance components from balanced data. Extending that method to unbalanced data began with the 1-way classification (Chapter 3) as in Cochran (1939) and Winsor and Clark (1940)—see Chapter 2. Extending it to higher-order classifications would nowadays appear to have been an obvious thing to do, and yet it seems to be that it was the Henderson (1953) paper that gave this extension its first major fillip—prompted, no doubt, by his interest in estimating variance components in a genetic setting where available data can be voluminous but severely unbalanced.

This chapter considers somewhat briefly the ANOVA method generally, as applicable to unbalanced data, and gives lengthy description of the three adaptations of ANOVA methodology suggested by Henderson (1953). Although those Henderson methods (as they have come to be known) are coming to be superseded by maximum likelihood (see Chapter 6) and other techniques, we know of no book that gives a detailed account of the Henderson methods, so this we proceed to do. For some readers this chapter may be mainly of historical interest. But it is important history, because for nigh on forty years the Henderson methods have been very widely used, in many cases on enormously large data sets. Moreover, they are methods that are likely to go on being used. This is because some researchers have solid confidence in understanding analysis of variance of balanced data and of expected mean squares derived therefrom, and feel that they can easily transfer that confidence to using the same concepts for unbalanced data. Furthermore, there is attraction in the fact that one of the Henderson methods is relatively easy to understand and to compute. In contrast, those same researchers can be quite apprehensive about maximum likelihood, for example. They might feel it is "too theoretic", and they might be overawed by the mathematics involved, by the lack of closed form expressions for estimators, and by the need for iterative techniques and sophisticated computing programs to carry out those techniques. In addition, of course, there are situations where the necessary computing power may not be available, and so resort has to be made to something more easily computed than maximum likelihood estimators, such as some kind of ANOVA-style estimates. We therefore deem it worthwhile to describe ANOVA methodology and especially the Henderson applications of it.

5.1. MODEL FORMULATION

a. Data

We take unbalanced data to be data in which there is not the same number of observations in every sub-most cell (see Section 1.2). For fixed effects models, emphatic distinction between all-cells-filled data and some-cells-empty data has been made by Searle (1987), but there seems to be less need for this distinction with mixed models. Nevertheless, in many applications where variance components are of interest the data frequently have empty cells, often with a very large percentage of the cells being empty (e.g., 70% empty). The problem of connectedness of the data (e.g., Searle 1987, Sec. 5.3) therefore raises its ugly head. It is important because for disconnected data (e.g., *loc cit.*, p. 157) certain calculation procedures are not appropriate when singular matrices occur where they would not do so with connected data.

The reader is assumed to be familiar with the $R(\cdot\,|\,\cdot)$ notation for reductions in sums of squares. For example

$$R(\boldsymbol{\alpha}\,|\,\mu) = R(\mu, \boldsymbol{\alpha}) - R(\mu)$$

is the difference between $R(\mu, \boldsymbol{\alpha})$, the reduction in sum of squares due to fitting $E(y_{ij}) = \mu + \alpha_i$ and $R(\mu)$, the reduction in sum of squares due to fitting, to the same data, the model $E(y_{ij}) = \mu$. $R(\boldsymbol{\alpha}\,|\,\mu)$ is thus often referred to as the sum of squares due to $\boldsymbol{\alpha}$ after μ. Lengthy discussion of $R(\cdot\,|\,\cdot)$ is given in Searle (1971, pp. 246–247; 1987, pp. 26–28).

In the 2-way crossed classification the sums of squares $R(\boldsymbol{\beta}\,|\,\mu)$ and $R(\boldsymbol{\beta}\,|\,\mu, \boldsymbol{\alpha})$ are equal for balanced data but not for unbalanced data. Using the model equation

$$E(y_{ijk}) = \mu + \alpha_i + \beta_j$$

for $i = 1, \ldots, a, j = 1, \ldots, b$ and $k = 1, \ldots, n_{ij}$ with $n_{ij} \geqslant 0$ when a cell has no data,

$$R(\boldsymbol{\beta}\,|\,\mu) = \sum_{i=1}^{a} n_{i.}(\bar{y}_{i..} - \bar{y}_{...})^2 .$$

But

$$R(\boldsymbol{\beta}\,|\,\mu, \alpha) = \mathbf{r}'\mathbf{T}^{-1}\mathbf{r} \quad \text{for } \mathbf{r} = \left\{_{\mathrm{c}}\; y_{.j.} - \sum_{i=1}^{a} n_{ij}\bar{y}_{i..} \right\}_{j=1}^{b-1}$$

where $\mathbf{T}$, symmetric of order $b-1$, has elements for $j \neq j' = 1, \ldots, b-1$ that are

$$t_{jj} = n_{.j} - \sum_{i=1}^{a} \frac{n_{ij}^2}{n_{i.}} \quad \text{and} \quad t_{jj'} = -\sum_{i=1}^{a} \frac{n_{ij}n_{ij'}}{n_{i.}} .$$

Details of these derivations are found, for example, in Searle (1971, p. 267; 1987, pp. 124–125), wherein the letter c is used in place of t.

b. A general model

The general model equation

$$\mathbf{y} = \mathbf{X}\boldsymbol{\beta} + \mathbf{Z}\mathbf{u} + \mathbf{e}$$

developed in Section 4.6 for balanced data can still be used with unbalanced data, although in the form

$$\mathbf{y} = \mathbf{X}\boldsymbol{\beta} + \mathbf{Z}_1\mathbf{u}_1 + \cdots + \mathbf{Z}_r\mathbf{u}_r + \mathbf{e} \tag{1}$$

the matrices $\mathbf{Z}_1, \ldots, \mathbf{Z}_r$ are no longer Kronecker products of identity matrices and summing vectors. Taking the $\mathbf{Z}$s as incidence matrices (with 0s and 1s as elements) they still have structure, but not such that it can be neatly formulated as with balanced data.

*-i. **Example 1: the 2-way crossed classification, random model.*** Suppose in the 2-way crossed classification that the numbers of observations in a set of data are as shown in Table 5.1. Then for the model equation

$$y_{ijk} = \mu + \alpha_i + \beta_j + \gamma_{ij} + e_{ijk}$$

with $i = 1, \ldots, a, j = 1, \ldots, b$ and $k = 1, \ldots, n_{ij}$, with $n_{ij} = 0$ when cell (i, j) has no data, the vector form

$$\mathbf{y} = \mu\mathbf{1} + \mathbf{Z}_1\boldsymbol{\alpha} + \mathbf{Z}_2\boldsymbol{\beta} + \mathbf{Z}_3\boldsymbol{\gamma} + \mathbf{e} \tag{2}$$

is

$$\mathbf{y} = \mathbf{1}_{21}\mu + \begin{bmatrix} \mathbf{1}_5 & \cdot & \cdot \\ \cdot & \mathbf{1}_9 & \cdot \\ \cdot & \cdot & \mathbf{1}_7 \end{bmatrix}\boldsymbol{\alpha} + \begin{bmatrix} \mathbf{1}_2 & \cdot & \cdot \\ \cdot & \mathbf{1}_3 & \cdot \\ \cdot & \mathbf{1}_1 & \cdot \\ \cdot & \cdot & \mathbf{1}_8 \\ \mathbf{1}_4 & \cdot & \cdot \\ \cdot & \mathbf{1}_3 & \cdot \end{bmatrix}\boldsymbol{\beta}$$

$$+ \begin{bmatrix} \mathbf{1}_2 & \cdot & \cdot & \cdot & \cdot & \cdot \\ \cdot & \mathbf{1}_3 & \cdot & \cdot & \cdot & \cdot \\ \cdot & \cdot & \mathbf{1}_1 & \cdot & \cdot & \cdot \\ \cdot & \cdot & \cdot & \mathbf{1}_8 & \cdot & \cdot \\ \cdot & \cdot & \cdot & \cdot & \mathbf{1}_4 & \cdot \\ \cdot & \cdot & \cdot & \cdot & \cdot & \mathbf{1}_3 \end{bmatrix}\boldsymbol{\gamma} + \mathbf{e} . \tag{3}$$

TABLE 5.1. VALUES OF n

	n_{ij}			$n_{i.}$
	2	3	—	5
	—	1	8	9
	4	3	—	7
$n_{.j}$	6	7	8	$21 = n_{..} = N$

Note that the matrix multiplying $\boldsymbol{\beta}$ can be partitioned into 3 sets of rows corresponding to the summing vectors in the matrix multiplying $\boldsymbol{\alpha}$. But in contrast to that matrix, which can be described quite generally as $\{_d \mathbf{1}_{n_{i.}}\}_{i=1}^{a}$, the matrix multiplying $\boldsymbol{\beta}$ has no general specification. This is because the order of the summing vectors in the coefficient of $\boldsymbol{\beta}$ depends on the numbers of observations in each row of Table 5.1 and the manner in which those numbers are spread across the columns of the table. For example, in Table 5.1 there are five observations in row 1, two of them being in column 1 and three in column 2. This gives rise to having $\mathbf{1}_2$ and $\mathbf{1}_3$ in columns 1 and 2 of the matrix multiplying $\boldsymbol{\beta}$ in (3). Unfortunately, in the presence of empty cells, i.e., with some-cells-empty data, there is no useful notation for this matrix, except that it is $\mathbf{Q}\{_d \mathbf{1}_{n_{.j}}\}_{i=1}^{a}$, where $\mathbf{Q}$ is a permutation matrix (and thus orthogonal) that is determined by the actual pattern of observations. As a result, a general form of (3) is

$$\mathbf{y} = \mu\mathbf{1}_N + \{_d \mathbf{1}_{n_{i.}}\}_{i=1}^{a}\boldsymbol{\alpha} + \mathbf{Q}\{_d \mathbf{1}_{n_{.j}}\}_{j=1}^{b}\boldsymbol{\beta} + \{_d \{_d \mathbf{1}_{n_{ij}}\}_{j=1}^{b}\}_{i=1}^{a}\boldsymbol{\gamma} + \mathbf{e}\,. \tag{4}$$

-ii. Dispersion matrix. For the random effects represented by the $\mathbf{u}_i$s in (1) we adopt the usual conventions of

$$E(\mathbf{u}_i) = \mathbf{0}, \quad \text{var}(\mathbf{u}_i) = \sigma_i^2\mathbf{I}_{q_i}, \tag{5a}$$

where q_i is the order of $\mathbf{u}_i$, and

$$\text{cov}(\mathbf{u}_i, \mathbf{u}_j') = \mathbf{0}, \quad \text{cov}(\mathbf{u}, \mathbf{e}') = \mathbf{0} \quad \text{and} \quad \text{var}(\mathbf{e}) = \sigma_e^2\mathbf{I}_N\,. \tag{5b}$$

Applying this to (1) gives

$$\mathbf{V} = \text{var}(\mathbf{y}) = \sum_{i=1}^{r} \mathbf{Z}_i\mathbf{Z}_i'\sigma_i^2 + \sigma_e^2\mathbf{I}_N\,. \tag{6}$$

This notation can be made more compact by defining

$$\mathbf{u}_0 = \mathbf{e}, \quad \sigma_0^2 = \sigma_e^2 \quad \text{and} \quad \mathbf{Z}_0 = \mathbf{I}_N\,. \tag{7}$$

Then we have

$$\mathbf{y} = \mathbf{X}\boldsymbol{\beta} + \sum_{i=0}^{r} \mathbf{Z}_i\mathbf{u}_i \quad \text{and} \quad \mathbf{V} = \sum_{i=0}^{r} \mathbf{Z}_i\mathbf{Z}_i'\sigma_i^2, \tag{8}$$

just as in (70) and (71) of Chapter 4. Considerable use is made of (7) and (8).

-iii. Example 1 (continued). From (4), on omitting the limits on i and j,

$$\mathbf{Z}_1 = \{_\mathrm{d}\, \mathbf{1}_{n_{i.}}\}, \quad \text{with } \mathbf{Z}_1\mathbf{Z}_1' = \{_\mathrm{d}\, \mathbf{J}_{n_{i.}}\}, \tag{9}$$

$$\mathbf{Z}_2 = \mathbf{Q}\{_\mathrm{d}\, \mathbf{1}_{n_{.j}}\}, \quad \text{with } \mathbf{Z}_2\mathbf{Z}_2' = \mathbf{Q}\{_\mathrm{d}\, \mathbf{J}_{n_{.j}}\}\mathbf{Q}', \tag{10}$$

$$\mathbf{Z}_3 = \{_\mathrm{d}\, \{_\mathrm{d}\, \mathbf{1}_{n_{ij}}\}_j\}_i \quad \text{with } \mathbf{Z}_3\mathbf{Z}_3' = \{_\mathrm{d}\, \{_\mathrm{d}\, \mathbf{J}_{n_{ij}}\}_j\}_i, \tag{11}$$

$$\mathbf{Z}_0 = \mathbf{I}_N, \quad \text{with } \mathbf{Z}_0\mathbf{Z}_0' = \mathbf{I}_N\,. \tag{12}$$

The $\mathbf{Z}_i$-matrices are used in the equation for $\mathbf{y}$, and the $\mathbf{Z}_i\mathbf{Z}_i'$-matrices occur in $\mathbf{V} = \mathrm{var}(\mathbf{y})$. Simplification of $\mathbf{Z}_2\mathbf{Z}_2'$ is not readily apparent.

5.2. ANOVA ESTIMATION

As developed for balanced data, ANOVA estimation is derived from equating analysis of variance sums of squares to their expected values.

a. Example 2—the 1-way random model, balanced data

From equations (52) and (53) of Section 3.5b we have

$$E\begin{bmatrix} \mathrm{SSE} \\ \mathrm{SSA} \end{bmatrix} = \begin{bmatrix} a(n-1) & 0 \\ a-1 & (a-1)n \end{bmatrix} \begin{bmatrix} \sigma_e^2 \\ \sigma_\alpha^2 \end{bmatrix}. \tag{13}$$

From this come the estimation equations

$$\begin{bmatrix} a(n-1) & 0 \\ a-1 & (a-1)n \end{bmatrix} \begin{bmatrix} \hat{\sigma}_e^2 \\ \hat{\sigma}_\alpha^2 \end{bmatrix} = \begin{bmatrix} \mathrm{SSE} \\ \mathrm{SSA} \end{bmatrix}. \tag{14}$$

These are, of course, easily solved in this simple case.

b. Estimation

-i. The general case. The principle of (13) and (14) is easily generalized. For a mixed model

$$\mathbf{y} = \mathbf{X}\boldsymbol{\beta} + \mathbf{Z}\mathbf{u} + \mathbf{e}$$

having r random factors suppose

$$\mathbf{s} = \{_\mathrm{c}\, s_i\}_{i=0}^{r} = \{_\mathrm{c}\, \mathbf{y}'\mathbf{A}_i\mathbf{y}\}_{i=0}^{r} \tag{15}$$

is a vector of $r + 1$ quadratic forms in $\mathbf{y}$ such that $\mathbf{A}_i$ is symmetric, i.e., $\mathbf{A}_i = \mathbf{A}_i'$ $\forall\ i$. Then, from Theorem S1 of Appendix S.5,

$$\begin{aligned} E(s_i) = E(\mathbf{y}'\mathbf{A}_i\mathbf{y}) &= \mathrm{tr}(\mathbf{A}_i\mathbf{V}) + E(\mathbf{y}')\mathbf{A}_i E(\mathbf{y}) \\ &= \mathrm{tr}(\mathbf{A}_i\mathbf{V}) + \boldsymbol{\beta}'\mathbf{X}'\mathbf{A}_i\mathbf{X}\boldsymbol{\beta}, \end{aligned} \tag{16}$$

since $E(\mathbf{y}) = \mathbf{X}\boldsymbol{\beta}$. This expectation will contain no terms in $\boldsymbol{\beta}$, the fixed effects, if $\mathbf{X}'\mathbf{A}_i\mathbf{X} = \mathbf{0}$. Thus, providing $\mathbf{A}_i = \mathbf{A}_i'$ is chosen so that $\mathbf{X}'\mathbf{A}_i\mathbf{X} = \mathbf{0}$, the expected value of $\mathbf{y}'\mathbf{A}_i\mathbf{y}$ contains only σ^2-terms and no fixed effects. A series of such quadratic forms can then be used in a generalization of the ANOVA method of estimating variance components, namely equating observed values of such forms to their expected values. With balanced data, $\mathbf{X}'\mathbf{A}_i\mathbf{X} = \mathbf{0}$ always holds for the sums of squares of the analysis of variance table. With unbalanced data it holds for some sums of squares and not others, and it depends on the model being used, i.e., on $\mathbf{X}$, as well as on $\mathbf{A}_i$. For the usual completely random models $\mathbf{X}'\mathbf{A}_i\mathbf{X} = \mathbf{0}$ reduces to $\mathbf{1}'\mathbf{A}_i\mathbf{1} = \mathbf{0}$ (all elements of $\mathbf{A}_i$ summing to zero) because $\mathbf{X}\boldsymbol{\beta}$ is then $\mathbf{X}\boldsymbol{\beta} = \mu\mathbf{1}$.

Then, because

$$E(s_i) = \operatorname{tr}(\mathbf{A}_i\mathbf{V}) = \operatorname{tr}\left[\mathbf{A}_i \sum_{j=0}^{r} \mathbf{Z}_j\mathbf{Z}'_j\sigma_j^2\right] = \sum_{j=0}^{r} \operatorname{tr}(\mathbf{Z}'_j\mathbf{A}_i\mathbf{Z}_j)\sigma_j^2,$$

$$E(\mathbf{s}) = \{_{\mathrm{m}}\, \operatorname{tr}(\mathbf{Z}'_j\mathbf{A}_i\mathbf{Z}_j)\}_{i,j}\{_{\mathrm{c}}\, \sigma_j^2\}_j, \tag{17}$$

which we write as

$$E(\mathbf{s}) = \mathbf{C}\boldsymbol{\sigma}^2, \quad \text{with } \mathbf{C} = \{_{\mathrm{m}}\, \operatorname{tr}(\mathbf{Z}'_j\mathbf{A}_i\mathbf{Z}_j)\}_{i,j} \text{ and } \boldsymbol{\sigma}^2 = \{_{\mathrm{c}}\, \sigma_j^2\}_{j=0}^{r}. \tag{18}$$

This immediately provides extension of the balanced data ANOVA method of estimation given in (29) and (30) of Chapter 4: from (18) equate the expected value of $\mathbf{s}$ to $\mathbf{s}$ and solve for the variance components. This gives

$$\mathbf{C}\hat{\boldsymbol{\sigma}}^2 = \mathbf{s}, \quad \text{or} \quad \hat{\boldsymbol{\sigma}}^2 = \mathbf{C}^{-1}\mathbf{s}, \tag{19}$$

providing $\mathbf{C}$ is non-singular. This procedure, which includes the balanced data case, of course, can be viewed in some sense as a special form of the method of moments.

Equation (19) is what is called the general ANOVA method of estimating variance components from unbalanced data: equate observed values of a set of quadratic forms to their expected values and solve for the variance components. The solutions are called ANOVA estimators of variance components. The only limitations on what one chooses for those quadratics is that their expectations contain only variance components; i.e., that $\mathbf{X}'\mathbf{A}_i\mathbf{X} = \mathbf{0}$. When symmetric $\mathbf{A}_i$ is non-negative definite, the condition $\mathbf{X}'\mathbf{A}_i\mathbf{X} = \mathbf{0}$ reduces to $\mathbf{A}_i\mathbf{X} = \mathbf{0}$.

-ii. Example 2 (continued). An example of (19) is (14):

$$\mathbf{C}\boldsymbol{\sigma}^2 = \mathbf{s} \quad \text{is} \quad \begin{bmatrix} a(n-1) & 0 \\ a-1 & (a-1)n \end{bmatrix}\begin{bmatrix} \hat{\sigma}_e^2 \\ \hat{\sigma}_\alpha^2 \end{bmatrix} = \begin{bmatrix} \mathrm{SSE} \\ \mathrm{SSA} \end{bmatrix},$$

$$\hat{\boldsymbol{\sigma}}^2 = \mathbf{C}^{-1}\mathbf{s} \quad \text{is} \quad \begin{bmatrix} \hat{\sigma}_e^2 \\ \hat{\sigma}_\alpha^2 \end{bmatrix} = \begin{bmatrix} \dfrac{1}{a(n-1)} & 0 \\ \dfrac{-1}{an(n-1)} & \dfrac{1}{(a-1)n} \end{bmatrix}\begin{bmatrix} \mathrm{SSE} \\ \mathrm{SSA} \end{bmatrix} = \begin{bmatrix} \mathrm{MSE} \\ (\mathrm{MSA} - \mathrm{MSE})/n \end{bmatrix},$$

as in (55) of Chapter 3.

-iii. Example 1 (continued). Although (18) applies for any choice of $\mathbf{A}_i$ that satisfies $\mathbf{X}'\mathbf{A}_i\mathbf{X} = \mathbf{0}$, when we decide to use some particular sum of squares as an element of $\mathbf{s}$ it is the nature of that sum of squares that determines its corresponding $\mathbf{A}_i$. We illustrate this for SSA, one of the sums of squares in the 2-way crossed classification. It is

$$\begin{aligned}\text{SSA} &= \sum_{i=1}^{a} n_{i\cdot}(\bar{y}_{i\cdot} - \bar{y}_{\cdot\cdot})^2 = \mathbf{y}'(\{_{\text{d}}\,\bar{\mathbf{J}}_{n_{i\cdot}}\} - \bar{\mathbf{J}}_N)\mathbf{y} \\ &= \mathbf{y}'\mathbf{A}\mathbf{y} \quad \text{for } \mathbf{A} = \{_{\text{d}}\,\bar{\mathbf{J}}_{n_{i\cdot}}\} - \bar{\mathbf{J}}_N\,.\end{aligned}$$

Using the random model, $E(\mathbf{y}) = \mu\mathbf{1}_N$ and $\mathbf{1}'_N\mathbf{A} = \mathbf{1}'_N - \mathbf{1}'_N = \mathbf{0}$. Therefore $E(\text{SSA})$ as an element of $E(\mathbf{s})$ of (17) is

$$\begin{aligned}E(\text{SSA}) &= [\text{tr}(\mathbf{Z}'_1\mathbf{A}\mathbf{Z}_1) \quad \text{tr}(\mathbf{Z}'_2\mathbf{A}\mathbf{Z}_2) \quad \text{tr}(\mathbf{Z}'_3\mathbf{A}\mathbf{Z}_3) \quad \text{tr}(\mathbf{Z}'_0\mathbf{A}\mathbf{Z}_0)]\begin{bmatrix}\sigma^2_\alpha \\ \sigma^2_\beta \\ \sigma^2_\gamma \\ \sigma^2_e\end{bmatrix} \\ &= \sigma^2_\alpha\,\text{tr}(\mathbf{A}\mathbf{Z}_1\mathbf{Z}'_1) + \sigma^2_\beta\,\text{tr}(\mathbf{A}\mathbf{Z}_2\mathbf{Z}'_2) + \sigma^2_\gamma\,\text{tr}(\mathbf{A}\mathbf{Z}_3\mathbf{Z}'_3) + \sigma^2_e\,\text{tr}(\mathbf{A}\mathbf{Z}_0\mathbf{Z}'_0)\,.\end{aligned}$$

On using $\mathbf{Z}_1$ from (9), the first term in this expression is

$$\begin{aligned}\sigma^2_\alpha\,\text{tr}(\mathbf{A}\mathbf{Z}_1\mathbf{Z}'_1) &= \sigma^2_\alpha\,\text{tr}[(\{_{\text{d}}\,\bar{\mathbf{J}}_{n_{i\cdot}}\} - \bar{\mathbf{J}}_N)\{_{\text{d}}\,\mathbf{J}_{n_{i\cdot}}\}] \\ &= \sigma^2_\alpha\,\text{tr}\left(\{_{\text{d}}\,\mathbf{J}_{n_{i\cdot}}\} - \frac{1}{N}\mathbf{1}_N\{_{\text{r}}\,n_{i\cdot}\mathbf{1}'_{n_{i\cdot}}\}\right) \\ &= \sigma^2_\alpha\left(N - \frac{\Sigma_i n^2_{i\cdot}}{N}\right).\end{aligned}$$

The second term, using $\mathbf{Z}_2$ from (10), is

$$\begin{aligned}\sigma^2_\beta\,\text{tr}(\mathbf{A}\mathbf{Z}_2\mathbf{Z}'_2) &= \sigma^2_\beta\,\text{tr}[\mathbf{Z}'_2(\{_{\text{d}}\,\bar{\mathbf{J}}_{n_{i\cdot}}\} - \bar{\mathbf{J}}_N)\mathbf{Z}_2] \\ &= \sigma^2_\beta\,\text{tr}\left[\mathbf{Z}'_2\left\{_{\text{d}}\,\frac{1}{n_{i\cdot}}\mathbf{1}_{n_{i\cdot}}\mathbf{1}'_{n_{i\cdot}}\right\}\mathbf{Z}_2 - \frac{1}{N}\mathbf{1}'_N\mathbf{Z}_2(\mathbf{1}'_N\mathbf{Z}_2)'\right].\end{aligned}$$

To simplify this, we use $\text{tr}(\mathbf{X}\mathbf{X}') = \text{sesq}(\mathbf{X})$, the sum of squares of elements of $\mathbf{X}$, from Appendix M.5. Then

$$\begin{aligned}\sigma^2_\beta\,\text{tr}(\mathbf{A}\mathbf{Z}_2\mathbf{Z}'_2) &= \sigma^2_\beta\left[\text{tr}\left(\mathbf{Z}'_2\left\{_{\text{d}}\,\frac{1}{\sqrt{n_{i\cdot}}}\mathbf{1}_{n_{i\cdot}}\right\}\left\{_{\text{d}}\,\frac{1}{\sqrt{n_{i\cdot}}}\mathbf{1}'_{n_{i\cdot}}\right\}\mathbf{Z}_2\right) - \frac{\text{sesq}(\mathbf{1}'_N\mathbf{Z}_2)}{N}\right] \\ &= \sigma^2_\beta\left[\text{sesq}\left(\left\{_{\text{d}}\,\frac{1}{\sqrt{n_{i\cdot}}}\mathbf{1}'_{n_{i\cdot}}\right\}\mathbf{Z}_2\right) - \frac{\Sigma_j n^2_{\cdot j}}{N}\right] \\ &= \sigma^2_\beta\left(\Sigma_i\frac{\Sigma_j n^2_{ij}}{n_{i\cdot}} - \frac{\Sigma_j n^2_{\cdot j}}{N}\right),\end{aligned}$$

the last expression's first term being evident from the nature of $\mathbf{Z}_2$ illustrated in (3). The complete expression for $E(\text{SSA})$ is shown in E 5.2. It is left for the reader to derive the terms other than those in σ_α^2 and σ_β^2.

Thus does (17) provide a mechanism for deriving the expected value of any set of quadratic forms $\mathbf{y}'\mathbf{A}_i\mathbf{y}$ having $\mathbf{A}_i = \mathbf{A}_i'$ and $\mathbf{X}'\mathbf{A}_i\mathbf{X} = \mathbf{0}$.

c. Unbiasedness

Turning to properties of $\hat{\boldsymbol{\sigma}}^2$, it is easily seen that $\hat{\boldsymbol{\sigma}}^2$ is unbiased since, from (19),

$$E(\hat{\boldsymbol{\sigma}}^2) = E(\mathbf{C}^{-1}\mathbf{s}) = \mathbf{C}^{-1}E(\mathbf{s}) = \mathbf{C}^{-1}\mathbf{C}\boldsymbol{\sigma}^2 = \boldsymbol{\sigma}^2 \,. \tag{20}$$

No matter what quadratic forms are used in elements of $\mathbf{s}$, so long as $\mathbf{C}$ of $E(\mathbf{s}) = \mathbf{C}\boldsymbol{\sigma}^2$ is non-singular, $\hat{\boldsymbol{\sigma}}^2 = \mathbf{C}^{-1}\mathbf{s}$ is an unbiased estimator of $\boldsymbol{\sigma}^2$.

In (18) and hence (19) we have implicitly assumed that $\mathbf{s}$ has as many elements as does $\boldsymbol{\sigma}^2$. Then $\mathbf{C}\hat{\boldsymbol{\sigma}}^2 = \mathbf{s}$ has as many equations as there are variance components, $\mathbf{C}$ is square, and $\hat{\boldsymbol{\sigma}}^2 = \mathbf{C}^{-1}\mathbf{s}$ provided $\mathbf{C}^{-1}$ exists. But $\mathbf{C}$ does not have to be square. $\mathbf{s}$ can have more elements than $\boldsymbol{\sigma}^2$. There are then more equations than variance components in $\mathbf{C}\hat{\boldsymbol{\sigma}}^2 = \mathbf{s}$, and the equations are unlikely to be consistent. However, providing $\mathbf{C}$ has full column rank, one could always adopt a least squares outlook and use $\tilde{\boldsymbol{\sigma}}^2 = (\mathbf{C}'\mathbf{C})^{-1}\mathbf{C}'\mathbf{s}$ as an unbiased estimator of $\boldsymbol{\sigma}^2$; it reduces to $\hat{\boldsymbol{\sigma}}^2 = \mathbf{C}^{-1}\mathbf{s}$ if $\mathbf{C}$ is square.

This unbiasedness arises without this method of estimation containing a word of how to choose what quadratic forms shall be used as elements—only that $\mathbf{X}'\mathbf{A}_i\mathbf{X} = \mathbf{0}$ be satisfied and that $\mathbf{C}^{-1}$ exist; and these are not severe limitations. But they provide no optimality characteristics of any sort for the resulting estimators. The only built-in property is that of unbiasedness.

In the context of designed experiments, where estimation of treatment contrasts, for example, is a prime consideration, unbiasedness may be a useful property. This is because we conceive of repeating the experiment, and unbiasedness means that over all conceivable repetitions the expected value of our estimated contrast will equal the true contrast: e.g., $E(\bar{y}_1 - \bar{y}_2) = \tau_1 - \tau_2$. But when estimating variance components this concept of repeating the data collection process may not be a practical feasibility. Many situations in which variance components estimates are sought involve very large amounts of data; e.g., a project having three million records at its disposal. Repetitions of the process by which such data were gathered may be simply impractical. In those circumstances unbiasedness may not be as useful a property as when estimating treatment contrasts from designed experiments. Nevertheless, we can still imagine conceptual repetitions of data collection and think of unbiasedness as being over those conceptual repetitions. However, since unbiasedness is the only property that is built into ANOVA estimators of variance components, it may be worthwhile to abandon it as a property in favor of other estimators that are not unbiased but which have better large-sample properties such as large-sample normality and efficiency, as do maximum likelihood estimators, for example. It is therefore useful to appreciate this situation as we discuss a variety of ANOVA estimators that are available, some of which have received widespread use in applications. ʳ d notice that in mentioning large-sample normality we are at once conceptualizing the idea of repeated sampling.

d. Sampling variances

Since ANOVA estimators are derived without reference to their variances, and yet are quadratic functions of the observations, it is natural to think of deriving sampling variances. We do this on the basis of assuming normality throughout.

-i. A general result. With $\mathbf{y} \sim \mathcal{N}(\mathbf{X}\boldsymbol{\beta}, \mathbf{V})$, we use $\text{var}(\mathbf{y}'\mathbf{A}\mathbf{y}) = 2\,\text{tr}(\mathbf{A}\mathbf{V})^2 + 4\boldsymbol{\mu}'\mathbf{A}\mathbf{V}\mathbf{A}\boldsymbol{\mu}$ of Theorem S4 in Appendix S.5, together with $\mathbf{X}'\mathbf{A}_i\mathbf{X} = \mathbf{0}$ and (19). With $\boldsymbol{\mu} = \mathbf{X}\boldsymbol{\beta}$ this gives

$$\text{var}(\mathbf{y}'\mathbf{A}\mathbf{y}) = 2\,\text{tr}(\mathbf{A}\mathbf{V})^2 \quad \text{and} \quad \text{cov}(\mathbf{y}'\mathbf{A}\mathbf{y}, \mathbf{y}'\mathbf{B}\mathbf{y}) = 2\,\text{tr}(\mathbf{A}\mathbf{V}\mathbf{B}\mathbf{V})\,.$$

Then the variance–covariance matrix of $\hat{\boldsymbol{\sigma}}^2 = \mathbf{C}^{-1}\mathbf{s}$ is

$$\begin{aligned}
\text{var}(\hat{\boldsymbol{\sigma}}^2) &= \mathbf{C}^{-1}\,\text{var}(\mathbf{s})\mathbf{C}^{-1'} \\
&= 2\mathbf{C}^{-1}\{_{\text{m}}\,\text{tr}(\mathbf{A}_i\mathbf{V}\mathbf{A}_{i'}\mathbf{V})\}_{i,i'}\mathbf{C}^{-1'} \\
&= 2\mathbf{C}^{-1}\left\{_{\text{m}}\,\text{tr}\left(\mathbf{A}_i \sum_{j=0}^{r} \mathbf{Z}_j\mathbf{Z}_j'\sigma_j^2\mathbf{A}_{i'} \sum_{j'=0}^{r} \mathbf{Z}_{j'}\mathbf{Z}_{j'}'\sigma_{j'}^2\right)\right\}_{i,i'}\mathbf{C}^{-1'} \qquad (21\text{a}) \\
&= 2\mathbf{C}^{-1}\{_{\text{m}}\,\boldsymbol{\sigma}^{2'}\{_{\text{m}}\,\text{tr}(\mathbf{A}_i\mathbf{Z}_j\mathbf{Z}_j'\mathbf{A}_{i'}\mathbf{Z}_{j'}\mathbf{Z}_{j'}')\}_{j,j'}\boldsymbol{\sigma}^2\}_{i,i'}\mathbf{C}^{-1'}\,. \qquad (21\text{b})
\end{aligned}$$

In (21b) the inner matrix is not symmetric and so no further useful simplifications seem readily available.

Since the derivation of (21b) from (21a) may not be clear to all readers, we demonstrate their equivalence for the case of $r = 2$. Let $\mathbf{W}$ denote the expression in braces in (21a). Then

$$\begin{aligned}
\mathbf{W} &= \left\{_{\text{m}}\,\text{tr}\left(\mathbf{A}_i \sum_{j=0}^{2} \mathbf{Z}_j\mathbf{Z}_j'\sigma_j^2\mathbf{A}_{i'} \sum_{j=0}^{2} \mathbf{Z}_{j'}\mathbf{Z}_{j'}'\sigma_{j'}^2\right)\right\}_{i,i'=0}^{2} \\
&= \{_{\text{m}}\,\text{tr}[\mathbf{A}_i(\mathbf{Z}_0\mathbf{Z}_0'\sigma_0^2 + \mathbf{Z}_1\mathbf{Z}_1'\sigma_1^2 + \mathbf{Z}_2\mathbf{Z}_2'\sigma_2^2) \\
&\qquad \times \mathbf{A}_{i'}(\mathbf{Z}_0\mathbf{Z}_0'\sigma_0^2 + \mathbf{Z}_1\mathbf{Z}_1'\sigma_1^2 + \mathbf{Z}_2\mathbf{Z}_2'\sigma_2^2)]\}_{i,i'} \\
&= \{_{\text{m}}\,\sigma_0^4\,\text{tr}(\mathbf{A}_i\mathbf{Z}_0\mathbf{Z}_0'\mathbf{A}_{i'}\mathbf{Z}_0\mathbf{Z}_0') + \sigma_1^4(\mathbf{A}_i\mathbf{Z}_1\mathbf{Z}_1'\mathbf{A}_{i'}\mathbf{Z}_1\mathbf{Z}_1') + \sigma_2^4\,\text{tr}(\mathbf{A}_i\mathbf{Z}_2\mathbf{Z}_2'\mathbf{A}_{i'}\mathbf{Z}_2\mathbf{Z}_2') \\
&\qquad + \sigma_0^2\sigma_1^2[\text{tr}(\mathbf{A}_i\mathbf{Z}_0\mathbf{Z}_0'\mathbf{A}_{i'}\mathbf{Z}_1\mathbf{Z}_1') + \text{tr}(\mathbf{A}_i\mathbf{Z}_1\mathbf{Z}_1'\mathbf{A}_{i'}\mathbf{Z}_0\mathbf{Z}_0')] \\
&\qquad + \sigma_0^2\sigma_2^2[\text{tr}(\mathbf{A}_i\mathbf{Z}_0\mathbf{Z}_0'\mathbf{A}_{i'}\mathbf{Z}_2\mathbf{Z}_2') + \text{tr}(\mathbf{A}_i\mathbf{Z}_2\mathbf{Z}_2'\mathbf{A}_{i'}\mathbf{Z}_0\mathbf{Z}_0')] \\
&\qquad + \sigma_1^2\sigma_2^2[\text{tr}(\mathbf{A}_i\mathbf{Z}_1\mathbf{Z}_1'\mathbf{A}_{i'}\mathbf{Z}_2\mathbf{Z}_2') + \text{tr}(\mathbf{A}_i\mathbf{Z}_2\mathbf{Z}_2'\mathbf{A}_{i'}\mathbf{Z}_1\mathbf{Z}_1')]\}_{i,i'}\,.
\end{aligned}$$

For the coefficient of $\sigma_0^2\sigma_1^2$ note that the two trace terms are equal:

$$\begin{aligned}
\text{tr}(\mathbf{A}_i\mathbf{Z}_0\mathbf{Z}_0'\mathbf{A}_{i'}\mathbf{Z}_1\mathbf{Z}_1') &= \text{tr}(\mathbf{A}_i\mathbf{Z}_0\mathbf{Z}_0'\mathbf{A}_{i'}\mathbf{Z}_1\mathbf{Z}_1')' = \text{tr}(\mathbf{Z}_1\mathbf{Z}_1'\mathbf{A}_{i'}\mathbf{Z}_0\mathbf{Z}_0'\mathbf{A}_i) \\
&= \text{tr}(\mathbf{A}_i\mathbf{Z}_1\mathbf{Z}_1'\mathbf{A}_{i'}\mathbf{Z}_0\mathbf{Z}_0')\,.
\end{aligned}$$

Hence

$$\mathbf{W} = \left\{ [\sigma_0^2 \quad \sigma_1^2 \quad \sigma_2^2] \right._{\mathrm{m}}$$

$$\times \begin{bmatrix} \mathrm{tr}(\mathbf{A}_i\mathbf{Z}_0\mathbf{Z}_0'\mathbf{A}_{i'}\mathbf{Z}_0\mathbf{Z}_0') & \mathrm{tr}(\mathbf{A}_i\mathbf{Z}_0\mathbf{Z}_0'\mathbf{A}_{i'}\mathbf{Z}_1\mathbf{Z}_1') & \mathrm{tr}(\mathbf{A}_i\mathbf{Z}_0\mathbf{Z}_0'\mathbf{A}_{i'}\mathbf{Z}_2\mathbf{Z}_2') \\ & \mathrm{tr}(\mathbf{A}_i\mathbf{Z}_1\mathbf{Z}_1'\mathbf{A}_{i'}\mathbf{Z}_1\mathbf{Z}_1') & \mathrm{tr}(\mathbf{A}_i\mathbf{Z}_1\mathbf{Z}_1'\mathbf{A}_{i'}\mathbf{Z}_2\mathbf{Z}_2') \\ \text{symmetric} & & \mathrm{tr}(\mathbf{A}_i\mathbf{Z}_2\mathbf{Z}_2'\mathbf{A}_{i'}\mathbf{Z}_2\mathbf{Z}_2') \end{bmatrix} \left.\begin{bmatrix} \sigma_0^2 \\ \sigma_1^2 \\ \sigma_2^2 \end{bmatrix}\right\},$$

for $i, i' = 0, 1, 2$. This is the essence of (21b).

Notice that in using (19) as a method of estimation, nothing has been said about what quadratic forms of the observations shall be used as elements of $\mathbf{s}$. Nothing. Results (19) for any quadratic forms $\mathbf{y}'\mathbf{A}_i\mathbf{y}$ having $\mathbf{X}'\mathbf{A}_i\mathbf{X} = 0$ lead to (20) and (21); and that methodology gives no guidance whatever as to what quadratics are optimal. Moreover, it is the quadratics that do get used as elements of $\mathbf{s}$ that determine $\mathbf{C}$ of (18) and (21). This all leads to an extremely difficult optimization problem, which is developed thoroughly in Malley (1986), wherein, building on results of Zyskind (1967) and Seely (1971), conditions are developed under which quadratic functions of data can be optimal estimates of variance components.

-ii. ***Example 2*** (continued). To illustrate the use of (21b), we use the 1-way classification random model, balanced data, giving some of the details here and leaving others to the reader as an exercise (E 5.1). Starting from (14),

$$\hat{\boldsymbol{\sigma}}^2 = \begin{bmatrix} \hat{\sigma}_e^2 \\ \hat{\sigma}_\alpha^2 \end{bmatrix} = \begin{bmatrix} a(n-1) & 0 \\ a-1 & (a-1)n \end{bmatrix}^{-1} \begin{bmatrix} \mathrm{SSE} \\ \mathrm{SSA} \end{bmatrix}$$

$$= \begin{bmatrix} \dfrac{1}{a(n-1)} & 0 \\ \dfrac{-1}{an(n-1)} & \dfrac{1}{(a-1)n} \end{bmatrix} \begin{bmatrix} \mathbf{y}'(\mathbf{I}_a \otimes \mathbf{I}_n - \mathbf{I}_a \otimes \bar{\mathbf{J}}_n)\mathbf{y} \\ \mathbf{y}'(\mathbf{I}_a \otimes \bar{\mathbf{J}}_n - \bar{\mathbf{J}}_a \otimes \bar{\mathbf{J}}_n)\mathbf{y} \end{bmatrix}$$

$$= \begin{bmatrix} \dfrac{1}{a(n-1)} & 0 \\ \dfrac{-1}{an(a-1)} & \dfrac{1}{(a-1)n} \end{bmatrix} \begin{bmatrix} \mathbf{y}'(\mathbf{I}_a \otimes \mathbf{C}_n)\mathbf{y} \\ \mathbf{y}'(\mathbf{C}_a \otimes \bar{\mathbf{J}}_n)\mathbf{y} \end{bmatrix} \quad \text{for } \mathbf{C}_n = \mathbf{I}_n - \bar{\mathbf{J}}_n\,.$$

Then for

$$\mathbf{C}^{-1} = \begin{bmatrix} \dfrac{1}{a(n-1)} & 0 \\ \dfrac{-1}{an(n-1)} & \dfrac{1}{(a-1)n} \end{bmatrix},$$

$$\hat{\boldsymbol{\sigma}}^2 = \mathbf{C}^{-1}\begin{bmatrix} \mathbf{y}'\mathbf{A}_0\mathbf{y} \\ \mathbf{y}'\mathbf{A}_1\mathbf{y} \end{bmatrix} \quad \text{with } \mathbf{A}_0 = \mathbf{I}_a \otimes \mathbf{C}_n \text{ and } \mathbf{A}_1 = \mathbf{C}_a \otimes \bar{\mathbf{J}}_n\,.$$

We then have (21b), with $\mathbf{V} = (\mathbf{I}_a \otimes \mathbf{I}_n)\sigma_e^2 + (\mathbf{I}_a \otimes \mathbf{J}_n)\sigma_\alpha^2$, as

$$\text{var}\begin{bmatrix} \hat{\sigma}_e^2 \\ \hat{\sigma}_\alpha^2 \end{bmatrix} = \mathbf{C}^{-1}\left\{ {}_{\mathbf{m}} \begin{bmatrix} \sigma_e^2 & \sigma_\alpha^2 \end{bmatrix} \begin{bmatrix} \text{tr}(\mathbf{A}_i\mathbf{I}\mathbf{A}_{i'}\mathbf{I}) & \text{tr}[\mathbf{A}_i\mathbf{I}\mathbf{A}_{i'}(\mathbf{I}_a \otimes \mathbf{J}_n)] \\ \text{sym} & \text{tr}[\mathbf{A}_i(\mathbf{I}_a \otimes \mathbf{J}_n)\mathbf{A}_{i'}(\mathbf{I}_a \otimes \mathbf{J}_n)] \end{bmatrix} \right.$$

$$\left. \times \begin{bmatrix} \sigma_e^2 \\ \sigma_\alpha^2 \end{bmatrix} \right\}_{i,i'=0}^{i,i'=1} \mathbf{C}^{-1} .$$

Write this as

$$\text{var}\begin{bmatrix} \hat{\sigma}_e^2 \\ \hat{\sigma}_\alpha^2 \end{bmatrix} = \mathbf{C}^{-1}\begin{bmatrix} t_{00} & t_{01} \\ t_{01} & t_{11} \end{bmatrix}\mathbf{C}^{-1} . \tag{22}$$

Then

$$t_{00} = \begin{bmatrix} \sigma_e^2 & \sigma_e^2 \end{bmatrix}\begin{bmatrix} \text{tr}(\mathbf{A}_0^2) & \text{tr}[\mathbf{A}_0^2(\mathbf{I}_a \otimes \mathbf{J}_n)] \\ \text{sym} & \text{tr}[\mathbf{A}_0(\mathbf{I}_a \otimes \mathbf{J}_n)]^2 \end{bmatrix}\begin{bmatrix} \sigma_e^2 \\ \sigma_\alpha^2 \end{bmatrix}$$

$$= \begin{bmatrix} \sigma_e^2 & \sigma_\alpha^2 \end{bmatrix}\begin{bmatrix} \text{tr}(\mathbf{I}_a \otimes \mathbf{C}_n)^2 & \text{tr}[(\mathbf{I}_a \otimes \mathbf{C}_n)^2(\mathbf{I}_a \otimes \mathbf{J}_n)] \\ \text{sym} & \text{tr}[(\mathbf{I}_a \otimes \mathbf{C}_n)(\mathbf{I}_a \otimes \mathbf{J}_n)] \end{bmatrix}\begin{bmatrix} \sigma_e^2 \\ \sigma_\alpha^2 \end{bmatrix},$$

and because $\mathbf{C}_n\mathbf{J}_n = \mathbf{0}$,

$$t_{00} = \begin{bmatrix} \sigma_e^2 & \sigma_e^2 \end{bmatrix}\begin{bmatrix} a(n-1) & 0 \\ 0 & 0 \end{bmatrix}\begin{bmatrix} \sigma_e^2 \\ \sigma_\alpha^2 \end{bmatrix} = a(n-1)\sigma_e^4 \quad \text{because} \quad \text{tr}(\mathbf{C}_n) = n - 1 .$$

It is left to the reader (E 5.1) to show that $t_{01} = 0$ and $t_{11} = (a-1)(\sigma_e^2 + n\sigma_\alpha^2)^2$, and hence that (22) reduces to

$$\text{var}\begin{bmatrix} \hat{\sigma}_e^2 \\ \hat{\sigma}_\alpha^2 \end{bmatrix} = \begin{bmatrix} \dfrac{2\sigma_e^4}{a(n-1)} & \dfrac{-2\sigma_e^4}{an(n-1)} \\ \text{sym} & \dfrac{1}{n^2}\left[\dfrac{2\sigma_e^4}{a(n-1)} + \dfrac{2(\sigma_e^2 + n\sigma_\alpha^2)^2}{a-1}\right] \end{bmatrix}, \tag{23}$$

which has the same results as (66), (68) and (71) of Section 3.5d-iii.

-iii. A direct approach. In each of the preceding examples we have seen that derivation of even one element of $E(\mathbf{s})$ or of $\text{var}(\hat{\sigma}^2)$ through using $\text{tr}(\mathbf{AV})$ or $2\text{tr}(\mathbf{AV})^2$, respectively, is usually very tedious. That is why for any particular model, e.g.,

$$y_{ijk} = \mu + \alpha_i + \beta_j + \gamma_{ij} + e_{ijk},$$

one often derives expected values and variances directly. For example, substituting the preceding model equation into $\bar{y}_{i..}$ and $y_{...}$ gives

$$E(\text{SSA}) = E \quad \Sigma_i n_{i\cdot}(\bar{y}_{i..} - y_{...})^2$$

$$= E \quad \Sigma_i n_{i\cdot}\left[\left(\alpha_i - \frac{\Sigma_i n_{i\cdot}\alpha_i}{N}\right) + \left(\frac{\Sigma_j n_{ij}\beta_j}{n_{i\cdot}} - \frac{\Sigma_j n_{\cdot j}\beta_j}{N}\right)\right.$$

$$\left. + \left(\frac{\Sigma_j n_{ij}\gamma_{ij}}{n_{i\cdot}} - \frac{\Sigma_i\Sigma_j n_{ij}\gamma_{ij}}{N}\right) + (\bar{e}_{i..} - \bar{e}_{...})\right]^2 .$$

Then, using properties of the random effects, stemming from (5), such as $E(\alpha_i^2) = \sigma_\alpha^2$, $E(\alpha_i\alpha_{i'}) = 0$ for $i \neq i'$, and so on, one can evaluate this expectation quite straightforwardly.

Similarly, deriving var(SSA) can be achieved a little more easily than using (21). Its original derivation was obtained (Searle, 1958) by writing $\text{SSA} = \Sigma_i n_i \bar{y}_{i..}^2 - N\bar{y}_{...}^2$ and obtaining the individual terms in

$$\text{var(SSA)} = \text{var}(\Sigma_i n_i \bar{y}_{i..})^2 + \text{var}(N\bar{y}_{...}^2) - 2\,\text{cov}(\Sigma_i n_i \bar{y}_{i..}^2, N\bar{y}_{...}^2)\,.$$

Thus by writing $T_A = \Sigma_i n_i \bar{y}_{i..}^2$, and $T_\mu = N\bar{y}_{...}^2$ for what may be called the uncorrected sums of squares, we have $\text{SSA} = T_A - T_\mu$ and

$$\text{var(SSA)} = \text{var}(T_A) + \text{var}(T_\mu) - 2\,\text{cov}(T_A, T_\mu)\,. \tag{24}$$

Using T_A and its natural extensions to other factors provides (see Appendix F) a reasonably economic procedure for deriving variances of SSA and its extensions, and of the covariances between these terms.

Variances of Ts and covariances between them were obtained directly from applying the expressions for var($\mathbf{y}'\mathbf{A}\mathbf{y}$) and cov($\mathbf{y}'\mathbf{A}\mathbf{y}, \mathbf{y}'\mathbf{B}\mathbf{y}$) from Appendix S.5. Although terms in μ occur in the Ts they do not occur in sums of squares like SSA. They were therefore ignored. Then although, as (21) shows, var(s_i) is a quadratic form in the σ^2s, the coefficients of the squares and products of the σ^2s turn out to be fairly complicated functions of the n_{ij}s. And derivation is tedious. We therefore omit the derivation and simply quote one result: excluding terms in μ, and under the normality assumption $\mathbf{y} \sim \mathcal{N}(\mu\mathbf{1}, \mathbf{V})$

$$\begin{aligned}\tfrac{1}{2}\,\text{var}(T_A) = {} & \sigma_\alpha^4 \Sigma_i n_{i.}^2 + \sigma_\beta^4\left(\sum_{i \neq i'}\sum \frac{(\Sigma_j n_{ij} n_{i'j})^2}{n_{i.} n_{i'.}} + \Sigma_i \frac{(\Sigma_j n_{ij}^2)^2}{n_{i.}^2}\right) \\ & + \sigma_\gamma^4\left(\Sigma_i \frac{(\Sigma_j n_{ij}^2)^2}{n_{i.}^2}\right) + a\sigma_e^4 + 2\sigma_\alpha^2(\sigma_\beta^2 + \sigma_\gamma^2)\Sigma_i\Sigma_j n_{ij}^2 \\ & + 2\sigma_\alpha^2\sigma_e^2 N + 2\sigma_\beta^2\sigma_\gamma^2 \Sigma_i \frac{(\Sigma_j n_{ij}^2)^2}{n_{i.}^2} \\ & + 2(\sigma_\beta^2 + \sigma_\gamma^2)\sigma_e^2 \Sigma_i \frac{\Sigma_j n_{ij}^2}{n_{i.}}\,.\end{aligned} \tag{25}$$

Of course, given a data set and a model, the $\mathbf{A}$ for each sum of squares (or quadratic form) expressed as $\mathbf{y}'\mathbf{A}\mathbf{y}$ is known numerically; and if $\mathbf{V}$ is known, or one is prepared to assign numerical values to the σ^2s and hence to $\mathbf{V}$, then one need not bother with algebraic forms of tr$(\mathbf{AV})^2$. Instead, with today's and tomorrow's supercomputers, one can calculate it directly. Nevertheless, there is value to having algebraic techniques available for when something other than numerical results are needed.

e. Unbiased estimation of sampling variances

Unbiased estimation of sampling variances (of estimated components), under normality assumptions, can be achieved for unbalanced data by a direct extension of the method used for mean squares in Appendix S.3b.

With $\hat{\boldsymbol{\sigma}}^2$ being the vector of all the variance components in a model, and $\hat{\boldsymbol{\sigma}}^2$ an unbiased ANOVA estimator, define

$$\mathbf{v} = \text{vech}[\text{var}(\hat{\boldsymbol{\sigma}}^2)] \quad \text{and} \quad \boldsymbol{\gamma} = \text{vech}(\boldsymbol{\sigma}^2\boldsymbol{\sigma}^{2\prime}),$$

where the matrix operator vech($\mathbf{X}$) is defined in Appendix M.7. Thus $\mathbf{v}$ is the vector of all variances of, and covariances between, the estimated components, and $\boldsymbol{\gamma}$ is the vector of all squares and products of the σ^2s.

What we seek is an unbiased estimator of $\mathbf{v} = \text{vech}[\text{var}(\hat{\boldsymbol{\sigma}}^2)]$. By (21), every element in $\text{var}(\hat{\boldsymbol{\sigma}}^2)$ is a linear combination of squares and products of σ^2s. Hence every element in $\mathbf{v} = \text{vech}[\text{var}(\hat{\boldsymbol{\sigma}}^2)]$ is a linear combination of elements in $\boldsymbol{\gamma} = \text{vech}(\boldsymbol{\sigma}^2\boldsymbol{\sigma}^{2\prime})$. Hence there is always a matrix, call it $\mathbf{B}$, having elements that are not functions of σ^2s, such that

$$\mathbf{v} = \mathbf{B}\boldsymbol{\gamma} . \tag{26}$$

Now ANOVA estimators are unbiased; and for any pair of unbiased estimators, $\hat{\sigma}_i^2$ and $\hat{\sigma}_j^2$,

$$E(\hat{\sigma}_i^4) = \text{var}(\hat{\sigma}_i^2) + \sigma_i^4 \quad \text{and} \quad E(\hat{\sigma}_i^2\hat{\sigma}_j^2) = \text{cov}(\hat{\sigma}_i^2, \hat{\sigma}_j^2) + \sigma_i^2\sigma_j^2 .$$

Therefore with $\hat{\boldsymbol{\gamma}} = \text{vech}(\hat{\boldsymbol{\sigma}}^2\hat{\boldsymbol{\sigma}}^{2\prime})$ being $\boldsymbol{\gamma}$ with each σ^2 replaced by the corresponding $\hat{\sigma}^2$,

$$E(\hat{\boldsymbol{\gamma}}) = \mathbf{v} + \boldsymbol{\gamma} = (\mathbf{I} + \mathbf{B})\boldsymbol{\gamma},$$

from (26). Hence $\hat{\boldsymbol{\gamma}}$ is an unbiased estimate of $(\mathbf{I} + \mathbf{B})\boldsymbol{\gamma}$. Thus $(\mathbf{I} + \mathbf{B})^{-1}\hat{\boldsymbol{\gamma}}$ is an unbiased estimator of $\boldsymbol{\gamma}$; and since (26) has $\mathbf{v} = \mathbf{B}\boldsymbol{\gamma}$,

$$\hat{\mathbf{v}} = \mathbf{B}(\mathbf{I} + \mathbf{B})^{-1}\hat{\boldsymbol{\gamma}} \tag{27}$$

is an unbiased estimator of $\mathbf{v}$. Thus for $\mathbf{B}$ defined by

$$\text{vech}[\text{var}(\hat{\boldsymbol{\sigma}}^2)] = \mathbf{B}\,\text{vech}(\boldsymbol{\sigma}^2\boldsymbol{\sigma}^{2\prime})$$

$$\mathbf{B}(\mathbf{I} + \mathbf{B})^{-1}\,\text{vech}(\hat{\boldsymbol{\sigma}}^2\hat{\boldsymbol{\sigma}}^{2\prime}) \quad \text{estimates} \quad \mathbf{v} = \text{vech}[\text{var}(\hat{\boldsymbol{\sigma}}^2)] \tag{28}$$

unbiasedly. Mahamunulu (1963) uses elements of this principle, although Ahrens (1965) derived (28). In passing, note that $\mathbf{B}(\mathbf{I} + \mathbf{B})^{-1} = (\mathbf{I} + \mathbf{B})^{-1}\mathbf{B}$.

Example 2 (continued). In the 1-way classification, with balanced data, we have, from writing (23) in the form of (28), that $\mathbf{B}$ is the 3×3 matrix in

$$\begin{bmatrix} \text{var}(\hat{\sigma}_e^2) \\ \text{var}(\hat{\sigma}_\alpha^2) \\ \text{var}(\hat{\sigma}_e^2, \hat{\sigma}_\alpha^2) \end{bmatrix} = \begin{bmatrix} \dfrac{2}{a(n-1)} & 0 & 0 \\ \dfrac{2(an-1)}{an^2(a-1)(n-1)} & \dfrac{2}{a-1} & \dfrac{4}{n(a-1)} \\ -\dfrac{2}{an(n-1)} & 0 & 0 \end{bmatrix} \begin{bmatrix} \sigma_e^4 \\ \sigma_\alpha^4 \\ \sigma_e^2\sigma_\alpha^2 \end{bmatrix} . \tag{29}$$

Then calculating $\mathbf{B}(\mathbf{I}+\mathbf{B})^{-1}$ gives (27) as

$$\begin{bmatrix} \hat{\text{var}}(\hat{\sigma}_e^2) \\ \hat{\text{var}}(\hat{\sigma}_\alpha^2) \\ \hat{\text{cov}}(\hat{\sigma}_e^2, \hat{\sigma}_\alpha^2) \end{bmatrix} = \begin{bmatrix} \dfrac{2}{a(n-1)+2} & 0 & 0 \\ \dfrac{2}{n^2}\left[\dfrac{1}{a+1} + \dfrac{1}{a(n-1)+2}\right] & \dfrac{2}{a+1} & \dfrac{4}{n(a+1)} \\ \dfrac{-2}{n[a(n-1)+2]} & 0 & 0 \end{bmatrix} \begin{bmatrix} \hat{\sigma}_e^4 \\ \hat{\sigma}_\alpha^4 \\ \hat{\sigma}_e^2\sigma_\alpha^4 \end{bmatrix},$$

which is equivalent to the expressions in (67), (70) and (72) of Chapter 3. It is left to the reader as E 5.3 to carry out the details, and as E 5.4 to do the same for unbalanced data.

5.3. HENDERSON'S METHOD I

The Henderson (1953) paper is a landmark in the estimation of variance components from unbalanced data. It established three different sets of quadratic forms that could be used for $\mathbf{s}$ in the ANOVA method of estimation of (19). All three sets are closely related to the sums of squares of analysis of variance calculations for unbalanced data: an extension to multivariate data is suggested by Wesolowska-Janczarek (1984). Although the methodology is the same with each set of quadratics (equate them to their expected values), the three uses of them have come to be known as Henderson's Methods I, II and III. In brief, Method I uses quadratic forms that are analogous to the sums of squares of balanced data; Method II is an adaptation of Method I that takes account of the fixed effects in the model; and Method III uses sums of squares from fitting whatever linear model (treated as a fixed effects model) is being used and submodels thereof. Henderson (1953) describes these methods without benefit of matrix notation. Searle (1968) reformulated the methods in matrix notation, generalized Method 2, and suggested it had no unique usage for any given set of data. But Henderson, Searle and Schaeffer (1974) show that this suggestion is wrong, and they also give simplified calculation procedures. We draw heavily on these papers in what follows. And in doing so, much of the description is in terms of the 2-way crossed classification, with interaction, random model and, of course, unbalanced data. This is the simplest case that provides opportunity for describing most, if not all, of the features of the Henderson methods.

a. The quadratic forms

Method I uses quadratic forms adapted directly from the sums of squares of the analysis of variance of balanced data. In some cases these are sums of squares and in others they are not. We use the 2-way crossed classification for illustration and begin with a familiar sum of squares used with balanced data; namely $bn\Sigma_i(\bar{y}_{i..} - \bar{y}_{...})^2$. The corresponding form for unbalanced data is $\text{SSA} = \Sigma_i n_{i.}(\bar{y}_{i..} - \bar{y}_{...})^2$, which can also be recognized as $R(\boldsymbol{\alpha} \mid \mu)$.

SSA can also be seen as a generalization of the balanced data sum of squares (to which it is equal, of course, when all $n_{ij} = n$), obtained by replacing bn by $n_{i.}$, in both cases the number of observations in $\bar{y}_{i..}$. Henderson's Method I uses SSA; and it also uses a quadratic form that comes from extending the interaction sum of squares, which for balanced data is

$$\Sigma_i\Sigma_j n(\bar{y}_{ij.} - \bar{y}_{i..} - \bar{y}_{.j.} - \bar{y}_{...})^2 = \Sigma_i\Sigma_j n\bar{y}_{ij.}^2 - \Sigma_i bn\bar{y}_{i..}^2 - \Sigma_j an\bar{y}_{.j.}^2 + abn\bar{y}_{...}^2 .$$

Unbalanced-data analogues of this equality are

$$\text{SSAB} = \Sigma_i\Sigma_j n_{ij}(\bar{y}_{ij.} - \bar{y}_{i..} - \bar{y}_{.j.} + \bar{y}_{...})^2$$

for the left-hand side and

$$\text{SSAB*} = \sum_i\sum_j \frac{y_{ij.}^2}{n_{ij}} - \sum_i \frac{y_{i..}^2}{n_{i.}} - \sum_j \frac{y_{.j.}^2}{n_{.j}} + \frac{y_{...}^2}{n_{..}}$$

for the right-hand side. But, despite what one might anticipate, SSAB and SSAB* are, in general, not equal. (They are equal, of course, when all $n_{ij} = n$.) But their difference is

$$\begin{aligned} \text{SSAB*} - \text{SSAB} &= -2\Sigma_i\Sigma_j n_{ij}(\bar{y}_{i..} - \bar{y}_{...})(\bar{y}_{.j.} - \bar{y}_{...}) \\ &= -2(\Sigma_i\Sigma_j n_{ij}\bar{y}_{i..}\bar{y}_{.j.} - N\bar{y}_{...}^2) . \end{aligned}$$

SSAB is clearly a sum of squares; it is therefore a positive semi-definite quadratic form. It can never be negative (for real values of y_{ijk}). In contrast SSAB* can be negative. For instance, with data of 2 rows and 2 columns

6	4	10
6, 42	12	60
54	16	70

$$\text{SSAB*} = \frac{6^2}{1} + \frac{4^2}{1} + \frac{48^2}{2} + \frac{12^2}{1} - \left(\frac{10^2}{2} + \frac{60^2}{3}\right) - \left(\frac{54^2}{3} + \frac{16^2}{2}\right) + \frac{70^2}{5} = -22 .$$

Thus SSAB*, although it is a quadratic form, is not non-negative, and so it is not a sum of squares. Nevertheless, this is what is used in Henderson's Method I.

The four quadratic forms that are used in Henderson's Method I are thus

$$\text{SSA} = \Sigma_i n_{i.}(\bar{y}_{i..} - \bar{y}_{...})^2,$$

$$\text{SSB} = \Sigma_j n_{.j}(\bar{y}_{.j.} - \bar{y}_{...})^2,$$

$$\text{SSAB*} = \Sigma_i\Sigma_j n_{ij}\bar{y}_{ij.}^2 - \Sigma_i n_{i.}\bar{y}_{i..}^2 - \Sigma_j n_{.j}\bar{y}_{.j.}^2 + N\bar{y}_{...}^2$$

and

$$\text{SSE} = \Sigma_i\Sigma_j\Sigma_k(y_{ijk} - \bar{y}_{ij.})^2 .$$

It is interesting to note that although these four expressions never occur all together in any traditional partitioning of the total sum of squares for the fixed effects model, they do add to that total,

$$\mathrm{SST_m} = \Sigma_i \Sigma_j \Sigma_k (y_{ijk} - \bar{y}_{...})^2 \,.$$

The four never occur together for at least two reasons: SSA is $R(\boldsymbol{\alpha} \mid \mu)$ and $\mathrm{SSB} = R(\boldsymbol{\beta} \mid \mu)$, and they never occur together; and SSAB* is not a sum of squares. Nevertheless, they do represent a partitioning of $\mathrm{SST_m}$.

b. Estimation

The estimation method is to find the expected value of each of the four quadratic forms and to equate those to the observed values of those forms—the values calculated from the data. This gives, for the random model, four linear equations in four variance components.

The tedious part is deriving the expected values. Two terms of $E(\mathrm{SSA})$ are shown in Section 5.2b-iii, and the complete expression for $E(\mathrm{SSA})$ is given in E 5.2. Expected values of SSB and SSAB* are derived in similar fashion; and, of course, $E(\mathrm{SSA}) = (N - s)\sigma_e^2$, where s is the number of filled cells. To simplify notation in the estimation equations, define

$$k_1 = \Sigma_i n_{i\cdot}^2, \quad k_2 = \Sigma_j n_{\cdot j}^2,$$
$$k_3 = \sum\nolimits_i \frac{\Sigma_j n_{ij}^2}{n_{i\cdot}}, \quad k_4 = \sum\nolimits_j \frac{\Sigma_i n_{ij}^2}{n_{\cdot j}}, \tag{30}$$
$$k_{23} = \Sigma_i \Sigma_j n_{ij}^2$$

and for any k_r define

$$k_r' = k_r / N \,.$$

Then for $N' = s - a - b + 1$ the estimation equations are

$$\begin{bmatrix} N - k_1' & k_3 - k_2' & k_3 - k_{23}' & a - 1 \\ k_4 - k_1' & N - k_2' & k_4 - k_{23}' & b - 1 \\ k_1' - k_4 & k_2' - k_3 & N - k_3 - k_4 + k_{23}' & N' \\ 0 & 0 & 0 & N - s \end{bmatrix} \begin{bmatrix} \hat{\sigma}_\alpha^2 \\ \hat{\sigma}_\beta^2 \\ \hat{\sigma}_\gamma^2 \\ \hat{\sigma}_e^2 \end{bmatrix} = \begin{bmatrix} \mathrm{SSA} \\ \mathrm{SSB} \\ \mathrm{SSAB^*} \\ \mathrm{SSE} \end{bmatrix}, \tag{31}$$

where the expected value of the right-hand side of (31) is the left-hand side with each $\hat{\sigma}^2$ replaced by the corresponding σ^2. That, through the ANOVA estimation principle of equating quadratic forms to their expected values, is the origin of (31).

Having the condition $\mathbf{X}'\mathbf{A}_i\mathbf{X} = \mathbf{0}$ in (16) is salient to ANOVA estimation of variance components. It takes the form $\mathbf{1}'\mathbf{A}_i\mathbf{1} = \mathbf{0}$ in random models. This condition for the terms in (31) has, for example, $\mathrm{SSA} = \mathbf{y}'\mathbf{A}_i\mathbf{y}$ for $\mathbf{A}_i = \{_\mathrm{d}\, \bar{\mathbf{J}}_{n_{i\cdot}}\} - \bar{\mathbf{J}}_N$ from Section 5.2b-iii, and so

$$\mathbf{1}'\mathbf{A}_i\mathbf{1} = \sum\nolimits_i \frac{n_{i\cdot}^2}{n_{i\cdot}} - \frac{N^2}{N} = N - N = 0 \,.$$

Solutions to (31) can, on defining

$$\mathbf{P} = \begin{bmatrix} N - k'_1 & k_3 - k'_2 & k_3 - k'_{23} \\ k_4 - k'_1 & N - k'_2 & k_4 - k'_{23} \\ k'_1 - k_4 & k'_2 - k_3 & N - k_3 - k_4 + k'_{23} \end{bmatrix}, \tag{32}$$

be expressed as

$$\hat{\sigma}_e^2 = \frac{\text{SSE}}{N - s} = \text{MSE}$$

and

$$\begin{bmatrix} \hat{\sigma}_\alpha^2 \\ \hat{\sigma}_\beta^2 \\ \hat{\sigma}_\gamma^2 \end{bmatrix} = \mathbf{P}^{-1} \begin{bmatrix} \text{SSA} - (a-1)\text{MSE} \\ \text{SSB} - (b-1)\text{MSE} \\ \text{SSAB*} - (N')\text{MSE} \end{bmatrix}. \tag{33}$$

These are the Henderson Method I estimators of the variance components in a 2-way cross-classification, random model.

c. Negative estimates

It is clear that $\hat{\sigma}_e^2 = \text{MSE}$ is always positive. But from (33) it is equally as clear that estimates of the other variance components are not necessarily non-negative. So here we are, back at the familiar problem of having an estimation method that does not preclude negative estimates. The reason is that nothing is built into the estimation method to ensure that negative estimates do not occur. This is true of all applications of ANOVA methodology.

d. Sampling variances

On the basis of assuming normality, namely $\mathbf{y} \sim \mathcal{N}(\mu\mathbf{1}, \mathbf{V})$ for the random model, expressions can be derived that lead to sampling variances of the Method I estimators. For $\hat{\sigma}_e^2$ we have $\text{SSE}/\sigma_e^2 \sim \chi^2_{N-s}$ and so

$$\text{var}(\hat{\sigma}_e^2) = \frac{2\sigma_e^4}{N - s}.$$

But for the other estimators the results are not so simple. Writing SSA, SSB and SSAB* in terms of uncorrected sums of squares, like $T_A = \Sigma_i n_{i\cdot} \bar{y}_{i\cdot\cdot}^2$, gives

$$\begin{bmatrix} \text{SSA} \\ \text{SSB} \\ \text{SSAB*} \end{bmatrix} = \begin{bmatrix} 1 & 0 & 0 & -1 \\ 0 & 1 & 0 & -1 \\ -1 & -1 & 1 & 1 \end{bmatrix} \mathbf{t} \quad \text{for } \mathbf{t} = \begin{bmatrix} T_A \\ T_B \\ T_{AB} \\ T_\mu \end{bmatrix}. \tag{34}$$

Using Theorem S4 of Appendix S.5 to show that MSE is independent of SSA, SSB and SSAB*, and on defining

$$\mathbf{H} = \begin{bmatrix} 1 & 0 & 0 & -1 \\ 0 & 1 & 0 & -1 \\ -1 & -1 & 1 & 1 \end{bmatrix} \quad \text{and} \quad \mathbf{f} = \begin{bmatrix} a-1 \\ b-1 \\ N' \end{bmatrix}, \tag{35}$$

we can, from (33), write

$$\text{var}\begin{bmatrix} \hat{\sigma}_\alpha^2 \\ \hat{\sigma}_\beta^2 \\ \hat{\sigma}_\gamma^2 \end{bmatrix} = \mathbf{P}^{-1}[\mathbf{H}\,\text{var}(\mathbf{t})\mathbf{H}' + \text{var}(\hat{\sigma}_e^2)\mathbf{ff}']\mathbf{P}^{-1\prime}\,. \tag{36}$$

Searle (1971, p. 482) gives the elements of var($\mathbf{t}$). There are ten of them, each a quadratic in σ_α^2, σ_β^2, σ_γ^2 and σ_e^2; and thus each of the ten different elements of var($\mathbf{t}$) is a linear combination of the ten squares and products of σ_α^2, σ_β^2, σ_γ^2 and σ_e^2. The coefficients of those squares and products are therefore set out in a 10×10 matrix. Those coefficients involve 28 different k-terms, of which but five are shown in (30). The full set of 28 is shown in Table F.1 of Appendix F. Also shown there are detailed formulae for three nested-classification random models, and for four forms of the 2-way crossed classification, embodying the double dichotomy of with and without interaction, and of random and mixed models. Searle (1971, pp. 491–493) also has details of Henderson I estimators for the 3-way crossed classification with all interactions, random model. That model has 8 variance components, $\mathbf{t}$ has 8 elements and so var($\mathbf{t}$) has 36 different elements, each involving the 36 squares and products of the 8 variance components. The required coefficients, developed by Blischke (1968) as a 36×36 matrix, are given in Searle (1971, pp. 494–514). Printing those twenty-one pages is only necessary once!

It is clear that expression (36) is not at all amenable to studying the behaviour of sampling variances of estimated variance components obtained by the Henderson Method I. And this is seen to be true for the specific models in Appendix F. Each element of (36) is a quadratic form of the unknown σ^2s. But the coefficients of the squares and products of the σ^2s are complicated functions of the numbers of observations in the cells and subclasses of the data. For example, in (36), the matrix $\mathbf{P}^{-1}$ has, by virtue of (30) and (32), elements that are in no way simple functions of the n_{ij}-values. This precludes any thought whatever of making analytical studies of the variance functions as to how they behave either for different sets of values for the σ^2s, or for different sets of n_{ij}-values. Even for the 1-way classification, as discussed in Section 3.6-iv, these kinds of studies were demonstrated as not being readily feasible. With two and more factors the intractability of expressions leading to sampling variances becomes increasingly aggravated.

e. A general coefficient for Method I

After substituting the model equation $y_{ijk} = \mu + \alpha_i + \beta_j + \gamma_{ij} + e_{ijk}$ into $\bar{y}_{i..}$ of $T_A = \Sigma_i n_i \bar{y}^2_{i..}$, it is not difficult to show that the coefficient of σ^2_β in $E(T_A)$, which we shall denote by $c[\sigma^2_\beta : E(T_A)]$ is

$$c[\sigma^2_\beta : E(T_A)] = \sum_{i=1}^{a} n_{i.} \frac{\sum_{j=1}^{b} n^2_{ij}}{n^2_{i.}} = \sum_{i=1}^{a} \frac{\sum_{j=1}^{b} n^2_{ij}}{n_{i.}} .$$

One of the minor advantages of Henderson's Method I is that the quadratic forms are all linear combinations of uncorrected sums of squares like Ts. Therefore, similar to $c[\sigma^2_\beta : E(T_A)]$, one can write down a general expression for the coefficient of any σ^2 in the expected value of any such T.

Suppose there are $n_{\cdot t(i)}$ observations in the tth level of the ith factor of a multi-factor model, their total being $y_{\cdot t(i)}$. Let q_i denote the number of levels of the ith factor that occur in the data. Then the T for the ith factor is

$$T_i = \sum_{t(i)=1}^{q_i} n_{\cdot t(i)} \bar{y}^2_{\cdot t(i)} .$$

Let $u_{s(j)}$ be the effect for the sth level of the jth random factor, and let $n_{\cdot t(i)s(j)}$ be the number of observations in the cell defined by the tth level of factor i and the sth level of factor j. Then, for r random factors (main effect factors, interactions or nested)

$$T_i = \sum_{t(i)=1}^{q_i} \frac{\left[n_{\cdot t(i)}\mu + \sum_{j=1}^{r} \sum_{s(j)=1}^{q_j} n_{\cdot t(i)s(j)} u_{s(j)} + e_{\cdot t(i)} \right]^2}{n_{\cdot t(i)}} .$$

Hence, similar to $c[\sigma^2_\beta : E(T_A)]$, the coefficient of σ^2_j in T_i is

$$c[\sigma^2_j : E(T_i)] = \sum_{t(i)=1}^{q_i} \frac{\sum_{s(j)=1}^{q_j} n^2_{\cdot t(i)s(j)}}{n_{\cdot t(i)}} .$$

It is easily shown that

$$c[\mu^2 : E(T_i)] = N \quad \text{and} \quad c[\sigma^2_e : E(T_i)] = q_i \sigma^2_e .$$

These results can be used as needed for the application of Henderson's Method I to any random model. The quadratic forms that are used are analogous to the analysis of variance sums of squares for balanced data, like those in the preceding sub-section b, for the 2-way crossed classification. They are formed as contrasts of the T_i-terms, of which (34) is an illustration.

As indicated in sub-section g that follows, Method I should not be used with mixed models. Hence the results of this sub-section being confined to random models is no restriction so far as Method I is concerned.

f. Synthesis

Synthesis is the name given by Hartley (1967) to his method of calculating coefficients of variance components in expected quadratic forms without needing to know the algebraic expressions for those coefficients. It operates as follows. Denote the sth column of $\mathbf{Z}_j$ in (1) by $\mathbf{z}_{sj}$. Then in any quadratic form $\mathbf{y}'\mathbf{A}\mathbf{y}$ the coefficient of σ_j^2 in $E(\mathbf{y}'\mathbf{A}\mathbf{y})$ is

$$c[\sigma_j^2 : E(\mathbf{y}'\mathbf{A}\mathbf{y})] = \Sigma_s \mathbf{z}'_{sj}\mathbf{A}\mathbf{z}_{sj} \,. \tag{37}$$

Observe that $\mathbf{z}'_{sj}\mathbf{A}\mathbf{z}_{sj}$ is the same quadratic form in $\mathbf{z}_{sj}$ as $\mathbf{y}'\mathbf{A}\mathbf{y}$ is in $\mathbf{y}$. Hence if $\mathbf{y}'\mathbf{A}\mathbf{y}$ is a sum of squares SS($\mathbf{y}$), the coefficient of σ_j^2 in $E[\text{SS}(\mathbf{y})]$ is obtained by summing SS($\mathbf{z}_{sj}$) for every column of $\mathbf{Z}_j$. Thus (37) represents a very general procedure for calculating the numerical coefficients in any particular case. Use each column $\mathbf{z}_{sj}$ of $\mathbf{Z}_j$ as data for the sum of squares and add over $s = 1, 2, \ldots$. Do this for each $j = 0, 1, \ldots, r$. It is feasible, but not so useful, for algebraic derivations, as illustrated in the following example.

Example 2 (continued). In the 1-way classification, random model, balanced data, $\mathbf{Z}_1$ corresponding to the random effect $\boldsymbol{\alpha}$ is $\mathbf{Z}_1 = \mathbf{I}_a \otimes \mathbf{1}_n$. And in $\text{SSA} = \Sigma_i n_i(\bar{y}_{i\cdot} - \bar{y}_{\cdot\cdot})^2$

$$c[\sigma_\alpha^2 : E(\text{SSA})] = \sum_{s=1}^{a} \text{SSA}(\mathbf{z}_{s1})$$

where SSA($\mathbf{z}_{s1}$) is SSA for the 1-way classification analysis of variance of the sth column of $\mathbf{Z}_1 = \mathbf{I}_a \otimes \mathbf{1}_n$ used as data. Thus, since $\text{SSA}(\mathbf{y}) = \Sigma_{i=1}^a n(\bar{y}_{i\cdot} - \bar{y}_{\cdot\cdot})^2$, and each column of $\mathbf{Z}_1$ has n unities and $an - n$ zeros,

$$\text{SSA}(\mathbf{z}_{s1}) = (an - n)\left(0 - \frac{n}{an}\right)^2 + n\left(1 - \frac{n}{an}\right)^2 = n\left(1 - \frac{1}{a}\right).$$

This is the case for all a columns of $\mathbf{Z}_1$ and so

$$c[\sigma_\alpha^2 : E(\text{SSA})] = \sum_{i=1}^{a} n\left(1 - \frac{1}{a}\right) = n(a - 1)$$

as one would expect (see Table 3.3).

The reason for (37) is almost self-evident:

$$E(\mathbf{y}'\mathbf{A}\mathbf{y}) = \text{tr}(\mathbf{A}\mathbf{V}) = \text{tr}\left(\mathbf{A} \sum_{j=0}^{r} \mathbf{Z}_j\mathbf{Z}'_j\sigma_j^2\right).$$

Therefore the coefficient of σ_j^2 is

$$c[\sigma_j^2 : E(\mathbf{y}'\mathbf{A}\mathbf{y})] = \text{tr}(\mathbf{A}\mathbf{Z}_j\mathbf{Z}'_j) = \text{tr}(\mathbf{Z}'_j\mathbf{A}\mathbf{Z}_j) = \sum_{s=1}^{q_j} \mathbf{z}'_{sj}\mathbf{A}\mathbf{z}_{sj} \,.$$

Something similar can also be done for sampling variances. See Hartley (1967).

g. Mixed models

At several points we have stated that Method I can be used only on random models. This is so because (with unbalanced data) any attempt at using Method I results in the expected mean squares containing functions of fixed effects that do not drop out as do the terms in μ^2. We illustrate this for the 2-way crossed classification.

Suppose the A-factor is a fixed effects factor. Then with $\mathbf{X}_1 = \{_d\, \mathbf{1}_{n_i}\}$, as is $\mathbf{Z}_1$ in the random effects model, the terms in μ and elements of $\boldsymbol{\alpha}$ that will occur in $E(\text{SSB})$ are $(\mu\mathbf{1} + \mathbf{X}_1\boldsymbol{\alpha})'\mathbf{B}(\mu\mathbf{1} + \mathbf{X}_1\boldsymbol{\alpha})$ for $\mathbf{B}$ defined by $\text{SSB} = \Sigma_j n_{.j}\bar{y}_{.j.}^2 - N\bar{y}_{...}^2 = \mathbf{y}'\mathbf{B}\mathbf{y}$. Motivated by $\text{SSA} = \mathbf{y}'\mathbf{A}_i\mathbf{y}$ for $\mathbf{A} = \{_d\, \bar{\mathbf{J}}_{n_{i.}}\} - \bar{\mathbf{J}}_N$, one might expect $\mathbf{B}$ to be $\{_d\, \bar{\mathbf{J}}_{n_{.j}}\} - \bar{\mathbf{J}}_N$. But it is not. The form of $\mathbf{A}$ depends on the fact that $\mathbf{y}$ has its elements ordered by j with i: if they were ordered by i within j then $\mathbf{B}$ would be as expected. But with elements of $\mathbf{y}$ ordered by i within j, the matrix $\mathbf{B}$ is, for $\mathbf{Q}$ being some permutation matrix,

$$\mathbf{B} = \mathbf{Q}[\{_d\, \bar{\mathbf{J}}_{n_{.j}}\} - \bar{\mathbf{J}}_N]\mathbf{Q}' = \mathbf{Q}\{_d\, \bar{\mathbf{J}}_{n_{.j}}\}\mathbf{Q}' - \bar{\mathbf{J}}_N .$$

Then the term in μ is

$$\mu\mathbf{1}'\mathbf{B}\mathbf{1}\mu = \mu^2(\mathbf{1}'\mathbf{Q}\{_d\, \bar{\mathbf{J}}_{n_{.j}}\}\mathbf{Q}'\mathbf{1} - \mathbf{1}'\bar{\mathbf{J}}_N\mathbf{1}) = \mu^2\left(\mathbf{1}'\{_d\, \bar{\mathbf{J}}_{n_{.j}}\}\mathbf{1} - \frac{N^2}{N}\right) = \mathbf{0} .$$

But using $\mathbf{B}$ for the term in αs in $E(\text{SSB})$ will be difficult because no specific form of the permutation matrix $\mathbf{Q}$ is known. Nevertheless, using $E(\text{SSB}) = E\ \Sigma_j n_{.j}(\bar{y}_{.j.} - \bar{y}_{...})^2$, it is easily shown that the term in αs in $E(\text{SSB})$ is

$$\theta_1 = \Sigma_j n_{.j}\left[\sum_i \alpha_i\left(\frac{n_{ij}}{n_{.j}} - \frac{n_{i.}}{n_{..}}\right)\right]^2 . \tag{38}$$

Similarly in

$$E(\text{SSAB*}) = E\left(\sum_i \sum_j \frac{y_{ij.}^2}{n_{ij}} - \sum_i \frac{y_{i..}^2}{n_{i.}} - \sum_j \frac{y_{.j.}^2}{n_{.j}} + \frac{y_{...}^2}{n_{..}}\right)$$

the term in the αs is

$$\theta_2 = \frac{(\Sigma_i n_{i.}\alpha_i)^2}{n_{..}} - \sum_j \frac{(\Sigma_i n_{ij}\alpha_i)^2}{n_{.j}} . \tag{39}$$

It can be noted that both θ_1 and θ_2 reduce to zero for balanced data; and that if the αs were random effects, θ_1 and θ_2 become the corresponding coefficients of $\hat{\sigma}_\alpha^2$ in (31), since the estimators in (31) are unbiased. See E 5.6.

The important feature of θ_1 and θ_2 in (38) and (39) is that they are functions of fixed effects that occur in expected values of quadratic forms. Thus in equating the observed values of those quadratic forms to their expectations we cannot simply solve for the variance components. The θs get in the way. Although μ drops out of the expected sums of squares, the other fixed effects do not. What this amounts to is that for mixed models $\mathbf{X}$ is not just $\mathbf{1}$, and whereas $\mathbf{1}'\mathbf{A}\mathbf{1} = \mathbf{0}$, which is $\mathbf{X}'\mathbf{A}_i\mathbf{X} = \mathbf{0}$ for $\mathbf{X} = \mathbf{1}$ for random models, the $\mathbf{X}$ of mixed models and the **A**-matrices of Method I are such that, for unbalanced data, $\mathbf{X}'\mathbf{A}\mathbf{X} \neq \mathbf{0}$; i.e., the condition for fixed effects to drop out of $E(\mathbf{y}'\mathbf{A}_i\mathbf{y})$ of (16) is not satisfied. Thus it is that with unbalanced data one cannot use Henderson's Method I for mixed models. It is suitable only for random models.

Because the arithmetic of Method 1 is the easiest of all methods, one can be tempted to use it on mixed models, even though one is then knowingly introducing error. The two ways of doing this are either (a) ignore the fixed effects, i.e., drop them from the mixed model entirely, or (b) treat the fixed effects as random and estimate variance components for them under that assumption. In either case the resulting variance components estimators for the true random effects are not unbiased.

h. Merits and demerits

The merits of Method I include the following.

(i) Computation is easy even for very large data sets. No matrices are involved except for one or two of order no more than the number of variance components—and that is usually a small number.

(ii) Estimators are unbiased.

(iii) In many cases unbiased estimators are available for the sampling dispersion matrix of the variance components estimators—assuming normality of the data.

(iv) For balanced data Method I simplifies to be identical to the (unique) ANOVA method.

Demerits include the following.

(i) The method does not preclude the possibility of negative estimates.

(ii) Under the usual normality assumptions, the probability density function of the estimators cannot be specified in closed form—save for that of the error variance, which is often proportional to a χ^2.

(iii) This method can be used only for random models. It cannot be used for mixed models.

i. A numerical example

We use the small, hypothetical data set of Table 5.2 to illustrate the calculations of Method I, for a 2-way crossed classification of 2 rows and 3 columns.

TABLE 5.2. DATA FOR A 2-WAY CROSSED CLASSIFICATION

	y_{ijk}			$y_{i..}$		$\bar{y}_{ij.}$			$\bar{y}_{i..}$		n_{ij}			$n_{i.}$
	7,9	6	2	24		8	6	2	6		2	1	1	4
	8	4,8	12	32		8	6	12	8		1	2	1	4
$y_{.j.}$	24	18	14	$56 = y_{...}$	$\bar{y}_{.j.}$	8	6	7	7	$n_{.j}$	3	3	2	$8 = n_{..}$

$$\begin{aligned}
\text{SSA} &= 4(6-7)^2 + 4(8-7)^2 &&= 8,\\
\text{SSB} &= 3(8-7)^2 + 3(6-7)^2 + 2(7-7)^2 &&= 6,\\
\text{SSAB*} &= 2(8^2) + 1(6^2) + 1(2^2) + 1(8^2) + 2(6^2) + 1(12^2) - [4(6^2) + 4(8^2)]\\
&\quad - [3(8^2) + 3(6^2) + 2(7^2)] + 8(7^2) = 42,\\
\text{SSE} &= 2 + 8 &&= 10\,.
\end{aligned}$$

From (29) and (30)

$$k_1 = 4^2 + 4^2 = 32, \quad k_2 = 3^2 + 3^2 + 2^2 = 22,$$

$$k_3 = \frac{2^2 + 1^2 + 1^2}{4} + \frac{1^2 + 2^2 + 1^2}{4} = 3, \quad k_4 = \frac{2^2 + 1^2}{3} + \frac{2^2 + 1^2}{3} + \frac{1^2 + 1^2}{2} = 4\tfrac{1}{3}$$

$$k_{23} = 2^2 + 1^2 + 1^2 + 1^2 + 2^2 + 1^2 = 12,$$

$$k'_1 = \frac{32}{8} = 4, \qquad k'_2 = \frac{22}{8} = 2\tfrac{3}{4}, \qquad k'_{23} = \frac{12}{8} = 1\tfrac{1}{2}\,.$$

Then (31) is

$$\begin{bmatrix} 8-4 & 3-2\frac{3}{4} & 3-1\frac{1}{2} & 2-1\\ 4\frac{1}{3}-4 & 8-2\frac{3}{4} & 4\frac{1}{3}-1\frac{1}{2} & 3-1\\ 4-4\frac{1}{3} & 2\frac{3}{4}-3 & 8-3-4\frac{1}{3}+1\frac{1}{2} & 6-2-3+1\\ 0 & 0 & 0 & 8-6 \end{bmatrix}\begin{bmatrix}\hat{\sigma}_\alpha^2\\ \hat{\sigma}_\beta^2\\ \hat{\sigma}_\gamma^2\\ \hat{\sigma}_e^2\end{bmatrix} = \begin{bmatrix}8\\6\\42\\10\end{bmatrix},$$

i.e.,

$$\begin{bmatrix} 4 & \frac{1}{4} & 1\frac{1}{2} & 1\\ \frac{1}{3} & 5\frac{1}{4} & 2\frac{5}{6} & 2\\ -\frac{1}{3} & -\frac{1}{4} & 2\frac{1}{6} & 2\\ 0 & 0 & 0 & 2 \end{bmatrix}\begin{bmatrix}\hat{\sigma}_\alpha^2\\ \hat{\sigma}_\beta^2\\ \hat{\sigma}_\gamma^2\\ \hat{\sigma}_e^2\end{bmatrix} = \begin{bmatrix}8\\6\\42\\10\end{bmatrix},$$

with solution

$$\hat{\sigma}_\alpha^2 = \frac{-454.6}{121} = -1.6909, \quad \hat{\sigma}_\beta^2 = \frac{-932}{121} = -7.7024,$$

$$\hat{\sigma}_\gamma^2 = \frac{1609.6}{121} = 13.3025, \quad \hat{\sigma}_e^2 = 5\,.$$

5.4. HENDERSON'S METHOD II

The purpose of Method II is to provide a method of estimation that retains the relatively easy arithmetic of Method I but which is usable for mixed models that contain a term $\mu\mathbf{1}$ for a general mean μ. It achieves this by adjusting the data (in some sense) for the fixed effects, and from the adjusted data the variance components are estimated by a variant of Method I. Method II can therefore be thought of as an adaptation of Method I that overcomes the deficiency of Method I that the need for having $\mathbf{X}'\mathbf{A}_i\mathbf{X} = \mathbf{0}$ in Method I makes it unavailable for mixed models; i.e., Method I cannot be used for mixed models. Method II can be used for mixed models, but only for those containing no interactions between fixed and random effects (see subsection f which follows). Method II involves adjusting the data in a manner that produces a vector of adjusted observations for which the linear model is a completely random model consisting of a general mean and all the random effects parts of the model for y—except for a transformation of the error term. Aside from that transformation, Method I is then used on those adjusted data based on that random model. This idea of adjusting records to get rid of fixed effects and then using a standard method on the adjusted records can nowadays, with benefit of hindsight, be viewed as a precursor of REML estimation (restricted maximum likelihood, see Chapter 6). That method adjusts data for the same reason, and then uses maximum likelihood on the adjusted data. Thus the general idea of Method II is straightforward; and the necessary calculations are mostly not difficult. But describing the underlying details and characteristics is. In this respect Method II is the most difficult of all three of the Henderson methods.

The general procedure is as follows. In the usual mixed model equation $\mathbf{y} = \mathbf{X}\boldsymbol{\beta} + \mathbf{Z}\mathbf{u} + \mathbf{e}$ separate out $\mu\mathbf{1}$ from the fixed effects and redefine $\mathscr{X}\boldsymbol{\beta}$ as excluding $\mu\mathbf{1}$ and then write the model equation as

$$\mathbf{y} = \mu\mathbf{1} + \mathscr{X}\boldsymbol{\beta} + \mathbf{Z}\mathbf{u} + \mathbf{e}\,. \tag{40}$$

Throughout this whole section we use this meaning of $\mathscr{X}\boldsymbol{\beta}$: it excludes $\mu\mathbf{1}$.

The general procedure of Method II is based on computing $\hat{\boldsymbol{\beta}} = \mathbf{L}\mathbf{y}$ for $\mathbf{L}$ chosen in such a way (as described in sub-section a that follows) that $\mathbf{y}_a = \mathbf{y} - \mathscr{X}\hat{\boldsymbol{\beta}}$ has a model equation on which it will be easy to use Method I. Thus $\hat{\boldsymbol{\beta}} = \mathbf{L}\mathbf{y}$ is chosen so that

$$\mathbf{y}_a = \mathbf{y} - \mathscr{X}\hat{\boldsymbol{\beta}} = \mu_0\mathbf{1} + \mathbf{Z}\mathbf{u} + \boldsymbol{\varepsilon}, \tag{41}$$

i.e., so that the model equation for $\mathbf{y}_a$ has the random effects in it in the same form as they are in $\mathbf{y}$, namely $\mathbf{Z}\mathbf{u}$. Then Method II consists of using Method I on $\mathbf{y}_a$. This is straightforward insofar as the random effects of $\mathbf{u}$ are concerned, because $\mathbf{Z}\mathbf{u}$ in $\mathbf{y}_a$ of (41) is the same as $\mathbf{Z}\mathbf{u}$ in $\mathbf{y}$ of (40)—and we know how to do Method I on $\mathbf{y}$. But account must be taken of the fact that $\boldsymbol{\varepsilon}$ of (41) is not $\mathbf{e}$ of (40) but is $\boldsymbol{\varepsilon} = (\mathbf{I} - \mathscr{X}\mathbf{L})\mathbf{e}$ for $\hat{\boldsymbol{\beta}} = \mathbf{L}\mathbf{y}$; also, μ_0 depends on $\mathbf{L}$, and is a scalar different from μ. But its actual value is of no importance. The question is "how is $\hat{\boldsymbol{\beta}}$ derived in order to achieve this?" Henderson (1953) shows this largely by means of an example; Searle (1968) gives a general description, which we follow here.

a. Estimating the fixed effects

No matter how $\hat{\boldsymbol{\beta}}$, as a linear function of the data, is calculated, it will be $\hat{\boldsymbol{\beta}} = \mathbf{L}\mathbf{y}$ for some $\mathbf{L}$. Then $\mathbf{y}_a = \mathbf{y} - \mathscr{X}\hat{\boldsymbol{\beta}} = (\mathbf{I} - \mathscr{X}\mathbf{L})\mathbf{y}$, for which the model equation is, using (40),

$$\mathbf{y}_a = \mu(\mathbf{I} - \mathscr{X}\mathbf{L})\mathbf{1} + (\mathbf{I} - \mathscr{X}\mathbf{L})\mathscr{X}\boldsymbol{\beta} + (\mathbf{I} - \mathscr{X}\mathbf{L})\mathbf{Z}\mathbf{u} + (\mathbf{I} - \mathscr{X}\mathbf{L})\mathbf{e}\,. \tag{42}$$

In wanting to choose $\mathbf{L}$ so that (42) reduces to (41), the easiest way of eliminating $\boldsymbol{\beta}$ from (42) would be to pick $\mathbf{L}$ so that the coefficient of $\boldsymbol{\beta}$ in (42) is null; i.e., so that $\mathscr{X} - \mathscr{X}\mathbf{L}\mathscr{X} = \mathbf{0}$. Such is achieved by having $\mathbf{L}$ as a generalized inverse of $\mathscr{X}$. This is what Searle (1971) calls a generalized Method II. But this is not necessarily successful for achieving our ends.

In addition to ridding (42) of $\boldsymbol{\beta}$ by having $\mathbf{L}$ be a generalized inverse of $\mathscr{X}$, we also want the coefficient of $\mathbf{u}$ to be the same in (42) as in (41), i.e., we need $\mathbf{Z} - \mathscr{X}\mathbf{L}\mathbf{Z} = \mathbf{Z}$, and hence $\mathscr{X}\mathbf{L}\mathbf{Z} = \mathbf{0}$. And to get the $\mu_0\mathbf{1}$ term in (41), we need $\mu(\mathbf{I} - \mathscr{X}\mathbf{L})\mathbf{1}$ to be of the form $\lambda\mathbf{1}$ for some scalar λ. This is achieved if the elements in each row of $\mathscr{X}\mathbf{L}$ all add to the same value, say δ_1; i.e., $\mathscr{X}\mathbf{L}\mathbf{1} = \delta_1\mathbf{1}$ for some δ_1. Furthermore, although it seems as if we also need $\mathscr{X} = \mathscr{X}\mathbf{L}\mathscr{X}$ as already discussed, we can in fact settle for $(\mathscr{X} - \mathscr{X}\mathbf{L}\mathscr{X})\boldsymbol{\beta}$ having the form $\delta_2\mathbf{1}$ for some scalar δ_2. For then, although $(\mathscr{X} - \mathscr{X}\mathbf{L}\mathscr{X})\boldsymbol{\beta}$ will not have disappeared from (42) through being a null vector, it will have effectively disappeared through having the form $\delta_2\mathbf{1}$ and ultimately being incorporated in $\mu_0\mathbf{1}$ of (41). This occurs if $\mathscr{X} - \mathscr{X}\mathbf{L}\mathscr{X}$ has all its rows the same, i.e., $\mathscr{X} - \mathscr{X}\mathbf{L}\mathscr{X} = \mathbf{1}\boldsymbol{\tau}'$. Thus the three conditions required for $\mathbf{L}$ are

(i) $\mathscr{X}\mathbf{L}\mathbf{Z} = \mathbf{0}$;

(ii) $\mathscr{X}\mathbf{L}$ having all row sums the same, i.e.,

$$\mathscr{X}\mathbf{L}\mathbf{1} = \delta_1\mathbf{1} \quad \text{for some scalar } \delta_1; \tag{43}$$

(iii) $\mathscr{X} - \mathscr{X}\mathbf{L}\mathscr{X}$ having all rows the same, i.e.,

$$\mathscr{X} - \mathscr{X}\mathbf{L}\mathscr{X} = \mathbf{1}\boldsymbol{\tau}' \quad \text{for some row vector } \boldsymbol{\tau}'\,.$$

With $\mathbf{L}$ satisfying (i), (ii) and (iii), we then have (42) reducing to

$$\mathbf{y}_a = \mu_0\mathbf{1} + \mathbf{Z}\mathbf{u} + (\mathbf{I} - \mathscr{X}\mathbf{L})\mathbf{e}, \tag{44}$$

where $\mu_0 = \mu - \delta_1 + \boldsymbol{\tau}'\boldsymbol{\beta}$.

b. Calculation

We now show details of how Henderson's method of choosing $\mathbf{L}$ of $\hat{\boldsymbol{\beta}} = \mathbf{L}\mathbf{y}$ satisfies (43). In doing so we are led to the calculation of estimates from Method II as follows.

(i) Use the model

$$E(\mathbf{y}) = \mu\mathbf{1} + \mathscr{X}\boldsymbol{\beta} + \mathbf{Z}\mathbf{u}$$

as if $\mathbf{u}$ were fixed effects and where $\mu \neq 0$. The normal equations would be

$$\begin{bmatrix} \mathbf{1}'\mathbf{1} & \mathbf{1}'\mathscr{X} & \mathbf{1}'\mathbf{Z} \\ \mathscr{X}'\mathbf{1} & \mathscr{X}'\mathscr{X} & \mathscr{X}'\mathbf{Z} \\ \mathbf{Z}'\mathbf{1} & \mathbf{Z}'\mathscr{X} & \mathbf{Z}'\mathbf{Z} \end{bmatrix} \begin{bmatrix} \hat{\mu} \\ \hat{\boldsymbol{\beta}} \\ \hat{\mathbf{u}} \end{bmatrix} = \begin{bmatrix} \mathbf{1}'\mathbf{y} \\ \mathscr{X}'\mathbf{y} \\ \mathbf{Z}'\mathbf{y} \end{bmatrix}. \tag{45}$$

(ii) Since these equations are always of less than full rank, simplify solving them by taking $\hat{\mu} = 0$. This reduces the equations to

$$\begin{bmatrix} \mathscr{X}'\mathscr{X} & \mathscr{X}'\mathbf{Z} \\ \mathbf{Z}'\mathscr{X} & \mathbf{Z}'\mathbf{Z} \end{bmatrix} \begin{bmatrix} \hat{\boldsymbol{\beta}} \\ \hat{\mathbf{u}} \end{bmatrix} = \begin{bmatrix} \mathscr{X}'\mathbf{y} \\ \mathbf{Z}'\mathbf{y} \end{bmatrix}.$$

These equations are usually not of full rank; their many solutions are obtainable by using generalized inverses of

$$\mathbf{C} = \begin{bmatrix} \mathscr{X}'\mathscr{X} & \mathscr{X}'\mathbf{Z} \\ \mathbf{Z}'\mathscr{X} & \mathbf{Z}'\mathbf{Z} \end{bmatrix}. \tag{46}$$

(iii) For Henderson's Method II a generalized inverse of $\mathbf{C}$ is chosen as follows (Searle, 1968, pp. 758–760):

> Strike out from $\mathbf{C}$ as many rows and corresponding columns as is necessary to leave a matrix of full rank (equal to the rank of $\mathbf{C}$). As many as possible of the struck-out rows and columns *must be* through $\mathscr{X}'\mathscr{X}$. (This is the *crux* of Henderson's Method II.)

Call the remaining full rank submatrix $\mathbf{B}$. It will consist of some rows and columns through $\mathscr{X}'\mathscr{X}$ and some through $\mathbf{Z}'\mathbf{Z}$. Within $\mathbf{C}$ replace $\mathbf{B}$ by $\mathbf{B}^{-1}$, element for element, and in the struck-out rows and columns put zeros. The result is a generalized inverse

$$\mathbf{C}^- = \begin{bmatrix} \mathbf{0} & \mathbf{0} & \mathbf{0} \\ \mathbf{0} & \mathbf{B}^{-1} & \mathbf{0} \\ \mathbf{0} & \mathbf{0} & \mathbf{0} \end{bmatrix} = \begin{bmatrix} \mathbf{P}_{11} & \mathbf{P}_{12} \\ \mathbf{P}_{21} & \mathbf{P}_{22} \end{bmatrix}, \tag{47}$$

where the partitioning into the $\mathbf{P}$-matrices is conformable with (46). Then

$$\hat{\boldsymbol{\beta}} = \mathbf{L}\mathbf{y} = [\mathbf{P}_{11} \quad \mathbf{P}_{12}] \begin{bmatrix} \mathscr{X}'\mathbf{y} \\ \mathbf{Z}'\mathbf{y} \end{bmatrix}. \tag{48}$$

(iv) Carry out Henderson's Method I on

$$\mathbf{y}_a = \mathbf{y} - \mathscr{X}\hat{\boldsymbol{\beta}} = \mu_0\mathbf{1} + \mathbf{Z}\mathbf{u} + (\mathbf{I} - \mathscr{X}\mathbf{L})\mathbf{e}.$$

Using $\mathbf{y}_a'\mathbf{A}\mathbf{y}_a$ for each $\mathbf{A}$ that would be used in applying Method I to $\mathbf{y}$ if there were no fixed effects, $E(\mathbf{y}_a'\mathbf{A}\mathbf{y}_a)$ will contain the same terms in the variance components as does $E(\mathbf{y}'\mathbf{A}\mathbf{y})$, except for terms in σ_e^2. This is because $\mathbf{Z}$ and $\mathbf{u}$ are the same in the model equation for $\mathbf{y}_a$ as in that for $\mathbf{y}$.

(v) The term in σ_e^2 in $E(\mathbf{y}_a'\mathbf{A}\mathbf{y}_a)$ is $(k_A + \delta_A)\sigma_e^2$, where $E(\mathbf{y}'\mathbf{A}\mathbf{y})$ contains $k_A\sigma_e^2$, and where δ_A is calculated as the trace of a matrix derived through the following steps.

Partition $\mathscr{X}$ and $\mathbf{Z}$ as

$$\mathscr{X} = [\mathscr{X}_1 \quad \mathscr{X}_2] \quad \text{and} \quad \mathbf{Z} = [\mathbf{Z}_1 \quad \mathbf{Z}_2] \tag{49}$$

so that in deriving $\mathbf{C}^-$ of (47) from $\mathbf{C}$ of (46) the rows and columns through $\mathscr{X}'\mathscr{X}$ of $\mathbf{C}$ that are deleted to obtain $\mathbf{B}$ correspond to the columns of $\mathscr{X}_1$ and of $\mathbf{Z}_2$, with $\mathscr{X}_1$ having as many such columns as possible.

On using (49) in (46), we then have $\mathbf{C}^-$ of (47) as

$$\mathbf{C}^- = \begin{bmatrix} \mathbf{0} & \mathbf{0} & \mathbf{0} & \mathbf{0} \\ \mathbf{0} & \multicolumn{2}{c}{\begin{bmatrix} \mathscr{X}_2'\mathscr{X}_2 & \mathscr{X}_2'\mathbf{Z}_1 \\ \mathbf{Z}_1'\mathscr{X}_2 & \mathbf{Z}_1'\mathbf{Z}_1 \end{bmatrix}^{-1}} & \mathbf{0} \\ \mathbf{0} & & & \mathbf{0} \\ \mathbf{0} & \mathbf{0} & \mathbf{0} & \mathbf{0} \end{bmatrix} = \begin{bmatrix} \mathbf{0} & \mathbf{0} & \mathbf{0} & \mathbf{0} \\ \mathbf{0} & \mathbf{Q}_{11} & \mathbf{Q}_{12} & \mathbf{0} \\ \mathbf{0} & \mathbf{Q}_{21} & \mathbf{Q}_{22} & \mathbf{0} \\ \mathbf{0} & \mathbf{0} & \mathbf{0} & \mathbf{0} \end{bmatrix} = \begin{bmatrix} \mathbf{P}_{11} & \mathbf{P}_{12} \\ \mathbf{P}_{21} & \mathbf{P}_{22} \end{bmatrix} \tag{50}$$

where the second equality in (50) defines the **Q**-matrices, and the third defines each $\mathbf{P}_{ij}$ as a matrix having $\mathbf{Q}_{ij}$ and three nulls as submatrices. Then $(k_A + \delta_A)\sigma_e^2$ of (v), wherein k_A is defined, has

$$\delta_A = \operatorname{tr} \mathbf{A}(\mathscr{X}_2\mathbf{Q}_{11}\mathscr{X}_2') . \tag{51}$$

The corresponding value of $\hat{\boldsymbol{\beta}}$ in (48) is

$$\hat{\boldsymbol{\beta}} = \begin{bmatrix} \hat{\boldsymbol{\beta}}_1 \\ \hat{\boldsymbol{\beta}}_2 \end{bmatrix} = \begin{bmatrix} \mathbf{0} \\ (\mathbf{Q}_{11}\mathscr{X}_2' + \mathbf{Q}_{12}\mathbf{Z}_1')\mathbf{y} \end{bmatrix}, \tag{52}$$

i.e.,

$$\mathbf{L} = \begin{bmatrix} \mathbf{0} \\ \mathbf{Q}_{11}\mathscr{X}_2' + \mathbf{Q}_{12}\mathbf{Z}_1' \end{bmatrix} . \tag{53}$$

c. Verification

We prove that **L** of (53) satisfies conditions (i), (ii) and (iii) of (43). The crux of the proof lies in properties of $[\mathscr{X} \;\; \mathbf{Z}]$ that arise from the manner in which rows and columns were deleted from **C** of (46) to obtain $\mathbf{C}^-$ in (50). With $\mathscr{X}_1$ of (49) defined as having as many columns as possible, the number of columns (deleted from **C**) in $\mathbf{Z}_2$ is as small as possible—the total number of columns in $\mathbf{Z}_2$ being dependent on the rank of $[\mathscr{X} \;\; \mathbf{Z}]$. A consequence of this is that among the columns of $\mathbf{Z}_1$ are those pertaining to *all* levels of one of the random factors. Our proof hangs entirely on this fact. Denote those columns pertaining to all levels of whichever random factor has all its levels represented in $\mathbf{Z}_1$ by $\mathbf{Z}_{1\mathrm{A}}$ and partition $\mathbf{Z}_1$ as

$$\mathbf{Z}_1 = [\mathbf{Z}_{11} \quad \mathbf{Z}_{1\mathrm{A}} \quad \mathbf{Z}_{13}], \tag{54}$$

where $\mathbf{Z}_{11}$ and/or $\mathbf{Z}_{13}$ may be dimensionless. Then, since in $[\mathscr{X} \;\; \mathbf{Z}]$ the sum of the columns pertaining to all levels of each factor is $\mathbf{1}_N$, we also have

$$\mathbf{Z}_{1\mathrm{A}}\mathbf{1} = \mathbf{1}_N . \tag{55}$$

With **L** of (53) we now prove (i), (ii) and (iii).

-i. The matrix $\mathscr{X}\mathbf{L}\mathbf{Z}$ is null. In $\mathbf{Z} = [\mathbf{Z}_1 \;\; \mathbf{Z}_2]$ each of the random factors that has (so to speak) some columns in $\mathbf{Z}_2$ also has some columns in $\mathbf{Z}_1$; and for each such factor adding its columns in $\mathbf{Z}_2$ and its columns in $\mathbf{Z}_1$ gives $\mathbf{1}_N$, which is, by (55), also the sum of columns in $\mathbf{Z}_{1\mathrm{A}}$. Therefore $\mathbf{Z}_2 = \mathbf{Z}_1\mathbf{K}$ for

some **K**, and so, on using **L** from (53),

$$\mathscr{X}\mathbf{LZ} = \mathscr{X}\mathbf{LZ}_1[\mathbf{I} \quad \mathbf{K}] = \mathscr{X}\begin{bmatrix} \mathbf{0} \\ \mathbf{Q}_{11}\mathscr{X}'_2\mathbf{Z}_1 + \mathbf{Q}_{12}\mathbf{Z}'_1\mathbf{Z}_1 \end{bmatrix}[\mathbf{I} \quad \mathbf{K}]$$

$$= \mathscr{X}\begin{bmatrix} \mathbf{0} \\ \mathbf{0} \end{bmatrix}[\mathbf{I} \quad \mathbf{K}] = \mathbf{0} \tag{56}$$

because, from the non-null submatrices of (50)

$$\mathbf{Q}_{11}\mathscr{X}'_2\mathscr{X}_2 + \mathbf{Q}_{12}\mathbf{Z}'_1\mathscr{X}_2 = \mathbf{I} \quad \text{and} \quad \mathbf{Q}_{11}\mathscr{X}'_2\mathbf{Z}_1 + \mathbf{Q}_{12}\mathbf{Z}'_1\mathbf{Z}_1 = \mathbf{0}\,. \tag{57}$$

-ii. ***Row sums of*** $\mathscr{X}\mathbf{L}$ ***are all the same.*** Because of (54) and (55),

$$[\mathscr{X}_1 \quad \mathscr{X}_2 \quad \mathbf{Z}_{11} \quad \mathbf{Z}_{1A} \quad \mathbf{Z}_{13} \quad \mathbf{Z}_2][\mathbf{0} \quad \mathbf{0} \quad \mathbf{0} \quad \mathbf{1}' \quad \mathbf{0} \quad \mathbf{0}]' = \mathbf{1}_N; \tag{58}$$

therefore, for $\mathbf{t}' = [\mathbf{0} \quad \mathbf{1}' \quad \mathbf{0}]$ conformable with $\mathbf{Z}_1 = [\mathbf{Z}_{11} \quad \mathbf{Z}_{1A} \quad \mathbf{Z}_{13}]$, with

$$\mathbf{Z}_1\mathbf{t} = \mathbf{1}, \tag{59}$$

$$[\mathscr{X}_1 \quad \mathscr{X}_2 \quad \mathbf{Z}_1 \quad \mathbf{Z}_2]\begin{bmatrix} \mathbf{0} \\ \mathbf{0} \\ \mathbf{t} \\ \mathbf{0} \end{bmatrix} = [\mathscr{X} \quad \mathbf{Z}]\begin{bmatrix} \mathbf{0} \\ \mathbf{0} \\ \mathbf{t} \\ \mathbf{0} \end{bmatrix} = \mathbf{1}_N\,. \tag{60}$$

Pre-multiplying both sides of the second equality in (60) by $[\mathscr{X} \quad \mathbf{Z}]'$ and extracting part of the result produces

$$\begin{bmatrix} \mathscr{X}'_2 \\ \mathbf{Z}'_1 \end{bmatrix}\mathbf{1}_N = \begin{bmatrix} \mathscr{X}'_2\mathscr{X}_2 & \mathscr{X}'_2\mathbf{Z}_1 \\ \mathbf{Z}'_1\mathscr{X}_2 & \mathbf{Z}'_1\mathbf{Z}_1 \end{bmatrix}\begin{bmatrix} \mathbf{0} \\ \mathbf{t} \end{bmatrix}. \tag{61}$$

Then, using (53), it can be seen that $\mathbf{L1}_N$ involves the left-hand side of (61), from which we get, with the aid of (57), $\mathbf{L1}_N = \mathbf{0}$. Therefore row sums of $\mathscr{X}\mathbf{L}$, which are elements of $\mathscr{X}\mathbf{L1}$, are zero, i.e., they are all the same.

-iii. ***All rows of*** $\mathscr{X} - \mathscr{X}\mathbf{L}\mathscr{X}$ ***are the same.*** From (49) and (53)

$$\mathscr{X} - \mathscr{X}\mathbf{L}\mathscr{X} = [\mathscr{X}_1 \quad \mathscr{X}_2] - [\mathscr{X}_1 \quad \mathscr{X}_2]\begin{bmatrix} \mathbf{0} \\ \mathbf{Q}_{11}\mathscr{X}'_2 + \mathbf{Q}_{12}\mathbf{Z}'_1 \end{bmatrix}[\mathscr{X}_1 \quad \mathscr{X}_2]$$

$$= [\mathscr{X}_1 \quad \mathscr{X}_2] - [\mathscr{X}_2(\mathbf{Q}_{11}\mathscr{X}'_2 + \mathbf{Q}_{12}\mathbf{Z}'_1)\mathscr{X}_1 \quad \mathscr{X}_2], \quad \text{using (57)},$$

$$= \left[\mathscr{X}_1 - \mathscr{X}_2[\mathbf{Q}_{11} \quad \mathbf{Q}_{12}]\begin{pmatrix} \mathscr{X}'_2 \\ \mathbf{Z}'_1 \end{pmatrix}\mathscr{X}_1 \quad \mathbf{0}\right]. \tag{62}$$

The reasoning used with $\mathbf{Z}_1$ and $\mathbf{Z}_2$ concerning sums of certain of their columns adding to $\mathbf{1}_N$ also applies to $\mathscr{X}_1$ and $\mathscr{X}_2$: the sum of certain columns in $\mathscr{X}_1$ and in $\mathscr{X}_2$ is also $\mathbf{1}_N$. But (58) and (60) give $\mathbf{Z}_{1A}\mathbf{1} = \mathbf{1}_N = \mathbf{Zt}$. Therefore

$$\mathscr{X}_1 = [\mathscr{X}_2 \quad \mathbf{Z}_1]\begin{bmatrix} \mathbf{S} \\ \mathbf{T} \end{bmatrix}, \tag{63}$$

where $\mathbf{S}$ is some matrix of elements that are each 0 or 1, and $\mathbf{T} = [\mathbf{0} \quad \mathbf{J} \quad \mathbf{0}]$. Therefore pre-multiplying (63) by $[\mathbf{Q}_{11} \quad \mathbf{Q}_{12}][\mathscr{X}_2 \quad \mathbf{Z}_1]'$ gives

$$[\mathbf{Q}_{11} \quad \mathbf{Q}_{12}]\begin{bmatrix} \mathscr{X}'_2 \\ \mathbf{Z}'_1 \end{bmatrix}\mathscr{X}_1 = [\mathbf{Q}_{11} \quad \mathbf{Q}_{12}]\begin{bmatrix} \mathscr{X}'_2 \\ \mathbf{Z}'_1 \end{bmatrix}[\mathscr{X}_2 \quad \mathbf{Z}_1]\begin{bmatrix} \mathbf{S} \\ \mathbf{T} \end{bmatrix},$$

$$= [\mathbf{I} \quad \mathbf{0}]\begin{bmatrix} \mathbf{S} \\ \mathbf{T} \end{bmatrix} = \mathbf{S} \tag{64}$$

from (57). Substituting (64) and (63) into (62) gives

$$\mathscr{X} - \mathscr{X}\mathbf{L}\mathscr{X} = \left[[\mathscr{X}_2 \quad \mathbf{Z}_1]\begin{pmatrix} \mathbf{S} \\ \mathbf{T} \end{pmatrix} - \mathscr{X}_2\mathbf{S} \quad \mathbf{0}\right] = [\mathbf{Z}_1\mathbf{T} \quad \mathbf{0}] . \tag{65}$$

But

$$\mathbf{Z}_1\mathbf{T} = [\mathbf{Z}_{11} \quad \mathbf{Z}_{1A} \quad \mathbf{Z}_{13}]\begin{bmatrix} \mathbf{0} \\ \mathbf{J} \\ \mathbf{0} \end{bmatrix} = \mathbf{Z}_{1A}\mathbf{J} = \mathbf{J}$$

by (55), and so (65) becomes $\mathscr{X} - \mathscr{X}\mathbf{L}\mathscr{X} = [\mathbf{J} \quad \mathbf{0}]$, which has all its rows the same.

d. Invariance

Since execution of Method II depends, as in (iii) following (46), upon deletion of rows and columns of $\mathbf{C}$ for deriving $\mathbf{C}^-$, it might be thought that the resulting variance components estimates would not be invariant to the manner in which this deletion is carried out (as suggested by Searle, 1971, p. 443). That is false, as proved by Henderson, Searle and Schaeffer (1974). We give their proof here.

-i. ***Rank properties.*** By the very choice of $\mathscr{X}_1$ and $\mathbf{Z}_2$, the rank of $[\mathscr{X}_2 \quad \mathbf{Z}_1]$ equals the number of its columns and it is the rank of $[\mathscr{X} \quad \mathbf{Z}]$; i.e., for $r(\mathscr{X})$ being the rank of $\mathscr{X}$,

$$r(\mathscr{X}_2 \quad \mathbf{Z}_1) = r(\mathscr{X}_2) + r(\mathbf{Z}_1) = r(\mathscr{X} \quad \mathbf{Z}) . \tag{66}$$

Furthermore, since $\mathbf{Z}_1$ has as many columns as possible, subject to (66),

$$r(\mathbf{Z}) = r(\mathbf{Z}_1) . \tag{67}$$

In addition, we confine attention to models wherein

$$r(\mathscr{X} \quad \mathbf{Z}) = r(\mathscr{X}) + r(\mathbf{Z}) - 1 . \tag{68}$$

This requirement excludes models that have interactions between fixed and random effects, but these are excluded, anyway, by other characteristics of Method II (see sub-section f which follows). Also excluded by (68) are models having any confounding between fixed and random effects.

Conditions (66)–(68) are used in proving that Method II is invariant to whatever solution $\hat{\boldsymbol{\beta}}$ is used in $\mathbf{y}_a = \mathbf{y} - \mathscr{X}\hat{\boldsymbol{\beta}}$ for applying Method II to $\mathbf{y}_a$. The proof depends upon the estimability of a certain function of $\boldsymbol{\beta}$, upon the relationship of one form of $\hat{\boldsymbol{\beta}}$ to another, and upon the quadratic forms used in Method I.

Since (67) is equivalent to $r(\mathbf{Z}_1) = r(\mathbf{Z}_1 \quad \mathbf{Z}_2)$, we have

$$\mathbf{Z}_2 = \mathbf{Z}_1\mathbf{H} \tag{69}$$

for some $\mathbf{H}$. Also, (59) is $\mathbf{Z}_1\mathbf{t} = \mathbf{1}_N$; therefore, because $(\mathscr{X}_2 \quad \mathbf{Z}_1)$ has full column rank, $\mathbf{1}$ and columns of $\mathscr{X}_2$ are linearly independent. Thus $r(\mathbf{1} \quad \mathscr{X}_2) = 1 + r(\mathscr{X}_2)$ and so from (66), (67) and (68)

$$\begin{aligned} r(\mathbf{1} \quad \mathscr{X}_2) &= 1 + r(\mathscr{X} \quad \mathbf{Z}) - r(\mathbf{Z}_1) = 1 + r(\mathscr{X} \quad \mathbf{Z}) - r(\mathbf{Z}) = 1 + r(\mathscr{X}) - 1 \\ &= r(\mathscr{X}) = r(\mathbf{1} \quad \mathscr{X}_1 \quad \mathscr{X}_2)\,. \end{aligned}$$

Hence $\mathscr{X}_1 = [\mathbf{1} \quad \mathscr{X}_2]\mathbf{R}$ for some matrix $\mathbf{R}$, which can be written as $\mathbf{R}' = [\mathbf{w} \quad \mathbf{W}']$ so that

$$\mathscr{X}_1 = \mathbf{1}\mathbf{w}' + \mathscr{X}_2\mathbf{W}, \tag{70}$$

for some row vector $\mathbf{w}'$ and matrix $\mathbf{W}$. Hence from $\mathbf{Z}_1\mathbf{t} = \mathbf{1}$ of (59)

$$\mathscr{X}_1 = \mathbf{Z}_1\mathbf{t}\mathbf{w}' + \mathscr{X}_2\mathbf{W}\,. \tag{71}$$

-ii. An estimable function. The model (40), $\mathbf{y} = \mu\mathbf{1} + \mathscr{X}\boldsymbol{\beta} + \mathbf{Z}\mathbf{u} + \mathbf{e}$, is now

$$\mathbf{y} = [\mathbf{1} \quad \mathscr{X}_1 \quad \mathbf{Z}_2]\begin{bmatrix}\mu \\ \boldsymbol{\beta}_1 \\ \mathbf{u}_2\end{bmatrix} + [\mathscr{X}_2 \quad \mathbf{Z}_1]\begin{bmatrix}\boldsymbol{\beta}_2 \\ \mathbf{u}_1\end{bmatrix} + \mathbf{e}\,.$$

On now using $\mathbf{Z}_1\mathbf{t} = \mathbf{1}_N$ of (59) together with (71) and (69) for $\mathscr{X}_1$ and $\mathbf{Z}_2$, respectively, this becomes

$$\begin{aligned} \mathbf{y} &= [\mathbf{Z}_1\mathbf{t} \quad \mathbf{Z}_1\mathbf{t}\mathbf{w}' + \mathscr{X}_2\mathbf{W} \quad \mathbf{Z}_1\mathbf{H}]\begin{bmatrix}\mu \\ \boldsymbol{\beta}_1 \\ \mathbf{u}_2\end{bmatrix} + [\mathscr{X}_2 \quad \mathbf{Z}_1]\begin{bmatrix}\boldsymbol{\beta}_2 \\ \mathbf{u}_1\end{bmatrix} + \mathbf{e} \\ &= [\mathscr{X}_2 \quad \mathbf{Z}_1]\begin{bmatrix}\mathbf{0} & \mathbf{W} & \mathbf{0} \\ \mathbf{t} & \mathbf{t}\mathbf{w}' & \mathbf{H}\end{bmatrix}\begin{bmatrix}\mu \\ \boldsymbol{\beta}_1 \\ \mathbf{u}_2\end{bmatrix} + [\mathscr{X}_2 \quad \mathbf{Z}_1]\begin{bmatrix}\boldsymbol{\beta}_2 \\ \mathbf{u}_1\end{bmatrix} + \mathbf{e} \\ &= [\mathscr{X}_2 \quad \mathbf{Z}_1]\begin{bmatrix}\mathbf{W}\boldsymbol{\beta}_1 + \boldsymbol{\beta}_2 \\ \mu\mathbf{t} + \mathbf{t}\mathbf{w}'\boldsymbol{\beta}_1 + \mathbf{u}_1 + \mathbf{H}\mathbf{u}_2\end{bmatrix} + \mathbf{e}\,. \end{aligned} \tag{72}$$

Since $[\mathscr{X}_2 \quad \mathbf{Z}_1]$ has full column rank, we conclude from (72) that

$$\mathbf{W}\boldsymbol{\beta}_1 + \boldsymbol{\beta}_2 \quad \text{is an estimable function}\,. \tag{73}$$

-iii. Two solutions. Suppose in the class of models satisfying (68) that $\boldsymbol{\beta}^0$ is *any* solution of the normal equations (45) for the model that treats the random effects as if they were fixed; and $\boldsymbol{\beta}^0$ is assumed to differ from $\hat{\boldsymbol{\beta}} = \mathbf{Ly}$. Then for

$$\begin{aligned} \mathbf{y}_a^0 &= \mathbf{y} - \mathscr{X}\boldsymbol{\beta}^0, \\ \mathbf{y}_a^0 - \mathbf{y}_a &= \mathbf{y} - \mathscr{X}\boldsymbol{\beta}^0 - (\mathbf{y} - \mathscr{X}\hat{\boldsymbol{\beta}}) \\ &= \mathscr{X}_1(\hat{\boldsymbol{\beta}}_1 - \boldsymbol{\beta}_1^0) + \mathscr{X}_2(\hat{\boldsymbol{\beta}}_2 - \boldsymbol{\beta}_2^0), \end{aligned}$$

and from (70) this is

$$\mathbf{y}_a^0 - \mathbf{y}_a = \mathbf{1}\mathbf{w}'(\hat{\boldsymbol{\beta}}_1 - \boldsymbol{\beta}_1^0) + \mathscr{X}_2[\mathbf{W}\hat{\boldsymbol{\beta}}_1 + \hat{\boldsymbol{\beta}}_2 - (\mathbf{W}\boldsymbol{\beta}_1^0 + \boldsymbol{\beta}_2^0)] \,. \tag{74}$$

But $\mathbf{W}\boldsymbol{\beta}_1 + \boldsymbol{\beta}_2$ is, by (73), estimable. Therefore $\mathbf{W}\hat{\boldsymbol{\beta}}_1 + \hat{\boldsymbol{\beta}}_2 = \mathbf{W}\boldsymbol{\beta}_1^0 + \boldsymbol{\beta}_2^0$ and so (74) reduces to

$$\mathbf{y}_a^0 = \mathbf{y}_a + \lambda\mathbf{1} \quad \text{for } \lambda = \mathbf{w}'(\hat{\boldsymbol{\beta}}_1 - \boldsymbol{\beta}_1^0) \,.$$

-iv. The quadratic forms. Method II is to use Method I on $\mathbf{y}_a$. But the quadratic forms of Method I, say $\mathbf{y}'\mathbf{A}\mathbf{y}$, have $\mathbf{A} = \mathbf{A}'$ and are such that $\mathbf{1}'\mathbf{A} = 0$. Therefore when those same quadratic forms are used on $\mathbf{y}_a$ and $\mathbf{y}_a^0$

$$\mathbf{y}_a^{0\prime}\mathbf{A}\mathbf{y}_a^0 = (\mathbf{y}_a' + \lambda\mathbf{1}')\mathbf{A}(\mathbf{y}_a + \lambda\mathbf{1}) = \mathbf{y}_a'\mathbf{A}\mathbf{y}_a \,.$$

Hence Method I on $\mathbf{y}_a^0$ calculates the same quadratic forms as on $\mathbf{y}_a$. But, of course, for the expected values of those Method I quadratic forms to have the same values as they do on $\mathbf{y}$ (other than σ_e^2-terms) the $\mathbf{y}_a$ that one uses must be of the form required for Method II, namely $\mathbf{y}_a = \mathbf{y} - \mathbf{Ly}$ for $\mathbf{L}$ of (53).

It can be noted in passing that because the only fixed effect term in the model equation for $\mathbf{y}_a$ is $\mu_0\mathbf{1}$, the condition that $\mathbf{y}_a'\mathbf{A}\mathbf{y}_a$ be suitable for ANOVA estimation, $\mathscr{X}'\mathbf{A}\mathscr{X} = \mathbf{0}$, is $\mathbf{1}'\mathbf{A}\mathbf{1} = \mathbf{0}$. This is satisfied because $\mathbf{A}$ is defined through $\mathbf{y}'\mathbf{A}\mathbf{y}$ being a Method I quadratic form, for which we know $\mathbf{1}'\mathbf{A}\mathbf{1} = \mathbf{0}$.

e. Coefficients of σ_e^2

Method II applied to $\mathbf{y} = \mu\mathbf{1} + \mathscr{X}\boldsymbol{\beta} + \mathbf{Zu} + \mathbf{e}$ is Method I used on $\mathbf{y}_a = \mu_0\mathbf{1} + \mathbf{Zu} + \boldsymbol{\varepsilon}$. This means calculating quadratic forms in $\mathbf{y}_a$ that are the same as those in $\mathbf{y}$ for Method I used on $\mathbf{y} = \mu\mathbf{1} + \mathbf{Zu} + \mathbf{e}$; for example, $\mathbf{y}'\mathbf{A}\mathbf{y}$, say. Then Method II equates $\mathbf{y}_a'\mathbf{A}\mathbf{y}_a$ to $E(\mathbf{y}_a'\mathbf{A}\mathbf{y}_a)$. Because $\mathbf{Zu}$ in $\mathbf{y}_a$ is the same as in $\mathbf{y}$, $E(\mathbf{y}_a'\mathbf{A}\mathbf{y}_a)$ is identical to $E(\mathbf{y}_a'\mathbf{A}\mathbf{y}_a)$ for the $\mathbf{y} = \mu\mathbf{1} + \mathbf{Zu} + \mathbf{e}$ model—except for the term in σ_e^2, since $\mathbf{y}$ contains $\mathbf{e}$ and $\mathbf{y}_a$ contains $\boldsymbol{\varepsilon} = (\mathbf{I} - \mathscr{X}\mathbf{L})\mathbf{e}$ of (44). Let the term in σ_e^2 in $E(\mathbf{y}'\mathbf{A}\mathbf{y})$ be $k_A\sigma_e^2$ and that in $E(\mathbf{y}_a'\mathbf{A}\mathbf{y}_a)$ be $(k_A + \delta_A)\sigma_e^2$. Then, since the variance components terms in $E(\mathbf{y}'\mathbf{A}\mathbf{y})$ come from $\text{tr}(\mathbf{AV})$ and because σ_e^2 occurs in $\mathbf{V}$ as $\sigma_e^2\mathbf{I}$, the σ_e^2 term in $E(\mathbf{y}'\mathbf{A}\mathbf{y})$ is

$$k_A\sigma_e^2 = \text{tr}(\mathbf{A}\sigma_e^2\mathbf{I}) = \sigma_e^2\,\text{tr}(\mathbf{A}) \,. \tag{75}$$

Similarly, with $\boldsymbol{\varepsilon} = (\mathbf{I} - \mathscr{X}\mathbf{L})\mathbf{e}$, the σ_e^2 term in $E(\mathbf{y}_a'\mathbf{A}\mathbf{y}_a)$ is

$$(k_A + \delta_A)\sigma_e^2 = \text{tr}\{\mathbf{A}\,\text{var}[(\mathbf{I} - \mathscr{X}\mathbf{L})\mathbf{e}]\} \,.$$

But from (49) and (53)

$$\mathscr{X}\mathbf{L} = \mathscr{X}_2\mathbf{U} \quad \text{for } \mathbf{U} = \mathbf{Q}_{11}\mathscr{X}_2' + \mathbf{Q}_{12}\mathbf{Z}_1' . \tag{76}$$

Hence

$$\begin{aligned} k_A + \delta_A &= \operatorname{tr}[\mathbf{A}(\mathbf{I} - \mathscr{X}_2\mathbf{U})(\mathbf{I} - \mathscr{X}_2\mathbf{U})'] \\ &= \operatorname{tr}[\mathbf{A}(\mathbf{I} - \mathbf{U}'\mathscr{X}_2' - \mathscr{X}_2\mathbf{U} + \mathscr{X}_2\mathbf{U}\mathbf{U}'\mathscr{X}_2')] \\ &= \operatorname{tr}\{\mathbf{A}[\mathbf{I} - \mathbf{U}'\mathscr{X}_2' - \mathscr{X}_2\mathbf{U} + \mathscr{X}_2\mathbf{U}(\mathscr{X}_2\mathbf{Q}_{11}' + \mathbf{Z}_1\mathbf{Q}_{12}')\mathscr{X}_2']\}, \end{aligned} \tag{77}$$

from (76). But (76) gives (57) as $\mathbf{U}\mathscr{X}_2 = \mathbf{I}$ and $\mathbf{U}\mathbf{Z}_1 = \mathbf{0}$. Using these and (75) in (77) gives

$$k_A + \delta_A = k_A - \operatorname{tr}[\mathbf{A}(\mathbf{U}'\mathscr{X}_2' + \mathscr{X}_2\mathbf{U} - \mathscr{X}_2\mathbf{Q}_{11}\mathscr{X}_2')] . \tag{78}$$

Any Method I quadratic form $\mathbf{y}_a'\mathbf{A}\mathbf{y}_a$ is, in fact, a quadratic form in the random factors' subclass totals—i.e., in $\mathbf{Z}'\mathbf{y}_a$. Therefore $\mathbf{y}_a'\mathbf{A}\mathbf{y}_a = \mathbf{y}_a'\mathbf{Z}\mathbf{M}\mathbf{Z}'\mathbf{y}_a$ for some $\mathbf{M}$. But $\mathbf{Z} = [\mathbf{Z}_1 \quad \mathbf{Z}_2]$ of (49) is, from (69), $\mathbf{Z} = \mathbf{Z}[\mathbf{I} \quad \mathbf{H}] = \mathbf{Z}_1\mathbf{F}$, say, for some $\mathbf{F}$. Therefore $\mathbf{y}_a'\mathbf{A}\mathbf{y}_a = \mathbf{y}_a'\mathbf{Z}_1\mathbf{F}\mathbf{M}\mathbf{F}'\mathbf{Z}_1'\mathbf{y}_a$. Hence for δ_A of (78)

$$\operatorname{tr}(\mathbf{A}\mathscr{X}_2\mathbf{U}) = \operatorname{tr}(\mathbf{Z}_1\mathbf{F}\mathbf{M}\mathbf{F}'\mathbf{Z}_1'\mathscr{X}_2\mathbf{U}) = \operatorname{tr}(\mathbf{U}\mathbf{Z}_1\mathbf{F}\mathbf{M}\mathbf{F}'\mathbf{Z}_1'\mathscr{X}_2) = 0,$$

because $\mathbf{U}\mathbf{Z}_1 = \mathbf{0}$, as precedes (78); i.e., $\operatorname{tr}(\mathbf{A}\mathscr{X}_2\mathbf{U}) = 0$. Therefore

$$0 = \operatorname{tr}(\mathbf{A}\mathscr{X}_2\mathbf{U}) = \operatorname{tr}(\mathbf{U}'\mathscr{X}_2'\mathbf{A}') = \operatorname{tr}(\mathbf{U}'\mathscr{X}_2'\mathbf{A}) = \operatorname{tr}(\mathbf{A}\mathbf{U}'\mathscr{X}_2') .$$

Hence (78) becomes $\delta_A = \operatorname{tr}(\mathbf{A}\mathscr{X}_2\mathbf{Q}_{11}\mathscr{X}_2')$ of (51).

f. No fixed-by-random interactions

It is a restrictive feature of Method II that it cannot be used on data from models that include interactions between fixed effects and random effects. This is so whether such interactions are defined as being random effects (which would be usual) or as being fixed effects. The reason that such interactions cannot be accommodated is that their existence is inconsistent with conditions (i)–(iii) of (43). This we now prove, taken from Searle (1968). But first an example, to illustrate the relationships that exist between $\mathscr{X}$ and $\mathbf{Z}$ when a model has interactions.

Example. Suppose for the 2-way crossed classification of two rows (factor A) and two columns (factor B) that the numbers of observations are

2	3	5
1	2	3
3	5	8

For these n_{ij}-values the model equation

$$y_{ijk} = \mu + \alpha_i + \beta_j + \gamma_{ij} + e_{ijk},$$

written as

$$\mathbf{y} = \mu\mathbf{1} + \mathscr{X}_A\boldsymbol{\alpha} + \mathscr{X}_B\boldsymbol{\beta} + \mathscr{X}_{AB}\boldsymbol{\gamma} + \mathbf{e},$$

has

$$\begin{matrix}[\mathbf{1} & \mathscr{X}_A & \mathscr{X}_B & \mathscr{X}_{AB}]\end{matrix}$$

$$\left[\begin{array}{c|cc|cc|cccc}
1 & 1 & \cdot & 1 & \cdot & 1 & \cdot & \cdot & \cdot \\
1 & 1 & \cdot & 1 & \cdot & 1 & \cdot & \cdot & \cdot \\
1 & 1 & \cdot & \cdot & 1 & \cdot & 1 & \cdot & \cdot \\
1 & 1 & \cdot & \cdot & 1 & \cdot & 1 & \cdot & \cdot \\
1 & 1 & \cdot & \cdot & 1 & \cdot & 1 & \cdot & \cdot \\
1 & \cdot & 1 & 1 & \cdot & \cdot & \cdot & 1 & \cdot \\
1 & \cdot & 1 & \cdot & 1 & \cdot & \cdot & \cdot & 1 \\
1 & \cdot & 1 & \cdot & 1 & \cdot & \cdot & \cdot & 1
\end{array}\right].$$

By inspection we see that $\mathscr{X}_A$ and $\mathscr{X}_B$ each have columns that are sums of columns of $\mathscr{X}_{AB}$. This is a direct consequence of there being A-by-B interactions in the model.

Suppose αs represent fixed effects and βs represent random effects. Then if the interactions are taken as random in the model equation $\mathbf{y} = \mu\mathbf{1} + \mathscr{X}\boldsymbol{\beta} + \mathbf{Zu} + \mathbf{e}$, we would have for the example $\mathscr{X} = \mathscr{X}_A$ and $\mathbf{Z} = [\mathscr{X}_B \quad \mathscr{X}_{AB}]$. Therefore some columns of $\mathscr{X}$ are sums of certain columns of $\mathbf{Z}$. This is true generally, whenever interactions of fixed effects factors with some random factors are taken as random effects. Then, apart from permuting columns of $\mathscr{X}$, we can partition $\mathscr{X}$ as

$$\mathscr{X} = [\mathscr{X}_1 \quad \mathscr{X}_2] \tag{79}$$

where $\mathscr{X}_2$ represents those columns of $\mathscr{X}$ that are sums of certain columns of $\mathbf{Z}$ (e.g., of $\mathscr{X}_{AB}$ in the example) and so we have

$$\mathscr{X}_2 = \mathbf{ZM} \quad \text{for some } \mathbf{M}\,. \tag{80}$$

Similarly, if those interactions are taken as fixed effects, $\mathbf{Z}$ can be partitioned as

$$\mathbf{Z} = [\mathbf{Z}_1 \quad \mathbf{Z}_2] \tag{81}$$

with

$$\mathbf{Z}_2 = \mathscr{X}\mathbf{K} \quad \text{for some } \mathbf{K}\,. \tag{82}$$

Note: $[\mathscr{X}_1 \quad \mathscr{X}_2]$ and $[\mathbf{Z}_1 \quad \mathbf{Z}_2)$ do *not* represent the same partitionings of $\mathscr{X}$ and $\mathbf{Z}$, respectively, as are used in (49).

Now we prove that interactions of this nature, be they taken as random or fixed, are inconsistent with conditions (i)–(iii) of (43). Suppose the interactions are taken as random. Then from (80)

$$\mathscr{X}\mathbf{L}\mathscr{X}_2 = \mathscr{X}\mathbf{LZM} = \mathbf{0}, \quad \text{from (i) of (43)}\,. \tag{83}$$

Therefore

$$\begin{aligned}\mathscr{X} - \mathscr{X}\mathbf{L}\mathscr{X} &= [\mathscr{X}_1 \quad \mathscr{X}_2] - \mathscr{X}\mathbf{L}[\mathscr{X}_1 \quad \mathscr{X}_2] \\ &= [(\mathscr{X}_1 - \mathscr{X}\mathbf{L}\mathscr{X}_1) \quad (\mathscr{X}_2 - \mathscr{X}\mathbf{L}\mathscr{X}_2)] \\ &= [(\mathscr{X}_1 - \mathscr{X}\mathbf{L}\mathscr{X}_1) \quad \mathscr{X}_2]\end{aligned}$$

and so by (ii) of (43) every row of $[(\mathscr{X}_1 - \mathscr{X}\mathbf{L}\mathscr{X}_1) \quad \mathscr{X}_2]$ is the same. But this means every row of $\mathscr{X}_2$ is the same—and that is unacceptable because $\mathscr{X}_2$ is the coefficient of a sub-vector of $\boldsymbol{\beta}$ (which does not include μ) in the model equation.

Now suppose the interactions are taken as fixed effects. Then condition (i) of (43) is $\mathscr{X}\mathbf{LZ} = \mathscr{X}\mathbf{L}[\mathbf{Z}_1 \quad \mathbf{Z}_2] = \mathbf{0}$ and so $\mathscr{X}\mathbf{LZ}_2 = \mathbf{0}$. Therefore using (82) gives $\mathscr{X}\mathbf{LZ}_2 = \mathscr{X}\mathbf{L}\mathscr{X}\mathbf{K} = \mathbf{0}$. This reduces, after post-multiplying (iii) of (43) by $\mathbf{K}$ to get $\mathscr{X}\mathbf{K} - \mathscr{X}\mathbf{L}\mathscr{X}\mathbf{K} = \mathbf{1}\boldsymbol{\tau}'\mathbf{K}$, to be $\mathscr{X}\mathbf{K} = \mathbf{1}\boldsymbol{\tau}'\mathbf{K}$, so that by (82) $\mathbf{Z}_2 = \mathbf{1}\boldsymbol{\tau}'\mathbf{K}$. But this means that every row of $\mathbf{Z}_2$ is the same—again an unacceptable situation. Thus data from models that include interactions between fixed and random factors cannot be used for estimating variance components by any method based on adjusted data $\mathbf{y}_a = \mathbf{y} - \mathscr{X}\hat{\boldsymbol{\beta}}$ for $\hat{\boldsymbol{\beta}} = \mathbf{Ly}$ where $\mathbf{L}$ satisfies (43). And Henderson's Method II is one such method. Thus, be they treated as fixed or random, interactions can be part of the model when Henderson's Method II is used only if they are interactions of fixed effects with each other, or of random effects with each other, and not of fixed effects with random effects.

g. Merits and demerits

Merits of Method II include the following.

(i) The inapplicability of Method I to mixed models is overcome, at least partially, by Method II: it can be used for mixed models that have no interactions between fixed and random factors.

(ii) Computation of $\mathbf{y}_a = \mathbf{y} - \mathbf{X}\hat{\boldsymbol{\beta}}$ requires care, but after that the computation is as easy as is that of Method I, save for coefficients of $\hat{\sigma}_e^2$ in the estimation equations.

(iii) Estimators are unbiased.

(iv) For balanced data Methods I and II are the same, and are the same as the ANOVA method for balanced data. (See E 5.18.)

Demerits include the following.

(i) Models with interactions between fixed and random factors cannot be analyzed using Method II.

(ii) No closed form expressions are available for sampling variances of estimators. They could be developed from those of Method I (see Appendix F) after taking account of $(k_A + \delta_A)\sigma_e^2$ discussed in the preceding sub-section e.

(iii) Negative estimates are possible.

5.5. HENDERSON'S METHOD III

Method III is based on borrowing sums of squares from the analysis of fixed effects models. The sums of squares used are the reductions in sums of squares due to fitting one model and various sub-models of it. We therefore begin a description of Method III with a brief summary of these sums of squares.

a. Borrowing from fixed effects models

-i. Reductions in sums of squares. In writing a general mixed model equation as $\mathbf{y} = \mathbf{X}\boldsymbol{\beta} + \mathbf{Z}\mathbf{u} + \mathbf{e}$ we clearly distinguish between fixed effects and random effects, representing them by $\boldsymbol{\beta}$ and $\mathbf{u}$, respectively. Suppose for the moment that we remove this distinction and combine $\boldsymbol{\beta}$ and $\mathbf{u}$ into a single vector $\mathbf{b}$ and write the model equation as

$$\mathbf{y} = \mathbf{W}\mathbf{b} + \mathbf{e} . \tag{84}$$

In this sub-section we consider (84) in its own right, forgetting that $\mathbf{b}$ contains both fixed and random effects. We do this because some of the sums of squares associated with fitting (84) as a fixed effects model and with fitting sub-models of that fixed effects model are the basis of Method III.

In fitting a fixed effects model having model equation (84) it is well known that the best linear unbiased estimator of $\mathbf{W}\mathbf{b}$ is (see Appendix M for the $\mathbf{A}^-$ and $\mathbf{A}^+$ notation)

$$\text{BLUE}(\mathbf{W}\mathbf{b}) = \mathbf{W}\mathbf{b}^0 = \mathbf{W}(\mathbf{W}'\mathbf{W})^-\mathbf{W}'\mathbf{y} = \mathbf{W}\mathbf{W}^+\mathbf{y} . \tag{85}$$

Then the residual error sum of squares after fitting the model is

$$\text{SSE} = (\mathbf{y} - \mathbf{W}\mathbf{b}^0)'(\mathbf{y} - \mathbf{W}\mathbf{b}^0) = \mathbf{y}'\mathbf{y} - \mathbf{y}'\mathbf{W}\mathbf{W}^+\mathbf{y} . \tag{86}$$

The partitioning of $\mathbf{y}'\mathbf{y}$ into two sums of squares SSE and $\mathbf{y}'\mathbf{W}\mathbf{W}^+\mathbf{y}$ (to be denoted by SSR) represented by (86) is summarized in Table 5.3. That table is, of course, the basis of the analysis of variance table for fitting (84). That analysis of variance usually includes calculating mean squares, which, on the basis of assuming normality in the form $\mathbf{y} \sim \mathcal{N}(\mathbf{W}\mathbf{b}, \sigma_e^2\mathbf{I}_N)$, then provide, through the F-statistic

$$F = \frac{\text{SSR}}{r(\mathbf{W})}\bigg/\frac{\text{SSE}}{N - r(\mathbf{W})},$$

a test of the hypothesis H: $\mathbf{b} = \mathbf{0}$. All this is for the fixed effects model. For estimating variance components for mixed models we concentrate attention on SSR.

SSR in Table 5.3 is seen to be the *reduction* in sum of squares due to fitting the model $\mathbf{y} = \mathbf{W}\mathbf{b} + \mathbf{e}$. We therefore denote it by $R(\mathbf{b})$ and so have

$$R(\mathbf{b}) = \mathbf{y}'\mathbf{W}(\mathbf{W}'\mathbf{W})^-\mathbf{W}'\mathbf{y} = \mathbf{y}'\mathbf{W}\mathbf{W}^+\mathbf{y} . \tag{87}$$

Method III is the ANOVA method of estimating variance components using quadratic forms based on $R(\mathbf{b})$.

TABLE 5.3. PARTITIONING THE TOTAL SUM OF SQUARES WHEN FITTING THE FIXED EFFECTS MODEL $\mathbf{y} = \mathbf{Wb} + \mathbf{e}$

Reduction due to fitting the model	$\text{SSR} = \mathbf{y}'\mathbf{WW}^+\mathbf{y}$
Residual error	$\text{SSE} = \mathbf{y}'\mathbf{y} - \mathbf{y}'\mathbf{WW}^+\mathbf{y}$
Total	$\text{SST} = \mathbf{y}'\mathbf{y}$

Consider partitioning $\mathbf{Wb}$ so that

$$E(\mathbf{y}) = \mathbf{W}_1\mathbf{b}_1 + \mathbf{W}_2\mathbf{b}_2, \tag{88}$$

with $R(\mathbf{b})$ now being denoted

$$R(\mathbf{b}_1, \mathbf{b}_2) = \mathbf{y}'[\mathbf{W}_1 \quad \mathbf{W}_2][\mathbf{W}_1 \quad \mathbf{W}_2]^+\mathbf{y}\,. \tag{89}$$

In fixed effects models we might want to compare the fitting of (88) with fitting

$$E(\mathbf{y}) = \mathbf{W}_1\mathbf{b}_1, \tag{90}$$

which has

$$R(\mathbf{b}_1) = \mathbf{y}'\mathbf{W}_1\mathbf{W}_1^+\mathbf{y}\,. \tag{91}$$

The comparison is based on the difference between the two reductions in sums of squares:

$$\begin{aligned} R(\mathbf{b}_2 \mid \mathbf{b}_1) &= R(\mathbf{b}_1, \mathbf{b}_2) - R(\mathbf{b}_1) \\ &= \mathbf{y}'[\mathbf{W}_1 \quad \mathbf{W}_2][\mathbf{W}_1 \quad \mathbf{W}_2]^+\mathbf{y} - \mathbf{y}'\mathbf{W}_1\mathbf{W}_1^+\mathbf{y}\,. \end{aligned} \tag{92}$$

Simplification of (92) comes from using $\mathbf{WW}^+ = \mathbf{W}(\mathbf{W}'\mathbf{W})^-\mathbf{W}$ and the generalized inverse of a partitioned $\mathbf{W}'\mathbf{W}$ as given in (21) of Appendix M.4c. This results in (92) reducing to

$$R(\mathbf{b}_2 \mid \mathbf{b}_1) = \mathbf{y}'\mathbf{M}_1\mathbf{W}_2(\mathbf{W}_2'\mathbf{M}_1\mathbf{W}_2)^-\mathbf{W}_2'\mathbf{M}_1\mathbf{y} \tag{93}$$

for

$$\mathbf{M}_1 = \mathbf{I} - \mathbf{W}_1(\mathbf{W}_1'\mathbf{W}_1)^-\mathbf{W}_1' = \mathbf{M}_1' = \mathbf{M}_1^2, \quad \text{with } \mathbf{M}_1\mathbf{W}_1 = \mathbf{0}\,. \tag{94}$$

It is sums of squares like (93) that are used in Method III. Although with fixed effects models such sums of squares are used in numerators of F-statistics, for which normality assumptions are required, no normality assumptions are implied when using those sums of squares in the Method III estimation procedure.

-ii. ***Expected sums of squares.*** Before specifically adapting (93) to the mixed models we are interested in (through writing $\mathbf{Wb}$ as $\mathbf{X\beta} + \mathbf{Zu}$), we consider a general formulation of $E[R(\mathbf{b}_2 \mid \mathbf{b}_1)]$ that illustrates an important property of Method III. It is not affected by having fixed effects in the model, as is Method I; nor is it affected by having fixed-by-random interactions, as is Method II. To do this, think of $\mathbf{b}$ as being any mixture of fixed and random effects, so that

without knowing which is which we can, in broad generality, have

$$\text{var}(\mathbf{b}) = E(\mathbf{bb}') - E(\mathbf{b})E(\mathbf{b}') \, . \tag{95}$$

For a sub-vector of $\mathbf{b}$ having elements that are the effects due to a random factor, the corresponding diagonal elements in (95) will be a variance component; and all other elements of (95) in the same rows and columns as those diagonal elements will be zero. This arises from the properties given in (5a) and (5b).

Without having to know or formulate which elements of $\mathbf{b}$ are fixed effects and which are random, the generalization (95) proves useful in considering what we need for any form of the ANOVA method of estimation, namely the expected value of a quadratic form $\mathbf{y}'\mathbf{Ay}$. With (95) we have

$$\mathbf{V} = \text{var}(\mathbf{y}) = \text{var}(\mathbf{Wb}) + \sigma_e^2 \mathbf{I}_N = \mathbf{W}\,\text{var}(\mathbf{b})\mathbf{W}' + \sigma_e^2 \mathbf{I}$$

and so

$$\begin{aligned} E(\mathbf{y}'\mathbf{Ay}) &= \text{tr}[\mathbf{AW}\,\text{var}(\mathbf{b})\mathbf{W}' + \mathbf{A}\sigma_e^2\mathbf{I}] + E(\mathbf{b}')\mathbf{W}'\mathbf{AW}E(\mathbf{b}) \\ &= \text{tr}[\mathbf{W}'\mathbf{AW}\,\text{var}(\mathbf{b}) + \mathbf{W}'\mathbf{AW}E(\mathbf{b})E(\mathbf{b}')] + \sigma_e^2\,\text{tr}(\mathbf{A}) \\ &= \text{tr}[\mathbf{W}'\mathbf{AW}E(\mathbf{bb}')] + \sigma_e^2\,\text{tr}(\mathbf{A}) \, . \end{aligned} \tag{96}$$

This, for the quadratic form $R(\mathbf{b}_2 \mid \mathbf{b}_1)$ of (93), where $\mathbf{W} = [\mathbf{W}_1 \quad \mathbf{W}_2]$, gives

$$\begin{aligned} E\,R(\mathbf{b}_2 \mid \mathbf{b}_1) &= \text{tr}\left\{\begin{bmatrix} \mathbf{W}_1' \\ \mathbf{W}_2' \end{bmatrix}[\mathbf{M}_1\mathbf{W}_2(\mathbf{W}_2'\mathbf{M}_1\mathbf{W}_2)^-\mathbf{W}_2'\mathbf{M}_1][\mathbf{W}_1 \quad \mathbf{W}_2]E(\mathbf{bb}')\right\} \\ &\quad + \sigma_e^2\,\text{tr}[\mathbf{M}_1\mathbf{W}_2(\mathbf{W}_2'\mathbf{M}_1\mathbf{W}_2)^-\mathbf{W}_2'\mathbf{M}_1] \, . \end{aligned} \tag{97}$$

Using $\mathbf{M}_1\mathbf{W}_1 = \mathbf{0}$ from (94), together with $\mathbf{b}' = [\mathbf{b}_1' \quad \mathbf{b}_2']$ and the idempotency of $\mathbf{MW}_2(\mathbf{W}_2'\mathbf{M}_1\mathbf{W}_2)^-\mathbf{W}_2\mathbf{M}_1$ reduces (97) to

$$E\,R(\mathbf{b}_2 \mid \mathbf{b}_1) = \text{tr}[\mathbf{W}_2'\mathbf{M}_1\mathbf{W}_2E(\mathbf{b}_2\mathbf{b}_2')] + \sigma_e^2(r_{[\mathbf{W}_1 \ \ \mathbf{W}_2]} - r_{\mathbf{W}_1}), \tag{98}$$

where the coefficient of σ_e^2 comes from (26) of Appendix M.4d; and $r_{\mathbf{W}_1} = r(\mathbf{W}_1)$, the rank of $\mathbf{W}_1$.

A notable feature of (98) is that, apart from σ_e^2, the only parameters of the model that are in (98) are those in $\mathbf{b}_2$. They occur in the form $E(\mathbf{b}_2\mathbf{b}_2')$. There is no occurrence in (98) of the parameters of $\mathbf{b}_1$ in any form. This means for Method III that expected sums of squares of the form $E\,R(\mathbf{b}_2 \mid \mathbf{b}_1)$ never involve $\mathbf{b}_1$. Therefore, so long as we formulate $\mathbf{b}_1$ to always include the fixed effects of our model, $E\,R(\mathbf{b}_2 \mid \mathbf{b}_1)$ never includes fixed effects. By this means, Method III avoids the deficiency of Method I being unsuitable for mixed models. And, by the general nature of $R(\mathbf{b}_2 \mid \mathbf{b}_1)$, there is no restriction, as there is with Method II, of having to do without fixed-by-random interactions.

b. Mixed models

We now revert to the mixed model having model equation $\mathbf{y} = \mathbf{X}\boldsymbol{\beta} + \mathbf{Zu} + \mathbf{e}$. For

$$\mathbf{M} = \mathbf{I} - \mathbf{XX}^+ = \mathbf{I} - \mathbf{X}(\mathbf{X}'\mathbf{X})^-\mathbf{X}' = \mathbf{M}' = \mathbf{M}^2 \quad \text{with } \mathbf{MX} = \mathbf{0}, \tag{99}$$

analogous to (94), we have from (98)

$$\begin{aligned} E\,R(\mathbf{u}\,|\,\boldsymbol{\beta}) &= \mathrm{tr}[\mathbf{Z}'\mathbf{M}\mathbf{Z}E(\mathbf{u}\mathbf{u}')] \quad + \sigma_e^2(r[\mathbf{X} \quad \mathbf{Z}] - r[\mathbf{X}]) \\ &= \mathrm{tr}[\mathbf{Z}'\mathbf{M}\mathbf{Z}\{_{\mathrm{d}}\,\sigma_i^2\mathbf{I}_{q_i}\}] + \sigma_e^2(r[\mathbf{X} \quad \mathbf{Z}] - r[\mathbf{X}]) \\ &= \Sigma_i\,\mathrm{tr}(\mathbf{Z}_i'\mathbf{M}\mathbf{Z}_i)\sigma_i^2 \quad + \sigma_e^2(r[\mathbf{X} \quad \mathbf{Z}] - r[\mathbf{X}])\,. \end{aligned} \tag{100}$$

In this we see at once that the fixed effects $\boldsymbol{\beta}$ do not occur at all.

Example 3. The 1-way classification, random model, with unbalanced data $y_{ij} = \mu + \alpha_i + e_{ij}$ has $\mathbf{X} = \mathbf{1}_N$, $\boldsymbol{\beta} = \mu$, $\mathbf{Z} = \{_{\mathrm{d}}\,\mathbf{1}_{n_i}\}$ and $\mathbf{u} = \boldsymbol{\alpha}$. Therefore with $\mathrm{SSA} = R(\mathbf{u}\,|\,\boldsymbol{\beta}) = R(\boldsymbol{\alpha}\,|\,\mu)$

$$\begin{aligned} E(\mathrm{SSA}) = E\,R(\boldsymbol{\alpha}\,|\,\mu) &= \mathrm{tr}[(\mathbf{I} - \bar{\mathbf{J}}_N)\sigma_\alpha^2\{_{\mathrm{d}}\,\mathbf{J}_{n_i}\}] + \sigma_e^2(r[\mathbf{1}_N \quad \{_{\mathrm{d}}\,\mathbf{J}_{n_i}\}] - r[\mathbf{1}_N]) \\ &= \sigma_\alpha^2\,\mathrm{tr}[\{_{\mathrm{d}}\,\mathbf{J}_{n_i}\} - N^{-1}\mathbf{1}_N\{_{\mathrm{r}}\,n_i\mathbf{1}_{n_i}'\}] + \sigma_e^2(a-1) \\ &= (N - \Sigma_i n_i^2/N)\sigma_\alpha^2 + (a-1)\sigma_e^2 \end{aligned}$$

as in Section 3.6a.

Now consider a slightly more general case, of just two random effects, represented by $\mathbf{u}_1$ and $\mathbf{u}_2$. Then (100) gives

$$E\,R(\mathbf{u}_1, \mathbf{u}_2\,|\,\boldsymbol{\beta}) = \mathrm{tr}[\mathbf{M}(\mathbf{Z}_1\mathbf{Z}_1'\sigma_1^2 + \mathbf{Z}_2\mathbf{Z}_2'\sigma_2^2)] + \sigma_e^2(r[\mathbf{X} \quad \mathbf{Z}] - r[\mathbf{X}])\,. \tag{101}$$

From (98) we can also obtain $ER(\mathbf{u}_2\,|\,\boldsymbol{\beta}, \mathbf{u}_1)$. It involves an $\mathbf{M}_1$ based on $[\mathbf{X} \quad \mathbf{Z}_1]$, and is in fact $\mathbf{I} - [\mathbf{X} \quad \mathbf{Z}_1][\mathbf{X} \quad \mathbf{Z}_1]^+$, by the nature of $\mathbf{M}_1$ in (94). Thus

$$\begin{aligned} E\,R(\mathbf{u}_2\,|\,\boldsymbol{\beta}, \mathbf{u}_1) &= \mathrm{tr}\{(\mathbf{I} - [\mathbf{X} \quad \mathbf{Z}_1][\mathbf{X} \quad \mathbf{Z}_1]^+)\mathbf{Z}_2\mathbf{Z}_2'\sigma_2^2\} \\ &\quad + \sigma_e^2(r[\mathbf{X} \quad \mathbf{Z}] - r[\mathbf{X} \quad \mathbf{Z}_1])\,. \end{aligned} \tag{102}$$

We also have, of course, that

$$E(\mathrm{SSE}) = E[\mathbf{y}'\mathbf{y} - R(\boldsymbol{\beta}, \mathbf{u}_1, \mathbf{u}_2)] = (N - r[\mathbf{X} \quad \mathbf{Z}])\sigma_e^2\,. \tag{103}$$

Equation (102) demonstrates a feature of Method III that can sometimes prove useful: just as each of $E\,R(\mathbf{u}_1, \mathbf{u}_2\,|\,\boldsymbol{\beta})$ and $E\,R(\mathbf{u}_2\,|\,\boldsymbol{\beta}, \mathbf{u}_1)$ involve no terms in elements of $\boldsymbol{\beta}$, arising from (98) involving no $\mathbf{b}_1$, so also does $E\,R(\mathbf{u}_2\,|\,\boldsymbol{\beta}, \mathbf{u}_1)$ of (102) not involve $\mathbf{u}_1$; it involves only $\mathbf{u}_2$, in the form of σ_2^2. This leads to being able to write Method III as a series of equations in the estimated components that can easily be solved progressively, first for σ_e^2 and then for one of the other components and then for another, and so on. They are, effectively, linear equations that have a triangular coefficient matrix. The estimation equations from (101)–(103) are an example:

$$\begin{aligned} R(\mathbf{u}_1, \mathbf{u}_2\,|\,\boldsymbol{\beta}) &= \mathrm{tr}(\mathbf{M}\mathbf{Z}_1\mathbf{Z}_1')\hat{\sigma}_1^2 + \mathrm{tr}(\mathbf{M}\mathbf{Z}_2\mathbf{Z}_2')\hat{\sigma}_2^2 + (r[\mathbf{X} \quad \mathbf{Z}] - r[\mathbf{X}])\hat{\sigma}_e^2, \\ R(\mathbf{u}_2\,|\,\boldsymbol{\beta}, \mathbf{u}_1) &= \lambda\hat{\sigma}_2^2 + (r[\mathbf{X} \quad \mathbf{Z}] - r[\mathbf{X} \quad \mathbf{Z}_1])\hat{\sigma}_e^2, \\ \mathbf{y}'\mathbf{y} - R(\beta, \mathbf{u}_1, \mathbf{u}_2) &= (N - r[\mathbf{X} \quad \mathbf{Z}])\hat{\sigma}_e^2 \end{aligned} \tag{104}$$

for

$$\lambda = \text{tr}\{(\mathbf{I} - [\mathbf{X} \quad \mathbf{Z}_1][\mathbf{X} \quad \mathbf{Z}_1]^+)\mathbf{Z}_2\mathbf{Z}_2'\}\ .$$

Note that $ER(\mathbf{b}_2 \mid \mathbf{b}_1)$ of (98) not involving $\mathbf{b}_1$ has an underlying condition that must not be overlooked: in deriving (98), $\mathbf{b}_1$ and $\mathbf{b}_2$ constitute *all* the parameters of the model. Suppose $R(\mathbf{b}_2 \mid \mathbf{b}_1)$ is such that $\mathbf{b}_1$ and $\mathbf{b}_2$ do not make up the whole model. Then there are more parameters in $\mathbf{b}$ than those in $\mathbf{b}_1$ and $\mathbf{b}_2$. Therefore the derivation of (97) from (96) would have to have $\mathbf{W}$ partitioned not just as $[\mathbf{W}_1 \quad \mathbf{W}_2]$ corresponding to $\mathbf{b}_1$ and $\mathbf{b}_2$ of $R(\mathbf{b}_2 \mid \mathbf{b}_1)$, but as $[\mathbf{W}_1 \quad \mathbf{W}_2 \quad \mathbf{W}_3]$, where $\mathbf{W}_3$ corresponds to $\mathbf{b}_3$, which contains the parameters in $\mathbf{b}$ that are not in $\mathbf{b}_1$ and $\mathbf{b}_2$. This would lead to (98) containing terms in $\mathbf{b}_3$ and $\mathbf{W}_3$ as well as $\mathbf{b}_2$ and $\mathbf{W}_2$, and the principle evident in (98) as it stands, that $E\,R(\mathbf{b}_2 \mid \mathbf{b}_1)$ involves only σ_e^2 and $\mathbf{b}_2$, would be negated. And the triangular nature of the estimating equations seen in (104) would be lost.

Although sums of squares defined in terms of only parts of a model do not, as just described, fit into the algorithm of (98), they can often be adapted so that (98) can be utilized. Consider $E\,R(\mathbf{u}_2 \mid \boldsymbol{\beta})$ for the model $E(\mathbf{y}) = \mathbf{X}\boldsymbol{\beta} + \mathbf{Z}_1\mathbf{u}_1 + \mathbf{Z}_2\mathbf{u}_2$. Because $\mathbf{u}_2$ and $\boldsymbol{\beta}$ of $R(\mathbf{u}_2 \mid \boldsymbol{\beta})$ do not constitute the whole model, consisting of $\boldsymbol{\beta}$, $\mathbf{u}_1$ and $\mathbf{u}_2$, we cannot use (98) to derive $E\,R(\mathbf{u}_2 \mid \boldsymbol{\beta})$. But

$$\begin{aligned} R(\mathbf{u}_2 \mid \boldsymbol{\beta}) &= R(\boldsymbol{\beta}, \mathbf{u}_2) - R(\boldsymbol{\beta}) \\ &= R(\boldsymbol{\beta}, \mathbf{u}_1, \mathbf{u}_2) - R(\boldsymbol{\beta}) - [R(\boldsymbol{\beta}, \mathbf{u}_1, \mathbf{u}_2) - R(\boldsymbol{\beta}, \mathbf{u}_2)] \\ &= R(\mathbf{u}_1, \mathbf{u}_2 \mid \boldsymbol{\beta}) - R(\mathbf{u}_1 \mid \boldsymbol{\beta}, \mathbf{u}_2), \end{aligned}$$

with (98) being applicable to each of these last two terms. In this way, reductions in sums of squares whose expected values cannot be obtained directly from (98) can be expressed as the differences between two reductions that can utilize (98).

Notice in (104) that in place of $R(\mathbf{u}_2 \mid \boldsymbol{\beta}, \mathbf{u}_1)$ it would be perfectly permissible to have $R(\mathbf{u}_1 \mid \boldsymbol{\beta}, \mathbf{u}_2)$—with a correspondingly different expected value. This means that there would then be two different sets of three sums of squares from which to estimate the three variance components: (104), and (104) with $R(\mathbf{u}_1 \mid \boldsymbol{\beta}, \mathbf{u}_2)$ in place of $R(\mathbf{u}_2 \mid \boldsymbol{\beta}, \mathbf{u}_1)$. There could be a third set: the last two equations of (104) together with $R(\mathbf{u}_1 \mid \boldsymbol{\beta}, \mathbf{u}_2)$. Herein lies one of the great deficiences of Method III; it gives no indication as to which of such different sets of equations is to be preferred.

c. A general result

The result given in (102) for $ER(\mathbf{u}_2, \boldsymbol{\beta} \mid \mathbf{u}_1)$, where each of $\mathbf{u}_1$ and $\mathbf{u}_2$ represent a single random factor, can be extended to where $\mathbf{u}_1$ and $\mathbf{u}_2$ each represent one or more random factors. In particular, partition $\mathbf{Z}_2\mathbf{u}_2$ as

$$\begin{aligned} \mathbf{Z}_2\mathbf{u}_2 &= \mathbf{Z}_{21}\mathbf{u}_{21} + \mathbf{Z}_{22}\mathbf{u}_{22} + \cdots + \mathbf{Z}_{2r_2}\mathbf{u}_{2r_2} \\ &= \sum_{i=1}^{r_2} \mathbf{Z}_{2i}\mathbf{u}_{2i}, \end{aligned}$$

where r_2 is the number of random factors having all their effects in $\mathbf{u}_2$, each $\mathbf{u}_{2i}$ being the effects for exactly one of those factors, with $\text{var}(\mathbf{u}_{2i}) = \sigma_{2i}^2 \mathbf{I}_{q_{2i}}$. Then with

$$\mathbf{y} = \mathbf{X}\boldsymbol{\beta} + \mathbf{Z}_1\mathbf{u}_1 + \sum_{i=1}^{r_2} \mathbf{Z}_{2i}\mathbf{u}_{2i} + \mathbf{e}$$

the extension of (102) is

$$E\,R(\mathbf{u}_2 \mid \boldsymbol{\beta}, \mathbf{u}_1) = \sum_{i=1}^{r_2} \sigma_{2i}^2 \,\text{tr}\{(\mathbf{I} - [\mathbf{X} \quad \mathbf{Z}_1][\mathbf{X} \quad \mathbf{Z}_1]^+)\mathbf{Z}_{2i}\mathbf{Z}'_{2i}\}$$
$$+ \sigma_e^2(r[\mathbf{X} \quad \mathbf{Z}_1 \quad \mathbf{Z}_2] - r([\mathbf{X} \quad \mathbf{Z}_1]) . \tag{105}$$

d. Sampling variances

In (21), for Method I, we established a general formula for $\text{var}(\hat{\boldsymbol{\sigma}}^2)$, knowing that in $\hat{\boldsymbol{\sigma}}^2 = \mathbf{C}^{-1}\mathbf{s}$ each element of $\mathbf{s}$ was of the form $\mathbf{y}'\mathbf{A}_i\mathbf{y}$, where $\mathbf{X}'\mathbf{A}_i\mathbf{X} = \mathbf{0}$, as in (16). We now show that this is also the case for Method III estimation, which means that (21) for $\text{var}(\hat{\boldsymbol{\sigma}}^2)$ can also be used for Method III.

In terms of the vector $\mathbf{s} = \{R(\cdot \mid \cdot)\}$ of reductions in sums of squares, we have, as usual for ANOVA estimation, $E(\mathbf{s}) = \mathbf{C}\boldsymbol{\sigma}^2$, giving $\mathbf{s} = \mathbf{C}\hat{\boldsymbol{\sigma}}^2$ for some $\mathbf{C}$ and so $\hat{\boldsymbol{\sigma}}^2 = \mathbf{C}^{-1}\mathbf{s}$. Denoting a typical element of $\mathbf{s}$ by $\mathbf{y}'\mathbf{A}_i\mathbf{y}$, we know from (98) that by $\mathbf{y}'\mathbf{A}_i\mathbf{y}$ being of the form $R(\cdot \mid \boldsymbol{\beta}, \cdot)$, its expected value contains no term in $\boldsymbol{\beta}$. Therefore, since in general $E(\mathbf{y}'\mathbf{A}\mathbf{y}) = \text{tr}(\mathbf{A}\mathbf{V}) + \boldsymbol{\beta}'\mathbf{X}'\mathbf{A}\mathbf{X}\boldsymbol{\beta}$, we know for $R(\cdot \mid \boldsymbol{\beta}, \cdot)$ that $\boldsymbol{\beta}'\mathbf{X}'\mathbf{A}\mathbf{X}\boldsymbol{\beta} = \mathbf{0} \quad \forall\, \boldsymbol{\beta}$ and hence $\mathbf{X}'\mathbf{A}\mathbf{X} = \mathbf{0}$; also, because $R(\cdot \mid \boldsymbol{\beta}, \cdot)$ is a sum of squares, $\mathbf{A}$ is real and n.n.d. and so $\mathbf{X}'\mathbf{A}\mathbf{X} = \mathbf{0}$ implies $\mathbf{A}\mathbf{X} = \mathbf{0}$.

More direct derivation of this result is achieved by writing $R(\cdot \mid \boldsymbol{\beta}, \cdot)$ as $R(\mathbf{u}_2 \mid \boldsymbol{\beta}, \mathbf{u}_1)$ for some $\mathbf{u}_1$ and $\mathbf{u}_2$. Then

$$R(\mathbf{u}_2 \mid \boldsymbol{\beta}, \mathbf{u}_1) = \mathbf{y}'\mathbf{M}_1\mathbf{Z}_2(\mathbf{Z}'_2\mathbf{M}_1\mathbf{Z}_2)^-\mathbf{Z}'_2\mathbf{M}_1\mathbf{y}, \tag{106}$$

where $\mathbf{M}_1$ of (94) is now

$$\mathbf{M}_1 = [\mathbf{X} \quad \mathbf{Z}_1]\left(\begin{bmatrix}\mathbf{X}' \\ \mathbf{Z}'_1\end{bmatrix}[\mathbf{X} \quad \mathbf{Z}_1]\right)^- \begin{bmatrix}\mathbf{X}' \\ \mathbf{Z}'_1\end{bmatrix}.$$

Recall that, in general $\mathbf{T}(\mathbf{T}'\mathbf{T})^-\mathbf{T}'\mathbf{T} = \mathbf{T}$. Using this with $\mathbf{T} = [\mathbf{X} \quad \mathbf{Z}_1]$ gives

$$\mathbf{M}_1\begin{bmatrix}\mathbf{X} \\ \mathbf{Z}_1\end{bmatrix} = \begin{bmatrix}\mathbf{X} \\ \mathbf{Z}_1\end{bmatrix} - \begin{bmatrix}\mathbf{X} \\ \mathbf{Z}_1\end{bmatrix} = \mathbf{0} .$$

Therefore $\mathbf{M}_1\mathbf{X} = \mathbf{0}$ and so writing (106) as $\mathbf{y}'\mathbf{A}\mathbf{y}$ gives $\mathbf{A}\mathbf{X} = \mathbf{M}_1\mathbf{Z}_2(\mathbf{Z}'_2\mathbf{M}_1\mathbf{Z}_2)^-\mathbf{Z}'_2\mathbf{M}_1\mathbf{X} = \mathbf{0}$. Hence the expected value and variance of $\mathbf{y}'\mathbf{M}_1\mathbf{T}\mathbf{M}_1\mathbf{y}$ for any $\mathbf{T}$ (and its covariance with any other quadratic form in $\mathbf{y}$) contains the term $\mathbf{T}\mathbf{M}_1\mathbf{X}\boldsymbol{\beta}$, which, with $\mathbf{M}_1\mathbf{X} = \mathbf{0}$, is null. Therefore, under normality, for $\mathbf{A}_i$ having the form $\mathbf{M}_1\mathbf{Z}_2(\mathbf{Z}'_2\mathbf{M}_1\mathbf{Z}_2)^-\mathbf{Z}'_2\mathbf{M}_1$,

$$\text{var}(\mathbf{y}'\mathbf{A}_i\mathbf{y}) = 2\,\text{tr}(\mathbf{A}_i\mathbf{V})^2 \quad \text{and} \quad \text{cov}(\mathbf{y}'\mathbf{A}_i\mathbf{y}, \mathbf{y}'\mathbf{A}_j\mathbf{y}) = 2\,\text{tr}(\mathbf{A}_i\mathbf{V}\mathbf{A}_j\mathbf{V}) .$$

Therefore, as in (21),

$$\text{var}(\hat{\boldsymbol{\sigma}}^2) = \mathbf{C}^{-1}\,\text{var}(\mathbf{s})\,\mathbf{C}^{-1\prime} = 2\mathbf{C}^{-1}\{_{\text{m}}\,\text{tr}(\mathbf{A}_i\mathbf{V}\mathbf{A}_j\mathbf{V})\}_{i,j=0}^{r}\mathbf{C}^{-1\prime}\,. \quad (107)$$

This is a succinct expression but its use is bedeviled with the usual complexities of a sampling dispersion matrix of estimated variance components. First, through its dependence on $\mathbf{V}$, (107) is in terms of the unknown components, $\boldsymbol{\sigma}^2$. Nevertheless, for any particular data set one can always determine $\mathbf{C}$, numerically, and then, for any pre-assigned value of $\boldsymbol{\sigma}^2$, say $\boldsymbol{\sigma}_0^2$, one can compute $\text{var}(\hat{\boldsymbol{\sigma}}^2)$ from (21) or (107). However, this does little for establishing closed form expressions for sampling variances and covariances of Method III estimators, which remains an intractable problem. Second, the numbers of observations in the cells and subclass totals of the data occur in (107) in very intractable ways, in the $\mathbf{A}$s and in the functions that are elements of $\mathbf{C}$. Third, as a result, studying the behavior of elements of (107) for changes in the number of observations is well nigh impossible, analytically—and arithmetic studies through simulation are fraught with the difficulties of all such studies: designing them in such a way as to be reasonably likely to be able to draw some conclusions.

e. Merits and demerits

(i) Method III is applicable to all mixed models; the restriction of having no interactions between fixed and random factors that applies in Method II does not apply to Method III.

(ii) Estimates are unbiased.

(iii) For balanced data Methods I, II and III are the same.

But demerits include the following.

(i) When there are two or more crossed random factors the method can be applied, for a given model, in more than one way; and there is no way, analytically, of deciding between one application and another (as illustrated in Sections 5.6a-ii and 5.6c-ii, which follow).

(ii) Computationally, the method can involve the inversion of large-sized matrices—of order equal to the number of levels of the effects in the model. This disadvantage will decline as today's computing power increases in speed and declines in cost (per arithmetic operation).

(iii) Sampling variances can be calculated arithmetically, through a series of matrix operations and with using estimated values for the variance components, but specific closed form expressions are not available.

5.6. METHOD III APPLIED TO THE 2-WAY CROSSED CLASSIFICATION

The 2-way crossed classification is the easiest case for illustrating some of the results of the preceding subsections. We begin with the no interaction, random model.

a. No interaction, random model

Taking the scalar form of the model equation

$$y_{ijk} = \mu + \alpha_i + \beta_j + e_{ijk},$$

its vector form is

$$\mathbf{y} = \mu \mathbf{1}_N + \mathbf{Z}_A \boldsymbol{\alpha} + \mathbf{Z}_B \boldsymbol{\beta} + \mathbf{e} \tag{108}$$

where

$$\mathbf{Z}_A = \{_{\mathrm{d}}\, \mathbf{1}_{n_{i.}}\}_{i=1}^{a} \quad \text{and} \quad \mathbf{Z}_B = \{_{\mathrm{c}}\, \{_{\mathrm{d}^*}\, \mathbf{1}_{n_{ij}}\}_{j=1}^{b}\}_{i=1}^{1} .$$

In $\mathbf{Z}_B$ the d* is described at the end of Appendix M.3: it means that for $n_{ij} = 0$ the symbol $\mathbf{1}_0$ has column position but no dimensions, i.e., no rows. Useful products are

$$\begin{gathered} \mathbf{1}'_N \mathbf{Z}_A = \{_{\mathrm{r}}\, n_{i.}\} \quad \text{and} \quad \mathbf{1}'_N \mathbf{Z}_B = \{_{\mathrm{r}}\, n_{.j}\}; \\ \mathbf{Z}_A \mathbf{Z}'_A = \{_{\mathrm{d}}\, \mathbf{J}_{N_{i.}}\}, \quad \mathbf{Z}'_A \mathbf{Z}_A = \{_{\mathrm{d}}\, n_{i.}\}, \quad \mathbf{Z}'_B \mathbf{Z}_B = \{_{\mathrm{d}}\, n_{.j}\} . \end{gathered} \tag{109}$$

-i. One set of sums of squares. One partitioning of the total sum of squares (corrected for the mean) that is used in the fixed effects model is

$$R(\boldsymbol{\alpha} \mid \mu) + R(\boldsymbol{\beta} \mid \mu, \boldsymbol{\alpha}) + \mathrm{SSE} = \mathrm{SST_m} \tag{110}$$

where

$$\mathrm{SSE} = \mathbf{y}'\mathbf{y} - R(\mu, \boldsymbol{\alpha}, \boldsymbol{\beta}) \quad \text{and} \quad \mathrm{SST_m} = \mathbf{y}'\mathbf{y} - N\bar{\mathbf{y}}^2 = \Sigma_i \Sigma_j \Sigma_k (y_{ijk} - \bar{y}_{...})^2 .$$

To derive expected values of the terms on the left-hand side of (110), we use (100) and (102), and to do so rewrite $R(\boldsymbol{\alpha} \mid \mu)$ in (110) as

$$R(\boldsymbol{\alpha} \mid \mu) = R(\boldsymbol{\alpha}, \boldsymbol{\beta} \mid \mu) - R(\boldsymbol{\beta} \mid \mu, \boldsymbol{\alpha}) . \tag{111}$$

Using (100) then gives

$$E\, R(\boldsymbol{\alpha}, \boldsymbol{\beta} \mid \mu) = \mathrm{tr}[\mathbf{M}(\mathbf{Z}_A \mathbf{Z}'_A \sigma^2_\alpha + \mathbf{Z}_B \mathbf{Z}'_B \sigma^2_\beta)] + \sigma^2_e[r(\mathbf{1} \quad \mathbf{Z}_A \quad \mathbf{Z}_B) - r(\mathbf{1})], \tag{112}$$

where $\mathbf{M}$ is

$$\mathbf{M} = \mathbf{I} - \mathbf{1}_N(\mathbf{1}'_N \mathbf{1}_N)^- \mathbf{1}_N = \mathbf{I} - \bar{\mathbf{J}}_N .$$

Hence

$$\begin{aligned} E\, R(\boldsymbol{\alpha}, \boldsymbol{\beta} \mid \mu) = {} & \sigma^2_\alpha \,\mathrm{tr}[(\mathbf{I} - \bar{\mathbf{J}}_N)\mathbf{Z}_A \mathbf{Z}'_A] + \sigma^2_\beta \,\mathrm{tr}[(\mathbf{I} - \bar{\mathbf{J}}_N)\mathbf{Z}_B \mathbf{Z}'_B] \\ & + \sigma^2_e(a + b - 1 - 1) . \end{aligned} \tag{113}$$

We now utilize (from Appendix M.6), for any matrix $\mathbf{T}$,

$$\mathrm{tr}(\mathbf{TT}') = \Sigma_i \Sigma_j t^2_{ij} = \mathrm{sesq}(\mathbf{T}), \tag{114}$$

where "sesq" is mnemonic for "sum of squares of elements". This reduces (113) to

$$E\,R(\boldsymbol{\alpha}, \boldsymbol{\beta} \mid \mu) = \sigma_\alpha^2\left[\operatorname{sesq}(\mathbf{Z}_A) - \frac{\operatorname{sesq}(\mathbf{1}'_N\mathbf{Z}_A)}{N}\right] + \sigma_\beta^2\left[\operatorname{sesq}(\mathbf{Z}_B) - \frac{\operatorname{sesq}(\mathbf{1}'_N\mathbf{Z}_B)}{N}\right]$$

$$+ (a + b - 2)\sigma_e^2$$

$$= \sigma_e^2\left(N - \frac{\Sigma_i n_{i.}^2}{N}\right)\sigma_\alpha^2 + \left(N - \frac{\Sigma_j n_{.j}^2}{N}\right)\sigma_\beta^2 + (a + b - 2)\sigma_e^2\,. \quad (115)$$

Similarly, using (102) with $\mathbf{M} = \mathbf{I} - [\mathbf{1} \quad \mathbf{X}_A][\mathbf{1} \quad \mathbf{X}_A]^+$ gives

$$E\,R(\boldsymbol{\beta} \mid \mu, \boldsymbol{\alpha}) = \operatorname{tr}(\mathbf{M}\mathbf{Z}_B\mathbf{Z}'_B\sigma_\beta^2) + \sigma_e^2[r(\mathbf{1} \quad \mathbf{Z}_A \quad \mathbf{Z}_B) - r(\mathbf{1} \quad \mathbf{Z}_A)]\,. \quad (116)$$

$$= \sigma_\beta^2\left(N - \sum_i \frac{\Sigma_j n_{ij}^2}{n_{i.}}\right) + (b - 1)\sigma_e^2\,. \quad (117)$$

And, of course

$$E(\mathrm{SSE}) = [N - r(\mathbf{1} \quad \mathbf{Z}_A \quad \mathbf{Z}_B)]\sigma_e^2 = (N - a - b + 1)\sigma_e^2\,. \quad (118)$$

It is left to the reader as E 5.7 to derive (117) from (116) and to explain (118).

On defining for any r, $k'_r = k_r/N$ with, as in (30),

$$k_1 = \Sigma_i n_{i.}^2, \qquad k_2 = \Sigma_j n_{.j}^2,$$

$$k_3 = \sum_i \frac{(\Sigma_j n_{ij}^2)}{n_{i.}}, \quad k_4 = \sum_j \frac{(\Sigma_i n_{ij}^2)}{n_{.j}} \quad (119)$$

and

$$k_{23} = \Sigma_i\Sigma_j n_{ij}^2,$$

the estimation equations can be written in convenient form. They come from using (115) and (118) with (111), together with (117) and (118), and are

$$\begin{aligned} R(\boldsymbol{\alpha} \mid \mu) &= (N - k'_1)\hat\sigma_\alpha^2 + (k_3 - k'_2)\hat\sigma_\beta^2 + (a - 1)\hat\sigma_e^2, \\ R(\boldsymbol{\beta} \mid \mu, \boldsymbol{\alpha}) &= \qquad\qquad\quad (N - k_3)\hat\sigma_\beta^2 + (b - 1)\hat\sigma_e^2, \\ \mathrm{SSE} &= \qquad\qquad\quad (N - a - b + 1)\hat\sigma_e^2\,. \end{aligned} \quad (120)$$

-ii. Three sets of sums of squares. The preceding results stem from (110), but that is only one of the two possible partitionings of $\mathrm{SST_m}$ that are used in the fixed effects model. The other is

$$R(\boldsymbol{\beta} \mid \mu) + R(\boldsymbol{\alpha} \mid \mu, \boldsymbol{\beta}) + \mathrm{SSE} = \mathrm{SST_m}\,. \quad (121)$$

By direct analogy (interchanging $\boldsymbol{\alpha}$ with $\boldsymbol{\beta}$, and i with j) with (120)—see E 5.7(d)—using expected values of the terms in (121) gives

$$\begin{aligned} R(\boldsymbol{\beta} \mid \mu) &= (k_4 - k'_1)\hat\sigma_\alpha^2 + (N - k'_2)\hat\sigma_\beta^2 + (b - 1)\hat\sigma_e^2, \\ R(\boldsymbol{\alpha} \mid \mu, \boldsymbol{\beta}) &= (N - k_4)\hat\sigma_\alpha^2 \qquad\qquad\quad + (a - 1)\hat\sigma_e^2, \\ \mathrm{SSE} &= \qquad\qquad\quad (N - a - b + 1)\hat\sigma_e^2\,. \end{aligned} \quad (122)$$

So this is a second set of equations that is available for estimating the σ^2s: it is not, of course, the same as (120).

Notation. It is convenient to define

$$h_1 = N - k_1' = N - \frac{\Sigma_i n_{i\cdot}^2}{N}, \quad h_2 = N - k_2' = N - \frac{\Sigma_j n_{\cdot j}^2}{N},$$

$$h_3 = N - k_{23}' = N - \frac{\Sigma_i \Sigma_j n_{ij}^2}{N}, \tag{123}$$

$$h_4 = N - k_3 = N - \sum_i \frac{\Sigma_j n_{ij}^2}{n_{i\cdot}}, \quad h_7 = N - k_4 = N - \sum_j \frac{\Sigma_i n_{ij}^2}{n_{\cdot j}}.$$

h_6 is defined in (137) and its calculation is described in Table F.3.

There are now two options for estimation: (120) or (122). Both have the same last equation, rewritten as (124a) below. Having thus obtained $\hat{\sigma}_e^2$, one could use the middle equations of (122) and (120) for calculating $\hat{\sigma}_\alpha^2$ and $\hat{\sigma}_\beta^2$, respectively: these are shown as (124b) and (124c). Doing this fails to utilize the first equations of (120) and (122). But a feature of them is that in each the sum of the first two equations is the same:

$$\begin{aligned} R(\boldsymbol{\alpha}, \boldsymbol{\beta} \mid \mu) &= R(\boldsymbol{\alpha} \mid \mu) + R(\boldsymbol{\beta} \mid \mu, \boldsymbol{\alpha}) \\ &= R(\boldsymbol{\beta} \mid \mu) + R(\boldsymbol{\alpha} \mid \mu, \boldsymbol{\beta}) \\ &= (N - k_1)\hat{\sigma}_\alpha^2 + (N - k_2')\hat{\sigma}_\beta^2 + (a + b - 2)\hat{\sigma}_e^2 \end{aligned}$$

which has been rewritten as estimation equation (124d).

$$\hat{\sigma}_e^2 = \frac{\text{SSE}}{N - a - b + 1} = \frac{\Sigma_i \Sigma_j \Sigma_k y_{ijk}^2 - R(\mu, \boldsymbol{\alpha}, \boldsymbol{\beta})}{N - a - b + 1}, \tag{124a}$$

$$\hat{\sigma}_\alpha^2 = \frac{R(\boldsymbol{\alpha} \mid \mu, \boldsymbol{\beta}) - (a-1)\hat{\sigma}_e^2}{h_7} = \frac{R(\mu, \boldsymbol{\alpha}, \boldsymbol{\beta}) - \Sigma_j n_{\cdot j}\bar{y}_{\cdot j \cdot}^2 - (a-1)\hat{\sigma}_e^2}{h_7}, \tag{124b}$$

$$\hat{\sigma}_\beta^2 = \frac{R(\boldsymbol{\beta} \mid \mu, \boldsymbol{\alpha}) - (b-1)\hat{\sigma}_e^2}{h_4} = \frac{R(\mu, \boldsymbol{\alpha}, \boldsymbol{\beta}) - \Sigma_i n_{i\cdot}\bar{y}_{i\cdot\cdot}^2 - (b-1)\hat{\sigma}_e^2}{h_4}, \tag{124c}$$

$$h_1\hat{\sigma}_\alpha^2 + h_2\hat{\sigma}_\beta^2 = R(\mu, \boldsymbol{\alpha}, \boldsymbol{\beta}) - N\bar{y}_{\cdot\cdot\cdot}^2 - (a + b - 2)\hat{\sigma}_e^2 . \tag{124d}$$

Therefore, since (124c) is the second equation of (120) and (124d) is the sum of the first and second, equations (120) are equivalent to (124a,c,d). Similarly, (122) is equivalent to (124a,b,d).

Equations (124a,c,d) come from the partitioning of SST_m shown in (110). Thus they are equivalent to using SAS Type I sums of squares, with the factors ordered *A*, *B*. Similarly, (124a,b,d) come from (121) and are equivalent to using SAS Type I sums of squares with the factors ordered *B*, *A*.

When $R(\boldsymbol{\beta} \mid \mu, \boldsymbol{\alpha})$ of (110) is used for fixed effects models, it has a different purpose from that of $R(\boldsymbol{\alpha} \mid \mu, \boldsymbol{\beta})$ in (128): e.g., $R(\boldsymbol{\beta} \mid \mu, \boldsymbol{\alpha})$ tests H: β_j *all equal*, and $R(\alpha \mid \mu, \boldsymbol{\beta})$ tests H: α_i *all equal*. Distinguishing between the utility of $R(\boldsymbol{\beta} \mid \mu, \boldsymbol{\alpha})$ and of $R(\boldsymbol{\alpha} \mid \mu, \boldsymbol{\beta})$ is therefore easy in the fixed effects model. But this distinction of purpose does not carry over to the use of these sums of squares in Method III. The method includes no way of deciding which of the two sets, a, c, d, or a, b, d of equations (124) should be used. Indeed, a third set of equations that includes (124a) is now apparent: Method III permits us to also use a, b, c. This is equivalent to the last two equations of each of (120) and (122), and so is the same as using SAS Type II sum of squares. Thus we have three possible ways of applying Method III to this case. They are set out in Table 5.4. That table identifies which equations a, b, c, or d of (124) can be put together to form a set of three equations for estimating the variance components σ_e^2, σ_α^2 and σ_β^2. It can also be interpreted as showing similarities between the three resulting sets of estimates. Thus $\hat{\sigma}_e^2$ is the same in all three options, $\hat{\sigma}_\alpha^2$ is the same in options 2 and 3 as obtained from (124b), and $\hat{\sigma}_\beta^2$ is the same in options 1 and 3, obtained from (124c).

It is the availability of more than one set of estimation equations $\hat{\boldsymbol{\sigma}}^2 = \mathbf{C}^{-1}\mathbf{s}$ that gives to Method III its unhappy characteristic of not always being uniquely defined for a given model. The method contains absolutely no criteria for deciding, for example, between options 1, 2 and 3 of Table 5.4. And in models with more than two crossed random factors there will be even more than three such sets of possible quadratic forms. Moreover, not only does the method itself provide no means for deciding between one option and another but, just as with trying to compare any forms of ANOVA estimation, the analytic intractability of sampling variances, for example, makes comparison on that basis effectively impossible. Numerical comparisons can be made, of course, but are fraught with all the usual difficulties already discussed.

TABLE 5.4. THREE OPTIONS FOR USING METHOD III ON THE 2-WAY CROSSED CLASSIFICATION, NO-INTERACTION, RANDOM MODEL

	Equations from (124)		
Estimate	Option 1	Option 2	Option 3
$\hat{\sigma}_e^2$	a	a	a
$\hat{\sigma}_\alpha^2$	d	b	b
$\hat{\sigma}_\beta^2$	c	d	c
Equivalent to			
Partitioning of SST_m	(110)	(121)	
Estimation equations	(120)	(122)	Last two of (120) and (122)
SAS sums of squares	Type I A, B	Type I B, A	Type II

Of course, one could always use the least-squares approach and, after arraying *all* the equations in the form $E(\mathbf{s}) = \mathbf{C}\boldsymbol{\sigma}^2$ use $\hat{\boldsymbol{\sigma}}^2 = (\mathbf{C}'\mathbf{C})^{-1}\mathbf{C}'\mathbf{s}$ as an unbiased estimator of $\boldsymbol{\sigma}^2$. (See Section 5.2c).

-iii. Calculation. The only difficult term to calculate in equations (124) is $R(\mu, \boldsymbol{\alpha}, \boldsymbol{\beta})$, the sum of squares due to fitting the no-interaction model having equation $E(y_{ijk}) = \mu + \alpha_i + \beta_j$. A computational method is given in Table F.3 of Appendix F. It is exactly the method given in Searle [1971, Chpt. 7, equation (26); 1987, Chpt. 5, equation (32); Chpt. 9, equation (99)].

-iv. Sampling Variances. Low (1964) derived sampling variances (under normality) for estimated σ^2s obtained by one of the three possible estimation options of Table 5.3, namely option 3, based on equations a, b and c of (124). Those sampling variances and covariances are shown in Appendix F.6e.

b. No interaction, mixed model

Suppose the βs in the model equation $y_{ijk} = \mu + \alpha_i + \beta_j + e_{ijk}$ are taken as fixed effects. Then the sums of squares are calculated exactly the same as in the random model of the preceding section. With the βs being fixed the only sum of squares having expected value that contains no βs is $R(\boldsymbol{\alpha} \mid \mu, \boldsymbol{\beta})$; and that expected value is precisely the same as in the random model. Therefore the estimation equations are the last two equations of (122), namely (124a,b):

$$\hat{\sigma}_e^2 = \frac{\Sigma_i\Sigma_j\Sigma_k y_{ijk}^2 - R(\mu, \boldsymbol{\alpha}, \boldsymbol{\beta})}{N - a - b + 1}$$

and

$$\hat{\sigma}_\alpha^2 = \frac{R(\mu, \boldsymbol{\alpha}, \boldsymbol{\beta}) - \Sigma_j n_{.j} \bar{y}_{.j.}^2 - (a-1)\hat{\sigma}_e^2}{h_7} .$$

These are the only equations for estimating σ_α^2 and σ_e^2 that Method III yields for this model. Hence this is a case where Method III is unique—in contrast to the random model case of Table 5.4, where it is not.

c. With interaction, random model

The with-interaction model has equation

$$y_{ijk} = \mu + \alpha_i + \beta_j + \gamma_{ij} + e_{ijk} .$$

Its vector form is thus

$$\mathbf{y} = \mu\mathbf{1}_N + \mathbf{Z}_A\boldsymbol{\alpha} + \mathbf{Z}_B\boldsymbol{\beta} + \mathbf{Z}_C\boldsymbol{\gamma} + \mathbf{e}, \tag{125}$$

exactly the same as in the no-interaction case of (108) except for the addition of $\mathbf{Z}_C\boldsymbol{\gamma}$ with

$$\mathbf{Z}_C = \{_d \{_d \mathbf{1}_{n_{ij}}\}_{j=1}^{a}\}_{j=1}^{b} \quad \text{for } n_{ij} \neq 0$$

which we write more simply as

$$\mathbf{Z}_C = \{_d \mathbf{1}_{n_{ij}}\} . \tag{126}$$

Useful products involving $\mathbf{Z}_C$ are

$$\mathbf{1}'_N\mathbf{Z}_C = \{_{\rm r}\, n_{ij}\}, \quad \mathbf{Z}_C\mathbf{Z}'_C = \{_{\rm d}\, \mathbf{J}_{n_{ij}}\} \quad \text{and} \quad \mathbf{Z}'_C\mathbf{Z}_C = \{_{\rm d}\, n_{ij}\} .$$

To take account of the possibility that some cells may contain no data, we define

$$s = \text{number of filled cells} . \tag{127}$$

*-i. **One set of sums of squares.*** Corresponding to (110) for the no-interaction case, one partitioning of $\mathrm{SST_m}$ used in the fixed effects model for the with-interaction case is

$$R(\boldsymbol{\alpha} \mid \mu) + R(\boldsymbol{\beta} \mid \mu, \boldsymbol{\alpha}) + R(\boldsymbol{\gamma} \mid \mu, \boldsymbol{\alpha}, \boldsymbol{\beta}) + \mathrm{SSE} = \mathrm{SST_m}, \tag{128}$$

where $\mathrm{SSE} = \mathbf{y}'\mathbf{y} - R(\mu, \boldsymbol{\alpha}, \boldsymbol{\beta}, \boldsymbol{\gamma})$ and $\mathrm{SST_m} = \Sigma_i\Sigma_j\Sigma_k(y_{ijk} - \bar{y}_{...})^2$. In order to use (100) and (102) for $R(\boldsymbol{\alpha} \mid \mu)$, we write it as

$$R(\boldsymbol{\alpha} \mid \mu) = R(\boldsymbol{\alpha}, \boldsymbol{\beta}, \boldsymbol{\gamma} \mid \mu) - R(\boldsymbol{\beta}, \boldsymbol{\gamma} \mid \mu, \boldsymbol{\alpha}) . \tag{129}$$

For this, similar to (112),

$$E\, R(\boldsymbol{\alpha}, \boldsymbol{\beta}, \boldsymbol{\gamma} \mid \mu) = \mathrm{tr}\{\mathbf{M}(\mathbf{Z}_A\mathbf{Z}'_A\sigma^2_\alpha + \mathbf{Z}_B\mathbf{Z}'_B\sigma^2_\beta + \mathbf{Z}_C\mathbf{Z}'_C\sigma^2_\gamma)\} + \sigma^2_e[r(\mathbf{1} \quad \mathbf{Z}) - r_1] \tag{130}$$

with, as following (112), $\mathbf{M} = \mathbf{I} - \bar{\mathbf{J}}_N$. Comparing this with (112), we conclude that the first two terms in (130) are the same as in (115); and the last two terms of (130) are

$$\begin{aligned}
\sigma^2_\gamma\, \mathrm{tr}(\mathbf{M}\mathbf{Z}_C\mathbf{Z}'_C) + \sigma^2_e[r(\mathbf{1} \quad \mathbf{Z}) - r_1] &= \sigma^2_\gamma\, \mathrm{tr}(\mathbf{Z}_C\mathbf{Z}'_C - \bar{\mathbf{J}}_N\mathbf{Z}_C\mathbf{Z}'_C) + \sigma^2_e(s-1) \\
&= \sigma^2_\gamma\left[\mathrm{sesq}(\mathbf{Z}_C) - \frac{\mathrm{sesq}(\mathbf{1}'_N\mathbf{Z}_C)}{N}\right] + \sigma^2_e(s-1) \\
&= \sigma^2_\gamma\left(N - \frac{\Sigma_i\Sigma_j n^2_{ij}}{N}\right) + \sigma^2_e(s-1) .
\end{aligned} \tag{131}$$

Therefore, with (131) and those first two terms of (115) used in (130),

$$\begin{aligned}
E\, R(\boldsymbol{\alpha}, \boldsymbol{\beta}, \boldsymbol{\gamma} \mid \mu) = {}& \sigma^2_\alpha\left(N - \frac{\Sigma_i n^2_{i.}}{N}\right) + \sigma^2_\beta\left(N - \frac{\Sigma_j n^2_{.j}}{N}\right) \\
&+ \sigma^2_\gamma\left(N - \frac{\Sigma_i\Sigma_j n^2_{ij}}{N}\right) + \sigma^2_e(s-1) .
\end{aligned} \tag{132}$$

Similarly, akin to (116),

$$E\, R(\boldsymbol{\beta}, \boldsymbol{\gamma} \mid \mu, \boldsymbol{\alpha}) = \mathrm{tr}[\mathbf{M}(\mathbf{Z}_B\mathbf{Z}'_B\sigma^2_\beta + \mathbf{Z}_C\mathbf{Z}'_C\sigma^2_\gamma)] + \sigma^2_e[r(\mathbf{1} \quad \mathbf{Z}) - r(\mathbf{1} \quad \mathbf{Z}_A)] \tag{133}$$

with the same $\mathbf{M}$ as used in (116). It is left to the reader to show that this reduces to

$$E\,R(\boldsymbol{\beta}, \boldsymbol{\gamma} \mid \mu, \boldsymbol{\alpha}) = \sigma_\beta^2\left(N - \sum_i \frac{\Sigma_j n_{ij}^2}{n_{i.}}\right) + \sigma_\gamma^2\left(N - \sum_i \frac{\Sigma_j n_{ij}^2}{n_{i.}}\right) + \sigma_e^2(s - a)\,. \tag{134}$$

The second term of (128) is

$$R(\boldsymbol{\beta} \mid \mu, \boldsymbol{\alpha}) = R(\boldsymbol{\beta}, \boldsymbol{\gamma} \mid \mu, \boldsymbol{\alpha}) - R(\boldsymbol{\gamma} \mid \mu, \boldsymbol{\alpha}, \boldsymbol{\beta})\,. \tag{135}$$

Its expected value is obtained from using (134) and $E\,R(\boldsymbol{\gamma} \mid \mu, \boldsymbol{\alpha}, \boldsymbol{\beta})$. But deriving the latter is difficult. From (105) it involves only two terms, σ_γ^2 and σ_e^2, with that in σ_e^2 being

$$\sigma_e^2[r(\mathbf{1} \quad \mathbf{Z}) - r(\mathbf{1} \quad \mathbf{Z}_A \quad \mathbf{Z}_B)] = \sigma_e^2(s - a - b + 1)\,.$$

Hence,

$$E\,R(\boldsymbol{\gamma} \mid \mu, \boldsymbol{\alpha}, \boldsymbol{\beta}) = h_6\sigma_\gamma^2 + \sigma_e^2(s - a - \ + 1), \tag{136}$$

where from (105)

$$h_6 = \mathrm{tr}\{(\mathbf{I} - [\mathbf{1} \quad \mathbf{Z}_A \quad \mathbf{Z}_B][\mathbf{1} \quad \mathbf{Z}_A \quad \mathbf{Z}_B]^+)\mathbf{Z}_C\mathbf{Z}_C'\}\,. \tag{137}$$

Unfortunately, a tractable form of h_6 is difficult to obtain—because $[\mathbf{1} \quad \mathbf{Z}_A \quad \mathbf{Z}_B][\mathbf{1} \quad \mathbf{Z}_A \quad \mathbf{Z}_B]^+$ has no tractable form. A procedure for calculating it is given in Table F.3 of Appendix F, taken from Searle and Henderson (1961). It would be nice to have something algebraically tractable rather than just that computing procedure.

On using (132) and (134) with (129); and (134) and (136) with (135); and (136) itself, together with $E(\mathrm{SSE}) = (N - s)\sigma_e^2$, we now have estimation equations similar to (120):

$$\begin{aligned}
R(\boldsymbol{\alpha} \mid \mu) &= (N - k_1')\hat{\sigma}_\alpha^2 + (k_3 - k_2')\hat{\sigma}_\beta^2 + (k_3 - k_{23}')\hat{\sigma}_\gamma^2 + (a - 1)\hat{\sigma}_e^2,\\
R(\boldsymbol{\beta} \mid \mu, \boldsymbol{\alpha}) &= (N - k_3)\hat{\sigma}_\beta^2 + (N - k_3 - h_6)\hat{\sigma}_\gamma^2 + (b - 1)\hat{\sigma}_e^2,\\
R(\boldsymbol{\gamma} \mid \mu, \boldsymbol{\alpha}, \boldsymbol{\beta}) &= h_6\hat{\sigma}_\gamma^2 + (s - a - b + 1)\hat{\sigma}_e^2,\\
\mathrm{SSE} &= (N - s)\hat{\sigma}_e^2\,.
\end{aligned} \tag{138}$$

-*ii*. ***Three sets of sums of squares.*** Just as (110) and (121) are two partitionings of $\mathrm{SST_m}$ for the no-interaction model, so are (128) and

$$R(\boldsymbol{\beta} \mid \mu) + R(\boldsymbol{\alpha} \mid \mu, \boldsymbol{\beta}) + R(\boldsymbol{\gamma} \mid \mu, \boldsymbol{\alpha}, \boldsymbol{\beta}) + \mathrm{SSE} = \mathrm{SST_m} \tag{139}$$

for the with-interaction model. Equation (139) is the with-interaction form of (121) just as (128) is of (110). Therefore, just as (122) is a second set of estimation equations analogous to (120), so is the analogy of (138) for the terms in (139), as follows—derivable from (138) by interchanging $\boldsymbol{\alpha}$ with $\boldsymbol{\beta}$ and i with j:

$$
\begin{aligned}
R(\boldsymbol{\beta} \mid \mu) &= (k_4' - k_1')\hat{\sigma}_\alpha^2 + (N - k_2')\hat{\sigma}_\beta^2 + (k_4 - k_{23}')\hat{\sigma}_\gamma^2 + (b-1)\hat{\sigma}_e^2,\\
R(\boldsymbol{\alpha} \mid \mu, \boldsymbol{\beta}) &= (N - k_4)\hat{\sigma}_\alpha^2 + (N - k_4 - h_6)\hat{\sigma}_\gamma^2 + (a-1)\hat{\sigma}_e^2,\\
R(\boldsymbol{\gamma} \mid \mu, \boldsymbol{\alpha}, \boldsymbol{\beta}) &= h_6\hat{\sigma}_\gamma^2 + (s - a - b + 1)\hat{\sigma}_e^2,\\
\text{SSE} &= (N - s)\hat{\sigma}_e^2 .
\end{aligned}
\tag{140}
$$

So, just as (122) is a set of equations second to (120) in the no-interaction case, so is (140) second to (138) in the with-interaction case.

Both sets of equations, (138) and (140), have the same last two equations and in each set the sum of the first two equations is the same:

$$
\begin{aligned}
R(\boldsymbol{\alpha}, \boldsymbol{\beta} \mid \mu) &= R(\boldsymbol{\alpha} \mid \mu) + R(\boldsymbol{\beta} \mid \mu, \boldsymbol{\alpha})\\
&= R(\boldsymbol{\beta} \mid \mu) + R(\boldsymbol{\alpha} \mid \mu, \boldsymbol{\beta})\\
&= (N - k_1')\hat{\sigma}_\alpha^2 + (N - k_2')\hat{\sigma}_\beta^2 + (N - k_{23}' - h_6)\hat{\sigma}_\gamma^2 + (a + b - 2)\hat{\sigma}_e^2 .
\end{aligned}
\tag{141}
$$

This leads, by exactly the same kind of reasoning as was used in deriving (124a,b,c,d) to having the following estimation equations for the with-interaction case—using (123).

$$
\hat{\sigma}_e^2 = \frac{\text{SSE}}{N - s} = \frac{\Sigma_i\Sigma_j\Sigma_k(y_{ijk} - \bar{y}_{ij.})^2}{N - s}, \tag{142a}
$$

$$
\begin{aligned}
\hat{\sigma}_\gamma^2 &= \frac{1}{h_6}[R(\boldsymbol{\gamma} \mid \mu, \boldsymbol{\alpha}, \boldsymbol{\beta}) - (s - a - b + 1)\hat{\sigma}_e^2]\\
&= \frac{1}{h_6}[\Sigma_i\Sigma_j n_{ij}\bar{y}_{ij.}^2 - R(\mu, \boldsymbol{\alpha}, \boldsymbol{\beta}) - (s - a - b + 1)\hat{\sigma}_e^2],
\end{aligned}
\tag{142b}
$$

$$
\begin{aligned}
\hat{\sigma}_\alpha^2 &= \frac{1}{h_7}[R(\boldsymbol{\alpha} \mid \mu, \boldsymbol{\beta}) - (h_7 - h_6)\hat{\sigma}_\gamma^2 - (a-1)\hat{\sigma}_e^2]\\
&= \frac{1}{h_7}[R(\mu, \boldsymbol{\alpha}, \boldsymbol{\beta}) - \Sigma_j n_{.j}\bar{y}_{.j.}^2 - (h_7 - h_6)\hat{\sigma}_\gamma^2 - (a-1)\hat{\sigma}_e^2]
\end{aligned}
$$

and on substituting for $h_6\hat{\sigma}_\gamma^2$ from (142b) this is

$$
\hat{\sigma}_\alpha^2 = \frac{1}{h_7}[\Sigma_i\Sigma_j n_{ij}\bar{y}_{ij.}^2 - \Sigma_j n_{.j}\bar{y}_{.j.}^2 - (s - b)\hat{\sigma}_e^2] - \hat{\sigma}_\gamma^2 . \tag{142c}
$$

Similarly

$$
\hat{\sigma}_\beta^2 = \frac{1}{h_4}[\Sigma_i\Sigma_j n_{ij}\bar{y}_{ij.}^2 - \Sigma_i n_{i.}\bar{y}_{i..}^2 - (s - a)\hat{\sigma}_e^2] - \hat{\sigma}_\gamma^2 . \tag{142d}
$$

Finally, from (141)

$$h_1\hat{\sigma}_\alpha^2 + h_2\hat{\sigma}_\beta^2 = R(\boldsymbol{\alpha}, \boldsymbol{\beta} \mid \mu) - (N - k'_{23} - h_6)\hat{\sigma}_\gamma^2 - (a + b - 2)\hat{\sigma}_e^2$$
$$= \Sigma_i\Sigma_j n_{ij}\bar{y}_{ij.}^2 - N\bar{y}_{..}^2 - h_3\hat{\sigma}_\gamma^2 - (s - 1)\hat{\sigma}_e^2 . \qquad (142e)$$

Equations (142a) and (142b) are the third and fourth equations of (138); (142d) is equivalent to the second; and (142e) is the sum of the first and second equations of (138). Therefore (142a,b,d,e) are equivalent to using (138). This in turn is equivalent to SAS Type I sums of squares when ordering the factors A, B and $A*B$. Similarly, (142a,b,c,e) are equivalent to (140), which is equivalent to SAS Type I sums of squares when ordering the factors B, A and $B*A$. Finally, Method III permits using (142a,b,c,d), which are equivalent to using $R(\boldsymbol{\alpha} \mid \boldsymbol{\beta}, \mu)$, $R(\boldsymbol{\beta} \mid \mu, \boldsymbol{\alpha})$, $R(\boldsymbol{\gamma} \mid \mu, \boldsymbol{\alpha}, \boldsymbol{\beta})$ and SSE—and so they are the SAS Type II sums of squares. Thus, as with the no-interaction model in equations (124), we again have three possible ways of applying Method III to the random model. They are set out in Table 5.5.

Again, it is the availability of more than one set of sums of squares, and hence more than one set of estimates, that characterizes Method III as being not always uniquely applicable to a set of data.

-iii. Calculation. As with the equations (124) for the no-interaction model, so also for (140) for the with-interaction model. The only difficult sum of squares is $R(\mu, \boldsymbol{\alpha}, \boldsymbol{\beta})$ with the additional difficult coefficient, h_6. Computing procedures for both of these are given in Table F.3.

-iv. Sampling variances. No specific expressions are known to be available for sampling variances and covariances of the estimates available from equations (142). A matrix formulation of the estimators could be used in Theorem S4 of

TABLE 5.5. THREE OPTIONS FOR USING METHOD III ON THE 2-WAY CROSSED CLASSIFICATION, WITH-INTERACTION, RANDOM MODEL

	Equations from (142)		
Estimate	Option 1	Option 2	Option 3
$\hat{\sigma}_e^2$	a	a	a
$\hat{\sigma}_\gamma^2$	b	b	b
$\hat{\sigma}_\alpha^2$	e	c	c
$\hat{\sigma}_\beta^2$	d	e	d
Equivalent to			
Partitioning of SST_m	(128)	(139)	
Estimation equations	(138)	(140)	Last three of (138) and (140)
SAS sums of squares	Type I $A,B, A*B$	Type I $B,A, B*A$	Type II

Appendix S.5, but it would not be at all tractable. It could yield computational procedures, no doubt, and they would involve quadratic forms of the unknown variance components. Rohde and Tallis (1969) have considered this approach.

d. With interaction, mixed model

As was done in the no-interaction model, suppose the β_js are fixed effects. Then since the only trio of equations (for estimating σ_e^2, σ_γ^2 and σ_α^2) that do not have β_js in them are (142a, b and c), these are the equations that are used. Therefore these are the estimation equations for the 2-way crossed classification, mixed model, with β_js being fixed effects. Das (1987) considers the special case of this model when $e_{ijk} \sim \mathcal{N}(0, \sigma_{ij}^2)$, i.e., having a different within-cell variance for each cell.

5.7. NESTED MODELS

It is difficult to make generalizations about the applicability of ANOVA estimation methods to mixed models that include nested factors. But, for completely nested models, those having no crossed factors at all, one or two general statements can be made. LaMotte (1972), for example, gives a general formulation of the dispersion matrix $\text{var}(\mathbf{y})$ applicable to any completely nested model.

For completely nested, random models the nested feature of such models makes the sequence of sums of squares for Method III self-evident, and that leads to Henderson's Methods I, II and III being all the same for these models; and they are the same as using the customary analysis of variance sums of squares. Details for the 1-way, 2-way and 3-way nested models are shown in Appendix F; also E 5.13.

For completely nested, mixed models, estimation of variance components is easy when all random factors are nested within fixed factors; by this is meant, for a 4-factor case, for example, that if the primary and secondary factors are fixed, and the tertiary factor nested within the secondary factor and the fourth factor nested within the tertiary one are both random, then the variance components for those two factors and for error are estimated from the three sums of squares for those three random contributions to the data. These will generally be the last three of the five sums of squares displayed in a partitioning of the total sums of squares. Thus if the factors are represented by A, B, C and D, with B nested within A, with C nested within B, and D nested with C, and with C and D being random, then the sums of squares that are the basis of the estimation equations are $R(\boldsymbol{\gamma} \mid \mu, \boldsymbol{\alpha}, \boldsymbol{\beta})$, $R(\boldsymbol{\delta} \mid \mu, \boldsymbol{\alpha}, \boldsymbol{\beta}, \boldsymbol{\gamma})$ and $R(\mu, \boldsymbol{\alpha}, \boldsymbol{\beta}, \boldsymbol{\gamma}, \boldsymbol{\delta})$.

Some of the papers that deal with nested models include, for example, Khattree and Naik (1990) who, for the 2-way nested model, derive locally best tests for H: $\sigma_\alpha^2 = 0$ and H: $\sigma_{\beta:\alpha}^2 = 0$ using partially balanced data. Burdick and Graybill (1985) deal with the same model, and data having the same number of observations in each sub-most cell, but unequal numbers of levels of the nested factor. They suggest an approximation for the distribution of a sum of squares and use that to develop an approximate confidence interval for the sum of the three variance components. And for the 3-way nested classification, random model, with unbalanced data, Tan and Cheng (1984) compare four different ratios of mean squares as statistics for testing H: $\sigma_\alpha^2 = 0$. In the case of the random effects being distributed in some manner other than normally Westfall (1986) provides conditions under which, for nested mixed models, the ANOVA estimators of variance components obtained from unbalanced data have an asymptotical multivariate normal distribution.

5.8. OTHER FORMS OF ANOVA ESTIMATION

Henderson's three methods are just particular ways of using ANOVA methods of estimation, just three different ways of choosing quadratic forms for the ANOVA procedure of $E(\mathbf{s}) = \mathbf{C}\boldsymbol{\sigma}^2$ giving $\hat{\boldsymbol{\sigma}}^2 = \mathbf{C}^{-1}\mathbf{s}$; indeed, in many cases, more than three ways because Method III can, as illustrated in Tables 5.4 and 5.5, provide more than one way.

There are, of course, other sets of quadratic forms that have been used in the ANOVA method. Two that have received some attention are the unweighted means analysis and the weighted squares of means analysis, which are now described for the 2-way crossed classification, each of which generalizes in a straightforward fashion. There is also the symmetric sums method of Koch (1967a, b, 1968), but since it has been little used in practice, it is not included here, and also the method of Hocking *et al.* (1989) discussed in section 11.2.

a. Unweighted means method: all cells filled

An easily calculated analysis of all-cells-filled data is to treat the observed cell means as observations and subject them to a balanced-data analysis of variance. This was suggested by Yates (1934) as providing approximate F-tests for fixed effects models. In the case of the 2-way crossed classification, random model the mean squares from that fixed model analysis of variance provide ANOVA estimators of variance components in the following manner.

Define

$$x_{ij} = \bar{y}_{ij.} = \frac{\sum_{k=1}^{n_{ij}} y_{ijk}}{n_{ij}} \quad \text{and} \quad n_h = \frac{1}{ab}\Sigma_i\Sigma_j\frac{1}{n_{ij}}, \tag{143}$$

with

$$\bar{x}_{i.} = \frac{\Sigma_j x_{ij}}{b}, \quad \bar{x}_{.j} = \frac{\Sigma_i x_{ij}}{a} \quad \text{and} \quad \bar{x}_{..} = \frac{\Sigma_i\Sigma_j x_{ij}}{ab}. \tag{144}$$

In doing so, take note that we are dealing *only* with data for which

$$n_{ij} > 0 \quad \forall\ i \text{ and } j\,.$$

Then the estimation equations are (using subscript u to denote "unweighted")

$$\begin{aligned}
\text{MSA}_\text{u} &= \frac{b}{a-1}\Sigma_i(\bar{x}_{i.} - \bar{x}_{..})^2 &&= b\hat{\sigma}_\alpha^2 + \hat{\sigma}_\gamma^2 + n_\text{h}\hat{\sigma}_e^2 \\
\text{MSB}_\text{u} &= \frac{a}{b-1}\Sigma_j(\bar{x}_{.j} - \bar{x}_{..})^2 &&= a\hat{\sigma}_\beta^2 + \hat{\sigma}_\gamma^2 + n_\text{h}\hat{\sigma}_e^2 \\
\text{MSAB}_\text{u} &= \frac{1}{(a-1)(b-1)}\Sigma_i\Sigma_j(x_{ij} - \bar{x}_{i.} - \bar{x}_{.j} - \bar{x}_{..})^2 = && \hat{\sigma}_\gamma^2 + n_\text{h}\hat{\sigma}_e^2 \\
\text{MSE} &= \frac{1}{N-ab}\Sigma_i\Sigma_j\Sigma_k(y_{ijk} - \bar{y}_{ij.})^2 &&= \hat{\sigma}_e^2
\end{aligned} \tag{145}$$

These arise from the expected values of the mean squares being the right-hand sides of (145) with σ^2s in place of $\hat{\sigma}^2$s. (See E 5.9.)

b. Weighted squares of means: all cells filled

A second analysis that Yates (1934) suggested for all-cells-filled data from fixed effects models is known as the weighted squares of means analysis. Its mean squares for the 2-way crossed classification, random model, used in the ANOVA method of estimating variance components yield estimation equations as follows.

Define

$$\begin{gathered}
w_i = 1\Big/\left(\frac{1}{b^2}\sum_j \frac{1}{n_{ij}}\right) \quad \text{and} \quad v_j = 1\Big/\left(\frac{1}{a^2}\sum_i \frac{1}{n_{ij}}\right), \\
\tilde{x}_I = \frac{\Sigma_i w_i \bar{x}_{i.}}{\Sigma_i w_i} \quad \text{and} \quad \tilde{x}_J = \frac{\Sigma_j v_j \bar{x}_{.j}}{\Sigma_j v_j},
\end{gathered} \tag{146}$$

where $\bar{x}_{i.}$ and $\bar{x}_{.j}$ are as in (144). Then the estimation equations are (with subscript w denoting "weighted")

$$\begin{aligned}
\text{MSA}_\text{w} &= \frac{1}{a-1}\Sigma_i w_i(\bar{x}_{i.} - \tilde{x}_I)^2 &&= \frac{1}{(a-1)b}\left(\Sigma_i w_i - \frac{\Sigma_i w_i^2}{\Sigma_i w_i}\right)(b\hat{\sigma}_\alpha^2 + \hat{\sigma}_\gamma^2) + \hat{\sigma}_e^2 \\
\text{MSB}_\text{w} &= \frac{1}{b-1}\Sigma_j v_j(\bar{x}_{.j} - \tilde{x}_J)^2 &&= \frac{1}{(b-1)a}\left(\Sigma_j v_j - \frac{\Sigma_j v_j^2}{\Sigma_j v_j}\right)(a\hat{\sigma}_\beta^2 + \hat{\sigma}_\gamma^2) + \hat{\sigma}_e^2 \\
\text{MSAB} &= \frac{1}{(a-1)(b-1)}R(\boldsymbol{\gamma} \mid \mu, \boldsymbol{\alpha}, \boldsymbol{\beta}) = && \frac{h_6}{(a-1)(b-1)}\hat{\sigma}_\gamma^2 + \hat{\sigma}_e^2 \\
\text{MSE} &= \frac{1}{N-ab}\Sigma_i\Sigma_j\Sigma_k(y_{ijk} - \bar{y}_{ij.})^2 = && \hat{\sigma}_e^2
\end{aligned} \tag{147}$$

Again, the right-hand sides of these equations, with σ^2s in place of $\hat{\sigma}^2$s, are expected values of the left-hand sides—in accord with the ANOVA method of estimation. (See E 5.9.)

It is noticeable in both (145) and (147) that $\hat{\sigma}_e^2 = \text{MSE}$; and (147) has $\hat{\sigma}_\gamma^2$ the same as in Henderson's Method III. [The equation for MSAB contains h_6, which is the same h_6 as used in equations (140)–(142), and which is defined in (137) and for which computing details are given in Table F.3 of Appendix F.] Although the sums of squares in (147) are those customarily recognized as constituting the weighted squares of means analysis, a variety of other weights can be used in place of w_i and v_j, as discussed by Gosslee and Lucas (1965).

The cell means $\bar{y}_{ij.}$ of (143) have also been used by Thomsen (1975) and by Khuri and Littell (1987) to establish tests of hypotheses that variance components are zero in the 2-way crossed classification, with interaction random model with unbalanced data, all cells filled.

5.9. COMPARING DIFFERENT FORMS OF ANOVA ESTIMATION

Applying the general ANOVA method of $E(\mathbf{s}) = \mathbf{C}\boldsymbol{\sigma}^2$ giving $\hat{\boldsymbol{\sigma}}^2 = \mathbf{C}^{-1}\mathbf{s}$ to the 2-way crossed classification, random model yielded five different sets of estimation equations: Henderson I, II and III, and unweighted means and weighted squares of means. Indeed, more than five because of the three forms of Method III—Table 5.5. This multiplicity of available quadratic forms is inherent in the general ANOVA method. So long as $\mathbf{1}'\mathbf{A}\mathbf{1} = \mathbf{0}$, the quadratic form $\mathbf{y}'\mathbf{A}\mathbf{y}$ can be an element of $\mathbf{s}$, for a random model. Any $r + 1$ such quadratic forms can be the elements of $\mathbf{s}$, where there are r random factors (main effect or interaction factors). Within this only slightly restricted confine ($\mathbf{1}'\mathbf{A}\mathbf{1} = 0$) there is an infinite number of sets of quadratic forms that can make up $\mathbf{s}$. They all have just one thing in common: they yield unbiased estimators for random models; as do the Henderson Method II quadratic forms for mixed models.

This property of unbiasedness might, however, be of questionable value. As a property of estimators it has been borrowed from fixed effects estimation; but in the context of variance component estimation it may not be appropriate. In estimating fixed effects the basis of desiring unbiasedness of estimators is the concept of repetition of data and associated estimates. The concept remains valid, but not its applicability for unbalanced data from random models—repeated data, perhaps, but not necessarily with the same pattern of unbalancedness or with the same set of (random) effects in the data. Replications of data are not, therefore, just replications of any existing data structure. This would be particularly so when considering the possibility of repeating the data collection of some of the very large data sets (e.g., 500,000 records) that get used for variance components estimation in animal breeding work with farm animals, such as dairy cows, beef animals and sheep. Under these circumstances mean unbiasedness may therefore no longer be pertinent, and replacing it with some other criterion might be worth considering. Modal unbiasedness is one possibility, suggested by Searle (1968, discussion), although Harville (1969b) doubts if modally unbiased estimators exist and questions the justification of such a criterion on decision-theoretic grounds. Nevertheless, as Kempthorne (1968) points out, mean unbiasedness in estimating fixed effects "...leads to residuals which do not contain systematic effects and is therefore valuable... and is fertile mathematically in that it reduces the class of candidate statistics (or estimates)." However, "...in the variance component problem it does not lead to a fertile smaller class of statistics."

All five modes of the ANOVA method that have been described reduce, for balanced data, to *the* ANOVA method in that case (e.g., E 5.18 and E 5.19), which has optimum properties of being minimum variance quadratic unbiased and minimum variance unbiased under normality. But for unbalanced data this reduction to an optimal balanced data situation and the unbiasedness of the resulting estimators are the only known properties of the methods. Otherwise, the quadratic forms involved in each method have been selected solely because they seemed "reasonable" in one way or another. The ANOVA methodology itself gives no guidance whatever as to which set of quadratic forms is, or might be, optimal in any sense. It includes no criteria for choosing one set of quadratic forms over any other. Moreover, the "reasonableness" of the quadratic forms in each case provides little or no comparison of any properties of the estimators that result from the different methods. Probably the simplest idea would be to compare sampling variances. Unfortunately this comparison soon becomes bogged down in algebraic complexity. Not only are the variances in any way tractable only if normality is assumed but also, just as with balanced data, the variances themselves are functions of the variance components. The complexity of the variances is evident in $\text{var}(\hat{\sigma}_\alpha^2)$ for the 1-way classification given in (102) of Chapter 3, where its behavior is briefly discussed. Yet that, apart from $\text{var}(\hat{\sigma}_e^2) = 2\sigma_e^4/(N - s)$, is the simplest example of a sampling variance (under normality assumptions) of an estimated variance component. But, as is apparent from (36), sampling variances in the 2-way crossed classification are considerably more complicated than in the 1-way case. Certainly, they are quadratic functions of the unknown variance components, but the coefficients multiplying the terms in the σ^2s are such that their behavior, and hence that of the sampling variance, for different sets of n_{ij}-values, cannot be studied algebraically. The functions of the n_{ij}-values are just too complicated. Moreover, the behavior depends upon what the values of the σ^2s are.

Two possibilities exist. One is for whatever particular data set is at hand. It will have a set of n_{ij}-values. We call that set an n-pattern, and then have

$$\text{var}(\hat{\boldsymbol{\sigma}}^2) = \mathbf{f}(\boldsymbol{\sigma}^2, n\text{-pattern}), \tag{148}$$

where $\mathbf{f}$ is a vector of elements that are quadratic forms of the σ^2s, with coefficients that are those complicated functions of the n-pattern. Now calculate $\text{var}(\hat{\boldsymbol{\sigma}}^2)$ for each of a range of values of $\boldsymbol{\sigma}^2$ around the estimate $\hat{\boldsymbol{\sigma}}^2$. Included in this would be $\hat{\boldsymbol{\sigma}}^2$ itself, used in the manner of (28) to get an unbiased estimator of $\text{var}(\hat{\boldsymbol{\sigma}}^2)$. This will give information about how changes in $\boldsymbol{\sigma}^2$ affect $\text{var}(\hat{\boldsymbol{\sigma}}^2)$—for that n-pattern. And one can do this for more than one estimator of $\boldsymbol{\sigma}^2$, and thus compare var($\hat{\boldsymbol{\sigma}}^2$) for one estimator with that of another—but only for the n-pattern of the data.

A second, and much more difficult, possibility is to try, arithmetically, to study the behavior of var($\hat{\boldsymbol{\sigma}}^2$) for different n-patterns. The difficulty is to decide what n-patterns to use. Whereas the arithmetic of calculating var($\hat{\boldsymbol{\sigma}}^2$) is relatively no longer time-consuming, the problem of what different n-patterns to choose still remains. One objective might be to see how var($\hat{\boldsymbol{\sigma}}^2$) [or even just var($\sigma_t^2$), where σ_t^2 is some element of $\boldsymbol{\sigma}^2$] behaves for different degrees of unbalancedness. But how can unbalancedness be categorized? What n-patterns will typify different degrees of unbalancedness? Even in the 1-way classification we saw in Section 3.6d-iv that var($\hat{\sigma}_\alpha^2$) was, for certain values of $\sigma_\alpha^2/\sigma_e^2$, bigger for the n-pattern (1, 1, 1, 11, 11) than for (1, 1, 1, 1, 21); and yet in some general sense the latter would usually be considered to represent greater unbalancedness than the former. This inconsistency with one's intuition about unbalancedness is occurring in a 1-way classification with but 25 observations in 5 groups. Contemplate how much more this may well arise with, say, 500 observations in 80 groups, and even more so with a 2-way crossed classification, wherein settling on n-patterns to use we have to decide on not only N but also on the $n_{i.}$- and $n_{.j}$-values, and the n_{ij}-values and the cells in which they will occur. Even in the most trite (and totally impractical) case of all, a 2-way crossed classification of but 2 rows and 2 columns with 8 observations and all cells filled, there are at least 11 distinguishably different n-patterns, as shown in Table 5.6. With something of even modest size, such as 50 rows and 80 columns, the number of n-patterns clearly becomes astronomically large. Categorizing them on some monotonic scale of unbalancedness seems quite impractical. And even if it were not, one would also need to select sets of values for $\boldsymbol{\sigma}^2$ and, using each set with each n-pattern, calculate var($\hat{\boldsymbol{\sigma}}^2$) of (148). The hope of matching those calculated values with unbalancedness in a manner than informs us about how unbalancedness affects var($\hat{\boldsymbol{\sigma}}^2$) seems unlikely to be fulfilled.

Despite the difficulties just described, some numerical comparisons have been reported in the literature. Kussmaul and Anderson (1967) studied a special case of the 2-way nested classification that makes it a particular form of the 1-way classification. A study of the latter by Anderson and Crump (1967) suggests that the unweighted means estimator of σ_α^2 appears, for very unbalanced data,

TABLE 5.6. ELEVEN DISTINGUISHABLE n-PATTERNS IN A 2-WAY CROSSED CLASSIFICATION OF 2 ROWS, 2 COLUMNS AND 8 OBSERVATIONS, WITH ALL CELLS FILLED

n-pattern										
1	2	3	4	5	6	7	8	9	10	11
2 2	2 2	1 3	1 3	1 4	4 1	3 2	2 3	4 1	4 1	1 5
2 2	1 3	3 1	1 3	1 2	1 2	1 2	1 2	1 2	2 1	1 1

to have larger variance than does the analysis of variance estimator for small values of $\rho = \sigma_\alpha^2/\sigma_e^2$, but that it has smaller variance for large ρ. The 2-way classification, interaction model has been studied by Bush and Anderson (1963) in terms of several cases of planned unbalancedness. With 6 rows and 6 columns in a 2-way crossed classification, three of the designs they used had filled cells (each with just one or two observations) either in an L-pattern in the 6×6 grid or in a diagonal band, more or less, across the grid. Designs such as these [and others, e.g., Anderson (1975)] were used to compare Henderson's Methods I and III and a weighted means application of the ANOVA method. Comparisons were made, by way of variances of the estimators, both of different designs as well as of different estimation procedures, over a range of values of the underlying variance components. For the designs used the general trend of the results is that, for values of the error variances much larger than the other components, the Method I estimators have smallest variance, but otherwise Method III estimators have. Later, Swallow and Searle (1978) and Swallow and Monahan (1984), in comparing ANOVA with other methods of estimation, use 13 different n-patterns for the 1-way classification and in doing so illustrate values of $\text{var}(\hat{\boldsymbol{\sigma}}^2)$ for a variety of values of $\sigma_\alpha^2/\sigma_e^2$.

Comparing the three Henderson methods is therefore virtually not feasible. Even with using a supercomputer so that vast arithmetic would be feasible, there is no assurance that the desired calculations, $\hat{\text{var}}(\hat{\boldsymbol{\sigma}}^2)$, say, could be displayed in a manner that would reveal any underlying patterns if indeed such patterns exist. For example, suppose in the 2-way crossed classification, we try planning to calculate $\hat{\text{var}}(\hat{\boldsymbol{\sigma}}^2)$ for a set of values of $\boldsymbol{\sigma}^2$s, and a set of n-patterns. How will one choose the set of $\boldsymbol{\sigma}^2$s? Certainly we could consider just $[\sigma_\alpha^2/\sigma_e^2 \quad \sigma_\beta^2/\sigma_e^2 \quad \sigma_\gamma^2/\sigma_e^2 \quad 1]$, but even this requires choosing triplets, and even for 4 different values of each ratio that gives 64 different triplets. How, one wonders, can the ultimate values of $\hat{\text{var}}(\hat{\boldsymbol{\sigma}}^2)$ be arrayed over those 64 triplets to yield information, if there is any, about how $\boldsymbol{\sigma}^2$ affects $\text{var}(\hat{\boldsymbol{\sigma}}^2)$? And the difficulty of this question is magnified greatly when one further considers choosing a set of n-patterns and looking at each of the 64 values of $\text{var}(\hat{\boldsymbol{\sigma}}^2)$, itself a 4×4 matrix, for each n-pattern. In choosing n-patterns one has such a large number of choices available: values for N, a, b and s; values for $n_{1\cdot}, \ldots, n_{a\cdot}$ and for $n_{\cdot 1}, \ldots, n_{\cdot b}$; values for the n_{ij}s, of which there are ab, with $ab - s$ of them having to be chosen as zero. So even for one set of values for N, a, b, s, $n_{i\cdot}$ and $n_{\cdot j}$, there will be a very large set of possible n-patterns. And although the computing of $\text{var}(\hat{\boldsymbol{\sigma}}^2)$ for the 64 sets of $\boldsymbol{\sigma}^2$ for each n-pattern is nowadays quite feasible, the big question is how can we relate those computed values of $\text{var}(\hat{\boldsymbol{\sigma}}^2)$ to the 64 values of $\boldsymbol{\sigma}^2$, and to the multitudinous n-patterns, so as to be able to draw conclusions about how $\text{var}(\hat{\boldsymbol{\sigma}}^2)$ is affected by different values of $\boldsymbol{\sigma}^2$ and by different degrees of what we implicitly think of as unbalancedness.

So maybe the only comparisons available are those stemming from the establishment of the methods—and they are not really very helpful. Method I commends itself because it is the obvious analog of the analysis of variance of balanced data, and it is easy to use; some of its terms are not sums of squares, and it gives biased estimators in mixed models. The generalized form of Henderson's Method II makes up for this deficiency, but his specific definition of it cannot be used when there are interactions between fixed and random effects. Method III uses sums of squares that have non-central χ^2-distributions in the fixed effects model, and it gives unbiased estimators in mixed models; but it can involve more quadratics than there are components to be estimated; and it can also involve inverting matrices of order equal to the number of random effects in the model. For data in which all subclasses are filled the analysis of means methods have the advantage of being easier to compute than Method III; the unweighted means analysis is especially easy. All of the methods reduce, for balanced data, to the analysis of variance method, and all of them can yield negative estimates. Little more than this can be said by way of comparing the methods.

5.10. ESTIMATING FIXED EFFECTS IN MIXED MODELS

The basic formulae for estimating $\mathbf{X\beta}$ in the model equation $\mathbf{y} = \mathbf{X\beta} + \mathbf{Zu} + \mathbf{e}$ are the same for unbalanced data as for balanced data of Section 4.8:

$$\text{OLSE}(\mathbf{X\beta}) = \mathbf{X}(\mathbf{X}'\mathbf{X})^{-}\mathbf{X}'\mathbf{y}$$

and (149)

$$\text{GLSE}(\mathbf{X\beta}) = \mathbf{X}(\mathbf{X}'\mathbf{V}^{-1}\mathbf{X})^{-}\mathbf{X}'\mathbf{V}^{-1}\mathbf{y}\,.$$

However, when data are unbalanced, these formulae are not necessarily equal, as they always are with balanced data and the customary mixed or random model [see the discussion following (113) in Section 4.9]; nor do they reduce to straightforward expressions for calculating estimates, as they do with balanced data, e.g., equations (109) and (112) in Chapter 4.

For completely random models, where μ is the only fixed effect, $\mathbf{X} = \mathbf{1}_N$ and

$$\text{OLSE}(\mu) = \bar{y} \quad \text{and} \quad \text{GLSE}(\mu) = \frac{\mathbf{1}'\mathbf{V}^{-1}\mathbf{y}}{\mathbf{1}'\mathbf{V}^{-1}\mathbf{1}}\,. \tag{150}$$

Otherwise, there are few other general simplifications of the expressions (149) except, of course, in the case of planned unbalancedness, such as discussed by Harville (1986).

The general inequality of the expressions in (149) prompts the question "What conditions on $\mathbf{X}$ and $\mathbf{V}$ will lead to GLSE($\mathbf{X\beta}$) equalling OLSE($\mathbf{X\beta}$)?" This is a question of some practical interest because equality of the two estimators means that GLSE can be calculated as the OLSE. And since GLSE utilizes variances of the random effects factors, which requires knowing or estimating those variances, being able to use OLSE, which does not require variances, is very advantageous. Thus answering the question "When does GLSE equal OLSE?" is of some importance and has engendered much research. The easiest answer is that GLSE = OLSE if and only if there is a matrix $\mathbf{F}$ such that

$\mathbf{VX} = \mathbf{XF}$. This is one of many equivalent answers given by Zyskind (1967) and reviewed by Puntanen and Styan (1989). Establishing this condition is relatively straightforward when $\mathbf{V}$ is nonsingular (see E 5.14). It is somewhat difficult when $\mathbf{V}$ is singular—see Puntanen and Styan (1989).

When the condition $\mathbf{VX} = \mathbf{XF}$ is not met, as is usually the case with unbalanced data, the GLSE is an estimator that has two optimal properties: unbiasedness and minimum variance. But its use requires knowing $\mathbf{V}$, and this is seldom the case. So something must be used in place of $\mathbf{V}$. An obvious choice is $\hat{\mathbf{V}} = \Sigma_{i=0}^{r} \mathbf{Z}_i\mathbf{Z}_i'\hat{\sigma}_i^2$ where $\hat{\sigma}_i^2$ is an estimate of σ_i^2. The difficulty, of course, lies in what estimator should be used as the basis for $\hat{\sigma}_i^2$. Kackar and Harville (1981) have shown that if the $\hat{\sigma}_i^2$ are calculated as even-valued functions of $\mathbf{y}$ [a function $s(\mathbf{y})$ is even if $s(\mathbf{y}) = s(-\mathbf{y})$ for all $\mathbf{y}$] and as translation-invariant [meaning that $s(\mathbf{y} + \mathbf{X\beta}) = s(\mathbf{y})$ for all $\mathbf{y}$ and $\boldsymbol{\beta}$] then

$$\text{GLSE}(\mathbf{X\beta}) = \mathbf{X}(\mathbf{X}'\hat{\mathbf{V}}^{-1}\mathbf{X})^{-}\mathbf{X}'\hat{\mathbf{V}}^{-1}\mathbf{y}$$

is an unbiased estimator of $\mathbf{X\beta}$. ANOVA estimators $\hat{\boldsymbol{\sigma}}^2$ satisfy these conditions (even and translation-invariant), and so do the ML and REML estimators, discussed in Chapter 6.

5.11. SUMMARY

Few details are given in this summary because details for any particular model are mostly somewhat voluminous. Appendix F contains detailed formulae for a variety of individual models; and this summary is mostly just a short list of topics. The table of contents has the complete list.

A general model for fixed β and random u

$$\mathbf{y} = \mathbf{X\beta} + \mathbf{Zu} + \mathbf{e}$$

$$= \mathbf{X\beta} + \sum_{i=1}^{r} \mathbf{Z}_i\mathbf{u}_i + \mathbf{e} \quad (1)$$

$$= \mathbf{X\beta} + \sum_{i=0}^{r} \mathbf{Z}_i\mathbf{u}_i; \quad (8)$$

$$\mathbf{V} = \text{var}(\mathbf{y}) = \sum_{i=0}^{r} \mathbf{Z}_i\mathbf{Z}_i'\sigma_i^2 . \quad (8)$$

Estimation

$$\mathbf{s} = \{_c\, \mathbf{y}'\mathbf{A}_i\mathbf{y}\}_{i=0}^{r}; \quad (15)$$

$$E(\mathbf{s}) = \{_m\, \text{tr}(\mathbf{Z}_j'\mathbf{A}_i\mathbf{Z}_j)\}_{i,j}\{_j\, \sigma_j^2\} \quad (17)$$

$$= \mathbf{C}\boldsymbol{\sigma}^2; \quad (18)$$

$$\hat{\boldsymbol{\sigma}}^2 = \mathbf{C}^{-1}\mathbf{s}; \quad (19)$$

$$E(\hat{\boldsymbol{\sigma}}^2) = \boldsymbol{\sigma}^2; \quad (20)$$

$$\text{var}(\hat{\boldsymbol{\sigma}}^2) = 2\mathbf{C}^{-1}\{_m\, \boldsymbol{\sigma}^{2\prime}\{_m\, \text{tr}(\mathbf{A}_i\mathbf{Z}_j\mathbf{Z}_j'\mathbf{A}_{i'}\mathbf{Z}_{j'}\mathbf{Z}_{j'}')\}_{j,j'}\boldsymbol{\sigma}^2\}_{i,i'}\mathbf{C}^{-1\prime} . \quad (21b)$$

Unbiased estimation of $\text{var}(\hat{\boldsymbol{\sigma}}^2)$

$$\mathbf{V} = \text{vech}[\text{var}(\hat{\boldsymbol{\sigma}}^2)] \quad \text{and} \quad \boldsymbol{\gamma} = \text{vech}(\boldsymbol{\sigma}^2\boldsymbol{\sigma}^{2\prime});$$

$$\text{var}(\hat{\boldsymbol{\sigma}}^2) = \mathbf{C}^{-1}\,\text{var}(\mathbf{s})\mathbf{C}^{-1\prime} \quad \Rightarrow \quad \mathbf{B} \text{ exist for } \mathbf{v} = \mathbf{B}\boldsymbol{\gamma}; \tag{26}$$

$$\hat{\mathbf{v}} = \mathbf{B}(\mathbf{I} + \mathbf{B})\,\text{vech}(\hat{\boldsymbol{\sigma}}^2\hat{\boldsymbol{\sigma}}^2) \text{ is unbiased for } \mathbf{v}\,. \tag{28}$$

Henderson's Methods I, II and III: Sections 5.3, 5.4 and 5.5

The 2-way cross classification: Sections 5.3, 5.6, and 5.8

Analysis of means methods: Section 5.9

5.12. EXERCISES

E 5.1. Show that in (22), $t_{01} = 0$ and $t_{11} = (a-1)(\sigma_e^2 + n\sigma_\alpha^2)^2$ and hence derive (23).

E 5.2. Show that $E(\text{SSA})$ of Section 5.2b-iii is

$$E(\text{SSA}) = \left(N - \frac{\Sigma_i n_{i.}^2}{N}\right)\sigma_\alpha^2 + \left(\sum_i \frac{\Sigma_j n_{ij}^2}{n_{i.}} - \frac{\Sigma_j n_{.j}^2}{N}\right)\sigma_\beta^2$$

$$+ \left(\sum_i \frac{\Sigma_j n_{ij}^2}{n_{i.}} - \frac{\Sigma_i\Sigma_j n_{ij}^2}{N}\right)\sigma_\gamma^2 + (a-1)\sigma_e^2\,.$$

E 5.3. Derive $\mathbf{B}(\mathbf{I} + \mathbf{B})^{-1}$ of (28) for the ANOVA estimators stemming from Tables 4.10 and 4.12.

E 5.4. Define $S_2 = \Sigma_i n_i^2$ and $S_3 = \Sigma_i n_i^3$,

$$k_1 = \frac{2}{N-a}, \quad k_2 = \frac{-2N(a-1)}{(N-a)(N^2 - S_2)}, \quad k_3 = \frac{2N^2(N-1)(a-1)}{(N^2 - S_2)^2(N-a)},$$

$$k_4 = \frac{4N}{N^2 - S_2}, \quad k_5 = \frac{2(N^2 S_2 + S_2^2 - 2NS_3)}{(N^2 - S_2)^2}\,.$$

Using Section 5.2e, derive unbiased estimators of the variances of, and covariance between, the ANOVA estimators of variance components [see (95), (96) and (102) of Chapter 3] in the 1-way classification with unbalanced data. The results are

$$\hat{\text{var}}(\hat{\sigma}_e^2) = \frac{2\hat{\sigma}_e^4}{N - a + 2}, \quad \hat{\text{cov}}(\hat{\sigma}_e^2, \hat{\sigma}_\alpha^2) = \frac{k_2}{k_1}\,\hat{\text{var}}(\hat{\sigma}_e^2)$$

and

$$\hat{\text{var}}(\hat{\sigma}_\alpha^2) = \frac{1}{1 + k_5}\left(\frac{k_3 - k_2 k_4}{1 + k_1}\,\hat{\sigma}_e^4 + k_4\hat{\sigma}_e^2\hat{\sigma}_\alpha^2 + k_5\hat{\sigma}_\alpha^4\right).$$

E 5.5. Prove that SSAB* − SSAB is as shown in Section 5.3a, and illustrate it for the data given there.

E 5.6. (a) Derive (38) and (39).
(b) Show that (38) and (39) are both zero for balanced data.
(c) Show for the random model that θ_r of (38) and (39) reduce to what one would expect from (31).

E 5.7. (a) Derive (117) from (116).
(b) Derive (118).
(c) Establish (120).
(d) Derive (122) from (120).

E 5.8. Derive (134) from (133).

E 5.9. (a) Derive the expected value of MSA_u, MSB_u and MSAB_u.
(b) Derive the expected value of MSA_w and MSB_w.

E 5.10. (a) From Tables F.1 and F.2 write down the variance of SSB.
(b) For balanced data show that it simplifies as expected.

E 5.11. Consider the following data from a 2-way classification of 2 rows and 2 columns:

Data

3,7	17	25
2	6,10	—

For a random model, with interaction, calculate
(a) estimated variance components using Henderson's Method I;
(b) the sampling variance of T_A used in (a);
(c) estimated variance components using all versions of Henderson's Method III.

For a mixed model, with fixed rows, without interaction, calculate
(d) estimated variance components using Henderson's Method II.

E 5.12. Repeat E 5.11 for the data set

7,9	6	2
8	4,8	12

E 5.13. The 2-way nested random model has model equation

$$y_{ijk} = \mu + \alpha_i + \beta_{ij} + e_{ijk}$$

for $i = 1, \ldots, a$, $j = 1, \ldots, b_i$ and $k = 1, \ldots, n_{ij}$. For convenience use the notation

$$\alpha \equiv \sigma_\alpha^2, \quad \beta \equiv \sigma_\beta^2 \quad \text{and} \quad e \equiv \sigma_e^2, \quad \theta = \alpha + \beta + e\,.$$

(a) For the following data write down $\mathbf{V} = \text{var}(\mathbf{y})$ *in extenso*:

Data

$i = 1$		$i = 2$		
$j = 1$	$j = 2$	$j = 1$	$j = 2$	$j = 3$
5	8	8	1	3
	9	10	2	7
	10		3	
			6	

(b) Explain why, in general,

$$\mathbf{V} = \{_{\text{d}}\ \alpha\mathbf{J}_{n_{i.}} + \beta\{_{\text{d}}\ \mathbf{J}_{n_{ij}}\}_{j=1}^{b_i} + e\mathbf{I}_{n_{i.}}\}_{i=1}^{a}\,.$$

(c) Derive

$$\mathbf{V}^{-1} = \left\{_{\text{d}}\ \frac{1}{e}\left\{_{\text{d}}\ \mathbf{I}_{n_{ij}} - \frac{\beta}{e + n_{ij}\beta}\mathbf{J}_{n_{ij}}\right\}_{j=1}^{b_i}\right.$$
$$\left. - \left\{_{\text{m}}\ \frac{1}{(e + n_{ij}\beta)(e + n_{ij'}\beta)}\mathbf{J}_{n_{ij}\times n_{ij'}}\right\}_{j,j'=1}^{b_i}\right\}_{i=1}^{a}\,.$$

(d) Verify $\mathbf{V}\mathbf{V}^{-1} = \mathbf{I}$ for $\mathbf{V}^{-1}$ of (c).

(e) Given $\text{SSA} = \Sigma_i n_{i.}(\bar{y}_{i..} - \bar{y}_{...})^2$, $\text{SSB:A} = \Sigma_i\Sigma_j n_{ij}(\bar{y}_{ij.} - \bar{y}_{i..})^2$ and $\text{SSE} = \Sigma_i\Sigma_j\Sigma_k(y_{ijk} - \bar{y}_{ij.})^2$, establish

$$E(\text{SSA}) = \left(N - \frac{\Sigma_i n_{i.}^2}{N}\right)\sigma_\alpha^2 + \left[\Sigma_i \frac{(\Sigma_j n_{ij}^2)}{n_{i.}} - \frac{\Sigma_i\Sigma_j n_{ij}^2}{N}\right]\sigma_\beta^2 + (a-1)\sigma_e^2,$$

$$E(\text{SSB:A}) = \left[N - \Sigma_i \frac{(\Sigma_j n_{ij}^2)}{n_{i.}}\right]\sigma_\beta^2 + (b_. - a)\sigma_e^2,$$

$$E(\text{SSE}) = (N - b_.)\sigma_e^2\,.$$

(f) Note in (e) that the coefficient of σ_α^2 can be written as

$$\Sigma_i n_{i.}^2\left(\frac{1}{n_{i.}} - \frac{1}{N}\right).$$

Express the other five coefficients in forms that involve differences between reciprocals, e.g.,

$$N - b_. = \Sigma_i\Sigma_j\Sigma_k 1^2\left(\frac{1}{1} - \frac{1}{n_{ij}}\right).$$

Conjecture corresponding results for the 3-way nested classification random model. [This is the formulation given by Ganguli (1941).]

E 5.14. (a) For non-singular $\mathbf{V}$ prove that $\mathbf{X}(\mathbf{X}'\mathbf{V}^{-1}\mathbf{X})^{-}\mathbf{X}'\mathbf{V}^{-1} = \mathbf{X}(\mathbf{X}'\mathbf{X})^{-}\mathbf{X}'$ if and only if $\mathbf{VX} = \mathbf{VF}$ for some $\mathbf{F}$.

(b) For the model $\mathbf{y} \sim \{\mu\mathbf{1}_N, \sigma^2[(1-\rho)\mathbf{I}_n + \rho\mathbf{J}_n]\}$ prove that $\text{GLSE}(\mu) = \text{OLSE}(\mu)$.

(c) For the 1-way classification, random model, balanced data, show that $\text{GLSE}(\mu) = \text{OLSE}(\mu)$. Why is this not the case for unbalanced data?

(d) Consider the usual 2-way nested classification, mixed model, having model equation $y_{ijk} = \mu + \alpha_i + \beta_{ij} + e_{ijk}$, with $i = 1, \ldots, a$, $j = 1, \ldots, b$ and $k = 1, \ldots, n$, where μ and the αs are fixed. Show that $\mathbf{F}$ exists such that $\mathbf{VX} = \mathbf{XF}$.

E 5.15. Suppose unbalanced data from a 2-way crossed classification have been wrongly analyzed using a computing routine for a 2-way nested classification. The user of the routine is so perplexed that the routine gets used again. As a result, there are now ANOVA variance components estimates on both a β-within-α basis and α-within-β. Show how they can be used to get Henderson Method I estimates for the 2-way crossed classification. (Use results shown in E 5.13.)

E 5.16. Equation (31) yields Henderson Method I estimators for the 2-way crossed classification, with interaction, random model.

(a) Describe what amendments have to be made to those equations to yield estimators for the no interaction form of the model.

(b) Carry out those amendments and show that they yield the estimators in Appendix F.6b.

E 5.17. The following data are to be considered coming from a 2-way crossed classification of three rows and three columns.

8,9	13	—
9	15	12
6	—	8

Estimate variance components using the Henderson methods, denoted I, II and III, for the

(a) no interaction, mixed model, using II and III;

(b) no interaction, random model, using I and III;

(c) with interaction, mixed model, using III;

(d) with interaction, random model, using I and III.

Note: Use rational fractions rather than decimals.

E 5.18. Using Appendix F.4b, demonstrate the validity of method (b) at the end of Section 5.3g.

E 5.19. For the 2-way classification having model equation

$$y_{ij} = \mu + \alpha_i + \beta_j + e_{ij}$$

for $i = 1, \ldots, a$ and $j = 1, \ldots, b$, with one observation in every cell, show that Method II is the same as Method I. Use $a = 3$ and $b = 2$ to illustrate parts of your derivation.

E 5.20. Show that (145) and (147) reduce to the usual ANOVA estimators for balanced data.

CHAPTER 6

MAXIMUM LIKELIHOOD (ML) AND RESTRICTED MAXIMUM LIKELIHOOD (REML)

ANOVA methods of estimating variance components described in preceding chapters have not required, for the actual derivation of estimators, any assumption of an underlying probability distribution for the data. All that has been needed is that the random effects and residual errors have finite first and second moments, and satisfy some mild correlation assumptions (e.g., Section 5.1b-ii). True, for balanced data (Chapter 4 and parts of Chapter 3) we have seen that making some normality assumptions (Sections 3.5d and 4.5) leads to being able to test certain hypotheses and to establish certain confidence intervals. And for unbalanced data (Chapter 5) those same normality assumptions provide (by means of Theorem S4 of Appendix S.5) the ability to provide expressions for, or that can lead to, computable forms of sampling variances of, and covariances among, variance components estimators (Sections 5.2d, 5.3d, 5.6a-iv and 5.6c-iv).

Abbreviations. We use ML acronymically for maximum likelihood and MLE for "maximum likelihood estimat--," with a variety of word endings, depending on context; e.g., estimate, estimates, estimator, estimation and so on.

In using the ML method of estimation we are turning to an old (e.g., Fisher, 1922), well-established and well-respected method of estimation that has a variety of optimality properties. For straightforward situations detailed description of these properties can be found in many mathematical statistics books (e.g., Casella and Berger, 1990, Chap. 7). Therefore we just use the method here, without detailing derivation of those general properties, the special applications of which, to variance components estimation, can be found in Hartley and Rao (1967), Anderson (1973) and Miller (1973, 1977).

In contrast to the ANOVA method of estimation, one of the basic requirements of ML estimation is that of having to assume an underlying probability distribution for the data. A natural choice is the normal distribution, the multivariate normal, of Appendix S.2. Normality is chosen not because it is necessarily appropriate for all the different kinds of data for which one might want to estimate variance components but, more practically, normality leads to mathematically tractable methodology—even for unbalanced data. We therefore refer the reader to Appendices S.3 and S.7 for brief accounts of certain features of the multivariate normal distribution and of the method of maximum likelihood estimation. With that as a base we proceed to show the derivation of MLEs of variance components. And in Section 6.6 we describe an amended form of ML estimation that we call restricted maximum likelihood (REML). It also goes by the names of residual maximum likelihood and marginal maximum likelihood.

6.1. THE MODEL AND LIKELIHOOD FUNCTION

We return to the linear model that is described in detail in Section 4.6a. Only its essential features are repeated here. $\mathbf{y}$, the $N \times 1$ vector of observations, is taken to have model equation

$$\mathbf{y} = \mathbf{X}\boldsymbol{\beta} + \mathbf{Zu} + \mathbf{e} \tag{1}$$

as in (58) of Chapter 4. The fixed effects occurring in $\mathbf{y}$ are represented by $\boldsymbol{\beta}$, and the random effects by $\mathbf{u}$, with $\mathbf{Zu}$ being partitioned as

$$\mathbf{Zu} = [\mathbf{Z}_1 \quad \dots \quad \mathbf{Z}_r]\begin{bmatrix} \mathbf{u}_1 \\ \vdots \\ \mathbf{u}_r \end{bmatrix} = \sum_{i=1}^{r} \mathbf{Z}_i\mathbf{u}_i, \tag{2}$$

where $\mathbf{u}_i$ is the vector, for random factor i, of the effects for all levels of that random factor (be it a main effect factor, a nested factor, or an interaction factor) occurring in the data. The number of such levels, and hence the order of $\mathbf{u}_i$, is denoted by q_i. In the customary random model the random effects represented by $\mathbf{u}_i$ have the properties

$$E(\mathbf{u}_i) = \mathbf{0} \quad \text{and} \quad \text{var}(\mathbf{u}_i) = \sigma_i^2\mathbf{I}_{q_i} \quad \forall\ i;$$

and

$$\text{cov}(\mathbf{u}_i, \mathbf{u}_h') = \mathbf{0} \quad \text{for } i \neq h\,. \tag{3}$$

Thus

$$\text{var}(\mathbf{u}) = \{_d\ \sigma_i^2\mathbf{I}_{q_i}\}_{i=1}^{r} \tag{4}$$

as in (67) of Chapter 4. Also,

$$E(\mathbf{e}) = \mathbf{0}, \quad \text{var}(\mathbf{e}) = \sigma_e^2\mathbf{I}_N \quad \text{and} \quad \text{cov}(\mathbf{u}_i, \mathbf{e}') = 0\ \forall\ i\,. \tag{5}$$

Using these assumptions in (1) leads to

$$E(\mathbf{y}) = \mathbf{X}\boldsymbol{\beta} \tag{6}$$

and

$$\mathbf{V} = \text{var}(\mathbf{y}) = \sum_{i=1}^{r} \mathbf{Z}_i\mathbf{Z}_i'\sigma_i^2 + \sigma_e^2\mathbf{I}_N . \tag{7}$$

A notational convenience is to define

$$\mathbf{u}_0 = \mathbf{e}, \quad q_0 = N \quad \text{and} \quad \mathbf{Z}_0 = \mathbf{I}_N . \tag{8}$$

This gives (1) and (7) as

$$\mathbf{y} = \mathbf{X}\boldsymbol{\beta} + \sum_{i=0}^{r} \mathbf{Z}_i\mathbf{u}_i \quad \text{and} \quad \mathbf{V} = \sum_{i=0}^{r} \mathbf{Z}_i\mathbf{Z}_i'\sigma_i^2 \tag{9}$$

as in (70) and (71) of Chapter 4.

In Appendix S.2 the density function of the vector of random variables

$$\mathbf{x} \sim \mathcal{N}(\boldsymbol{\mu}, \mathbf{V}) \tag{10}$$

is given as

$$f(\mathbf{x}) = \frac{e^{-\frac{1}{2}(\mathbf{x}-\boldsymbol{\mu})'\mathbf{V}^{-1}(\mathbf{x}-\boldsymbol{\mu})}}{(2\pi)^{\frac{1}{2}N}|\mathbf{V}|^{\frac{1}{2}}} . \tag{11}$$

For our data vector $\mathbf{y} \sim \mathcal{N}_N(\mathbf{X}\boldsymbol{\beta}, \mathbf{V})$, and the function corresponding to (11), viewed as a function of the parameters $\boldsymbol{\beta}$ and $\mathbf{V}$, is called the likelihood function

$$L = L(\boldsymbol{\beta}, \mathbf{V}\,|\,\mathbf{y}) = \frac{e^{-\frac{1}{2}(\mathbf{y}-\mathbf{X}\boldsymbol{\beta})'\mathbf{V}^{-1}(\mathbf{y}-\mathbf{X}\boldsymbol{\beta})}}{(2\pi)^{\frac{1}{2}N}|\mathbf{V}|^{\frac{1}{2}}}, \tag{12}$$

similar to (103) of Chapter 3 and (85) of Chapter 4.

6.2. THE ML ESTIMATION EQUATIONS

Maximum likelihood estimation uses as estimators of $\boldsymbol{\beta}$ and $\mathbf{V}$ those values of $\boldsymbol{\beta}$ and $\mathbf{V}$ that maximize the likelihood L of (12). More accurately, we maximize L with respect to $\boldsymbol{\beta}$ and to $\boldsymbol{\sigma}^2 = [\sigma_0^2 \quad \sigma_1^2 \quad \dots \quad \sigma_r^2]$, the latter being used in $\mathbf{V}$ as in (7) and (10).

a. A direct derivation

Maximizing L can be achieved by maximizing the logarithm of L of (12), which shall be denoted by l:

$$l = \log L = -\tfrac{1}{2}N \log 2\pi - \tfrac{1}{2}\log|\mathbf{V}| - \tfrac{1}{2}(\mathbf{y}-\mathbf{X}\boldsymbol{\beta})'\mathbf{V}^{-1}(\mathbf{y}-\mathbf{X}\boldsymbol{\beta}) . \tag{13}$$

To maximize l, we differentiate (13), first with respect to $\boldsymbol{\beta}$, using Appendix M.7d, which yields

$$l_{\boldsymbol{\beta}} = \frac{\partial l}{\partial \boldsymbol{\beta}} = \mathbf{X}'\mathbf{V}^{-1}\mathbf{y} - \mathbf{X}'\mathbf{V}^{-1}\mathbf{X}\boldsymbol{\beta} . \tag{14}$$

Second, differentiating (13) with respect to σ_i^2 using Appendices M.7e and f, together with

$$\frac{\partial \mathbf{V}}{\partial \sigma_i^2} = \mathbf{Z}_i \mathbf{Z}_i', \tag{15}$$

gives, for $i = 0, 1, \ldots, r$,

$$l_{\sigma_i^2} = \frac{\partial l}{\partial \sigma_i^2} = -\tfrac{1}{2}\operatorname{tr}(\mathbf{V}^{-1}\mathbf{Z}_i\mathbf{Z}_i') + \tfrac{1}{2}(\mathbf{y} - \mathbf{X}\boldsymbol{\beta})'\mathbf{V}^{-1}\mathbf{Z}_i\mathbf{Z}_i'\mathbf{V}^{-1}(\mathbf{y} - \mathbf{X}\boldsymbol{\beta})\,. \tag{16}$$

A general principle for maximizing l with respect to $\boldsymbol{\beta}$ and the σ_i^2 is to equate (14) and (16) to zero and solve the resulting equations for $\boldsymbol{\beta}$ and the σ_i^2s. In general, values of $\boldsymbol{\beta}$ and $\boldsymbol{\sigma}^2$ that maximize l of (13) are solutions to equating (14) and (16) to zero. But these solutions are not necessarily the maximum likelihood estimators of $\boldsymbol{\beta}$ and the σ_i^2s, merely candidates. Completing the maximization demands checking second derivatives (see Section 6.3c), and also demands checking the likelihood function on the boundary of the parameter space, since the maximization must be confined to the parameter space. In many situations this confinement is not a restrictive requirement. For example, equating $l_{\boldsymbol{\beta}}$ of (14) to $\mathbf{0}$ gives, denoting a solution to $\boldsymbol{\beta}$ by $\boldsymbol{\beta}^0$,

$$\mathbf{X}'\mathbf{V}^{-1}\mathbf{X}\boldsymbol{\beta}^0 = \mathbf{X}'\mathbf{V}^{-1}\mathbf{y}, \tag{17}$$

which yields the MLE $\mathbf{X}\boldsymbol{\beta}^0$ of $\mathbf{X}\boldsymbol{\beta}$ when $\mathbf{V}$ is known. Since for a typical element of $\boldsymbol{\beta}$, say β_k, the parameter space is usually $-\infty < \beta_k < \infty$, there is no concern in solving (17) as to whether elements of $\boldsymbol{\beta}^0$ are positive, negative or zero. But being unconcerned for solution values vis-à-vis $\boldsymbol{\beta}^0$ does not carry over to solutions for elements of $\boldsymbol{\sigma}^2$ obtained from equating (16) to zero for $i = 0, 1, \ldots, r$. This is because the parameter space for the variance components in the linear model described in (1)–(10) is

$$\sigma_e^2 > 0 \quad \text{and} \quad \sigma_i^2 \geqslant 0 \quad \text{for } i = 1, \ldots, r\,. \tag{18}$$

Therefore, if $\tilde{\sigma}_e^2$ and $\tilde{\sigma}_i^2$ are to be maximum likelihood estimators, they must satisfy $\tilde{\sigma}_e^2 > 0$ and $\tilde{\sigma}_i^2 \geqslant 0$, conditions similar to (18). Denote the solutions for $\boldsymbol{\sigma}^2$ to the equations $\partial l/\partial\boldsymbol{\beta} = \mathbf{0}$ and $\{_c\, \partial l/\sigma_i^2\}_{i=0}^{r} = \mathbf{0}$ by $\dot{\boldsymbol{\sigma}}^2 = \{_c\, \dot{\sigma}_i^2\}_{i=0}^{r}$. Then $\dot{\boldsymbol{\sigma}}^2$ is the MLE of $\boldsymbol{\sigma}^2$ only when

$$\dot{\sigma}_e^2 > 0 \quad \text{and} \quad \dot{\sigma}_i^2 \geqslant 0 \quad \text{for } i = 1, \ldots, r; \tag{19}$$

i.e., provided (19) is satisfied, the ML estimator is $\tilde{\boldsymbol{\sigma}}^2 = \dot{\boldsymbol{\sigma}}^2$.

When (19) is not satisfied (usually by one or more $\dot{\sigma}_i^2$ being negative), the ML procedure is the extension of that described in Section 3.7a known as pooling the minimum violator, which often results in replacing any negative value with a zero, which is tantamount to dropping the corresponding factor from the model. That extension is described in Herbach (1959) and Thompson (1962). After applying it, one then uses the model so reduced to re-estimate $\boldsymbol{\sigma}^2$, obtaining a new $\dot{\boldsymbol{\sigma}}^2$ and applying (19) again.

Notation. $\dot{\mathbf{V}}$ and $\tilde{\mathbf{V}}$ are $\mathbf{V}$ with $\dot{\boldsymbol{\sigma}}^2$ and $\tilde{\boldsymbol{\sigma}}^2$ used, respectively, in place of $\boldsymbol{\sigma}^2$.

The ML solutions are solutions to equating (14) and (16) to zero:

$$\mathbf{X}'\dot{\mathbf{V}}^{-1}\mathbf{X}\dot{\boldsymbol{\beta}} = \mathbf{X}'\dot{\mathbf{V}}^{-1}\mathbf{y} \tag{20}$$

and

$$\operatorname{tr}(\dot{\mathbf{V}}^{-1}\mathbf{Z}_i\mathbf{Z}_i') = (\mathbf{y} - \mathbf{X}\dot{\boldsymbol{\beta}})'\dot{\mathbf{V}}^{-1}\mathbf{Z}_i\mathbf{Z}_i'\dot{\mathbf{V}}^{-1}(\mathbf{y} - \mathbf{X}\dot{\boldsymbol{\beta}}) \quad \text{for } i = 0, 1, \ldots r\,. \tag{21}$$

An algebraically simpler expression for (21) is derived by defining

$$\mathbf{P} = \mathbf{V}^{-1} - \mathbf{V}^{-1}\mathbf{X}(\mathbf{X}'\mathbf{V}^{-1}\mathbf{X})^{-}\mathbf{X}'\mathbf{V}^{-1}\,. \tag{22}$$

Then from (20) it is clear that for $\dot{\mathbf{P}}$ being $\mathbf{P}$ with $\mathbf{V}$ replaced by $\dot{\mathbf{V}}$

$$\dot{\mathbf{V}}^{-1}(\mathbf{y} - \mathbf{X}\dot{\boldsymbol{\beta}}) = \dot{\mathbf{P}}\mathbf{y}, \tag{23}$$

so that the ML equations (20) and (21) are

$$\mathbf{X}'\dot{\mathbf{V}}^{-1}\mathbf{X}\dot{\boldsymbol{\beta}} = \mathbf{X}'\dot{\mathbf{V}}^{-1}\mathbf{y} \tag{24}$$

and

$$\{_c \operatorname{tr}(\dot{\mathbf{V}}^{-1}\mathbf{Z}_i\mathbf{Z}_i')\}_{i=0}^{r} = \{_c\, \mathbf{y}'\dot{\mathbf{P}}\mathbf{Z}_i\mathbf{Z}_i'\dot{\mathbf{P}}\mathbf{y}\}_{i=0}^{r}\,. \tag{25}$$

Before deriving alternative expressions for these equations we should notice two features of them that are important. First, (24) is similar to but not the same as the equation $\mathbf{X}'\mathbf{V}^{-1}\mathbf{X}\boldsymbol{\beta}^0 = \mathbf{X}'\mathbf{V}^{-1}\mathbf{y}$ that yields BLUE($\mathbf{X}\boldsymbol{\beta}$). It is not the same because (24) uses $\dot{\mathbf{V}}$ where the equation for the BLUE uses $\mathbf{V}$. Second, equations (25) are nonlinear in the variance components. Elements of $\mathbf{V}$ are linear in the σ_i^2s, but $\dot{\mathbf{V}}$ occurs in (21) only in the form $\dot{\mathbf{V}}^{-1}$, once in each element of the left-hand side of (21) and twice in each right-hand element of (21), together with its occurrences in $\dot{\boldsymbol{\beta}}$. Thus equations (21), or equivalently (25), are complicated polynomial functions of the variance components, an illustration of which is evident in (134) and (135) of Chapter 3. Hence (except in what turns out to be a very few cases of balanced data) closed form expressions for the solutions of (25) cannot be obtained. Therefore on a case-by-case basis, for each individual data set, solutions to (25) have to be obtained numerically, usually by iteration. All the problems that this entails are in the ken of the numerical analyst. They are mentioned briefly in Section 6.5, and considered again in Chapter 8.

b. An alternative form

The left-hand side of the ML equation (25) is a vector of elements

$$\begin{aligned}
\operatorname{tr}(\dot{\mathbf{V}}^{-1}\mathbf{Z}_i\mathbf{Z}_i') &= \operatorname{tr}(\dot{\mathbf{V}}^{-1}\mathbf{Z}_i\mathbf{Z}_i'\dot{\mathbf{V}}^{-1}\dot{\mathbf{V}}) \\
&= \operatorname{tr}\left(\dot{\mathbf{V}}^{-1}\mathbf{Z}_i\mathbf{Z}_i'\dot{\mathbf{V}}^{-1}\sum_{j=0}^{r}\mathbf{Z}_j\mathbf{Z}_j\dot{\sigma}_j^2\right) \\
&= \sum_{j=0}^{r}\operatorname{tr}(\dot{\mathbf{V}}^{-1}\mathbf{Z}_i\mathbf{Z}_i'\dot{\mathbf{V}}^{-1}\mathbf{Z}_j\mathbf{Z}_j')\dot{\sigma}_j^2 = \sum_{j=0}^{r}\operatorname{sesq}(\mathbf{Z}_i'\dot{\mathbf{V}}^{-1}\mathbf{Z}_j)\dot{\sigma}_j^2 \\
&= \{_r \operatorname{tr}(\dot{\mathbf{V}}^{-1}\mathbf{Z}_i\mathbf{Z}_i'\dot{\mathbf{V}}^{-1}\mathbf{Z}_j\mathbf{Z}_j')\}_{j=0}^{r}\dot{\boldsymbol{\sigma}}^2 = \{_r \operatorname{sesq}(\mathbf{Z}_i'\dot{\mathbf{V}}^{-1}\mathbf{Z}_j)\}_{j=1}^{r}\dot{\boldsymbol{\sigma}}^2,
\end{aligned} \tag{26}$$

where, as in Appendix M.6, sesq(**A**) is the sum of squares of elements of the matrix **A**. Therefore (25) can be written as

$$\{_{\mathrm{m}}\ \mathrm{tr}(\dot{\mathbf{V}}^{-1}\mathbf{Z}_i\mathbf{Z}_i'\dot{\mathbf{V}}^{-1}\mathbf{Z}_j\mathbf{Z}_j')\}_{i,j=0}^{r}\dot{\boldsymbol{\sigma}}^2 = \{_{\mathrm{c}}\ \mathbf{y}'\dot{\mathbf{P}}\mathbf{Z}_i\mathbf{Z}_i'\dot{\mathbf{P}}\mathbf{y}\}_{i=0}^{r} \tag{27a}$$

or, equivalently, as

$$\{_{\mathrm{m}}\ \mathrm{sesq}(\mathbf{Z}_i'\dot{\mathbf{V}}^{-1}\mathbf{Z}_j)\}_{i,j=0}^{r}\dot{\boldsymbol{\sigma}}^2 = \{_{\mathrm{c}}\ \mathrm{sesq}(\mathbf{Z}_i'\dot{\mathbf{P}}\mathbf{y})\}_{i=0}^{r}\ . \tag{27b}$$

A cautionary note: equations (27) might seem to be linear in elements of $\dot{\boldsymbol{\sigma}}^2$, the variance components estimates, but they are not. Those estimates also occur in $\dot{\mathbf{V}}^{-1}$, which is involved in the left-hand side of (27) and, through $\dot{\mathbf{P}}$, in the right-hand side also. Thus the equations are non-linear in the elements of $\dot{\boldsymbol{\sigma}}^2$.

The form of equations (27) is not necessarily optimum for computing purposes, but it is useful for illustrating how an iterative procedure for obtaining a solution could be set up: use a set of starting values for $\dot{\boldsymbol{\sigma}}^2$ in $\dot{\mathbf{V}}^{-1}$ and $\dot{\mathbf{P}}$, so that (27) is then linear in $\dot{\boldsymbol{\sigma}}^2$ and is easily solved for the next value of $\dot{\boldsymbol{\sigma}}^2$.

c. The Hartley–Rao form

Hartley and Rao (1967) formulate the likelihood function in terms of **H** defined by

$$\mathbf{V} = \mathbf{H}\sigma_e^2 \quad \text{with} \quad \mathbf{V}^{-1} = \mathbf{H}^{-1}/\sigma_e^2\ . \tag{28}$$

Thus **H** has exactly the same form as **V** except that where σ_e^2 occurs in **V** there is a 1 in **H**, and where there is σ_i^2 in **V** there is $\gamma_i = \sigma_i^2/\sigma_e^2$ in **H**, for $i = 1,\ldots, r$. This means that in the Hartley–Rao formulation of ML estimation it is $\boldsymbol{\beta}$, σ_e^2 and $\gamma_1,\ldots,\gamma_r$ that are the parameters—in particular, γ_i instead of σ_i^2, for $i = 1,\ldots,r$. This leads to a separate equation for $\dot{\sigma}_e^2$ rather than having it be included in (27).

We derive, for (28), the estimation equations of Hartley and Rao, using (25) as the starting point. (The basis of being able to make this derivation, i.e., of deriving estimation equations for σ_e^2 and γ_i from equations for $\boldsymbol{\sigma}^2$, is the chain rule used in Appendix S.7d.) First, (25) for $i = 0$ is

$$\mathrm{tr}(\dot{\mathbf{V}}^{-1}) = \mathbf{y}'\dot{\mathbf{P}}^2\mathbf{y}, \tag{29}$$

since $i = 0$ corresponds to $\sigma_0^2 = \sigma_e^2$ with $\mathbf{Z}_0 = \mathbf{I}$. But using (23) gives (29) as

$$\mathrm{tr}(\dot{\mathbf{V}}^{-1}) = (\mathbf{y} - \mathbf{X}\dot{\boldsymbol{\beta}})'\dot{\mathbf{V}}^{-2}(\mathbf{y} - \mathbf{X}\dot{\boldsymbol{\beta}}),$$

which, with (28), is

$$\mathrm{tr}(\dot{\mathbf{H}}^{-1}) = (\mathbf{y} - \mathbf{X}\dot{\boldsymbol{\beta}})'\dot{\mathbf{H}}^{-2}(\mathbf{y} - \mathbf{X}\dot{\boldsymbol{\beta}})/\dot{\sigma}_e^2\ .$$

Therefore

$$\dot{\sigma}_e^2 = \frac{(\mathbf{y} - \mathbf{X}\dot{\boldsymbol{\beta}})'\dot{\mathbf{H}}^{-2}(\mathbf{y} - \mathbf{X}\dot{\boldsymbol{\beta}})}{\mathrm{tr}(\dot{\mathbf{H}}^{-1})}\ . \tag{30}$$

Now consider equations (25) for $i = 1, 2,\ldots, r$ (excluding $i = 0$):

$$\mathrm{tr}(\dot{\mathbf{V}}^{-1}\mathbf{Z}_i\mathbf{Z}_i') = \mathbf{y}'\dot{\mathbf{P}}\mathbf{Z}_i\mathbf{Z}_i'\dot{\mathbf{P}}\mathbf{y}\ . \tag{31}$$

Multiply (31) by $\dot{\sigma}_i^2$ and sum over $i = 1, 2, \ldots, r$:

$$\operatorname{tr}\left(\dot{\mathbf{V}}^{-1} \sum_{i=1}^{r} \mathbf{Z}_i\mathbf{Z}_i\dot{\sigma}_i^2\right) = \mathbf{y}'\dot{\mathbf{P}}\left(\sum_{i=1}^{r} \mathbf{Z}_i\mathbf{Z}_i'\dot{\sigma}_i^2\right)\dot{\mathbf{P}}\mathbf{y}\,. \tag{32}$$

This becomes, on using (23), (28) and

$$\sum_{i=1}^{r} \mathbf{Z}_i\mathbf{Z}_i'\sigma_i^2 = \mathbf{V} - \sigma_e^2\mathbf{I} = (\mathbf{H} - \mathbf{I})\sigma_e^2, \tag{33}$$

$$\operatorname{tr}[\dot{\mathbf{H}}^{-1}\dot{\sigma}_e^{-2}(\dot{\mathbf{H}} - \mathbf{I})\dot{\sigma}_e^2] = (\mathbf{y} - \mathbf{X}\dot{\boldsymbol{\beta}})'\dot{\mathbf{H}}^{-1}\dot{\sigma}_e^{-2}(\dot{\mathbf{H}} - \mathbf{I})\dot{\sigma}_e^2\dot{\mathbf{H}}^{-1}\dot{\sigma}_e^{-2}(\mathbf{y} - \mathbf{X}\dot{\boldsymbol{\beta}}),$$

which is equivalent to

$$N\dot{\sigma}_e^2 = (\mathbf{y} - \mathbf{X}\dot{\boldsymbol{\beta}})'\dot{\mathbf{H}}^{-1}(\mathbf{y} - \mathbf{X}\dot{\boldsymbol{\beta}}) + \operatorname{tr}(\dot{\mathbf{H}}^{-1})\left[\dot{\sigma}_e^2 - \frac{(\mathbf{y} - \mathbf{X}\dot{\boldsymbol{\beta}})'\dot{\mathbf{H}}^{-2}(\mathbf{y} - \mathbf{X}\dot{\boldsymbol{\beta}})}{\operatorname{tr}(\dot{\mathbf{H}}^{-1})}\right].$$

But, by (30), the term in the square brackets is zero. Hence, on repeating (30),

$$\dot{\sigma}_e^2 = \frac{(\mathbf{y} - \mathbf{X}\dot{\boldsymbol{\beta}})'\dot{\mathbf{H}}^{-2}(\mathbf{y} - \mathbf{X}\dot{\boldsymbol{\beta}})}{\operatorname{tr}(\dot{\mathbf{H}}^{-1})} = \frac{(\mathbf{y} - \mathbf{X}\dot{\boldsymbol{\beta}})'\dot{\mathbf{H}}^{-1}(\mathbf{y} - \mathbf{X}\dot{\boldsymbol{\beta}})}{N}\,. \tag{34}$$

The equations that Hartley and Rao then have are

$$\mathbf{X}'\dot{\mathbf{V}}^{-1}\mathbf{X}\dot{\boldsymbol{\beta}} = \mathbf{X}'\dot{\mathbf{V}}^{-1}\mathbf{y},$$

$$\dot{\sigma}_e^2 = (\mathbf{y} - \mathbf{X}\dot{\boldsymbol{\beta}})'\dot{\mathbf{H}}^{-1}(\mathbf{y} - \mathbf{X}\dot{\boldsymbol{\beta}})/N, \tag{35a}$$

and

$$\{_c \operatorname{tr}(\dot{\mathbf{H}}^{-1}\mathbf{Z}_i\mathbf{Z}_i')\}_{i=1}^{r} = \{_c (\mathbf{y} - \mathbf{X}\dot{\boldsymbol{\beta}})'\dot{\mathbf{H}}^{-1}\mathbf{Z}_i\mathbf{Z}_i'\dot{\mathbf{H}}^{-1}(\mathbf{y} - \mathbf{X}\dot{\boldsymbol{\beta}})/\dot{\sigma}_e^2\}_{i=1}^{r}\,. \tag{35b}$$

In (35b) it is not σ_i^2 that is estimated but $\gamma_i = \sigma_i^2/\sigma_e^2$ for $i = 1, \ldots, r$; and, because of (35a) and the fact that the γ_is are ratios, iterative solution of (35a) and (35b) may, in fact, be easier than of (25).

6.3. ASYMPTOTIC DISPERSION MATRICES FOR ML ESTIMATORS

As indicated in Appendix S.7, one of the attractive features of ML estimation is that the large-sample, asymptotic dispersion matrix of the estimators is always available. It is the inverse of what is called the information matrix, $\operatorname{var}(\tilde{\boldsymbol{\theta}}) \simeq [\mathbf{I}(\boldsymbol{\theta})]^{-1}$. We now develop $\mathbf{I}(\boldsymbol{\theta})$ for $\boldsymbol{\beta}$, $\boldsymbol{\sigma}^2$, and for $[\sigma_0^2 \;\; \boldsymbol{\gamma}']'$ where $\boldsymbol{\gamma}$ has elements $\gamma_i = \sigma_i^2/\sigma_0^2$ for $i = 1, \ldots, r$, of Hartley and Rao (1967).

a. For variance components

In (14) we used the symbol $l_{\boldsymbol{\beta}}$ for $\partial l/\partial\boldsymbol{\beta}$. This is extended to using $l_{\boldsymbol{\beta\beta}}$ for $\partial^2 l/\partial\boldsymbol{\beta}\,\partial\boldsymbol{\beta}'$ and $l_{\boldsymbol{\beta}\boldsymbol{\sigma}^2}$ for $\partial^2 l/\partial\boldsymbol{\beta}\,\partial\boldsymbol{\sigma}^{2\prime}$. Then, from Appendix S.7c,

$$\mathbf{I}\begin{bmatrix}\boldsymbol{\beta} \\ \boldsymbol{\sigma}^2\end{bmatrix} = -E\begin{bmatrix} l_{\boldsymbol{\beta\beta}} & l_{\boldsymbol{\beta}\boldsymbol{\sigma}^2} \\ l_{\boldsymbol{\sigma}^2\boldsymbol{\beta}} & l_{\boldsymbol{\sigma}^2\boldsymbol{\sigma}^2}\end{bmatrix}.$$

Since in (14)

$$l_{\boldsymbol{\beta}} = \mathbf{X}'\mathbf{V}^{-1}\mathbf{y} - \mathbf{X}'\mathbf{V}^{-1}\mathbf{X}\boldsymbol{\beta},$$
$$l_{\boldsymbol{\beta}\boldsymbol{\beta}} = -\mathbf{X}'\mathbf{V}^{-1}\mathbf{X} \quad \text{and} \quad l_{\boldsymbol{\beta}\sigma_i^2} = -\mathbf{X}'\mathbf{V}^{-1}\mathbf{Z}_i\mathbf{Z}_i'\mathbf{V}^{-1}(\mathbf{y} - \mathbf{X}\boldsymbol{\beta}). \tag{36}$$

And with (16) being

$$l_{\sigma_i^2} = -\tfrac{1}{2}\operatorname{tr}(\mathbf{V}^{-1}\mathbf{Z}_i\mathbf{Z}_i') + \tfrac{1}{2}(\mathbf{y} - \mathbf{X}\boldsymbol{\beta})'\mathbf{V}^{-1}\mathbf{Z}_i\mathbf{Z}_i'\mathbf{V}^{-1}(\mathbf{y} - \mathbf{X}\boldsymbol{\beta})$$

we get, with Appendix M.7e giving

$$\frac{\partial \mathbf{V}^{-1}}{\partial \sigma_i^2} = -\mathbf{V}^{-1}\frac{\partial \mathbf{V}}{\partial \sigma_i^2}\mathbf{V}^{-1} = -\mathbf{V}^{-1}\mathbf{Z}_i\mathbf{Z}_i'\mathbf{V}^{-1},$$

$$\begin{aligned} l_{\sigma_i^2\sigma_j^2} &= \tfrac{1}{2}\operatorname{tr}(\mathbf{V}^{-1}\mathbf{Z}_j\mathbf{Z}_j'\mathbf{V}^{-1}\mathbf{Z}_i\mathbf{Z}_i') - \tfrac{1}{2}(\mathbf{y} - \mathbf{X}\boldsymbol{\beta})'\mathbf{V}^{-1}\mathbf{Z}_j\mathbf{Z}_j'\mathbf{V}^{-1}\mathbf{Z}_i\mathbf{Z}_i'\mathbf{V}^{-1}(\mathbf{y} - \mathbf{X}\boldsymbol{\beta}) \\ &\quad - \tfrac{1}{2}(\mathbf{y} - \mathbf{X}\boldsymbol{\beta})'\mathbf{V}^{-1}\mathbf{Z}_i\mathbf{Z}_i'\mathbf{V}^{-1}\mathbf{Z}_j\mathbf{Z}_j'\mathbf{V}^{-1}(\mathbf{y} - \mathbf{X}\boldsymbol{\beta}) \\ &= \tfrac{1}{2}\operatorname{tr}(\mathbf{V}^{-1}\mathbf{Z}_i\mathbf{Z}_i'\mathbf{V}^{-1}\mathbf{Z}_j\mathbf{Z}_j') - (\mathbf{y} - \mathbf{X}\boldsymbol{\beta})'\mathbf{V}^{-1}\mathbf{Z}_i\mathbf{Z}_i'\mathbf{V}^{-1}\mathbf{Z}_j\mathbf{Z}_j'\mathbf{V}^{-1}(\mathbf{y} - \mathbf{X}\boldsymbol{\beta}). \end{aligned} \tag{37}$$

In taking expected values of (36) and (37) we use $E(\mathbf{y}) = \mathbf{X}\boldsymbol{\beta}$ and hence $E(\mathbf{y} - \mathbf{X}\boldsymbol{\beta}) = \mathbf{0}$, and

$$E\,(\mathbf{y} - \mathbf{X}\boldsymbol{\beta})'\mathbf{T}(\mathbf{y} - \mathbf{X}\boldsymbol{\beta}) = \operatorname{tr}(\mathbf{T}\mathbf{V}) \quad \text{for non-stochastic } \mathbf{T}$$

gives

$$-E\, l_{\boldsymbol{\beta}\boldsymbol{\beta}} = E(\mathbf{X}'\mathbf{V}^{-1}\mathbf{X}) = \mathbf{X}'\mathbf{V}^{-1}\mathbf{X},$$
$$-E\, l_{\boldsymbol{\beta}\sigma_i^2} = \mathbf{X}'\mathbf{V}^{-1}\mathbf{Z}_i\mathbf{Z}_i'\mathbf{V}^{-1}E(\mathbf{y} - \mathbf{X}\boldsymbol{\beta}) = \mathbf{0}$$

and

$$\begin{aligned} -E\, l_{\sigma_i^2\sigma_j^2} &= -\tfrac{1}{2}\operatorname{tr}(\mathbf{V}^{-1}\mathbf{Z}_i\mathbf{Z}_i'\mathbf{V}^{-1}\mathbf{Z}_j\mathbf{Z}_j') + \operatorname{tr}(\mathbf{V}^{-1}\mathbf{Z}_i\mathbf{Z}_i'\mathbf{V}^{-1}\mathbf{Z}_j\mathbf{Z}_j'\mathbf{V}^{-1}\mathbf{V}) \\ &= \tfrac{1}{2}\operatorname{tr}(\mathbf{V}^{-1}\mathbf{Z}_i\mathbf{Z}_i'\mathbf{V}^{-1}\mathbf{Z}_j\mathbf{Z}_j'). \end{aligned}$$

Therefore the information matrix is

$$\mathbf{I}\begin{bmatrix} \boldsymbol{\beta} \\ \boldsymbol{\sigma}^2 \end{bmatrix} = \begin{bmatrix} \mathbf{X}'\mathbf{V}^{-1}\mathbf{X} & \mathbf{0} \\ \mathbf{0} & \tfrac{1}{2}\{_{m} \operatorname{tr}(\mathbf{V}^{-1}\mathbf{Z}_i\mathbf{Z}_i'\mathbf{V}^{-1}\mathbf{Z}_j\mathbf{Z}_j')\}_{i,j=0}^{r} \end{bmatrix}. \tag{38}$$

Therefore, asymptotically,

$$\begin{aligned} \operatorname{var}\begin{bmatrix} \tilde{\boldsymbol{\beta}} \\ \tilde{\boldsymbol{\sigma}}^2 \end{bmatrix} &\simeq \left(\mathbf{I}\begin{bmatrix} \boldsymbol{\beta} \\ \boldsymbol{\sigma}^2 \end{bmatrix}\right)^{-1} \\ &= \begin{bmatrix} (\mathbf{X}'\mathbf{V}^{-1}\mathbf{X})^{-1} & \mathbf{0} \\ \mathbf{0} & 2[\{_{m} \operatorname{tr}(\mathbf{V}^{-1}\mathbf{Z}_i\mathbf{Z}_i'\mathbf{V}^{-1}\mathbf{Z}_j\mathbf{Z}_j')\}_{i,j=0}^{r}]^{-1} \end{bmatrix}, \end{aligned} \tag{39}$$

as in Searle (1970). Hence, in the limit

$$\operatorname{var}(\tilde{\boldsymbol{\beta}}) \to (\mathbf{X}'\mathbf{V}^{-1}\mathbf{X})^{-1}, \tag{40}$$

$$\operatorname{cov}(\tilde{\boldsymbol{\beta}}, \tilde{\boldsymbol{\sigma}}^2) \to \mathbf{0}, \tag{41}$$

$$\operatorname{var}(\tilde{\boldsymbol{\sigma}}^2) \to 2[\{_{m} \operatorname{tr}(\mathbf{V}^{-1}\mathbf{Z}_i\mathbf{Z}_i'\mathbf{V}^{-1}\mathbf{Z}_j\mathbf{Z}_j')\}_{i,j=0}^{r}]^{-1} \tag{42a}$$

$$= 2[\{_{m} \operatorname{sesq}(\mathbf{Z}_i'\mathbf{V}^{-1}\mathbf{Z}_j)\}_{i,j=0}^{r}]^{-1}. \tag{42b}$$

In these expressions $\tilde{\boldsymbol{\sigma}}^2$ denotes the vector of solutions to the maximum likelihood equations when those solutions satisfy the non-negativity requirements of (19) and so are therefore ML estimators. Similarly, $\tilde{\boldsymbol{\beta}}$ is the solution vector from $\mathbf{X}'\tilde{\mathbf{V}}^{-1}\mathbf{X}\tilde{\boldsymbol{\beta}} = \mathbf{X}'\tilde{\mathbf{V}}^{-1}\mathbf{y}$ when using the true ML estimator $\tilde{\boldsymbol{\sigma}}^2$ in deriving $\tilde{\mathbf{V}}$ and not just any solution vector. Of course, making this distinction in (39) is really unnecessary, because although ML estimators are not generally unbiased, they are consistent. This means that in the limit (as sample size tends to infinity) the estimators converge to the parameter values; and since (39) and its sequels are only true in the limit, there is no problem about the solution $\tilde{\boldsymbol{\sigma}}^2$ not being the MLE. Nevertheless, we persist with this notation to be emphatic about distinguishing between solutions and MLEs. Furthermore, although (40)–(42) are exact only in the limit, they are results that provide some information about sampling variances even for finite-sized data sets. Even though the ML estimators are not unbiased, use of (40)–(42) with $\boldsymbol{\sigma}^2$ replaced with $\tilde{\boldsymbol{\sigma}}^2$ may lead, with small-sized samples of data, to under-estimation of variances of the ML estimators. Nevertheless, calculated values of (40)–(42) using this replacement are to be found in much of today's computer package output for ML estimation, and so will undoubtedly gain ever-increasing use. Even in the limit there are, of course, difficult questions as to what is meant by "sample size tending to infinity" in mixed models. For example, in a 2-way crossed classification what does that phrase mean with regard to the numbers of levels of each factor, the numbers of empty cells, the number of observations per filled cell and the total number of observations. Both Hartley-Rao (1967) and Miller (1977) give consideration to this kind of question. Finally, note that (39) is also the Cramer–Rao lower bound for the variance-covariance matrix of unbiased estimators. [See Casella and Berger (1990, Theorem 7.3.1).]

b. For ratios of components

The Hartley–Rao equations (35) lead to direct estimation of σ_0^2 and of

$$\boldsymbol{\gamma} = \{_c\,\gamma_i\}_{i=1}^{r} \quad \text{for } \gamma_i = \sigma_i^2/\sigma_e^2 \tag{43}$$

rather than of $\boldsymbol{\sigma}^2 = \{\sigma_i^2\}$ for $i = 0, \ldots, r$. On defining $\gamma_0 = \sigma_0^2 = \sigma_e^2$ we have the relationship between these two sets of parameters as

$$\begin{bmatrix} \gamma_0 \\ \boldsymbol{\gamma} \end{bmatrix} = \begin{bmatrix} 1 & \mathbf{0} \\ \mathbf{0} & \mathbf{I}_r/\sigma_e^2 \end{bmatrix} \boldsymbol{\sigma}^2 . \tag{44}$$

The information matrix for γ_0 and $\boldsymbol{\gamma}$ is then obtained from (44) by using the theorem in Appendix S.7d. Through writing

$$\boldsymbol{\theta} = \boldsymbol{\sigma}^2 \quad \text{and} \quad \boldsymbol{\Delta} = \begin{bmatrix} \gamma_0 \\ \boldsymbol{\gamma} \end{bmatrix}, \tag{45}$$

the theorem is

$$\mathbf{I}(\boldsymbol{\Delta}) = \mathbf{S}\mathbf{I}(\boldsymbol{\theta})\mathbf{S}' \quad \text{for } \mathbf{S} = \left\{\frac{\partial \theta_i}{\partial \Delta_j}\right\}. \tag{46}$$

Therefore, from (44),

$$\mathbf{S} = \left\{_{\mathrm{m}} \frac{\partial \sigma_i^2}{\partial \gamma_j}\right\}_{i,j=0}^{r} = \begin{bmatrix} 1 & \mathbf{0} \\ \boldsymbol{\gamma} & \gamma_0 \mathbf{I} \end{bmatrix} \tag{47}$$

and so, with (38) yielding

$$\mathbf{I}(\boldsymbol{\sigma}^2) = \tfrac{1}{2}\{_{\mathrm{m}}\, \mathrm{tr}(\mathbf{V}^{-1}\mathbf{Z}_i\mathbf{Z}_i'\mathbf{V}^{-1}\mathbf{Z}_j\mathbf{Z}_j')\}_{i,j=0}^{r} \tag{48}$$

$$\mathbf{I}\begin{bmatrix} \sigma_e^2 \\ \boldsymbol{\gamma} \end{bmatrix} = \begin{bmatrix} 1 & \boldsymbol{\gamma}' \\ \mathbf{0} & \gamma_0 \mathbf{I} \end{bmatrix} \tfrac{1}{2}\{_{\mathrm{m}}\, \mathrm{tr}(\mathbf{V}^{-1}\mathbf{Z}_i\mathbf{Z}_i'\mathbf{V}^{-1}\mathbf{Z}_j\mathbf{Z}_j')\} \begin{bmatrix} 1 & \mathbf{0} \\ \boldsymbol{\gamma} & \gamma_0 \mathbf{I} \end{bmatrix}. \tag{49}$$

For $i, j = 1, 2, \ldots, r$ this is

$$\mathbf{I}\begin{bmatrix} \sigma_e^2 \\ \boldsymbol{\gamma} \end{bmatrix} = \begin{bmatrix} 1 & \boldsymbol{\gamma}' \\ \mathbf{0} & \sigma_e^2 \mathbf{I} \end{bmatrix} \tfrac{1}{2} \begin{bmatrix} \mathrm{tr}(\mathbf{V}^{-2}) & \{_{\mathrm{r}}\, \mathrm{tr}(\mathbf{V}^{-2}\mathbf{Z}_i\mathbf{Z}_i')\} \\ \{_{\mathrm{c}}\, \mathrm{tr}(\mathbf{V}^{-2}\mathbf{Z}_i\mathbf{Z}_i')\} & \{_{\mathrm{m}}\, \mathrm{tr}(\mathbf{V}^{-1}\mathbf{Z}_i\mathbf{Z}_i'\mathbf{V}^{-1}\mathbf{Z}_j\mathbf{Z}_j')\} \end{bmatrix} \begin{bmatrix} 1 & \mathbf{0} \\ \boldsymbol{\gamma} & \sigma_e^2 \mathbf{I} \end{bmatrix}, \tag{50}$$

which ultimately reduces (see E 6.6) to

$$\mathbf{I}\begin{bmatrix} \sigma_e^2 \\ \boldsymbol{\gamma} = \left\{_{\mathrm{c}} \dfrac{\sigma_i^2}{\sigma_e^2}\right\} \end{bmatrix} = \frac{1}{2}\begin{bmatrix} \dfrac{N}{\sigma_e^4} & \left\{_{\mathrm{r}} \dfrac{\mathrm{tr}(\mathbf{H}^{-1}\mathbf{Z}_i\mathbf{Z}_i')}{\sigma_e^2}\right\} \\ \left\{_{\mathrm{c}} \dfrac{\mathrm{tr}(\mathbf{H}^{-1}\mathbf{Z}_i\mathbf{Z}_i')}{\sigma_e^2}\right\} & \{_{\mathrm{m}}\, \mathrm{tr}(\mathbf{H}^{-1}\mathbf{Z}_i\mathbf{Z}_i'\mathbf{H}^{-1}\mathbf{Z}_j\mathbf{Z}_j')\} \end{bmatrix}. \tag{51}$$

Further reduction seems difficult.

c. Maximum?

The matrix of second derivatives, known as the Hessian, is

$$\mathbf{Q} = \begin{bmatrix} l_{\boldsymbol{\beta}\boldsymbol{\beta}} & l_{\boldsymbol{\beta}\boldsymbol{\sigma}^2} \\ l_{\boldsymbol{\sigma}^2\boldsymbol{\beta}} & l_{\boldsymbol{\sigma}^2\boldsymbol{\sigma}^2} \end{bmatrix},$$

and from (35), (36) and (37) this is

$$\mathbf{Q} = -\begin{bmatrix} \mathbf{X}'\mathbf{V}^{-1}\mathbf{X} & \{_{\mathrm{r}}\, \mathbf{X}'\mathbf{V}^{-1}\mathbf{Z}_i\mathbf{Z}_i'\mathbf{P}\mathbf{y}\}_{i=0}^{r} \\ \mathrm{sym} & \tfrac{1}{2}\{_{\mathrm{m}} -\mathrm{tr}(\mathbf{V}^{-1}\mathbf{Z}_i\mathbf{Z}_i'\mathbf{V}^{-1}\mathbf{Z}_j\mathbf{Z}_j') + 2\mathbf{y}'\mathbf{P}\mathbf{Z}_i\mathbf{Z}_i'\mathbf{V}^{-1}\mathbf{Z}_j\mathbf{Z}_j'\mathbf{P}\mathbf{y}\}_{i,j=0}^{r} \end{bmatrix}.$$

By standard results in advanced calculus (e.g., Buck, 1978, p. 426), $\mathbf{Q}$ will be negative definite when evaluated at $\boldsymbol{\beta} = \tilde{\boldsymbol{\beta}}$ and $\boldsymbol{\sigma} = \tilde{\boldsymbol{\sigma}}^2$ or at any local maximum of the log likelihood, so long as the maximum is in the interior of the parameter space. It is easy to see that $\mathbf{Q}$ need not always be negative definite; or even negative definite at all points that satisfy the likelihood equations (20) and (21). Consider Figure 8.1, which exhibits a log likelihood surface with two local maxima and a saddlepoint. At the saddlepoint, equations (20) and (21) will be satisfied but the Hessian will be indefinite.

6.4. SOME REMARKS ON COMPUTING

Equations (24) and (25), or either of the alternative forms for (25), namely (27) or (35), are clearly nonlinear in the elements of $\dot{\boldsymbol{\sigma}}^2$, and solutions are usually obtained by numerical iteration. This raises all kinds of questions in numerical analysis, such as the following.

(i) What method of iteration is the best to use for these equations?

(ii) Does the choice of iterative method depend on the form of the equations used, (25), (27) or (35)? Or are there other forms that are even more suitable?

(iii) Clearly, (27) is the most succint and easily understood form of the estimation equations, but is it the best?

(iv) Is convergence of the iteration always assured?

(v) If convergence is achieved, can we be sure that it is at a value that corresponds to a global maximum of the likelihood and not just a local maximum?

(vi) Does the value of $\boldsymbol{\sigma}^2$ chosen as an initial value for starting the iteration affect the value at which convergence is achieved?

(vii) If so, is there any particular set of starting values that will always yield a value at convergence that corresponds to the global maximum of the likelihood?

(viii) What is the cost, in terms of computer time and/or money to do the necessary computing? [We might note in passing that as of March, 1990, the time required for inverting matrices on Cornell's supercomputer was quoted for matrices of order 1000, 2000 and 9000 (an upper limit) as being approximately 17 seconds, 2 minutes and 2 hours, respectively.]

(ix) The matrix $\mathbf{V}$ is, by definition, always non-negative definite; and usually positive definite. The latter has been assumed. What, therefore, is to be done numerically if, at some step in the iteration, the calculated $\boldsymbol{\sigma}^2$ is such that the calculated $\mathbf{V}$ is not positive definite?

(x) More seriously, what is to be done if the calculated $\mathbf{V}$ is singular? [Harville (1977) addresses this concern.]

(xi) Since ML estimators, as distinct from just solutions to the estimation equations, must satisfy the conditions (19) that $\tilde{\sigma}_e^2 > 0$ and $\tilde{\sigma}_i^2 \geqslant 0$ for $i = 1, \ldots, r$, these conditions must be taken into account in computer programs that are used for solving the ML equations to obtain ML estimators. Customarily, any $\tilde{\sigma}_i^2$ that is computed as a negative value is put equal to zero—an action which can sometimes be interpreted as altering the model being used. It also raises the further difficulty of having a computer program which, for any $\tilde{\sigma}_i^2$ that has been put equal to zero after some iteration, enables that $\tilde{\sigma}_i^2$ to come back into the calculations again at some later iteration if it were then to be positive. Conditions of this nature are considered in such papers as Hemmerle and

Hartley (1973) and Jennrich and Sampson (1976). Maybe replacing the negative solution by a small possible number, e.g., 0.5, would be better—as is done in some packages; and the use of algorithms for solving nonlinear equations, adapted by constraints such as $\sigma_i^2 \geqslant 0$ is also a possibility.

Clearly, these difficulties are not necessarily overcome in any easy manner when building a computer package for estimating variance components by maximum likelihood. It is a job for the expert, with a sound appreciation of numerical analysis. Computer packages designed by those who are amateur in this regard are usually to be deemed suspect. A more detailed discussion of computing variance components estimates is given in Chapter 8.

6.5. ML RESULTS FOR 2-WAY CROSSED CLASSIFICATION, BALANCED DATA

Section 4.7b contains ML solutions and estimators for a variety of balanced data cases, and Section 4.7d displays the ML equations for two cases of the 2-way crossed classification random model, with and without interaction. For the with-interaction case we show here the details of deriving certain parts of those equations—and leave it to the reader (E 6.8) to derive the others.

a. 2-way crossed, random model, with interaction

The scalar form of the model equation is

$$y_{ijk} = \mu + \alpha_i + \beta_j + \gamma_{ij} + e_{ijk} \tag{52}$$

for $i = 1,\ldots, a$, $j = 1,\ldots, b$ and $k = 1,\ldots, n$. The vector form is

$$\mathbf{y} = \mu \mathbf{1}_{abn} + \mathbf{Z}_1 \boldsymbol{\alpha} + \mathbf{Z}_2 \boldsymbol{\beta} + \mathbf{Z}_3 \boldsymbol{\gamma} + \mathbf{Z}_0 \mathbf{e}$$

with

$$\begin{aligned}
\mathbf{Z}_0 &= \mathbf{I}_a \otimes \mathbf{I}_b \otimes \mathbf{I}_n = \mathbf{I}, & \mathbf{Z}_0\mathbf{Z}_0' &= \mathbf{I}_a \otimes \mathbf{I}_b \otimes \mathbf{I}_n,\\
\mathbf{Z}_1 &= \mathbf{I}_a \otimes \mathbf{1}_b \otimes \mathbf{1}_n, & \mathbf{Z}_1\mathbf{Z}_1' &= \mathbf{I}_a \otimes \mathbf{J}_b \otimes \mathbf{J}_n,\\
\mathbf{Z}_2 &= \mathbf{1}_a \otimes \mathbf{I}_b \otimes \mathbf{1}_n, & \mathbf{Z}_2\mathbf{Z}_2' &= \mathbf{J}_a \otimes \mathbf{I}_b \otimes \mathbf{J}_n,\\
\mathbf{Z}_3 &= \mathbf{I}_a \otimes \mathbf{I}_b \otimes \mathbf{1}_n, & \mathbf{Z}_3\mathbf{Z}_3' &= \mathbf{I}_a \otimes \mathbf{I}_b \otimes \mathbf{J}_n
\end{aligned} \tag{53}$$

and

$$\mathbf{V} = \mathbf{Z}_1\mathbf{Z}_1'\sigma_\alpha^2 + \mathbf{Z}_2\mathbf{Z}_2'\sigma_\beta^2 + \mathbf{Z}_3\mathbf{Z}_3'\sigma_\gamma^2 + \mathbf{Z}_0\mathbf{Z}_0'\sigma_e^2 \,. \tag{54}$$

-i. Notation. The ML equations turn out to be quite complicated. Relative simplicity is achieved by relying on some substitutional notation. First, the familiar sums of squares:

$$\begin{aligned}
\text{SSA} &= \Sigma_i bn(\bar{y}_{i..} - \bar{y}_{...})^2,\\
\text{SSB} &= \Sigma_j an(\bar{y}_{.j.} - \bar{y}_{...})^2,\\
\text{SSAB} &= \Sigma_i\Sigma_j n(\bar{y}_{ij.} - \bar{y}_{i..} - \bar{y}_{.j.} + \bar{y}_{...})^2,\\
\text{SSE} &= \Sigma_i\Sigma_j\Sigma_k(y_{ijk} - \bar{y}_{ij.})^2 \,.
\end{aligned} \tag{55}$$

Next, for notational simplicity, define

$$\begin{aligned} \theta_0 &= \sigma_e^2, & \theta_{11} &= \sigma_e^2 + n\sigma_\gamma^2 + bn\sigma_\alpha^2, \\ \theta_1 &= \sigma_e^2 + n\sigma_\gamma^2, & \theta_{12} &= \sigma_e^2 + n\sigma_\gamma^2 + an\sigma_\beta^2 \end{aligned} \tag{56}$$

and

$$\theta_4 = \sigma_e^2 + n\sigma_\gamma^2 + bn\sigma_\alpha^2 + an\sigma_\beta^2 = \theta_{11} + \theta_{12} - \theta_1 . \tag{57}$$

-ii. ***Inverse of* V.** Using results of Henderson and Searle (1979), start with

$$\begin{aligned} \mathbf{V} = {} & \sigma_\alpha^2(\mathbf{J}_a^0 \otimes \mathbf{J}_b^1 \otimes \mathbf{J}_n^1) + \sigma_\beta^2(\mathbf{J}_a^1 \otimes \mathbf{J}_b^0 \otimes \mathbf{J}_n^1) + \sigma_\gamma^2(\mathbf{J}_a^0 \otimes \mathbf{J}_b^0 \otimes \mathbf{J}_n^1) \\ & + \sigma_e^2(\mathbf{J}_a^0 \otimes \mathbf{J}_b^0 \otimes \mathbf{J}_n^0) \end{aligned} \tag{58}$$

from (80) of Chapter 4. Then with $\boldsymbol{\theta}$ being $\boldsymbol{\theta}' = [\sigma_e^2 \quad \sigma_\gamma^2 \quad 0 \quad \sigma_\alpha^2 \quad 0 \quad \sigma_\beta^2 \quad 0 \quad 0]$, the $\boldsymbol{\lambda} = \mathbf{T}\boldsymbol{\theta}$ of equation (81) in Section 4.6e-ii is

$$\boldsymbol{\lambda} = \begin{bmatrix} 1 & 0 \\ 1 & a \end{bmatrix} \otimes \begin{bmatrix} 1 & 0 \\ 1 & b \end{bmatrix} \otimes \begin{bmatrix} 1 & 0 \\ 1 & n \end{bmatrix} \boldsymbol{\theta}$$

$$= \begin{bmatrix} 1 & \cdot & \cdot & \cdot & \cdot & \cdot & \cdot & \cdot \\ 1 & n & \cdot & \cdot & \cdot & \cdot & \cdot & \cdot \\ 1 & \cdot & b & \cdot & \cdot & \cdot & \cdot & \cdot \\ 1 & n & b & bn & \cdot & \cdot & \cdot & \cdot \\ 1 & \cdot & \cdot & \cdot & a & \cdot & \cdot & \cdot \\ 1 & n & \cdot & \cdot & a & an & \cdot & \cdot \\ 1 & \cdot & b & \cdot & a & \cdot & ab & \cdot \\ 1 & n & b & bn & a & an & ab & abn \end{bmatrix} \begin{bmatrix} \sigma_e^2 \\ \sigma_\gamma^2 \\ 0 \\ \sigma_\alpha^2 \\ 0 \\ \sigma_\beta^2 \\ 0 \\ 0 \end{bmatrix} = \begin{bmatrix} \sigma_e^2 \\ \sigma_e^2 + n\sigma_\gamma^2 \\ \sigma_e^2 \\ \sigma_e^2 + n\sigma_\gamma^2 + bn\sigma_\alpha^2 \\ \sigma_e^2 \\ \sigma_e^2 + n\sigma_\gamma^2 + an\sigma_\beta^2 \\ \sigma_e^2 \\ \sigma_e^2 + n\sigma_\gamma^2 + bn\sigma_\alpha^2 + an\sigma_\beta^2 \end{bmatrix} = \begin{bmatrix} \theta_0 \\ \theta_1 \\ \theta_0 \\ \theta_{11} \\ \theta_0 \\ \theta_{12} \\ \theta_0 \\ \theta_4 \end{bmatrix}.$$

Hence with $\mathbf{v}$ having elements that are reciprocals of those of $\boldsymbol{\lambda}$, the equation $\boldsymbol{\tau} = \mathbf{T}^{-1}\mathbf{v}$ of Section 4.6e-ii is

$$\boldsymbol{\tau} = \frac{1}{abn}\begin{bmatrix} a & 0 \\ -1 & 1 \end{bmatrix} \otimes \begin{bmatrix} b & 0 \\ -1 & 1 \end{bmatrix} \otimes \begin{bmatrix} n & 0 \\ -1 & 1 \end{bmatrix} \mathbf{v}$$

$$= \frac{1}{abn} \begin{bmatrix} abn & \cdot & \cdot & \cdot & \cdot & \cdot & \cdot & \cdot \\ -ab & ab & \cdot & \cdot & \cdot & \cdot & \cdot & \cdot \\ -an & \cdot & an & \cdot & \cdot & \cdot & \cdot & \cdot \\ a & -a & -a & a & \cdot & \cdot & \cdot & \cdot \\ -bn & \cdot & \cdot & \cdot & bn & \cdot & \cdot & \cdot \\ b & -b & \cdot & \cdot & -b & b & \cdot & \cdot \\ n & \cdot & -n & \cdot & -n & \cdot & n & \cdot \\ -1 & 1 & 1 & -1 & 1 & -1 & -1 & 1 \end{bmatrix} \begin{bmatrix} 1/\theta_0 \\ 1/\theta_1 \\ 1/\theta_0 \\ 1/\theta_{11} \\ 1/\theta_0 \\ 1/\theta_{12} \\ 1/\theta_0 \\ 1/\theta_4 \end{bmatrix} = \begin{bmatrix} \tau_{000} \\ \tau_{001} \\ 0 \\ \tau_{011} \\ 0 \\ \tau_{101} \\ 0 \\ \tau_{111} \end{bmatrix}. \tag{59}$$

Thus

$$\tau_{000} = \frac{1}{\theta_0}, \qquad \tau_{001} = \frac{1}{n}\left(\frac{1}{\theta_1} - \frac{1}{\theta_0}\right),$$

$$\tau_{011} = \frac{1}{bn}\left(\frac{1}{\theta_{11}} - \frac{1}{\theta_1}\right), \quad \tau_{101} = \frac{1}{an}\left(\frac{1}{\theta_{12}} - \frac{1}{\theta_1}\right), \tag{60}$$

$$\tau_{111} = \frac{1}{abn}\left(\frac{1}{\theta_1} - \frac{1}{\theta_{11}} - \frac{1}{\theta_{12}} + \frac{1}{\theta_4}\right).$$

Therefore (86) of Chapter 4 gives

$$\mathbf{V}^{-1} = \tau_{000}(\mathbf{I}_a \otimes \mathbf{I}_b \otimes \mathbf{I}_n) + \tau_{001}(\mathbf{I}_a \otimes \mathbf{I}_b \otimes \mathbf{J}_n)$$
$$+ \tau_{011}(\mathbf{I}_a \otimes \mathbf{J}_b \otimes \mathbf{J}_n) + \tau_{101}(\mathbf{J}_a \otimes \mathbf{I}_b \otimes \mathbf{J}_n) + \tau_{111}(\mathbf{J}_a \otimes \mathbf{J}_b \otimes \mathbf{J}_n),$$

and on replacing τs by θs, as in (60), this reduces to

$$\mathbf{V}^{-1} = \theta_0^{-1}(\mathbf{I}_a \otimes \mathbf{I}_b \otimes \mathbf{C}_n) + \theta_1^{-1}(\mathbf{C}_a \otimes \mathbf{C}_b \otimes \bar{\mathbf{J}}_n) + \theta_{11}^{-1}(\mathbf{C}_a \otimes \bar{\mathbf{J}}_b \otimes \bar{\mathbf{J}}_n)$$
$$+ \theta_{12}^{-1}(\bar{\mathbf{J}}_a \otimes \mathbf{C}_b \otimes \bar{\mathbf{J}}_n) + \theta_4^{-1}(\bar{\mathbf{J}}_a \otimes \bar{\mathbf{J}}_b \otimes \bar{\mathbf{J}}_n), \tag{61}$$

where, for example, $\mathbf{C}_a$ is the centering matrix $\mathbf{C}_a = \mathbf{I} - \bar{\mathbf{J}}_a$. This form for $\mathbf{V}^{-1}$ makes multiplication with $\mathbf{Z}_i\mathbf{Z}_i'$ very easy because any time that a $\mathbf{J}$ or $\bar{\mathbf{J}}$ multiplies a $\mathbf{C}$ in a Kronecker product, the resulting Kronecker product is null.

-iii. The estimation equations. The form of the general estimation equations that we use is

$$\{_c \operatorname{tr}(\mathbf{V}^{-1}\mathbf{Z}_i\mathbf{Z}_i')\}_{i=0}^{r} = \{_c \mathbf{y}'\mathbf{P}\mathbf{Z}_i\mathbf{Z}_i'P\mathbf{y}\}_{i=0}^{r} = \{_c \operatorname{sesq}(\mathbf{Z}_i'\mathbf{P}\mathbf{y})\}_{i=0}^{r}. \tag{62}$$

For simplifying the left-hand side of (62) it is useful to note that

$$\operatorname{tr}(\mathbf{A} \otimes \mathbf{B}) = \operatorname{tr}(\mathbf{A})\operatorname{tr}(\mathbf{B}), \quad \operatorname{tr}(\bar{\mathbf{J}}_a) = 1 \quad \text{and} \quad \operatorname{tr}(\mathbf{C}_a) = a - 1\,. \tag{63}$$

Simplification of equations (62) is demonstrated for just some terms. Others are left for the reader.

First, the left-hand side of (62) for $i = 0$, for which $\mathbf{Z}_i = \mathbf{Z}_0 = \mathbf{I}$, is

$$\operatorname{tr}(\mathbf{V}^{-1}\mathbf{Z}_0\mathbf{Z}_0') = \operatorname{tr}(\mathbf{V}^{-1})$$
$$= \frac{ab(n-1)}{\theta_0} + \frac{(a-1)(b-1)}{\theta_1} + \frac{a-1}{\theta_{11}} + \frac{b-1}{\theta_{12}} + \frac{1}{\theta_4}. \tag{64}$$

And for $i = 1$ it is

$$\operatorname{tr}(\mathbf{V}^{-1}\mathbf{Z}_1\mathbf{Z}_1') = \operatorname{tr}[\mathbf{V}^{-1}(\mathbf{I}_a \otimes \mathbf{J}_b \otimes \mathbf{J}_n)],$$

which, because of the property, $\mathbf{C}_a\mathbf{J}_a = \mathbf{0}$, and using $\bar{\mathbf{J}}_b\mathbf{J}_b = \mathbf{J}_b$, is

$$\operatorname{tr}(\mathbf{V}^{-1}\mathbf{Z}_1\mathbf{Z}_1') = \theta_{11}^{-1}\operatorname{tr}(\mathbf{C}_a \otimes \mathbf{J}_b \otimes \mathbf{J}_n) + \theta_4^{-1}(\bar{\mathbf{J}}_a \otimes \mathbf{J}_b \otimes \mathbf{J}_n)$$
$$= \frac{bn(a-1)}{\theta_{11}} + \frac{bn}{\theta_4}. \tag{65}$$

The right-hand side of (62) requires

$$\mathbf{P} = \mathbf{V}^{-1} - \mathbf{V}^{-1}\mathbf{X}(\mathbf{X}'\mathbf{V}^{-1}\mathbf{X})^{-}\mathbf{X}'\mathbf{V}^{-1}, \quad \text{where } \mathbf{X} = \mathbf{1}_{abn} = \mathbf{1}_a \otimes \mathbf{1}_b \otimes \mathbf{1}_n \,.$$

Then with $\mathbf{V}^{-1}$ of (61), $\mathbf{X}'\mathbf{V}^{-1} = \mathbf{1}'\mathbf{V}^{-1}$, which, after using the property, $\mathbf{1}'_a\mathbf{C}_a = \mathbf{0}$, is

$$\mathbf{X}'\mathbf{V}^{-1} = \theta_4^{-1}\mathbf{1}'_{abn}(\bar{\mathbf{J}}_a \otimes \bar{\mathbf{J}}_b \otimes \bar{\mathbf{J}}_n) = \theta_4^{-1}\mathbf{1}'_{abn} \,.$$

Hence

$$\begin{aligned}
\mathbf{P} &= \mathbf{V}^{-1} - \theta_4^{-1}\mathbf{1}_{abn}(\theta_4^{-1}\mathbf{1}'_{abn}\mathbf{1}_{abn})^{-}\theta_4^{-1}\mathbf{1}'_{abn} \\
&= \mathbf{V}^{-1} - \theta_4^{-1}\bar{\mathbf{J}}_{abn} = \mathbf{V}^{-1} - \theta_4^{-1}(\bar{\mathbf{J}}_a \otimes \bar{\mathbf{J}}_b \otimes \bar{\mathbf{J}}_n) \\
&= \theta_0^{-1}(\mathbf{I}_a \otimes \mathbf{I}_b \otimes \mathbf{C}_n) + \theta_1^{-1}(\mathbf{C}_a \otimes \mathbf{C}_b \otimes \bar{\mathbf{J}}_n) \\
&\quad + \theta_{11}^{-1}(\mathbf{C}_a \otimes \bar{\mathbf{J}}_b \otimes \bar{\mathbf{J}}_n) + \theta_{12}^{-1}(\bar{\mathbf{J}}_a \otimes \mathbf{C}_b \otimes \bar{\mathbf{J}}_n) \,.
\end{aligned} \tag{66}$$

Therefore for $i = 1$ the right-hand side of (62) is

$$\begin{aligned}
\mathbf{y}'\mathbf{P}\mathbf{Z}_1\mathbf{Z}'_1\mathbf{P}\mathbf{y} &= \text{sesq}(\mathbf{Z}'_i\mathbf{P}\mathbf{y}) \\
&= \text{sesq}[(\mathbf{I}_a \otimes \mathbf{1}'_b \otimes \mathbf{1}'_n)\mathbf{P}\mathbf{y}]; \quad \text{and using (66) gives} \qquad (67) \\
&= \text{sesq}[\theta_{11}^{-1}(\mathbf{C}_a \otimes \mathbf{1}'_b \otimes \mathbf{1}'_n)\mathbf{y}] = \theta_{11}^{-2}\,\text{sesq}\{[(\mathbf{I}_a - \bar{\mathbf{J}}_a) \otimes \mathbf{1}'_{bn}]\mathbf{y}\} \\
&= \theta_{11}^{-2}\,\text{sesq}[(\{_d\,\mathbf{1}'_{bn}\}_{i=1}^{a} - \tfrac{1}{a}\mathbf{J}_{a \times abn})\mathbf{y}] \\
&= \theta_{11}^{-2}\,\text{sesq}(\{_c\, y_{i..} - bn\bar{y}_{...}\}_{i=1}^{a}) \\
&= \frac{bn\text{SSA}}{\theta_{11}^2},
\end{aligned} \tag{68}$$

Hence on equating (65) and (68),

$$\frac{1}{\theta_4} + \frac{a-1}{\theta_{11}} = \frac{\text{SSA}}{\theta_{11}^2},$$

and this is equation (62) for $i = 1$, as in the second equation of (90) in Chapter 4.

As another illustration of the right-hand side of (62), consider its value for $i = 3$, for which $\mathbf{Z}_i = \mathbf{Z}_3 = \mathbf{1}_a \otimes \mathbf{I}_b \otimes \mathbf{1}_n$, as in (53):

$$\begin{aligned}
\mathbf{y}'\mathbf{P}\mathbf{Z}_3\mathbf{Z}'_3\mathbf{P}\mathbf{y} &= \text{sesq}(\mathbf{Z}'_3\mathbf{P}\mathbf{y}) \\
&= \text{sesq}[(\mathbf{I}_a \otimes \mathbf{I}_b \otimes \mathbf{1}'_n)\mathbf{P}\mathbf{y}] \\
&= \text{sesq}[\theta_1^{-1}(\mathbf{C}_a \otimes \mathbf{C}_b \otimes \mathbf{1}'_n)\mathbf{y} + \theta_{11}^{-1}(\mathbf{C}_a \otimes \bar{\mathbf{J}}_b \otimes \mathbf{1}'_n)\mathbf{y} \\
&\quad + \theta_{12}^{-1}(\bar{\mathbf{J}}_a \otimes \mathbf{C}_b \otimes \mathbf{1}'_n)\mathbf{y}] \,.
\end{aligned} \tag{69}$$

In general, note that when $\mathbf{u}'\mathbf{v} = 0$

$$\text{sesq}(\mathbf{u} + \mathbf{v}) = \text{sesq}(\mathbf{u}) + \text{sesq}(\mathbf{v}) + 2\mathbf{u}'\mathbf{v} = \text{sesq}(\mathbf{u}) + \text{sesq}(\mathbf{v}) \,. \tag{70}$$

And for the terms in (69) the inner product of any one with another is zero—because of the occurrence of products such as $\mathbf{C}_b\bar{\mathbf{J}}_b = \mathbf{0}$ in the resulting products of Kronecker products. Therefore (69) is

$$\mathbf{y}'\mathbf{P}\mathbf{Z}_3\mathbf{Z}_3'\mathbf{P}\mathbf{y} = \theta_{12}^{-2}\,\text{sesq}[(\mathbf{C}_a \otimes \mathbf{C}_b \otimes \mathbf{1}_n')\mathbf{y}] + \theta_{11}^{-2}\,\text{sesq}[(\mathbf{C}_a \otimes \bar{\mathbf{J}}_b \otimes \mathbf{1}_n')\mathbf{y}] + \theta_{12}^{-2}\,\text{sesq}[(\bar{\mathbf{J}}_a \otimes \mathbf{C}_b \otimes \mathbf{1}_n')\mathbf{y}]\,. \tag{71}$$

We deal with each term separately.

$$\begin{aligned}\theta_1^{-2}\,&\text{sesq}[(\mathbf{C}_a \otimes \mathbf{C}_b \otimes \mathbf{1}_n')\mathbf{y}]\\ &= \theta_1^{-2}\,\text{sesq}\{[(\mathbf{I}_{ab} - \mathbf{I}_a \otimes \bar{\mathbf{J}}_b - \bar{\mathbf{J}}_a \otimes \mathbf{I}_b + \bar{\mathbf{J}}_{ab}) \otimes \mathbf{1}_n']\mathbf{y}\}\\ &= \theta_1^{-2}\,\text{sesq}[(\{_{\text{d}}\,\mathbf{1}_n'\}_{i=1,j=1}^{a\quad b} - \{_{\text{d}}\,\mathbf{J}_{b\times bn}/b\}_{i=1}^{a}\\ &\quad - \{_{\text{m}}\,\{_{\text{d}}\,\mathbf{1}_n'/a\}_{j=1}^{b}\}_{i,i'=1}^{a} + \mathbf{J}_{ab\times abn}/ab)\mathbf{y}]\\ &= \theta_1^{-2}\,\text{sesq}[\{_{\text{c}}\,y_{ij\cdot} - n\bar{y}_{i\cdot\cdot} - n\bar{y}_{\cdot j\cdot} + n\bar{y}_{\cdots}\}_{i=1,j=1}^{a\quad b}] = \frac{n\text{SSAB}}{\theta_1^2}\,,\end{aligned} \tag{72}$$

$$\begin{aligned}\theta_{11}^{-2}\,&\text{sesq}[(\mathbf{C}_a \otimes \bar{\mathbf{J}}_b \otimes \mathbf{1}_n')\mathbf{y}]\\ &= \theta_{11}^{-2}\,\text{sesq}[(\mathbf{I}_a \otimes \mathbf{1}_b/b \otimes 1)(\mathbf{C}_a \otimes \mathbf{1}_b' \otimes \mathbf{1}_n')\mathbf{y}]\\ &= \theta_{11}^{-2}b^{-2}\,\text{sesq}[\{_{\text{d}}\,\mathbf{1}_b\}_{i=1}^{a}\{_{\text{c}}\,y_{i\cdot\cdot} - bn\bar{y}_{\cdots}\}] = \frac{n\text{SSA}}{\theta_{11}^2}\,.\end{aligned} \tag{73}$$

Similarly,

$$\theta_{12}^{-2}\,\text{sesq}[(\bar{\mathbf{J}}_a \otimes \mathbf{C}_b \otimes \mathbf{1}_n')\mathbf{y}] = \frac{b\text{SSB}}{\theta_{12}^2}\,. \tag{74}$$

Therefore substituting (72)–(74) into (71) gives

$$\mathbf{y}'\mathbf{P}\mathbf{Z}_3\mathbf{Z}_3'\mathbf{P}\mathbf{y} = \frac{n\text{SSAB}}{\theta_1^2} + \frac{n\text{SSA}}{\theta_{11}^2} + \frac{n\text{SSB}}{\theta_{12}^2}\,. \tag{75}$$

Simplifying the remaining terms of (62), namely the left-hand side for $i = 2$ and 3, and the right-hand side for $i = 0$ and 2, is left to the reader (E 6.8).

-iv. ***Information matrix.*** Even though no closed form exists for the variance component estimators, their information matrix can be obtained. From (38), a typical element of the information matrix is

$$\text{tr}(\mathbf{V}^{-1}\mathbf{Z}_i\mathbf{Z}_i'\mathbf{V}^{-1}\mathbf{Z}_j\mathbf{Z}_j') = \text{sesq}(\mathbf{Z}_i'\mathbf{V}^{-1}\mathbf{Z}_j)\,. \tag{76}$$

This is evaluated using general results such as

$$\text{sesq}(\mathbf{C}_a) = a\left(1 - \frac{1}{a}\right)^2 + (a^2 - a)\frac{1}{a^2} = a - 1,$$

$$\text{sesq}(\mathbf{A} + \mathbf{B}) = \text{sesq}(\mathbf{A}) + \text{sesq}(\mathbf{B}) \quad \text{when } \mathbf{AB} = \mathbf{0},$$

and

$$\text{sesq}(\mathbf{A} \otimes \mathbf{B}) = \text{sesq}\{_{\text{m}}\, a_{ij}\mathbf{B}\}_{i,j} = \Sigma_i \Sigma_j a_{ij}^2 \,\text{sesq}(\mathbf{B}) = \text{sesq}(\mathbf{A})\text{sesq}(\mathbf{B})\,.$$

We derive four cases of (76) for (38) and leave the other six for the reader (E 6.9). These terms are derived using (61).

$$\begin{aligned}\text{sesq}(\mathbf{Z}_0'\mathbf{V}^{-1}\mathbf{Z}_0) &= \text{sesq}(\mathbf{V}^{-1})\\ &= \frac{ab(n-1)}{\theta_0^2} + \frac{(a-1)(b-1)}{\theta_1^2} + \frac{a-1}{\theta_{11}^2} + \frac{b-1}{\theta_{12}^2}\\ &\quad + \frac{1}{\theta_4^2} = t_{ee}, \quad \text{say}\,.\end{aligned}$$

$$\begin{aligned}\text{sesq}(\mathbf{Z}_0'\mathbf{V}^{-1}\mathbf{Z}_1) &= \text{sesq}[\mathbf{V}^{-1}(\mathbf{I}_a \otimes \mathbf{1}_b \otimes \mathbf{1}_n)]\\ &= \text{sesq}[\theta_{11}^{-1}(\mathbf{C}_a \otimes \mathbf{1}_b \otimes \mathbf{1}_n) + \theta_4^{-1}(\bar{\mathbf{J}}_a \otimes \mathbf{1}_b \otimes \mathbf{1}_n)]\\ &= \frac{(a-1)bn}{\theta_{11}^2} + \frac{bn}{\theta_4^2} = \frac{t_{\alpha\alpha}}{bn}, \quad \text{say}\,.\end{aligned}$$

$$\begin{aligned}\text{sesq}(\mathbf{Z}_0'\mathbf{V}^{-1}\mathbf{Z}_2) &= \text{sesq}[\mathbf{V}^{-1}(\mathbf{1}_b \otimes \mathbf{I}_b \otimes \mathbf{1}_n)]\\ &= \text{sesq}[\theta_{12}^{-1}(\mathbf{1}_a \otimes \mathbf{C}_b \otimes \mathbf{1}_n) + \theta_4^{-1}(\mathbf{1}_a \otimes \bar{\mathbf{J}}_b \otimes \mathbf{1}_n)]\\ &= \frac{a(b-1)n}{\theta_{12}^2} + \frac{an}{\theta_4^2} = \frac{t_{\beta\beta}}{an}, \quad \text{say}\,.\end{aligned}$$

$$\begin{aligned}\text{sesq}(\mathbf{Z}_0'\mathbf{V}^{-1}\mathbf{Z}_3) &= \text{sesq}[\mathbf{V}^{-1}(\mathbf{I}_a \otimes \mathbf{I}_b \otimes \mathbf{1}_n)]\\ &= \text{sesq}[\theta_1^{-1}(\mathbf{C}_a \otimes \mathbf{C}_b \otimes \mathbf{1}_n) + \theta_{11}^{-1}(\mathbf{C}_a \otimes \bar{\mathbf{J}}_b \otimes \mathbf{1}_n)\\ &\quad + \theta_{12}^{-1}(\bar{\mathbf{J}}_a \otimes \mathbf{C}_b \otimes \mathbf{1}_n) + \theta_4^{-1}(\bar{\mathbf{J}}_a \otimes \bar{\mathbf{J}}_b \otimes \mathbf{1}_n)]\\ &= \frac{(a-1)(b-1)n}{\theta_1^2} + \frac{(a-1)n}{\theta_{11}^2} + \frac{(b-1)n}{\theta_{12}^2} + \frac{n}{\theta_4^2}\\ &= \frac{t_{\gamma\gamma}}{n}, \quad \text{say}\,.\end{aligned}$$

The information matrix then turns out to be

$$\mathbf{I}\begin{bmatrix}\tilde{\sigma}_e^2\\ \tilde{\sigma}_\alpha^2\\ \tilde{\sigma}_\beta^2\\ \tilde{\sigma}_\gamma^2\end{bmatrix} = \frac{1}{2}\begin{bmatrix} t_{ee} & t_{\alpha\alpha}/bn & t_{\beta\beta}/an & t_{\gamma\gamma}/n\\ & t_{\alpha\alpha} & abn^2/\theta_4^2 & t_{\alpha\alpha}/b\\ \text{symmetric} & & t_{\beta\beta} & t_{\beta\beta}/a\\ & & & t_{\gamma\gamma}\end{bmatrix} \tag{77}$$

for

$$t_{ee} = \frac{ab(n-1)}{\theta_0^2} + \frac{t_{\gamma\gamma}}{n^2}, \quad t_{\alpha\alpha} = b^2n^2\left(\frac{a-1}{\theta_{11}^2} + \frac{1}{\theta_4^2}\right),$$

$$t_{\beta\beta} = a^2n^2\left(\frac{b-1}{\theta_{12}^2} + \frac{1}{\theta_4^2}\right) \text{ and } t_{\gamma\gamma} = n^2\left[\frac{(a-1)(b-1)}{\theta_1^2} + \frac{a-1}{\theta_{11}^2} + \frac{b-1}{\theta_{12}^2} + \frac{1}{\theta_4^2}\right]. \tag{78}$$

And the dispersion matrix $\text{var}(\boldsymbol{\sigma}^2)$ is the inverse of (77), as in (94) of Chapter 4.

b. 2-way crossed, random model, no interaction

The no-interaction model is easily derived from the with-interaction model by putting $\boldsymbol{\gamma} = \mathbf{0}$ and $\mathbf{Z}_3 = \mathbf{0}$, and adapting the θs of (56) and (57) as follows:

$$\theta_1 = \theta_0 = \sigma_e^2, \quad \theta_{11} = \sigma_e^2 + bn\sigma_\alpha^2, \quad \theta_{12} = \sigma_e^2 + an\sigma_\beta^2$$

and (79)

$$\theta_4 = \sigma_e^2 + bn\sigma_\alpha^2 + an\sigma_\beta^2 = \theta_{11} + \theta_{12} - \theta_0 .$$

This reduces the estimation equations (90) of Chapter 4, for the with-interaction model, to be (92) of Chapter 4, for the no-interaction model.

The information matrix will be that for the with-interaction case, with its last row and column deleted and with the θs defined as in (79): with

$$\varphi = \frac{abn - a - b + 1}{\theta_0^2} + \frac{a-1}{\theta_{11}^2} + \frac{b-1}{\theta_{12}^2} + \frac{1}{\theta_4^2},$$

$$\mathbf{I}\begin{bmatrix} \tilde{\sigma}_e^2 \\ \tilde{\sigma}_\alpha^2 \\ \tilde{\sigma}_\beta^2 \end{bmatrix} = \begin{bmatrix} \varphi & bn\left(\frac{a-1}{\theta_{11}^2} + \frac{1}{\theta_4^2}\right) & an\left(\frac{b-1}{\theta_{12}^2} + \frac{1}{\theta_4^2}\right) \\ & b^2n^2\left(\frac{a-1}{\theta_{11}^2} + \frac{1}{\theta_4^2}\right) & \frac{abn^2}{\theta_4^2} \\ \text{symmetric} & & a^2n^2\left(\frac{b-1}{\theta_{12}^2} + \frac{1}{\theta_4^2}\right) \end{bmatrix}. \tag{80}$$

This leads to (96) of Chapter 4.

6.6. RESTRICTED MAXIMUM LIKELIHOOD (REML)

A property of ML estimation is that in estimating variance components it takes no account of the degrees of freedom that are involved in estimating fixed effects. For example, when data are a simple random sample, $x_1, \ldots, x_n$, identically and independently distributed $\mathcal{N}(\mu, \sigma^2)$, the unbiased ANOVA estimator of σ^2 is $\hat{\sigma}^2 = \Sigma_i(x_i - \bar{x})^2/(n-1)$; but the MLE is $\tilde{\sigma}^2 = \Sigma_i(x_i - \bar{x})^2/n$. Likewise in the 1-way classification random model (e.g., Table 4.9), the ML solution for $\dot{\sigma}_\alpha^2$ is $\dot{\sigma}_\alpha^2 = (\text{SSA}/a - \text{MSE})/n$, wherein we might expect the denominator a in SSA/a to be $a - 1$ as it is in the ANOVA estimator

$\hat{\sigma}_\alpha^2 = (\text{MSA} - \text{MSE})/n$. Thus $E(\dot{\sigma}_\alpha^2) = (1 - 1/a)\sigma_\alpha^2 - \sigma_e^2/an$ and so $\dot{\sigma}_\alpha^2$ is biased; whereas $E(\hat{\sigma}_\alpha^2) = \sigma_\alpha^2$ and so $\hat{\sigma}_\alpha^2$ is unbiased. Thus, although ANOVA estimators have the attractive property under normality of being minimum variance unbiased, ML estimators do not. (In particular, they are not even unbiased.) Nor do ML solutions, if used as estimators. Even for balanced data, neither ML estimators nor ML solutions are the same as ANOVA estimators. Thus the minimum variance property is not applicable to ML estimation; we return to this property in Chapter 11.

The feature of ML not taking account of the degrees of freedom used for estimating fixed effects when estimating variance components is overcome by what has come to be known as restricted (or, more usually in Europe, residual) maximum likelihood (REML) estimation. First developed for certain balanced data situations by Anderson and Bancroft (1952) and Russell and Bradley (1958), it was extended by W.A. Thompson (1962) to balanced data in general and by Patterson and R. Thompson (1971, 1974) to mixed models generally. It has received all manner of descriptions in the literature, ranging from consideration of negative estimates to "maximizing that part of the likelihood which is invariant to the fixed effects" [e.g., Thompson (1962); and also Harville (1977, p. 325), who additionally suggests it is a method that is marginally sufficient for $\boldsymbol{\sigma}^2$ "in the sense described by Sprott (1975)"]. Whatever description is preferred, a basic idea of restricted maximum likelihood (REML) estimation is that of estimating variance components based on residuals calculated after fitting by ordinary least squares just the fixed effects part of the model. REML estimation can also be viewed as maximizing a marginal likelihood—as described in Section 9.3d.

a. Linear combinations of observations

Rather than using $\mathbf{y}$ (the data vector) directly, REML is based on linear combinations of elements of $\mathbf{y}$, chosen in such a way that those combinations do not contain any fixed effects, no matter what their value. These linear combinations turn out to be equivalent to residuals obtained after fitting the fixed effects. This results from starting with a set of values $\mathbf{k}'\mathbf{y}$ where vectors $\mathbf{k}'$ are chosen so that $\mathbf{k}'\mathbf{y} = \mathbf{k}'\mathbf{X}\boldsymbol{\beta} + \mathbf{k}'\mathbf{Z}\mathbf{u}$ contains no term in $\boldsymbol{\beta}$, i.e., so that

$$\mathbf{k}'\mathbf{X}\boldsymbol{\beta} = \mathbf{0} \quad \forall\ \boldsymbol{\beta}\,. \tag{81}$$

Hence

$$\mathbf{k}'\mathbf{X} = \mathbf{0}\,. \tag{82}$$

Therefore, from Appendix M.4e, the form of $\mathbf{k}$ must be $\mathbf{k}' = \mathbf{c}'(\mathbf{I} - \mathbf{X}\mathbf{X}^-)$ or

$$\mathbf{k}' = \mathbf{c}'[\mathbf{I} - \mathbf{X}(\mathbf{X}'\mathbf{X})^-\mathbf{X}'] = \mathbf{c}'(\mathbf{I} - \mathbf{X}\mathbf{X}^+) = \mathbf{c}'\mathbf{M} \tag{83}$$

for any $\mathbf{c}'$ and where $\mathbf{M}$ is defined as

$$\mathbf{M} = \mathbf{I} - \mathbf{X}(\mathbf{X}'\mathbf{X})^-\mathbf{X}' = \mathbf{I} - \mathbf{X}\mathbf{X}^+\,. \tag{84}$$

Harville (1977) refers to $\mathbf{k}'\mathbf{y}$ for $\mathbf{k}'$ of this nature as being an "error contrast": its expected value is zero:

$$E(\mathbf{k}'\mathbf{y}) = \mathbf{k}'\mathbf{X}\boldsymbol{\beta} = \mathbf{0}\,.$$

The number of linearly independent error contrasts depends on $\mathbf{X}$: for $\mathbf{X}$ of order $N \times p$ and rank r equation (81) is satisfied by only $N - r$ linearly independent values of $\mathbf{k}'$. Thus, in using a set of linearly independent vectors $\mathbf{k}'$ as rows of $\mathbf{K}'$ we confine attention to

$$\mathbf{K}'\mathbf{y} \quad \text{for } \mathbf{K}' = \mathbf{TM}, \tag{85}$$

where $\mathbf{K}'$ and $\mathbf{T}$ have full row rank $N - r$. (There is clearly no point in having more than $N - r$ vectors $\mathbf{k}'$ because some of them will then be linear combinations of others, as will the corresponding values $\mathbf{k}'\mathbf{y}$.)

b. The REML equations

With $\mathbf{y} \sim \mathcal{N}(\mathbf{X}\boldsymbol{\beta}, \mathbf{V})$ we have, for $\mathbf{K}'\mathbf{X} = \mathbf{0}$,

$$\mathbf{K}'\mathbf{y} \sim \mathcal{N}(\mathbf{0}, \mathbf{K}'\mathbf{V}\mathbf{K})\,.$$

The REML equations can therefore be derived from the ML equations of (25), namely

$$\{_c\, \mathrm{tr}(\dot{\mathbf{V}}^{-1}\mathbf{Z}_i\mathbf{Z}_i')\}_{i=0}^{r} = \{_c\, \mathbf{y}'\dot{\mathbf{P}}\mathbf{Z}_i\mathbf{Z}_i'\dot{\mathbf{P}}\mathbf{y}\}_{i=0}^{r}, \tag{86}$$

by making suitable replacements:

$$\text{replace} \quad \begin{matrix} \mathbf{y} \text{ by } \mathbf{K}'\mathbf{y} \\ \mathbf{X} \text{ by } \mathbf{K}'\mathbf{X} = \mathbf{0} \end{matrix} \quad \text{and} \quad \begin{matrix} \mathbf{Z} \text{ by } \mathbf{K}'\mathbf{Z} \\ \mathbf{V} \text{ by } \mathbf{K}'\mathbf{V}\mathbf{K}\,. \end{matrix}$$

Then (86) becomes

$$\{_c\, \mathrm{tr}[(\mathbf{K}'\dot{\mathbf{V}}\mathbf{K})^{-1}\mathbf{K}'\mathbf{Z}_i\mathbf{Z}_i'\mathbf{K}]\}_{i=0}^{r} = \{_c\, \mathbf{y}'\mathbf{K}(\mathbf{K}'\dot{\mathbf{V}}\mathbf{K})^{-1}\mathbf{K}'\mathbf{Z}_i\mathbf{Z}_i'\mathbf{K}(\mathbf{K}'\dot{\mathbf{V}}\mathbf{K})^{-1}\mathbf{K}'\mathbf{y}\}_{i=0}^{r}\,. \tag{87}$$

With

$$\mathbf{P} = \mathbf{V}^{-1} - \mathbf{V}^{-1}\mathbf{X}(\mathbf{X}'\mathbf{V}^{-1}\mathbf{X})^{-}\mathbf{X}'\mathbf{V}^{-1} = \mathbf{K}(\mathbf{K}'\mathbf{V}\mathbf{K})^{-1}\mathbf{K}' \tag{88}$$

from Appendix M.4f, (87) reduces to

$$\{_c\, \mathrm{tr}(\dot{\mathbf{P}}\mathbf{Z}_i\mathbf{Z}_i')\}_{i=0}^{r} = \{_c\, \mathbf{y}'\dot{\mathbf{P}}\mathbf{Z}_i\mathbf{Z}_i'\dot{\mathbf{P}}\mathbf{y}\}_{i=0}^{r}\,. \tag{89}$$

These are the REML equations. Comparison with the ML equations of (86) reveals that they have the same right-hand side as the ML equations: and the left-hand sides are the same except that the $\mathbf{P}$ in REML replaces $\mathbf{V}^{-1}$ of ML.

c. An alternative form

Through direct multiplication, it is easily established that $\mathbf{PVP} = \mathbf{P}$. Hence in the left-hand side of (89) we can use the identity

$$\mathrm{tr}(\mathbf{P}\mathbf{Z}_i\mathbf{Z}_i') = \mathrm{tr}(\mathbf{PVP}\mathbf{Z}_i\mathbf{Z}_i') = \mathrm{tr}(\mathbf{P}\mathbf{Z}_i\mathbf{Z}_i'\mathbf{P}\sum_{j=0}^{r}\mathbf{Z}_j\mathbf{Z}_j'\sigma_j^2) = \sum_{j=0}^{r}\mathrm{tr}(\mathbf{P}\mathbf{Z}_i\mathbf{Z}_i'\mathbf{P}\mathbf{Z}_j\mathbf{Z}_j')\sigma_j^2\,.$$

Thus, similar to (27), the REML equations can be put in the form

$$\{_{\rm m}\,\mathrm{tr}(\mathbf{Z}_i'\dot{\mathbf{P}}\mathbf{Z}_j\mathbf{Z}_j'\dot{\mathbf{P}}\mathbf{Z}_i)\}_{i,j=0}^{r}\dot{\boldsymbol{\sigma}}^2 = \{_{\rm c}\,\mathbf{y}'\dot{\mathbf{P}}\mathbf{Z}_i\mathbf{Z}_i'\dot{\mathbf{P}}\mathbf{y}\}_{i=0}^{r}\,. \tag{90}$$

Whatever form of the REML equations are used, the comments made in Section 6.5 about computing iterative solutions of the ML equations apply equally as well to those REML equations, (89) or (90). For a particular class of models often appropriate to dairy breeding data, Smith and Graser (1986) describe some computational simplifications for calculating REML estimates. This is extended by Graser and Smith (1987) to avoid matrix inversion, using instead a one-dimensional search involving just the variance part of the log likelihood. A suggestion from Giesbrecht and Burns (1985) is to use only two iterations of the REML equations.

d. Invariance to choice of error contrasts

It is clear from (88)–(90) that the REML equations (90) do not contain **K**. It occurs only through its relationship to **P** in (88), although **P**, as defined in (22), does not involve **K**. Therefore the REML equations are invariant to whatever set of error contrasts are chosen as $\mathbf{K}'\mathbf{y}$ so long as $\mathbf{K}'$ is of full row rank $N - r_{\mathbf{X}}$ with $\mathbf{K}'\mathbf{X} = \mathbf{0}$. We can also observe this directly, from the likelihood of $\mathbf{K}'\mathbf{y}$ (see E 6.11).

e. The information matrix

With $L_{\rm R}$ being the likelihood function of $\mathbf{K}'\mathbf{y}$ define

$$l_{\rm R} = \log L_{\rm R} = -\tfrac{1}{2}(N-r)\log 2\pi - \tfrac{1}{2}\log|\mathbf{K}'\mathbf{V}\mathbf{K}| - \tfrac{1}{2}\mathbf{y}'\mathbf{K}(\mathbf{K}'\mathbf{V}\mathbf{K})^{-1}\mathbf{K}'\mathbf{y}\,.$$

Then, using Appendix M.4f in the form

$$\begin{aligned}\frac{\partial\mathbf{P}}{\partial\sigma_i^2} &= \frac{\partial}{\partial\sigma_i^2}\mathbf{K}(\mathbf{K}'\mathbf{V}\mathbf{K})^{-1}\mathbf{K}'\\ &= -\mathbf{K}(\mathbf{K}'\mathbf{V}\mathbf{K})^{-1}\frac{\partial\mathbf{K}\mathbf{V}\mathbf{K}'}{\partial\sigma_i^2}(\mathbf{K}'\mathbf{V}\mathbf{K})^{-1}\mathbf{K}'\\ &= -\mathbf{K}(\mathbf{K}'\mathbf{V}\mathbf{K})^{-1}\mathbf{K}'\frac{\partial\mathbf{V}}{\partial\sigma_i^2}\mathbf{K}(\mathbf{K}'\mathbf{V}\mathbf{K})^{-1}\mathbf{K}'\\ &= -\mathbf{P}\frac{\partial\mathbf{V}}{\partial\sigma_i^2}\mathbf{P} = -\mathbf{P}\mathbf{Z}_i\mathbf{Z}_i'\mathbf{P},\end{aligned}\tag{91}$$

$$\begin{aligned}\frac{\partial l_{\rm R}}{\partial\sigma_i^2} &= -\tfrac{1}{2}\,\mathrm{tr}[(\mathbf{K}'\mathbf{V}\mathbf{K})^{-1}\mathbf{K}'\mathbf{Z}_i\mathbf{Z}_i'\mathbf{K}] - \tfrac{1}{2}\mathbf{y}'(-1)\mathbf{P}\mathbf{Z}_i\mathbf{Z}_i'\mathbf{P}\mathbf{y}\\ &= -\tfrac{1}{2}\,\mathrm{tr}(\mathbf{P}\mathbf{Z}_i\mathbf{Z}_i') + \tfrac{1}{2}\mathbf{y}'\mathbf{P}\mathbf{Z}_i\mathbf{Z}_i'\mathbf{P}\mathbf{y}\,.\end{aligned}\tag{92}$$

For the information matrix we need second derivatives of $l_{\rm R}$:

$$\begin{aligned}\frac{\partial^2 l_{\rm R}}{\partial\sigma_i^2\,\partial\sigma_j^2} &= \tfrac{1}{2}\,\mathrm{tr}(\mathbf{P}\mathbf{Z}_j\mathbf{Z}_j'\mathbf{P}\mathbf{Z}_i\mathbf{Z}_i') - \tfrac{1}{2}\mathbf{y}'\mathbf{P}\mathbf{Z}_j\mathbf{Z}_j'\mathbf{P}\mathbf{Z}_i\mathbf{Z}_i'\mathbf{P}\mathbf{y} - \tfrac{1}{2}\mathbf{y}'\mathbf{P}\mathbf{Z}_i\mathbf{Z}_i'\mathbf{P}\mathbf{Z}_j\mathbf{Z}_j'\mathbf{P}\mathbf{y}\\ &= \tfrac{1}{2}\,\mathrm{tr}(\mathbf{P}\mathbf{Z}_j\mathbf{Z}_j'\mathbf{P}\mathbf{Z}_i\mathbf{Z}_i') - \mathbf{y}'\mathbf{P}\mathbf{Z}_j\mathbf{Z}_j'\mathbf{P}\mathbf{Z}_i\mathbf{Z}_i'\mathbf{P}\mathbf{y}\,.\end{aligned}\tag{93}$$

Therefore, on using $E(\mathbf{y}'\mathbf{A}\mathbf{y}) = \text{tr}(\mathbf{AV}) + \boldsymbol{\beta}'\mathbf{X}'\mathbf{AX}\boldsymbol{\beta}$ from Theorem S1 of Appendix S.5,

$$-E\left(\frac{\partial^2 l_R}{\partial\sigma_i^2\,\sigma_j^2}\right) = -\tfrac{1}{2}\,\text{tr}(\mathbf{PZ}_j\mathbf{Z}_j'\mathbf{PZ}_i\mathbf{Z}_i') + \text{tr}(\mathbf{PZ}_j\mathbf{Z}_j'\mathbf{PZ}_i\mathbf{Z}_i'\mathbf{PV}) - \boldsymbol{\beta}'\mathbf{X}'\mathbf{PZ}_j\mathbf{Z}_j'\mathbf{PZ}_i\mathbf{Z}_i'\mathbf{PX}\boldsymbol{\beta}$$

$$= -\tfrac{1}{2}\,\text{tr}(\mathbf{PZ}_j\mathbf{Z}_j'\mathbf{PZ}_i\mathbf{Z}_i') + \text{tr}(\mathbf{Z}_j\mathbf{Z}_j'\mathbf{PZ}_i\mathbf{Z}_i'\mathbf{PVP}) + \mathbf{0}, \quad \text{because } \mathbf{PX} = \mathbf{0},$$

$$= \tfrac{1}{2}\,\text{tr}(\mathbf{PZ}_j\mathbf{Z}_j'\mathbf{PZ}_i\mathbf{Z}_i'), \quad \text{because } \mathbf{PVP} = \mathbf{P}\,. \tag{94}$$

Hence, denoting REML estimations by $\tilde{\sigma}^2_{\text{REML}}$, we have in the limit

$$\text{var}(\tilde{\sigma}^2_{\text{REML}}) \simeq 2[\{_m\ \text{tr}(\mathbf{PZ}_i\mathbf{Z}_i'\mathbf{PZ}_j\mathbf{Z}_j')\}_{i,j=0}^{r}]^{-1}$$

$$\simeq 2[\{_m\ \text{sesq}(\mathbf{Z}_i'\mathbf{PZ}_j)\}_{i,j=0}^{r}]^{-1}, \tag{95}$$

exactly the same as (42) for ML, except for ML there is $\mathbf{V}^{-1}$ where here we have $\mathbf{P}$.

f. Balanced data

Solutions of REML equations, for all cases of balanced data from mixed models, are the same as ANOVA estimators—and this result is true whether normality is assumed or not. That is, if one ignores normality but nevertheless solves equations (90), which are the REML equations under normality, the solutions are identical to ANOVA estimators—for balanced data from all mixed models. Those solutions are, of course, not REML estimators unless one assumes normality *and* takes into account the non-negativity requirement of the maximum likelihood method of estimating variance components.

Several authors give lip service to this result, either in the form of a simple statement of it, or with specific examples: e.g., Patterson and Thompson (1971), Corbeil and Searle (1976b) and Harville (1977). Detailed (and necessarily lengthy) proof that REML solutions are ANOVA estimators, without relying on normality, is given in Anderson (1978, pp. 97–104).

g. Using cell means models for fixed effects

Suppose in the mixed model $\mathbf{y} = \mathbf{X}\boldsymbol{\beta} + \mathbf{Zu} + \mathbf{e}$ that the fixed effects $\boldsymbol{\beta}$ are all taken to be cell means of the sub-most cells of the fixed effects factors. Then $\mathbf{X}\boldsymbol{\beta}$ will have the form

$$\mathbf{X}\boldsymbol{\mu} = \{_d\ \mathbf{1}_{n_t}\}_{t=1}^{s}\boldsymbol{\mu} \tag{96}$$

where $\boldsymbol{\mu}$ is of order s, the number of filled sub-most cells of the fixed effects factors, with the tth such cell having n_t observations and the tth element of $\boldsymbol{\mu}$ being the population mean μ_t for that cell. Then, since the form of $\mathbf{X}$ in (96) gives $\mathbf{X}'\mathbf{X} = \{_d\ n_t\}_{t=1}^{s}$, the form of $\mathbf{M}$ is

$$\mathbf{M} = \mathbf{I} - \mathbf{X}(\mathbf{X}'\mathbf{X})^-\mathbf{X}' = \{_d\ \mathbf{I}_{n_t} - \bar{\mathbf{J}}_{n_t}\}_{t=1}^{s}\,. \tag{97}$$

Under these circumstances, one form of $\mathbf{K}'$, as described by Corbeil and Searle (1976a), is $\mathbf{M}$ with its n_1th, $(n_1 + n_2)$th, $\ldots$, $(n_1 + n_2 + \cdots + n_s)$th rows deleted.

Thus $\mathbf{K}'$ is $\mathbf{M}$ after deleting the last row of each submatrix on the diagonal of $\mathbf{M}$; and so, by reference to (97), we gather that

$$\mathbf{K}' = \{_d \, \mathbf{I}_{v_t} - \mathbf{J}_{v_t}/n_t \quad -\mathbf{1}_{v_t}/n_t\}_{t=1}^{s} \quad \text{for } v_t = n - 1 \,. \tag{98}$$

It is easily shown that $\mathbf{K}(\mathbf{K}'\mathbf{K})^{-1}\mathbf{K}' = \mathbf{M}$.

6.7. ESTIMATING FIXED EFFECTS IN MIXED MODELS

a. ML

With ML estimation the equations for the fixed effects are

$$\mathbf{X}'\tilde{\mathbf{V}}^{-1}\mathbf{X}\tilde{\boldsymbol{\beta}} = \mathbf{X}'\tilde{\mathbf{V}}^{-1}\mathbf{y}$$

for $\tilde{\mathbf{V}}$ being the MLE of $\mathbf{V}$, i.e., $\tilde{\mathbf{V}} = \Sigma_{i=1}^{t} \mathbf{Z}_i\mathbf{Z}_i\tilde{\sigma}_i^2$. Hence the MLE of $\mathbf{X}\boldsymbol{\beta}$ is

$$\text{MLE}(\mathbf{X}\boldsymbol{\beta}) = \mathbf{X}(\mathbf{X}'\tilde{\mathbf{V}}^{-1}\mathbf{X})^{-}\mathbf{X}'\tilde{\mathbf{V}}^{-1}\mathbf{y}, \tag{99}$$

and its asymptotic dispersion matrix is

$$\text{var}[\text{MLE}(\mathbf{X}\boldsymbol{\beta})] = \mathbf{X}(\mathbf{X}'\tilde{\mathbf{V}}^{-1}\mathbf{X})^{-}\mathbf{X}' \,. \tag{100}$$

b. REML

REML estimation includes no procedure for estimating fixed effects. However, it would seem to be reasonable to use (99) and (100) with $\tilde{\mathbf{V}}$ being $\Sigma_{i=0}^{r} \mathbf{Z}_i\mathbf{Z}_i'\tilde{\sigma}_{i,\text{R}}^2$ where $\tilde{\sigma}_{i,\text{R}}^2$ is the REML estimate of σ_i^2. This is similar to empirical Bayes estimation discussed in Section 9.3c.

6.8. ML OR REML?

It is our considered opinion that for unbalanced data each of ML and REML are to be preferred over any ANOVA method. This is because the maximum likelihood principle that is behind ML and REML is known to have useful properties: consistency and asymptotic normality of the estimators; and the asymptotic sampling dispersion matrix of the estimators is also known. This provides some opportunity for establishing confidence intervals and testing hypotheses about parameters. In contrast, ANOVA estimators have only unbiasedness as an established property; and their sampling dispersion matrices are often very difficult to derive. True, the ML and REML estimators are based on assuming normality of the data, but in many circumstances that assumption is unlikely to be seriously wrong. And of course, the asymptotic variance–covariance properties are valid only in the large-sample sense, and for small or modest-sized samples this may somewhat nullify their usefulness. Nevertheless, these properties seem to us to be sufficiently reliable for us to have more faith in ML and REML than in the ANOVA methods, for which we often have no means for making a rational decision between one ANOVA method and another. Maximum likelihood, however, is firmly established as a respected method of estimation. Initially, ML and REML were impractical because of their computing requirements, but this impracticality is now fast disappearing with the rapid development of bigger and faster computers. Adequate software is probably a more limiting factor than adequate hardware.

As to the question "ML or REML?" there is probably no hard and fast answer. Both have the same merits of being based on the maximum likelihood principle—and they have the same demerit of computability requirements. ML provides estimators of fixed effects, whereas REML, of itself, does not. But with balanced data REML solutions are identical to ANOVA estimators which have optimal minimum variance properties—and to many users this is a sufficiently comforting feature of REML that they prefer it over ML.

6.9. SUMMARY

Model: Section 6.1

$$\mathbf{y} = \mathbf{X}\boldsymbol{\beta} + \sum_{i=0}^{r} \mathbf{Z}_i\mathbf{u}_i; \quad \mathbf{V} = \sum_{i=0}^{r} \mathbf{Z}_i\mathbf{Z}_i'\sigma_i^2 . \tag{9}$$

Likelihood—under normality: Section 6.1

$$L = \frac{e^{-\frac{1}{2}(\mathbf{y}-\mathbf{X}\boldsymbol{\beta})'\mathbf{V}^{-1}(\mathbf{y}-\mathbf{X}\boldsymbol{\beta})}}{(2\pi)^{\frac{1}{2}N}|\mathbf{V}|^{\frac{1}{2}}} . \tag{12}$$

ML equations: Section 6.2

$$\mathbf{X}'\dot{\mathbf{V}}^{-1}\mathbf{X}\dot{\boldsymbol{\beta}} = \mathbf{X}'\dot{\mathbf{V}}^{-1}\mathbf{y}, \tag{20}$$

and, for $i = 0, 1, \ldots, r$,

$$\operatorname{tr}(\dot{\mathbf{V}}^{-1}\mathbf{Z}_i\mathbf{Z}_i') = (\mathbf{y} - \mathbf{X}\dot{\boldsymbol{\beta}})'\dot{\mathbf{V}}^{-1}\mathbf{Z}_i\mathbf{Z}_i'\dot{\mathbf{V}}^{-1}(\mathbf{y} - \mathbf{X}\dot{\boldsymbol{\beta}}) . \tag{21}$$

Alternatively, for

$$\mathbf{P} = \mathbf{V}^{-1} - \mathbf{V}^{-1}\mathbf{X}(\mathbf{X}'\mathbf{V}^{-1}\mathbf{X})^{-}\mathbf{X}'\mathbf{V}^{-1} \tag{22}$$

$$\{_{c} \operatorname{tr}(\dot{\mathbf{V}}^{-1}\mathbf{Z}_i\mathbf{Z}_i')\}_{i=0}^{r} = \{_{c}\, \mathbf{y}'\dot{\mathbf{P}}\mathbf{Z}_i\mathbf{Z}_i'\dot{\mathbf{P}}\mathbf{y}\}_{i=0}^{r} \tag{25}$$

or

$$\{_{m} \operatorname{sesq}(\mathbf{Z}_i'\dot{\mathbf{V}}^{-1}\mathbf{Z}_j)\}_{i,j=0}^{r}\dot{\boldsymbol{\sigma}}^2 = \{_{c} \operatorname{sesq}(\mathbf{Z}_i'\dot{\mathbf{P}}\mathbf{y})\}_{i=1}^{r}, \tag{27b}$$

where

$$\operatorname{sesq}(\mathbf{A}) = \operatorname{sesq}\{a_{pq}\} = \Sigma_p\Sigma_q a_{pq}^2 .$$

Asymptotic variances: Section 6.3

$$\operatorname{var}(\tilde{\boldsymbol{\beta}}) \simeq (\mathbf{X}'\mathbf{V}^{-1}\mathbf{X})^{-1}, \tag{40}$$

$$\text{var}(\tilde{\boldsymbol{\sigma}}) \simeq 2[\{_{\text{m}}\, \text{tr}(\mathbf{V}^{-1}\mathbf{Z}_i\mathbf{Z}_i'\mathbf{V}^{-1}\mathbf{Z}_j\mathbf{Z}_j')\}_{i,j=0}^{r}]^{-1} . \tag{42}$$

Comments on computing: Section 6.4

Deriving some balanced data results: Section 6.5

$\mathbf{V}$ and $\mathbf{V}^{-1}$ for the 2-way classification . (58),(61)

Restricted maximum likelihood (REML): Section 6.6

For $\mathbf{K}'\mathbf{X} = \mathbf{0}$, with $\mathbf{K}'_{r \times N}$ of rank $r = r_{\mathbf{X}}$

$$\mathbf{P} = \mathbf{V}^{-1} - \mathbf{V}^{-1}\mathbf{X}(\mathbf{X}'\mathbf{V}^{-1}\mathbf{X})^{-}\mathbf{X}'\mathbf{V}^{-1} = \mathbf{K}(\mathbf{K}'\mathbf{V}\mathbf{K})^{-1}\mathbf{K}' . \tag{88}$$

Estimation equations

$$\{_{\text{m}}\, \text{tr}(\mathbf{Z}_i\dot{\mathbf{P}}\mathbf{Z}_j\mathbf{Z}_j'\dot{\mathbf{P}}\mathbf{Z}_i)\}_{i,j=0}^{r} = \{_{\text{c}}\, \mathbf{y}'\dot{\mathbf{P}}\mathbf{Z}_i\mathbf{Z}_i'\dot{\mathbf{P}}\mathbf{y}\}_{i=0}^{r} . \tag{90}$$

Estimating fixed effects: Section 6.7

ML or REML?: Section 6.8

6.10. EXERCISES

E 6.1. Derive (20) and (21), showing all details

E 6.2. Use equation (20) to show that in any random model the MLE of μ, when the data are balanced, is $\bar{y}$, the grand mean of the data.

E 6.3. Use equation (21), or one of its equivalent forms, to derive the ML solutions of variance components for balanced data from the following models:

(a) the 1-way classification (see Table 4.8);

(b) the 2-way nested classification (see Table 4.10);

(c) the 2-way crossed classification, without interaction (see Table 4.12).

E 6.4. Using

$$\mathbf{Q} = \mathbf{H}^{-1} - \mathbf{H}^{-1}\mathbf{X}(\mathbf{X}'\mathbf{H}^{-1}\mathbf{X})^{-}\mathbf{X}'\mathbf{H}^{-1} = \mathbf{P}\sigma_e^2$$

for $\mathbf{P}$ of (22), recast (35b) in a form akin to (27).

E 6.5. Use the alternative form of $\mathbf{I}(\boldsymbol{\theta})$ given at the end of Appendix S.7c to derive (38).

E 6.6. Reduce (50) to (51).

E 6.7. Use equation (20) and equation (21), or one of its equivalent forms, to derive (133), (134) and (135) of Chapter 3.

E 6.8. Simplify the left-hand side of (62) for $i = 2$ and 3 and the right-hand side for $i = 0$ and 2, and along with the other simplifications shown in Section 6.6a, derive equations (90) of Chapter 4.

E 6.9. Derive the six elements of the information matrix not derived in Section 6.5a–iv.

E 6.10. For the 1-way classification, random model derive the REML equations from (89) or (90)

(a) for unbalanced data;

(b) for balanced data.

Notation: Use e to represent σ_e^2, and α for σ_α^2, and λ_i for $e + n_i\alpha$.

E 6.11. (a) Write down the likelihood of $\mathbf{K}'\mathbf{y}$ for $\mathbf{K}'\mathbf{X} = \mathbf{0}$.

(b) Why does the numerator of that likelihood not involve $\mathbf{K}$?

(c) On defining $\Pi(\mathbf{A})$ as the product of the non-zero eigenroots of a square matrix $\mathbf{A}$, prove that $\Pi(\mathbf{AB}) = \Pi(\mathbf{BA})$ when $\mathbf{AB}$ and $\mathbf{BA}$ both exist.

(d) Use (c) to prove that $\log|\mathbf{K}'\mathbf{VK}| = \log|\mathbf{K}'\mathbf{K}| - \log\Pi(\mathbf{P})$.

(e) Explain why the matrix $\mathbf{K}$ plays no role in maximizing the likelihood in (a), and therefore REML estimation is invariant to $\mathbf{K}$.

E 6.12. Derive (95) from (42) using the replacements that follow (86).

CHAPTER 7

PREDICTION OF RANDOM VARIABLES

7.1. INTRODUCTION

Consider measuring intelligence in humans. Each of us has some level of intelligence, usually quantified as IQ. It can never be measured exactly. As a substitute, we have test scores, which are used for putting a value to an individual's IQ. An example of this is introduced in Section 3.4. It leads to the problem "Exactly how are the test scores to be used?", a problem that is addressed very directly in the following textbook exercise taken from Mood (1950, p. 164, exercise 23). With important changes it is also to be found in Mood and Graybill (1963, p. 195, exercise 32), and in Mood, Graybill and Boes (1974, p. 370, exercise 52).

> 23. Suppose intelligence quotients for students in a particular age group are normally distributed about a mean of 100 with standard deviation 15. The IQ, say x_1, of a particular student is to be estimated by a test on which he scores 130. It is further given that test scores are normally distributed about the true IQ as a mean with standard deviation 5. What is the maximum-likelihood estimate of the student's IQ? (The answer is not 130.)

This exercise, with its tantalizing last sentence, played a prominent role in initially motivating C. R. Henderson in his lifelong contributions (e.g., 1948, 1963, 1973a,b, 1975) to the problem of estimating genetic merit of dairy cattle. That and the estimation of IQ represent the classic prediction problem of predicting the unobservable realized value of a random effect that is part of a mixed model.

One way of solving Mood's problem is achieved by starting with the model equation

$$y_{ij} = \mu + \alpha_i + e_{ij}$$

for the jth test score of the ith person, where $\mu + \alpha_i$ is that ith person's true IQ. We first operate conditional on the value of α_i for the particular person who has been given label i. But in thinking about people in general, that particular person is really just a random person: and α_i is, accordingly, simply a realized (but unobservable) value of a random effect—the effect on test score of the intelligence level of the ith randomly chosen person. Therefore we treat α_i as random and have IQ and score, namely $\mu + \alpha_i$ and y_{ij}, jointly distributed with bivariate normal density:

$$\begin{bmatrix} \text{IQ} \\ \text{score} \end{bmatrix} = \begin{bmatrix} \mu + \alpha_i \\ y_{ij} \end{bmatrix} \sim \mathcal{N}\left(\begin{bmatrix} 100 \\ 100 \end{bmatrix}, \begin{bmatrix} 15^2 & 15^2 \\ 15^2 & 15^2 + 5^2 \end{bmatrix} \right).$$

From this, using (iv) of Appendix S.2, the conditional mean of $\mu + \alpha_i$ given $y_{ij} = 130$, namely $E(\mu + \alpha_i \mid y_{ij} = 130)$, is

$$E(\mu + \alpha_i \mid y_{ij} = 130) = 100 + \frac{15^2}{15^2 + 5^2}(130 - 100) = 127 .$$

This is what is used to quantify the student's IQ. It shows how one can obtain a reasonable answer to Mood's exercise other than 130, as alluded to in the last sentence of the exercise.

Note that although Mood's (1950) question asks for a maximum likelihood estimate of the student's IQ, we have used just the conditional mean. This is because, once we confine ourselves to a particular student having a test score of 130, we are then in the conditional situation of being interested only in quantifying $\mu + \alpha_i$ conditional on $y_{ij} = 130$. And under these circumstances the conditional mean, $E(\mu + \alpha_i \mid y_{ij} = 130)$, is what we use as a predictor of $[(\mu + \alpha_i) \mid y_{ij} = 130]$, namely of $\mu + \alpha_i$ given that $y_{ij} = 130$. The connection with the maximum likelihood estimation of Mood's question is that under the normality assumptions given in the question, the conditional variable $(\mu + \alpha_i) \mid y_{ij} = 130$ is indeed normally distributed with mean 127. Then, whilst taking $\tilde{\mu} + \tilde{\alpha}_i = 127$ as the predictor of $[(\mu + \alpha_i) \mid y_{ij} = 130]$ it is not, in the strictest sense, a maximum likelihood estimator; but it does maximize the density function of $[(\mu + \alpha_i) \mid y_{ij} = 130]$.

An interesting feature of the question in the Mood (1950) book is that in its later forms in the 1963 and 1974 editions the "What is the maximum likelihood estimate?" question is replaced by a "What is the Bayes estimator?" type of question. With the general topic of Bayes estimation being dealt with in Chapter 9, we here simply note that the above predictor, $E[(\mu + \alpha_i) \mid y_{ij} = 130] = 127$, is indeed the same as is derived using the results given in that chapter. In particular, the derivation is

$$\begin{aligned} E[(\mu + \alpha_i) \mid y_{ij} = 130] &= \mu + E(\alpha_i \mid y_{ij} = 130) \\ &= \mu + n\sigma_\alpha^2 \left(\frac{\bar{y}_{i\cdot} - \bar{y}_{\cdot\cdot}}{\sigma_e^2 + n\sigma_\alpha^2} + \frac{\bar{y}_{\cdot\cdot} - \mu}{\sigma_e^2 + n\sigma_\alpha^2 + an\sigma_\mu^2} \right), \end{aligned}$$

from (54) of Chapter 9. And with $\mu = 100$, $n = 1$, $\bar{y}_{i.} = 130 = \bar{y}_{..}$, $\sigma_e^2 = 5^2$, $\sigma_\alpha^2 = 15^2$ and $\sigma_\mu^2 = 0$, which are the characteristics of Mood's question, this becomes

$$E(\mu + \alpha_i \mid y_{ij} = 130) = 100 + \frac{15^2}{15^2 + 5^2}(130 - 100) = 127 .$$

There are many situations similar to that of the student's IQ, of wanting to quantify the realization of an unobservable random variable. A biological example is that of predicting the genetic merit of a dairy bull from the milk yields of his daughters. A non-biological example is that of predicting instrument bias in micrometers selected randomly from a manufacturer's lot, using the micrometers to measure ball-bearing diameters. And an example in psychology is that just considered: predicting a person's intelligence from IQ scores. In all of these we have a vector of observations on some random variables from which we wish to predict the value of some other random variable (or variables) that cannot be observed.

A statement of the general problem is easy. Suppose **U** and **Y** are jointly distributed vectors of random variables, with those in **Y** being observable but those in **U** not being observable. The problem is to predict **U** from some realized, observed value of **Y**, say **y**. Usually **Y** contains more elements than **U**, and indeed **U** is often scalar. In the IQ example **U** is the scalar, unknowable true value of a person's intelligence, and **y** is the vector of test scores.

Three methods of prediction are of interest: best prediction (BP), best linear prediction (BLP), and mixed model prediction, which leads to what is now called best, linear, unbiased prediction (BLUP). Of these three methods of prediction, BP is available when we know all the parameters of the joint distribution of **U** and **Y**; i.e., when we know $f(\mathbf{y}, \mathbf{u})$. BLP and BLUP are methods that are best in situations when we know some of the parameters of $f(\mathbf{y}, \mathbf{u})$ but not all of them. For BLP only first and second moments are assumed known, and for BLUP second, but not first, moments are assumed known. In each case the less that is assumed known, the more restrictive are the resulting predictors.

The description that follows is strongly influenced by the work of C. R. Henderson, who for thirty years sustained the interest of one of us (S.R.S.) in the prediction problem in the context of animal breeding. In particular, the opening paragraphs of Henderson (1973a) have been of especial assistance in preparing this account of prediction.

Notation: WARNING. In contrast to the notation of (70) in Chapter 4 and of (19) in Chapter 6, the vector **u** no longer includes **e**. Rather than define an adorned **Z** and **u**, such as $\breve{\mathbf{Z}}$ and $\breve{\mathbf{u}}$, to represent **Z** without $\mathbf{Z}_0 = \mathbf{I}$ and **u** without $\mathbf{u}_0 = \mathbf{e}$, we simply emphasize that in this chapter

$$\mathbf{Zu} = \mathbf{Z}_1\mathbf{u}_1 + \mathbf{Z}_2\mathbf{u}_2 + \cdots + \mathbf{Z}_r\mathbf{u}_r,$$

and the model equation is

$$\mathbf{y} = \mathbf{X}\boldsymbol{\beta} + \mathbf{Zu} + \mathbf{e}$$

with

$$\mathbf{V} = \text{var}(\mathbf{y}) = \mathbf{ZDZ}' + \sigma_e^2\mathbf{I}$$

for

$$\mathbf{D} = \text{var}(\mathbf{u}); \text{ and with } \mathbf{C} = \text{cov}(\mathbf{y}, \mathbf{u}') = \mathbf{ZD}.$$

The reason for excluding $\mathbf{u}_0 = \mathbf{e}$ from $\mathbf{u}$ is that we are interested in predicting elements of $\mathbf{u}$ that are random effects, but not those that are residual errors.

7.2. BEST PREDICTION (BP)

Suppose that $\mathbf{U}$ is scalar, U. When $f(u, \mathbf{y})$ is the joint density function of the random variables U and $\mathbf{Y}$ at the point $u, \mathbf{y}$ then with the predictor being denoted by $\tilde{u}$ the mean square error of prediction is

$$E\,(\tilde{u} - u)^2 = \iint (\tilde{u} - u)^2 f(u, \mathbf{y})\, d\mathbf{y}\, du, \tag{1}$$

where E represents expectation. A generalization of this to a vector of random variables $\mathbf{u}$ is

$$E\,(\tilde{\mathbf{u}} - \mathbf{u})'\mathbf{A}(\tilde{\mathbf{u}} - \mathbf{u}) = \iint (\tilde{\mathbf{u}} - \mathbf{u})'\mathbf{A}(\tilde{\mathbf{u}} - \mathbf{u}) f(\mathbf{u}, \mathbf{y})\, d\mathbf{y}\, d\mathbf{u}, \tag{2}$$

where $\mathbf{A}$ is any positive definite symmetric matrix. Clearly, for $\mathbf{A}$ being scalar and unity (2) is identical to (1). In passing, note that decomposition of the error of prediction, $\tilde{u} - u$, is discussed at length by Harville (1985) for four different states of knowledge.

a. The best predictor

Our criterion for deriving a predictor is minimum mean square, i.e., we minimize (2). The result is what we call the *best predictor*. Note that "best" here means minimum mean square error of prediction, which is different from the usual meaning of "best" being minimum variance. Because variance is variability around a fixed value and because u in (1) is a random variable, (1) is not the definition of the variance of u. Thus, whereas as an estimation criterion we use minimum variance for estimating a parameter, we use minimum mean square for predicting the realized value of a random variable. Thus from minimizing (2) we get

$$\text{best predictor: } \tilde{\mathbf{u}} = \text{BP}(\mathbf{u}) = E(\mathbf{u} \mid \mathbf{y}); \tag{3}$$

i.e., the best predictor of $\mathbf{u}$ is the conditional mean of $\mathbf{u}$ given $\mathbf{y}$.

Noteworthy features of this result are: (i) it holds for all probability density functions $f(\mathbf{u}, \mathbf{y})$, and (ii) it does not depend on the positive definite symmetric matrix $\mathbf{A}$.

Verification of (3). In the mean square on the left-hand side of (2), to $\tilde{\mathbf{u}} - \mathbf{u}$ add and subtract $E(\mathbf{u} \mid \mathbf{y})$, which, for convenience, will be denoted $\mathbf{u}_0$; i.e., with

$$\mathbf{u}_0 \equiv E(\mathbf{u} \mid \mathbf{y}),$$

$$\begin{aligned} E\,(\tilde{\mathbf{u}} - \mathbf{u})'\mathbf{A}(\tilde{\mathbf{u}} - \mathbf{u}) &= E\,(\tilde{\mathbf{u}} - \mathbf{u}_0 + \mathbf{u}_0 - \mathbf{u})'\mathbf{A}(\tilde{\mathbf{u}} - \mathbf{u}_0 + \mathbf{u}_0 - \mathbf{u}) \\ &= E\,(\tilde{\mathbf{u}} - \mathbf{u}_0)'\mathbf{A}(\tilde{\mathbf{u}} - \mathbf{u}_0) + 2E\,(\tilde{\mathbf{u}} - \mathbf{u}_0)'\mathbf{A}(\mathbf{u}_0 - \mathbf{u}) \\ &\quad + E\,(\mathbf{u}_0 - \mathbf{u})'\mathbf{A}(\mathbf{u}_0 - \mathbf{u})\,. \end{aligned}$$

To choose a $\tilde{\mathbf{u}}$ that minimizes this, note that the last term, $E\,(\mathbf{u}_0 - \mathbf{u})'\mathbf{A}(\mathbf{u}_0 - \mathbf{u})$, does not involve $\tilde{\mathbf{u}}$. And in the second term,

$$\begin{aligned} E\,(\tilde{\mathbf{u}} - \mathbf{u}_0)'\mathbf{A}(\mathbf{u}_0 - \mathbf{u}) &= E_y\{E_u[(\tilde{\mathbf{u}} - \mathbf{u}_0)'\mathbf{A}(\mathbf{u}_0 - \mathbf{u}) \mid \mathbf{y}]\} \\ &= E_y\{(\tilde{\mathbf{u}} - \mathbf{u}_0)'\mathbf{A}(\mathbf{u}_0 - \mathbf{u}_0)\} = \mathbf{0} \end{aligned}$$

since, given $\mathbf{y}$, only $\mathbf{u}|\mathbf{y}$ is not fixed with $E_u(\mathbf{u}|\mathbf{y}) = \mathbf{u}_0$. Therefore

$$E\,(\tilde{\mathbf{u}} - \mathbf{u})'\mathbf{A}(\tilde{\mathbf{u}} - \mathbf{u}) = E\,(\tilde{\mathbf{u}} - \mathbf{u}_0)'\mathbf{A}(\tilde{\mathbf{u}} - \mathbf{u}_0) + \text{terms without } \tilde{\mathbf{u}}\,.$$

Since $E\,(\tilde{\mathbf{u}} - \mathbf{u}_0)'\mathbf{A}(\tilde{\mathbf{u}} - \mathbf{u}_0)$ must be non-negative it is minimized by choosing $\tilde{\mathbf{u}} = \mathbf{u}_0$; i.e., the best predictor is $\tilde{\mathbf{u}} = E(\mathbf{u} \mid \mathbf{y})$.

Three features of this derivation merit comment. First, adding and subtracting $E(\mathbf{u} \mid \mathbf{y})$ is simply centering about the conditional mean, and this is often a useful methodological step. Second, the cross-product term is merely a covariance and the centering often reduces it to zero. Third, the final step illustrates that the problem of predicting a random variable is simply that of estimating its conditional mean.

b. Mean and variance properties

First and second moments of the best predictor are important. They are discussed in Cochran (1951) and in Rao (1965, pp. 79 and 220–222) for the case of scalar U.

First, the best predictor is unbiased for sampling over $\mathbf{y}$: for $E_{\mathbf{y}}$ representing expectation over $\mathbf{y}$

$$E_{\mathbf{y}}(\tilde{\mathbf{u}}) = E_{\mathbf{y}}[E_{\mathbf{u}|\mathbf{y}}(\mathbf{u} \mid \mathbf{y})] = E(\mathbf{u}), \tag{4}$$

as in $E(g)$ of Appendix S.1. Note that the meaning of unbiasedness here is that the expected value of the predictor equals that of the random variable for which it is a predictor. This differs from the usual meaning of unbiasedness as defined in the statistical literature when estimating a parameter. In that case unbiasedness means that the expected value of (estimator minus parameter) is zero; e.g., $E(\hat{\beta} - \beta) = 0$, where β is a constant. With prediction, unbiasedness means that the expected value of (predictor minus random variable) is zero; e.g., $E(\tilde{\mathbf{u}} - \mathbf{u}) = \mathbf{0}$ where $\mathbf{u}$ is a random variable. The former gives $E(\hat{\beta}) = \beta$, whereas the latter gives $E(\tilde{\mathbf{u}}) = E(\mathbf{u})$.

Second, prediction errors $\tilde{\mathbf{u}} - \mathbf{u}$ have a variance–covariance matrix that is the mean value, over sampling on $\mathbf{y}$, of that of $\mathbf{u} \mid \mathbf{y}$:

$$\text{var}(\tilde{\mathbf{u}} - \mathbf{u}) = E_{\mathbf{y}}[\text{var}(\mathbf{u} \mid \mathbf{y})]\,. \tag{5}$$

Also,

$$\text{cov}(\tilde{\mathbf{u}}, \mathbf{u}') = \text{var}(\tilde{\mathbf{u}}) \quad \text{and} \quad \text{cov}(\tilde{\mathbf{u}}, \mathbf{y}') = \text{cov}(\mathbf{u}, \mathbf{y}') . \tag{6}$$

Verification of (5) comes from using var(g) at the end of Appendix S.1, namely

$$\text{var}(g) = E_y[\text{var}(g \mid y)] + \text{var}_y[E(g \mid y)] .$$

With $\tilde{\mathbf{u}} - \mathbf{u}$ used for g this gives

$$\begin{aligned}\text{var}(\tilde{\mathbf{u}} - \mathbf{u}) &= E_{\mathbf{y}}\{\text{var}[(\tilde{\mathbf{u}} - \mathbf{u}) \mid \mathbf{y}]\} + \text{var}_{\mathbf{y}}\{E[(\tilde{\mathbf{u}} - \mathbf{u}) \mid \mathbf{y}]\}\\ &= E_{\mathbf{y}}\{\text{var}[(\tilde{\mathbf{u}} - \mathbf{u}) \mid \mathbf{y}\} + \text{var}_{\mathbf{u}}(\mathbf{0}), \quad \text{because } E(\tilde{\mathbf{u}} \mid \mathbf{y}) = \tilde{\mathbf{u}} = E(\mathbf{u} \mid \mathbf{y})\\ &= E_{\mathbf{y}}[\text{var}(\mathbf{u} \mid \mathbf{y})],\end{aligned}$$

which is (5).

The two results in (6) are established by using cov(g, h) developed in Appendix S.1:

$$\text{cov}(g, h) = E_y[\text{cov}(g \mid y,\ h \mid y)] + \text{cov}_y[E(g \mid y),\ E(h \mid y)] .$$

With $g = \tilde{\mathbf{u}}$ and $h = \mathbf{u}'$ this is

$$\text{cov}(\tilde{\mathbf{u}}, \mathbf{u}') = E_{\mathbf{y}}[\text{cov}(\tilde{\mathbf{u}} \mid \mathbf{y},\ \mathbf{u}' \mid \mathbf{y})] + \text{cov}_{\mathbf{y}}[E(\tilde{\mathbf{u}} \mid \mathbf{y}),\ E(\mathbf{u}' \mid \mathbf{y})] .$$

Note that the first term involves the covariance of $\mathbf{u} \mid \mathbf{y}$ with its mean $\tilde{\mathbf{u}} = E(\mathbf{u} \mid \mathbf{y})$. It is therefore zero. Hence

$$\text{cov}(\tilde{\mathbf{u}}, \mathbf{u}') = \text{cov}_{\mathbf{y}}(\tilde{\mathbf{u}}, \tilde{\mathbf{u}}') = \text{var}(\tilde{\mathbf{u}}),$$

which is the first result in (6). Likewise, for the second result we start with

$$\text{cov}(\mathbf{u}, \mathbf{y}') = E_{\mathbf{y}}[\text{cov}(\mathbf{u} \mid \mathbf{y},\ \mathbf{y}' \mid \mathbf{y})] + \text{cov}_{\mathbf{y}}[E(\mathbf{u} \mid \mathbf{y}),\ E(\mathbf{y}' \mid \mathbf{y})] .$$

In the first term the covariance is of $\mathbf{u} \mid \mathbf{y}$ with the constant $\mathbf{y}' \mid \mathbf{y}$. Therefore it is zero, and so

$$\text{cov}(\mathbf{u}, \mathbf{y}') = \text{cov}_{\mathbf{y}}(\tilde{\mathbf{u}}, \mathbf{y}') .$$

Thus is (6) established.

c. Two properties of the best predictor of a scalar

For scalar u there are two further properties of interest. The first is that the correlation between u and any predictor of it that is a function of $\mathbf{y}$ is maximum for the best predictor, that maximum value being

$$\rho(\tilde{u}, u) = \sigma_{\tilde{u}}/\sigma_u . \tag{7}$$

Second, selecting any upper fraction of the population on the basis of values of $\tilde{u}$ ensures that

$$\text{for that selected proportion,} \quad E(u) \text{ is maximized} . \tag{8}$$

-i. ***Maximizing a correlation.*** The correlation between $\tilde{u}$ and u is, using (6),

$$\rho(\tilde{u}, u) = \frac{\text{cov}(\tilde{u}, u)}{\sigma_{\tilde{u}}\sigma_u} = \frac{\sigma_{\tilde{u}}}{\sigma_u},$$

which is (7). Now consider some function of **y**, say f, as a predictor of u different from $\tilde{u}$. Then

$$\begin{aligned}
\text{cov}(f, u) &= E\{[f - E(f)][u - E(u)]\}, \quad \text{by definition,} \\
&= E\{[f - E(f)][u - \tilde{u} + \tilde{u} - E(\tilde{u})]\}, \quad \text{because } E(\tilde{u}) = E(u), \\
&= E\{[f - E(f)](u - \tilde{u})\} + \text{cov}(f, \tilde{u}) \\
&= \text{cov}(f, \tilde{u}) + E_{\mathbf{y}}E_{u|\mathbf{y}}(\{[f - E(f)](u - \tilde{u})\} \mid \mathbf{y}) \\
&= \text{cov}(f, \tilde{u}) + E_{\mathbf{y}}\{[f - E(f)]E[(u - \tilde{u}) \mid \mathbf{y}]\},
\end{aligned}$$

because f being a function of **y** means that $f \mid \mathbf{y}$ is constant with respect to the E-operator. And then, because $E(u \mid \mathbf{y}) = \tilde{u}$, this becomes

$$\begin{aligned}
\text{cov}(f, u) &= \text{cov}(f, \tilde{u}) + E_{\mathbf{y}}\{[f - E(f)](\tilde{u} - \tilde{u})\} \\
&= \text{cov}(f, \tilde{u}) .
\end{aligned}$$

Hence

$$\rho^2(f, u) = \frac{\text{cov}^2(f, u)}{\sigma_f^2\sigma_u^2} = \frac{\text{cov}^2(f, \tilde{u})}{\sigma_f^2\sigma_{\tilde{u}}^2}\,\frac{\sigma_{\tilde{u}}^2}{\sigma_u^2} = \rho^2(f, \tilde{u})\rho^2(\tilde{u}, u) .$$

The maximum over all f is when $\rho^2(f, \tilde{u}) = 1$, i.e., $f = \tilde{u}$. Hence (7) is the maximum $\rho(\tilde{u}, u)$. This proof follows Rao (1973, p. 265–266).

-ii. ***Maximizing the mean of a selected proportion.*** Begin by contemplating the selection of a proportion α of the population of u-values, using **y** in some way as the basis of selection. We want to make the selection such that for given α the value of $E(u)$ is maximized. Hence we want a region of values of **y**, R say, such that

$$\int_R f(\mathbf{y})\, d\mathbf{y} \leqslant \alpha \quad \text{and} \quad \int_R \int_{-\infty}^{\infty} uf(u, \mathbf{y})\, du\, d\mathbf{y} \text{ is maximized .}$$

The latter is equivalent to maximizing

$$\int_R E(u \mid \mathbf{y}) f(\mathbf{y})\, d\mathbf{y},$$

which is equivalent to maximizing $E(u)$ for the selected proportion. By a generalization of the Neyman–Pearson Lemma [e.g., Cochran (1951), Rao (1973, Sec. 7a.2) or Casella and Berger (1990, p. 372)], this maximum is attained by the set of values of **y** for which $\tilde{u} = E(u \mid \mathbf{y}) \geqslant k$, where k has the one-to-one relationship to α

$$\int_R f(\mathbf{y})\,d\mathbf{y} = \alpha \quad \text{for} \quad R = \{\mathbf{y} : E(u \mid \mathbf{y}) \geqslant k\} .$$

Use this to determine k from α. Then selecting all those observational units for which $E(u \mid \mathbf{y}) = \tilde{u} \geqslant k$ yields a sample from the upper α-fraction of the population of unobservable u-values in which $E(u)$ is maximized.

Practitioners, in using estimates in place of the population parameters in this procedure, may often be found using it not exactly as specified. After choosing a value for α they might, for simplification, avoid determining k from α and selecting on the basis of $\tilde{u} > k$ but, instead, simply select the upper α-fraction of $\tilde{u}$-values. For small values of α (say .10 or less) and for a large number of elements in $\mathbf{u}$ (100 or more, say) this simplified practice might not be seriously different from the procedure as specified.

d. Normality

It is to be emphasized that $\tilde{\mathbf{u}} = E(\mathbf{u} \mid \mathbf{y})$ is a random variable, being a function of $\mathbf{y}$ and unknown parameters. Thus the problem of estimating the best predictor $\tilde{\mathbf{u}}$ remains, and demands some knowledge of the joint density $f(\mathbf{u} \mid \mathbf{y})$. Should this be normal,

$$\begin{bmatrix} \mathbf{u} \\ \mathbf{y} \end{bmatrix} \sim \mathcal{N}\left(\begin{bmatrix} \boldsymbol{\mu}_U \\ \boldsymbol{\mu}_Y \end{bmatrix}, \begin{bmatrix} \mathbf{D} & \mathbf{C} \\ \mathbf{C}' & \mathbf{V} \end{bmatrix} \right), \tag{9}$$

then with $\mathbf{C} = \mathbf{DZ}'$ as in Section 7.1, and using Appendix S.3,

$$\tilde{\mathbf{u}} = E(\mathbf{u} \mid \mathbf{y}) = \boldsymbol{\mu}_U + \mathbf{CV}^{-1}(\mathbf{y} - \boldsymbol{\mu}_Y) . \tag{10}$$

Properties (5)–(8) of $\tilde{\mathbf{u}}$ still hold. In (5) we now have from (9) that $\text{var}(\mathbf{u} \mid \mathbf{y}) = \mathbf{D} - \mathbf{CV}^{-1}\mathbf{C}'$, so that in (5)

$$\text{var}(\tilde{\mathbf{u}} - \mathbf{u}) = \mathbf{D} - \mathbf{CV}^{-1}\mathbf{C}' . \tag{11}$$

And using (10) in (6) gives

$$\text{cov}(\tilde{\mathbf{u}}, \mathbf{u}') = \text{var}(\tilde{\mathbf{u}}) = \mathbf{CV}^{-1}\mathbf{C}', \quad \text{and hence} \quad \rho(\tilde{u}_i, u_i) = \sqrt{\frac{\mathbf{c}_i'\mathbf{V}^{-1}\mathbf{c}_i}{\sigma^2_{u_i}}} \tag{12}$$

where $\mathbf{c}_i'$ is the ith row of $\mathbf{C}$.

The estimation problem is clearly visible in these results. The predictor is given in (10) but it and its succeeding properties cannot be estimated without having values for, or estimating, the four parameters $\boldsymbol{\mu}_U$, $\boldsymbol{\mu}_Y$, $\mathbf{C}$ and $\mathbf{V}$.

7.3. BEST LINEAR PREDICTION (BLP)

a. BLP(u)

The best predictor (3) is not necessarily linear in $\mathbf{y}$. Suppose attention is now confined to predictors of $\mathbf{u}$ that *are* linear in $\mathbf{y}$, of the form

$$\tilde{\mathbf{u}} = \mathbf{a} + \mathbf{By} \tag{13}$$

for some vector **a** and matrix **B**. Minimizing (2) for $\tilde{\mathbf{u}}$ of (13), in order to obtain the best linear predictor, leads (without any assumption of normality) to

$$\text{BLP}(\mathbf{u}) = \tilde{\mathbf{u}} = \boldsymbol{\mu}_U + \mathbf{C}\mathbf{V}^{-1}(\mathbf{y} - \boldsymbol{\mu}_Y), \tag{14}$$

where $\boldsymbol{\mu}_U$, $\boldsymbol{\mu}_Y$, $\mathbf{C}$ and $\mathbf{V}$ are as defined in (9) but without assuming normality as there.

An immediate observation on (14) is that it is identical to (10). This shows that the best linear predictor (14), derivation of which demands no knowledge of the form of $f(\mathbf{u}, \mathbf{y})$, is *identical* to the best predictor under normality, (10). Properties (11) and (12) therefore apply equally to (14) as to (10). And, of course, BLP(**u**) is unbiased, in the sense described following (4), namely that $E(\tilde{\mathbf{u}}) = E(\mathbf{u})$. Problems of estimation of the unknown parameters in (14) still remain.

b. Example

To illustrate (14) we use the 1-way classification random model of Chapter 3. It has model equation $y_{ij} = \mu + \alpha_i + e_{ij}$, or equivalently

$$\mathbf{y} = \mu \mathbf{1}_N + \mathbf{Z}\boldsymbol{\alpha} + \mathbf{e},$$

where

$$\mathbf{Z} = \{_{\text{d}}\, \mathbf{1}_{n_i}\}, \quad \mathbf{u} = \boldsymbol{\alpha},$$

$$\boldsymbol{\mu}_Y = \mu \mathbf{1}_N, \quad \boldsymbol{\mu}_U = \mathbf{0},$$

$$\mathbf{V} = \text{var}(\mathbf{y}) = \{_{\text{d}}\, \sigma_\alpha^2 \mathbf{J}_{n_i} + \sigma_e^2 \mathbf{I}_{n_i}\},$$

as in (81) of Section 3.6b, and

$$\begin{aligned}\mathbf{C} = \text{cov}(\mathbf{u}, \mathbf{y}') = \text{cov}(\mathbf{u}, \mathbf{u}'\mathbf{Z}') = [\text{var}(\mathbf{u})]\mathbf{Z}' &= \sigma_\alpha^2 \mathbf{I}_a \{_{\text{d}}\, \mathbf{1}'_{n_i}\} \\ &= \{_{\text{d}}\, \sigma_\alpha^2 \mathbf{1}'_{n_i}\}\,.\end{aligned}$$

Also

$$\mathbf{V}^{-1} = \left\{_{\text{d}}\, \frac{1}{\sigma_e^2}\left(\mathbf{I}_{n_i} - \frac{\sigma_\alpha^2}{\sigma_e^2 + n_i\sigma_\alpha^2}\mathbf{J}_{n_i}\right)\right\},$$

as immediately precedes (104) of Section 3.7. Using these expressions in (14) gives

$$\begin{aligned}\tilde{\mathbf{u}} = \tilde{\boldsymbol{\alpha}} &= \mathbf{0} + \{_{\text{d}}\, \sigma_\alpha^2 \mathbf{1}'_{n_i}\}\left\{_{\text{d}}\, \frac{1}{\sigma_e^2}\left(\mathbf{I}_{n_i} - \frac{\sigma_\alpha^2}{\sigma_e^2 + n_i\sigma_\alpha^2}\mathbf{J}_{n_i}\right)\right\}(\mathbf{y} - \mu\mathbf{1}_N) \\ &= \left\{_{\text{d}}\, \frac{\sigma_\alpha^2}{\sigma_e^2 + n_i\sigma_\alpha^2}\mathbf{1}'_{n_i}\right\}(\mathbf{y} - \mu\mathbf{1}_N)\,.\end{aligned} \tag{15}$$

Hence

$$\text{BLP}(\alpha_i) = \frac{n_i\sigma_\alpha^2}{\sigma_e^2 + n_i\sigma_\alpha^2}(\bar{y}_{i\cdot} - \mu), \tag{16}$$

as in (40) of Section 3.4.

It is to be noticed that $\text{BLP}(\alpha_i)$ of (16) involves not only the unknown mean μ but also the unknown variance components, as is evident in the general result (14) where they occur in $\mathbf{C}$ and $\mathbf{V}^{-1}$. Hence, in order to use $\text{BLP}(\alpha_i)$ in practice, one must have estimates of those variance components. Then, on using $\hat{\mu}$ in place of μ, and on replacing each σ^2 in $\text{BLP}(\alpha_i)$ by its estimate, $\hat{\sigma}^2$ say, one has

$$\text{an estimate of BLP }(\alpha_i)\text{ is } \frac{n_i\hat{\sigma}_\alpha^2}{\hat{\sigma}_e^2 + n_i\hat{\sigma}_\alpha^2}(\bar{y}_{i\cdot} - \hat{\mu}),$$

which is no longer a best linear predictor; indeed, it is not even linear in $\mathbf{y}$.

c. Derivation

To derive $\tilde{\mathbf{u}} = \text{BLP}(\mathbf{u})$ of (14), which, it is to be emphasized, needs no assumption of normality, we use $\tilde{\mathbf{u}} = \mathbf{a} + \mathbf{By}$ and minimize, for positive definite symmetric $\mathbf{A}$, the generalized mean squared error of prediction used in (2), namely

$$\begin{aligned} q &= E\,(\tilde{\mathbf{u}} - \mathbf{u})'\mathbf{A}(\tilde{\mathbf{u}} - \mathbf{u}) \\ &= E\,(\mathbf{a} + \mathbf{By} - \mathbf{u})'\mathbf{A}(\mathbf{a} + \mathbf{By} - \mathbf{u})\,. \end{aligned} \tag{17}$$

Using $\tilde{\mathbf{u}} = \mathbf{a} + \mathbf{By}$ ensures that $\mathbf{u}$ is linear in $\mathbf{y}$, and since from (3) we know that the best predictor is $E(\mathbf{u}\,|\,\mathbf{y})$, we now want $\mathbf{a} + \mathbf{By} = E(\mathbf{u}\,|\,\mathbf{y})$. Under normality, this means, from (10), that

$$\tilde{\mathbf{u}} = \mathbf{a} + \mathbf{By} = \boldsymbol{\mu}_U + \mathbf{CV}^{-1}(\mathbf{y} - \boldsymbol{\mu}_Y)\,. \tag{18}$$

However, although $E(\mathbf{u}\,|\,\mathbf{y})$ under normality is always linear in $\mathbf{y}$, we do not know its form under other distributions. Therefore, to derive $\tilde{\mathbf{u}}$ in a more general framework, we begin with q of (17) and have, after defining

$$\begin{aligned} \mathbf{t} &\equiv \mathbf{By} - \mathbf{u}, \\ q &= E\,(\mathbf{a} + \mathbf{t})'\mathbf{A}(\mathbf{a} + \mathbf{t}) \\ &= \mathbf{a}'\mathbf{Aa} + 2\mathbf{a}'\mathbf{A}E(\mathbf{t}) + E(\mathbf{t}'\mathbf{At}) \\ &= [\mathbf{a} + E(\mathbf{t})]'\mathbf{A}[\mathbf{a} + E(\mathbf{t})] - E(\mathbf{t}')\mathbf{A}E(\mathbf{t}) + E(\mathbf{t}'\mathbf{At})\,. \end{aligned} \tag{19}$$

Clearly, this is minimized with respect to $\mathbf{a}$ by choosing

$$\mathbf{a} = -E(\mathbf{t}) = -E(\mathbf{By} - \mathbf{u}) = \boldsymbol{\mu}_U - \mathbf{B}\boldsymbol{\mu}_Y\,. \tag{20}$$

This makes the first term of (19) zero, and so minimizing the other two terms of (19) with respect to $\mathbf{B}$ involves minimizing

$$\begin{aligned} q_1 &= q - [\mathbf{a} + E(\mathbf{t})]'\mathbf{A}[\mathbf{a} + E(\mathbf{t})] \\ &= -E(\mathbf{t}')\mathbf{A}E(\mathbf{t}) + E(\mathbf{t}'\mathbf{At}) \\ &= \text{tr}[\mathbf{A}\,\text{var}(\mathbf{t})], \quad \text{using Theorem S4 of Appendix S.5,} \\ &= \text{tr}[\mathbf{A}\,\text{var}(\mathbf{By} - \mathbf{u})] \\ &= \text{tr}[\mathbf{A}(\mathbf{BVB}' + \mathbf{D} - \mathbf{BC}' - \mathbf{CB}')]\,. \end{aligned}$$

Since tr(**AD**) does not involve **B**, we minimize

$$q_2 = \text{tr}[\mathbf{A}(\mathbf{BVB}' - \mathbf{BC}' - \mathbf{CB}')]$$
$$= \text{tr}[\mathbf{A}(\mathbf{B} - \mathbf{CV}^{-1})\mathbf{V}(\mathbf{B} - \mathbf{CV}^{-1})' - \mathbf{ACV}^{-1}\mathbf{C}'] . \tag{21}$$

Then, because $\mathbf{CV}^{-1}\mathbf{C}'$ does not involve **B**, minimizing q_2 with respect to **B** is achieved by taking

$$\mathbf{B} = \mathbf{CV}^{-1} . \tag{22}$$

This, together with (20), gives

$$\tilde{\mathbf{u}} = \mathbf{a} + \mathbf{By} = \boldsymbol{\mu}_U + \mathbf{CV}^{-1}(\mathbf{y} - \boldsymbol{\mu}_Y), \tag{23}$$

as in (14). And it is identical to $E(\mathbf{u} \mid \mathbf{y})$ under normality.

Note that $\tilde{\mathbf{u}}$ of (23) does not depend on the **A** introduced for generality in (2) and used in (17). But the mean squared error of prediction does involve **A**:

$$E\ \ [\boldsymbol{\mu}_U + \mathbf{CV}^{-1}(\mathbf{y} - \boldsymbol{\mu}_Y) - \mathbf{u}]'\mathbf{A}[\boldsymbol{\mu}_U + \mathbf{CV}^{-1}(\mathbf{y} - \boldsymbol{\mu}_Y) - \mathbf{u}]$$
$$= E\ \ [\mathbf{CV}^{-1}(\mathbf{y} - \boldsymbol{\mu}_Y) - (\mathbf{u} - \boldsymbol{\mu}_U)]'\mathbf{A}[\mathbf{CV}^{-1}(\mathbf{y} - \boldsymbol{\mu}_Y) - (\mathbf{u} - \boldsymbol{\mu}_U)]$$
$$= \text{tr}(\mathbf{A}\{\mathbf{CV}^{-1}E[(\mathbf{y} - \boldsymbol{\mu}_Y)(\mathbf{y} - \boldsymbol{\mu}_Y)']\mathbf{V}^{-1}\mathbf{C}' + E\ \ (\mathbf{u} - \boldsymbol{\mu}_U)(\mathbf{u} - \boldsymbol{\mu}_U)'$$
$$- 2\mathbf{CV}^{-1}E[(\mathbf{y} - \boldsymbol{\mu}_Y)(\mathbf{u} - \boldsymbol{\mu}_U)']\})$$
$$= \text{tr}[\mathbf{A}(\mathbf{CV}^{-1}\mathbf{C}' + \mathbf{D} - 2\mathbf{CV}^{-1}\mathbf{C}')]$$
$$= \text{tr}[\mathbf{A}(\mathbf{D} - \mathbf{CV}^{-1}\mathbf{C}')] .$$

d. Ranking

In establishing (8), that selection on the basis of the best predictor $\tilde{u}$ maximizes $E(u)$ of the selected proportion of the population, Cochran's (1951) development implicitly relies on each scalar $\tilde{u}$ having the same variance and being derived from a y that is independent of other ys. Sampling is over repeated samples of u (scalar) and **y**. However, these conditions are not met for the elements of $\tilde{\mathbf{u}}$ derived in (10). Each such element is derived from the whole vector **y**, their variances are not equal, and the elements of **y** used in one element of $\tilde{\mathbf{u}}$ are not necessarily independent of those used for another element of $\tilde{\mathbf{u}}$. Maximizing the probability of correctly ranking individuals on the basis of elements in $\tilde{\mathbf{u}}$ is therefore not assured. In place of this there is a property about pairwise ranking.

Having predicted the (unobservable) realized values of the random variables in the data, a salient problem that is often of great importance is this: How does the ranking on predicted values compare with the ranking on the true (realized but unobservable) values? Henderson (1963) and Searle (1974) show, under certain conditions (including normality), that the probability that predictors of u_i and u_j have the same pairwise ranking as u_i and u_j is maximized when those predictors are elements of BLUP(**u**) of (23); i.e., the probability $P(\tilde{u}_i - \tilde{u}_j \gtrless 0 \mid u_i - u_j \gtrless 0)$ is maximized. Portnoy (1982) extends this to the case of the usual components of variance model, for which he shows that ranking all the u_is of $\mathbf{u}$ in the same order as the $\tilde{u}_i$s (the best linear predictors) rank themselves does maximize the probability of correctly ranking the u_is. He does, however, go on to show that in models more general than variance components models, there can be predictors that lead to higher values of this probability than do the best linear predictors, which are elements of the vector $\text{BLP}(\mathbf{u}) = \boldsymbol{\mu}_U + \mathbf{C}\mathbf{V}^{-1}(\mathbf{y} - \boldsymbol{\mu}_Y)$.

7.4. MIXED MODEL PREDICTION (BLUP)

The preceding discussion is concerned with the prediction of random variables. Through maximizing the probability of correct ranking, the predictors are appropriate values upon which to base selection; e.g., in genetics, selecting the animals with highest predictions to be parents of the next generation. Since we are concerned here with the prediction (and selection) of random variables, the procedure might be called Model II prediction corresponding to Model II, the random effects model, in analysis of variance. In this connection Lehman (1961) has discussed Model I prediction, corresponding to the fixed effects model. Consideration is now given to mixed model prediction, corresponding to mixed models in which some factors are fixed and others are random.

a. Combining fixed and random effects

The model we use for $\mathbf{y}$ is the familiar

$$\mathbf{y} = \mathbf{X}\boldsymbol{\beta} + \mathbf{Z}\mathbf{u} + \mathbf{e} \tag{24}$$

for $\boldsymbol{\beta}$ being the vector of fixed effects and with $\mathbf{u}$ excluding $\mathbf{e}$, just as at the end of Section 7.1. Then, with

$$E(\mathbf{u}) = \mathbf{0}, \tag{25}$$

we consider the problem of predicting

$$\mathbf{w} = \mathbf{L}'\boldsymbol{\beta} + \mathbf{u} \tag{26}$$

for some known matrix $\mathbf{L}'$, such that $\mathbf{L}'\boldsymbol{\beta}$ is estimable; i.e., $\mathbf{L}' = \mathbf{T}'\mathbf{X}$ for some $\mathbf{T}'$. Since $\mathbf{w}$ involves both fixed effects and random variables, there might be debate as to whether we should "estimate" $\mathbf{w}$ or "predict" $\mathbf{w}$. We will "predict" $\mathbf{w}$, and will choose $\tilde{\mathbf{w}}$ as a predictor to have three properties:

$$\text{"best" in the sense of (2):} \quad \text{minimizing } E(\tilde{\mathbf{w}} - \mathbf{w})'\mathbf{A}(\tilde{\mathbf{w}} - \mathbf{w}), \tag{27}$$

$$\text{linear in } \mathbf{y}: \quad \tilde{\mathbf{w}} = \mathbf{a} + \mathbf{B}\mathbf{y}, \quad \text{with } \mathbf{a} \text{ and } \mathbf{B} \text{ not involving } \boldsymbol{\beta}, \tag{28}$$

$$\text{unbiased:} \quad E(\tilde{\mathbf{w}}) = E(\mathbf{w}). \tag{29}$$

The resulting predictor is a best linear unbiased predictor (BLUP). Note that unbiasedness is now a criterion of the prediction procedure and not just a by-product of it as in Section 7.2. Introducing it as a criterion arises from the presence of $\boldsymbol{\beta}$ in (26).

It is clear from (25) and (26) that $E(\mathbf{w}) = \mathbf{L}'\boldsymbol{\beta}$. We then have

$$\begin{bmatrix} \mathbf{w} \\ \mathbf{y} \end{bmatrix} \sim \left\{ \begin{bmatrix} \mathbf{L}'\boldsymbol{\beta} \\ \mathbf{X}\boldsymbol{\beta} \end{bmatrix}, \begin{bmatrix} \mathbf{D} & \mathbf{C} \\ \mathbf{C}' & \mathbf{V} \end{bmatrix} \right\} \tag{30}$$

similar to (9), although without yet assuming normality. The unbiasedness required of $\tilde{\mathbf{w}}$ in (29) demands that $\mathbf{a} + \mathbf{BX}\boldsymbol{\beta} = \mathbf{L}'\boldsymbol{\beta}$ for all $\boldsymbol{\beta}$, and if $\mathbf{a}$ is not to depend on $\boldsymbol{\beta}$ then $\mathbf{a} = \mathbf{0}$ and $\mathbf{BX} = \mathbf{L}'$. Consequently, the predictor is $\tilde{\mathbf{w}} = \mathbf{By}$, and in $\mathbf{w} = \mathbf{L}'\boldsymbol{\beta} + \mathbf{u}$ the term $\mathbf{L}'\boldsymbol{\beta}$ is an estimable function of $\boldsymbol{\beta}$ in the model $E(\mathbf{y}) = \mathbf{X}\boldsymbol{\beta}$. This limits the form of $\mathbf{L}'$ in $\mathbf{w}$, but it is obviously a reasonable limitation, and the predictor is called BLUP, the best linear unbiased predictor. It is, as we show in sub-section c that follows,

$$\text{BLUP}(\mathbf{w}) = \tilde{\mathbf{w}} = \mathbf{L}'\boldsymbol{\beta}^0 + \mathbf{CV}^{-1}(\mathbf{y} - \mathbf{X}\boldsymbol{\beta}^0) \tag{31}$$

with, as in Appendix S.2,

$$\text{BLUE}(\mathbf{X}\boldsymbol{\beta}) = \mathbf{X}\boldsymbol{\beta}^0 = \mathbf{X}(\mathbf{X}'\mathbf{V}^{-1}\mathbf{X})^{-}\mathbf{X}'\mathbf{V}^{-1}\mathbf{y}\,. \tag{32}$$

Recall (10): the best predictor under normality [and in (14) the best linear predictor] is

$$\tilde{\mathbf{u}} = \boldsymbol{\mu}_U + \mathbf{CV}^{-1}(\mathbf{y} - \boldsymbol{\mu}_Y)\,.$$

When $\boldsymbol{\mu}_U = \mathbf{0}$, and $\boldsymbol{\mu}_Y = E(\mathbf{y}) = \mathbf{X}\boldsymbol{\beta}$, then $\tilde{\mathbf{u}} = \mathbf{CV}^{-1}(\mathbf{y} - \mathbf{X}\boldsymbol{\beta})$, which is the same as the second term of (31) except that (31) has $\mathbf{X}\boldsymbol{\beta}^0$ where $\tilde{\mathbf{u}}$ has $\mathbf{X}\boldsymbol{\beta}$. Moreover, $\tilde{\mathbf{w}}$ has $\mathbf{L}'\boldsymbol{\beta}^0$ where $\mathbf{w}$ has $\mathbf{L}'\boldsymbol{\beta}$. Thus in the predictor $\tilde{\mathbf{w}}$ of $\mathbf{w} = \mathbf{L}'\boldsymbol{\beta} + \mathbf{u}$, there are two parts: $\mathbf{L}'\boldsymbol{\beta}^0$, the BLUE of $\mathbf{L}'\boldsymbol{\beta}$, and $\mathbf{CV}^{-1}(\mathbf{y} - \mathbf{X}\boldsymbol{\beta}^0)$, the best linear predictor of $\mathbf{u}$ when $E(\mathbf{u}) = 0$ and with $\mathbf{X}\boldsymbol{\beta}$ therein replaced by its BLUE, $\mathbf{X}\boldsymbol{\beta}^0$ of (32). To emphasize this we write the predictor as

$$\tilde{\mathbf{w}} = \mathbf{L}'\boldsymbol{\beta}^0 + \mathbf{u}^0 \quad \text{on defining} \quad \mathbf{u}^0 = \text{BLUP}(\mathbf{u}) = \mathbf{CV}^{-1}(\mathbf{y} - \mathbf{X}\boldsymbol{\beta}^0)\,. \tag{33}$$

$\tilde{\mathbf{w}}$ is thus the sum of the BLUE of $\mathbf{L}'\boldsymbol{\beta}$ and the BLUP of $\mathbf{u}$, using $\boldsymbol{\beta}^0$. Result (33) is given in Henderson (1973a) and that part of it not involving $\mathbf{L}'\boldsymbol{\beta}$ is also in Henderson (1963) in a slightly different context. Broad generalizations are considered by Harville (1976).

b. Example

We continue with the example used for illustrating BLP($\mathbf{u}$) in Section 7.3b, namely the 1-way classification random model of Chapter 3. Suppose we take $\mathbf{w}$ as

$$\mathbf{w} = \mu\mathbf{1}_a + \boldsymbol{\alpha}\,.$$

Then

$$\mathbf{L}' = \mathbf{1}_a, \quad \boldsymbol{\beta} = \mu \quad \text{and} \quad \mathbf{X} = \mathbf{1}_N$$

and (31) is

$$\text{BLUP}(\mathbf{w}) = \mu^0\mathbf{1}_a + \mathbf{CV}^{-1}(\mathbf{y} - \mu^0\mathbf{1}_N)\,. \tag{34}$$

$\mathbf{C}$ and $\mathbf{V}^{-1}$ in (34) are exactly as in the example of Section 7.3b, and μ^0 is, from (32), given by

$$\mathbf{1}_N\mu^0 = \mathbf{1}_N(\mathbf{1}'\mathbf{V}^{-1}\mathbf{1})^{-}\mathbf{1}'\mathbf{V}^{-1}\mathbf{y}\,.$$

This reduces, as shown in (34) of Section 3.3, to

$$\mu^0 = \frac{\displaystyle\sum_{i=1}^{a} \frac{n_i\bar{y}_{i\cdot}}{\sigma_e^2 + n_i\sigma_\alpha^2}}{\displaystyle\sum_{i=1}^{a} \frac{n_i}{\sigma_e^2 + n_i\sigma_\alpha^2}}\,.$$

Thus, after substituting into (34) for $\mathbf{CV}^{-1}$ implicit in (15),

$$\text{BLUP}(\mathbf{w}) = \mu^0\mathbf{1}_a + \left\{_{\text{d}}\; \frac{\sigma_\alpha^2}{\sigma_e^2 + n_i\sigma_\alpha^2}\,\mathbf{1}'_{n_i}\right\}(\mathbf{y} - \mu^0\mathbf{1}_N),$$

which gives

$$\text{BLUP}(w_i) = \text{BLUP}(\mu + \alpha_i) = \mu^0 + \frac{n_i\sigma_\alpha^2}{\sigma_e^2 + n_i\sigma_\alpha^2}(\bar{y}_{i\cdot} - \mu^0)\,.$$

This is the same as (42) of Section 3.4. Again, we notice that, in order to use this in practice, estimates of σ_e^2 and σ_α^2 must be available. In contrast to BLP(α_i) of (16), which involves μ, we see that BLUP$(\mu + \alpha_i)$ also uses μ^0; i.e., BLUP includes estimation of the fixed effects whereas BLP does not.

c. Derivation of BLUP

We start with $\mathbf{w} = \mathbf{L}'\boldsymbol{\beta} + \mathbf{u}$ of (26) and the distributional properties of (30); also with $\tilde{\mathbf{w}} = \mathbf{By}$ with $\mathbf{BX} = \mathbf{L}'$ as discussed following (30). Then we choose $\mathbf{B}$ by minimizing, with respect to elements of $\mathbf{B}$ and subject to $\mathbf{BX} = \mathbf{L}'$, the mean squared error of prediction,

$$\begin{aligned} q^* &= E\,(\tilde{\mathbf{w}} - \mathbf{w})'\mathbf{A}(\tilde{\mathbf{w}} - \mathbf{w}) \\ &= E\,(-\mathbf{L}'\boldsymbol{\beta} + \mathbf{By} - \mathbf{u})'\mathbf{A}(-\mathbf{L}'\boldsymbol{\beta} + \mathbf{By} - \mathbf{u})\,. \end{aligned} \tag{35}$$

This is just q of (19) but with $\mathbf{a}$ of (19) replaced by $-\mathbf{L}'\boldsymbol{\beta}$ in (35). Conveniently, this form of $\mathbf{a}$ is exactly as prescribed in (20) for minimizing (19), namely $\mathbf{a} = \boldsymbol{\mu}_U - \mathbf{B}\boldsymbol{\mu}_Y$ with $\boldsymbol{\mu}_U = \mathbf{0}$ and $\boldsymbol{\mu}_Y = \mathbf{X}\boldsymbol{\beta}$, so that $\mathbf{a} = \mathbf{0} - \mathbf{BX}\boldsymbol{\beta} = -\mathbf{L}'\boldsymbol{\beta}$. Hence, just as minimizing (19) was reduced to minimizing (21), so here, too, with (35), except that it is to be done under the condition $\mathbf{BX} = \mathbf{L}'$.

Thus, from (21), we seek to minimize, with respect to $\mathbf{B}$,

$$q_1^* = \text{tr}[\mathbf{A}(\mathbf{BVB}' - \mathbf{BC}' - \mathbf{CB}')], \quad \text{subject to } \mathbf{BX} = \mathbf{L}'\,. \tag{36}$$

To do this, define

$$\mathbf{T} = \mathbf{X}(\mathbf{X}'\mathbf{V}^{-1}\mathbf{X})^{-}\mathbf{X}'\mathbf{V}^{-1} = \mathbf{T}^2 \quad \text{and} \quad \mathbf{Q} = \mathbf{I} - \mathbf{T} = \mathbf{Q}^2, \tag{37}$$

observing that

$$\mathbf{TVT}' = \mathbf{TV} = \mathbf{VT}' \quad \text{and} \quad \mathbf{QVQ}' = \mathbf{QV} = \mathbf{VQ}'\,. \tag{38}$$

Further, note that

$$\mathbf{B} = \mathbf{BT} + \mathbf{BQ}, \tag{39}$$

in which

$$\mathbf{BT} = \mathbf{BX}(\mathbf{X}'\mathbf{V}^{-1}\mathbf{X})^{-}\mathbf{X}'\mathbf{V}^{-1} = \mathbf{L}'(\mathbf{X}'\mathbf{V}^{-1}\mathbf{X})^{-}\mathbf{X}'\mathbf{V}^{-1} . \tag{40}$$

In (40) we see that **BT** does not involve **B**. Therefore minimizing q_1^* subject to $\mathbf{BX} = \mathbf{L}'$ is achieved by substituting into q_1^* of (36) the **B** of (39) after ignoring **BT**; i.e., by replacing **B** in q_1^* with **BQ** and then minimizing with respect to elements of **B**:

$$q_2^* = \text{tr}[\mathbf{A}(\mathbf{BQVQ}'\mathbf{B}' - \mathbf{BQC}' - \mathbf{CQ}'\mathbf{B}')] .$$

But on using (38) it will be found that

$$q_2^* = \text{tr}[\mathbf{A}(\mathbf{BQ} - \mathbf{CV}^{-1}\mathbf{Q})\mathbf{V}(\mathbf{BQ} - \mathbf{CV}^{-1}\mathbf{Q})' - \mathbf{ACV}^{-1}\mathbf{QVQ}'\mathbf{V}^{-1}\mathbf{C}'] .$$

Therefore, since q_2^* is just q_2 of (21) with **B** replaced by **Q** and $\mathbf{CV}^{-1}$ replaced by $\mathbf{CV}^{-1}\mathbf{Q}$, and because just as minimizing q_2 yielded $\mathbf{B} = \mathbf{CV}^{-1}$ of (22), so now the minimization of q_2^* yields

$$\mathbf{BQ} = \mathbf{CV}^{-1}\mathbf{Q} . \tag{41}$$

Substituting this and (40) into (39) and using (37) gives

$$\mathbf{B} = \mathbf{L}'\mathbf{X}(\mathbf{X}'\mathbf{V}^{-1}\mathbf{X})^{-}\mathbf{X}'\mathbf{V}^{-1} + \mathbf{CV}^{-1}[\mathbf{I} - \mathbf{X}(\mathbf{X}'\mathbf{V}^{-1}\mathbf{X})^{-}\mathbf{X}'\mathbf{V}^{-1}] .$$

Hence

$$\tilde{\mathbf{w}} = \mathbf{By}$$

is, using $\mathbf{X}\boldsymbol{\beta}^0$ of (32),

$$\tilde{\mathbf{w}} = \text{BLUP}(\mathbf{w}) = \mathbf{L}'\boldsymbol{\beta}^0 + \mathbf{CV}^{-1}(\mathbf{y} - \mathbf{X}\boldsymbol{\beta}^0),$$

which is (31).

d. Variances and covariances

A variety of results leading to $\text{var}(\tilde{\mathbf{w}})$ is available:

$$\text{var}(\mathbf{L}'\boldsymbol{\beta}^0) = \mathbf{L}'(\mathbf{X}'\mathbf{V}^{-1}\mathbf{X})^{-}\mathbf{L}, \tag{42a}$$

$$\text{var}(\mathbf{u}^0) = \mathbf{CV}^{-1}\mathbf{C}' - \mathbf{CV}^{-1}\mathbf{X}(\mathbf{X}'\mathbf{V}^{-1}\mathbf{X})^{-}\mathbf{X}'\mathbf{V}^{-1}\mathbf{C}', \tag{42b}$$

$$\text{cov}(\mathbf{L}'\boldsymbol{\beta}^0, \mathbf{u}^{0\prime}) = \mathbf{0}, \tag{42c}$$

$$\text{var}(\tilde{\mathbf{w}}) = \text{var}(\mathbf{L}'\boldsymbol{\beta}^0) + \text{var}(\mathbf{u}^0), \tag{42d}$$

$$\text{cov}(\mathbf{u}^0, \mathbf{u}') = \text{var}(\mathbf{u}^0), \tag{42e}$$

$$\text{var}(\mathbf{u}^0 - \mathbf{u}) = \mathbf{D} - \text{var}(\mathbf{u}^0), \tag{42f}$$

$$\text{cov}(\mathbf{L}\boldsymbol{\beta}^0, \mathbf{u}') = \mathbf{L}'(\mathbf{X}'\mathbf{V}^{-1}\mathbf{X})^{-}\mathbf{X}'\mathbf{V}^{-1}\mathbf{C}, \tag{42g}$$

$$\text{var}(\tilde{\mathbf{w}} - \mathbf{w}) = \text{var}(\mathbf{L}'\boldsymbol{\beta}^0) + \text{var}(\mathbf{u}^0 - \mathbf{u}) - \text{cov}(\mathbf{L}'\boldsymbol{\beta}^0, \mathbf{u}') - \text{cov}(\mathbf{u}, \boldsymbol{\beta}^{0\prime}\mathbf{L}) . \tag{42h}$$

Derivation of these results is left to the reader (see E 7.1).

e. Normality

All of the preceding results involve no assumption of normality. On introducing that assumption, as in (9), with $\boldsymbol{\mu}_U = \mathbf{0}$ and $\boldsymbol{\mu}_Y = \mathbf{X}\boldsymbol{\beta}$, we have

$$\begin{bmatrix} \mathbf{w} = \mathbf{L}'\boldsymbol{\beta} + \mathbf{u} \\ \mathbf{y} \end{bmatrix} \sim \mathcal{N}\left(\begin{bmatrix} \mathbf{L}'\boldsymbol{\beta} \\ \mathbf{X}\boldsymbol{\beta} \end{bmatrix}, \begin{bmatrix} \mathbf{D} & \mathbf{C} \\ \mathbf{C}' & \mathbf{V} \end{bmatrix}\right). \tag{43}$$

Then $\mathbf{X}\boldsymbol{\beta}^0$ is the maximum likelihood (as well as the best linear unbiased) estimator of $\mathbf{X}\boldsymbol{\beta}$, for $\mathbf{V}$ assumed known, and since from (43), using Appendix S.3,

$$E(\mathbf{w} \mid \mathbf{y}) = \mathbf{L}'\boldsymbol{\beta} + \mathbf{C}\mathbf{V}^{-1}(\mathbf{y} - \mathbf{X}\boldsymbol{\beta}),$$

it follows that for $\mathbf{V}$ known, $\tilde{\mathbf{w}}$ of (33) is the maximum likelihood estimator of $E(\mathbf{w} \mid \mathbf{y})$. Furthermore, with $\mathbf{u}^0 = \mathbf{C}\mathbf{V}^{-1}(\mathbf{y} - \mathbf{X}\boldsymbol{\beta}^0)$, $\mathbf{u}$ and $\mathbf{u}^0$ are normally distributed with zero means and because of (42b) and (42c)

$$E(\mathbf{u} \mid \mathbf{u}^0) = \text{cov}(\mathbf{u}^0, \mathbf{u}')\,[\text{var}(\mathbf{u}^0)]^{-1}\mathbf{u}^0 = \mathbf{u}^0$$

and

$$\begin{aligned} \text{var}(\mathbf{u} \mid \mathbf{u}^0) &= \mathbf{V} - \text{cov}(\mathbf{u}^0, \mathbf{u}')\,[\text{var}(\mathbf{u}^0)]^{-1}\,\text{cov}(\mathbf{u}^0, \mathbf{u}') \\ &= \mathbf{D} - \text{var}(\mathbf{u}^0) \\ &= \text{var}(\mathbf{u}^0 - \mathbf{u}), \quad \text{as in (42f)}\,. \end{aligned}$$

And, of course, as has already been shown, the elements of $\mathbf{u}^0$ have the property of maximizing the probability of correct rankings. But this property does not hold for elements of $\tilde{\mathbf{w}}$, unless $E(\mathbf{w}) = \mathbf{L}'\boldsymbol{\beta}$ is of the form $\theta\mathbf{1}$, for some scalar θ.

7.5. OTHER DERIVATIONS OF BLUP

The literature contains a number of other derivations of $\mathbf{u}^0 = \mathbf{C}\mathbf{V}^{-1}(\mathbf{y} - \mathbf{X}\boldsymbol{\beta}^0)$ of (33), some of which are now outlined, with details left for the reader in E 7.7, E 7.10 and E 7.11.

a. A two-stage derivation

Bulmer (1980, pp. 208–209) suggests a two-stage approach to the problem of predicting $\mathbf{u}$: first, form a vector of the data $\mathbf{y}$ corrected for the fixed effects (in the genetic context, corrected for the environmental effects):

$$\mathbf{y}_c = \mathbf{y} - \mathbf{X}\boldsymbol{\beta}^0, \tag{44}$$

where $\mathbf{X}\boldsymbol{\beta}^0$ is as in (32). Then, under normality assumptions, $\mathbf{u}$ is predicted by the intuitively appealing regression estimator $E(\mathbf{u} \mid \mathbf{y}_c)$,

$$\tilde{\mathbf{u}} = \text{cov}(\mathbf{u}, \mathbf{y}_c')\,[\text{var}(\mathbf{y}_c)]^{-1}\mathbf{y}_c, \tag{45}$$

which is well-known to be optimal among predictors based on $\mathbf{y}_c$.

Gianola and Goffinet (1982) show that Bulmer's $\tilde{\mathbf{u}}$ is identical to Henderson's $\mathbf{u}^0$ of (33), and include a discussion by Bulmer in which the equivalence is gladly acknowledged. A shorter verification than theirs of this identity, along with the extension of replacing $[\text{var}(\mathbf{y}_c)]^{-1}$ in (45) by $[\text{var}(\mathbf{y}_c)]^{-}$, since $\text{var}(\mathbf{y}_c)$ is singular, uses the matrix $\mathbf{P} = \mathbf{V}^{-1} - \mathbf{V}^{-1}\mathbf{X}(\mathbf{X}'\mathbf{V}^{-1}\mathbf{X})^{-}\mathbf{X}'\mathbf{V}^{-1}$ of (22) in Chapter 6, with $\mathbf{PVP} = \mathbf{P}$. Then $\mathbf{y}_c = \mathbf{VPy}$ and so $\text{var}(\mathbf{y}_c) = \mathbf{VPV}$ and $\text{cov}(\mathbf{u}, \mathbf{y}'_c) = \mathbf{DZ}'\mathbf{PV}$. Then, as can be easily shown (E 7.10), $\tilde{\mathbf{u}}$ of (45) reduces to $\mathbf{u}^0$ of (33).

b. A direct derivation assuming linearity

We confine attention to a predictor of $\mathbf{u}$ that is linear in $\mathbf{y}$, say $\tilde{\mathbf{u}} = \mathbf{a} + \mathbf{By}$; and that is unbiased, in the sense of $E(\tilde{\mathbf{u}}) = E(\mathbf{u})$. Then with $E(\mathbf{u}) = \mathbf{0}$ we have $E(\mathbf{a} + \mathbf{By}) = \mathbf{0}$ and so $\mathbf{a} = \mathbf{0}$ and $\mathbf{BX\boldsymbol{\beta}} = \mathbf{0}$. The latter is true for all $\boldsymbol{\beta}$ only if $\mathbf{BX} = \mathbf{0}$. But this implies $\mathbf{B} = \mathbf{K}(\mathbf{I} - \mathbf{XX}^+)$, for arbitrary $\mathbf{K}$. More than that, though, $\mathbf{BX} = \mathbf{0}$ is also $\mathbf{BV}^{\frac{1}{2}}\mathbf{V}^{-\frac{1}{2}}\mathbf{X} = \mathbf{0}$ and so we can also take $\mathbf{BV}^{\frac{1}{2}} = \mathbf{K}[\mathbf{I} - \mathbf{V}^{-\frac{1}{2}}\mathbf{X}(\mathbf{V}^{-\frac{1}{2}}\mathbf{X})^+]$, which reduces (E 7.11) to $\mathbf{B} = \mathbf{SVP}$, for arbitrary $\mathbf{S}$.

Having $\tilde{\mathbf{u}} = \mathbf{a} + \mathbf{By}$ and $\mathbf{a} = \mathbf{0}$ gives $\tilde{\mathbf{u}} = \mathbf{By}$. With $\mathbf{B} = \mathbf{SVP}$ for arbitrary $\mathbf{S}$, unbiasedness of $\tilde{\mathbf{u}}$ is assured, so in order for $\tilde{\mathbf{u}} = \mathbf{By}$ to be BLUP($\mathbf{u}$) we must also choose $\mathbf{B} = \mathbf{SVP}$ such that $\mathbf{By}$ is BLP($\mathbf{u}$). But with

$$\begin{bmatrix} \mathbf{u} \\ \mathbf{y} \end{bmatrix} \sim \mathcal{N}\left(\begin{bmatrix} \boldsymbol{\mu}_U \\ \boldsymbol{\mu}_Y \end{bmatrix}, \begin{bmatrix} \mathbf{D} & \mathbf{C} \\ \mathbf{C}' & \mathbf{V} \end{bmatrix}\right)$$

of (9), we have from (14)

$$\text{BLP}(\mathbf{u}) = \boldsymbol{\mu}_U + \mathbf{CV}^{-1}(\mathbf{y} - \boldsymbol{\mu}_Y) .$$

Therefore if $\mathbf{By}$ is to be used for deriving BLP($\mathbf{u}$), it will, with $\boldsymbol{\mu}_U = \mathbf{0}$ and $\mathbf{BX\boldsymbol{\beta}} = \mathbf{0}$, be

$$\text{BLP}(\mathbf{u}) = \mathbf{CB}'(\mathbf{BVB}')^{-}\mathbf{By} .$$

But we want $\mathbf{B}$ to be such that this is $\mathbf{By}$. Therefore, with $\mathbf{C} = \text{cov}(\mathbf{u}, \mathbf{y}') = \mathbf{DZ}'$, because $\mathbf{y} = \mathbf{X\boldsymbol{\beta}} + \mathbf{Zu} + \mathbf{e}$, we want

$$\mathbf{CB}'(\mathbf{BVB}')^{-}\mathbf{By} = \mathbf{By} .$$

Since $\mathbf{B} = \mathbf{SVP}$ for arbitrary $\mathbf{S}$, we find that this equality holds if $\mathbf{S} = \mathbf{DZ}'\mathbf{V}^{-1}$. This gives

$$\tilde{\mathbf{u}} = \mathbf{By} = \mathbf{SVPy} = \mathbf{DZ}'\mathbf{V}^{-1}\mathbf{VPy} = \mathbf{DZ}'\mathbf{V}^{-1}(\mathbf{y} - \mathbf{X\boldsymbol{\beta}}^0) .$$

Thus $\tilde{\mathbf{u}}$ is obtained in this fashion is BLUP($\mathbf{u}$) of (33).

c. Partitioning y into two parts

It is of interest to observe that $\mathbf{y}$ can be partitioned into two uncorrelated parts, one of which yields BLUE($\mathbf{X\boldsymbol{\beta}}$) and the other yields BLUP($\mathbf{u}$). To do this write $\mathbf{y}$ as

$$\mathbf{y} = \mathbf{y} - \mathbf{X\boldsymbol{\beta}}^0 + \mathbf{X\boldsymbol{\beta}}^0 = \mathbf{VPy} + (\mathbf{I} - \mathbf{VP})\mathbf{y} . \qquad (46)$$

Then, whereas $\mathbf{VPy}$ is the basis for $\tilde{\mathbf{u}} = \mathbf{DZ'V}^{-1}(\mathbf{VPy})$, the remaining portion of $\mathbf{y}$, namely

$$\mathbf{y} - \mathbf{VPy} = (\mathbf{I} - \mathbf{VP})\mathbf{X\beta} + (\mathbf{I} - \mathbf{VP})(\mathbf{Zu} + \mathbf{e}), \tag{47}$$

which is uncorrelated with $\mathbf{VPy}$, yields a generalized least squares estimator of $\mathbf{X\beta}$ that is $\mathbf{X\beta}^0$ (see E 7.7). Extensions to partitioning prediction error are considered by Harville (1985).

d. A Bayes estimator

In developing (7) of Appendix S.6 we have

$$\pi(p \mid \mathbf{y}) = \frac{f(\mathbf{y}, p)}{f(\mathbf{y})} \tag{48}$$

as the posterior density of p, the parameter associated with the vector $\mathbf{y}$. Adapted to our mixed model

$$\mathbf{y} = \mathbf{X\beta} + \mathbf{Zu} + \mathbf{e}, \tag{49}$$

we have

$$\pi(\mathbf{u} \mid \mathbf{y}) = \frac{f(\mathbf{y}, \mathbf{u})}{f(\mathbf{y})} = \frac{f(\mathbf{y} \mid \mathbf{u})f(\mathbf{u})}{f(\mathbf{y})}.$$

A Bayes estimator of $\mathbf{u}$ is then $E(\mathbf{u} \mid \mathbf{y})$. But, under normality assumptions

$$\begin{bmatrix} \mathbf{u} \\ \mathbf{y} \end{bmatrix} \sim \mathcal{N}\left(\begin{bmatrix} \mathbf{0} \\ \mathbf{X\beta} \end{bmatrix} \begin{bmatrix} \mathbf{D} & \mathbf{DZ'} \\ \mathbf{ZD} & \mathbf{V} \end{bmatrix}\right)$$

we know that

$$\mathbf{u} \mid \mathbf{y} \sim \mathcal{N}[\mathbf{0} + \mathbf{DZ'V}^{-1}(\mathbf{y} - \mathbf{X\beta}), \mathbf{D} - \mathbf{DZ'V}^{-1}\mathbf{ZD}]. \tag{50}$$

Thus a Bayes estimator of $\mathbf{u}$ is $\mathbf{DZ'V}^{-1}(\mathbf{y} - \mathbf{X\beta})$, namely the best linear predictor.

7.6. HENDERSON'S MIXED MODEL EQUATIONS (MME)

a. Derivation

Henderson, in Henderson *et al.* (1959), developed a set of equations that simultaneously yield BLUE($\mathbf{X\beta}$) and BLUP($\mathbf{u}$). They have come to be known as the mixed model equations (MME). They were derived by maximizing the joint density of $\mathbf{y}$ and $\mathbf{u}$, which is, for var($\mathbf{e}$) = $\mathbf{R}$ and $\mathbf{D}$ of order q.

$$\begin{aligned} f(\mathbf{y}, \mathbf{u}) &= f(\mathbf{y} \mid \mathbf{u})f(\mathbf{u}) \\ &= \frac{\exp\{-\frac{1}{2}[(\mathbf{y} - \mathbf{X\beta} - \mathbf{Zu})'\mathbf{R}^{-1}(\mathbf{y} - \mathbf{X\beta} - \mathbf{Zu}) + \mathbf{u}'\mathbf{D}^{-1}\mathbf{u}]\}}{(2\pi)^{\frac{1}{2}(N+q.)}|\mathbf{R}|^{\frac{1}{2}}|\mathbf{D}|^{\frac{1}{2}}}. \end{aligned} \tag{51}$$

Equating to zero the partial derivatives of (51) with respect to elements first of $\boldsymbol{\beta}$ and then of $\mathbf{u}$ gives, using $\tilde{\boldsymbol{\beta}}$ and $\tilde{\mathbf{u}}$ to denote the solutions,

$$\mathbf{X}'\mathbf{R}^{-1}\mathbf{X}\tilde{\boldsymbol{\beta}} + \mathbf{X}'\mathbf{R}^{-1}\mathbf{Z}\tilde{\mathbf{u}} = \mathbf{X}'\mathbf{R}^{-1}\mathbf{y}$$

and (52)

$$\mathbf{Z}'\mathbf{R}^{-1}\mathbf{X}\tilde{\boldsymbol{\beta}} + (\mathbf{Z}'\mathbf{R}^{-1}\mathbf{Z} + \mathbf{D}^{-1})\tilde{\mathbf{u}} = \mathbf{Z}'\mathbf{R}^{-1}\mathbf{y}\,.$$

These are the mixed model equations, written more compactly as

$$\begin{bmatrix} \mathbf{X}'\mathbf{R}^{-1}\mathbf{X} & \mathbf{X}'\mathbf{R}^{-1}\mathbf{Z} \\ \mathbf{Z}'\mathbf{R}^{-1}\mathbf{X} & \mathbf{Z}'\mathbf{R}^{-1}\mathbf{Z} + \mathbf{D}^{-1} \end{bmatrix} \begin{bmatrix} \tilde{\boldsymbol{\beta}} \\ \tilde{\mathbf{u}} \end{bmatrix} = \begin{bmatrix} \mathbf{X}'\mathbf{R}^{-1}\mathbf{y} \\ \mathbf{Z}'\mathbf{R}^{-1}\mathbf{y} \end{bmatrix}. \tag{53}$$

Their form is worthy of note: without the $\mathbf{D}^{-1}$ in the lower right-hand submatrix of the matrix on the left, they would be the ML equations for the model (49) treated as if $\mathbf{u}$ represented fixed effects, rather than random effects.

The form of equations (53) is similar to that resulting from various diagonal augmentation strategies [Piegorsch and Casella (1989) give a history]. Many such strategies arose from numerical, not statistical, considerations (e.g., Levenberg, 1944; Marquardt, 1963; Moré, 1977) designed to provide stable numerical procedures involving ill-conditioned matrices.

A statistical strategy leading to equations like (53) is that of incorporating prior knowledge into analysis of data, i.e., Bayes estimation. Although Henderson's derivation of (53) was essentially classical in nature, it yields the same results as formal Bayes analysis, some details of which are shown in Chapter 9. This analysis dates back to at least Durbin (1953), mostly in the context of the fixed effects, and not the mixed, linear model. An outgrowth of this statistical approach, combined with numerical advantages, led to procedures such as ridge regression (e.g., Hoerl and Kennard, 1970; Hoerl, 1985) and hierarchical Bayes estimation (Lindley and Smith, 1972) and to a variety of other applications.

b. Solutions

Equation (53) has solutions that are even more noteworthy than its form. First, substituting for $\tilde{\mathbf{u}}$ from the second equation of (53) into its first gives

$$\mathbf{X}'\mathbf{R}^{-1}\mathbf{X}\tilde{\boldsymbol{\beta}} + \mathbf{X}'\mathbf{R}^{-1}\mathbf{Z}(\mathbf{Z}'\mathbf{R}^{-1}\mathbf{Z} + \mathbf{D}^{-1})^{-1}\mathbf{Z}'\mathbf{R}^{-1}(\mathbf{y} - \mathbf{X}\tilde{\boldsymbol{\beta}}) = \mathbf{X}'\mathbf{R}^{-1}\mathbf{y},$$

which is

$$\mathbf{X}'\mathbf{B}\mathbf{X}\tilde{\boldsymbol{\beta}} = \mathbf{X}'\mathbf{B}\mathbf{y} \tag{54}$$

for

$$\mathbf{B} = \mathbf{R}^{-1} - \mathbf{R}^{-1}\mathbf{Z}(\mathbf{Z}'\mathbf{R}^{-1}\mathbf{Z} + \mathbf{D}^{-1})^{-1}\mathbf{Z}'\mathbf{R}^{-1} = \mathbf{V}^{-1}\,. \tag{55}$$

The equality $\mathbf{B} = \mathbf{V}^{-1}$ in (55) comes from (28) of Appendix M.5. It gives (54) as $\mathbf{X}'\mathbf{V}^{-1}\mathbf{X}\tilde{\boldsymbol{\beta}} = \mathbf{X}'\mathbf{V}^{-1}\mathbf{y}$, which is the GLS equation for $\boldsymbol{\beta}$, and it yields $\mathbf{X}\tilde{\boldsymbol{\beta}} = \mathbf{X}\boldsymbol{\beta}^0$ of (32).

Second, $\tilde{\mathbf{u}}$ of (53) is BLUP($\mathbf{u}$) of (33). We see this from rewriting the second equation of (53) as

$$\tilde{\mathbf{u}} = (\mathbf{Z}'\mathbf{R}^{-1}\mathbf{Z} + \mathbf{D}^{-1})^{-1}\mathbf{Z}'\mathbf{R}^{-1}(\mathbf{y} - \mathbf{X}\tilde{\boldsymbol{\beta}}),$$

which, on using the identity $(\mathbf{D}^{-1} + \mathbf{Z}'\mathbf{R}^{-1}\mathbf{Z})^{-1}\mathbf{Z}'\mathbf{R}^{-1} = \mathbf{D}\mathbf{Z}'\mathbf{V}^{-1}$, is $\tilde{\mathbf{u}} = \mathbf{D}\mathbf{Z}'\mathbf{V}^{-1}(\mathbf{y} - \mathbf{X}\tilde{\boldsymbol{\beta}})$; and this, since $\mathbf{C} = \text{cov}(\mathbf{u}, \mathbf{y}') = \mathbf{D}\mathbf{Z}'$, is $\mathbf{u}^0$ as in (33).

Equations (53) not only represent a procedure for calculating a $\boldsymbol{\beta}^0$ and $\tilde{\mathbf{u}}$, but they also are computationally more economic than the GLS equations that lead to $\mathbf{X}\boldsymbol{\beta}^0$. Those equations require inversion of $\mathbf{V}$ of order N. But the MMEs of (53) need inversion of a matrix of order only $p + q.$, the total number of levels of fixed and random effects in the data. And this number is usually much smaller than the N, the number of observations. True, (53) does require inversion both of $\mathbf{R}$ and $\mathbf{D}$, but these are often diagonal, e.g., $\mathbf{R}$ having σ_e^2 for every diagonal element and $\mathbf{D}$ having q_i diagonal elements σ_i^2 for $i = 1, \ldots, r$. This makes those inversions easy.

An interesting aspect of the mixed model equations is that elements of them can be used for setting up iterative procedures for calculating solutions to the maximum likelihood (ML) and the restricted maximum likelihood (REML) equations for estimating variance components (see Harville, 1977). Those same elements can also be used for calculating the information matrices for ML and REML estimation. We now show a derivation of these relationships of the MMEs to ML and REML. Unfortunately the algebra is tedious—but is detailed here to save readers from becoming embroiled in their own attempts at establishing the results. Even more detail is available in Searle (1979).

In the development of the MMEs of (53) and their solution (54) and (55) the variance matrices var($\mathbf{u}$) = $\mathbf{D}$ and var($\mathbf{e}$) = $\mathbf{R}$ are perfectly general. But we now confine them to their special forms of the traditional components model, as in (4) and (5) of Section 6.1, namely

$$\mathbf{R} = \sigma_e^2\mathbf{I}_N \quad \text{and} \quad \mathbf{D} = \{_d\, \sigma_i^2\mathbf{I}_{q_i}\}_{i=1}^{r}\ .$$

We then define two further matrices, to be denoted $\mathbf{W}$ and $\mathbf{F}_{ii}$, with $q. = \Sigma_{i=1}^{r} q_i$ being the order of both. The first is

$$\mathbf{W} = (\mathbf{I} + \mathbf{Z}'\mathbf{R}^{-1}\mathbf{Z}\mathbf{D})^{-1} = \{\mathbf{W}_{ij}\}_{i,j=1}^{r}, \quad \text{with } \mathbf{W}^{-1} - \mathbf{I}_{q.} = \mathbf{Z}'\mathbf{R}^{-1}\mathbf{Z}\mathbf{D}\ . \tag{56}$$

The second is a variant of $\mathbf{D}$:

$\mathbf{F}_{ii}$ is $\mathbf{D}$ with unity in place of σ_i^2 and zero in place of σ_j^2 for $j \neq i$. (57)

Then note that

$$\mathbf{D}\mathbf{F}_{ii} = \sigma_i^2\mathbf{F}_{ii} \quad \text{and so} \quad \mathbf{F}_{ii} = \mathbf{D}\mathbf{F}_{ii}/\sigma_i^2\ . \tag{58}$$

Example. For $q_1 = 2$, $q_2 = 3$ and $q_3 = 4$

$$\mathbf{F}_{22} = \begin{bmatrix} \mathbf{0}_{2\times 2} & \mathbf{0} & \mathbf{0} \\ \mathbf{0} & \mathbf{I}_{3\times 3} & \mathbf{0} \\ \mathbf{0} & \mathbf{0} & \mathbf{0}_{4\times 4} \end{bmatrix}.$$

c. Calculations for ML estimation

-i. The estimation equations. The ML equations for the variance components are written in (25) of Section 6.2a as

$$\{_c \operatorname{tr}(\dot{\mathbf{V}}^{-1}\mathbf{Z}_i\mathbf{Z}_i')\}_{i=0}^{r} = \{_c \mathbf{y}'\dot{\mathbf{P}}\mathbf{Z}_i\mathbf{Z}_i'\dot{\mathbf{P}}\mathbf{y}\}_{i=0}^{r} . \tag{59}$$

These are the equations we need to solve for the elements of $\dot{\boldsymbol{\sigma}}^2$ that occur in elements of $\dot{\mathbf{V}}$ and $\dot{\mathbf{P}}$. For notational convenience we here drop the dots, and then show how each element of the two vectors of (59) can be written in terms of the solutions of the MMEs, (53). First, in (59) separate out the elements corresponding to $i = 0$ and use (34) of Chapter 6 with $\mathbf{H}^{-1}$ replaced by $\sigma_e^2\mathbf{V}^{-1}$: thus, for $i = 0$,

$$\begin{aligned}\sigma_e^2 &= \sigma_e^2(\mathbf{y} - \mathbf{X}\boldsymbol{\beta}^0)'\mathbf{V}^{-1}(\mathbf{y} - \mathbf{X}\boldsymbol{\beta}^0)/N \\ &= \sigma_e^2\mathbf{y}'\mathbf{V}^{-1}(\mathbf{y} - \mathbf{X}\boldsymbol{\beta}^0)/N \end{aligned} \tag{60}$$

because $\mathbf{X}'\mathbf{V}^{-1}(\mathbf{y} - \mathbf{X}\boldsymbol{\beta}^0) = \mathbf{X}'\mathbf{V}^{-1}\mathbf{y} - \mathbf{X}'\mathbf{V}^{-1}\mathbf{X}\boldsymbol{\beta}^0 = \mathbf{0}$. It is apparent in (60) that σ_e^2 could be factored out of both sides, but leaving it there simplifies subsequent results. With $\mathbf{V}^{-1}$ of (55)

$$\begin{aligned}\mathbf{V}^{-1}(\mathbf{y} - \mathbf{X}\boldsymbol{\beta}^0) &= [\mathbf{R}^{-1} - \mathbf{R}^{-1}\mathbf{Z}(\mathbf{D}^{-1} + \mathbf{Z}'\mathbf{R}^{-1}\mathbf{Z})^{-1}\mathbf{Z}'\mathbf{R}^{-1}](\mathbf{y} - \mathbf{X}\boldsymbol{\beta}^0) \\ &= \mathbf{R}^{-1}(\mathbf{y} - \mathbf{X}\boldsymbol{\beta}^0) - \mathbf{R}^{-1}\mathbf{Z}(\mathbf{D}^{-1} + \mathbf{Z}'\mathbf{R}^{-1}\mathbf{Z})^{-1} \\ &\quad \times (\mathbf{Z}'\mathbf{R}^{-1}\mathbf{y} - \mathbf{Z}'\mathbf{R}^{-1}\mathbf{X}\boldsymbol{\beta}^0) \\ &= \mathbf{R}^{-1}(\mathbf{y} - \mathbf{X}\boldsymbol{\beta}^0) - \mathbf{R}^{-1}\mathbf{Z}\tilde{\mathbf{u}}, \quad \text{from (53)}, \\ &= \mathbf{R}^{-1}(\mathbf{y} - \mathbf{X}\boldsymbol{\beta}^0 - \mathbf{Z}\tilde{\mathbf{u}}) . \end{aligned} \tag{61}$$

Therefore, using (61) in (60), together with $\mathbf{R} = \sigma_e^2\mathbf{I}$, gives

$$\sigma_e^2 = \sigma_e^2\mathbf{y}'\mathbf{R}^{-1}(\mathbf{y} - \mathbf{X}\boldsymbol{\beta}^0 - \mathbf{Z}\tilde{\mathbf{u}})/N = \mathbf{y}'(\mathbf{y} - \mathbf{X}\boldsymbol{\beta}^0 - \mathbf{Z}\tilde{\mathbf{u}})/N .$$

Therefore, with superscript (m) denoting computed values after m rounds of iteration, we have

$$\tilde{\sigma}_e^{2(m+1)} = \mathbf{y}'(\mathbf{y} - \mathbf{X}\boldsymbol{\beta}^{0(m)} - \mathbf{Z}\tilde{\mathbf{u}}^{(m)})/N . \tag{62}$$

Having, for $i = 0$ in (59) arrived at (62), we now consider

$$\operatorname{tr}(\mathbf{V}^{-1}\mathbf{Z}_i\mathbf{Z}_i') = \mathbf{y}'\mathbf{P}\mathbf{Z}_i\mathbf{Z}_i'\mathbf{P}\mathbf{y} \quad \text{for } i = 1, \ldots, r . \tag{63}$$

First, from (55)

$$\begin{aligned}\mathbf{V}^{-1} &= \mathbf{R}^{-1} - \mathbf{R}^{-1}\mathbf{Z}(\mathbf{Z}'\mathbf{R}^{-1}\mathbf{Z} + \mathbf{D}^{-1})^{-1}\mathbf{Z}'\mathbf{R}^{-1} \\ &= \mathbf{R}^{-1} - \mathbf{R}^{-1}\mathbf{Z}\mathbf{D}(\mathbf{I} + \mathbf{Z}'\mathbf{R}^{-1}\mathbf{Z}\mathbf{D})^{-1}\mathbf{Z}'\mathbf{R}^{-1} \\ &= \mathbf{R}^{-1} - \mathbf{R}^{-1}\mathbf{Z}\mathbf{D}\mathbf{W}\mathbf{Z}'\mathbf{R}^{-1}, \quad \text{using (56)} . \end{aligned} \tag{64}$$

Hence, using $\mathbf{F}_{ii}$, the left-hand side of (63) is

$$\begin{aligned}
\mathrm{tr}(\mathbf{V}^{-1}\mathbf{Z}_i\mathbf{Z}_i') &= \mathrm{tr}(\mathbf{V}^{-1}\mathbf{Z}\mathbf{F}_{ii}\mathbf{Z}'), \quad \text{by the nature of } \mathbf{F}_{ii} \text{ in (57)},\\
&= \mathrm{tr}(\mathbf{Z}'\mathbf{V}^{-1}\mathbf{Z}\mathbf{F}_{ii})\\
&= \mathrm{tr}[\mathbf{Z}'(\mathbf{R}^{-1} - \mathbf{R}^{-1}\mathbf{Z}\mathbf{D}\mathbf{W}\mathbf{Z}'\mathbf{R}^{-1})\mathbf{Z}\mathbf{F}_{ii}], \quad \text{using (64)},\\
&= \mathrm{tr}[(\mathbf{Z}'\mathbf{R}^{-1}\mathbf{Z}\mathbf{D} - \mathbf{Z}'\mathbf{R}^{-1}\mathbf{Z}\mathbf{D}\mathbf{W}\mathbf{Z}'\mathbf{R}^{-1}\mathbf{Z}\mathbf{D})\mathbf{F}_{ii}/\sigma_i^2], \quad \text{using (58)},\\
&= \mathrm{tr}\{[\mathbf{W}^{-1} - \mathbf{I} - (\mathbf{W}^{-1} - \mathbf{I})\mathbf{W}(\mathbf{W}^{-1} - \mathbf{I})]\mathbf{F}_{ii}/\sigma_i^2\}, \quad \text{using (56)},\\
&= \mathrm{tr}[(\mathbf{I} - \mathbf{W})\mathbf{F}_{ii}/\sigma_i^2] \qquad (65)\\
&= \frac{q_i - \mathrm{tr}(\mathbf{W}_{ii})}{\sigma_i^2}, \quad \text{by the nature of } \mathbf{F}_{ii}, \qquad (66)
\end{aligned}$$

where $\mathbf{W}_{ii}$ is the (i, i)th submatrix of $\mathbf{W}$. And the right-hand side of (63) is

$$\mathbf{y}'\mathbf{P}\mathbf{Z}_i\mathbf{Z}_i'\mathbf{P}\mathbf{y} = \mathbf{y}'\mathbf{P}\mathbf{Z}\mathbf{F}_{ii}\mathbf{Z}'\mathbf{P}\mathbf{y} = \mathbf{y}'\mathbf{P}\mathbf{Z}\mathbf{D}\mathbf{F}_{ii}\mathbf{D}\mathbf{Z}'\mathbf{P}\mathbf{y}/\sigma_i^4,$$

from (58). But

$$\mathbf{D}\mathbf{Z}'\mathbf{P}\mathbf{y} = \mathbf{D}\mathbf{Z}'\mathbf{V}^{-1}\mathbf{V}\mathbf{P}\mathbf{y} = \mathbf{D}\mathbf{Z}'\mathbf{V}^{-1}(\mathbf{y} - \mathbf{X}\boldsymbol{\beta}^0) = \tilde{\mathbf{u}},$$

as in (33) with $\mathbf{C} = \mathbf{D}\mathbf{Z}'$. Hence

$$\mathbf{y}'\mathbf{P}\mathbf{Z}_i\mathbf{Z}_i'\mathbf{P}\mathbf{y} = \frac{\tilde{\mathbf{u}}\mathbf{F}_{ii}\tilde{\mathbf{u}}}{\sigma_i^4} = \frac{\tilde{\mathbf{u}}_i'\tilde{\mathbf{u}}_i}{\sigma_i^4}. \qquad (67)$$

Using (66) and (67) in (63) therefore gives (63) as

$$\frac{q_i - \mathrm{tr}(\mathbf{W}_{ii})}{\sigma_i^2} = \frac{\tilde{\mathbf{u}}_i'\tilde{\mathbf{u}}_i}{\sigma_i^4}.$$

This is the result first noted in Patterson and Thompson (1971) and Henderson (1973a). From it come two different iterative procedures:

$$\sigma_i^{2(m+1)} = \frac{\tilde{\mathbf{u}}_i'^{(m)}\tilde{\mathbf{u}}_i^{(m)} + \sigma_i^{2(m)}\,\mathrm{tr}[\mathbf{W}_{ii}^{(m)}]}{q_i} \qquad (68a)$$

or

$$\sigma_i^{2(m+1)} = \frac{\tilde{\mathbf{u}}_i'^{(m)}\tilde{\mathbf{u}}_i^{(m)}}{q_i - \mathrm{tr}[\mathbf{W}_{ii}^{(m)}]}, \qquad (68b)$$

each along with (62)

$$\sigma_e^{2(m+1)} = \mathbf{y}'[(\mathbf{y} - \mathbf{X}\boldsymbol{\beta}^{0(m)} - \mathbf{Z}\tilde{\mathbf{u}}^{(m)})]/N.$$

Comments

(i) $\mathbf{W} = (\mathbf{I} + \mathbf{Z}'\mathbf{R}^{-1}\mathbf{Z}\mathbf{D})^{-1}$ does exist because, on using Appendix M.5, the determinant of $\mathbf{I} + \mathbf{Z}'\mathbf{R}^{-1}\mathbf{Z}\mathbf{D}$ is $|\mathbf{R}^{-1}|\,|\mathbf{V}| \neq 0$.

(ii) $\text{tr}(\mathbf{W}_{ii}) > \mathbf{0}$. This is so because

$$\mathbf{D}^{\frac{1}{2}}\mathbf{W}^{-1} = (\mathbf{I} + \mathbf{D}^{\frac{1}{2}}\mathbf{Z}'\mathbf{R}^{-1}\mathbf{Z}\mathbf{D}^{\frac{1}{2}})\mathbf{D}^{\frac{1}{2}},$$

which implies

$$\mathbf{D}^{\frac{1}{2}}\mathbf{W} = (\mathbf{I} + \mathbf{D}^{\frac{1}{2}}\mathbf{Z}'\mathbf{R}^{-1}\mathbf{Z}\mathbf{D}^{\frac{1}{2}})^{-1}\mathbf{D}^{\frac{1}{2}} = (\mathbf{I} + \mathbf{T}'\mathbf{T})^{-1}\mathbf{D}^{\frac{1}{2}}, \tag{69}$$

for $\mathbf{T} = \mathbf{R}^{-\frac{1}{2}}\mathbf{Z}\mathbf{D}^{\frac{1}{2}}$. And $\mathbf{I} + \mathbf{T}'\mathbf{T}$ is positive definite. Its diagonal elements are positive, as also are those of its inverse. Therefore, since $\mathbf{D}^{\frac{1}{2}}$ is diagonal, the diagonal elements of $\mathbf{W}$ in (69) are positive. Hence $\text{tr}(\mathbf{W}_{ii}) > 0$.

(iii) $q_i - \text{tr}(\mathbf{W}_{ii}) \geqslant 0$. From (66)

$$\frac{q_i - \text{tr}(\mathbf{W}_{ii})}{\sigma_i^2} = \text{tr}(\mathbf{V}^{-1}\mathbf{Z}_i\mathbf{Z}_i') = \text{tr}(\mathbf{Z}_i'\mathbf{V}^{-1}\mathbf{V}\mathbf{V}^{-1}\mathbf{Z}_i) = \text{var}(\mathbf{Z}_i'\mathbf{V}^{-1}\mathbf{y}) \geqslant 0\,.$$

These results, noted by Harville (1977), indicate that as iterative procedures both pairs of equations, (62) and (68a), and (62) and (68b), always yield positive values of $\tilde{\sigma}_i^2$. And these results are for population values of the σ^2 parameters that are to be such that $\mathbf{V}$ is positive definite. But if for a computed iterate of $\tilde{\boldsymbol{\sigma}}^2$ the resulting $\tilde{\mathbf{V}}$ is not positive definite then these positive conditions may not be upheld.

-ii. The information matrix. From (38) of Section 6.3, the information matrix is

$$\tfrac{1}{2}\{_m\, \text{tr}(\mathbf{V}^{-1}\mathbf{Z}_i\mathbf{Z}_i'\mathbf{V}^{-1}\mathbf{Z}_j\mathbf{Z}_j')\}_{i,j=0}^{r}$$

$$= \frac{1}{2}\begin{bmatrix} \text{tr}(\mathbf{V}^{-2}) & \{_r\, \text{tr}(\mathbf{V}^{-1}\mathbf{Z}_i\mathbf{Z}_i'\mathbf{V}^{-1})\}_{i=1}^{r} \\ \{_c\, \text{tr}(\mathbf{V}^{-1}\mathbf{Z}_i\mathbf{Z}_i'\mathbf{V}^{-1})\}_{i=1}^{r} & \{_m\, \text{tr}(\mathbf{V}^{-1}\mathbf{Z}_i\mathbf{Z}_i'\mathbf{V}^{-1}\mathbf{Z}_j\mathbf{Z}_j')\}_{i,j=1}^{r} \end{bmatrix}. \tag{70}$$

The three different trace terms in (70) are now considered, one at a time. First, in the (1, 1) position of (70)

$$\begin{aligned} \text{tr}(\mathbf{V}^{-2}) &= \text{tr}(\mathbf{R}^{-1} - \mathbf{R}^{-1}\mathbf{Z}\mathbf{D}\mathbf{W}\mathbf{Z}'\mathbf{R}^{-1})^2, \quad \text{using (64)}, \\ &= \text{tr}(\mathbf{R}^{-2} - \mathbf{R}^{-2}\mathbf{Z}\mathbf{D}\mathbf{W}\mathbf{Z}'\mathbf{R}^{-1} - \mathbf{R}^{-1}\mathbf{Z}\mathbf{D}\mathbf{W}\mathbf{Z}'\mathbf{R}^{-2} \\ &\qquad - \mathbf{R}^{-1}\mathbf{Z}\mathbf{D}\mathbf{W}\mathbf{Z}'\mathbf{R}^{-2}\mathbf{Z}\mathbf{D}\mathbf{W}\mathbf{Z}'\mathbf{R}^{-1}) \\ &= [\text{tr}(\mathbf{I}_N) - \text{tr}(\mathbf{W}\mathbf{Z}'\mathbf{R}^{-1}\mathbf{Z}\mathbf{D} + \mathbf{W}\mathbf{Z}'\mathbf{R}^{-1}\mathbf{Z}\mathbf{D} \\ &\qquad - \mathbf{W}\mathbf{Z}'\mathbf{R}^{-1}\mathbf{Z}\mathbf{D}\mathbf{W}\mathbf{Z}'\mathbf{R}^{-1}\mathbf{Z}\mathbf{D})]/\sigma_e^4, \end{aligned}$$

after using $\mathbf{R} = \sigma_e^2\mathbf{I}$ and the trace property $\text{tr}(\mathbf{AB}) = \text{tr}(\mathbf{BA})$. Then, with (56) yielding $\mathbf{W}\mathbf{Z}'\mathbf{R}^{-1}\mathbf{Z}\mathbf{D} = \mathbf{I}_{q.} - \mathbf{W}$

$$\begin{aligned} \text{tr}(\mathbf{V}^{-2}) &= [\text{tr}(\mathbf{I}_N) - \text{tr}(\mathbf{I}_{q.} - \mathbf{W} + \mathbf{I}_{q.} - \mathbf{W} - \mathbf{I}_{q.} + 2\mathbf{W} - \mathbf{W}^2)]/\sigma_e^4 \\ &= \frac{N - q. + \text{tr}(\mathbf{W}^2)}{\sigma_e^4}\,. \end{aligned} \tag{71}$$

Second, an element of the row vector in the (1, 2) position of (70) is

$$\text{tr}(\mathbf{V}^{-1}\mathbf{Z}_i\mathbf{Z}_i'\mathbf{V}^{-1}) = \text{tr}(\mathbf{V}^{-1}\mathbf{Z}\mathbf{F}_{ii}\mathbf{Z}'\mathbf{V}^{-1}),$$

where, from using (64) and then (56),

$$\mathbf{Z'V^{-1}} = \mathbf{Z'R^{-1}} - \mathbf{Z'R^{-1}ZDWZ'R^{-1}} = \mathbf{Z'R^{-1}} - (\mathbf{W^{-1}} - \mathbf{I})\mathbf{WZ'R^{-1}} = \mathbf{WZ'R^{-1}}\,.$$

Therefore

$$\begin{aligned}\operatorname{tr}(\mathbf{V}^{-1}\mathbf{Z}_i\mathbf{Z}_i'\mathbf{V}^{-1}) &= \operatorname{tr}(\mathbf{R}^{-1}\mathbf{ZW'F}_{ii}\mathbf{WZ'R}^{-1})\\ &= \operatorname{tr}(\mathbf{F}_{ii}\mathbf{WZ'R}^{-1}\mathbf{ZW'})/\sigma_e^2, \quad \text{because } \mathbf{R} = \sigma_e^2\mathbf{I},\\ &= \operatorname{tr}[\mathbf{F}_{ii}\mathbf{W}(\mathbf{W}^{-1} - \mathbf{I})\mathbf{D}^{-1}\mathbf{W'}]/\sigma_e^2, \quad \text{using (56)},\\ &= \operatorname{tr}[\mathbf{F}_{ii}(\mathbf{D}^{-1}\mathbf{W'} - \mathbf{WD}^{-1}\mathbf{W'})]/\sigma_e^2\,.\end{aligned}$$

But from (56) it is easily shown that $\mathbf{D}^{-1}\mathbf{W'} = \mathbf{WD}^{-1}$. Therefore

$$\begin{aligned}\operatorname{tr}(\mathbf{V}^{-1}\mathbf{Z}_i\mathbf{Z}_i'\mathbf{V}^{-1}) &= \operatorname{tr}[\mathbf{F}_{ii}(\mathbf{D}^{-1}\mathbf{W'} - \mathbf{D}^{-1}\mathbf{W'}^2)]/\sigma_e^2\\ &= \frac{\operatorname{tr}(\mathbf{W}_{ii}) - \sum_{k=1}^{r}\operatorname{tr}(\mathbf{W}_{ik}\mathbf{W}_{ki})}{\sigma_e^2\sigma_i^2}, \quad \text{using (58)}, \qquad (72)\end{aligned}$$

where $\mathbf{W}_{ik}$ is the (i, k)th submatrix of $\mathbf{W}$. Finally, an element of the matrix in the (2, 2) position of (70) is

$$\begin{aligned}\operatorname{tr}(\mathbf{V}^{-1}\mathbf{Z}_i\mathbf{Z}_i'\mathbf{V}^{-1}\mathbf{Z}_j\mathbf{Z}_j') &= \operatorname{tr}(\mathbf{V}^{-1}\mathbf{ZF}_{ii}\mathbf{Z'V}^{-1}\mathbf{ZF}_{jj}\mathbf{Z'}), \quad \text{by the nature of } F_{ii},\\ &= \operatorname{tr}(\mathbf{Z'V}^{-1}\mathbf{ZF}_{ii}\mathbf{Z'V}^{-1}\mathbf{ZF}_{jj}),\\ &= \operatorname{tr}[(\mathbf{I} - \mathbf{W})\mathbf{F}_{ii}(\mathbf{I} - \mathbf{W})\mathbf{F}_{jj}]/\sigma_i^2\sigma_j^2, \quad \text{see (65)},\\ &= \operatorname{tr}(\mathbf{F}_{ii}\mathbf{F}_{jj} - \mathbf{F}_{ii}\mathbf{WF}_{jj} - \mathbf{WF}_{ii}\mathbf{F}_{jj} + \mathbf{WF}_{ii}\mathbf{WF}_{jj})/\sigma_i^2\sigma_j^2\,.\end{aligned}$$

For $i \neq j$ this is

$$0 - 0 - 0 + \operatorname{tr}(\mathbf{W}_{ij}\mathbf{W}_{ji})/\sigma_i^2\sigma_j^2 = \operatorname{tr}(\mathbf{W}_{ij}\mathbf{W}_{ji})/\sigma_i^2\sigma_j^2 \qquad (73)$$

and for $i = j$ it is

$$\operatorname{tr}(\mathbf{F}_{ii} - 2\mathbf{F}_{ii}\mathbf{WF}_{ii} + \mathbf{F}_{ii}\mathbf{WF}_{ii}\mathbf{W})/\sigma_i^4 = [q_i - 2\operatorname{tr}(\mathbf{W}_{ii}) + \operatorname{tr}(\mathbf{W}_{ii}^2)]/\sigma_i^4\,. \qquad (74)$$

Hence on substituting (71)–(74) into (70) the information matrix is

$$I(\tilde{\boldsymbol{\sigma}}^2_{\mathrm{ML}}) = \frac{1}{2}\begin{bmatrix} \dfrac{N - q. + \operatorname{tr}(\mathbf{W}^2)}{\sigma_e^4} & \left\{_{\mathrm{r}}\ \dfrac{\operatorname{tr}(\mathbf{W}_{ii}) - \sum_{k=1}^{r}\operatorname{tr}(\mathbf{W}_{ik}\mathbf{W}_{ki})}{\sigma_e^2\sigma_i^2}\right\}_{i=1}^{r} \\ \text{sym} & \left\{_{\mathrm{m}}\ \dfrac{\delta_{ij}[q_i - 2\operatorname{tr}(\mathbf{W}_{ii})] + \operatorname{tr}(\mathbf{W}_{ij}\mathbf{W}_{ji})}{\sigma_i^2\sigma_j^2}\right\}_{i,j=1}^{r}\end{bmatrix}, \qquad (75)$$

where δ_{ij} is the Kronecker delta, with $\delta_{ii} = 1$ and $\delta_{ij} = 0$ for $i \neq j$.

d. Calculations for REML estimation

-i. The estimation equations. In equation (89) of Chapter 6 the REML estimation equations are given as

$$\{_c\, \text{tr}(\mathbf{PZ}_i\mathbf{Z}_i')\}_{i=0}^{r} = \{_c\, \mathbf{y}'\mathbf{PZ}_i\mathbf{Z}_i'\mathbf{Py}\}_{i=0}^{r} \,. \tag{76}$$

This is equivalent to

$$\text{tr}(\mathbf{P}) = \mathbf{y}'\mathbf{P}^2\mathbf{y} \quad \text{for } i = 0 \tag{77}$$

and

$$\text{tr}(\mathbf{PZ}_i\mathbf{Z}_i') = \mathbf{y}'\mathbf{PZ}_i\mathbf{Z}_i'\mathbf{Py} \quad \text{for } i = 1, 2, \ldots, r\,. \tag{78}$$

We deal first with (77). Multiplying both sides of (78) by σ_i^2, and summing over $i = 1, \ldots, r$ gives

$$\text{tr}\left(\mathbf{P} \sum_{i=1}^{r} \sigma_i^2 \mathbf{Z}_i\mathbf{Z}_i'\right) = \mathbf{y}'\mathbf{P} \sum_{i=1}^{r} \sigma_i^2\mathbf{Z}_i\mathbf{Z}_i'\mathbf{Py}\,.$$

Using $\mathbf{V} = \Sigma_{i=1}^{r} \sigma_i^2 \mathbf{Z}_i\mathbf{Z}_i' + \sigma_e^2\mathbf{I}$ reduces this to

$$\text{tr}[\mathbf{P}(\mathbf{V} - \sigma_e^2\mathbf{I})] = \mathbf{y}'\mathbf{P}(\mathbf{V} - \sigma_e^2\mathbf{I})\mathbf{Py},$$

$$\text{tr}(\mathbf{PV}) - \sigma_e^2\, \text{tr}(\mathbf{P}) = \mathbf{y}'\mathbf{Py} - \sigma_e^2\mathbf{y}'\mathbf{P}^2\mathbf{y}, \quad \text{since } \mathbf{PVP} = \mathbf{P}, \tag{79}$$

$$\text{tr}(\mathbf{PV}) = \mathbf{y}'\mathbf{Py}, \quad \text{on using (77)}\,. \tag{80}$$

Therefore instead of (77) for $i = 0$ in (76) we use (80), which becomes

$$\text{tr}[\mathbf{I} - \mathbf{V}^{-1}\mathbf{X}(\mathbf{X}'\mathbf{V}^{-1}\mathbf{X})^{-}\mathbf{X}'] = \mathbf{y}'\mathbf{PVV}^{-1}\mathbf{VPy},$$

$$N - r_{\mathbf{X}} = (\mathbf{y} - \mathbf{X}\boldsymbol{\beta}^0)'\mathbf{V}^{-1}(\mathbf{y} - \mathbf{X}\boldsymbol{\beta}^0) \tag{81}$$

$$= \mathbf{y}'(\mathbf{y} - \mathbf{X}\boldsymbol{\beta}^0 - \mathbf{Z}\tilde{\mathbf{u}})/\sigma_e^2, \tag{82}$$

where the left-hand side of (81) arises from $\mathbf{I} - \mathbf{V}^{-1}\mathbf{X}(\mathbf{X}'\mathbf{V}^{-1}\mathbf{X})^{-}\mathbf{X}'$ being idempotent and hence having its trace equal to its rank. And the right-hand side of (82) comes from (60) and (61). Therefore (82) gives

$$\hat{\sigma}_e^2 = \frac{\mathbf{y}'(\mathbf{y} - \mathbf{X}\boldsymbol{\beta}^0 - \mathbf{Z}\tilde{\mathbf{u}})}{N - r_{\mathbf{X}}}\,. \tag{83}$$

To deal with (78) we need a lemma, based on observing that since $\mathbf{V} = \mathbf{ZDZ}' + \mathbf{R}$, putting $\mathbf{Z} = \mathbf{0}$ in

$$\mathbf{P} = \mathbf{V}^{-1} - \mathbf{V}^{-1}\mathbf{X}(\mathbf{X}'\mathbf{V}^{-1}\mathbf{X})^{-}\mathbf{X}'\mathbf{V}^{-1} \tag{84}$$

yields what shall be denoted as $\mathbf{S}$:

$$\mathbf{S} = \mathbf{R}^{-1} - \mathbf{R}^{-1}\mathbf{X}(\mathbf{X}'\mathbf{R}^{-1}\mathbf{X})^{-}\mathbf{X}'\mathbf{R}^{-1}\,. \tag{85}$$

Then we have the following lemma.

Lemma

$$\mathbf{P} = \mathbf{S} - \mathbf{SZ}(\mathbf{D}^{-1} + \mathbf{Z}'\mathbf{SZ})^{-1}\mathbf{Z}'\mathbf{S}\,.$$

Proof. For $\mathbf{K}'$ of full row rank $N - r_{\mathbf{X}}$ and $\mathbf{K}'\mathbf{X} = \mathbf{0}$ it is shown in Appendix M.4f that

$$\mathbf{P} = \mathbf{K}(\mathbf{K}'\mathbf{V}\mathbf{K})^{-1}\mathbf{K}' .$$

Therefore, in this equation, as in deriving (85), put $\mathbf{Z} = \mathbf{0}$ and get

$$\mathbf{S} = \mathbf{K}(\mathbf{K}'\mathbf{R}\mathbf{K})^{-1}\mathbf{K}' . \tag{86}$$

But also

$$\begin{aligned}\mathbf{P} &= \mathbf{K}[\mathbf{K}'(\mathbf{ZDZ}' + \mathbf{R})\mathbf{K}]^{-1}\mathbf{K}' = \mathbf{K}(\mathbf{K}'\mathbf{RK} + \mathbf{K}'\mathbf{ZDZ}'\mathbf{K})^{-1}\mathbf{K}' \\ &= \mathbf{K}\{(\mathbf{K}'\mathbf{RK})^{-1} - (\mathbf{K}'\mathbf{RK})^{-1}\mathbf{K}'\mathbf{ZD}[\mathbf{I} + \mathbf{Z}'\mathbf{K}(\mathbf{K}'\mathbf{RK})^{-1}\mathbf{K}'\mathbf{ZD}]^{-1} \\ &\quad \times \mathbf{Z}'\mathbf{K}(\mathbf{K}'\mathbf{RK})^{-1}\}\mathbf{K}', \quad \text{using (28) of Appendix M,} \\ &= \mathbf{S} - \mathbf{SZD}(\mathbf{I} + \mathbf{Z}'\mathbf{SZD})^{-1}\mathbf{Z}'\mathbf{S}, \quad \text{using (86),} \\ &= \mathbf{S} - \mathbf{SZ}(\mathbf{D}^{-1} + \mathbf{Z}'\mathbf{SZ})^{-1}\mathbf{Z}'\mathbf{S} .\end{aligned} \tag{87}$$

Q.E.D.

Note that this result has a form similar to

$$\mathbf{V}^{-1} = \mathbf{R}^{-1} - \mathbf{R}^{-1}\mathbf{Z}(\mathbf{D}^{-1} + \mathbf{Z}'\mathbf{R}^{-1}\mathbf{Z})^{-1}\mathbf{Z}'\mathbf{R}^{-1} . \tag{88}$$

$\mathbf{P}$ of (87) is $\mathbf{V}^{-1}$ of (88) with $\mathbf{R}^{-1}$ replaced by $\mathbf{S}$.

We return to (78). Its left-hand side is $\text{tr}(\mathbf{P}\mathbf{Z}_i\mathbf{Z}_i')$. In (66) we derived

$$\text{tr}(\mathbf{V}^{-1}\mathbf{Z}_i\mathbf{Z}_i') = [q_i - \text{tr}(\mathbf{W}_{ii})]/\sigma_i^2$$

for $\mathbf{W}_{ii}$ being the (i, i)th submatrix of $\mathbf{W} = (\mathbf{I} + \mathbf{Z}'\mathbf{R}^{-1}\mathbf{ZD})^{-1}$. Therefore with $\mathbf{P}$ of (87) being the same as $\mathbf{V}^{-1}$ of (80) only with $\mathbf{R}^{-1}$ replaced by $\mathbf{S}$, we immediately have

$$\text{tr}(\mathbf{P}\mathbf{Z}_i\mathbf{Z}_i') = [q_i - \text{tr}(\mathbf{T}_{ii})]/\sigma_i^2 \tag{89}$$

for $\mathbf{T}_{ii}$ being the (i, i)th submatrix of $\mathbf{W}$ with $\mathbf{R}^{-1}$ replaced by $\mathbf{S}$, namely

$$\mathbf{T} = (\mathbf{I} + \mathbf{Z}'\mathbf{SZD})^{-1} . \tag{90}$$

Thus the left-hand side of (78) is (89). And the right-hand side of (78) is identical to that of (63) and therefore equals (67). Thus the REML equation (78) is the same as (68) only with $\mathbf{T}_{ii}$ in place of $\mathbf{W}_{ii}$, as is evident on comparing (66) and (89). Therefore, with this replacement of $\mathbf{T}$ for $\mathbf{W}$ we can immediately rewrite (68a) and (68b) in the form of iteration equations as

$$\sigma_i^{2(m+1)} = \frac{\tilde{\mathbf{u}}_i'^{(m)}\tilde{\mathbf{u}}_i^{(m)} + \sigma_i^{2(m)}\,\text{tr}(\mathbf{T}_{ii}^{(m)})}{q_i} \tag{91a}$$

or

$$\sigma_i^{2(m+1)} = \frac{\tilde{\mathbf{u}}_i'^{(m)}\tilde{\mathbf{u}}_i^{(m)}}{q_i - \text{tr}\,(\mathbf{T}_{ii}^{(m)})} \tag{91b}$$

along with using (83) as

$$\hat{\sigma}_e^{2(m+1)} = \frac{\mathbf{y}'(\mathbf{y} - \mathbf{X}\boldsymbol{\beta}^{0(m)} - \mathbf{Z}\tilde{\mathbf{u}}^{(m)})}{N - r_{\mathbf{X}}} \,. \tag{92}$$

-ii. The information matrix. The same replacement of **T** for **W** in (75) gives

$$\mathbf{I}(\hat{\sigma}^2_{\text{REML}}) = \frac{1}{2}\begin{bmatrix} \dfrac{N - q. + \operatorname{tr}(\mathbf{T}^2)}{\sigma_e^4} & \left\{{}_{\text{r}} \dfrac{\operatorname{tr}(\mathbf{T}_{ii}) - \sum_{k=1}^{r} \operatorname{tr}(\mathbf{T}_{ik}\mathbf{T}_{ki})}{\sigma_e^2\sigma_i^2}\right\}_{i=1}^{r} \\ \text{sym} & \left\{{}_{\text{m}} \dfrac{\delta_{ij}[q_i - 2\operatorname{tr}(\mathbf{T}_{ii})] + \operatorname{tr}(\mathbf{T}_{ij}\mathbf{T}_{ji})}{\sigma_i^2\sigma_j^2}\right\}_{i,j=1}^{r} \end{bmatrix} . \tag{93}$$

e. Iterative procedures summarized

-i. Adapting the MME. The part that **W** of (56) plays in the MME of (53) is that on defining **v** by means of

$$\mathbf{D}\mathbf{v} = \mathbf{u}, \tag{94}$$

equations (53) can be written as

$$\begin{bmatrix} \mathbf{X}'\mathbf{R}^{-1}\mathbf{X} & \mathbf{X}'\mathbf{R}^{-1}\mathbf{Z}\mathbf{D} \\ \mathbf{Z}'\mathbf{R}^{-1}\mathbf{X} & \mathbf{I} + \mathbf{Z}'\mathbf{R}^{-1}\mathbf{Z}\mathbf{D} \end{bmatrix}\begin{bmatrix} \tilde{\boldsymbol{\beta}} \\ \tilde{\mathbf{v}} \end{bmatrix} = \begin{bmatrix} \mathbf{X}'\mathbf{R}^{-1}\mathbf{y} \\ \mathbf{Z}'\mathbf{R}^{-1}\mathbf{y} \end{bmatrix} . \tag{95}$$

Even though the matrix on the left-hand side is not symmetric, there is a computational advantage to these equations over (53). Suppose at some intermediate round of a numerical (iterative) procedure for calculating variance components estimates that one of the calculated σ^2-values is zero (which it can be, under ML estimation). Using that zero in **D** makes **D** singular, and then the $\mathbf{D}^{-1}$ occurring in (53) does not exist. But $\mathbf{W} = (\mathbf{I} + \mathbf{Z}'\mathbf{R}^{-1}\mathbf{Z}\mathbf{D})^{-1}$ always exists, as pointed out by Harville (1977).

-ii. Using the MME. Using $\mathbf{R} = \sigma_e^2\mathbf{I}$, from (56), (95) and (94)

$$\mathbf{W}^{(m)} = (\sigma_e^{2(m)}\mathbf{I} + \mathbf{Z}'\mathbf{Z}\mathbf{D}^{(m)})^{-1}\sigma_e^{2(m)} \,. \tag{96}$$

$$\begin{bmatrix} \mathbf{X}'\mathbf{X} & \mathbf{X}'\mathbf{Z}\mathbf{D}^{(m)} \\ \mathbf{Z}'\mathbf{X} & \mathbf{W}^{(m)} \end{bmatrix}\begin{bmatrix} \boldsymbol{\beta}^{(m)} \\ \mathbf{v}^{(m)} \end{bmatrix} = \begin{bmatrix} \mathbf{X}'\mathbf{y} \\ \mathbf{Z}'\mathbf{y} \end{bmatrix}, \tag{97}$$

$$\mathbf{u}^{(m)} = \mathbf{D}^{(m)}\mathbf{v}^{(m)}, \tag{98}$$

where symbols $\boldsymbol{\beta}^{0(m)}$ and $\hat{\mathbf{u}}^{(m)}$ of (92) replaced here by $\boldsymbol{\beta}^{(m)}$ and $\mathbf{u}^{(m)}$ emphasizes that the latter are simply computed values.

-iii. Iterating for ML

$$\sigma_e^{2(m+1)} = \mathbf{y}'(\mathbf{y} - \mathbf{X}\boldsymbol{\beta}^{(m)} - \mathbf{Z}\mathbf{u}^{(m)})/N, \tag{62}$$

$$\sigma_i^{2(m+1)} = \frac{\mathbf{u}_i'^{(m)}\mathbf{u}_i^{(m)} + \sigma_i^{2(m)}\operatorname{tr}(\mathbf{W}_{ii}^{(m)})}{q_i}, \tag{68a}$$

$$\sigma_i^{2(m+1)} = \frac{\mathbf{u}_i'^{(m)}\mathbf{u}_i^{(m)}}{q_i - \operatorname{tr}(\mathbf{W}_{ii}^{(m)})}. \tag{68b}$$

Procedure

1. Decide on starting values $\sigma_e^{2(0)}$ and $\sigma_i^{2(0)}$, and set $m = 0$.
2. Calculate (96), solve (97) for $\boldsymbol{\beta}^{(m)}$ and $\mathbf{v}^{(m)}$, and calculate (98).
3. Calculate (62) and *either* (68a) or (68b).

4a. If convergence is reached for the σ^2s, set $\tilde{\boldsymbol{\sigma}}^2 = \boldsymbol{\sigma}^{2(m+1)}$. Denote the resulting calculated terms as $\tilde{\mathbf{W}} = \mathbf{W}^{(m+1)}$, $\tilde{\boldsymbol{\beta}} = \boldsymbol{\beta}^{(m+1)}$, $\tilde{\mathbf{v}} = \mathbf{v}^{(m+1)}$ and $\tilde{\mathbf{u}} = \mathbf{u}^{(m+1)}$. Use $\tilde{\boldsymbol{\sigma}}^2$ and $\tilde{\mathbf{W}}$ to calculate $\mathbf{I}(\tilde{\boldsymbol{\sigma}}^2)$ from (75).

4b. If convergence is not reached, increase m by unity and return to step 2. At each repeat of step 3 use whichever of (68a) or (68b) was used on the first occasion.

-iv. Iterating for REML. From (85) and (90)

$$\mathbf{T}^{(m)} = \left\{\mathbf{I} + \frac{\mathbf{Z}'[\mathbf{I} - \mathbf{X}(\mathbf{X}'\mathbf{X})^-\mathbf{X}']\mathbf{Z}\mathbf{D}^{(m)}}{\sigma_e^{2(m)}}\right\}^{-1}, \tag{99}$$

$$\sigma_e^{2(m+1)} = \frac{\mathbf{y}'(\mathbf{y} - \mathbf{X}\boldsymbol{\beta}^{(m)} - \mathbf{Z}\mathbf{u}^{(m)})}{N - r_{\mathbf{X}}}, \tag{92}$$

$$\sigma_i^{2(m+1)} = \frac{\mathbf{u}_i'^{(m)}\mathbf{u}_i^{(m)} + \sigma_i^{2(m)}\operatorname{tr}(\mathbf{T}_{ii}^{(m)})}{q_i}, \tag{91a}$$

$$\sigma_i^{2(m+1)} = \frac{\mathbf{u}_i'^{(m)}\mathbf{u}_i^{(m)}}{q_i - \operatorname{tr}(\mathbf{T}_{ii}^{(m)})}. \tag{91b}$$

Procedure

1. Decide on starting values $\sigma_e^{2(0)}$ and $\sigma_i^{2(0)}$ and set $m = 0$.
2. As for ML, calculate (96), solve (97) for $\boldsymbol{\beta}^{(m)}$ and $\mathbf{v}^{(m)}$, and calculate (98).
3. Calculate (99).
4. Calculate (92) and *either* (91a) or (91b).

5a. If convergence is reached for the σ^2s, set $\hat{\boldsymbol{\sigma}}^2 = \boldsymbol{\sigma}^{2(m+1)}$, and do steps 2 and 3 using $\boldsymbol{\sigma}^{2(m+1)}$. Denote the resulting calculated terms as $\tilde{\mathbf{W}} = \mathbf{W}^{(m+1)}$, $\tilde{\boldsymbol{\beta}} = \boldsymbol{\beta}^{(m+1)}$, $\tilde{\mathbf{v}} = \mathbf{v}^{(m+1)}$, $\tilde{\mathbf{u}} = \mathbf{u}^{(m+1)}$ and $\tilde{\mathbf{T}} = \mathbf{T}^{(m+1)}$. Use $\tilde{\boldsymbol{\sigma}}^2$ and $\tilde{\mathbf{T}}$ to calculate $\mathbf{I}(\hat{\boldsymbol{\sigma}}^2)$ from (93).

5b. If convergence is not reached, increase m by one, and return to step 2. At each repeat of step 4 use whichever of (91a) or (91b) was used on the first occasion.

f. A summary

Henderson's mixed model equations (53) are a convenient device for simultaneously calculating BLUE($\mathbf{X\beta}$) and BLUP($\mathbf{u}$). Elements of those equations [equations (62) and (68)] can also be used to iteratively calculate ML estimates of variance components and their information matrix, (75). Equations (91), (92) and (93) show companion results for REML estimation.

7.7. SUMMARY

Model

$$\mathbf{y} = \mathbf{X\beta} + \mathbf{Zu} + \mathbf{e}\,.$$

Best predictor: Sections 7.2a, b, c

$$\tilde{\mathbf{u}} = E(\mathbf{u}\,|\,\mathbf{y}); \tag{3}$$

$$E_y(\tilde{\mathbf{u}}) = E(\tilde{\mathbf{u}}); \tag{4}$$

$$\text{var}(\tilde{\mathbf{u}} - \mathbf{u}) = E_y[\text{var}(\mathbf{u}\,|\,\mathbf{y})]; \tag{5}$$

$$\text{cov}(\tilde{\mathbf{u}}, \mathbf{u}') = \text{var}(\tilde{\mathbf{u}}), \quad \text{cov}(\tilde{\mathbf{u}}, \mathbf{y}') = \text{cov}(\mathbf{u}, \mathbf{y}'); \tag{6}$$

$$\rho(\tilde{u}, u) \text{ is maximized}; \quad \rho(\tilde{u}, u) = \sigma_{\tilde{u}}/\sigma_u\,. \tag{7}$$

Under normality: Section 7.2d

$$\tilde{\mathbf{u}} = E(\mathbf{u}\,|\,\mathbf{y}) = \boldsymbol{\mu}_U + \mathbf{CV}^{-1}(\mathbf{y} - \boldsymbol{\mu}_Y); \tag{10}$$

$$\text{var}(\tilde{\mathbf{u}} - \mathbf{u}) = \mathbf{D} - \mathbf{CV}^{-1}\mathbf{C}'; \tag{11}$$

$$\begin{aligned} \text{cov}(\tilde{\mathbf{u}}, \mathbf{u}') = \text{var}(\tilde{\mathbf{u}}) &= \mathbf{CV}^{-1}\mathbf{C} \\ \rho(\tilde{u}_i, u_i) &= \sqrt{\mathbf{c}_i'\mathbf{V}^{-1}\mathbf{c}_i/\sigma_{u_i}^2}\,. \end{aligned} \tag{12}$$

Best linear predictor: Section 7.3a

$$\text{BLP}(\mathbf{u}) = \boldsymbol{\mu}_U + \mathbf{CV}^{-1}(\mathbf{y} - \boldsymbol{\mu}_Y)\,. \tag{14}$$

Mixed model prediction: Section 7.4

$$\mathbf{w} = \mathbf{L}'\boldsymbol{\beta} + \mathbf{u}; \tag{26}$$

$$\tilde{\mathbf{w}} = \text{BLUP}(\mathbf{L}'\boldsymbol{\beta} + \mathbf{u}) = \mathbf{L}'\boldsymbol{\beta}^0 + \mathbf{CV}^{-1}(\mathbf{y} - \mathbf{X}\boldsymbol{\beta}^0) \tag{31}$$

$$= \mathbf{L}'\boldsymbol{\beta}^0 + \mathbf{u}^0, \tag{33}$$

with

$$\mathbf{X}\boldsymbol{\beta}^0 = \mathbf{X}(\mathbf{X}'\mathbf{V}^{-1}\mathbf{X})^-\mathbf{X}'\mathbf{V}^{-1}\mathbf{y} \tag{32}$$

and

$$\mathbf{u}^0 = \mathbf{CV}^{-1}(\mathbf{y} - \mathbf{X}\boldsymbol{\beta}^0); \tag{33}$$

$$\text{var}(\mathbf{L}'\boldsymbol{\beta}^0) = \mathbf{L}'(\mathbf{X}'\mathbf{V}^{-1}\mathbf{X})^{-}\mathbf{L}, \tag{42a}$$

$$\text{var}(\mathbf{u}^0) = \mathbf{CV}^{-1}\mathbf{C}' - \mathbf{CV}^{-1}\mathbf{X}(\mathbf{X}'\mathbf{V}^{-1}\mathbf{X})^{-}\mathbf{X}'\mathbf{V}^{-1}\mathbf{C}, \tag{42b}$$

$$\text{cov}(\mathbf{L}'\boldsymbol{\beta}^0, \mathbf{u}^{0\prime}) = \mathbf{0}, \tag{42c}$$

$$\text{var}(\tilde{\mathbf{w}}) = \text{var}(\mathbf{L}'\boldsymbol{\beta}^0) + \text{var}(\mathbf{u}^0), \tag{42d}$$

$$\text{cov}(\mathbf{u}^0, \mathbf{u}') = \text{var}(\mathbf{u}^0), \tag{42e}$$

$$\text{var}(\mathbf{u}^0 - \mathbf{u}) = \mathbf{D} - \text{var}(\mathbf{u}^0), \tag{42f}$$

$$\text{cov}(\mathbf{L}'\boldsymbol{\beta}^0, \mathbf{u}) = \mathbf{L}'(\mathbf{X}'\mathbf{V}^{-1}\mathbf{X})^{-}\mathbf{X}'\mathbf{V}^{-1}\mathbf{C}', \tag{42g}$$

$$\text{var}(\tilde{\mathbf{w}} - \mathbf{w}) = \text{var}(\mathbf{L}'\boldsymbol{\beta}^0) + \text{var}(\mathbf{u}^0 - \mathbf{u}) - \text{cov}(\mathbf{L}'\boldsymbol{\beta}^0, \mathbf{u}) - \text{cov}(\mathbf{u}, \boldsymbol{\beta}^{0\prime}\mathbf{L})\,. \tag{42h}$$

Other derivations

Two-stage derivation: Section 7.5a (45)

A direct derivation assuming linearity: Section 7.5b.

Partitioning **y** into two parts: Section 7.5c (46)

A Bayes estimator: Section 7.5d (48)(50)

Henderson's mixed model equations: Section 7.6 (53)

Used in ML: Section 7.6c

Equations (68)

Information matrix (75)

Used in REML: Section 7.6d

Equations (91)

Information matrix (93)

7.8. EXERCISES

E 7.1. Verify (42a) through (42h).

E 7.2. For the conditions of Section 7.6 derive inverses of

$$\mathbf{V}, \quad \text{var}\begin{bmatrix}\mathbf{u}\\ \mathbf{y}\end{bmatrix} \quad \text{and} \quad \mathbf{C} = \begin{bmatrix}\mathbf{X}'\mathbf{R}^{-1}\mathbf{X} & \mathbf{X}'\mathbf{R}^{-1}\mathbf{Z}\\ \mathbf{Z}'\mathbf{R}^{-1}\mathbf{X} & \mathbf{Z}'\mathbf{R}^{-1}\mathbf{Z} + \mathbf{D}\end{bmatrix}.$$

[Use (27) and (28b) of Appendix M.5.]

E 7.3. Based on E 7.2 with $\mathbf{W} = [\mathbf{X} \quad \mathbf{Z}]$ and $\mathbf{P}$ of (84), show that $\mathbf{WC}^{-1}\mathbf{W}' = \mathbf{R} - \mathbf{RPR}$.

E 7.4. Prove that $(\mathbf{D}^{-1} + \mathbf{Z}'\mathbf{R}^{-1}\mathbf{Z})^{-1} = \mathbf{D} - \mathbf{DZ}'\mathbf{V}^{-1}\mathbf{ZD}$, in three different ways.

E 7.5. [Henderson (1977).] Suppose in addition to $\mathbf{u}$ of the model equation $\mathbf{y} = \mathbf{X}\boldsymbol{\beta} + \mathbf{Zu} + \mathbf{e}$ that we have $\mathbf{v}$, another vector of random effects, of the same nature as $\mathbf{u}$ but with no observations on them. Suppose $E(\mathbf{v}) = \mathbf{0}$, $\text{var}(\mathbf{v}) = \mathbf{T}$ and $\text{cov}(\mathbf{u}, \mathbf{v}') = \mathbf{H}$.

(a) What is the best linear predictor of $\mathbf{v}$?

(b) What is BLUP($\mathbf{v}$)?

(c) Show how to calculate BLUP($\mathbf{v}$) from BLUP($\mathbf{u}$).

(d) On defining

$$\begin{bmatrix} \mathbf{Q}_{11} & \mathbf{Q}_{12} \\ \mathbf{Q}_{12} & \mathbf{Q}_{22} \end{bmatrix} = \begin{bmatrix} \mathbf{D} & \mathbf{H} \\ \mathbf{H}' & \mathbf{T} \end{bmatrix}^{-1},$$

show that

$$\begin{bmatrix} \mathbf{X}'\mathbf{R}^{-1}\mathbf{X} & \mathbf{X}'\mathbf{R}^{-1}\mathbf{Z} & \mathbf{0} \\ \mathbf{Z}'\mathbf{R}^{-1}\mathbf{X} & \mathbf{Z}'\mathbf{R}^{-1}\mathbf{Z} + \mathbf{Q}_{11} & \mathbf{Q}_{12} \\ \mathbf{0} & \mathbf{Q}'_{12} & \mathbf{Q}_{22} \end{bmatrix} \begin{bmatrix} \boldsymbol{\beta}^0 \\ \mathbf{u}^0 \\ \mathbf{v}^0 \end{bmatrix} = \begin{bmatrix} \mathbf{X}'\mathbf{R}^{-1}\mathbf{y} \\ \mathbf{Z}'\mathbf{R}^{-1}\mathbf{y} \\ \mathbf{0} \end{bmatrix}$$

yields BLUP($\mathbf{v}$) and $\boldsymbol{\beta}^0$ and $\mathbf{u}^0$ of (32) and (33).

E 7.6. Using the usual normality assumptions for the terms of the model equation $\mathbf{y} = \mathbf{X}\boldsymbol{\beta} + \mathbf{Zu} + \mathbf{e}$, derive and simplify $L(\mathbf{y}) - f(\mathbf{y}, \mathbf{u})$, where $L(\mathbf{y})$ is the likelihood of $\mathbf{y}$ and $f(\mathbf{y}, \mathbf{u})$ is given in (51).

E 7.7. (a) Show that $(\mathbf{I} - \mathbf{VP})\mathbf{y}$ and $\mathbf{VPy}$ are uncorrelated.

(b) Show that $\text{var}(\mathbf{y} - \mathbf{VPy})$ is $\mathbf{V} - \mathbf{VPV}$ with generalized inverse $\mathbf{V}^{-1} - \mathbf{P}$.

(c) Derive the GLSE of $\mathbf{X}\boldsymbol{\beta}$ from (47).

E 7.8. The model equation for the 2-way crossed classification can be taken as $y_{ij} = \mu + \alpha_i + \beta_j + e_{ij}$ for $i = 1, \ldots, a$ and $j = 1, \ldots, b$. If the α_is are taken as random with dispersion matrix $\sigma_\alpha^2 \mathbf{I}_a$, and if the β_js are taken as fixed, establish the following results.

(a) $$\mathbf{V}^{-1} = \mathbf{I}_a \otimes \frac{1}{\sigma_e^2}(\mathbf{I}_b - \lambda \mathbf{J}_b) \quad \text{for } \lambda = \frac{\sigma_\alpha^2}{\sigma_e^2 + b\sigma_\alpha^2}.$$

(b) $$(\mathbf{X}'\mathbf{V}^{-1}\mathbf{X})^- = \sigma_e^2 \begin{bmatrix} 0 & \mathbf{0} \\ \mathbf{0} & \frac{1}{a}\left(\mathbf{I}_b + \frac{b\lambda}{1 - b\lambda}\mathbf{J}_b\right) \end{bmatrix}.$$

(c) $\tilde{\beta}_j = \bar{y}_{.j}$.

(d) $\tilde{\alpha}_i = b\lambda(\bar{y}_{i.} - \bar{y}_{..})$.

E 7.9. For the mixed model $\mathbf{y} = \mathbf{X}\boldsymbol{\beta} + \mathbf{Zu} + \mathbf{e}$ of this chapter derive the variances of $\text{BLP}(\mathbf{u}) - \mathbf{u}$, of $\text{BLUP}(\mathbf{u}) - \mathbf{u}$ and of $\text{BLUP}(\mathbf{u}) - \text{BLP}(\mathbf{u})$, and show a linear relationship between those variances.

E 7.10. Show the details of reducing (45) to (34), as outlined in Section 7.5a.

E 7.11. Prove

(a) $(\mathbf{D}^{-1} + \mathbf{Z}'\mathbf{R}^{-1}\mathbf{Z})^{-1}\mathbf{Z}'\mathbf{R}^{-1} = \mathbf{DZ}'\mathbf{V}^{-1}$;

(b) $|\mathbf{I} + \mathbf{Z}'\mathbf{R}^{-1}\mathbf{ZD}| = |\mathbf{V}|/|\mathbf{R}|$.

E 7.12. Define $\tilde{\mathbf{e}} = \mathbf{RV}^{-1}(\mathbf{y} - \mathbf{X}\boldsymbol{\beta}^0)$ for $\mathbf{V} = \mathbf{ZDZ}' + \mathbf{R}$ and $\mathbf{X}\boldsymbol{\beta}^0$ of (32). Prove that $\tilde{\mathbf{e}} = \mathbf{y} - \mathbf{X}\boldsymbol{\beta}^0 - \mathbf{Z}\tilde{\mathbf{u}}$.

E 7.13. (a) Show that q^* of (35) can be expressed as

$$q^* = \text{tr}\left(\begin{bmatrix} \mathbf{A} & -\mathbf{AB} \\ -\mathbf{B}'\mathbf{A} & \mathbf{B}'\mathbf{AB} \end{bmatrix}\begin{bmatrix} \mathbf{D} & \mathbf{C} \\ \mathbf{C}' & \mathbf{V} \end{bmatrix}\right).$$

(b) Minimize q^* with respect to $\mathbf{B}$, subject to $\mathbf{BX} = \mathbf{L}'$, using a Lagrange multiplier matrix $\mathbf{T}$ to take account of the condition $\mathbf{BX} = \mathbf{L}'$.

(c) For $\mathbf{B}$ obtained in (b) show that $\mathbf{By}$ is $\tilde{\mathbf{w}}$ of (31).

CHAPTER 8

COMPUTING ML AND REML ESTIMATES

8.1. INTRODUCTION

Maximum likelihood and restricted maximum likelihood estimation of variance components in Chapter 6 produce, in general, no analytical expressions for the estimators. Indeed, in only some cases of balanced data (see Section 4.8e) are there analytical expressions for variance component estimators. Furthermore, equating to zero the first derivatives of the log likelihood leads to nonlinear equations, e.g., equations (25), (27a), (89) and (90) of Chapter 6. Even solving these complicated systems of nonlinear equations is an oversimplification of the situation, since the log likelihood must be maximized within the parameter space. If the maximum occurs on the boundary of the parameter space then the derivative of the log likelihood is unlikely to equal zero at the maximum and the systems of equations no longer apply. In this chapter we review considerations in computing ML and REML estimates and outline some of the numerical techniques used for calculating the estimates.

The starting point for the numerical methods is the log likelihood rather than its first derivative or the equations of Chapter 6 just referred to. We use the log likelihood rather than the likelihood because it takes a simpler form and is numerically more tractable. We operate on the log likelihood itself rather than use derivative equations for two reasons. First, when techniques are based just on setting the first derivatives equal to zero, it is not possible to distinguish saddlepoints and minima from maxima. All three have first derivatives equal to zero. Second, progression towards a maximum can be checked by making sure that whatever iterative procedure is being used does increase the value of the log likelihood at each iteration. Checking that the first derivative is decreasing in absolute value towards zero is only peripherally related to increases in the log likelihood.

Trying to maximize the log likelihood or restricted log likelihood is, in general, a very difficult numerical problem. It, like the first derivatives of the log likelihood, is a complicated, nonlinear function of the parameters. The likelihood contains the inverse of an $N \times N$ matrix for ML and an $(N-p) \times (N-p)$ matrix for REML, a difficult computational problem in its own right, especially for large data sets. In addition, the maximization problem is a constrained maximization problem: even in the simplest situation of estimating variance components, the variances must be constrained to be not less than zero; and maxima can occur on the boundary of the parameter space. For models with covariances between the random effects, the constraints become even more complicated. Finally, the log likelihood surface can have local maxima. So, for a particular data set, even if an algorithm has converged to a local maximum, there is no guarantee that it is a global maximum.

Figure 8.1 illustrates some of the difficulties that can occur. It shows the log likelihood surface for a two-way crossed classification, random model, with no interaction. The error variance has been set equal to unity, so that the surface is only a function of σ_1^2 and σ_2^2. The surface exhibits two local maxima at (0.1, 5.5) and (6.2, 0.1). And each maximum is near one of the boundaries. These are some of the features of likelihoods that make it very difficult for numerical algorithms to reliably find ML or REML estimates.

How then do numerical algorithms attempt to maximize the log likelihood or restricted likelihood? We can first note that the problem is primarily one of estimating the variance components and not the fixed effects. For example, in Chapter 6 equation (25) does not involve the fixed effects and, given a solution

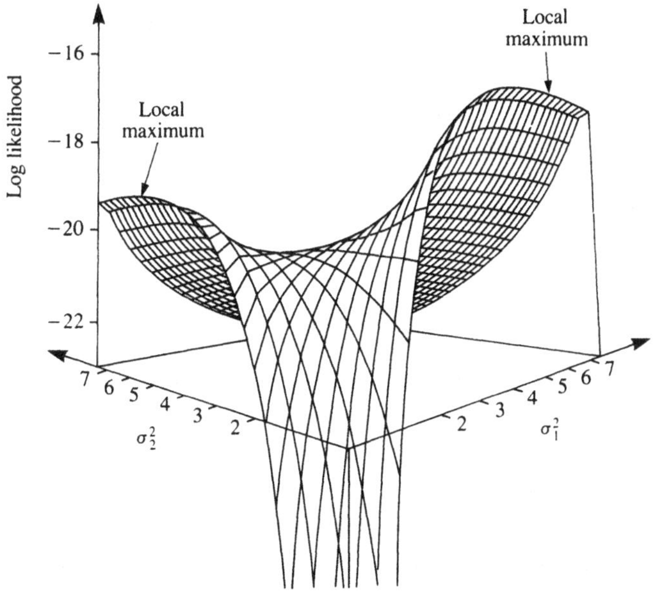

Figure 8.1. Log likelihood for a 2-way crossed classification, random model, with no interaction and $\sigma_e^2 = 1.0$.

to (25), it is a simple matter (weighted least squares) to solve equation (24) for the ML estimator of $\boldsymbol{\beta}$. Furthermore, equation (34), part of the Hartley–Rao form of the equations, shows that it is also a simple matter to solve for the error component given the others. Thus the difficult portion of maximizing the log likelihood involves r parameters, the variances of the r random effects factors, not including $\mathbf{e}$.

A simple approach would be to numerically evaluate the log likelihood or restricted likelihood for a fine grid of values of the variance components. The values giving rise to the largest value of the log likelihood would then be a close approximation to the ML or REML estimate. This is feasible for $r = 1$ or 2 but becomes unwieldy and time-consuming when the number of variance components is much larger than that. This is how Figure 8.1 was generated. To generate a grid with, say, 40 values per axis requires 40^r total evaluations of the log likelihood.

Numerical analysts have extensively studied the general problem of maximization of nonlinear functions and a number of methods have been proposed. The consensus is that iterative methods which use information about the first and/or second derivatives of the function to be maximized tend to perform best (Bard, 1974, p. 118; Gill, Murray and Wright, 1981, p. 93). Such an approach is acceptable as long as the derivatives are relatively easy to calculate. Statisticians have also developed other routines that exploit the special features that are characteristic of likelihood functions.

Iterative methods all have a common structure. They must be provided with starting values. Beginning with the starting values, they have a rule for getting the next value in the iteration and a rule for deciding when to stop iterating and declare the current value to be (in our case) the ML or REML estimate. For doing this there are at least three characteristics that a good iterative method should have. It should converge to a global maximum from a wide range of starting values, at each iteration it should be relatively quick to compute, and it should converge in relatively few iterations. Unfortunately, no known techniques guarantee convergence to a global maximum from arbitrary starting values.

8.2. ITERATIVE METHODS BASED ON DERIVATIVES

We first consider iterative methods that are explicitly based on the derivatives of the log likelihood. These are called gradient methods in the numerical analysis literature.

a. The basis of the methods

The heart of any iterative method is the rule it uses to find the next estimate given the current one. Essentially two decisions need to be made. In which direction will the next estimate be in relation to the current one, and how far will the next estimate be from the current one? These are termed the *step direction* and *step size* of the method.

A logical choice of direction would seem to be the direction in which the log likelihood increases the fastest near the current estimate, and a logical step size would be to choose the step size as large as possible while still having the likelihood increase. It can easily be shown (E 8.1) that the direction of fastest increase is the vector of first derivatives of the log likelihood evaluated at the current estimate. This is called the *gradient* of the log likelihood function. The method defined by this choice of step size and direction is called the *method of steepest ascent*. Unfortunately it performs very poorly in practice (Bard, 1974, p. 88), tending to require many iterations before it converges.

Nevertheless, the gradient information is what is generally used to define the step direction, but the direction chosen is usually something other than that of steepest ascent. Before describing better methods, we will consider a general class of methods that use the gradient to define the step direction. The step direction is usually modified by a multiplier matrix $\mathbf{M}$ so that the direction actually chosen is $\mathbf{M}\nabla l$, where ∇l is the gradient of the log likelihood function. If $\boldsymbol{\theta}^{(m)}$ represents the value of the estimate of the parameter vector $\boldsymbol{\theta}$ at the mth iteration then we can represent the step by the equation

$$\boldsymbol{\theta}^{(m+1)} = \boldsymbol{\theta}^{(m)} + s^{(m)}\mathbf{M}^{(m)} \left.\frac{\partial l}{\partial \boldsymbol{\theta}}\right|_{\boldsymbol{\theta}^{(m)}} \tag{1}$$

In this equation $s^{(m)}$, a scalar, is the step size for the mth step, $\mathbf{M}^{(m)}$ is the modifier matrix for the mth step and $\partial l/\partial \boldsymbol{\theta}|_{\boldsymbol{\theta}^{(m)}}$ is the gradient calculated at $\boldsymbol{\theta} = \boldsymbol{\theta}^{(m)}$. Bard (1974, p. 86) shows that as long as the matrix $\mathbf{M}^{(m)}$ is positive definite and the iterations have not yielded a maximum log likelihood, there is some step size $s^{(m)}$ for which the likelihood will increase. The method of steepest ascent is (1) with $\mathbf{M}^{(m)} = \mathbf{I}$, i.e., $\boldsymbol{\theta}^{(m+1)} = \boldsymbol{\theta}^{(m)} + s^{(m)}\, \partial l/\partial \boldsymbol{\theta}|_{\boldsymbol{\theta}^{(m)}}$, where $s^{(m)}$ is chosen as the largest value such that the likelihood continues to increase in the direction $\partial l/\partial \boldsymbol{\theta}|_{\boldsymbol{\theta}^{(m)}}$.

More specifically $s^{(m)}$ is chosen by searching along the line that goes through the point $\boldsymbol{\theta}^{(m)}$ in the direction $\partial l/\partial \boldsymbol{\theta}|_{\boldsymbol{\theta}^{(m)}}$ until a maximum or approximate maximum is found. This can be achieved in a variety of ways. One simple method is to first find an interval in which the maximum lies and then bisect it to find a subinterval. This process is repeated to achieve the desired accuracy. Other line search methods are detailed in Kennedy and Gentle (1980, Sec. 10.1.4).

b. The Newton–Raphson and Marquardt methods

A method commonly used for maximizing nonlinear functions is the Newton–Raphson method. Suppose the function f that we are trying to maximize is quadratic in $\boldsymbol{\theta}$:

$$f(\boldsymbol{\theta}) = \mathbf{a} + \mathbf{b}'\boldsymbol{\theta} - \tfrac{1}{2}\boldsymbol{\theta}'\mathbf{C}\boldsymbol{\theta},$$

where $\mathbf{C}$ is positive definite. $f(\boldsymbol{\theta})$ has gradient equal to $\mathbf{b} - \mathbf{C}\boldsymbol{\theta}$, the matrix of second derivatives of $f(\boldsymbol{\theta})$ with respect to $\boldsymbol{\theta}$ is $-\mathbf{C}$, and the global (and only) maximum of $f(\boldsymbol{\theta})$ is at $\boldsymbol{\theta} = \mathbf{C}^{-1}\mathbf{b}$. If we try to maximize $f(\boldsymbol{\theta})$ iteratively and want to arrive at the maximum in a single step, no matter from where we start, then, on setting (1) equal to $\mathbf{C}^{-1}\mathbf{b}$ and substituting for $\partial l/\partial \boldsymbol{\theta}$ we would need

$$\mathbf{C}^{-1}\mathbf{b} = \boldsymbol{\theta}^{(m+1)} = \boldsymbol{\theta}^{(m)} + s^{(m)}\mathbf{M}^{(m)}(\mathbf{b} - \mathbf{C}\boldsymbol{\theta}^{(m)})$$

to hold for all possible $\boldsymbol{\theta}^{(m)}$. That can be achieved using $s^{(m)} = 1$ and $\mathbf{M}^{(m)} = \mathbf{C}^{-1}$. Since the matrix of second derivatives of a function f is called the Hessian of that function, we denote it by $\mathbf{H}$ and so have in the above, quadratic, case $-\mathbf{C} = \mathbf{H}$. Therefore the iteration step, which would always get us to the maximum in a single step, would be

$$\begin{aligned}\boldsymbol{\theta}^{(m+1)} &= \boldsymbol{\theta}^{(m)} + \mathbf{C}^{-1}(\mathbf{b} - \mathbf{C}\boldsymbol{\theta}^{(m)}) \\ &= \boldsymbol{\theta}^{(m)} - \mathbf{H}^{-1}\nabla f^{(m)} .\end{aligned}$$

Of course, if the function were quadratic in $\boldsymbol{\theta}$, the iterative technique would not be needed. Nevertheless, this idea is applied by assuming a more complicated function can be approximated by a quadratic function and using the preceding iteration idea in the form

$$\boldsymbol{\theta}^{(m+1)} = \boldsymbol{\theta}^{(m)} - (\mathbf{H}^{(m)})^{-1}\nabla f^{(m)}, \tag{2}$$

where $\mathbf{H}^{(m)}$ and $\nabla f^{(m)}$ indicate the Hessian and gradient vector, respectively, with $\boldsymbol{\theta}$ replaced by $\boldsymbol{\theta}^{(m)}$. This is the Newton–Raphson method. One drawback of it is that it requires computation of the first and second derivatives of the log likelihood function.

A compromise between the Newton–Raphson and steepest ascent methods has been suggested by Marquardt (1963). He suggests the iteration

$$\boldsymbol{\theta}^{(m+1)} = \boldsymbol{\theta}^{(m)} - (\mathbf{H}^{(m)} + \tau^{(m)}\mathbf{I})^{-1}\nabla f^{(m)},$$

where $\tau^{(m)}$ is a scalar that partially determines the step size and $\mathbf{I}$ is the identity matrix. If $\tau^{(m)}$ is small, the procedure approximates Newton–Raphson. If $\tau^{(m)}$ is large, a small step is taken in approximately the direction of steepest ascent. The usual recommendations for choosing $\tau^{(m)}$ are to choose progressively smaller $\tau^{(m)}$ as long as the $\boldsymbol{\theta}^{(m)}$ increases the value of the function to be maximized (use steps more and more like Newton–Raphson). If $\boldsymbol{\theta}^{(m)}$ fails to increase the function to be maximized then progressively larger values of $\tau^{(m)}$ are used until it does increase (take a short step in a direction near steepest ascent).

In the variance components estimation problem, letting $\boldsymbol{\theta}$ denote all the parameters to be estimated, i.e., $\boldsymbol{\theta}' = [\boldsymbol{\beta}' \quad \boldsymbol{\sigma}^{2\prime}]$ for ML and $\boldsymbol{\theta}' = \boldsymbol{\sigma}^{2\prime}$ for REML, the Newton–Raphson iterations would be (2) with $f^{(m)}$ replaced by $l^{(m)}$:

$$\boldsymbol{\theta}^{(m+1)} = \boldsymbol{\theta}^{(m)} - (\mathbf{H}^{(m)})^{-1}\left.\frac{\partial l}{\partial \boldsymbol{\theta}}\right|_{\boldsymbol{\theta}^{(m)}},$$

with the entries in the Hessian $\mathbf{H}$ given by (36) and (37) of Chapter 6 for ML and by (93) of Chapter 6 for REML. In both cases $\mathbf{H}^{(m)}$ is found by replacing $\mathbf{V}$ and $\boldsymbol{\beta}$ in $\mathbf{H}$ by $\Sigma_i \mathbf{Z}_i \mathbf{Z}_i' \sigma_i^{2(m)}$ and $\boldsymbol{\beta}^{(m)}$. And $\partial l/\partial \boldsymbol{\theta}|_{\boldsymbol{\theta}^{(m)}}$ is found by similar replacements in Chapter 6: in (14) and (16) for ML and (92) for REML. Alternatively, the Hartley–Rao form of the likelihood (Section 6.2c) could be used with corresponding changes in the derivatives.

c. Method of scoring

To avoid the heavy computational burden of the second-derivative matrix, another method that has been used is the method of scoring, in which $-\mathbf{H}^{-1}$ is replaced by the inverse of the information matrix. That is, the Hessian is replaced by its expected value (see Appendix S.8c). An advantage of this is that the information matrix is often easier to compute than the Hessian. Comparing (36) and (37) with (38)—in Chapter 6—shows, for example, that large sections of the information matrix are zero, whereas the corresponding second derivatives are not. Jennrich and Sampson (1986) report that the method of scoring is also more robust to poor starting values than is the Newton–Raphson method. They recommend an iterative algorithm which starts by using scoring for the first few steps and then switches to Newton–Raphson.

The method of scoring thus uses an iteration scheme defined by

$$\boldsymbol{\theta}^{(m+1)} = \boldsymbol{\theta}^{(m)} + [\mathbf{I}(\boldsymbol{\theta}^{(m)})]^{-1} \left.\frac{\partial l}{\partial \boldsymbol{\theta}}\right|_{\boldsymbol{\theta}^{(m)}},$$

where $\mathbf{I}(\boldsymbol{\theta}^{(m)})$ is the information matrix calculated using $\boldsymbol{\theta} = \boldsymbol{\theta}^{(m)}$. In Chapter 6 $[\mathbf{I}(\boldsymbol{\theta})]^{-1}$ is given in (38) for ML and (94) for REML. Some details for the method of scoring for the 1-way random model are given in Section 8.5 that follows.

d. Quasi-Newton methods

A collection of popular methods for the maximization of nonlinear functions is known as quasi-Newton methods. These are similar to Newton–Raphson but they have the advantage of not requiring the calculation of second derivatives. In quasi-Newton algorithms the Hessian in (2) is replaced by an approximation which only requires the first derivatives. From a specified beginning matrix (often the identity matrix), updates are made to this approximate Hessian with simple-to-calculate matrices that have rank 2. For more details on quasi-Newton methods see Kennedy and Gentle (1980, Sec. 10.2.3).

e. Obtaining starting values

Any iterative technique needs a starting value and, in view of the difficult functions encountered as likelihoods, having starting values close to the values corresponding to the global maximum of the likelihood improves the chances of converging to a global maximum. For the fixed effects, a logical starting value would be any solution $\mathbf{X}\boldsymbol{\beta}^{(0)} = \mathbf{X}(\mathbf{X}'\mathbf{X})^{-}\mathbf{X}'\mathbf{y}$ to the ordinary least squares equations, since it is unbiased even when the elements of $\mathbf{y}$ are correlated and does not require knowledge of the values of the variance components. For the variance components any of the easy-to-calculate estimates, e.g., from ANOVA estimators, could be used as starting values for the ML or REML iterations. If any of the estimates are negative or zero, they need to be modified. Laird, Lange and Stram (1987) and Jennrich and Schluchter (1986) give suggestions for starting values for some special cases.

f. Termination rules

There is no general consensus as to when iterative methods should be stopped and the current values declared to be ML or REML estimates. Various suggestions include the following:

(i) stop when changes in the log likelihood are small;
(ii) stop when changes in the current parameter values are small;
(iii) stop when the values of the gradient are small;
(iv) combinations of the above.

Kennedy and Gentle (1980, p. 438) and Bard (1974, p. 114) recommend Marquardt's (1963) idea of using suggestion (ii) above, to stop when

$$\max_i |\theta_i^{(m+1)} - \theta_i^{(m)}| \leqslant \varepsilon_1(|\theta_i^{(m)}| + \varepsilon_2), \tag{3}$$

where $\varepsilon_1 = 10^{-4}$ and $\varepsilon_2 = 10^{-3}$. However, they make the point that some algorithms tend to stall temporarily before reaching the maximum, and a safer alternative would be to require (3) to hold over several iterations rather than just a single one. They also note that methods based on the gradients are often subject to rounding error. Combinations are typically implemented by requiring a sequence of iterations to satisfy more than one stopping criterion. Lindstrom and Bates (1988), quoting the method of Bates and Watts (1981), note that none of the above methods are actually checks on whether the estimates have converged to a maximum and they suggest an alternate method based on comparing the size of the numerical variability to the radius of an asymptotic confidence ellipse.

g. Incorporation of non-negativity constraints

As noted earlier, maximization with respect to the variance components is a constrained maximization problem, since the variances must at least be non-negative. It is possible that an iteration will give a negative value, which is unfortunate for a positive parameter. Jennrich and Schluchter (1986) use "step-halving" methods: if the step length in the iteration will yield a negative value then a new step is attempted using half the length. If that step gives all positive values then it is used to calculate the iteration, otherwise the step size is halved again, and so on. Callanan and Harville (1989) recommend "active-constraint" methodologies (see Gill, Murray and Wright, 1981, Sec. 5.2). Techniques such as the EM algorithm (see Section 8.3) automatically keep iterations in the parameter space.

h. Easing the computational burden

There are many techniques that can be applied to reduce the amount of computation necessary for ML or REML iterative methods. Harville (1977), Jennrich and Sampson (1986) and Hemmerle and Hartley (1973) give matrix identities that greatly reduce the size of the matrices that must be manipulated.

Lindstrom and Bates (1988) give a number of details on matrix decompositions that can be exploited to speed iterations. And in Section 7.6, where we describe what is widely known as Henderson's mixed model equations, we give details of how parts of these equations can be used to develop iterative procedures for calculating ML and REML estimators. A method of computing REML estimators (through using MINQUE, see Section 11.3) that avoids matrix inversion is given by Giesbrecht and Burrows (1978) for nested models with extensive details for unbalanced data from the 3-way nested model.

8.3. THE EM ALGORITHM

a. A general formulation

An iterative algorithm for calculating ML or REML estimates that differs from those like the Newton–Raphson or scoring methods is the EM algorithm. Its name stands for expectation–maximization, and it is so named because it alternates between calculating conditional expected values and maximizing simplified likelihoods. The EM algorithm only generates estimates and does not give variance estimates as a byproduct, as do the Newton–Raphson and methods of scoring. To obtain variance estimates extra computations must be performed (e.g. Louis, 1982).

The EM algorithm was designed to be used for maximum likelihood estimation for situations in which augmenting the data set leads to a simpler problem. The key to applying the EM algorithm is therefore the decision as to what to treat as the complete (augmented) data. The actual data set is typically called the incomplete data in application of the EM algorithm. Thus it is that for variance components estimation we think of the incomplete data as being the observed data $\mathbf{y}$ and the complete data as being $\mathbf{y}$ and the unobservable random effects $\mathbf{u}_i$ $(i = 1, 2, \ldots, r)$ of the usual mixed model described in Section 6.1, where the random effects are all uncorrelated.

The reason this is convenient is that if we knew the realized values of the unobservable random effects then we would estimate their variance with the average of their squared values (they are known to have zero mean). More explicitly, for a vector $\mathbf{u}_i$ of q_i random effects we would form

$$\hat{\sigma}_i^2 = \mathbf{u}_i'\mathbf{u}_i/q_i \tag{4}$$

to calculate the maximum likelihood estimates (under normality) based on the complete data. Being maximum likelihood estimators they are functions of the sufficient statistics of the complete (augmented) data.

However, in real life we do not know the realized values of the random effects. But the EM algorithm gives us a way to calculate values to use in place of those realized random effects in order to effect this estimation scheme. Starting with initial guesses for the parameters, we calculate the conditional expected values of the sufficient statistics of the complete data, $\mathbf{u}_i'\mathbf{u}_i$, given the incomplete data, $\mathbf{y}$. These conditional expected values are then used in place of the sufficient statistics in (4) to form improved estimates of the parameters. This is the maximization step, since (4) represents maximum likelihood estimation for the complete data. We can then use the new $\hat{\sigma}^2$-estimates to re-calculate the conditional expected value, and so on. This gives an iterative scheme that is used until it converges. Convergence is guaranteed under relatively unrestricted conditions (Dempster, Laird and Rubin, 1977; Wu, 1983).

An important feature of the EM algorithm is that, since it is performing maximum likelihood estimation for the complete data, the iterations will always remain in the parameter space. This is evident in (4).

b. Distributional derivations needed for the EM algorithm

As outlined, the EM algorithm is based on being able to calculate expected values conditional on the incomplete (observed) data $\mathbf{y}$. For this we need the joint distribution of $\mathbf{y}$ and $\mathbf{u} = [\mathbf{u}_1' \quad \mathbf{u}_2' \quad \dots \quad \mathbf{u}_r']'$. In taking the model equation for $\mathbf{y}$ as in (9) of Chapter 6,

$$\begin{aligned}\mathbf{y} &= \mathbf{X}\boldsymbol{\beta} + \sum_{i=1}^{r} \mathbf{Z}_i\mathbf{u}_i + \mathbf{e}\\ &= \mathbf{X}\boldsymbol{\beta} + \sum_{i=0}^{r} \mathbf{Z}_i\mathbf{u}_i\end{aligned}$$

with $\mathbf{u}_0 = \mathbf{e}$, $q_0 = N$ and $\mathbf{Z}_0 = \mathbf{I}_n$, as in (8) of that same chapter, we have

$$\text{cov}(\mathbf{y}, \mathbf{u}_j') = \text{cov}\left(\mathbf{X}\boldsymbol{\beta} + \sum_{i=1}^{r} \mathbf{Z}_i\mathbf{u}_i, \mathbf{u}_j'\right) = \mathbf{Z}_j \,\text{cov}(\mathbf{u}_j, \mathbf{u}_j') = \sigma_j^2 \mathbf{Z}_j \,.$$

Then, with

$$\mathbf{V} = \text{var}(\mathbf{y}) = \sum_{i=0}^{r} \mathbf{Z}_i\mathbf{Z}_i'\sigma_i^2 = \sum_{i=1}^{r} \mathbf{Z}_i\mathbf{Z}_i'\sigma_i^2 + \sigma_0^2\mathbf{I}_N, \tag{5}$$

the joint distribution of $\mathbf{y}$ and $\mathbf{u}_1, \mathbf{u}_2, \dots, \mathbf{u}_r$ is $\mathcal{N}(\boldsymbol{\mu}, \boldsymbol{\Sigma})$, where

$$\boldsymbol{\mu} = \begin{bmatrix} \mathbf{X}\boldsymbol{\beta} \\ \mathbf{0} \\ \vdots \\ \vdots \\ \mathbf{0} \end{bmatrix} \quad \text{and} \quad \boldsymbol{\Sigma} = \begin{bmatrix} \mathbf{V} & \{_r\, \sigma_i^2\mathbf{Z}_i\}_{i=1}^{r} \\ \{_c\, \sigma_i^2\mathbf{Z}_i'\}_{i=1}^{r} & \{_d\, \sigma_i^2\mathbf{I}_{q_i}\}_{i=1}^{r} \end{bmatrix}. \tag{6}$$

This gives a density function of

$$f_{\mathbf{y},\mathbf{u}_1,\mathbf{u}_2,\dots,\mathbf{u}_r}(\mathbf{y}, \mathbf{u}_1, \mathbf{u}_2, \dots, \mathbf{u}_r) = (2\pi)^{-\frac{1}{2}\Sigma_{i=0}^{r} q_i}|\boldsymbol{\Sigma}|^{-\frac{1}{2}} \exp(-\tfrac{1}{2}Q), \tag{7}$$

where

$$Q = [(\mathbf{y} - \mathbf{X}\boldsymbol{\beta})' \quad \mathbf{u}_1' \quad \mathbf{u}_2' \quad \dots \quad \mathbf{u}_r']\boldsymbol{\Sigma}^{-1} \begin{bmatrix} \mathbf{y} - \mathbf{X}\boldsymbol{\beta} \\ \mathbf{u}_1 \\ \vdots \\ \mathbf{u}_r \end{bmatrix}. \tag{8}$$

To simplify (7) in terms of the variance components we first use the standard result from Appendix M.5

$$\begin{vmatrix} \mathbf{A} & \mathbf{B} \\ \mathbf{C} & \mathbf{D} \end{vmatrix} = |\mathbf{D}|\,|\mathbf{A} - \mathbf{B}\mathbf{D}^{-1}\mathbf{C}|$$

to derive (with $i = 1, 2, \ldots, r$), from (6)

$$\begin{aligned}
|\mathbf{\Sigma}| &= |\{_{\rm d}\, \sigma_i^2 \mathbf{I}_{q_i}\}|\,|\mathbf{V} - \{_{\rm r}\, \sigma_i^2 \mathbf{Z}_i\}\{_{\rm d}\, (\sigma_i^2)^{-1} \mathbf{I}_{q_i}\}\{_{\rm c}\, \sigma_i^2 \mathbf{Z}_i'\}| \\
&= \prod_{i=1}^{r} (\sigma_i^2)^{q_i} |\mathbf{V} - \Sigma_i \sigma_i^2 \mathbf{Z}_i \mathbf{Z}_i'| \\
&= \prod_{i=1}^{r} (\sigma_i^2)^{q_i} |\sigma_0^2 \mathbf{I}_N|, \quad \text{from (5)}, \\
&= \prod_{i=1}^{r} (\sigma_i^2)^{q_i} (\sigma_0^2)^N \\
&= \prod_{i=0}^{r} (\sigma_i^2)^{q_i} \,. \qquad (9)
\end{aligned}$$

And, because $\mathbf{V} - \Sigma_{i=1}^r \sigma_i^2 \mathbf{Z}_i \mathbf{Z}_i' = \sigma_0^2 \mathbf{I}_N$ and

$$\{_{\rm d}\, \sigma_i^{-2} \mathbf{I}_{q_i}\}\{_{\rm c}\, \sigma_i^2 \mathbf{Z}_i'\} = \{_{\rm c}\, \mathbf{Z}_i'\},$$

the nonsingular analog of (22) in Appendix M.4c gives the inverse of (6) as

$$\mathbf{\Sigma}^{-1} = \begin{bmatrix} \mathbf{0} & \mathbf{0} \\ \mathbf{0} & \{_{\rm d}\, \sigma_i^{-2} \mathbf{I}_{q_i}\} \end{bmatrix} + \begin{bmatrix} \mathbf{I} \\ -\{_{\rm c}\, \mathbf{Z}_i'\} \end{bmatrix} \sigma_0^{-2} \mathbf{I}_N [\mathbf{I} \quad -\{_{\rm r}\, \mathbf{Z}_i\}] \,. \qquad (10)$$

Hence the log likelihood based on the complete data, $\mathbf{y}, \mathbf{u}_1, \mathbf{u}_2, \ldots, \mathbf{u}_r$, is, from substituting (8), (9) and (10) into (7), and taking logs

$$\begin{aligned}
l &= -\tfrac{1}{2}\left(\sum_{i=0}^{r} q_i\right) \ln 2\pi - \tfrac{1}{2} \sum_{i=0}^{r} q_i \ln \sigma_i^2 - \tfrac{1}{2} \sum_{i=1}^{r} \frac{\mathbf{u}_i'\mathbf{u}_i}{\sigma_i^2} \\
&\quad - \tfrac{1}{2}(\mathbf{y} - \mathbf{X}\boldsymbol{\beta} - \sum_1^r \mathbf{Z}_i \mathbf{u}_i)'(\mathbf{y} - \mathbf{X}\boldsymbol{\beta} - \sum_1^r \mathbf{Z}_i \mathbf{u}_i)/\sigma_0^2 \\
&= -\tfrac{1}{2}\left(\sum_{i=0}^{r} q_i\right) \ln 2\pi - \tfrac{1}{2} \sum_{i=0}^{r} q_i \ln \sigma_i^2 - \tfrac{1}{2} \sum_{i=0}^{r} \frac{\mathbf{u}_i'\mathbf{u}_i}{\sigma_i^2}
\end{aligned}$$

because $\mathbf{y} - \mathbf{X}\boldsymbol{\beta} - \Sigma_{i=1}^r \mathbf{Z}_i \mathbf{u}_i = \mathbf{e} = \mathbf{u}_0$.

From this it is a simple matter to derive the ML estimators based on the complete data, namely $\mathbf{y}$ and the $\mathbf{u}_i$s, as

$$\hat{\sigma}_i^2 = \mathbf{u}_i'\mathbf{u}_i / q_i, \quad i = 0, 1, 2, \ldots, r, \qquad (11)$$

and

$$\mathbf{X}\hat{\boldsymbol{\beta}} = \mathbf{X}(\mathbf{X}'\mathbf{X})^{-}\mathbf{X}'\left(\mathbf{y} - \sum_{i=1}^{r} \mathbf{Z}_i \mathbf{u}_i\right). \qquad (12)$$

The estimator for σ_i^2 in (11) is, of course, the same as in (4); and the estimator for $\mathbf{X\beta}$ in (12) is equivalent to subtracting the random effects other than $\mathbf{e}$ from $\mathbf{y}$ and then applying ordinary least squares.

To finalize the iteration steps for the EM algorithm we need the conditional expected values of $\mathbf{u}_i'\mathbf{u}_i$ and $\mathbf{y} - \Sigma_{i=1}^r \mathbf{Z}_i\mathbf{u}_i$ (the sufficient statistics) given $\mathbf{y}$. These are straightforward using standard multivariate normal results and (6). The conditional distributions of the $\mathbf{u}_i$ given $\mathbf{y}$ are, from (iv) in Appendix S.3,

$$\mathbf{u}_i \mid \mathbf{y} \sim \mathcal{N}[\sigma_i^2\mathbf{Z}_i'\mathbf{V}^{-1}(\mathbf{y} - \mathbf{X\beta}),\ \sigma_i^2\mathbf{I}_{q_i} - \sigma_i^4\mathbf{Z}_i'\mathbf{V}^{-1}\mathbf{Z}_i],$$

so that

$$E(\mathbf{u}_i \mid \mathbf{y}) = \sigma_i^2\mathbf{Z}_i'\mathbf{V}^{-1}(\mathbf{y} - \mathbf{X\beta})\,. \tag{13}$$

Therefore, on using Theorem S1 in Appendix S.5,

$$E(\mathbf{u}_i'\mathbf{u}_i \mid \mathbf{y}) = \sigma_i^4(\mathbf{y} - \mathbf{X\beta})'\mathbf{V}^{-1}\mathbf{Z}_i\mathbf{Z}_i'\mathbf{V}^{-1}(\mathbf{y} - \mathbf{X\beta}) + \operatorname{tr}(\sigma_i^2\mathbf{I}_{q_i} - \sigma_i^4\mathbf{Z}_i'\mathbf{V}^{-1}\mathbf{Z}_i)\,. \tag{14}$$

c. EM algorithm for ML estimation (Version 1)

We can now make a formal statement of the EM algorithm for maximum likelihood estimation. In the statement of the algorithm superscripts in parentheses indicate either current values of parameters or functions of current values of parameters. For example, $\sigma_i^{2(m)}$ is the computed value of σ_i^2 after the mth round of iteration and $\mathbf{V}^{(m)}$ is $\mathbf{V}$ with $\sigma_i^{2(m)}$ in place of σ_i^2 for $i = 0, \ldots, r$.

Step 0. Decide on starting values $\boldsymbol{\beta}^{(0)}$ and $\boldsymbol{\sigma}^{2(0)}$. Set $m = 0$.

Step 1 (E-step). Calculate from (14) the conditional expected value of the sufficient statistics. Label them

$$\begin{aligned}\hat{t}_i^{(m)} &= E(\mathbf{u}_i'\mathbf{u}_i \mid \mathbf{y})|_{\boldsymbol{\beta} = \boldsymbol{\beta}^{(m)} \text{ and } \boldsymbol{\sigma}^2 = \boldsymbol{\sigma}^{2(m)}}\\ &= \sigma_i^{4(m)}(\mathbf{y} - \mathbf{X\beta}^{(m)})'(\mathbf{V}^{(m)})^{-1}\mathbf{Z}_i\mathbf{Z}_i'(\mathbf{V}^{(m)})^{-1}(\mathbf{y} - \mathbf{X\beta}^{(m)})\\ &\quad + \operatorname{tr}[\sigma_i^{2(m)}\mathbf{I}_{q_i} - \sigma_i^{4(m)}\mathbf{Z}_i'(\mathbf{V}^{(m)})^{-1}\mathbf{Z}_i]\,.\end{aligned} \tag{15}$$

And, for (12), using (13),

$$\begin{aligned}\hat{\mathbf{s}}^{(m)} &= E(\mathbf{y} - \sum_{i=1}^{r} \mathbf{Z}_i\mathbf{u}_i \mid \mathbf{y})|_{\boldsymbol{\beta} = \boldsymbol{\beta}^{(m)} \text{ and } \boldsymbol{\sigma}^2 = \boldsymbol{\sigma}^{2(m)}}\\ &= \mathbf{y} - \sum_{i=1}^{r} \mathbf{Z}_i\mathbf{Z}_i'\sigma_i^{2(m)}(\mathbf{V}^{(m)})^{-1}(\mathbf{y} - \mathbf{X\beta}^{(m)})\\ &= \mathbf{y} - (\mathbf{V}^{(m)} - \sigma_0^{2(m)}\mathbf{I})(\mathbf{V}^{(m)})^{-1}(\mathbf{y} - \mathbf{X\beta}^{(m)})\\ &= \mathbf{X\beta}^{(m)} + \sigma_0^{2(m)}(\mathbf{V}^{(m)})^{-1}(\mathbf{y} - \mathbf{X\beta}^{(m)})\,.\end{aligned} \tag{16}$$

Step 2 (M-step). Maximize the likelihood of the complete data, based on (12)

$$\sigma_i^{2(m+1)} = \hat{t}_i^{(m)}/q_i, \quad i = 0, 1, 2, \ldots, r, \tag{17}$$

and

$$\mathbf{X}\boldsymbol{\beta}^{(m+1)} = \mathbf{X}(\mathbf{X}'\mathbf{X})^{-}\mathbf{X}'\hat{\mathbf{s}}^{(m)}\,. \tag{18}$$

Step 3. If convergence is reached, set $\tilde{\boldsymbol{\sigma}}^2 = \boldsymbol{\sigma}^{2(m+1)}$ and $\tilde{\boldsymbol{\beta}} = \boldsymbol{\beta}^{(m+1)}$; otherwise increase m by unity and return to step 1.

d. EM algorithm for ML estimation (Version 2)

Previously, in equation (99) of Chapter 6, we set out the result that, given MLEs of the variance components used in $\mathbf{V}$ to yield $\tilde{\mathbf{V}}$, then MLE($\mathbf{X}\boldsymbol{\beta}$) is $\mathbf{X}(\mathbf{X}'\tilde{\mathbf{V}}^{-1}\mathbf{X})^{-}\mathbf{X}'\tilde{\mathbf{V}}^{-1}\mathbf{y}$. Laird (1982) suggests not calculating $\hat{\mathbf{s}}^{(m)}$ and $\boldsymbol{\beta}^{(m)}$ in the iterations of the EM algorithm but only calculating $\tilde{\boldsymbol{\beta}}$ at the end of iterating to streamline the calculations. The rationale for this is as follows. If $\mathbf{X}\boldsymbol{\beta}^{(m)}$ were to be the same form as $\mathbf{X}\tilde{\boldsymbol{\beta}}$, say

$$\mathbf{X}\boldsymbol{\beta}_*^{(m)} = \mathbf{X}[\mathbf{X}'(\mathbf{V}^{(m)})^{-1}\mathbf{X}]^{-}\mathbf{X}'(\mathbf{V}^{(m)})^{-1}\mathbf{y},$$

then in (15) and (16) we would have $(\mathbf{V}^{(m)})^{-1}(\mathbf{y} - \mathbf{X}\boldsymbol{\beta}_*^{(m)})$, which is $\mathbf{P}^{(m)}\mathbf{y}$, where, based on (22) and (23) of Chapter 6,

$$\mathbf{P}^{(m)} = (\mathbf{V}^{(m)})^{-1} - (\mathbf{V}^{(m)})^{-1}\mathbf{X}[\mathbf{X}'(\mathbf{V}^{(m)})^{-1}\mathbf{X}]^{-}\mathbf{X}'(\mathbf{V}^{(m)})^{-1}\,.$$

Then $\boldsymbol{\beta}^{(m)}$ would no longer be explicitly needed in the iterations. Since $\mathbf{X}\boldsymbol{\beta}^{(m)}$ is not, in general, of the form $\mathbf{X}\boldsymbol{\beta}_*^{(m)}$, replacing $(\mathbf{V}^{(m)})^{-1}(\mathbf{y} - \mathbf{X}\boldsymbol{\beta}^{(m)})$ by $\mathbf{P}^{(m)}\mathbf{y}$ in the algorithm is, strictly speaking, not EM. The differences are slight and probably do not affect convergence properties of the algorithm. Nevertheless, this replacement of $(\mathbf{V}^{(m)})^{-1}(\mathbf{y} - \mathbf{X}\boldsymbol{\beta}^{(m)})$ by $\mathbf{P}^{(m)}\mathbf{y}$ gives what we call a second version of the EM algorithm. This modified version of the EM algorithm avoids a poor property of Version 1 (see E 8.10).

Step 0. Obtain a starting value $\boldsymbol{\sigma}^{2(0)}$. Set $m = 0$.

Step 1 (E-step). Calculate from (14) the conditional expected value of the sufficient statistics. Label them $\hat{t}_i^{(m)}$ for $i = 0, 1, \ldots, r$:

$$\begin{aligned}\hat{t}_i^{(m)} &= E(\mathbf{u}_i'\mathbf{u}_i \mid \mathbf{y})|_{\boldsymbol{\sigma}^2 = \boldsymbol{\sigma}^{2(m)}} \\ &= \sigma_i^{4(m)}\mathbf{y}'\mathbf{P}^{(m)}\mathbf{Z}_i\mathbf{Z}_i'\mathbf{P}^{(m)}\mathbf{y} + \mathrm{tr}[\sigma_i^{2(m)}\mathbf{I}_{q_i} - \sigma_i^{4(m)}\mathbf{Z}_i'(\mathbf{V}^{(m)})^{-1}\mathbf{Z}_i]\,. \end{aligned} \tag{19}$$

Step 2 (M-step). Maximize the likelihood of the complete data, based on (12).

$$\sigma_i^{2(m+1)} = \hat{t}_i^{(m)}/q_i \quad \text{for } i = 0, 1, \ldots, r\,. \tag{20}$$

Step 3. If convergence is reached, set $\tilde{\boldsymbol{\sigma}}^2 = \boldsymbol{\sigma}^{2(m+1)}$ and $\mathbf{X}\tilde{\boldsymbol{\beta}} = \mathbf{X}(\mathbf{X}'\tilde{\mathbf{V}}^{-1}\mathbf{X})^{-}\mathbf{X}'\tilde{\mathbf{V}}^{-1}\mathbf{y}$; otherwise increase m by unity and return to Step 1.

As pointed out by Laird (1982), this is the same as an algorithm that had previously been proposed on an *ad hoc* basis by Henderson (1973a) [see Harville (1977), equations (61) and (62)].

e. Equivalence of the EM algorithm to the ML equations

We now show how the steps of the EM algorithm are related to the ML equations of (24) and (25) in Chapter 6. To do so, consider the situation when iteration has ended. At that point, to some designated degree of approximation, $\tilde{\boldsymbol{\beta}} = \boldsymbol{\beta}^{(m+1)} = \boldsymbol{\beta}^{(m)}$ and $\tilde{\boldsymbol{\sigma}}^2 = \boldsymbol{\sigma}^{2(m+1)} = \boldsymbol{\sigma}^{2(m)}$. Using these equalities, and substituting (16) into (18), then gives

$$\mathbf{X}\tilde{\boldsymbol{\beta}} = \mathbf{X}(\mathbf{X}'\mathbf{X})^{-}\mathbf{X}'[\mathbf{X}\tilde{\boldsymbol{\beta}} + \tilde{\sigma}_0^2\tilde{\mathbf{V}}^{-1}(\mathbf{y} - \mathbf{X}\tilde{\boldsymbol{\beta}})],$$

which reduces to

$$\mathbf{X}(\mathbf{X}'\mathbf{X})^{-}\mathbf{X}'\tilde{\mathbf{V}}^{-1}\mathbf{X}\tilde{\boldsymbol{\beta}} = \mathbf{X}(\mathbf{X}'\mathbf{X})^{-}\mathbf{X}'\tilde{\mathbf{V}}^{-1}\mathbf{y}\,.$$

Pre-multiplication of this by $\mathbf{X}'$ gives

$$\mathbf{X}'\tilde{\mathbf{V}}^{-1}\mathbf{X}\tilde{\boldsymbol{\beta}} = \mathbf{X}'\tilde{\mathbf{V}}^{-1}\mathbf{y},$$

which is (24) of Chapter 6. Likewise, substituting (19) into (20) gives

$$q_i\tilde{\sigma}_i^2 = \tilde{\sigma}_i^4\mathbf{y}'\tilde{\mathbf{P}}\mathbf{Z}_i\mathbf{Z}_i'\tilde{\mathbf{P}}\mathbf{y} + q_i\tilde{\sigma}_i^2 - \tilde{\sigma}_i^4\,\mathrm{tr}(\mathbf{Z}_i\tilde{\mathbf{V}}^{-1}\mathbf{Z}_i'),$$

which, so long as $\tilde{\sigma}_i^2 \neq 0$, reduces to

$$\mathrm{tr}(\tilde{\mathbf{V}}^{-1}\mathbf{Z}_i'\mathbf{Z}_i) = \mathbf{y}'\tilde{\mathbf{P}}\mathbf{Z}_i\mathbf{Z}_i'\tilde{\mathbf{P}}\mathbf{y},$$

which is a typical term in the ML equation (25) of Chapter 6.

f. EM algorithm for REML estimation

Section 6.6b shows, for $\mathbf{K}'$ of maximum full row rank $N - p$ such that $\mathbf{K}'\mathbf{X} = \mathbf{0}$, how

$$\begin{aligned} &\text{replacing } \mathbf{y} \text{ by } \mathbf{K}'\mathbf{y},\\ &\text{replacing } \mathbf{X} \text{ by } \mathbf{K}'\mathbf{X} = \mathbf{0},\\ &\text{replacing } \mathbf{Z} \text{ by } \mathbf{K}'\mathbf{Z}\\ \text{and } &\text{replacing } \mathbf{V} \text{ by } \mathbf{K}'\mathbf{V}\mathbf{K} \end{aligned} \tag{21}$$

leads to being able to derive the REML equations (89) from the ML equations (86). Part of the derivation involves the identity developed in Appendix M.4f that

$$\mathbf{P} = \mathbf{K}(\mathbf{K}'\mathbf{V}\mathbf{K})^{-1}\mathbf{K}' \tag{22}$$

for $\mathbf{P}$ defined as

$$\mathbf{P} = \mathbf{V}^{-1} - \mathbf{V}^{-1}\mathbf{X}(\mathbf{X}'\mathbf{V}^{-1}\mathbf{X})^{-}\mathbf{X}'\mathbf{V}^{-1}\,. \tag{23}$$

We can use the same replacements (21) and identity (22) to derive the EM algorithm for REML from that (Version 2) for ML. In doing so, note that (21) applied to $\mathbf{Z}_i'\mathbf{P}\mathbf{y}$ causes it to become $\mathbf{Z}_i'\mathbf{K}(\mathbf{K}'\mathbf{V}\mathbf{K})^{-1}\mathbf{K}'\mathbf{y}$ which, by (22), is $\mathbf{Z}_i'\mathbf{P}\mathbf{y}$; i.e., the replacements (21) cause no change in $\mathbf{Z}_i'\mathbf{P}\mathbf{y}$. In this way the EM algorithm for REML is as follows.

Step 0. Obtain a starting value $\boldsymbol{\sigma}^{2(0)}$. Set $m = 0$.

Step 1 (E-step). From (19) for $i = 0, 1, \ldots, r$ calculate

$$\begin{aligned}\hat{t}_i^{(m)} &= \sigma_i^{4(m)}(\mathbf{K}'\mathbf{y})'(\mathbf{K}'\mathbf{V}^{(m)}\mathbf{K})^{-1}\mathbf{K}'\mathbf{Z}_i\mathbf{Z}_i'\mathbf{K}(\mathbf{K}'\mathbf{V}^{(m)}\mathbf{K})^{-1}\mathbf{K}'\mathbf{y} \\ &\quad + \operatorname{tr}[\sigma_i^{2(m)}\mathbf{I}_{q_i} - \sigma_i^{4(m)}\mathbf{Z}_i'\mathbf{K}(\mathbf{K}'\mathbf{V}^{(m)}\mathbf{K})^{-1}\mathbf{K}'\mathbf{Z}_i] \\ &= \sigma_i^{4(m)}\mathbf{y}'\mathbf{P}^{(m)}\mathbf{Z}_i\mathbf{Z}_i'\mathbf{P}^{(m)}\mathbf{y} + \operatorname{tr}[\sigma_i^{2(m)}\mathbf{I}_{q_i} - \sigma_i^{4(m)}\mathbf{Z}_i'\mathbf{P}^{(m)}\mathbf{Z}_i] \,. \qquad (24)\end{aligned}$$

Step 2 (M-step). Maximize the likelihood of the complete data, using (20) for $i = 0, 1, \ldots, r$:

$$\sigma_i^{2(m+1)} = \hat{t}_i^{(m)}/q_i \,.$$

Step 3. If convergence is reached, set $\tilde{\boldsymbol{\sigma}}^2 = \boldsymbol{\sigma}^{2(m+1)}$; otherwise increase m by one and return to Step 1.

Thus, if for ML we use Version 2 of the EM algorithm, as in (19) and (20), then the only difference between that and EM for REML is the appearance of $\mathbf{P}$ in the trace term in (24) rather than $\mathbf{V}^{-1}$ in (19). Since, by the nature of $\mathbf{P}$ in (23), $\operatorname{tr}(\mathbf{Z}_i\mathbf{P}^{(m)}\mathbf{Z}_i') \leqslant \operatorname{tr}[\mathbf{Z}_i(\mathbf{V}^{(m)})^{-1}\mathbf{Z}_i']$, use of (24) leads to a larger value of $\sigma_i^{2(m)}$ than does (18), for $i = 0, 1, \ldots, r$.

g. A Bayesian justification for REML

The EM algorithm for REML estimation can also be derived using the connection identified in (24) and (25) of Chapter 9. If we regard the fixed effects as random effects with distribution given by $\boldsymbol{\beta} \sim \mathcal{N}(\boldsymbol{\beta}_0, \mathbf{B})$ then the conditional distribution of $\mathbf{u}_i$ given $\mathbf{y}$ is

$$\mathbf{u}_i \mid \mathbf{y} \sim \mathcal{N}[\sigma_i^2\mathbf{Z}_i'\mathbf{W}^{-1}(\mathbf{y} - \mathbf{X}\boldsymbol{\beta}_0),\ \sigma_i^2\mathbf{I}_{q_i} - \sigma_i^4\mathbf{Z}_i'\mathbf{W}^{-1}\mathbf{Z}_i],$$

where $\mathbf{W} = \mathbf{X}\mathbf{B}\mathbf{X}' + \mathbf{V}$. To obtain the EM iterations for REML we need the limiting distribution as $\mathbf{B}^{-1} \to \mathbf{0}$. For that we need the following.

Proposition. If $\mathbf{W} = \mathbf{X}\mathbf{B}\mathbf{X}' + \mathbf{V}$ then

$$\lim_{\mathbf{B}^{-1}\to\mathbf{0}} \mathbf{W}^{-1} = \mathbf{P} = \mathbf{V}^{-1} - \mathbf{V}^{-1}\mathbf{X}(\mathbf{X}'\mathbf{V}^{-1}\mathbf{X})^{-1}\mathbf{X}'\mathbf{V}^{-1} \,.$$

Proof. First note that

$$(\mathbf{V} + \mathbf{X}\mathbf{B}\mathbf{X}')^{-1}\mathbf{X}\mathbf{B} = \mathbf{V}^{-1}\mathbf{X}(\mathbf{B}^{-1} + \mathbf{X}'\mathbf{V}^{-1}\mathbf{X})^{-1},$$

which follows easily from the identity

$$\mathbf{X}\mathbf{B}(\mathbf{B}^{-1} + \mathbf{X}'\mathbf{V}^{-1}\mathbf{X}) = (\mathbf{V} + \mathbf{X}\mathbf{B}\mathbf{X}')\mathbf{V}^{-1}\mathbf{X} \,.$$

Then

$$\begin{aligned}
\mathbf{W}^{-1} &= (\mathbf{V} + \mathbf{XBX}')^{-1} \\
&= (\mathbf{V} + \mathbf{XBX}')^{-1}[(\mathbf{V} + \mathbf{XBX}')\mathbf{V}^{-1} - \mathbf{XBX}'\mathbf{V}^{-1}] \\
&\qquad\qquad \text{(matrix in square brackets is identity)} \\
&= \mathbf{V}^{-1} - (\mathbf{V} + \mathbf{XBX}')^{-1}\mathbf{XBX}'\mathbf{V}^{-1} \\
&= \mathbf{V}^{-1} - \mathbf{V}^{-1}\mathbf{X}(\mathbf{B}^{-1} + \mathbf{X}'\mathbf{V}^{-1}\mathbf{X})^{-1}\mathbf{X}'\mathbf{V}^{-1}\,. \\
&\to \mathbf{V}^{-1} - \mathbf{V}^{-1}\mathbf{X}(\mathbf{X}'\mathbf{V}^{-1}\mathbf{X})^{-1}\mathbf{X}'\mathbf{V}^{-1} = \mathbf{P} \quad \text{as } \mathbf{B}^{-1} \to \mathbf{0}\,. \qquad \text{Q.E.D.}
\end{aligned}$$

Using the Proposition and $\mathbf{PX} = \mathbf{0}$, the limiting distribution of $\mathbf{u}_i \mid \mathbf{y}$ as $\mathbf{B}^{-1} \to \mathbf{0}$ is

$$\mathbf{u}_i \mid \mathbf{y} \sim \mathcal{N}(\sigma_i^2 \mathbf{Z}_i'\mathbf{Py},\ \sigma_i^2 \mathbf{I}_{q_i} - \sigma_i^4 \mathbf{Z}_i'\mathbf{PZ}_i)\,.$$

As with the ML iterations, for REML the ML estimates based on the complete data are given by $\hat{\sigma}_i^2 = \mathbf{u}_i'\mathbf{u}_i/q_i$ for $i = 0, 1, 2, \ldots, r$. We therefore have

$$E(\mathbf{u}_i'\mathbf{u}_i \mid \mathbf{y})|_{\boldsymbol{\sigma}^2 = \boldsymbol{\sigma}^{2(m)}} = \sigma_i^{4(m)} \mathbf{y}'\mathbf{P}^{(m)}\mathbf{Z}_i\mathbf{Z}_i'\mathbf{P}^{(m)}\mathbf{y} + \operatorname{tr}(\sigma_i^{2(m)}\mathbf{I}_{q_i} - \sigma_i^{4(m)}\mathbf{Z}_i'\mathbf{P}^{(m)}\mathbf{Z}_i), \qquad i = 0, 1, \ldots, r,$$

which gives the EM algorithm (24) for REML estimation.

h. Non-zero correlations among the u_is

We have seen that the calculations for the EM algorithm revolve around the sufficient statistics for the complete data likelihood. If the model for $\mathbf{y}$ is different from the model having equation $\mathbf{y} = \mathbf{X}\boldsymbol{\beta} + \Sigma_{i=0}^{r} \mathbf{Z}_i\mathbf{u}_i$ (i.e., having non-zero correlations among the $\mathbf{u}_i$s) then the sufficient statistics will also be different and the EM algorithm will take a different form. Dempster, Rubin and Tsutakawa (1981) give some details for covariance components estimation. For some patterned variance–covariance matrices Andrade and Helms (1984) give results, and for longitudinal data models Laird and Ware (1982), Laird, Lange and Stram (1987) and Jennrich and Schluchter (1986) give specifics.

8.4. GENERAL METHODS THAT CONVERGE RAPIDLY FOR BALANCED DATA

The iterative algorithms identified so far (Newton–Raphson, Marquardt, scoring, quasi-Newton and EM) are all general purpose algorithms that can be applied whether the data are balanced or not. This generality comes at a price, since the algorithms do not take advantage of the simplicity that arises with balanced data. A consequence is that with balanced data, the aforementioned algorithms will still require a number of iterations to converge, perhaps even more than with unbalanced data. To address this concern Anderson (1973), Thompson and Meyer (1986) and more recently Callanan and Harville (1989) have devised generally applicable algorithms, which, when applied to some balanced data situations for which exact, analytic solutions exist, yield those exact solutions in a single iteration.

Callanan and Harville (1989) studied a number of the traditional algorithms, which they "linearized" to improve their convergence properties with balanced data. Their goal was to improve the convergence properties of the algorithms in general for use with balanced or unbalanced data. Under limited evaluation, they preferred a linearized version of the Newton–Raphson algorithm.

8.5. POOLING ESTIMATORS FROM SUBSETS OF A LARGE DATA SET

The advent of supercomputers is rapidly reducing the computational effort required for what have heretofore been tasks of unimaginable magnitude, e.g., inverting a matrix of order 1000 in 17 seconds; or of 2000 in 2 minutes. Nevertheless, there will be occasions when, depending on local computing facilities and the size of one's data set, iterative solution of the maximum likelihood equations for the whole data set will be effectively impractical. In such cases, a method from Babb (1986) provides opportunity for dividing the large data set into disjoint data sets (to be small enough to do maximum likelihood on each one) and pooling the ML estimators from the subsets. Since estimators from one data subset are not necessarily independent of those from another, simple averaging of the subset estimators can be improved upon in an approximate generalized least squares fashion. Details are as follows.

First, for each set of data that the complete data have been divided into, calculate $\tilde{\boldsymbol{\sigma}}^2$ from the ML equations (27) of Chapter 6. For the pth data set the model equation will be

$$\mathbf{y}_p = \mathbf{X}_p\boldsymbol{\beta}_p + \sum_{i=0}^{r} \mathbf{Z}_{ip}\mathbf{u}_{ip}, \tag{25}$$

with subscript p denoting the pth data set. Also, $\mathbf{u}_{ip}$ is not necessarily the same as $\mathbf{u}_i$, because $\mathbf{u}_{ip}$ will have as elements only those of $\mathbf{u}_i$ that actually occur in the pth data set. Then, on writing

$$\mathbf{A}_p = \{_m\, \mathrm{tr}(\mathbf{Z}_{ip}\mathbf{Z}'_{ip}\tilde{\mathbf{V}}_p^{-1}\mathbf{Z}_{jp}\mathbf{Z}'_{jp}\tilde{\mathbf{V}}_p^{-1})\}_{i,j=0}^{r} \tag{26}$$

and

$$\mathbf{f}_p = \{_c\, \mathbf{y}'_p\tilde{\mathbf{P}}_p\mathbf{Z}_{ip}\mathbf{Z}'_{ip}\tilde{\mathbf{P}}_p\mathbf{y}_p\}_{i,j=0}^{r}, \tag{27}$$

the ML equations (27) of Chapter 6 yield, on convergence and ignoring negativity constraints,

$$\tilde{\boldsymbol{\sigma}}_p^2 = \mathbf{A}_p^{-1}\mathbf{f}_p\,. \tag{28}$$

Second, suppose there are s data sets. An initial value for the estimator of $\boldsymbol{\sigma}^2$ from the whole data set is

$$\hat{\boldsymbol{\sigma}}_0^2 = \{_c\, \hat{\sigma}_{i(0)}^2\}_{i=0}^{r} = \sum_{p=1}^{s} \tilde{\boldsymbol{\sigma}}_p^2/s\,. \tag{29}$$

With this, calculate the following three matrices for each data set; i.e., for $p = 1, 2, \ldots, s$:

$$\mathbf{V}_{pp(0)} = \text{var}(\mathbf{y}_t), \quad \text{calculated at } \hat{\boldsymbol{\sigma}}_0^2$$

$$= \sum_{i=0}^{r} \mathbf{Z}_{ip}\mathbf{Z}'_{ip}\hat{\sigma}^2_{i(0)} \tag{30}$$

$$\mathbf{A}_{p(0)} = \{_{\text{m}}\, \text{tr}(\mathbf{Z}_{ip}\mathbf{Z}'_{ip}\mathbf{V}^{-1}_{pp(0)}\mathbf{Z}_{jp}\mathbf{Z}'_{jp}\mathbf{V}^{-1}_{pp(0)})\}_{i,j=0}^{r} \tag{31}$$

$$\mathbf{P}_{p(0)} = \mathbf{V}^{-1}_{pp(0)} - \mathbf{V}^{-1}_{pp(0)}\mathbf{X}_p(\mathbf{X}'_p\mathbf{V}^{-1}_{pp(0)}\mathbf{X}_p)^{-}\mathbf{X}'_p\mathbf{V}_{pp(0)} \,. \tag{32}$$

Third, we rewrite the model equation (25) for $\mathbf{y}_p$ not just in terms of $\mathbf{u}_{ip}$, the random effects that are in the pth data set, but in terms of $\mathbf{u}_i$, the random effects that are in the whole data set:

$$\mathbf{y}_p = \mathbf{X}_p\boldsymbol{\beta}_p + \sum_{i=0}^{r} \dot{\mathbf{Z}}_{ip}\mathbf{u}_i \,. \tag{33}$$

As a result, $\dot{\mathbf{Z}}_{ip}$ has a column of zeros corresponding to each element of $\mathbf{u}_i$ that is not in $\mathbf{u}_{ip}$. But it is this formulation that permits having an expression for the covariance of $\mathbf{y}_p$ with $\mathbf{y}_q$:

$$\mathbf{V}_{pq(0)} = \text{cov}(\mathbf{y}_p, \mathbf{y}'_q), \quad \text{calculated at } \hat{\boldsymbol{\sigma}}_0^2,$$

$$= \sum_{i=1}^{r} \dot{\mathbf{Z}}_{ip}\dot{\mathbf{Z}}'_{iq}\hat{\sigma}^2_{i(0)} \quad \text{for } p \neq q = 1, \ldots, s \,. \tag{34}$$

Note that the summation here excludes $\sigma_0^2 = \sigma_e^2$, because no two data sets have any error terms in common. Moreover, in any particular case, many of the terms in (34) will be zero; indeed for many pairs of data sets (34) itself will be zero. For example, when the random effects in a model constitute a two-way classification, (34) will seldom involve the interaction variance component, for the same kind of reason that it never contains σ_e^2; and only for data sets that have levels of the A-factor in common can it have σ_A^2, and so on.

Fourth, we are going to combine the estimators $\hat{\boldsymbol{\sigma}}_p^2$ from (28) for the individual data sets by weighted least squares, making use of an approximation to the sampling variance–covariance matrix of the vector

$$[\tilde{\boldsymbol{\sigma}}_1^{2\prime} \quad \tilde{\boldsymbol{\sigma}}_2^{2\prime} \quad \ldots \quad \tilde{\boldsymbol{\sigma}}_s^{2\prime}]' \,.$$

This is done as follows. In each equation (28) for $p = 1, \ldots, s$, we treat $\mathbf{A}_p$ as non-random and $\mathbf{f}_p$ as random. Then, since

$$\text{cov}(\mathbf{f}_p, \mathbf{f}'_q) = \{_{\text{m}}\, \text{cov}(\mathbf{y}'_p\mathbf{P}_p\mathbf{Z}_{ip}\mathbf{Z}'_{ip}\mathbf{P}_p\mathbf{y}_p,\; \mathbf{y}'_q\mathbf{P}_q\mathbf{Z}_{iq}\mathbf{Z}'_{iq}\mathbf{P}_q\mathbf{y}_q)\}_{i,j=0}^{r},$$

we calculate this at $\hat{\boldsymbol{\sigma}}_0^2$, calling it $\mathbf{G}_{pq(0)}$:

$$\mathbf{G}_{pq(0)} = 2\{_{\text{m}}\, \text{tr}[\mathbf{P}_{p(0)}\mathbf{Z}_{ip}\mathbf{Z}'_{ip}\mathbf{P}_{p(0)}\mathbf{V}_{pq(0)}\mathbf{P}_{q(0)}\mathbf{Z}_{iq}\mathbf{Z}'_{iq}\mathbf{P}_{q(0)}\mathbf{V}_{qp(0)}]\}_{i,j=0}^{r}$$

$$= 2\{_{\text{m}}\, \text{tr}[\mathbf{E}_{pq,ij(0)}(\mathbf{E}_{pq,ij(0)})']\}_{i,j=0}^{r} \tag{35}$$

for $\mathbf{E}_{pq,ij(0)} = \mathbf{Z}'_{ip}\mathbf{P}_{p(0)}\mathbf{V}_{pq(0)}\mathbf{P}_{q(0)}\mathbf{Z}_{iq}$. Then assemble these matrices into a single matrix as

$$\mathbf{G}_{(0)} = \text{var}\begin{bmatrix}\mathbf{f}_1\\ \vdots \\ \mathbf{f}_s\end{bmatrix} = \{_\text{m}\, \mathbf{G}_{pq(0)}\}_{p,q=1}^{s} \tag{36}$$

and define

$$[\mathbf{G}_{(0)}]^{-1} = \{_\text{m}\, \mathbf{G}_{(0)}^{p,q}\}_{p,q=1}^{s} \,. \tag{37}$$

Thus, on treating $\mathbf{A}_p$ and $\mathbf{A}_q$ as non-random, the covariance of $\tilde{\boldsymbol{\sigma}}_p^2$ and $\tilde{\boldsymbol{\sigma}}_q^2$ is, when calculated at $\hat{\boldsymbol{\sigma}}_0^2$,

$$\mathbf{C}_{pq(0)} = \mathbf{A}_{p(0)}^{-1}\mathbf{G}_{pq(0)}\mathbf{A}_{q(0)}^{-1};$$

and

$$\mathbf{C}_{(0)} = \{_\text{m}\, \mathbf{C}_{pq(0)}\}_{p,q=1}^{s} = \{_\text{d}\, \mathbf{A}_{p(0)}^{-1}\}_{p=1}^{s}\{_\text{m}\, \mathbf{G}_{pq(0)}\}_{p,q=1}^{s}\{_\text{d}\, \mathbf{A}_{p(0)}^{-1}\}_{p=1}^{s} \,.$$

Now the vector being estimated by every $\tilde{\boldsymbol{\sigma}}_p^2$ is $\boldsymbol{\sigma}^2$. Therefore the vector estimated by $\{_\text{c}\, \tilde{\boldsymbol{\sigma}}_p^2\}_{p=1}^{s}$ is $\{_\text{c}\, \boldsymbol{\sigma}^2\}_{p=1}^{s} = \mathbf{1}_s \otimes \boldsymbol{\sigma}^2 = (\mathbf{1}_s \otimes \mathbf{I}_{r+1})\boldsymbol{\sigma}^2$. We therefore estimate $\boldsymbol{\sigma}^2$ by pooling the $\tilde{\boldsymbol{\sigma}}_p^2$ vectors in a weighted least squares fashion, using $\mathbf{C}_{(0)}^{pq}$ as the (p, q)'th submatrix of $[\mathbf{C}_{(0)}]^{-1}$.

$$\begin{aligned}
\hat{\boldsymbol{\sigma}}_{(1)}^2 &= [(\mathbf{1}_s \otimes \mathbf{I}_{r+1})'\mathbf{C}_{(0)}^{-1}(\mathbf{1}_s \otimes \mathbf{I}_{r+1})]^{-1}(\mathbf{1}'_s \otimes \mathbf{I}_{r+1})\mathbf{C}_{(0)}^{-1}\{_\text{c}\, \tilde{\boldsymbol{\sigma}}_p^2\}_{p=1}^{s}\\
&= \left(\sum_{p=1}^{s}\sum_{q=1}^{s}\mathbf{C}_{(0)}^{pq}\right)^{-1}\sum_{p=1}^{s}\left(\sum_{q=1}^{s}\mathbf{C}_{(0)}^{pq}\right)\tilde{\boldsymbol{\sigma}}_p^2\\
&= \left(\sum_{p=1}^{s}\sum_{q=1}^{s}\mathbf{A}_{p(0)}\mathbf{G}_{(0)}^{pq}\mathbf{A}_{q(0)}\right)^{-1}\sum_{p=1}^{s}\mathbf{A}_{p(0)}\left(\sum_{q=1}^{s}\mathbf{G}_{(0)}^{pq}\mathbf{A}_{q(0)}\right)\tilde{\boldsymbol{\sigma}}_p^2 \,.
\end{aligned} \tag{38}$$

One now iterates this process, using $\hat{\boldsymbol{\sigma}}_{(1)}^2$ in place of $\tilde{\boldsymbol{\sigma}}_{(0)}^2$ in equations (30)–(32) and then in (34)–(38). More study is needed to evaluate this procedure.

8.6. EXAMPLE: THE 1-WAY RANDOM MODEL

For the 1-way classification, random model we show details for ML estimation for two iterative algorithms: EM (Version 1) and the method of scoring. In doing so we use the model equation

$$y_{ij} = \mu + \alpha_i + e_{ij}$$

for $i = 1, \ldots, a$ and $j = 1, \ldots, n_i$ with vector equivalence

$$\mathbf{y} = \mathbf{X}\boldsymbol{\beta} + \mathbf{Z}_1\mathbf{u}_1 + \mathbf{Z}_0\mathbf{u}_0,$$

where

$$\mathbf{X} = \mathbf{1}_N, \quad \boldsymbol{\beta} = \mu, \quad \mathbf{Z}_1 = \{_\text{d}\, \mathbf{1}_{n_i}\}, \quad \mathbf{u}_1 = \boldsymbol{\alpha}, \quad q_1 = a, \quad \mathbf{Z}_0 = \mathbf{I}_N \quad \text{and} \quad \mathbf{u}_0 = \mathbf{e} \,. \tag{39}$$

Also

$$\operatorname{var}(\mathbf{y}) = \mathbf{V} = \{_{\mathrm{d}}\, \sigma_e^2 \mathbf{I}_{n_i} + \sigma_\alpha^2 \mathbf{J}_{n_i}\} \quad \text{and} \quad \mathbf{V}^{-1} = \frac{1}{\sigma_e^2}\left\{_{\mathrm{d}}\ \mathbf{I}_{n_i} - \frac{\sigma_\alpha^2}{\sigma_e^2 + n_i\sigma_\alpha^2}\mathbf{J}_{n_i}\right\}. \tag{40}$$

Other useful notations taken from Chapter 3 are

$$\lambda_i = \sigma_e^2 + n_i\sigma_\alpha^2, \quad \tau = \frac{\sigma_\alpha^2}{\sigma_e^2} \quad \text{and} \quad \frac{n_i}{\lambda_i} = \frac{1}{\sigma_e^2(\tau + 1/n_i)}. \tag{41}$$

a. The EM algorithm (Version 1)

Using (39), (40) and (41) in (14), it can be shown (E 8.8) that

$$E(\mathbf{u}_i'\mathbf{u}_i|\mathbf{y}) = \tau^2 \sum_{i=1}^{a} \frac{(\bar{y}_{i\cdot} - \mu)^2}{(\tau + 1/n_i)^2} + a\sigma_\alpha^2 - \sigma_\alpha^2\tau \sum_{i=1}^{a} \frac{1}{\tau + 1/n_i} \tag{42}$$

and

$$E(\mathbf{u}_0'\mathbf{u}_0 \,|\, \mathbf{y}) = \sum_{i=1}^{a}\sum_{j=1}^{n_i} (y_{ij} - \mu)^2 - \tau \sum_{i=1}^{a} \frac{(\bar{y}_{i\cdot} - \mu)^2}{(\tau + 1/n_i)^2}(2 + n_i\tau) + \sigma_e^2 \sum_{i=1}^{a} \frac{\tau}{\tau + 1/n_i}. \tag{43}$$

And likewise, using (13),

$$E(\mathbf{y} - \mathbf{Z}_1\mathbf{u}_1 \,|\, \mathbf{y}) = \left\{_{\mathrm{c}} \left\{_{\mathrm{c}}\ y_{ij} - \frac{\tau(\bar{y}_{i\cdot} - \mu)}{\tau + 1/n_i}\right\}_{j=1}^{n_i}\right\}_{i=1}^{a}. \tag{44}$$

The iteration procedure then comes from using first (42) and then (43) as the right-hand side of (15), which is then put into (17) to give

$$a\sigma_\alpha^{2(m+1)} = \tau^{2(m)}\sum\nolimits_i \frac{(\bar{y}_{i\cdot} - \mu^{(m)})^2}{(\tau^{(m)} + 1/n_i)^2} + a\sigma_\alpha^{2(m)} - \sigma_\alpha^{2(m)}\tau^{(m)}\sum\nolimits_i \frac{1}{\tau^{(m)} + 1/n_i} \tag{45}$$

and

$$N\sigma_e^{2(m+1)} = \sum_i\sum_j (y_{ij} - \mu^{(m)})^2 - \tau^{(m)}\sum\nolimits_i \frac{(\bar{y}_{i\cdot} - \mu^{(m)})^2}{(\tau^{(m)} + 1/n_i)^2}(2 + n_i\tau^{(m)})$$
$$+ \sigma_e^{2(m)}\sum\nolimits_i \frac{\tau^{(m)}}{\tau^{(m)} + 1/n_i}. \tag{46}$$

Then, using (44) for $\hat{\mathbf{s}}^{(m)}$ in (16) and putting that in (18) gives [dropping the first $\mathbf{X}$ on each side of (18)]

$$\mu^{(m+1)} = \frac{1}{N}\mathbf{1}_N'\left\{_{\mathrm{c}} \left\{_{\mathrm{c}}\ y_{ij} - \frac{\tau^{(m)}(\bar{y}_{i\cdot} - \mu^{(m)})}{\tau^{(m)} + 1/n_i}\right\}_{j=1}^{n_i}\right\}_{i=1}^{a}, \tag{47}$$

which (see E 8.9) reduces to

$$\mu^{(m+1)} = \frac{1}{N}\sum_{i=1}^{a}\frac{\bar{y}_{i.} + n_i\tau^{(m)}\mu^{(m)}}{\tau^{(m)} + 1/n_i}, \tag{48}$$

and this in turn can also be written as

$$\mu^{(m+1)} = \frac{1}{N}\sum_{i=1}^{a}\frac{n_i(\bar{y}_{i.}/\sigma_\alpha^{2(m)} + n_i\mu^{(m)}/\sigma_e^{2(m)})}{1/\sigma_\alpha^{2(m)} + n_i/\sigma_e^{2(m)}}.$$

This shows that $\mu^{(m+1)}$ is the weighted average of weighted averages of $\bar{y}_{i.}$ and $\mu^{(m)}$.

One carries out the iterations by first setting $m = 0$, choosing starting values $\mu^{(0)}$, $\sigma_\alpha^{2(0)}$ and $\sigma_e^{2(0)}$, and using them in (45), (46) and (47) to get $\mu^{(1)}$, $\sigma_\alpha^{2(1)}$ and $\sigma_e^{2(1)}$. The latter are then used again in (45), (46) and (47) with $m = 1;\ldots$, and so on, until satisfactory convergence is attained.

We note in passing that when the iteration ceases, (48) will be $\tilde{\mu}$ and (see E 8.9) it reduces to

$$\tilde{\mu} = \sum_i \frac{\bar{y}_{i.}}{\tilde{\sigma}_\alpha^2 + \tilde{\sigma}_e^2/n_i}\Big/\sum_i \frac{1}{\tilde{\sigma}_\alpha^2 + \tilde{\sigma}_e^2/n_i} = \sum_i \frac{\bar{y}_{i.}}{\tilde{\text{var}}(\bar{y}_{i.})}\Big/\sum_i \frac{1}{\tilde{\text{var}}(\bar{y}_{i.})} \tag{49}$$

where $\tilde{\text{var}}(\bar{y}_{i.}) = \tilde{\sigma}_\alpha^2 + \tilde{\sigma}_e^2/n_i$ is the MLE of $\text{var}(\bar{y}_{i.})$. Moreover, (49) is GLSE(μ) with $\tilde{\sigma}_\alpha^2$ and $\tilde{\sigma}_e^2$ in place of σ_α^2 and σ_e^2; and, of course, for balanced data $\tilde{\mu} = \bar{y}_{..}$.

Note that for balanced data (48) reduces (see E 8.10) to

$$\mu^{(m+1)} = \frac{\bar{y}_{..}}{n\tau^{(m)} + 1} + \left(1 - \frac{1}{n\tau^{(m)} + 1}\right)\mu^{(m)} \tag{50}$$

which is not $\bar{y}_{..}$; although it does, of course, reduce to $\tilde{\mu} = \bar{y}_{..}$ when iteration has ended. This is for Version 1 of the EM algorithm. In contrast, though, we can show that Version 2 gives every $\mu^{(m)}$ as $\bar{y}_{..}$.

In Version 2 of the EM algorithm step 3 has

$$\mathbf{X}\tilde{\boldsymbol{\beta}} = \mathbf{X}(\mathbf{X}'\tilde{\mathbf{V}}^{-1}\mathbf{X})^{-}\mathbf{X}'\tilde{\mathbf{V}}^{-1}\mathbf{y}\,. \tag{51}$$

Using $\mathbf{X}$ and $\mathbf{V}^{-1}$ of (39) and (40), this reduces to

$$\tilde{\mu} = \sum_i \frac{\bar{y}_{i.}}{\tilde{\text{var}}(\bar{y}_{i.})}\Big/\sum_i \frac{1}{\tilde{\text{var}}(\bar{y}_{i.})} \tag{52}$$

just as in (49). So, for the $(m+1)$th step it is

$$\mu^{(m+1)} = \sum_i \frac{\bar{y}_{i.}}{\text{var}^{(m)}(\bar{y}_{i.})}\Big/\sum_i \frac{1}{\text{var}^{(m)}(\bar{y}_{i.})}.$$

Then for balanced data $\text{var}(\bar{y}_{i.}) = \sigma_\alpha^2 + \sigma_e^2/n$, the same for all i, and so both $\tilde{\mu}$ and $\mu^{(m+1)}$ reduce to

$$\tilde{\mu} = \mu^{(m+1)} = \bar{y}_{..};$$

i.e., for every iteration Version 2 yields $\mu^{(m)} = \bar{y}_{..}$ for balanced data.

b. The method of scoring algorithm

The general iteration equation for the method of scoring is the equation in Section 8.2c:

$$\boldsymbol{\theta}^{(m+1)} = \boldsymbol{\theta}^{(m)} - [\mathbf{I}(\boldsymbol{\theta}^{(m)})]^{-1} \left.\frac{\partial l}{\partial \boldsymbol{\theta}}\right|_{\boldsymbol{\theta}=\boldsymbol{\theta}^{(m)}}$$

For the 1-way classification this is

$$\begin{bmatrix} \mu^{(m+1)} \\ \sigma_e^{2(m+1)} \\ \sigma_\alpha^{2(m+1)} \end{bmatrix} = \begin{bmatrix} \mu^{(m)} \\ \sigma_e^{2(m)} \\ \sigma_\alpha^{2(m)} \end{bmatrix} + \left[\mathbf{I}\begin{pmatrix} \mu \\ \sigma_e^2 \\ \sigma_\alpha^2 \end{pmatrix}\right]^{-1} \left.\begin{bmatrix} l_\mu \\ l_{\sigma_e^2} \\ l_{\sigma_\alpha^2} \end{bmatrix}\right|_{\mu=\mu^{(m)},\, \sigma_e^2=\sigma_e^{2(m)},\, \sigma_\alpha^2=\sigma_\alpha^{2(m)}}$$

where the information matrix $\mathbf{I}(\boldsymbol{\theta})$ and the first derivatives are available from Chapter 3, in equations (138) and (132), respectively. Hence, for $\lambda_i = \sigma_e^2 + n_i\sigma_\alpha^2$, using (138) of Chapter 3,

$$\begin{bmatrix} \mu^{(m+1)} \\ \sigma_e^{2(m+1)} \\ \sigma_\alpha^{2(m+1)} \end{bmatrix} = \begin{bmatrix} \mu^{(m)} \\ \sigma_e^{2(m)} \\ \sigma_\alpha^{2(m)} \end{bmatrix} + \frac{1}{2}\begin{bmatrix} 2\sum_i \dfrac{n_i}{\lambda_i^{(m)}} & 0 & 0 \\ 0 & \dfrac{N-q}{\sigma_e^{2(m)}} + \sum_i \dfrac{1}{(\lambda_i^{(m)})^2} & \sum_i \dfrac{n_i}{(\lambda_i^{(m)})^2} \\ 0 & \sum_i \dfrac{n_i}{(\lambda_i^{(m)})^2} & \sum_i \dfrac{n_i^2}{(\lambda_i^{(m)})^2} \end{bmatrix}^{-1} \begin{bmatrix} l_{\mu^{(m)}} \\ l_{\sigma_e^{2(m)}} \\ l_{\sigma_i^{2(m)}} \end{bmatrix},$$

where, from (132) of Chapter 3,

$$\begin{bmatrix} l_{\mu^{(m)}} \\ l_{\sigma_e^{2(m)}} \\ l_{\sigma_i^{2(m)}} \end{bmatrix} = \begin{bmatrix} \sum_i \dfrac{n_i(\bar{y}_{i\cdot} - \mu^{(m)})}{\lambda_i^{(m)}} \\ \dfrac{-(N-a)}{2\sigma_e^{2(m)}} - \dfrac{1}{2}\sum_i \dfrac{1}{\lambda_i^{(m)}} + \dfrac{\text{SSE}}{(\sigma_e^{2(m)})^2} + \sum_i \dfrac{n_i(\bar{y}_{i\cdot} - \mu^{(m)})}{2(\lambda_i^{(m)})^2} \\ -\dfrac{1}{2}\sum_i \dfrac{n_i}{\lambda_i^{(m)}} + \sum_i \dfrac{n_i^2(\bar{y}_{i\cdot} - \mu^{(m)})^2)}{2(\lambda_i^{(m)})^2} \end{bmatrix}.$$

It is clear that the iterations for σ_α^2 and σ_e^2 are somewhat complicated so we will not detail them here. However, the iteration for μ is given by

$$\mu^{(m+1)} = \mu^{(m)} + \frac{\sum_i \dfrac{n_i(\bar{y}_{i\cdot} - \mu^{(m)})}{\sigma_e^{2(m)} + n_i\sigma_\alpha^{2(m)}}}{\sum_i \dfrac{n_i}{\sigma_e^{2(m)} + n_i\sigma_\alpha^{2(m)}}}. \tag{53}$$

Again, when iteration has ceased, this reduces to (49), and thence to $\tilde{\mu} = \bar{y}_{..}$ for balanced data. When the data are balanced, (53) will always yield $\mu^{(m)} = \bar{y}_{..}$ for $m > 0$. Version 1 of the EM algorithm does not have this property, but Version 2, which uses $\mathbf{P}^{(m)}\mathbf{y}$ in place of $\mathbf{V}^{(m-1)}(\mathbf{y} - \mathbf{X}\boldsymbol{\beta}^{(m)})$, does. Figure 8.2 shows, as squares

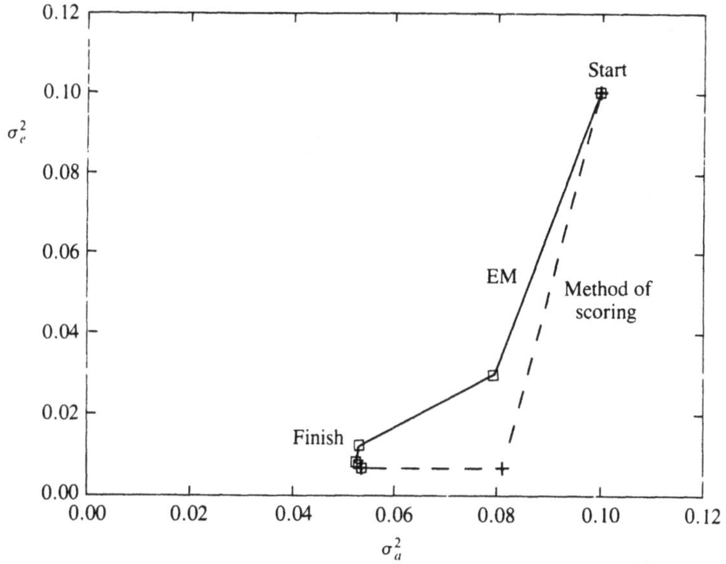

Figure 8.2. Convergence of two methods of solving the likelihood equations for the turnip data of Snedecor and Cochran (1989, p. 238), a 1-way classification with balanced data, $a = n = 4$: EM algorithm and method of scoring.

and plusses joined by lines, the iterations of the method of scoring and of Version 2 of the EM algorithm for the turnip data of Snedecor and Cochran (1989, p. 238). The data set is balanced with $a = 4 = n$.

While this is a very simple, balanced data example, it nevertheless illustrates several of the points made in this chapter. It shows that the EM algorithm requires a large number of steps to converge even though it is very close to the maximum after only two steps. It shows that even though the data are balanced, the iterative methods do not converge in a single step, and it illustrates the ideas of step direction and step size.

8.7. DISCUSSION

a. Computing packages

We have outlined in this chapter some of the methods available for computing ML and REML estimates. There are myriad difficulties involved in actually implementing these methods including, but not limited to, stability of numerical methods applied to the matrices involved, methods of avoiding the inversion of large matrices and the details of diagnosing convergence or non-convergence of the algorithms. All of these matters have to be attended to satisfactorily when designing and writing computing packages—and this is not necessarily an easy task. It is a job for an expert, who must have a sound appreciation of numerical analysis. Computer packages designed by those who are amateur in this regard can therefore usually be deemed suspect. That said, we now make but a comment or two on some of the widely-available commercial packages, and on some of the recommendations given in the recent literature.

The computing packages SAS, GENSTAT 5 and BMDP all have (as of 1988) procedures for calculating variance components estimates in mixed models. In SAS the procedure is VARCOMP. It calculates ANOVA estimators (based on Henderson's Method III) from the SAS Type I sums of squares. It also calculates MIVQUE(0) estimates—as described in Section 11.3g. For ML and REML estimation it uses the W-transform (Hemmerle and Hartley, 1973) to reduce the computational burden and a modified Newton–Raphson method that protects against the value of the objective function going in the wrong direction. Iteration commences with the MIVQUE(0) estimates, and convergence is assumed to have been achieved when the objective function ($\log_e|\mathbf{V}|$ for ML and $\log_e|\mathbf{K}'\mathbf{V}\mathbf{K}|$ for REML) changes by no more than 10^{-8}. GENSTAT 5 contains a REML routine developed by H.D. Patterson of Edinburgh, Scotland.

The BMDP package has several programs that compute variance components estimates. From BMDP (1988) we find that program 3V is the primary one for ML and REML estimation from unbalanced data of any mixed model. Its iterative procedure (p. 1182) starts with the Fisher scoring method, and when the change in the log likelihood becomes less than unity the algorithm changes to Newton–Raphson. Starting values (p. 1182) are $\hat{\sigma}_e^{2(0)} = (\mathbf{y}'\mathbf{y} - N\bar{y}^2)/(N-1)$ and $\hat{\sigma}_i^{2(0)} = 0$ for $i = 1, \ldots, r$, and convergence is assumed (p. 1042) when the relative change in log likelihood is less than 10^{-p} for a user-supplied p, the default being $p = 8$. Program 8V also calculates variance components estimates, but is confined to balanced data, although for almost any kind of mixed model (p. 1115). For a mixed model with repeated measures program 5V is preferred (p. 1026).

b. Evaluation of algorithms

Several recent research papers evaluate algorithms for variance components estimation (Dempster, Selwyn, Patel and Roth, 1984; Jennrich and Schluchter, 1986; Laird, Lange and Stram, 1988; Lindstrom and Bates, 1988). While there is no consensus on the best method, some general conclusions seem to be as follows.

1. The Newton–Raphson method often converges in the fewest iterations, followed by the scoring method and then the EM algorithm. In some cases the EM algorithm requires a very large number of iterations. The individual iterations tend to be slightly shorter for the EM algorithm, but this depends greatly on the details of the programming.

2. The robustness of the methods to their starting values (ability to converge given poor starting values) is the reverse of the rate of convergence. The EM algorithm is better than Newton–Raphson.

3. The EM algorithm automatically takes care of inequality constraints imposed by the parameter space. Other algorithms need specialized programming to incorporate constraints.

4. Newton–Raphson and scoring generate an estimated, asymptotic variance–covariance matrix for the estimates as a part of their calculations. At the end of the EM iterations, special programming [perhaps a single step of Newton–Raphson or use of the results of Louis (1982)] needs to be employed to calculate asymptotic standard errors.

8.8. SUMMARY

Iterative methods of maximizing the likelihood begin with an initial estimate $\boldsymbol{\theta}^{(0)\prime}$ of $\boldsymbol{\theta}' = (\boldsymbol{\beta}', \boldsymbol{\sigma}^{2\prime})$ and then proceed by calculating new estimates, $\boldsymbol{\theta}^{(m+1)}$, $m = 0, 1, 2, \ldots$. Iteration continues until the estimates converge (Section 8.2f). The iterations for the various methods are as follows.

Newton–Raphson

$$\boldsymbol{\theta}^{(m+1)} = \boldsymbol{\theta}^{(m)} - (\mathbf{H}^{(m)})^{-1} \left. \frac{\partial l}{\partial \boldsymbol{\theta}} \right|_{\boldsymbol{\theta}^{(m)}},$$

where $\mathbf{H}^{(m)}$ is the Hessian of l evaluated at $\boldsymbol{\theta}^{(m)}$ [equations (36) and (38) for ML and equations (108) for REML of Chapter 6] and $\partial l/\partial \boldsymbol{\theta}$ is given in (14) and (16) for ML and (106) for REML of Chapter 6.

Scoring

$$\boldsymbol{\theta}^{(m+1)} = \boldsymbol{\theta}^{(m)} + [\mathbf{I}(\theta^{(m)})]^{-1} \left. \frac{\partial l}{\partial \boldsymbol{\theta}} \right|_{\boldsymbol{\theta}^{(m)}},$$

where $\mathbf{I}(\boldsymbol{\theta}^{(m)})$ is the information matrix evaluated at $\boldsymbol{\theta}^{(m)}$ [equation (38) for ML and equation (108) for REML of Chapter 6].

EM algorithm version 1 (ML)

$$\mathbf{X}\boldsymbol{\beta}^{(m+1)} = \mathbf{X}\boldsymbol{\beta}^{(m)} + \sigma_0^{2(m)}\mathbf{X}(\mathbf{X}'\mathbf{X})^{-}\mathbf{X}'(\mathbf{V}^{(m)})^{-1}(\mathbf{y} - \mathbf{X}\boldsymbol{\beta}^{(m)}),$$

$$q_i\sigma_i^{2(m+1)} = \sigma_i^{4(m)}(\mathbf{y} - \mathbf{X}\boldsymbol{\beta}^{(m)})'(\mathbf{V}^{(m)})^{-1}\mathbf{Z}_i\mathbf{Z}_i'(\mathbf{V}^{(m)})^{-1}(\mathbf{y} - \mathbf{X}\boldsymbol{\beta}^{(m)}) + \operatorname{tr}[\sigma_i^{2(m)}\mathbf{I}_{q_i} - \sigma_i^{4(m)}\mathbf{Z}_i'(\mathbf{V}^{(m)})^{-1}\mathbf{Z}_i], \quad \text{for } i = 0, 1, 2, \ldots, r,$$

where $\mathbf{V}^{(m)}$ is var(y) evaluated at $\boldsymbol{\theta}^{(m)}$.

EM algorithm version 2 (ML)

$$q_i\sigma_i^{2(m+1)} = \sigma_i^{4(m)}\mathbf{y}'\mathbf{P}^{(m)}\mathbf{Z}_i\mathbf{Z}_i'\mathbf{P}^{(m)} + \operatorname{tr}[\sigma_i^{2(m)}\mathbf{I}_{q_i} - \sigma_i^{4(m)}\mathbf{Z}_i'(\mathbf{V}^{(m)})^{-1}\mathbf{Z}_i], \quad \text{for } i = 0, 1, 2, \ldots, r.$$

At convergence set $\mathbf{X}\tilde{\boldsymbol{\beta}} = \mathbf{X}(\mathbf{X}'\tilde{\mathbf{V}}^{-1}\mathbf{X}')^{-1}\mathbf{X}'\tilde{\mathbf{V}}^{-1}\mathbf{y}$.

EM algorithm version 2 (REML)

$$q_i\sigma_i^{2(m+1)} = \sigma_i^{4(m)}\mathbf{y}'\mathbf{P}^{(m)}\mathbf{Z}_i\mathbf{Z}_i'\mathbf{P}^{(m)}\mathbf{y} + \operatorname{tr}[\sigma_i^{2(m)}\mathbf{I}_{q_i} - \sigma_i^{4(m)}\mathbf{Z}_i'\mathbf{P}^{(m)}\mathbf{Z}_i], \quad \text{for } i = 0, 1, 2, \ldots, r.$$

8.9. EXERCISES

E 8.1. Show that the direction that a function $f(\boldsymbol{\theta})$ increases most rapidly at $\boldsymbol{\theta} = \boldsymbol{\theta}^*$ is given by $\nabla f(\boldsymbol{\theta}^*)$.

E 8.2. Show that $\mathbf{X}\boldsymbol{\beta}^0 = \mathbf{X}(\mathbf{X}'\mathbf{X})^-\mathbf{X}'\mathbf{y}$ is an unbiased estimator of $\mathbf{X}\boldsymbol{\beta}$ under the model $\mathbf{y} = \mathbf{X}\boldsymbol{\beta} + \mathbf{Z}\mathbf{u} + \mathbf{e}$ where $E(\mathbf{u}) = \mathbf{0}$ and $E(\mathbf{e}) = \mathbf{0}$.

E 8.3. If ε_2 in (3) is set equal to zero instead of 10^{-3} then the stopping rule is to stop when the relative change in all the parameter values is less than $10^{-4} = .01\%$. Why is this unsatisfactory?

E 8.4. Derive the log likelihood and MLEs for the complete data likelihood

$$l = -\frac{1}{2}\left(\sum_{i=0}^{r} q_i\right)\log 2\pi - \frac{1}{2}\sum_{i=0}^{r} q_i \log\sigma_i^2 - \frac{1}{2}\sum_{i=0}^{r}\frac{\mathbf{u}_i'\mathbf{u}_i}{\sigma_i^2}\,.$$

E 8.5. Derive (12) as the GLSE obtained from $\mathbf{y} - \sum_{i=1}^{r}\mathbf{Z}_i\mathbf{u}_i$.

E 8.6. Derive (13) and (14).

E 8.7. Show that for $\mathbf{P} = \mathbf{V}^{-1} - \mathbf{V}^{-1}\mathbf{X}(\mathbf{X}'\mathbf{V}^{-1}\mathbf{X})^-\mathbf{X}'\mathbf{V}^{-1}$

$$\operatorname{tr}(\mathbf{Z}_i'\mathbf{P}\mathbf{Z}_i) \leqslant \operatorname{tr}(\mathbf{Z}_i'\mathbf{V}^{-1}\mathbf{Z}_i)\,.$$

E 8.8. Derive (42), (43) and (44) from (13) and (14).

E 8.9. (a) Derive (47) from (16), (18) and (44).

(b) Show that (47) simplifies to (48).

(c) Show that after iteration has ended

$$\tilde{\mu} = \frac{\Sigma_i[\bar{y}_{i\cdot}/\tilde{\text{var}}(\bar{y}_{i\cdot})]}{\Sigma_i[1/\tilde{\text{var}}(\bar{y}_{i\cdot})]},$$

where $\tilde{\text{var}}(\bar{y}_{i\cdot})$ is the MLE of $\text{var}(\bar{y}_{i\cdot})$.

E 8.10. For Version 1 of the EM algorithm for ML in the balanced 1-way classification, random model, show that the iterations for μ are of the form

$$\mu^{(m+1)} = (1 - \eta^{(m)})\bar{y}_{\cdot\cdot} + \eta^{(m)}\mu^{(m)},$$

where

$$\eta^{(m)} = \frac{n\sigma_\alpha^{2(m)}}{\sigma_e^{2(m)} + n\sigma_\alpha^{2(m)}} < 1\,.$$

What happens to the iterations if $\sigma_e^{2(m)} \approx 0$?

CHAPTER 9

HIERARCHICAL MODELS AND BAYESIAN ESTIMATION

In this chapter a slightly different approach to analysis of the mixed model is explored, an approach that is arrived at through an amalgamation of many views. Although the idea of modeling in a hierarchy has a distinct Bayesian flavor, the purpose of hierarchical modeling goes beyond Bayesian analysis. For example, hierarchical techniques can help both our understanding of models and our estimation and interpretation of them. In particular, we will see that only a few simple ideas are necessary to arrive at some broad estimation principles.

9.1. BASIC PRINCIPLES

a. Introduction

A hierarchical model is one that is specified in stages, with each stage building upon another. The advantage of building a model in stages is that each stage can be relatively simple and easy to understand, while the entire model may be rather complicated. Thus, sophisticated models may be built by layering together relatively straightforward pieces.

Bayesian methods are strongly tied to hierarchical models. Recall from Section 3.9 that Bayes estimation is based on calculating a *posterior distribution*, which arises from combining a *prior distribution* with the sampling distribution. (These ideas are covered in Appendix S.6.) The specification of the sampling distribution and the prior distribution is an example of a hierarchical model. For example, if we observe a random variable X with distribution $f(x \mid \theta)$, and suppose that θ has a (prior) distribution $\pi(\theta)$, then

$$\begin{aligned} X \mid \theta &\sim f(x \mid \theta), \\ \theta &\sim \pi(\theta) \end{aligned} \tag{1}$$

is a hierarchical specification. Here we have two levels in the hierarchy: the first level deals with the distribution of the variable X (here the data) conditional on θ, and the second level deals with the distribution of the variable θ (here the parameter), on which the distribution of X depends. The hierarchical specification can continue. For example, if the distribution of θ depended on another variable λ, that is $\theta \sim \pi(\theta \mid \lambda)$, we could then specify a distribution for the variable λ in the third level of the hierarchy. In most cases three (or fewer) levels in a hierarchy will suffice, but the theory knows no limit.

Once a hierarchical model, such as (1), is specified, we can use the hierarchy to derive estimators using Bayesian methodology. Note that this *estimation method* can be used no matter how the hierarchy is arrived at. For example, to estimate θ from the model in (1), we could use its posterior distribution $\pi(\theta \mid x)$, given by

$$\pi(\theta \mid x) = \frac{f(x \mid \theta)\pi(\theta)}{\int f(x \mid \theta)\pi(\theta)\, d\theta}, \tag{2}$$

where $f(x \mid \theta)\pi(\theta)$ is the *joint distribution* of X and θ, and $\int f(x \mid \theta)\pi(\theta)\, d\theta = m(x)$ is the *marginal distribution* of X. Of course, other estimation methods (e.g., maximum likelihood) can be used in a hierarchy such as (1). For now, however, we will concentrate on the Bayesian estimation techniques that are natural for the specified hierarchy.

From the posterior distribution we could obtain a posterior mean, posterior variance, or any other parameter associated with a distribution. A common choice for a point estimate of θ is the posterior mean $E(\theta \mid x)$, given by

$$E(\theta \mid x) = \frac{\int \theta f(x \mid \theta)\pi(\theta)\, d\theta}{\int f(x \mid \theta)\pi(\theta)\, d\theta}. \tag{3}$$

Estimating the parameter θ by the mean of its posterior distributions seems quite reasonable, and is also justifiable on more formal grounds. If we assume that our penalty for misjudging θ is measured by squared error loss then the posterior mean is an optimal estimator of θ.

b. Simple examples

Section 3.9 contains an example of Bayesian estimation using a hierarchical model for a normal variance. For $[x_1 \ \ldots \ x_n]' \sim \mathcal{N}(\mu\mathbf{1}, \sigma^2\mathbf{I})$ we estimate σ^2 under the following hierarchical model:

$$\begin{aligned} s^2 \mid \sigma^2 &\sim f(s^2 \mid \sigma^2) = \sigma^2\chi^2_{n-1}/(n-1) \\ \sigma^2 &\sim \pi(\sigma^2) = (\sigma^2)^{-3}e^{-1/\sigma^2}. \end{aligned} \tag{4}$$

Applying formula (2), we can derive the posterior distribution of σ^2, or we could use formula (3) to obtain a point estimate of σ^2, $E(\sigma^2 \mid s^2)$, given by

$$E(\sigma^2 \mid s^2) = \frac{s^2 + 2/(n-1)}{1 + 2/(n-1)} = \frac{n-1}{n+1}s^2 + \frac{2}{n+1}(1). \tag{5}$$

The second expression in (5) shows that the posterior mean is a weighted average, an occurrence that will often happen. Note that $E(\sigma^2 \mid s^2)$ is a weighted average of s^2, the sample estimator, and of 1, the prior mean from $\pi(\sigma^2)$, with weights that are dependent on the sample size. As the sample size increases, more weight is given to the sample estimator. (More generally, the weights reflect the relative variance of the sample and prior information.)

Whether a hierarchical model is considered a Bayesian model depends on the interpretation of the prior distribution. There is a subtle difference here between Bayesian estimation and Bayesian modeling. Bayesian estimation leads to equations like (2) and (3), and can be used with *any* hierarchy. Bayesian modeling, a branch of hierarchical modeling, arises when the second (or third) level of a hierarchy reflects some prior (subjective) belief. If the distribution $\pi(\sigma^2)$ of (4) reflects a prior belief then (4) specifies a Bayesian hierarchical model. If $\pi(\sigma^2)$ is derived through some other means (as in the following example), model (4) remains a hierarchical model, but is not a Bayesian hierarchical model.

As an example of a hierarchical model that is not Bayesian, consider the following classical model for insect populations. An insect lays a number of eggs, Λ, according to a Poisson distribution with parameter λ. Each egg can either hatch or not, and if it hatches it survives with probability p. The interest is in estimating the number of surviving insects.

To specify this as a hierarchy, let X denote the number of survivors from a batch of Λ eggs. We can then write

$$\begin{aligned} X \mid \Lambda &\sim \text{binomial}(\Lambda, p), \\ \Lambda &\sim \text{Poisson}(\lambda)\,. \end{aligned} \tag{6}$$

Neither of the distributions specified in (6) came from a subjective belief, but rather can be attributed to the structure of the problem. Therefore this hierarchy does not specify a Bayesian model. To derive an expression for the number of survivors, however, we could use Bayesian methods, or more generally the calculus of probabilities. The conditional expectation of X given Λ is $E(X \mid \Lambda) = p\Lambda$, which can be estimated using the marginal distribution of X, a Poisson$(p\lambda)$. See E 9.1.

c. The mixed model hierarchy

The general mixed model equation, for a data vector $\mathbf{y}$, has been written as

$$\mathbf{y} = \mathbf{X}\boldsymbol{\beta} + \mathbf{Z}\mathbf{u} + \mathbf{e} \tag{7}$$

[as in (58) of Chapter 4], where $\boldsymbol{\beta}$ is an unknown, fixed parameter and $\mathbf{u}$ is an unknown, random variable. The matrices $\mathbf{X}$ and $\mathbf{Z}$ are considered fixed and known, and $\mathbf{e}$ is an unknown random vector. In the classical approach to analysis of data using a mixed model the distinctions of fixed versus random, known versus unknown, parameter versus statistic are all important. These classifications dictate the type of estimation and inference that is possible. When analyzing the mixed model (or any model) using a hierarchical approach, however, it only matters whether a specified quantity is *observable* or *unobservable*.

In equation (7) $\mathbf{y}$, $\mathbf{X}$ and $\mathbf{Z}$ are observable (because we see their values), while $\boldsymbol{\beta}$, $\mathbf{u}$ and $\mathbf{e}$ are unobservable (because we do not see their values). No further classifications are necessary. In particular, both fixed effects and random effects are handled within the same general framework. In hierarchical modeling we treat $\boldsymbol{\beta}$, $\mathbf{u}$ and all variance components in the same way: they are unobservable.

For modeling the hierarchy the distribution of $\mathbf{e}$ gives the *sampling distribution* which, in classical statistics, is the distribution of the data conditional on all parameters. The distribution of $\boldsymbol{\beta}$ and $\mathbf{u}$ gives the *prior distribution*. In a hierarchical model the first stage is always the sampling distribution, with prior distributions relegated to other stages. The mixed model of equation (7) is interpreted as a conditional ordinary (fixed) linear model in the following way. For fixed (but unknown) values of $\boldsymbol{\beta}$ and $\mathbf{u}$ we would have a usual linear model in equation (7). But these pieces can vary, so we model them in a hierarchy, that is, we put distributions on them. Formally we can write the first level of the hierarchy as

1. Given $\mathbf{u} = \mathbf{u}_0$ and $\boldsymbol{\beta} = \boldsymbol{\beta}_0$, we have

$$\mathbf{y} = \mathbf{X}\boldsymbol{\beta}_0 + \mathbf{Z}\mathbf{u}_0 + \mathbf{e}, \tag{8}$$

where $\mathbf{e}$ is the sampling error, $\mathbf{e} \sim f_{\mathbf{e}}(\cdot)$. Thus, the first level of the hierarchy is an ordinary fixed linear model. The second level of the hierarchy specifies the distribution of the unobservables $\mathbf{u}$ and $\boldsymbol{\beta}$. We write

2.
$$(\mathbf{u}, \boldsymbol{\beta}) \sim \mathbf{f}_{\mathbf{u},\boldsymbol{\beta}}(\cdot\,,\cdot), \tag{9}$$

where $f_{\mathbf{u},\boldsymbol{\beta}}(\cdot\,,\cdot)$ is a joint probability distribution on the unobservables $\mathbf{u}$ and $\boldsymbol{\beta}$.

Expressions (8) and (9) completely specify the model, with (8) giving a fixed effects model for $\boldsymbol{\beta}$ and $\mathbf{u}$, and (9) giving the hierarchical component. The variance components are parameters (unobservable) of the distributions of $\mathbf{e}$, $\boldsymbol{\beta}$ and $\mathbf{u}$. As such, they are modeled at a lower level of the hierarchy than $\boldsymbol{\beta}$ and $\mathbf{u}$. As we will see in Section 9.2, this leads to some rather straightforward estimation schemes for these variance components.

Building on our Bayesian estimation principles, we can state a broad estimation principle for unobservables (an estimation principle for observables is not needed!). An unobservable is estimated using the distribution obtained by conditioning on all observables and integrating over all other unobservables. This principle is a logical generalization of calculating conditional expectations, and is applicable in models of any complexity. The practice of integrating out the unobservables that are not of interest will always yield estimates that are functions only of the data (observables).

d. The normal hierarchy

In this subsection we illustrate a most popular and useful hierarchical model, the normal hierarchy, which we will use extensively throughout this chapter. Here we will only establish some notation, and discuss some general principles.

Details of estimation are left to Sections 9.2 and 9.3. For a comprehensive treatment of hierarchical linear models the reader is referred to Lindley and Smith (1972).

We now specify that $f_{\boldsymbol{\beta}}$, $f_{\mathbf{u}}$ and $f_{\mathbf{e}}$, described in (8) and (9), be normal distributions. Thus

$$\boldsymbol{\beta} \sim \mathcal{N}(\boldsymbol{\beta}_0, \mathbf{B}), \quad \mathbf{u} \sim \mathcal{N}(\mathbf{0}, \mathbf{D}), \quad \mathbf{e} \sim \mathcal{N}(\mathbf{0}, \mathbf{R}), \tag{10}$$

with $\boldsymbol{\beta}$, $\mathbf{u}$ and $\mathbf{e}$ being independent, where the zero means for $\mathbf{u}$ and $\mathbf{e}$ are taken without loss of generality. [This is illustrated following (15) of Chapter 1.] Also, although we are using the same symbol $\mathbf{D} = \text{var}(\mathbf{u})$ as in (67) of Chapter 4, the matrix $\mathbf{D}$ is not restricted to be diagonal here.

To illustrate some conditional and unconditional moments of $\mathbf{y}$, we get

conditional on $\boldsymbol{\beta}$:

$$E(\mathbf{y} \mid \boldsymbol{\beta}) = \mathbf{X}\boldsymbol{\beta} \quad \text{and} \quad \text{var}(\mathbf{y} \mid \boldsymbol{\beta}) = \mathbf{ZDZ}' + \mathbf{R} = \mathbf{V}; \tag{11}$$

unconditional on $\boldsymbol{\beta}$:

$$E(\mathbf{y}) = \mathbf{X}\boldsymbol{\beta}_0 \quad \text{and} \quad \text{var}(\mathbf{y}) = \mathbf{XBX}' + \mathbf{ZDZ}' + \mathbf{R} = \mathbf{XBX}' + \mathbf{V}; \tag{12}$$

where the results in (11) are reminiscent of (59) and (69) of Chapter 4. These equations show that we can place the usual treatment of the mixed model within the framework of a hierarchical model.

The variance components in $\mathbf{B}$, $\mathbf{D}$ and $\mathbf{R}$ can formally be modeled by adding another layer to the hierarchy. This can be done by expanding the hierarchy of (8) and (9) to

1. Given $\mathbf{u}, \boldsymbol{\beta}, \mathbf{R}$,

$$\mathbf{y} \sim \mathcal{N}(\mathbf{X}\boldsymbol{\beta} + \mathbf{Zu}, \mathbf{R});$$

2. Given $\boldsymbol{\beta}_0, \mathbf{B}, \mathbf{D}$, (13)

$$\boldsymbol{\beta} \sim \mathcal{N}(\boldsymbol{\beta}_0, \mathbf{B}), \quad \mathbf{u} \sim \mathcal{N}(\mathbf{0}, \mathbf{D});$$

3. $(\mathbf{B}, \mathbf{D}, \mathbf{R}) \sim \Pi_{\mathbf{B},\mathbf{D},\mathbf{R}}(\cdot, \cdot, \cdot)$.

However, this level of modeling is usually not done, and the hierarchy of (8) and (9) is used instead.

e. Point estimator of variance or variance of point estimator?

For the most part, we are concerned with point estimation of variances, a strategy that is different from (and perhaps easier than) estimating the variance of a point estimator. To illustrate these differences, consider estimation, in the mixed model, of $\boldsymbol{\beta}$, $\mathbf{V}$ and the variance of our estimate of $\boldsymbol{\beta}$.

If the matrix $\mathbf{V}$ were known, we could estimate $\boldsymbol{\beta}$ by $\hat{\boldsymbol{\beta}}_{\mathbf{V}}$, the BLUE, and also calculate its variance $\text{var}(\hat{\boldsymbol{\beta}}_{\mathbf{V}})$ by

$$\hat{\boldsymbol{\beta}}_{\mathbf{V}} = (\mathbf{X}'\mathbf{V}^{-1}\mathbf{X})^{-1}\mathbf{X}'\mathbf{V}^{-1}\mathbf{y} \text{ and } \text{var}(\hat{\boldsymbol{\beta}}_{\mathbf{V}}) = (\mathbf{X}'\mathbf{V}^{-1}\mathbf{X})^{-1}, \quad \text{respectively} . \tag{14}$$

With $\mathbf{V}$ unknown, a common practice is to replace $\mathbf{V}$ by an estimate $\hat{\mathbf{V}}$ in both expressions in (14). Even though $\hat{\mathbf{V}}$ may be a good point estimator of $\mathbf{V}$ (a good point estimator of a variance), and the estimate $\hat{\boldsymbol{\beta}}_{\hat{\mathbf{V}}}$ can sometimes be reasonable for the estimate of $\boldsymbol{\beta}$, it turns out that $(\mathbf{X}'\hat{\mathbf{V}}^{-1}\mathbf{X})^{-1}$ is not a reasonable estimate of $\text{var}(\hat{\boldsymbol{\beta}}_{\hat{\mathbf{V}}})$.

The variance estimate $(\mathbf{X}'\hat{\mathbf{V}}^{-1}\mathbf{X})^{-1}$ implicitly treates $\hat{\mathbf{V}}$ as known, and does not take into account the variation in $\hat{\mathbf{V}}$ as an estimate of $\mathbf{V}$. For this reason $(\mathbf{X}'\hat{\mathbf{V}}^{-1}\mathbf{X})^{-1}$ will be an underestimate of the true variance of the estimate $\hat{\boldsymbol{\beta}}_{\hat{\mathbf{V}}} = (\mathbf{X}'\hat{\mathbf{V}}^{-1}\mathbf{X})^{-1}\mathbf{X}'\hat{\mathbf{V}}^{-1}\mathbf{y}$. This problem, of treating $\hat{\mathbf{V}}$ as fixed, is also a shortcoming of the estimate of $\boldsymbol{\beta}$. However, Kackar and Harville (1981) show that, under mild conditions on $\hat{\mathbf{V}}$,

$$E(\hat{\boldsymbol{\beta}}_{\hat{\mathbf{V}}}) = E(\hat{\boldsymbol{\beta}}_{\mathbf{V}}) = \boldsymbol{\beta}\,. \tag{15}$$

Thus, replacing $\mathbf{V}$ by $\hat{\mathbf{V}}$ in $\hat{\boldsymbol{\beta}}_{\mathbf{V}}$ of (14) leads to an estimator of $\boldsymbol{\beta}$ that is consistent but has larger variance than $\hat{\boldsymbol{\beta}}_{\mathbf{V}}$. Although $(\mathbf{X}'\hat{\mathbf{V}}^{-1}\mathbf{X})^{-1}$ is a consistent estimator of $\text{var}(\hat{\boldsymbol{\beta}}_{\mathbf{V}}) = \text{var}[(\mathbf{X}'\mathbf{V}^{-1}\mathbf{X})^{-1}\mathbf{X}'\mathbf{V}^{-1}\mathbf{y}]$, it is an underestimate of $\text{var}(\hat{\boldsymbol{\beta}}_{\hat{\mathbf{V}}}) = \text{var}[(\mathbf{X}'\hat{\mathbf{V}}^{-1}\mathbf{X})^{-1}\mathbf{X}'\hat{\mathbf{V}}^{-1}\mathbf{y}]$. This also can be seen by applying the variance identity in Appendix M. We have

$$\begin{aligned}\text{var}(\hat{\boldsymbol{\beta}}_{\hat{\mathbf{V}}}) &= \text{var}[(\mathbf{X}'\hat{\mathbf{V}}^{-1}\mathbf{X})^{-1}\mathbf{X}'\hat{\mathbf{V}}^{-1}\mathbf{y}] \\ &= E[\text{var}(\hat{\boldsymbol{\beta}}_{\mathbf{V}} \mid \mathbf{V} = \hat{\mathbf{V}})] + \text{var}[E(\hat{\boldsymbol{\beta}}_{\mathbf{V}} \mid \mathbf{V} = \hat{\mathbf{V}})]\,.\end{aligned} \tag{16}$$

If (15) holds then the second term above is $\text{var}[E(\hat{\boldsymbol{\beta}}_{\mathbf{V}} \mid \mathbf{V} = \hat{\mathbf{V}})] = \text{var}(\boldsymbol{\beta}) = \mathbf{0}$, and so

$$\begin{aligned}\text{var}(\hat{\boldsymbol{\beta}}_{\hat{\mathbf{V}}}) &= E[\text{var}(\hat{\boldsymbol{\beta}}_{\mathbf{V}} \mid \mathbf{V} = \hat{\mathbf{V}}] \\ &= E[(\mathbf{X}'\hat{\mathbf{V}}^{-1}\mathbf{X})^{-1}\mathbf{X}'\hat{\mathbf{V}}^{-1}\,\text{var}(\mathbf{y} \mid \mathbf{V} = \hat{\mathbf{V}})\,\hat{\mathbf{V}}^{-1}\mathbf{X}(\mathbf{X}'\hat{\mathbf{V}}^{-1}\mathbf{X})^{-1}]\,.\end{aligned}$$

If $\text{var}(\mathbf{y} \mid \mathbf{V} = \hat{\mathbf{V}}) \approx \hat{\mathbf{V}}$ then $\text{var}(\hat{\boldsymbol{\beta}}_{\hat{\mathbf{V}}}) \approx E(\mathbf{X}'\hat{\mathbf{V}}^{-1}\mathbf{X})^{-1}$, and thus $(\mathbf{X}'\hat{\mathbf{V}}^{-1}\mathbf{X})^{-1}$ would be a reasonable estimate of $\text{var}(\hat{\boldsymbol{\beta}}_{\hat{\mathbf{V}}})$. However, assuming $\text{var}(\mathbf{y} \mid \mathbf{V} = \hat{\mathbf{V}}) \approx \hat{\mathbf{V}}$ is not always justified.

Thus, we see that the point estimator $\hat{\mathbf{V}}$ is a reasonable estimator of $\mathbf{V}$, that we can replace $\mathbf{V}$ by $\hat{\mathbf{V}}$ and obtain what we have called $\hat{\boldsymbol{\beta}}_{\hat{\mathbf{V}}}$, which is a reasonable estimator of $\boldsymbol{\beta}$, but it is more difficult to obtain a good estimator of $\text{var}(\hat{\boldsymbol{\beta}}_{\hat{\mathbf{V}}})$. Since we are mainly concerned with point estimation of variance components like $\mathbf{V}$, these problems do not affect us greatly here. However, they do appear when we deal with estimation of fixed and random effects, and must always be accounted for.

Later in this chapter we deal with this "variance underestimation" problem using a Bayesian approach in the hierarchical model. There are also a number of classical methods available to correct this variance underestimation problem, but unfortunately they can be difficult to implement. One that immediately comes to mind is to expand the estimate $\hat{\mathbf{V}}$ in a Taylor series around $\mathbf{V}$, and use the expansion to correct the underestimation problem. Another technique is to calculate a *bootstrap* estimate of variance [see, e.g., Efron (1982), and Laird and Louis (1987)]. Both of these methods may lead to implementation difficulties: the Taylor series may be an extremely involved calculation, while the bootstrap may require enormous computing power.

9.2. VARIANCE ESTIMATION IN THE NORMAL HIERARCHY

a. Formal hierarchical estimation

To estimate a variance component, we can proceed formally as outlined in (1)–(3), and derive a posterior distribution and calculate a posterior mean. For a specific example consider estimation of **D** in the hierarchy (13).

To proceed, we first obtain the posterior distribution of **D** given **y**. Formally, using (13) and keeping track of parameters, we use the laws of probability to write

$$\pi(\mathbf{D} \mid \mathbf{y}) = \frac{\int \cdots \int f(\mathbf{y} \mid \boldsymbol{\beta}, \mathbf{u}, \boldsymbol{\beta}_0, \mathbf{B}, \mathbf{D}, \mathbf{R})\pi(\boldsymbol{\beta} \mid \boldsymbol{\beta}_0, \mathbf{B})\pi(\mathbf{u} \mid \mathbf{D})\pi(\boldsymbol{\beta}_0)\pi(\mathbf{B}, \mathbf{D}, \mathbf{R})\, d\boldsymbol{\beta}\, d\mathbf{u}\, d\boldsymbol{\beta}_0\, d\mathbf{B}\, d\mathbf{R}}{\int \cdots \int f(\mathbf{y} \mid \boldsymbol{\beta}, \mathbf{u}, \boldsymbol{\beta}_0, \mathbf{B}, \mathbf{D}, \mathbf{R})\pi(\boldsymbol{\beta} \mid \boldsymbol{\beta}_0, \mathbf{B})\pi(\mathbf{u} \mid \mathbf{D})\pi(\boldsymbol{\beta}_0)\pi(\mathbf{B}, \mathbf{D}, \mathbf{R})\, d\boldsymbol{\beta}\, d\mathbf{u}\, d\boldsymbol{\beta}_0\, d\mathbf{B}\, d\mathbf{R}\, d\mathbf{D}}, \tag{17}$$

where $\pi(\boldsymbol{\beta}_0)$ is a prior distribution for $\boldsymbol{\beta}_0$, and $\pi(\mathbf{B}, \mathbf{D}, \mathbf{R})$ represents the prior distribution of the variance components.

Although (17) reflects a straightforward derivation of a density, this calculation can be extremely difficult to carry out. In particular, a numerical evaluation would involve high-dimension integrations, which can be quite tricky and demanding of computer time. Moreover, the choice of the prior density for the variances is non-trivial, as naïve choices (e.g., independent conjugate priors) can lead to difficult calculations. Such difficulties arise even in the 1-way model, as noted first by Hill (1965) and Tiao and Tan (1965, 1966). [A particularly readable account of estimation methods can be found in Gianola and Fernando (1986).] However, recent advances in hierarchical computing methods, particularly the use of Gibbs sampling techniques (Gelfand and Smith, 1990; Gelfand *et al.*, 1990) show great promise for alleviating the computational burden.

Here we will concentrate on variance component estimation strategies that are both conceptually easier to understand and computationally simpler. These techniques can be thought of as approximations to the results of the calculations in (17), where simplified forms of prior densities are employed. As we will see, these resulting strategies are closely related to likelihood-based methods.

b. Likelihood methods

Both maximum likelihood (ML) and restricted maximum likelihood (REML) estimates, as discussed in Chapter 6, can be obtained through a hierarchical model. In this section we describe the relationship between ML, REML and Bayes estimation in hierarchical models. In particular, REML is a special case of marginal likelihood, and is equivalent to Bayes estimation with a non-informative prior.

Before we describe the connection to likelihood methods, the relationship between the likelihood function and the densities specified in a hierarchy must be clarified. Thus far, a likelihood function has been defined only for a given normal distribution, as in (103) of Chapter 3, (85) of Chapter 4 and (12) of Chapter 6. Moreover, although it will turn out to be straightforward, it is not immediately clear how a likelihood function is to be defined with a specification like

$$\begin{gathered}\mathbf{y} \mid \boldsymbol{\beta}, \mathbf{u}, \mathbf{R} \sim \mathcal{N}(\mathbf{X}\boldsymbol{\beta} + \mathbf{Z}\mathbf{u}, \mathbf{R}), \\ \boldsymbol{\beta} \sim \mathcal{N}(\boldsymbol{\beta}_0, \mathbf{B}), \quad \mathbf{u} \sim \mathcal{N}(\mathbf{0}, \mathbf{D}), \\ \boldsymbol{\beta}, \mathbf{u} \text{ independent} .\end{gathered} \tag{18}$$

In a hierarchy like (18) the sampling density of $\mathbf{y}$ (the density that describes the variation in repeated sampling) is the marginal density of $\mathbf{y}$. Thus, the likelihood function associated with (18) is the one that is derived from the marginal distribution of $\mathbf{y}$. To be specific, we state the following definition.

Definition. For the hierarchical model

$$\begin{gathered}\mathbf{y} \sim f(\mathbf{y} \mid \boldsymbol{\beta}, \mathbf{u}, \mathbf{R}), \\ \boldsymbol{\beta} \sim f_{\boldsymbol{\beta}}(\boldsymbol{\beta} \mid \boldsymbol{\beta}_0, \mathbf{B}), \quad \mathbf{u} \sim f_{\mathbf{u}}(\mathbf{u} \mid \mathbf{D}),\end{gathered} \tag{19}$$

where $f(\mathbf{y} \mid \boldsymbol{\beta}, \mathbf{u}, \mathbf{R})$ is the sample density and $f_{\boldsymbol{\beta}}(\boldsymbol{\beta} \mid \boldsymbol{\beta}_0, \mathbf{B})$ and $f_{\mathbf{u}}(\mathbf{u} \mid \mathbf{D})$ are the densities of the parameters (unobservable quantities), the *likelihood function for the hierarchical model* (sometimes called the *full likelihood*) is given by

$$L(\boldsymbol{\beta}_0, \mathbf{B}, \mathbf{D}, \mathbf{R} \mid \mathbf{y}) = \iint f(\mathbf{y} \mid \boldsymbol{\beta}, \mathbf{u}, \mathbf{R}) f_{\boldsymbol{\beta}}(\boldsymbol{\beta} \mid \boldsymbol{\beta}_0, \mathbf{B}) f_{\mathbf{u}}(\mathbf{u} \mid \mathbf{D})\, d\boldsymbol{\beta}\, d\mathbf{u} . \tag{20}$$

Variations in either the hierarchical specification or in the densities in the hierarchy lead to different likelihoods. For example, for the normal mixed model (7), the likelihood function [as in (12) of Chapter 6], is

$$L(\boldsymbol{\beta}, \mathbf{V} \mid \mathbf{y}) = \frac{1}{(2\pi)^{\frac{1}{2}N} |\mathbf{V}|^{\frac{1}{2}}} \exp[-\tfrac{1}{2}(\mathbf{y} - \mathbf{X}\boldsymbol{\beta})' \mathbf{V}^{-1} (\mathbf{y} - \mathbf{X}\boldsymbol{\beta})], \tag{21}$$

where $\mathbf{V} = \mathbf{Z}\mathbf{D}\mathbf{Z}' + \mathbf{R}$. To obtain this likelihood from (20), we use a hierarchical specification with a *point-mass* prior density for $\boldsymbol{\beta}$. (A point-mass prior is a density that concentrates all mass on one point.) This is equivalent to leaving the specification of $f_{\boldsymbol{\beta}}(\boldsymbol{\beta})$ out of the hierarchy, and writing

$$\begin{gathered}\mathbf{y} \mid \boldsymbol{\beta}, \mathbf{u} \sim \mathcal{N}(\mathbf{X}\boldsymbol{\beta} + \mathbf{Z}\mathbf{u}, \mathbf{R}), \\ \mathbf{u} \mid \mathbf{D} \sim \mathcal{N}(\mathbf{0}, \mathbf{D}),\end{gathered} \tag{22}$$

which leads to the marginal density (or likelihood function)

$$\begin{aligned} L(\boldsymbol{\beta}, \mathbf{D}, \mathbf{R} \mid \mathbf{y}) &= \int L(\boldsymbol{\beta}, \mathbf{u}, \mathbf{D}, \mathbf{R} \mid \mathbf{y})\, d\mathbf{u} \\ &= \frac{1}{(2\pi)^{\frac{1}{2}N} |\mathbf{V}|^{\frac{1}{2}}} \exp[-\tfrac{1}{2}(\mathbf{y} - \mathbf{X}\boldsymbol{\beta})' \mathbf{V}^{-1} (\mathbf{y} - \mathbf{X}\boldsymbol{\beta})], \end{aligned} \tag{23}$$

which is the likelihood function in (21). Thus, ordinary maximum likelihood estimation is estimation that is conditional on the value of $\boldsymbol{\beta}$, and has the value of $\mathbf{u}$ integrated out, according to the hierarchy in (22).

Therefore ordinary maximum likelihood estimation (ML) can be derived from a hierarchy where the value of $\boldsymbol{\beta}$ is taken to be a fixed, unknown constant. In contrast, restricted maximum likelihood estimation (REML) can be derived from a hierarchy where $\boldsymbol{\beta}$ is integrated out using a non-informative, or flat, prior. Start from a hierarchical specification

$$\begin{gathered}\mathbf{y} \mid \boldsymbol{\beta}, \mathbf{u} \sim \mathcal{N}(\mathbf{X}\boldsymbol{\beta} + \mathbf{Z}\mathbf{u}, \mathbf{R}), \\ \boldsymbol{\beta} \sim \text{uniform}(-\infty, \infty), \quad \mathbf{u} \mid \mathbf{D} \sim \mathcal{N}(\mathbf{0}, \mathbf{D}),\end{gathered} \tag{24}$$

where uniform$(-\infty, \infty)$ is interpreted as the "density" $f_{\boldsymbol{\beta}}(\boldsymbol{\beta}) = 1$, and $\boldsymbol{\beta}$ and $\mathbf{u}$ are independent. Then integrate out $\boldsymbol{\beta}$ and $\mathbf{u}$ to obtain

$$\begin{aligned}L(\mathbf{D}, \mathbf{R} \mid \mathbf{y}) &= \iint L(\boldsymbol{\beta}, \mathbf{u}, \mathbf{D}, \mathbf{R} \mid \mathbf{y})\, d\mathbf{u}\, d\boldsymbol{\beta} \\ &= \frac{1}{(2\pi)^{\frac{1}{2}(N-r)}|\mathbf{K}'\mathbf{V}\mathbf{K}|^{\frac{1}{2}}} \exp[-\tfrac{1}{2}\mathbf{y}'\mathbf{K}(\mathbf{K}'\mathbf{V}\mathbf{K})^{-1}\mathbf{K}'\mathbf{y}],\end{aligned} \tag{25}$$

where r is the rank of $\mathbf{X}$ and $\mathbf{K}$ is any $N \times (N - r)$ matrix of rank $N - r$ that satisfies $\mathbf{K}'\mathbf{X} = \mathbf{0}$.

As we will see, the integration in (25) is an equivalent way of deriving the REML likelihood given in Section 6.7. The appearance of the matrix $\mathbf{K}$ results from the fact that the REML likelihood is based on data of smaller dimension than the full likelihood, and is related to a projection matrix for this new space through the identity $\mathbf{K}(\mathbf{K}'\mathbf{V}\mathbf{K})^{-1}\mathbf{K}' = \mathbf{V}^{-1} - \mathbf{V}^{-1}\mathbf{X}(\mathbf{X}'\mathbf{V}^{-1}\mathbf{X})^{-}\mathbf{X}'\mathbf{V}^{-1}$ (see Appendix M.4f). The likelihood in (25) is the REML likelihood, and thus restricted maximum likelihood estimation is estimation that has the values of both $\boldsymbol{\beta}$ and $\mathbf{u}$ integrated out. A non-informative prior is used for $\boldsymbol{\beta}$, and the usual normal prior for $\mathbf{u}$.

The ordinary maximum likelihood function is equal to (20) if either $\boldsymbol{\beta}$ is a fixed, unknown constant, or if the density of $\boldsymbol{\beta}$ is a point-mass density. This type of likelihood is also sometimes called a *conditional likelihood.* The restricted maximum likelihood function is equal to (20) if the density of $\boldsymbol{\beta}$ is uniform$(-\infty, \infty)$, or if the density of $\boldsymbol{\beta}$ is omitted from the hierarchy. This type of likelihood is also sometimes called a *marginal likelihood.*

Since REML plays such an important role in variance component estimation, derivation of the likelihood (25) will now be given. Although the derivation is somewhat involved, and the appearance of $\mathbf{K}$ may seem mysterious, it is really quite straightforward. Starting from the hierarchy (24), the likelihood function is

$$\begin{aligned}L(\boldsymbol{\beta}, \mathbf{u}, \mathbf{D}, \mathbf{R} \mid \mathbf{y}) = {}& \frac{1}{(2\pi)^{\frac{1}{2}N}|\mathbf{R}|^{\frac{1}{2}}} \exp\{-\tfrac{1}{2}[\mathbf{y} - (\mathbf{X}\boldsymbol{\beta} + \mathbf{Z}\mathbf{u})]'\mathbf{R}^{-1}[\mathbf{y} - (\mathbf{X}\boldsymbol{\beta} + \mathbf{Z}\mathbf{u})]\} \\ & \times \frac{1}{(2\pi)^{\frac{1}{2}q}|\mathbf{D}|^{\frac{1}{2}}} \exp(-\tfrac{1}{2}\mathbf{u}'\mathbf{D}^{-1}\mathbf{u}),\end{aligned}$$

where $q \equiv q_.$, the order of $\mathbf{D}$. By expanding the exponent we get

$$L(\boldsymbol{\beta}, \mathbf{u}, \mathbf{D}, \mathbf{R} \mid \mathbf{y}) = \frac{1}{(2\pi)^{\frac{1}{2}N}|\mathbf{R}|^{\frac{1}{2}}} \frac{1}{(2\pi)^{\frac{1}{2}q}|\mathbf{D}|^{\frac{1}{2}}}$$
$$\times \exp\{-\tfrac{1}{2}(\mathbf{y} - \mathbf{X}\boldsymbol{\beta})'\mathbf{R}^{-1}(\mathbf{y} - \mathbf{X}\boldsymbol{\beta}) - \tfrac{1}{2}\mathbf{u}'[\mathbf{D}^{-1} + \mathbf{Z}'\mathbf{R}^{-1}\mathbf{Z}]\mathbf{u}$$
$$+ \mathbf{u}'\mathbf{Z}'\mathbf{R}^{-1}(\mathbf{y} - \mathbf{X}\boldsymbol{\beta})\}\ .$$

Let $\mathbf{A} = \mathbf{D}^{-1} + \mathbf{Z}'\mathbf{R}^{-1}\mathbf{Z}$, and complete the square in the exponent (in $\mathbf{u}$) to get

$$L(\boldsymbol{\beta}, \mathbf{u}, \mathbf{D}, \mathbf{R} \mid \mathbf{y}) = \frac{1}{(2\pi)^{\frac{1}{2}N}|\mathbf{R}|^{\frac{1}{2}}} \frac{1}{(2\pi)^{\frac{1}{2}q}|\mathbf{D}|^{\frac{1}{2}}}$$
$$\times \exp\{-\tfrac{1}{2}(\mathbf{y} - \mathbf{X}\boldsymbol{\beta})'\mathbf{R}^{-1}(\mathbf{y} - \mathbf{X}\boldsymbol{\beta})$$
$$- \tfrac{1}{2}[\mathbf{u} - \mathbf{A}^{-1}\mathbf{Z}'\mathbf{R}^{-1}(\mathbf{y} - \mathbf{X}\boldsymbol{\beta})]'\mathbf{A}[\mathbf{u} - \mathbf{A}^{-1}\mathbf{Z}'\mathbf{R}^{-1}(\mathbf{y} - \mathbf{X}\boldsymbol{\beta})]$$
$$- \tfrac{1}{2}(\mathbf{y} - \mathbf{X}\boldsymbol{\beta})'\mathbf{R}^{-1}\mathbf{Z}\mathbf{A}^{-1}\mathbf{Z}'\mathbf{R}^{-1}(\mathbf{y} - \mathbf{X}\boldsymbol{\beta})\}\ .$$

Combining terms, we see that in the quadratic form in $\mathbf{y} - \mathbf{X}\boldsymbol{\beta}$ the matrix is $\mathbf{R}^{-1} - \mathbf{R}^{-1}\mathbf{Z}\mathbf{A}^{-1}\mathbf{Z}'\mathbf{R}^{-1} = (\mathbf{Z}\mathbf{D}\mathbf{Z}' + \mathbf{R})^{-1} = \mathbf{V}^{-1}$ (see E 9.12). Thus

$$L(\boldsymbol{\beta}, \mathbf{u}, \mathbf{D}, \mathbf{R} \mid \mathbf{y}) = \frac{1}{(2\pi)^{\frac{1}{2}N}|\mathbf{R}|^{\frac{1}{2}}} \frac{1}{(2\pi)^{\frac{1}{2}q}|\mathbf{D}|^{\frac{1}{2}}}$$
$$\times \exp\{-\tfrac{1}{2}(\mathbf{y} - \mathbf{X}\boldsymbol{\beta})'\mathbf{V}^{-1}(\mathbf{y} - \mathbf{X}\boldsymbol{\beta})$$
$$- \tfrac{1}{2}[\mathbf{u} - \mathbf{A}^{-1}\mathbf{Z}'\mathbf{R}^{-1}(\mathbf{y} - \mathbf{X}\boldsymbol{\beta})]'\mathbf{A}[\mathbf{u} - \mathbf{A}^{-1}\mathbf{Z}'\mathbf{R}^{-1}(\mathbf{y} - \mathbf{X}\boldsymbol{\beta})]\}\ .$$

We are now ready to carry out the first integration in (25). Using the properties of the multivariate normal, we have

$$\int \exp\{-\tfrac{1}{2}[\mathbf{u} - \mathbf{A}^{-1}\mathbf{Z}'\mathbf{R}^{-1}(\mathbf{y} - \mathbf{X}\boldsymbol{\beta})]'\mathbf{A}[\mathbf{u} - \mathbf{A}^{-1}\mathbf{Z}'\mathbf{R}^{-1}(\mathbf{y} - \mathbf{X}\boldsymbol{\beta})]\}\, d\mathbf{u} = (2\pi)^{\frac{1}{2}q}|\mathbf{A}|^{-\frac{1}{2}},$$

and thus

$$\int L(\boldsymbol{\beta}, \mathbf{u}, \mathbf{D}, \mathbf{R} \mid \mathbf{y})\, d\mathbf{u} = \frac{1}{(2\pi)^{\frac{1}{2}N}|\mathbf{R}|^{\frac{1}{2}}|\mathbf{D}|^{\frac{1}{2}}|\mathbf{A}|^{\frac{1}{2}}} \exp[-\tfrac{1}{2}(\mathbf{y} - \mathbf{X}\boldsymbol{\beta})'\mathbf{V}^{-1}(\mathbf{y} - \mathbf{X}\boldsymbol{\beta})]$$
$$= \frac{1}{(2\pi)^{\frac{1}{2}N}|\mathbf{V}|^{\frac{1}{2}}} \exp[-\tfrac{1}{2}(\mathbf{y} - \mathbf{X}\boldsymbol{\beta})'\mathbf{V}^{-1}(\mathbf{y} - \mathbf{X}\boldsymbol{\beta})]$$
$$= L(\boldsymbol{\beta}, \mathbf{V} \mid \mathbf{y}), \text{ as in (21)},$$

where we used the identity $|\mathbf{R}||\mathbf{D}||\mathbf{A}| = |\mathbf{V}|$ (see E 9.12). Next, we factor the exponent in $L(\boldsymbol{\beta}, \mathbf{V} \mid \mathbf{y})$ as

$$(\mathbf{y} - \mathbf{X}\boldsymbol{\beta})'\mathbf{V}^{-1}(\mathbf{y} - \mathbf{X}\boldsymbol{\beta}) = \mathbf{y}'[\mathbf{V}^{-1} - \mathbf{V}^{-1}\mathbf{X}(\mathbf{X}'\mathbf{V}^{-1}\mathbf{X})^{-1}\mathbf{X}'\mathbf{V}^{-1}]\mathbf{y}$$
$$+ (\boldsymbol{\beta} - \hat{\boldsymbol{\beta}})'\mathbf{X}'\mathbf{V}^{-1}\mathbf{X}(\boldsymbol{\beta} - \hat{\boldsymbol{\beta}}),$$

where $\hat{\boldsymbol{\beta}} = (\mathbf{X}'\mathbf{V}^{-1}\mathbf{X})^{-1}\mathbf{X}'\mathbf{V}^{-1}\mathbf{y}$. Using this factorization, we integrate over $\boldsymbol{\beta}$ to obtain

$$\int L(\boldsymbol{\beta}, \mathbf{V} \mid \mathbf{y})\, d\boldsymbol{\beta} = \frac{(2\pi)^{\frac{1}{2}r}|\mathbf{X}'\mathbf{V}^{-1}\mathbf{X}|^{-\frac{1}{2}}}{(2\pi)^{\frac{1}{2}N}|\mathbf{V}|^{\frac{1}{2}}} \times \exp\{-\tfrac{1}{2}\mathbf{y}'[\mathbf{V}^{-1} - \mathbf{V}^{-1}\mathbf{X}(\mathbf{X}'\mathbf{V}^{-1}\mathbf{X})^{-1}\mathbf{X}'\mathbf{V}^{-1}]\mathbf{y}\},$$

which is equal to (25), as $|\mathbf{V}| = |\mathbf{K}'\mathbf{V}\mathbf{K}|\,|\mathbf{X}'\mathbf{V}^{-1}\mathbf{X}|$ (E 9.12) and $\mathbf{K}(\mathbf{K}'\mathbf{V}\mathbf{K})^{-1}\mathbf{K}' = \mathbf{V}^{-1} - \mathbf{V}^{-1}\mathbf{X}(\mathbf{X}'\mathbf{V}^{-1}\mathbf{X})^{-1}\mathbf{X}'\mathbf{V}^{-1}$, as in Appendix M.4f.

c. Empirical Bayes estimation

The term "empirical Bayes" is non-precise, and thus has many different interpretations. Here, we will be quite specific about our definition. Empirical Bayes estimation will refer to using a marginal distribution to estimate parameters in a hierarchical model, and substituting these estimates for their corresponding parameters in a formal Bayes estimator.

-i. General strategies. We outline an empirical Bayes estimation principle which, when used in conjunction with a hierarchical model, will lead to empirical Bayes estimates of any desired parameter. To obtain the empirical Bayes estimate of a particular parameter $\boldsymbol{\tau}$:

(i) Specify, for $\boldsymbol{\tau}$, a distribution $\pi(\boldsymbol{\tau} \mid \boldsymbol{\eta})$, where $\boldsymbol{\eta}$ represents the parameters of the distribution of $\boldsymbol{\tau}$, sometimes known as *hyperparameters*.

(ii) Calculate the formal Bayes posterior of $\boldsymbol{\tau}$,

$$\pi(\boldsymbol{\tau} \mid \mathbf{y}, \boldsymbol{\eta}) = \frac{f(\mathbf{y} \mid \boldsymbol{\tau}, \boldsymbol{\eta})\pi(\boldsymbol{\tau} \mid \boldsymbol{\eta})}{\int f(\mathbf{y} \mid \boldsymbol{\tau}, \boldsymbol{\eta})\pi(\boldsymbol{\tau} \mid \boldsymbol{\eta})\, d\boldsymbol{\tau}}, \tag{26}$$

and use it to estimate $\boldsymbol{\tau}$, for example by using $\hat{\boldsymbol{\tau}} = E(\boldsymbol{\tau} \mid \mathbf{y}, \boldsymbol{\eta})$.

(iii) Calculate estimates $\hat{\boldsymbol{\eta}}$ of any unknown (hyper)parameters from the marginal distribution

$$m(\mathbf{y} \mid \boldsymbol{\eta}) = \int f(\mathbf{y} \mid \boldsymbol{\tau}, \boldsymbol{\eta})\pi(\boldsymbol{\tau} \mid \boldsymbol{\eta})\, d\boldsymbol{\tau}, \tag{27}$$

for example, by using ML on (27). Finally, produce the empirical Bayes estimate of $\boldsymbol{\tau}$ by substituting $\hat{\boldsymbol{\eta}}$ for $\boldsymbol{\eta}$ in $E(\boldsymbol{\tau} \mid \mathbf{y}, \boldsymbol{\eta})$ to give $\hat{\boldsymbol{\tau}} = E(\boldsymbol{\tau} \mid \mathbf{y}, \hat{\boldsymbol{\eta}})$.

-ii. Estimation. We outline an empirical Bayes estimation strategy, using (26), for obtaining estimates of the variances $\mathbf{D}$ and $\mathbf{R}$ in the hierarchy of (22). Then the connection to both ML and REML estimates of $\mathbf{D}$ and $\mathbf{R}$ will be made clear.

Empirical Bayes estimation of D and R in the hierarchy (22)

Step 1 [equivalent to (i) and (ii) above]. Calculate the posterior distribution

$$\pi(\mathbf{D}, \mathbf{R} \mid \mathbf{y}, \boldsymbol{\eta}) = \frac{f(\mathbf{y} \mid \mathbf{D}, \mathbf{R})\pi(\mathbf{D}, \mathbf{R} \mid \boldsymbol{\eta})}{\iint f(\mathbf{y} \mid \mathbf{D}, \mathbf{R})\pi(\mathbf{D}, \mathbf{R} \mid \boldsymbol{\eta})\, d\mathbf{D}\, d\mathbf{R}}, \tag{28}$$

where $f(\mathbf{y} \mid \mathbf{D}, \mathbf{R})$ is the distribution of $\mathbf{y}$ and $\pi(\mathbf{D}, \mathbf{R} \mid \boldsymbol{\eta})$ is a prior distribution on $\mathbf{D}$ and $\mathbf{R}$, with $\boldsymbol{\eta}$ being a hyperparameter (possibly vector-valued).

Step 2 [equivalent to (iii) above]. Estimate $\boldsymbol{\eta}$ with $\hat{\boldsymbol{\eta}}$, obtained from the marginal distribution of $\mathbf{y}$, the denominator of (28), using maximum likelihood on

$$m(\mathbf{y} \mid \boldsymbol{\eta}) = \iint f(\mathbf{y} \mid \mathbf{D}, \mathbf{R})\pi(\mathbf{D}, \mathbf{R} \mid \boldsymbol{\eta})\, d\mathbf{D}\, d\mathbf{R}, \tag{29}$$

the marginal distribution of $\mathbf{y}$.

Step 3. Obtain empirical Bayes estimates of $\mathbf{D}$ and $\mathbf{R}$ by substituting $\hat{\boldsymbol{\eta}}$ for $\boldsymbol{\eta}$ in the Bayes estimators obtained from the posterior (28) in Step 1.

Summary. Empirical Bayes estimation of $\mathbf{D}$ and $\mathbf{R}$ is accomplished by performing maximum likelihood on $\pi(\mathbf{D}, \mathbf{R} \mid \mathbf{y}, \boldsymbol{\eta})$, considering $\boldsymbol{\eta}$ fixed and known, then estimating $\boldsymbol{\eta}$ using maximum likelihood on $m(\mathbf{y} \mid \boldsymbol{\eta})$.

*-iii. **Connections with likelihood.*** To see the connection between ML, REML and empirical Bayes, first notice that from the hierarchy (22) the marginal distribution of $\mathbf{y}$ is exactly equal to the full likelihood function as given after the Definition, in equation (21). Now assume that the distribution of $\boldsymbol{\beta}$ does not depend on any unknown hyperparameters $\boldsymbol{\beta}_0$ and $\mathbf{B}$. Upon performing the integration of $\mathbf{u}$ in (20), we have

$$\begin{aligned} f(\mathbf{y} \mid \mathbf{D}, \mathbf{R}) = L(\mathbf{D}, \mathbf{R} \mid \mathbf{y}) &= \int L(\boldsymbol{\beta}, \mathbf{D}, \mathbf{R} \mid \mathbf{y}) f_{\boldsymbol{\beta}}(\boldsymbol{\beta})\, d\boldsymbol{\beta} \qquad (30) \\ &= \text{likelihood for ML, using the hierarchy in (22)} \\ &= \text{likelihood for REML, using the hierarchy in (24)}\,. \end{aligned}$$

Now, to complete the likelihood–empirical Bayes connection, we must identify $\pi(\mathbf{D}, \mathbf{R} \mid \mathbf{y}, \boldsymbol{\beta})$ of (28) with $f(\mathbf{y} \mid \mathbf{D}, \mathbf{R})$ of (30). However, this is easy, for we see that (28) is the same as (30) only if $\pi(\mathbf{D}, \mathbf{R} \mid \boldsymbol{\eta}) = 1$, that is, both ML and REML exactly correspond to empirical Bayes estimation using a flat (non-informative) prior distribution for $\mathbf{D}$ and $\mathbf{R}$. Thus, from (28),

$$\begin{aligned} \pi(\mathbf{D}, \mathbf{R} \mid \mathbf{y}, \boldsymbol{\eta}) &= \frac{f(\mathbf{y} \mid \mathbf{D}, \mathbf{R})\pi(\mathbf{D}, \mathbf{R} \mid \boldsymbol{\eta})}{\iint f(\mathbf{y} \mid \mathbf{D}, \mathbf{R})\pi(\mathbf{D}, \mathbf{R} \mid \boldsymbol{\eta})\, d\mathbf{D}\, d\mathbf{R}} \\ &= \frac{f(\mathbf{y} \mid \mathbf{D}, \mathbf{R})}{\iint f(\mathbf{y} \mid \mathbf{D}, \mathbf{R})\, d\mathbf{D}\, d\mathbf{R}} \quad [\pi(\mathbf{D}, \mathbf{R} \mid \boldsymbol{\eta}) = 1] \qquad (31) \\ &= \begin{cases} L(\boldsymbol{\beta}, \mathbf{D}, \mathbf{R} \mid \mathbf{y}) & [\text{likelihood for ML, using a hierarchy such as (22)}] \\ L(\mathbf{D}, \mathbf{R} \mid \mathbf{y}) & [\text{likelihood for REML, using a hierarchy such as (24)}] \end{cases} \end{aligned}$$

where we have implicitly assumed that $\iint f(\mathbf{y} \mid \mathbf{D}, \mathbf{R})\, d\mathbf{D}\, d\mathbf{R} = 1$, which is needed for the equalities in (31). In fact, we only need assume that $\iint f(\mathbf{y} \mid \mathbf{D}, \mathbf{R})\, d\mathbf{D}\, d\mathbf{R} < \infty$, and replace the "=" in (31) with "$\propto$" (proportional to), and all likelihood estimates will remain the same. If $\iint f(\mathbf{y} \mid \mathbf{D}, \mathbf{R})\, d\mathbf{D}\, d\mathbf{R} = \infty$, which is a distinct possibility once we leave the normal case, then this argument will not establish a likelihood–empirical Bayes connection.

This implementation of the empirical Bayes strategy yields point estimators for **D** and **R**, that is, point estimators of the variance components, not estimates of the variance of other estimators. Thus, the $\hat{\mathbf{D}}$ and $\hat{\mathbf{R}}$ resulting from (28) and (29), or its likelihood variations, are point estimates of **D** and **R**, and should not be used in estimates of var(**y**) or var($\hat{\mathbf{u}}$).

When $\pi(\mathbf{D}, \mathbf{R} \mid \boldsymbol{\eta}) = 1$, or in general when $\pi(\mathbf{D}, \mathbf{R})$ does not depend on any unknown hyperparameters, then part (iii) of the empirical Bayes estimation strategy is unnecessary as there are no more parameters to estimate. There are few easy-to-use alternative prior densities that would keep computations from getting out of hand. One alternative that is feasible is the *Wishart distribution*, the multivariate analog of a chi-squared distribution. [Anderson (1984) has a full treatment of the Wishart distribution.] The density is given by

$$f(\mathbf{T} \mid n, \boldsymbol{\Sigma}) = \frac{|\mathbf{T}|^{\frac{1}{2}(n-p-1)} \exp[-\frac{1}{2}\operatorname{tr}(\boldsymbol{\Sigma}^{-1}\mathbf{T})]}{2^{\frac{1}{2}np}\pi^{4p(p-1)}|\boldsymbol{\Sigma}|^{\frac{1}{2}n} \prod_{i=1}^{p} \Gamma(\frac{1}{2}[n+1-i])}, \tag{32}$$

where **T** is a $p \times p$ matrix, n and $\boldsymbol{\Sigma}$ are parameters, and both **T** and $\boldsymbol{\Sigma}$ are positive definite. Use of the Wishart distribution to implement empirical Bayes estimation in a hierarchy such as (22) or (24) will not yield closed form solutions, but is computationally feasible. If we take separate independent priors on **D** and **R**, such that $\mathbf{D}^{-1} \sim$ Wishart and $\mathbf{R}^{-1} \sim$ Wishart, this is as close as we can come to a joint conjugate prior (in the sense that a Wishart is conjugate for estimating a single variance matrix). Note that such a set-up is a direct multivariate analog of the univariate variance estimation described in Sections 3.9 and 9.1b.

9.3. ESTIMATION OF EFFECTS

a. Hierarchical estimation

In the mixed model

$$\mathbf{Y} = \mathbf{X}\boldsymbol{\beta} + \mathbf{Z}\mathbf{u} + \mathbf{e}, \tag{33}$$

we have, thus far in this chapter, concentrated on estimating the variance components **B**, **D** and **R**. However, there are many situations in which estimation of the effects $\boldsymbol{\beta}$ and **u** is also of interest. (Notice that, since a hierarchical model does not distinguish between fixed and random effects, neither do we. It will be seen that hierarchical strategies for estimation of $\boldsymbol{\beta}$ are the same as those for estimation of **u**.)

In hierarchical modeling the effects $\boldsymbol{\beta}$ and **u** are treated similarly, in that both parameters have their prior distributions, and no distinction is made between fixed and random effects. It is in the estimation of the effects $\boldsymbol{\beta}$ and $\mathbf{u}$ where the power of the hierarchical model is really seen. For example, the hierarchical model (22), which is equivalent to the classical mixed models, leads to a likelihood function that does not contain $\mathbf{u}$. Hence, straightforward likelihood estimation of $\mathbf{u}$ cannot be done. Using hierarchical models, however, we will see that such estimation is straightforward.

The estimation is straightforward in that we will only employ the principle outlined in Section 9.1a. That is, any parameter (unobservable) will be estimated using its posterior distribution. In particular, we will calculate its posterior expectation, as in (3). Thus, in the next two subsections we are concerned with calculating the posterior distributions of $\boldsymbol{\beta}$ and $\mathbf{u}$.

It is interesting to note that using the hierarchy to obtain estimates of $\boldsymbol{\beta}$ and $\mathbf{u}$ can be viewed as a generalization of the BLUP methodology. Indeed, our estimates of $\boldsymbol{\beta}$ and $\mathbf{u}$ will reduce to the BLUE and BLUP in special cases. The original derivations done by C.R. Henderson to obtain BLUP estimates were, in effect, derivations based on a hierarchical model.

Although the derivations we will be doing are straightforward in that the steps to be taken are clearly laid out, these steps may often require a large computational or analytical effort. To simplify matters somewhat, when calculating posterior distributions of $\boldsymbol{\beta}$ and $\mathbf{u}$, we make the (very common) assumption that the variance components are known. In the final steps, when estimators are derived, we indicate how to use estimates of the variance components to substitute for the assumed known quantities. As discussed in Section 9.1e, this strategy is acceptable for point estimation, but not for estimation of the variance of point estimators. This point is further dealt with in Section 9.3dii.

-i. ***Estimation of*** $\boldsymbol{\beta}$. For estimation of $\boldsymbol{\beta}$ we use the distribution $f(\boldsymbol{\beta} \mid \mathbf{y}, \mathbf{B}, \mathbf{D}, \mathbf{R})$. Formally, suppressing the dependence on the dispersion matrices, and using results (1) and (2) of Appendix S.6 (see E 9.7),

$$f_{\boldsymbol{\beta}}(\boldsymbol{\beta} \mid \mathbf{y}) = \frac{\int f(\mathbf{y} \mid \boldsymbol{\beta}, \mathbf{u}) f_{\boldsymbol{\beta}}(\boldsymbol{\beta}) f_{\mathbf{u}}(\mathbf{u})\, d\mathbf{u}}{\iint f(\mathbf{y} \mid \boldsymbol{\beta}, \mathbf{u}) f_{\boldsymbol{\beta}}(\boldsymbol{\beta}) f_{\mathbf{u}}(\mathbf{u})\, d\mathbf{u}\, d\boldsymbol{\beta}}, \tag{34}$$

where

$$\begin{gathered} \mathbf{y} \mid \boldsymbol{\beta}, \mathbf{u}, \mathbf{R} \sim \mathcal{N}(\mathbf{X}\boldsymbol{\beta} + \mathbf{Z}\mathbf{u}, \mathbf{R}), \\ \boldsymbol{\beta} \sim \mathcal{N}(\boldsymbol{\beta}_0, \mathbf{B}), \quad \text{and} \quad \mathbf{u} \sim \mathcal{N}(\mathbf{0}, \mathbf{D}) \,. \end{gathered} \tag{35}$$

Although simplification of $f_{\boldsymbol{\beta}}(\boldsymbol{\beta} \mid \mathbf{y})$ of (34) is involved, it is straightforward. The integrand of (34) is given by

$$\begin{aligned} f(\mathbf{y} \mid \boldsymbol{\beta}, \mathbf{u}) f_{\boldsymbol{\beta}}(\boldsymbol{\beta}) f_{\mathbf{u}}(\mathbf{u}) &= \frac{1}{(2\pi)^{\frac{1}{2}N} |\mathbf{R}|^{\frac{1}{2}}} \exp\{-\tfrac{1}{2}[\mathbf{y} - (\mathbf{X}\boldsymbol{\beta} + \mathbf{Z}\mathbf{u})]'\mathbf{R}^{-1}[\mathbf{y} - (\mathbf{X}\boldsymbol{\beta} + \mathbf{Z}\mathbf{u})]\} \\ &\times \frac{1}{(2\pi)^{\frac{1}{2}p} |\mathbf{B}|^{\frac{1}{2}}} \exp\{-\tfrac{1}{2}(\boldsymbol{\beta} - \boldsymbol{\beta}_0)'\mathbf{B}^{-1}(\boldsymbol{\beta} - \boldsymbol{\beta}_0)\} \\ &\times \frac{1}{(2\pi)^{\frac{1}{2}q} |\mathbf{D}|^{\frac{1}{2}}} \exp(-\tfrac{1}{2}\mathbf{u}'\mathbf{D}^{-1}\mathbf{u}), \end{aligned} \tag{36}$$

where N, p and q are the dimensions of $\mathbf{y}$, $\boldsymbol{\beta}$ and $\mathbf{u}$, respectively. This joint density is a product of normal densities, and is now factored to yield the desired conditional density. That is, we now factor the joint density in the order

$$f(\mathbf{y}, \boldsymbol{\beta}, \mathbf{u}) = f(\mathbf{y} \mid \boldsymbol{\beta}, \mathbf{u}) f_{\boldsymbol{\beta}}(\boldsymbol{\beta}) f_{\mathbf{u}}(\mathbf{u}) = f_{\mathbf{u}}(\mathbf{u} \mid \boldsymbol{\beta}, \mathbf{y}) f_{\boldsymbol{\beta}}(\boldsymbol{\beta} \mid \mathbf{y}) f(\mathbf{y}), \tag{37}$$

because $f_{\boldsymbol{\beta}}(\boldsymbol{\beta} \mid \mathbf{y})$ is the density of interest. To obtain this decomposition requires some algebraic work, in particular repeated use of the operation of "completing the square." [A useful identity is $\mathbf{x}'\mathbf{G}\mathbf{x} - 2\mathbf{x}'\mathbf{H}\mathbf{y} + \mathbf{y}'\mathbf{H}'\mathbf{G}^{-1}\mathbf{H}\mathbf{y} = (\mathbf{x} - \mathbf{G}^{-1}\mathbf{H}\mathbf{y})'\mathbf{G}(\mathbf{x} - \mathbf{G}^{-1}\mathbf{H}\mathbf{y})$.] After these manipulations, we obtain

$$\begin{aligned} f(\mathbf{y}, \boldsymbol{\beta}, \mathbf{u}) &= f(\mathbf{y} \mid \boldsymbol{\beta}, \mathbf{u}) f_{\boldsymbol{\beta}}(\boldsymbol{\beta}) f_{\mathbf{u}}(\mathbf{u}) = f_{\mathbf{u}}(\mathbf{u} \mid \boldsymbol{\beta}, \mathbf{y}) f_{\boldsymbol{\beta}}(\boldsymbol{\beta} \mid \mathbf{y}) f(\mathbf{y}) \\ &= \frac{\exp\{-\frac{1}{2}[\mathbf{u} - E(\mathbf{u} \mid \boldsymbol{\beta}, \mathbf{y})]'\mathbf{A}^{-1}[\mathbf{u} - E(\mathbf{u} \mid \boldsymbol{\beta}, \mathbf{y})]\}}{(2\pi)^{\frac{1}{2}q}|\mathbf{A}|^{\frac{1}{2}}} \\ &\quad \times \frac{\exp\{-\frac{1}{2}[\boldsymbol{\beta} - E(\boldsymbol{\beta} \mid \mathbf{y})]'\mathbf{C}[\boldsymbol{\beta} - E(\boldsymbol{\beta} \mid \mathbf{y})]\}}{(2\pi)^{\frac{1}{2}p}|\mathbf{C}^{-1}|^{\frac{1}{2}}} \\ &\quad \times \frac{\exp\{-\frac{1}{2}[\mathbf{y} - E(\mathbf{y})]'(\mathbf{L} - \mathbf{L}\mathbf{X}\mathbf{C}^{-1}\mathbf{X}'\mathbf{L})[\mathbf{y} - E(\mathbf{y})]\}}{(2\pi)^{\frac{1}{2}N}|\mathbf{L} - \mathbf{L}\mathbf{X}\mathbf{C}^{-1}\mathbf{X}'\mathbf{L}|^{\frac{1}{2}}}, \end{aligned} \tag{38}$$

where

$$\begin{aligned} \mathbf{A} &= \mathbf{D}^{-1} + \mathbf{Z}'\mathbf{R}^{-1}\mathbf{Z}, \\ \mathbf{L} &= \mathbf{R}^{-1} - \mathbf{R}^{-1}\mathbf{Z}\mathbf{A}^{-1}\mathbf{Z}'\mathbf{R}^{-1} = (\mathbf{Z}\mathbf{D}\mathbf{Z}' + \mathbf{R})^{-1} = \mathbf{V}^{-1}, \\ \mathbf{C} &= \mathbf{X}'\mathbf{V}^{-1}\mathbf{X} + \mathbf{B}^{-1}, \\ E(\mathbf{u} \mid \boldsymbol{\beta}, \mathbf{y}) &= \mathbf{A}^{-1}\mathbf{Z}'\mathbf{R}^{-1}(\mathbf{y} - \mathbf{X}\boldsymbol{\beta}), \\ E(\boldsymbol{\beta} \mid \mathbf{y}) &= \mathbf{C}^{-1}(\mathbf{X}'\mathbf{V}^{-1}\mathbf{y} + \mathbf{B}^{-1}\boldsymbol{\beta}_0), \\ E(\mathbf{y}) &= (\mathbf{L} - \mathbf{L}\mathbf{X}\mathbf{C}^{-1}\mathbf{X}'\mathbf{L})^{-1}\mathbf{L}'\mathbf{X}\mathbf{C}^{-1}\mathbf{B}^{-1}\boldsymbol{\beta}_0. \end{aligned} \tag{39}$$

A number of matrix identities can be applied to the expressions in (39) to derive alternate, perhaps more familiar forms for these estimators. In particular, using the identities of E 9.12, we have from (39)

$$\begin{aligned} E(\mathbf{u} \mid \boldsymbol{\beta}, \mathbf{y}) &= \mathbf{D}\mathbf{Z}'\mathbf{V}^{-1}(\mathbf{y} - \mathbf{X}\boldsymbol{\beta}), \\ E(\boldsymbol{\beta} \mid \mathbf{y}) &= (\mathbf{X}'\mathbf{V}^{-1}\mathbf{X} + \mathbf{B}^{-1})^{-1}(\mathbf{X}'\mathbf{V}^{-1}\mathbf{y} + \mathbf{B}^{-1}\boldsymbol{\beta}_0), \\ E(\mathbf{y}) &= \mathbf{X}\boldsymbol{\beta}_0 \,. \end{aligned} \tag{40}$$

The factorization in (37) gives the conditional distribution that we are interested in. Formally, we can complete the required integrations of (34) or, informally, read the answers from (38) knowing that, in this case, all distributions are normal (Appendix S). We have

$$\boldsymbol{\beta} \mid \mathbf{y} \sim \mathcal{N}[E(\boldsymbol{\beta} \mid \mathbf{y}), \mathbf{C}^{-1}], \tag{41}$$

where $E(\boldsymbol{\beta} \mid \mathbf{y})$ and $\mathbf{C}$ are given in (39) and (40).

Using (41), we can obtain estimates of $\boldsymbol{\beta}$. For example, a possible point estimate of $\boldsymbol{\beta}$ is $E(\boldsymbol{\beta} \mid \mathbf{y})$, given in (39), and an estimate of dispersion might be taken as $\text{var}(\boldsymbol{\beta} \mid \mathbf{y}) = \mathbf{C}^{-1}$. Note, however, that $\mathbf{C}^{-1}$ is not the variance of $E(\boldsymbol{\beta} \mid \mathbf{y})$. This variance is the same when using either a Bayesian or frequentist approach, and is commonly given as

$$\text{var}[E(\boldsymbol{\beta} \mid \mathbf{y})] = \begin{cases} \mathbf{C}^{-1}(\mathbf{X}'\mathbf{V}^{-1}\mathbf{X})\mathbf{C}^{-1} & \text{(Bayesian)}, \\ \mathbf{C}^{-1} - \mathbf{C}^{-1}\mathbf{B}^{-1}\mathbf{C}^{-1} & \text{(frequentist)}. \end{cases} \quad (42)$$

We note that expression (43) could form a basis for variance estimation with unknown variance components. Exercise E 9.9 explores the relationship between $E(\boldsymbol{\beta} \mid \mathbf{y})$ and the best linear unbiased estimator of $\boldsymbol{\beta}$.

These hierarchical estimates, based on the posterior distribution, require specification of $\boldsymbol{\beta}_0$ and $\mathbf{B}$, quantities that an experimenter may be reluctant to specify [although Angers (1987) details some robust versions of these estimates]. Sometimes an experimenter will choose values for $\boldsymbol{\beta}_0$ and $\mathbf{B}$ that (seemingly) impart no prior information, using a so-called non-informative prior. In this situation a non-informative prior would specify $\mathbf{B}^{-1} = \mathbf{0}$. Substituting this value into (39) and (41), noting that $\boldsymbol{\beta}_0$ vanishes, we obtain

$$\hat{\boldsymbol{\beta}} = E(\boldsymbol{\beta} \mid \mathbf{y}) = (\mathbf{X}'\mathbf{V}^{-1}\mathbf{X})^{-1}\mathbf{X}'\mathbf{V}^{-1}\mathbf{y}, \quad \text{var}(\boldsymbol{\beta} \mid \mathbf{y}) = (\mathbf{X}'\mathbf{V}^{-1}\mathbf{X})^{-1}, \quad (43)$$

the generalized least squares estimate of $\boldsymbol{\beta}$ and its variance, Note also that $\mathbf{X}\hat{\boldsymbol{\beta}}$ is the BLUE (best linear unbiased estimator) of $\mathbf{X}\boldsymbol{\beta}$ (Appendix S.2).

The assumption that the prior variance matrix $\mathbf{B}$ satisfies $\mathbf{B}^{-1} = \mathbf{0}$ is often equated with $\mathbf{B}$ satisfying $\mathbf{B} = \infty$, although this equivalence can be slippery (see E 9.8). If we assume that this equivalence holds then the generalized least squares estimate of (43) can be viewed as a posterior estimate obtained from prior information with infinite variance, and in that sense the prior is non-informative. This can also be interpreted in the reverse way. If there is any reasonable prior knowledge (where "reasonable" means that our prior variance is smaller than infinity) then we should be using an estimate other than a generalized least squares estimate.

***-ii. Estimation of* u.** For estimation and inference about $\mathbf{u}$ we similarly use $f_{\mathbf{u}}(\mathbf{u} \mid \mathbf{y})$. We cannot use the decomposition in (37) for inference about $\mathbf{u}$ because this would give us $f_{\mathbf{u}}(\mathbf{u} \mid \boldsymbol{\beta}, \mathbf{y})$. This is unsatisfactory since it requires knowing $\boldsymbol{\beta}$. Hence, we decompose the density in an alternate way, writing

$$f(\mathbf{y} \mid \boldsymbol{\beta}, \mathbf{u}) f_{\boldsymbol{\beta}}(\boldsymbol{\beta}) f_{\mathbf{u}}(\mathbf{u}) = f_{\boldsymbol{\beta}}(\boldsymbol{\beta} \mid \mathbf{y}, \mathbf{u}) f_{\mathbf{u}}(\mathbf{u} \mid \mathbf{y}) f(\mathbf{y}). \quad (44)$$

To derive $f_{\mathbf{u}}(\mathbf{u} \mid \mathbf{y})$, operate as in (36)–(39), with the end result being

$$\begin{aligned} f(\mathbf{y} \mid \boldsymbol{\beta}, \mathbf{u}) f_{\boldsymbol{\beta}}(\boldsymbol{\beta}) f_{\mathbf{u}}(\mathbf{u}) &= f_{\boldsymbol{\beta}}(\boldsymbol{\beta} \mid \mathbf{y}, \mathbf{u}) f_{\mathbf{u}}(\mathbf{u} \mid \mathbf{y}) f(\mathbf{y}) \\ &= \frac{\exp\{-\tfrac{1}{2}[\boldsymbol{\beta} - E(\boldsymbol{\beta} \mid \mathbf{u}, \mathbf{y})]'\mathscr{A}^{-1}[\boldsymbol{\beta} - \mathbf{E}(\boldsymbol{\beta} \mid \mathbf{u}, \mathbf{y})]\}}{(2\pi)^{\frac{1}{2}p}|\mathscr{A}|^{\frac{1}{2}}} \\ &\quad \times \frac{\exp\{-\tfrac{1}{2}[\mathbf{u} - E(\mathbf{u} \mid \mathbf{y})]'\mathscr{C}[\mathbf{u} - E(\mathbf{u} \mid \mathbf{y})]\}}{(2\pi)^{\frac{1}{2}q}|\mathscr{C}|^{-\frac{1}{2}}} \\ &\quad \times \frac{\exp\{-\tfrac{1}{2}[\mathbf{y} - E(\mathbf{y})]'(\mathbf{L} - \mathbf{LXC}^{-1}\mathbf{X}'\mathbf{L})[\mathbf{y} - E(\mathbf{y})]\}}{(2\pi)^{\frac{1}{2}N}|\mathbf{L} - \mathbf{LXC}^{-1}\mathbf{X}'\mathbf{L}|^{-\frac{1}{2}}}, \end{aligned} \quad (45)$$

where we define, analogous to (39),

$$\begin{aligned}
\mathscr{A} &= \mathbf{B}^{-1} + \mathbf{X}'\mathbf{R}^{-1}\mathbf{X}, \\
\mathscr{L} &= \mathbf{R}^{-1} - \mathbf{R}^{-1}\mathbf{X}\mathscr{A}^{-1}\mathbf{X}'\mathbf{R}^{-1} = (\mathbf{XBX}' + \mathbf{R})^{-1}, \\
\mathscr{C} &= \mathbf{Z}'\mathscr{L}\mathbf{Z} + \mathbf{D}^{-1}, \\
E(\boldsymbol{\beta} \mid \mathbf{u}, \mathbf{y}) &= \mathscr{A}^{-1}[\mathbf{X}'\mathbf{R}^{-1}(\mathbf{y} - \mathbf{Zu}) + \mathbf{B}^{-1}\boldsymbol{\beta}_0], \\
E(\mathbf{u} \mid \mathbf{y}) &= \mathscr{C}^{-1}\mathbf{Z}'\mathscr{L}(\mathbf{y} - \mathbf{X}\boldsymbol{\beta}_0).
\end{aligned} \tag{46}$$

Of course, the marginal distribution of $\mathbf{y}$ remains the same as in (38). An exact correspondence between $E(\boldsymbol{\beta} \mid \mathbf{u}, \mathbf{y})$ and $E(\mathbf{u} \mid \mathbf{y})$ of (46), and $E(\mathbf{u} \mid \boldsymbol{\beta}, \mathbf{y})$ and $E(\boldsymbol{\beta} \mid \mathbf{y})$ of (40) exists, but is not immediately apparent. This is because we have assumed the prior mean of $\mathbf{u}$ to be $\mathbf{0}$, while the prior mean of $\boldsymbol{\beta}$, $\boldsymbol{\beta}_0$ is not necessarily $\mathbf{0}$. The case with $\mathbf{u}$ having a non-zero prior mean is treated in the next subsection. The distribution of interest, $f_{\mathbf{u}}(\mathbf{u} \mid \mathbf{y})$, is given by

$$\mathbf{u} \mid \mathbf{y} \sim \mathscr{N}[E(\mathbf{u} \mid \mathbf{y}), \mathscr{C}^{-1}], \tag{47}$$

and we again could use the posterior expectation $E(\mathbf{u} \mid \mathbf{y})$, as a point estimate of $\mathbf{u}$, with $\text{var}(\mathbf{u} \mid \mathbf{y}) = \mathscr{C}^{-1}$. Again considering the special case of $\mathbf{B}^{-1} = 0$ (interpreted as $\mathbf{B} = \infty$), which led to the generalized least squares estimate of (43), we obtain,

$$E(\mathbf{u} \mid \mathbf{y}) = \mathbf{DZ}'\mathbf{V}^{-1}(\mathbf{y} - \mathbf{X}\hat{\boldsymbol{\beta}}), \tag{48}$$

where $\hat{\boldsymbol{\beta}} = (\mathbf{X}'\mathbf{V}^{-1}\mathbf{X})^{-1}\mathbf{X}'\mathbf{V}^{-1}\mathbf{y}$ and $\mathbf{V} = \mathbf{ZDZ}' + \mathbf{R}$. This is the BLUP (best linear unbiased predictor) of $\mathbf{u}$, discussed in Section 3.4 and in Chapter 7. The details of this derivation are in the next subsection.

b. An alternative derivation

The derivation of $f_{\boldsymbol{\beta}}(\boldsymbol{\beta} \mid \mathbf{y})$ of (34) and (41) and $f_{\mathbf{u}}(\mathbf{u} \mid \mathbf{y})$ of (47) was done in a general fashion, without exploiting some of the particular properties of the normal distribution. If we take advantage of those properties, especially facts about conditional distributions derived from multivariate normal distributions (Appendix S.2), we can simplify some derivations in the normal hierarchical mixed model. Of course, by taking advantage of the normal distribution, our results will not generalize to other distributions as easily.

-i. Exploiting the multivariate normal structure. With $\mathbf{y} = \mathbf{X}\boldsymbol{\beta} + \mathbf{Zu} + \mathbf{e}$ of (33), it follows from the normal distributions of (10) that $\text{cov}(\mathbf{y}, \boldsymbol{\beta}') = \mathbf{XB}$ and $\text{cov}(\mathbf{y}, \mathbf{u}') = \mathbf{ZD}$. Hence, using (33) and (10), the joint distribution of $\boldsymbol{\beta}$, $\mathbf{u}$ and $\mathbf{y}$ is given by

$$\begin{bmatrix} \boldsymbol{\beta} \\ \mathbf{u} \\ \mathbf{y} \end{bmatrix} \sim \mathscr{N}\left(\begin{bmatrix} \boldsymbol{\beta}_0 \\ \mathbf{u}_0 \\ \mathbf{X}\boldsymbol{\beta}_0 + \mathbf{Zu}_0 \end{bmatrix}, \begin{bmatrix} \mathbf{B} & \mathbf{0} & \mathbf{BX}' \\ \mathbf{0} & \mathbf{D} & \mathbf{DZ}' \\ \mathbf{XB} & \mathbf{ZD} & \mathbf{XBX}' + \mathbf{ZDZ}' + \mathbf{R} \end{bmatrix} \right). \tag{49}$$

Previously we specified $\mathbf{u}_0 = \mathbf{0}$, as in (10), but here retaining it as (potentially non-zero) $\mathbf{u}_0$ is subsequently helpful. Note also that (49) is a direct consequence of a hierarchical model such as (18).

From (49) we can read off the joint distribution of $[\boldsymbol{\beta}' \quad \mathbf{y}']'$, and obtain the conditional distribution of $\boldsymbol{\beta} \mid \mathbf{y}$. Using the conditional distribution from (iv) of Appendix S.3, which is

$$\mathbf{x}_1 \mid \mathbf{x}_2 \sim \mathcal{N}[\boldsymbol{\mu}_1 + \mathbf{V}_{12}\mathbf{V}_{22}^{-1}(\mathbf{x}_2 - \boldsymbol{\mu}_2), \quad \mathbf{V}_{11} - \mathbf{V}_{12}\mathbf{V}_{22}^{-1}\mathbf{V}_{21}], \tag{50}$$

applied to the joint distribution of $[\boldsymbol{\beta}' \quad \mathbf{y}']'$, we obtain

$$\boldsymbol{\beta} \mid \mathbf{y} \sim \mathcal{N}[E(\boldsymbol{\beta} \mid \mathbf{y}), \quad \text{var}(\boldsymbol{\beta} \mid \mathbf{y})] \tag{51}$$

with (as established in E 9.11)

$$E(\boldsymbol{\beta} \mid \mathbf{y}) = \boldsymbol{\beta}_0 + \mathbf{BX}'(\mathbf{XBX}' + \mathbf{V})^{-1}(\mathbf{y} - \mathbf{X}\boldsymbol{\beta}_0 - \mathbf{Zu}_0)$$

and (52)

$$\text{var}(\boldsymbol{\beta} \mid \mathbf{y}) = \mathbf{B} - \mathbf{BX}'(\mathbf{XBX}' + \mathbf{V})^{-1}\mathbf{XB}\,.$$

The apparent difference between the expressions in (52) and those in (39) and (40) can be explained using the identities of E 9.12. The resulting alternative expressions are

$$E(\boldsymbol{\beta} \mid \mathbf{y}) = \mathbf{C}^{-1}[\mathbf{X}'\mathbf{L}(\mathbf{y} - \mathbf{Zu}_0) + \mathbf{B}^{-1}\boldsymbol{\beta}_0] \quad \text{and} \quad \text{var}(\boldsymbol{\beta} \mid \mathbf{y}) = \mathbf{C}^{-1}\,, \tag{53}$$

Now, if we substitute $\mathbf{u}_0 = \mathbf{0}$, we obtain, analogous to (41),

$$\boldsymbol{\beta} \mid \mathbf{y} \sim \mathcal{N}[\mathbf{C}^{-1}(\mathbf{X}'\mathbf{Ly} + \mathbf{B}^{-1}\boldsymbol{\beta}_0), \mathbf{C}^{-1}]\,. \tag{54}$$

Using similar methods, including (27) of Appendix M.5, we can also derive (see E 9.10)

$$E(\mathbf{u} \mid \boldsymbol{\beta}, \mathbf{y}) = \mathbf{u}_0 + \mathbf{A}^{-1}\mathbf{Z}'\mathbf{R}^{-1}(\mathbf{y} - \mathbf{X}\boldsymbol{\beta} - \mathbf{Zu}_0) \quad \text{and} \quad \text{var}(\mathbf{u} \mid \boldsymbol{\beta}, \mathbf{y}) = \mathbf{A}^{-1} \tag{55}$$

and, substituting $\mathbf{u}_0 = \mathbf{0}$,

$$\mathbf{u} \mid \boldsymbol{\beta}, \mathbf{y} \sim \mathcal{N}[\mathbf{A}^{-1}\mathbf{Z}'\mathbf{R}^{-1}(\mathbf{y} - \mathbf{X}\boldsymbol{\beta}), \mathbf{A}^{-1}]\,. \tag{56}$$

To derive the analogous expressions for $\mathbf{u}$, that is, $E(\mathbf{u} \mid \mathbf{y})$ and $\text{var}(\mathbf{u} \mid \mathbf{y})$, we can use a simple set of notation interchanges, exploiting the symmetry of the model specification. Then, with these interchanges, we can immediately write down the parameters of the normal distributions of $\mathbf{u} \mid \mathbf{y}$ and $\boldsymbol{\beta} \mid \mathbf{u}, \mathbf{y}$. This is done by interchanging $\boldsymbol{\beta}$ and $\mathbf{u}$, $\boldsymbol{\beta}_0$ and $\mathbf{u}_0$, $\mathbf{B}$ and $\mathbf{D}$, and $\mathbf{X}$ and $\mathbf{Z}$. (This is the reason for carrying out these calculations with $\mathbf{u}_0 \neq \mathbf{0}$.) To do this, recall from (46) that

$$\mathscr{A} = \mathbf{B}^{-1} + \mathbf{X}'\mathbf{R}^{-1}\mathbf{X}, \quad \mathscr{L}^{-1} = \mathbf{R} + \mathbf{XBX}', \quad \mathscr{C} = \mathbf{D}^{-1} + \mathbf{Z}'\mathscr{L}\mathbf{Z}\,. \tag{57}$$

and hence, analogous to (54),

$$\mathbf{u} \mid \mathbf{y} \sim \mathcal{N}(\mathscr{C}^{-1}[\mathbf{Z}'\mathscr{L}(\mathbf{y} - \mathbf{X}\boldsymbol{\beta}_0) + \mathbf{D}^{-1}\mathbf{u}_0], \quad \mathscr{C}^{-1}),$$

which, with $\mathbf{u}_0 = \mathbf{0}$, is

$$\mathbf{u} \mid \mathbf{y} \sim \mathcal{N}[\mathscr{C}^{-1}\mathbf{Z}'\mathscr{L}(\mathbf{y} - \mathbf{X}\boldsymbol{\beta}_0), \quad \mathscr{C}^{-1}]\,. \tag{58}$$

Likewise making the interchanges in (55) gives

$$\begin{aligned} E(\boldsymbol{\beta} \mid \mathbf{u}, \mathbf{y}) &= \boldsymbol{\beta}_0 + \mathscr{A}^{-1}\mathbf{X}'\mathbf{R}^{-1}(\mathbf{y} - \mathbf{Z}\mathbf{u} - \mathbf{X}\boldsymbol{\beta}_0) \\ &= \mathscr{A}^{-1}[\mathbf{X}'\mathbf{R}^{-1}(\mathbf{y} - \mathbf{Z}\mathbf{u}) + \mathbf{B}^{-1}\boldsymbol{\beta}_0] \end{aligned} \tag{59}$$

and

$$\boldsymbol{\beta} \mid \mathbf{u}, \mathbf{y} \sim \mathcal{N}[\mathscr{A}^{-1}[\mathbf{X}'\mathbf{R}^{-1}(\mathbf{y} - \mathbf{Z}\mathbf{u}) + \mathbf{B}^{-1}\boldsymbol{\beta}_0], \quad \mathscr{A}^{-1}]\,. \tag{60}$$

-ii. Relationship to BLUP. As noted before, if $\mathbf{B} = \infty$ then $E(\mathbf{u} \mid \mathbf{y})$ becomes the BLUP of $\mathbf{u}$, as given in (48). That expression can also be derived by starting with (52) for $E(\boldsymbol{\beta} \mid \mathbf{y})$ and by making the notation interchanges noted just prior to (57), to get

$$E(\mathbf{u} \mid \mathbf{y}) = \mathbf{u}_0 + \mathbf{D}\mathbf{Z}'(\mathbf{X}\mathbf{B}\mathbf{X}' + \mathbf{Z}\mathbf{D}\mathbf{Z}' + \mathbf{R})^{-1}(\mathbf{y} - \mathbf{X}\boldsymbol{\beta}_0 - \mathbf{Z}\mathbf{u}_0). \tag{61}$$

Thus, for $\mathbf{u}_0 = \mathbf{0}$ and recalling that $\mathbf{V} = \mathbf{Z}\mathbf{D}\mathbf{Z}' + \mathbf{R}$, it follows from E 9.13 that as $\mathbf{B} \to \infty$ we obtain

$$\begin{aligned} E(\mathbf{u} \mid \mathbf{y}) &= \mathbf{D}\mathbf{Z}'(\mathbf{V} + \mathbf{X}\mathbf{B}\mathbf{X}')^{-1}(\mathbf{y} - \mathbf{X}\boldsymbol{\beta}_0) \\ &= \mathbf{D}\mathbf{Z}'\mathbf{V}^{-1}[\mathbf{y} - \mathbf{X}(\mathbf{X}'\mathbf{V}^{-1}\mathbf{X})^{-1}\mathbf{X}'\mathbf{V}^{-1}\mathbf{y}]\,. \end{aligned}$$

Therefore, on defining $\mathbf{X}\hat{\boldsymbol{\beta}} = \mathbf{X}(\mathbf{X}'\mathbf{V}^{-1}\mathbf{X})^{-1}\mathbf{X}'\mathbf{V}^{-1}\mathbf{y}$, we again have

$$E(\mathbf{u} \mid \mathbf{y}) = \mathbf{D}\mathbf{Z}'\mathbf{V}^{-1}(\mathbf{y} - \mathbf{X}\hat{\boldsymbol{\beta}}), \tag{62}$$

the BLUP of $\mathbf{u}$. [See (33) in Section 7.4a.] This connection of Bayes estimation to BLUP has also been demonstrated by Dempfle (1977).

c. The 1-way classification, random model

To illustrate both the formal method and the difficulties that might be encountered in implementation, we give some details for the 1-way classification model, having the familiar model equation

$$y_{ij} = \mu + \alpha_i + e_{ij} \quad \text{for } i = 1, \ldots, a \text{ and } j = 1, \ldots, n,$$

or, in matrix form,

$$\mathbf{y} = (\mathbf{1}_a \otimes \mathbf{1}_n)\mu + (\mathbf{I}_a \otimes \mathbf{1}_n)\boldsymbol{\alpha} + \mathbf{e}, \tag{63}$$

as in equation (23) of Chapter 3. Although we assume balanced data here, similar (but more involved) derivations can be carried out for unbalanced data.

To identify the general matrices of this chapter with the more specific forms here, write

$$\begin{aligned} &\mathbf{X} = \mathbf{1}_a \otimes \mathbf{1}_n, \quad \mathbf{B} = \text{var}(\mu) = \sigma_\mu^2, \\ &\mathbf{Z} = \mathbf{I}_a \otimes \mathbf{1}_n, \quad \mathbf{D} = \text{var}(\boldsymbol{\alpha}) = \sigma_\alpha^2\mathbf{I}_a, \\ &\mathbf{R} = \text{var}(\mathbf{e}) = \sigma_e^2\mathbf{I}_{an} = \sigma_e^2(\mathbf{I}_a \otimes \mathbf{I}_n), \end{aligned} \tag{64}$$

and

$$\mathbf{V} = \text{var}(\mathbf{y}) = \mathbf{ZDZ}' + \mathbf{R} = \mathbf{I}_a \otimes (\sigma_\alpha^2 \mathbf{J}_n + \sigma_e^2 \mathbf{I}_n) = \mathbf{I}_a \otimes \mathbf{V}_0,$$

where $\mathbf{V}_0 = \sigma_\alpha^2 \mathbf{J}_n + \sigma_e^2 \mathbf{I}_n$. Then

$$\mathbf{V}^{-1} = (\mathbf{I}_a \otimes \mathbf{V}_0)^{-1} = \mathbf{I}_a \otimes \mathbf{V}_0^{-1},$$

with (65)

$$\mathbf{V}_0^{-1} = \frac{1}{\sigma_e^2}\left(\mathbf{I}_n - \frac{\sigma_\alpha^2}{\sigma_e^2 + n\sigma_\alpha^2}\mathbf{J}_n\right).$$

The prior distributions taken for the parameters are

$$\mu \sim \mathcal{N}(\mu_0, \sigma_\mu^2), \quad \boldsymbol{\alpha} \sim \mathcal{N}(\mathbf{0}, \sigma_\alpha^2 \mathbf{I}), \tag{66}$$

which for $\boldsymbol{\alpha}$ is the same distribution assumed in the classical treatment of the random model (e.g., in Chapter 3).

If these assumptions are written in the hierarchical form of (35), we have

$$\begin{aligned}
\mathbf{y} \mid \mu, \boldsymbol{\alpha}, \sigma_e^2 &\sim \mathcal{N}[\mathbf{1}_{an}\mu + (\mathbf{I}_a \otimes \mathbf{1}_n)\boldsymbol{\alpha}, \quad \sigma_e^2 \mathbf{I}_{an}],\\
\mu &\sim \mathcal{N}[\mu_0, \quad \sigma_\mu^2],\\
\boldsymbol{\alpha} &\sim \mathcal{N}[\mathbf{0}, \quad \sigma_\alpha^2 \mathbf{I}_a],
\end{aligned} \tag{67}$$

and μ and $\boldsymbol{\alpha}$ are independent.

-i. ***Estimation of $\boldsymbol{\mu}$.*** Applying the arguments leading to (41), we find that the distribution of the fixed effects conditional on the data, that is, $f(\mu \mid \mathbf{y})$, is given by

$$\mu \mid \mathbf{y} \sim \mathcal{N}[\mathbf{C}^{-1}(\mathbf{1}'\mathbf{L}\mathbf{y} + \mu_0/\sigma_\mu^2), \quad \mathbf{C}^{-1}], \tag{68}$$

where

$$\begin{aligned}
\mathbf{C} &= \mathbf{1}'\mathbf{L}\mathbf{1} + 1/\sigma_\mu^2, && \text{from (39) and (64)},\\
&= \mathbf{1}'\mathbf{V}^{-1}\mathbf{1} + 1/\sigma_\mu^2, && \text{using } \mathbf{L} = \mathbf{V}^{-1} \text{ of (39)},\\
&= \mathbf{1}'(\mathbf{I}_a \otimes \mathbf{V}_0^{-1})\mathbf{1} + 1/\sigma_\mu^2, && \text{from (65)}, \qquad (69)\\
&= a\mathbf{1}'\mathbf{V}_0^{-1}\mathbf{1} + 1/\sigma_\mu^2\\
&= \frac{a}{\sigma_e^2}\left(n - \frac{n^2\sigma_\alpha^2}{\sigma_e^2 + n\sigma_\alpha^2}\right) + \frac{1}{\sigma_\mu^2}, && \text{from (65)},\\
&= \frac{an}{\sigma_e^2 + n\sigma_\alpha^2} + \frac{1}{\sigma_\mu^2}. && (70)
\end{aligned}$$

Hence from (68)

$$E(\mu \mid \mathbf{y}) = \mathbf{C}^{-1}\left(\mathbf{1}'\mathbf{L}\mathbf{y} + \frac{\mu_0}{\sigma_\mu^2}\right)$$

$$= \left(\frac{1}{\sigma_\mu^2} + \frac{an}{\sigma_e^2 + n\sigma_\alpha^2}\right)^{-1}\left[(\mathbf{1}_a' \otimes \mathbf{1}_n')(\mathbf{I}_a \otimes \mathbf{V}_0^{-1})\mathbf{y} + \frac{\mu_0}{\sigma_\mu^2}\right]$$

$$= \left(\frac{1}{\sigma_\mu^2} + \frac{an}{\sigma_e^2 + n\sigma_\alpha^2}\right)^{-1}\left[\mathbf{1}_a' \otimes \frac{1}{\sigma_e^2}\left(\mathbf{1}_n' - \frac{n\sigma_\alpha^2}{\sigma_e^2 + n\sigma_\alpha^2}\mathbf{1}_n'\right)\mathbf{y} + \frac{\mu_0}{\sigma_\mu^2}\right] \quad (71)$$

$$= \left(\frac{1}{\sigma_\mu^2} + \frac{an}{\sigma_e^2 + n\sigma_\alpha^2}\right)^{-1}\left[\frac{1}{\sigma_e^2}\left(1 - \frac{n\sigma_\alpha^2}{\sigma_e^2 + n\sigma_\alpha^2}\right)\mathbf{1}_{an}'\mathbf{y} + \frac{\mu_0}{\sigma_\mu^2}\right]$$

$$= \left(\frac{1}{\sigma_\mu^2} + \frac{an}{\sigma_e^2 + n\sigma_\alpha^2}\right)^{-1}\left(\frac{an}{\sigma_e^2 + n\sigma_\alpha^2}\bar{y}_{..} + \frac{\mu_0}{\sigma_\mu^2}\right).$$

Notice that the expression for $E(\mu \mid \mathbf{y})$ is a weighted mean of $\bar{y}_{..}$ (the average of all observations) and μ_0, the prior mean of μ. From (68) we also get a posterior variance of μ,

$$\text{var}(\mu \mid \mathbf{y}) = \mathbf{C}^{-1} = \left(\frac{1}{\sigma_\mu^2} + \frac{an}{\sigma_e^2 + n\sigma_\alpha^2}\right)^{-1} = \frac{\sigma_\mu^2(\sigma_e^2 + n\sigma_\alpha^2)}{an\sigma_\mu^2 + (\sigma_e^2 + n\sigma_\alpha^2)}. \quad (72)$$

Setting $\sigma_\mu^2 = \infty$, which is a special case of setting $\mathbf{B} = \infty$ as in (43), is often interpreted as complete uncertainty about μ. Then (71) and (72) reduce to

$$E(\mu \mid \mathbf{y}) = \bar{y}_{..} \quad \text{and} \quad \text{var}(\mu \mid \mathbf{y}) = \frac{\sigma_e^2}{an} + \frac{\sigma_\alpha^2}{a} = \text{var}(\bar{y}_{..}), \quad (73)$$

which has been derived previously in Section 3.3 where $\hat{\mu} = \bar{y}_{...}$. This also illustrates that the Bayesian quantity $\text{var}(\mu \mid \mathbf{y})$ can agree with the classical quantity $\text{var}(\bar{y}_{..})$ when an improper prior is used.

-ii. ***Estimation of*** $\boldsymbol{\alpha}$. Turning now to the prediction of $\boldsymbol{\alpha}$, the random effects, analogous to (47), the posterior distribution of $\boldsymbol{\alpha} \mid \mathbf{y}$ is also normal:

$$\boldsymbol{\alpha} \mid \mathbf{y} \sim \mathcal{N}[E(\boldsymbol{\alpha} \mid \mathbf{y}), \mathscr{C}^{-1}]. \quad (74)$$

Substituting into (58), we have

$$E(\boldsymbol{\alpha} \mid \mathbf{y}) = \mathscr{C}^{-1}(\mathbf{I}_a \otimes \mathbf{1}_n)'\frac{1}{\sigma_e^2}\left(\mathbf{I}_{an} - \frac{\sigma_\mu^2}{\sigma_e^2 + an\sigma_\mu^2}\mathbf{J}_{an}\right)(\mathbf{y} - \mu_0\mathbf{1}_{an}), \quad (75)$$

where, from (46) and (64),

$$\mathscr{C} = \frac{\sigma_e^2 + n\sigma_\alpha^2}{\sigma_e^2\sigma_\alpha^2}\mathbf{I}_a - \frac{n^2\sigma_\mu^2}{\sigma_e^2(\sigma_e^2 + an\sigma_\mu^2)}\mathbf{J}_a$$

and so (76)

$$\mathscr{C}^{-1} = \frac{\sigma_e^2\sigma_\alpha^2}{\sigma_e^2 + n\sigma_\alpha^2}\left[\mathbf{I}_a + \frac{n^2\sigma_\alpha^2\sigma_\mu^2}{\sigma_e^2(\sigma_e^2 + n\sigma_\alpha^2 + an\sigma_\mu^2)}\mathbf{J}_a\right].$$

Substituting these values into (75), after some algebra we have

$$E(\boldsymbol{\alpha}\,|\,\mathbf{y}) = \mathscr{C}^{-1}(\mathbf{I}_a \otimes \mathbf{1}_n)'\frac{1}{\sigma_e^2}\left(\mathbf{y} - \frac{an\sigma_\mu^2\bar{y}_{..}}{\sigma_e^2 + an\sigma_\mu^2}\mathbf{1}_{an} - \frac{\sigma_e^2\mu_0}{\sigma_e^2 + an\sigma_\mu^2}\mathbf{1}_{an}\right)$$

$$= \frac{n\sigma_\alpha^2}{\sigma_e^2 + n\sigma_\alpha^2}\left[(\bar{\mathbf{y}} - \mu_0\mathbf{1}_a) - \frac{an\sigma_\mu^2}{\sigma_e^2 + n\sigma_\alpha^2 + an\sigma_\mu^2}(\bar{y}_{..} - \mu_0)\mathbf{1}_a\right], \quad (77)$$

where $\bar{y}_{..} = \Sigma_{i,j}y_{ij}/an$. Note also that $(\mathbf{I}_a \otimes \mathbf{1}_n)'\mathbf{y} = n\bar{\mathbf{y}}$, where $\bar{\mathbf{y}} = [\bar{y}_{1.} \;\; \cdots \;\; \bar{y}_{a.}]'$ and $\bar{y}_{i.} = \Sigma_j y_{ij}/n$. From (74) we also see that $\text{var}(\boldsymbol{\alpha}\,|\,\mathbf{y}) = \mathscr{C}^{-1}$, and hence for the individual effects

$$\text{var}(\alpha_i\,|\,\mathbf{y}) = \frac{\sigma_e^2\sigma_\alpha^2}{\sigma_e^2 + n\sigma_\alpha^2}\left[1 + \frac{n^2\sigma_\alpha^2\sigma_\mu^2}{\sigma_e^2(\sigma_e^2 + n\sigma_\alpha^2 + an\sigma_\mu^2)}\right]$$

and (78)

$$\text{cov}(\alpha_i, \alpha_j\,|\,\mathbf{y}) = \frac{n^2\sigma_\alpha^4\sigma_\mu^2}{(\sigma_e^2 + n\sigma_\alpha^2)(\sigma_e^2 + n\sigma_\alpha^2 + an\sigma_\mu^2)}.$$

If we again take $\sigma_\mu^2 = \infty$ then from (77) and (78)

$$E(\alpha_i\,|\,\mathbf{y}) = \frac{n\sigma_\alpha^2}{\sigma_e^2 + n\sigma_\alpha^2}(\bar{y}_{i.} - \bar{y}_{..}), \quad \text{var}(\alpha_i\,|\,\mathbf{y}) = \frac{\sigma_\alpha^2}{\sigma_e^2 + n\sigma_\alpha^2}\,\frac{a\sigma_e^2 + n\sigma_\alpha^2}{a}$$

and (79)

$$\text{cov}(\alpha_i, \alpha_j\,|\,\mathbf{y}) = \frac{n\sigma_\alpha^4}{a(\sigma_e^2 + n\sigma_\alpha^2)}.$$

If, instead of a 1-way random model, a 1-way fixed model were hypothesized, this can also be handled within the hierarchical framework. Recall that classical fixed effects are modelled as having infinite variances. We can handle the fixed effects case by first deriving all necessary quantities using $\sigma_\alpha^2 < \infty$, and then letting it tend to infinity. Letting $\sigma_\alpha^2 \to \infty$ in (79) gives

$$E(\alpha_i\,|\,\mathbf{y}) = \bar{y}_{i.} - \bar{y}_{..}\,. \quad (80)$$

Thus, the usual fixed effects ANOVA can also be fitted into the hierarchical model. [Note that the assumption $E(\boldsymbol{\alpha}) = 0$, implicit in (67), alleviates any overparameterization problem.]

Finally, it is interesting to note what happens if we use the distribution of $\boldsymbol{\alpha} \mid \mu, \mathbf{y}$ to make inferences about $\boldsymbol{\alpha}$, in particular calculating $E(\boldsymbol{\alpha} \mid \mu, \mathbf{y})$ for a point estimate of $\boldsymbol{\alpha}$. [Admittedly, this would be mostly of theoretical concern since $E(\boldsymbol{\alpha} \mid \mu, \mathbf{y})$ would depend on μ, thus making it useless for inference unless μ is known.] Using (39) and (64), we derive

$$E(\boldsymbol{\alpha} \mid \mu, \mathbf{y}) = \frac{n\sigma_\alpha^2}{\sigma_e^2 + n\sigma_\alpha^2}(\bar{\mathbf{y}} - \mu\mathbf{1}_a),$$

which has already been derived in a classical manner [Chapter 3, equation (40)]. See E 9.14.

d. Empirical Bayes estimation

The empirical Bayes strategy outlined in Section 9.2c can be applied directly to the estimation of the effects $\boldsymbol{\beta}$ and $\mathbf{u}$. In fact, the application here is quite straightforward.

As illustration of the empirical Bayes principle, consider the estimation of $\mathbf{u}$ in the mixed model. The posterior distribution is given in (47), and keeping track of all parameters we would write, using (46),

$$E(\mathbf{u} \mid \mathbf{y}, \mathbf{B}, \mathbf{D}, \mathbf{R}, \boldsymbol{\beta}_0) = \mathscr{C}^{-1}\mathbf{Z}'\mathscr{L}(\mathbf{y} - \mathbf{X}\boldsymbol{\beta}_0)\,. \tag{81}$$

where $\mathscr{C}$ and $\mathscr{L}$ are given in (46). To estimate the unknown parameters implicit in (81), namely the variance matrices $\mathbf{B}$, $\mathbf{D}$, $\mathbf{R}$ and the prior mean $\boldsymbol{\beta}_0$, we obtain the marginal distribution for part (iii) of the empirical Bayes strategy of Section 9.2c–i (keeping track of the unknown parameters). Starting from the hierarchy in (35), the marginal distribution is given by

$$m(\mathbf{y} \mid \mathbf{B}, \mathbf{D}, \mathbf{R}, \boldsymbol{\beta}_0) = \int\left[\int f(\mathbf{y} \mid \boldsymbol{\beta}, \mathbf{u}, \mathbf{R}) f_{\boldsymbol{\beta}}(\boldsymbol{\beta} \mid \boldsymbol{\beta}_0, \mathbf{B}) f_{\mathbf{u}}(\mathbf{u} \mid \mathbf{D})\, d\boldsymbol{\beta}\right] d\mathbf{u}\,. \tag{82}$$

Notice that the integration over $\boldsymbol{\beta}$ is necessary to carry out part (iii); that is, the resulting marginal distribution must only depend on the *unknown parameters of interest*, and thus any other unknown parameter must be integrated out (a process known as *marginalization*). Although the parameter $\boldsymbol{\beta}$ is often of interest, in the estimation of $\mathbf{u}$ we treat it as a nuisance parameter and integrate it out.

Because of the factorization already performed in (45), the integration in (82) is easy to perform. In fact, $m(\mathbf{y} \mid \mathbf{B}, \mathbf{D}, \mathbf{R}, \boldsymbol{\beta}_0)$ is itself a likelihood function, a *marginal likelihood*, and we write

$$m(\mathbf{y} \mid \mathbf{B}, \mathbf{D}, \mathbf{R}, \boldsymbol{\beta}_0) = L(\mathbf{B}, \mathbf{D}, \mathbf{R}, \boldsymbol{\beta}_0 \mid \mathbf{y})\,. \tag{83}$$

Maximum likelihood can now be done on (83), and estimates for $\mathbf{B}$, $\mathbf{D}$, $\mathbf{R}$ and $\boldsymbol{\beta}_0$ can be found. These marginal maximum likelihood estimates $\hat{\mathbf{B}}$, $\hat{\mathbf{D}}$, $\hat{\mathbf{R}}$ and $\hat{\boldsymbol{\beta}}_0$ can be substituted into (81) to obtain an empirical Bayes estimate of $\mathbf{u}$. Furthermore, the estimates $\hat{\mathbf{B}}$, $\hat{\mathbf{D}}$ and $\hat{\mathbf{R}}$ can be used as point estimates of $\mathbf{B}$, $\mathbf{D}$ and $\mathbf{R}$. (Exercise E 9.3 shows how to obtain an empirical Bayes estimate of $\boldsymbol{\beta}$.)

-i. The 1-way classification. A special case of the above estimation is the 1-way random model, as detailed in Section 9.3c. From (77) and (78) formal

Bayes estimators of α_i and $\text{var}(\alpha_i)$ are

$$E(\alpha_i \mid \mathbf{y}, \mu_0, \sigma_e^2, \sigma_\mu^2, \sigma_\alpha^2) = \frac{n\sigma_\alpha^2}{\sigma_e^2 + n\sigma_\alpha^2}\left[\bar{y}_{i\cdot} - \mu_0 - \frac{an^2\sigma_\mu^2}{\sigma_e^2 + n\sigma_\alpha^2 + an\sigma_\mu^2}(\bar{y}_{\cdot\cdot} - \mu_0)\right] \tag{84}$$

and

$$\text{var}(\alpha_i \mid \mathbf{y}, \mu_0, \sigma_e^2, \sigma_\mu^2, \sigma_\alpha^2) = \frac{\sigma_e^2\sigma_\alpha^2}{\sigma_e^2 + n\sigma_\alpha^2}\left[1 + \frac{n^2\sigma_\alpha^2\sigma_\mu^2}{\sigma_e^2(\sigma_e^2 + n\sigma_\alpha^2 + an\sigma_\mu^2)}\right]. \tag{85}$$

To estimate the unknown parameters σ_e^2, σ_α^2, σ_μ^2 and μ_0 using the empirical Bayes strategy, we must obtain the marginal distribution of part (iii) of Section 9.3c–i, as parts (i) and (ii) are already implicit in (85). From (67) the likelihood is

$$\begin{aligned} &L(\mu, \mu_0, \sigma_e^2, \sigma_\mu^2, \sigma_\alpha^2 \mid \mathbf{y}) \\ &= f(\mathbf{y} \mid \mu, \alpha, \sigma_e^2) f(\mu \mid \mu_0, \sigma_\mu^2) f(\boldsymbol{\alpha} \mid \sigma_\alpha^2) \\ &= \frac{1}{(2\pi\sigma_e^2)^{\frac{1}{2}an}} \exp\left(-\frac{an}{2\sigma_e^2}\{\mathbf{y} - [\mathbf{1}_{an}\mu + (\mathbf{I}_a \otimes \mathbf{1}_n)\boldsymbol{\alpha}]\}'\{\mathbf{y} - [\mathbf{1}_{an}\mu + (\mathbf{I}_a \otimes \mathbf{1}_n)\boldsymbol{\alpha}]\}\right) \quad (86) \\ &\quad \times \frac{1}{(2\pi\sigma_\mu^2)^{\frac{1}{2}}} \exp\left[-\frac{1}{2\sigma_\mu^2}(\mu - \mu_0)^2\right] \times \frac{1}{(2\pi\sigma_\alpha^2)^{\frac{1}{2}a}} \exp\left(-\frac{1}{2\sigma_\alpha^2}\boldsymbol{\alpha}'\boldsymbol{\alpha}\right), \end{aligned}$$

and the appropriate marginal likelihood for part (iii) of Section 9.2c–i [equation (27)] is

$$L(\mu_0, \sigma_e^2, \sigma_\mu^2, \sigma_\alpha^2 \mid \mathbf{y}) = m(\mathbf{y} \mid \mu_0, \sigma_e^2, \sigma_\mu^2, \sigma_\alpha^2) = \iint L(\mu, \mu_0, \sigma_e^2, \sigma_\mu^2, \sigma_\alpha^2 \mid \mathbf{y})\, d\mu\, d\boldsymbol{\alpha}\,. \tag{87}$$

From this marginal likelihood we can obtain estimates $\tilde{\mu}_0$, $\tilde{\sigma}_e^2$, $\tilde{\sigma}_\mu^2$, $\tilde{\sigma}_\alpha^2$ and produce empirical Bayes point estimates $E(\alpha_i \mid \mathbf{y}, \tilde{\mu}_0, \tilde{\sigma}_e^2, \tilde{\sigma}_\mu^2, \tilde{\sigma}_\alpha^2)$ and $\text{var}(\alpha_i \mid \mathbf{y}, \tilde{\mu}_0, \tilde{\sigma}_e^2, \tilde{\sigma}_\mu^2, \tilde{\sigma}_\alpha^2)$. Realize, once again, that $\text{var}(\alpha_i \mid \mathbf{y}, \tilde{\mu}_0, \tilde{\sigma}_e^2, \tilde{\sigma}_\mu^2, \tilde{\sigma}_\alpha^2)$ is a point estimate of $\text{var}(\alpha_i \mid \mathbf{y}, \mu_0, \sigma_e^2, \sigma_\mu^2, \sigma_\alpha^2)$ and not an estimate of the variance of $E(\alpha_i \mid \mathbf{y}, \tilde{\mu}_0, \tilde{\sigma}_e^2, \tilde{\sigma}_\mu^2, \tilde{\sigma}_\alpha^2)$, nor of σ_α^2.

The likelihood in (87) is actually a straightforward calculation, and follows directly from applying the general decomposition (45) to the 1-way classification. See E 9.20.

-ii. Cautions. As in the classical approach, unknown variances pose no problem for point estimation in the hierarchical model using conditional expectations. For example, analogous to the above classical situation, from (39), if matrices $\mathbf{V}$, $\mathbf{C}$ and $\mathbf{B}$ all contain unknown variance components then the point estimator of $\boldsymbol{\beta}$,

$$E(\boldsymbol{\beta} \mid \mathbf{y}) = \mathbf{C}^{-1}(\mathbf{X}'\mathbf{V}^{-1}\mathbf{y} + \mathbf{B}^{-1}\boldsymbol{\beta}_0), \tag{88}$$

can be modified by replacing the variance components with estimates to obtain

$$\hat{E}(\boldsymbol{\beta} \mid \mathbf{y}) = \hat{\mathbf{C}}^{-1}(\mathbf{X}'\hat{\mathbf{V}}^{-1}\mathbf{y} + \hat{\mathbf{B}}^{-1}\boldsymbol{\beta}_0) \tag{89}$$

as an estimate of $E(\boldsymbol{\beta} \mid \mathbf{y})$.

The variance of the estimate in (89), $\text{var}[\hat{E}(\boldsymbol{\beta} \mid \mathbf{y})]$, is not straightforward to derive, since it involves estimates of $\mathbf{C}$, $\mathbf{V}$ and $\mathbf{B}$. Instead, we use the easy-to-obtain posterior variance of $\boldsymbol{\beta}$ to approximate it. The exact conditions under which this is justified are not known; however, the approximation tends to work well in practice. See Steffey and Kass (1991) for a discussion. From (39) and (41) we have an expression for $\text{var}(\boldsymbol{\beta} \mid \mathbf{y})$, and a convenient approximation to $\text{var}[E(\boldsymbol{\beta} \mid \mathbf{y})]$ is

$$\text{var}[E(\boldsymbol{\beta} \mid \mathbf{y})] \approx (\mathbf{X}'\mathbf{V}^{-1}\mathbf{X} + \mathbf{B}^{-1})^{-1}, \tag{90}$$

and a natural analog to (89) would be (also see E 9.21)

$$\hat{\text{var}}(\boldsymbol{\beta} \mid \mathbf{y}) = (\mathbf{X}'\hat{\mathbf{V}}^{-1}\mathbf{X} + \hat{\mathbf{B}}^{-1})^{-1}. \tag{91}$$

This straightforward substitution for $\mathbf{V}$ and $\mathbf{B}$, however, is reasonable only as an estimate of $\text{var}[E(\boldsymbol{\beta} \mid \mathbf{y})]$, and not as an estimate of $\text{var}[\hat{E}(\boldsymbol{\beta} \mid \mathbf{y})]$. Unfortunately, this latter quantity is usually the one of interest, and using (91) as its estimate may result in underestimation, which would lead to overly short confidence intervals. This is the same problem as before, that (91) does not take into account the variance of the estimates $\hat{\mathbf{V}}$ and $\hat{\mathbf{B}}$ that we substituted into (90) in place of $\mathbf{V}$ and $\mathbf{B}$, and rather treats them as constants.

In the hierarchical model we can see this more clearly, as long as we are careful to keep track of conditioning variables. The variance in (90) is, formally,

$$\text{var}(\boldsymbol{\beta} \mid \mathbf{y}, \mathbf{R}, \mathbf{B}, \mathbf{D}) = (\mathbf{X}'\mathbf{V}^{-1}\mathbf{X} + \mathbf{B}^{-1})^{-1}, \tag{92}$$

since it is derived conditional on the knowledge of $\mathbf{R}$, $\mathbf{B}$ and $\mathbf{D}$. The matrices $\mathbf{R}$ and $\mathbf{D}$ are used to obtain L. Continuing in this way, we write (91) as

$$\text{var}(\boldsymbol{\beta} \mid \mathbf{y}, \mathbf{R} = \hat{\mathbf{R}}, \mathbf{B} = \hat{\mathbf{B}}, \mathbf{D} = \hat{\mathbf{D}}) = (\mathbf{X}'\hat{\mathbf{V}}^{-1}\mathbf{X} + \hat{\mathbf{B}}^{-1})^{-1}, \tag{93}$$

which only can be used as a point estimate of the variance in (92). However, since the variances $\mathbf{R}$, $\mathbf{B}$ and $\mathbf{D}$ are unknown, the variance estimate of (91), to be useful, must be unconditional on $\mathbf{R}$, $\mathbf{B}$ and $\mathbf{D}$. A standard derivation (see Appendix S.1) gives the identity

$$\text{var}(\boldsymbol{\beta} \mid \mathbf{y}) = E[\text{var}(\boldsymbol{\beta} \mid \mathbf{y}, \mathbf{R}, \mathbf{B}, \mathbf{D})] + \text{var}[E(\boldsymbol{\beta} \mid \mathbf{y}, \mathbf{R}, \mathbf{B}, \mathbf{D})], \tag{94}$$

which involves integrating over the joint distribution of $\mathbf{R}$, $\mathbf{B}$ and $\mathbf{D}$. Now we see that (93) gives an estimate only of the first piece in (94), and the second piece is not dealt with. This is why using (93) as a variance estimate results in underestimation. Note also that this discussion applies equally to estimation of $\text{var}(\mathbf{u} \mid \mathbf{y})$, or the variance of any estimator.

The shortcoming of the "substitution principle" for estimating the variance of an estimator can also be seen by investigating the equivalence of (94) to the substitution in (92) and (93). More formally, we ask "When does $\text{var}(\boldsymbol{\beta} \mid \mathbf{y}) = \text{var}[\boldsymbol{\beta} \mid \mathbf{y}, \mathbf{R}, \mathbf{B}, \mathbf{D})$?"—a question that can be answered by derivations similar to those in Section 9.1e. Formally, we have

$$\begin{aligned}\text{var}(\boldsymbol{\beta} \mid \mathbf{y}) &= E[\text{var}(\boldsymbol{\beta} \mid \mathbf{y}, \mathbf{R}, \mathbf{B}, \mathbf{D})] + \text{var}[E(\boldsymbol{\beta} \mid \mathbf{y}, \mathbf{R}, \mathbf{B}, \mathbf{D})] \\ &\approx E[\text{var}(\boldsymbol{\beta} \mid \mathbf{y}, \mathbf{R}, \mathbf{B}, \mathbf{D})] \\ &\approx \text{var}(\boldsymbol{\beta} \mid \mathbf{y}, \mathbf{R}, \mathbf{B}, \mathbf{D})\end{aligned} \tag{95}$$

only if $\text{var}[E(\boldsymbol{\beta} \mid \mathbf{y}, \mathbf{R}, \mathbf{B}, \mathbf{D})] \approx \mathbf{0}$.

When this type of substitution is used, as in (92) and (93), we are using it with the values $\mathbf{R} = \hat{\mathbf{R}}$, $\mathbf{B} = \hat{\mathbf{B}}$, $\mathbf{D} = \hat{\mathbf{D}}$. Thus, we are implicitly assuming that the variances of $\hat{\mathbf{R}}$, $\hat{\mathbf{B}}$ and $\hat{\mathbf{D}}$ are negligible, which is, of course, false. It is this assumption that makes estimates such as (93) an underestimate of variance. Notice that the actual size of $\text{var}[E(\boldsymbol{\beta} \mid \mathbf{y}, \mathbf{R}, \mathbf{B}, \mathbf{D})]$ gives an indication of how much we are underestimating the variance using a direct substitution. If this term really is close to zero, then we will not be doing too badly.

Kackar and Harville (1981) address this problem. Working in the classical mixed model [equivalent to the hierarchy (22)], they show that if the variance component estimates are even, translation-invariant functions of $\mathbf{y}$ then the expected value of point estimators remain unchanged when variance estimates are substituted for known variances. In the model addressed here this implies that the estimators (88) and (89) have the same expected value, that is

$$E[E(\boldsymbol{\beta} \mid \mathbf{y})] = E[\hat{E}(\boldsymbol{\beta} \mid \mathbf{y})], \tag{96}$$

where the outer expectation is over the sampling distribution of $\mathbf{y}$. Such a property gives us some hope that $\text{var}[E(\boldsymbol{\beta} \mid \mathbf{y}, \mathbf{R}, \mathbf{B}, \mathbf{D})] \approx \mathbf{0}$, but of course, this is not a proven fact. Kackar and Harville (1984) go on to investigate various approximations to the variance. In general, it is probably wise not to assume $\text{var}[E(\boldsymbol{\beta} \mid \mathbf{y}, \mathbf{R}, \mathbf{B}, \mathbf{D})] \approx \mathbf{0}$, and use a more sophisticated variance approximation.

In the progression from (88) to (89), where we are dealing with an expected value, not a variance, this problem does not occur. Again, if we keep track of the conditioning variables, we have

$$E(\boldsymbol{\beta} \mid \mathbf{y}, \mathbf{R}, \mathbf{B}, \mathbf{D}) = \mathbf{C}^{-1}(\mathbf{X}'\mathbf{V}^{-1}\mathbf{y} + \mathbf{B}^{-1}\boldsymbol{\beta}_0) \tag{97}$$

and

$$\hat{E}(\boldsymbol{\beta} \mid \mathbf{y}, \mathbf{R} = \hat{\mathbf{R}}, \mathbf{B} = \hat{\mathbf{B}}, \mathbf{D} = \hat{\mathbf{D}}) = \hat{\mathbf{C}}^{-1}(\mathbf{X}'\hat{\mathbf{V}}^{-1}\mathbf{y} + \hat{\mathbf{B}}^{-1}\boldsymbol{\beta}_0)\,. \tag{98}$$

Applying E 9.15,

$$E(\boldsymbol{\beta} \mid \mathbf{y}) = E[E(\boldsymbol{\beta} \mid \mathbf{y}, \mathbf{R}, \mathbf{B}, \mathbf{D})], \tag{99}$$

so we can use (98) as an estimate of the entire quantity in (99), and the problems of (92)–(94) do not arise.

The substitution illustrated in (90), (91) and (97), (98) will work for deriving a *point estimator of the variance of* $\boldsymbol{\beta}$, that is, a point estimator of the quantity in (92). Thus, the moral of the story is that substitution of estimates for parameters can be reasonable if we are estimating means (even means of quantities that are variances), but is unreasonable if we are dealing with variances. This is because the calculation of the variance must take two pieces into account, and substitution will usually neglect one of these pieces.

*-iii. **Variance approximations.*** There has been much research aimed at obtaining approximations of the variance pieces in (94). Some examples, of different approaches, are Morris (1983), who gave one of the first approximations, and Kackar and Harville (1984). Here, we will outline a more recent strategy given by Kass and Steffey (1989). Recall from Section 9.3d–ii that when calculating a variance it should be obtained unconditional on all parameters other than the one of interest.

Although the Kass–Steffey strategy is in its infancy, and its worth can only be judged against time, it provides an easy-to-calculate approximation based on reasonable statistical assumptions. We illustrate the Kass–Steffey approximation first for a general hierarchy (as in E 9.15), and then give some details for the normal mixed model. For the hierarchical specification

$$\begin{aligned}\mathbf{X} \mid \boldsymbol{\theta}, \boldsymbol{\lambda} &\sim f(\mathbf{x} \mid \boldsymbol{\theta}, \boldsymbol{\lambda}),\\ \boldsymbol{\theta} \mid \boldsymbol{\lambda} &\sim \pi_{\boldsymbol{\theta}}(\boldsymbol{\theta} \mid \boldsymbol{\lambda}),\\ \boldsymbol{\lambda} &\sim \pi_{\boldsymbol{\lambda}}(\boldsymbol{\lambda})\end{aligned} \tag{100}$$

the variance of any function $g(\boldsymbol{\theta})$ is given by

$$\operatorname{var}[g(\boldsymbol{\theta}) \mid \mathbf{x}] = E\{\operatorname{var}[g(\boldsymbol{\theta}) \mid \mathbf{x}, \boldsymbol{\lambda}]\} + \operatorname{var}\{E[g(\boldsymbol{\theta}) \mid \mathbf{x}, \boldsymbol{\lambda}]\}, \tag{101}$$

where the right-hand side calculations of expectation and variance are done using the density

$$\pi(\boldsymbol{\lambda} \mid \mathbf{x}) = \frac{\iint f(\mathbf{x} \mid \boldsymbol{\theta}, \boldsymbol{\lambda}) \pi_{\boldsymbol{\theta}}(\boldsymbol{\theta} \mid \boldsymbol{\lambda}) \pi_{\boldsymbol{\lambda}}(\boldsymbol{\lambda})\, d\boldsymbol{\theta}}{\iint f(\mathbf{x} \mid \boldsymbol{\theta}, \boldsymbol{\lambda}) \pi_{\boldsymbol{\theta}}(\boldsymbol{\theta} \mid \boldsymbol{\lambda}) \pi_{\boldsymbol{\lambda}}(\boldsymbol{\lambda})\, d\boldsymbol{\theta}\, d\boldsymbol{\lambda}}. \tag{102}$$

In an empirical Bayes analysis, however, we would not specify $\pi_{\boldsymbol{\lambda}}(\boldsymbol{\lambda})$ but instead estimate $\boldsymbol{\lambda}$ from the marginal likelihood

$$L(\boldsymbol{\lambda} \mid \mathbf{x}) = \int f(\mathbf{x} \mid \boldsymbol{\theta}, \boldsymbol{\lambda}) \pi_{\boldsymbol{\theta}}(\boldsymbol{\theta} \mid \boldsymbol{\lambda})\, d\boldsymbol{\theta}\,. \tag{103}$$

Substitution of $\tilde{\boldsymbol{\lambda}}$, the MLE of $\boldsymbol{\lambda}$ from (103), into (101) may cause underestimation, but Kass and Steffey have *first-order approximations*

$$\begin{aligned}E[g(\theta_i) \mid \mathbf{x}] &\approx E[g(\theta_i) \mid \mathbf{x}, \tilde{\boldsymbol{\lambda}}],\\ \operatorname{var}[g(\theta_i) \mid \mathbf{x}] &\approx \operatorname{var}[g(\theta_i) \mid \mathbf{x}, \tilde{\boldsymbol{\lambda}}] + \Sigma_{j,h} \tilde{\sigma}_{jh} \tilde{\delta}_j \tilde{\delta}_h,\end{aligned} \tag{104}$$

where $\tilde{\sigma}_{jh}$ is the (j, h) element of the inverse negative Hessian of $l(\boldsymbol{\lambda} \mid \mathbf{x}) = \log L(\boldsymbol{\lambda} \mid \mathbf{x})$

$$\tilde{\boldsymbol{\Sigma}} = \{\tilde{\sigma}_{jh}\} = \left\{ -\frac{\partial^2}{\partial \lambda_j\, \partial \lambda_h} l(\boldsymbol{\lambda} \mid \mathbf{x}) \bigg|_{\boldsymbol{\lambda} = \tilde{\boldsymbol{\lambda}}} \right\}^{-1} \tag{105}$$

and

$$\tilde{\delta}_j = \frac{\partial}{\partial \lambda_j} E[g(\theta_i) \mid \mathbf{x}, \boldsymbol{\lambda}]\bigg|_{\boldsymbol{\lambda}=\tilde{\boldsymbol{\lambda}}} .$$

The results of Kass and Steffey are actually more general than reported here. It is possible to use the more general $\pi(\boldsymbol{\lambda} \mid \mathbf{x})$ in (105) instead of $L(\boldsymbol{\lambda} \mid \mathbf{x})$. Notice that $L(\boldsymbol{\lambda} \mid \mathbf{x})$ is equal to $\pi(\boldsymbol{\lambda} \mid \mathbf{x})$ when a uniform prior is put on $\boldsymbol{\lambda}$ [assuming the integral in (102) remains finite]. Thus, in (105) we have calculations made from the marginal likelihood, where we could otherwise have done them from a general marginal distribution.

For a mixed model-type hierarchy (but with general distributions),

$$\begin{gathered} \mathbf{y} \mid \boldsymbol{\beta}, \mathbf{u}, \mathbf{R} \sim f(\mathbf{y} \mid \boldsymbol{\beta}, \mathbf{u}, \mathbf{R}), \\ \boldsymbol{\beta} \mid \boldsymbol{\beta}_0, \mathbf{B} \sim f_{\boldsymbol{\beta}}(\boldsymbol{\beta} \mid \boldsymbol{\beta}_0, \mathbf{B}), \quad \mathbf{u} \mid \mathbf{D} \sim f_{\mathbf{u}}(\mathbf{u} \mid \mathbf{D}), \end{gathered} \tag{106}$$

the variance of $\mathbf{u}$ (for example) could be derived as

$$\operatorname{var}(\mathbf{u} \mid \mathbf{y}) = \int [\mathbf{u} - E(\mathbf{u} \mid \mathbf{y})]^2 f_{\mathbf{u}}(\mathbf{u} \mid \mathbf{y})\, d\mathbf{u}, \quad \text{where } E(\mathbf{u} \mid \mathbf{y}) = \int \mathbf{u} f_{\mathbf{u}}(\mathbf{u} \mid \mathbf{y})\, d\mathbf{u}, \tag{107}$$

and, keeping track of all parameters,

$$f_{\mathbf{u}}(\mathbf{u} \mid \mathbf{y}) = \int [f(\mathbf{u} \mid \mathbf{y}, \boldsymbol{\beta}_0, \mathbf{R}, \mathbf{B}, \mathbf{D})] \pi(\boldsymbol{\beta}_0, \mathbf{R}, \mathbf{B}, \mathbf{D})\, d\boldsymbol{\beta}_0\, d\mathbf{R}\, d\mathbf{B}\, d\mathbf{D} . \tag{108}$$

The density in square brackets is the posterior distribution of $\mathbf{u}$ from the hierarchy (106), and $\pi(\boldsymbol{\beta}_0, \mathbf{R}, \mathbf{B}, \mathbf{D})$ is a prior distribution on the other parameters. It is this distribution that gives the second piece in an expression such as (101), and is an aim of these approximations. As in (101), the variance of $\mathbf{u}$ can be written

$$\operatorname{var}(\mathbf{u} \mid \mathbf{y}) = E[\operatorname{var}(\mathbf{u} \mid \mathbf{y}, \boldsymbol{\beta}_0, \mathbf{R}, \mathbf{B}, \mathbf{D})] + \operatorname{var}[E(\mathbf{u} \mid \mathbf{y}, \boldsymbol{\beta}_0, \mathbf{R}, \mathbf{B}, \mathbf{D})] .$$

For notational convenience write $\boldsymbol{\eta}$ for the vector of elements of $\boldsymbol{\beta}_0$, $\mathbf{R}$, $\mathbf{B}$ and $\mathbf{D}$. Then we can write for element u_i of $\mathbf{u}$

$$\operatorname{var}(u_i \mid \mathbf{y}) = E[\operatorname{var}(u_i \mid \mathbf{y}, \boldsymbol{\eta})] + \operatorname{var}[E(u_i \mid \mathbf{y}, \boldsymbol{\eta})], \tag{109}$$

with approximations

$$\begin{gathered} E[\operatorname{var}(u_i \mid \mathbf{y}, \boldsymbol{\eta})] \approx \operatorname{var}(u_i \mid \mathbf{y}, \tilde{\boldsymbol{\eta}}) \\ \operatorname{var}[E(u_i \mid \mathbf{y}, \tilde{\boldsymbol{\eta}})] \approx \Sigma_{j,h} \tilde{\sigma}_{jh} \tilde{\delta}_{ij} \tilde{\delta}_{ih}, \\ \tilde{\boldsymbol{\Sigma}} = \left[-\frac{\partial^2}{\partial \boldsymbol{\eta} \partial \boldsymbol{\eta}'} l(\boldsymbol{\eta} \mid \mathbf{y})\bigg|_{\boldsymbol{\eta}=\tilde{\boldsymbol{\eta}}} \right]^{-1}, \quad \tilde{\delta}_{ij} = \frac{\partial}{\partial \eta_j} E\{u_i \mid \mathbf{y}, \boldsymbol{\eta})\}\bigg|_{\boldsymbol{\eta}=\tilde{\boldsymbol{\eta}}} \end{gathered} \tag{110}$$

and

$$L(\boldsymbol{\eta} \mid \mathbf{y}) = \int f(\mathbf{y}, \boldsymbol{\beta}, \mathbf{u} \mid \boldsymbol{\eta})\, d\boldsymbol{\beta}\, d\mathbf{u} = \iint f(\mathbf{y} \mid \boldsymbol{\beta}, \mathbf{u}, \mathbf{R}) f_{\boldsymbol{\beta}}(\boldsymbol{\beta} \mid \boldsymbol{\beta}_0, \mathbf{B}) f_{\mathbf{u}}(\mathbf{u} \mid \mathbf{D})\, d\boldsymbol{\beta}\, d\mathbf{u} .$$

Specializing even further, consider the normal hierarchy

$$\begin{gathered} \mathbf{y} \mid \boldsymbol{\beta}, \mathbf{u}, \mathbf{R} \sim \mathcal{N}(\mathbf{X}\boldsymbol{\beta} + \mathbf{Z}\mathbf{u}, \mathbf{R}), \\ \boldsymbol{\beta} \mid \mathbf{B} \sim \mathcal{N}(\mathbf{0}, \mathbf{B}), \quad \mathbf{u} \mid \mathbf{D} \sim \mathcal{N}(\mathbf{0}, \mathbf{D}) . \end{gathered} \tag{111}$$

An empirical Bayes estimate of $\mathbf{u}$, along with an approximate estimate of variance, can be obtained in the following way. Using the notation $\boldsymbol{\eta}$ for the vector of elements of $\mathbf{R}$, $\mathbf{B}$ and $\mathbf{D}$, from (38) and (39),

$$E(\mathbf{u} \mid \boldsymbol{\beta}, \mathbf{y}, \boldsymbol{\eta}) = \mathbf{A}^{-1}\mathbf{Z}'\mathbf{R}^{-1}(\mathbf{y} - \mathbf{X}\boldsymbol{\beta}), \quad \text{var}(\mathbf{u} \mid \boldsymbol{\beta}, \mathbf{y}, \boldsymbol{\eta}) = \mathbf{A}^{-1}, \tag{112}$$

where $\mathbf{A} = \mathbf{D}^{-1} + \mathbf{Z}'\mathbf{R}^{-1}\mathbf{Z}$. An empirical Bayes point estimate of $\mathbf{u}$ is

$$E(\mathbf{u} \mid \tilde{\boldsymbol{\beta}}, \mathbf{y}, \tilde{\boldsymbol{\eta}}) = \tilde{\mathbf{A}}^{-1}\mathbf{Z}'\tilde{\mathbf{R}}^{-1}(\mathbf{y} - \mathbf{X}\tilde{\boldsymbol{\beta}}), \tag{113}$$

where the estimates are MLEs from the likelihood

$$L(\boldsymbol{\beta}, \mathbf{B}, \mathbf{D}, \mathbf{R} \mid \mathbf{y}) = \int f(\mathbf{y} \mid \boldsymbol{\beta}, \mathbf{u}, \mathbf{R}) f_{\boldsymbol{\beta}}(\boldsymbol{\beta} \mid \mathbf{B}) f_{\mathbf{u}}(\mathbf{u} \mid \mathbf{D})\, d\mathbf{u}, \tag{114}$$

which is equal to the joint distribution of $\boldsymbol{\beta}$ and $\mathbf{y}$, and is given by the last two terms in (38). The variance is then estimated [detailed calculations are in Kass and Steffey (1986), and are similar to calculations given in Harville (1977)] with

$$\text{var}(\mathbf{u} \mid \boldsymbol{\beta}, \mathbf{y}, \boldsymbol{\eta}) \approx \text{var}(\mathbf{u} \mid \tilde{\boldsymbol{\beta}}, \mathbf{y}, \tilde{\boldsymbol{\eta}}) + \tilde{\text{var}}[E(\mathbf{u} \mid \boldsymbol{\beta}, \mathbf{y}, \boldsymbol{\eta})] = \tilde{\mathbf{A}}^{-1} + \hat{\boldsymbol{\delta}}'\tilde{\boldsymbol{\Sigma}}\hat{\boldsymbol{\delta}}$$

where (115)

$$\tilde{\sigma}_{jh} = -\frac{\partial^2}{\partial\eta_j\partial\eta_h} l(\boldsymbol{\eta} \mid \mathbf{y})\Bigg|_{\boldsymbol{\eta}=\tilde{\boldsymbol{\eta}}}, \quad \text{and} \quad \tilde{\boldsymbol{\delta}}_j = \frac{\partial}{\partial\eta_j} E(\mathbf{u} \mid \boldsymbol{\beta}, \mathbf{y}, \boldsymbol{\eta})\Bigg|_{\boldsymbol{\eta}=\tilde{\boldsymbol{\eta}}}.$$

Note that here we started with the distribution of $\mathbf{u} \mid \boldsymbol{\beta}, \mathbf{y}$, which gave us the expressions in (112) for the posterior expectation and variance. We could also have started with the distribution of $\mathbf{u} \mid \mathbf{y}$, with $\boldsymbol{\beta}$ integrated out. This would have led to different estimates. At present, there are no definite criteria for preferring one strategy over another. The different strategies, perhaps, lie at the heart of a Bayes/empirical Bayes choice. Exercises E 9.16–E 9.19 contain some complementary situations, and E 9.20 specializes to the 1-way classification.

9.4. OTHER TYPES OF HIERARCHIES

In this section we apply some of our hierarchical modeling and estimation strategy to hierarchies that fall outside of the linear model/normal case, illustrated with two hierarchies that are also treated in Chapter 10. The general techniques illustrated here are applicable to other nonlinear hierarchies, and are all examples of a *generalized linear model*; see McCullagh and Nelder (1983).

We will examine some empirical Bayes estimation strategies which have been used, for example, by Leonard (1975) and Laird (1978). These are only some of the strategies that are being used in the generalized linear model. Moreover, empirical Bayes estimation strategies can be adapted to even more complicated models than here, as is done in DuCrocq *et al.*, (1988a,b). There, mixed model ideas are applied to proportional hazards models, and empirical Bayes techniques are used to estimate parameters.

The generalized linear model, with a general *link function* (see Section 10.4), can also be analyzed in some detail using hierarchical models. Albert (1988) does this using a formal hierarchical Bayes analysis, and Piegorsch and Casella (1990) do it with empirical Bayes methodology.

a. A beta–binomial hierarchy

As an example, consider the "beta–binomial" hierarchy, which is described in some detail in Chapter 10. Although this model has some shortcomings, it represents a reasonable place to start, as it allows some explicit calculations (which does not often occur outside of the normal case). A version of the beta–binomial hierarchy can be described by writing

$$\begin{aligned} y_{ijk} \mid p_{ij} &\sim \text{Bernoulli}(p_{ij}), \quad \text{independent}, \\ p_{ij} &\sim \text{beta}(\alpha_i, \beta_i), \quad \text{independent}, \end{aligned} \tag{116}$$

for $i = 1, \ldots, a, j = 1, \ldots, b_i, k = 1, \ldots, n_{ij}$.

Here there are a groups, and subject j in group i has success probability p_{ij}. Estimation centers on p_{ij} and $\text{var}(p_{ij})$. Such a model might arise in an animal breeding experiment in the following way. Suppose that in herd i there are b_i cows to be artificially inseminated. For cow j in herd i the artificial insemination process might be thought of as a Bernoulli trial, with success probability p_{ij}. If n_{ij} trials are to be carried out on cow j then p_{ij} represents the success rate of calving of that cow (and may be confounded with other factors, e.g., the technicians). The second stage of the hierarchy models variation over animals within herds. Estimation of both p_{ij} and $\text{var}(p_{ij})$ is of interest. (See E 9.24 for a similar model.)

One shortcoming of the beta–binomial model is the problem that, unlike the linear/normal hierarchy, there is no unambiguously defined variance component. This problem is discussed in some detail in Section 10.3. As it turns out, estimation of $\text{var}(p_{ij})$ is a good compromise.

To estimate p_{ij} and $\text{var}(p_{ij})$, we first their obtain posterior expectations. Based on (116), we can derive $E(p_{ij} \mid \mathbf{y}, \alpha_i, \beta_i)$ and $\text{var}(p_{ij} \mid \mathbf{y}, \alpha_i, \beta_i)$, where $\mathbf{y} = \{y_{ijk}\}$. From first principles

$$\pi(p_{ij} \mid \mathbf{y}, \alpha_i, \beta_i) = \pi(p_{ij} \mid t, \alpha_i, \beta_i), \quad \left[\text{using sufficiency and } t = y_{ij\cdot} = \sum_{k=1}^{n_{ij}} y_{ijk}\right]$$

$$= \frac{\binom{n_{ij}}{t} p^t (1-p)^{n_{ij}-t} \dfrac{\Gamma(\alpha_i + \beta_i)}{\Gamma(\alpha_i)\Gamma(\beta_i)} p^{\alpha_i}{}_{-1}(1-p)^{\beta_i - 1}}{\displaystyle\int \binom{n_{ij}}{t} p^t (1-p)^{n_{ij}-t} \frac{\Gamma(\alpha_i + \beta_i)}{\Gamma(\alpha_i)\Gamma(\beta_i)} p^{\alpha_i - 1}(1-p)^{\beta_i - 1}\, dp} \quad [\text{equation (2)}] \tag{117}$$

Completing the integration in (117), we obtain the posterior distribution

$$\pi(p \mid t, \alpha, \beta) = \frac{\Gamma(n + \alpha + \beta)}{\Gamma(\alpha + t)\Gamma(n - t + \beta)} p^{t + \alpha - 1}(1-p)^{n - t + \beta - 1}. \tag{118}$$

which is a beta distribution with parameters $t + \alpha$ and $n - t + \beta$. Using the formulae for the mean and variance of a beta distribution [Appendix S, (9) and (10)], or calculating directly from (118), we obtain

$$E(p \mid t, \alpha, \beta) = \frac{t + \alpha}{n + \alpha + \beta} \quad \text{and} \quad \operatorname{var}(p \mid t, \alpha, \beta) = \frac{(t + \alpha)(n - t + \beta)}{(n + \alpha + \beta + 1)(n + \alpha + \beta)^2}, \tag{119}$$

the posterior mean and variance of p. Note that the posterior mean is a weighted average of the prior mean, $\alpha/(\alpha + \beta)$, and the sample mean, t/n, (as in the normal case), namely,

$$E(p \mid t, \alpha, \beta) = \left(\frac{n}{n + \alpha + \beta}\right)\frac{t}{n} + \left(\frac{\alpha + \beta}{n + \alpha + \beta}\right)\frac{\alpha}{\alpha + \beta}. \tag{120}$$

The weights are functions of the prior parameters and n, the sample size, with estimates from larger samples getting more weight.

To estimate $E(p \mid t, \alpha, \beta)$, we can use $E(p \mid t, \hat{\alpha}, \hat{\beta})$, which we know to be a reasonable estimator. To estimate α and β, the prior parameters, we use the marginal distribution given by

$$\begin{aligned} m(t \mid \alpha, \beta) &= \int f(t \mid p)\pi(p \mid \alpha, \beta)\, dp \\ &= \int \binom{n}{t} p^t (1 - p)^{n-t} \frac{\Gamma(\alpha + \beta)}{\Gamma(\alpha)\Gamma(\beta)} p^{\alpha - 1}(1 - p)^{\beta - 1}\, dp \qquad (121) \\ &= \binom{n}{t} \frac{\Gamma(\alpha + \beta)}{\Gamma(\alpha)\Gamma(\beta)} \frac{\Gamma(\alpha + t)\Gamma(n - t + \beta)}{\Gamma(\alpha + \beta + t)}, \end{aligned}$$

a *beta–binomial distribution*. The density $m(t \mid \alpha, \beta)$ forms a basis for estimating α and β; however, the constructive use of (121) requires multiple values (observations) on t. Otherwise, estimation of α and β will not gain anything—the estimates will be confounded with those of p. For example, using (121), the marginal mean of t is

$$E(t \mid \alpha, \beta) = n\frac{\alpha}{\alpha + \beta}, \tag{122}$$

while the fact that $t \sim \text{binomial}(n, p)$ yields

$$E(t \mid p) = np\,. \tag{123}$$

With only one observation t it would be impossible to estimate p separately from $\alpha/(\alpha + \beta)$. This is because our estimate of p is a "within" group estimate, while that of $\alpha/(\alpha + \beta)$ is a "between" group estimate. We can only estimate both quantities distinctly if we have multiple groups, which we have in a model such as (116).

Recall that $t = y_{ij\cdot}$ and, from (116) and (121), the marginal distribution of $y_{ij\cdot}$ is

$$m(y_{ij\cdot} \mid \alpha, \beta) = \binom{n_{ij}}{y_{ij\cdot}} \frac{\Gamma(\alpha + \beta)}{\Gamma(\alpha)\Gamma(\beta)} \frac{\Gamma(\alpha + y_{ij\cdot})\Gamma(n_{ij} - y_{ij\cdot} + \beta)}{\Gamma(\alpha + \beta + y_{ij\cdot})}, \qquad (124)$$

the beta–binomial distribution. Marginally, for each i, the $y_{ij\cdot}$s are identically distributed with parameters α_i and β_i, and these parameters can then be estimated. Thus, even though the sampling distributions of the $y_{ij\cdot}$s depend on the p_{ij}s, so that the $y_{ij\cdot}$s do not have identical sampling distributions, when the p_{ij}s are integrated out to obtain the marginal distribution for fixed i, the $y_{ij\cdot}$s become identically distributed.

From (124) we can obtain the marginal likelihood of the data. However, this likelihood factors so that we can look at the ith piece separately. Thus, the likelihood for α_i and β_i is

$$L(\alpha_i, \beta_i \mid y_{i1\cdot}, \ldots, y_{ib_i\cdot}) = \prod_{j=1}^{b_i} \binom{n_{ij}}{y_{ij\cdot}} \frac{\Gamma(\alpha_i + \beta_i)}{\Gamma(\alpha_i)\Gamma(\beta_i)} \frac{\Gamma(\alpha_i + y_{ij\cdot})\Gamma(n_{ij} - y_{ij\cdot} + \beta_i)}{\Gamma(\alpha_i + \beta_i + y_{ij\cdot})}, \qquad (125)$$

for $i = 1, \ldots, a$.

Estimation of each α_i and β_i can now proceed using (125). Two simple estimation methods come to mind: maximum likelihood and method of moments. Although maximum likelihood is preferred, there are (as usual) no closed-form expressions for the estimators. However, if we define

$$\bar{y}_i = \frac{1}{b_i} \sum_{j=1}^{b_i} y_{ij\cdot} \quad \text{and} \quad s_i^2 = \frac{1}{b_i - 1} \sum_{j=1}^{b_i} (y_{ij\cdot} - \bar{y}_i)^2 \qquad (126)$$

then the method of moments estimators of α_i and β_i are given by

$$\hat{\alpha}_i = \bar{y}_i \frac{\bar{y}_i(1 - \bar{y}_i) - s_i^2}{s_i^2 - n_i\bar{y}_i(1 - \bar{y}_i)} \quad \text{and} \quad \hat{\beta}_i = (1 - \bar{y}_i)\frac{\bar{y}_i(1 - \bar{y}_i) - s_i^2}{s_i^2 - n_i\bar{y}_i(1 - \bar{y}_i)}, \qquad (127)$$

for the case $n_{ij} = n_i$. Details are left to E 9.22.

Maximizing the likelihood for each i yields the marginal likelihood estimates of α_i and β_i, $\tilde{\alpha}_i$ and $\tilde{\beta}_i$. Substituting in (119) gives an estimate of p_{ij},

$$\begin{aligned} E(p_{ij} \mid y_{ij\cdot}, \tilde{\alpha}_i, \tilde{\beta}_i) &= \frac{y_{ij\cdot} + \tilde{\alpha}_i}{n_{ij} + \tilde{\alpha}_i + \tilde{\beta}_i} \\ &= \left(\frac{n_{ij}}{n_{ij} + \tilde{\alpha}_i + \tilde{\beta}_i}\right)\frac{y_{ij\cdot}}{n_{ij}} + \left(\frac{\tilde{\alpha}_i + \tilde{\beta}_i}{n_{ij} + \tilde{\alpha}_i + \tilde{\beta}_i}\right)\frac{\tilde{\alpha}_i}{\tilde{\alpha}_i + \tilde{\beta}_i}; \end{aligned}$$

and for the variance,

$$\operatorname{var}(p_{ij} \mid y_{ij\cdot}, \tilde{\alpha}_i, \tilde{\beta}_i) = \frac{(y_{ij\cdot} + \tilde{\alpha}_i)(n_{ij} - y_{ij\cdot} + \tilde{\beta}_i)}{(n_{ij} + \tilde{\alpha}_i + \tilde{\beta}_i + 1)(n_{ij} + \tilde{\alpha}_i + \tilde{\beta}_i)^2}, \qquad (128)$$

an underestimate of the true variance of our estimate of p_{ij}. Equation (128) estimates only the part of the variance given by the first piece in (101), and ignores the second term. One way to estimate the second term is by specifying a distribution for α_i and β_i, $\pi(\alpha_i, \beta_i)$. Given such a distribution, we could write

$$\operatorname{var}[E(p_{ij} \mid y_{ij\cdot}, \alpha_i, \beta_i)]$$
$$= \int \{E(p_{ij} \mid y_{ij\cdot}, \alpha_i, \beta_i) - E[E(p_{ij} \mid y_{ij\cdot}, \alpha_i, \beta_i)]\}^2 \pi(\alpha_i, \beta_i)\, d\alpha_i\, d\beta_i, \quad (129)$$

which would depend only on $y_{ij\cdot}$ and n_{ij}, and could be used in the variance estimate. This represents the type of formal hierarchical estimation discussed in Section 9.2a, where the hyperparameters are integrated out, leaving us with the marginal (unconditional) variance.

Another way to estimate the second variance piece in (101), in fact to estimate the entire variance, is to again apply the approximations of Kass and Steffey (1989). Using the approximation, calculate

$$\hat{\operatorname{var}}(p_{ij} \mid y_{ij\cdot}, n_{ij}) = \hat{\operatorname{var}}(p_{ij} \mid y_{ij\cdot}, \tilde{\alpha}_i, \tilde{\beta}_i) + \hat{\operatorname{var}}[E(p_{ij}) \mid y_{ij\cdot}, n_{ij}], \quad (130)$$

where the first part of the right-hand side is given in (128), and the second part is an approximation of (129) given by

$$\hat{\operatorname{var}}[E(p_{ij}) \mid y_{ij\cdot}, n_{ij}] = \Sigma_{\alpha,\beta} \hat{\sigma}_{\alpha\beta} \hat{\delta}_\alpha \hat{\delta}_\beta \,. \quad (131)$$

On suppressing the subscript i,

$$\hat{\mathbf{\Sigma}} = \{\hat{\sigma}_{\alpha\beta}\} = \begin{bmatrix} \dfrac{-\partial^2 l}{\partial\alpha^2} & \dfrac{-\partial^2 l}{\partial\alpha\,\partial\beta} \\[2ex] \dfrac{-\partial^2 l}{\partial\alpha\,\partial\beta} & \dfrac{-\partial^2 l}{\partial\beta^2} \end{bmatrix}^{-1}, \quad (132)$$

where l is the logarithm of the likelihood in (125), and

$$\begin{aligned} \hat{\delta}_\alpha &= \frac{\partial}{\partial\alpha_i} E(p_{ij} \mid y_{ij\cdot}, \alpha_i, \beta_i)\Bigg|_{\alpha_i=\tilde{\alpha}_i, \beta_i=\tilde{\beta}_i} = \frac{n_{ij} + \tilde{\beta}_i - y_{ij\cdot}}{(n_{ij} + \tilde{\alpha}_i + \tilde{\beta}_i)^2}, \\ \hat{\delta}_\beta &= \frac{\partial}{\partial\beta_i} E(p_{ij} \mid y_{ij\cdot}, \alpha_i, \beta_i)\Bigg|_{\alpha_i=\tilde{\alpha}_i, \beta_i=\tilde{\beta}_i} = \frac{-(y_{ij\cdot} + \tilde{\alpha}_i)}{(n_{ij} + \tilde{\alpha}_i + \tilde{\beta}_i)^2}\,. \end{aligned} \quad (133)$$

Defining $\hat{\boldsymbol{\delta}} = [\hat{\delta}_\alpha \quad \hat{\delta}_\beta]'$, using (131)–(133) we can write

$$\hat{\operatorname{var}}[E(p_{ij}) \mid y_{ij\cdot}, n_{ij}] = \hat{\boldsymbol{\delta}}'\hat{\mathbf{\Sigma}}\hat{\boldsymbol{\delta}}, \quad (134)$$

and combining (128)–(134) our empirical Bayes variance estimate is

$$\hat{\operatorname{var}}(p_{ij} \mid y_{ij\cdot}, n_{ij}) = \frac{(y_{ij\cdot} + \tilde{\alpha}_i)(n_{ij} - y_{ij\cdot} + \tilde{\beta}_i)}{(n_{ij} + \tilde{\alpha}_i + \tilde{\beta}_i + 1)(n_{ij} + \tilde{\alpha}_i + \tilde{\beta}_i)^2} + \hat{\boldsymbol{\delta}}'\hat{\mathbf{\Sigma}}\hat{\boldsymbol{\delta}}\,. \quad (135)$$

b. A generalized linear model

Analogous to, but more flexible than, the beta–binomial hierarchy is a special case of the generalized linear model, the logit–normal hierarchy. Although this hierarchy uses normal distributions, it is decidedly nonlinear, having a logit *link function* (see Section 10.4). The hierarchy is given by

$$
\begin{gathered}
y_i \mid p_i \sim \text{Bernoulli}(p_i), \quad i = 1, \ldots, n, \\
p_i = E(y_i \mid \boldsymbol{\beta}, \mathbf{u}), \\
\text{logit}(p_i) = \log\left(\frac{p_i}{1 - p_i}\right) = \mathbf{x}_i'\boldsymbol{\beta} + \mathbf{z}_i'\mathbf{u}, \qquad (136) \\
\boldsymbol{\beta} \sim \mathcal{N}(\boldsymbol{\beta}_0, \mathbf{B}), \quad \mathbf{u} \sim \mathcal{N}(\mathbf{0}, \mathbf{D}), \\
\boldsymbol{\beta}, \mathbf{u} \text{ independent}
\end{gathered}
$$

where $\mathbf{x}_i'$ and $\mathbf{z}_i'$ are the ith rows of $\mathbf{X}$ and $\mathbf{Z}$, respectively.

To illustrate estimation in this hierarchy, first obtain the sample density using the logit relationship between p_i, $\boldsymbol{\beta}$ and $\mathbf{u}$ given in (136). We have

$$
\begin{aligned}
f(\mathbf{y} \mid \boldsymbol{\beta}, \mathbf{u}) &= \prod_{i=1}^{n} f(y_i \mid \boldsymbol{\beta}, \mathbf{u}) = \prod_{i=1}^{n} p_i^{y_i}(1 - p_i)^{1 - y_i} \\
&= \prod_{i=1}^{n} \left[\frac{\exp(\mathbf{x}_i'\boldsymbol{\beta} + \mathbf{z}_i'\mathbf{u})}{1 + \exp(\mathbf{x}_i'\boldsymbol{\beta} + \mathbf{z}_i'\mathbf{u})}\right]^{y_i} \left[1 - \frac{\exp(\mathbf{x}_i'\boldsymbol{\beta} + \mathbf{z}_i'\mathbf{u})}{1 + \exp(\mathbf{x}_i'\boldsymbol{\beta} + \mathbf{z}_i'\mathbf{u})}\right]^{1 - y_i} \\
&= \prod_{i=1}^{n} \frac{\exp[y_i(\mathbf{x}_i'\boldsymbol{\beta} + \mathbf{z}_i'\mathbf{u})]}{1 + \exp(\mathbf{x}_i'\boldsymbol{\beta} + \mathbf{z}_i'\mathbf{u})}. \qquad (137)
\end{aligned}
$$

Now we can write the full likelihood for the hierarchy of (136) as

$$
\begin{aligned}
L(\boldsymbol{\beta}_0, \mathbf{B}, \mathbf{D} \mid \mathbf{y}) &= \iint f(\mathbf{y} \mid \boldsymbol{\beta}, \mathbf{u}) f_{\boldsymbol{\beta}}(\boldsymbol{\beta} \mid \boldsymbol{\beta}_0, \mathbf{B}) f_{\mathbf{u}}(\mathbf{u} \mid \mathbf{D})\, d\boldsymbol{\beta}\, d\mathbf{u} \\
&= \iint \left\{\prod_{i=1}^{n} \frac{\exp[y_i(\mathbf{x}_i'\boldsymbol{\beta} + \mathbf{z}_i'\mathbf{u})]}{1 + \exp(\mathbf{x}_i'\boldsymbol{\beta} + \mathbf{z}_i'\mathbf{u})}\right\} f_{\boldsymbol{\beta}}(\boldsymbol{\beta} \mid \boldsymbol{\beta}_0, \mathbf{B}) f_{\mathbf{u}}(\mathbf{u} \mid \mathbf{D})\, d\boldsymbol{\beta}\, d\mathbf{u},
\end{aligned} \qquad (138)
$$

where the densities of $\boldsymbol{\beta}$ and $\mathbf{u}$ are the normal densities given in (136).

To obtain estimates of $\boldsymbol{\beta}_0$, $\mathbf{B}$ and $\mathbf{D}$, the likelihood in (138) is now maximized. This cannot be accomplished in closed form, but a numerical solution may be obtainable. Maximization of the likelihood in (138) will yield a solution using normal prior densities on both $\boldsymbol{\beta}$ and $\mathbf{u}$, which, as we saw previously, does not correspond to the usual notion of REML or ML (but may be desirable in its own right). Connections with REML and ML are straightforward, as discussed in Section 9.2c–iii, and are obtained by specifying different forms of prior distribution of $\boldsymbol{\beta}$.

If $\boldsymbol{\beta}$ is given a flat (non-informative) prior, a uniform$(-\infty, \infty)$, then we obtain the analog of a REML likelihood,

$$
L(\mathbf{D} \mid \mathbf{y}) = \iint \left\{\prod_{i=1}^{n} \frac{\exp[y_i(\mathbf{x}_i'\boldsymbol{\beta} + \mathbf{z}_i'\mathbf{u})]}{1 + \exp(\mathbf{x}_i'\boldsymbol{\beta} + \mathbf{z}_i'\mathbf{u})}\right\} f_{\mathbf{u}}(\mathbf{u} \mid \mathbf{D})\, d\boldsymbol{\beta}\, d\mathbf{u}, \qquad (139)
$$

which, when maximized, gives a REML estimate of $\mathbf{D}$. The ordinary ML estimates would come from using a point-mass prior density on $\boldsymbol{\beta}$, producing a likelihood

$$L(\boldsymbol{\beta}, \mathbf{D} \mid \mathbf{y}) = \int \left\{ \prod_{i=1}^{n} \frac{\exp[y_i(\mathbf{x}_i'\boldsymbol{\beta} + \mathbf{z}_i'\mathbf{u})]}{1 + \exp(\mathbf{x}_i'\boldsymbol{\beta} + \mathbf{z}_i'\mathbf{u})} \right\} f_{\mathbf{u}}(\mathbf{u} \mid \mathbf{D})\, d\mathbf{u}, \tag{140}$$

which can be maximized to produce estimates of $\boldsymbol{\beta}$ and $\mathbf{D}$. [See Stiratelli, Laird, and Ware (1984) for more details.]

When point estimates of the variances are obtained, from any of (138)–(140), $\mathbf{u}$ and $\boldsymbol{\beta}$ can also be estimated. For example, using (139) to estimate $\mathbf{D}$, we would (if we could) calculate

$$E(\mathbf{u} \mid \hat{\mathbf{D}}, \mathbf{y}) = \frac{\int \mathbf{u} f(\mathbf{y} \mid \mathbf{u}) f_{\mathbf{u}}(\mathbf{u} \mid \hat{\mathbf{D}})\, d\mathbf{u}}{\int f(\mathbf{y} \mid \mathbf{u}) f_{\mathbf{u}}(\mathbf{u} \mid \hat{\mathbf{D}})\, d\mathbf{u}}$$

$$= \frac{\int \mathbf{u} \left\{ \int \prod_{i=1}^{n} \frac{\exp[y_i(\mathbf{x}_i'\boldsymbol{\beta} + \mathbf{z}_i'\mathbf{u})}{1 + \exp(\mathbf{x}_i'\boldsymbol{\beta} + \mathbf{z}_i'\mathbf{u})}\, d\boldsymbol{\beta} \right\} f_{\mathbf{u}}(\mathbf{u} \mid \hat{\mathbf{D}})\, d\mathbf{u}}{\int \left\{ \int \prod_{i=1}^{n} \frac{\exp[y_i(\mathbf{x}_i'\boldsymbol{\beta} + \mathbf{z}_i'\mathbf{u})]}{1 + \exp(\mathbf{x}_i'\boldsymbol{\beta} + \mathbf{z}_i'\mathbf{u})}\, d\boldsymbol{\beta} \right\} f_{\mathbf{u}}(\mathbf{u} \mid \hat{\mathbf{D}})\, d\mathbf{u}}, \tag{141}$$

where $f_{\mathbf{u}}(\mathbf{u} \mid \mathbf{D})$ is the $\mathcal{N}(\mathbf{0}, \mathbf{D})$ density of (136). As can be seen, this is a difficult calculation, and could be quite time-consuming. What is often done, however, is to estimate $E(\mathbf{u} \mid \mathbf{D}, \mathbf{y})$ with the posterior mode of the distribution of $\mathbf{u} \mid \mathbf{D}, \mathbf{y}$, with $\mathbf{D}$ estimated by $\hat{\mathbf{D}}$. To do this, we only need to work with the numerator of (141), and maximize

$$\left\{ \int \prod_{i=1}^{n} \frac{\exp[y_i(\mathbf{x}_i'\boldsymbol{\beta} + \mathbf{z}_i'\mathbf{u})]}{1 + \exp(\mathbf{x}_i'\boldsymbol{\beta} + \mathbf{z}_i'\mathbf{u})}\, d\boldsymbol{\beta} \right\} f_{\mathbf{u}}(\mathbf{u} \mid \hat{\mathbf{D}})$$

$$= \left\{ \int \prod_{i=1}^{n} \frac{\exp[y_i(\mathbf{x}_i'\boldsymbol{\beta} + \mathbf{z}_i'\mathbf{u})]}{1 + \exp(\mathbf{x}_i'\boldsymbol{\beta} + \mathbf{z}_i'\mathbf{u})}\, d\boldsymbol{\beta} \right\} \frac{e^{-\frac{1}{2}\mathbf{u}'\hat{\mathbf{D}}^{-1}\mathbf{u}}}{(2\pi)^{\frac{1}{2}q} |\hat{\mathbf{D}}|^{\frac{1}{2}}} \tag{142}$$

as a function of $\mathbf{u}$. A similar strategy, based on (140), can be used to estimate $\boldsymbol{\beta}$ (see E 9.27). Similar models are treated by Foulley *et al.* in Gianola and Hammond (1990).

9.5. PRACTICAL CONSIDERATIONS IN HIERARCHICAL MODELING

a. Computational problems

Much of the estimation methodology outlined in this chapter requires either the evaluation or approximation of integrals. Furthermore, in many practical problems these integrals can be of very high dimension. This evaluation can be a problem, since high-dimensional integration can be a computational problem. [For example, in Section 9.4b a posterior mode is suggested as an alternative to a posterior mean. This substitutes a maximization for an integration. Smith

(1983) discusses when these might be equivalent, an equivalence that will occur when empirical and formal hierarchical Bayes estimation yield the same answers.]

If integration is to be avoided, there are numerous alternative methods available for doing computations, many of which have seen great improvements in recent years. Approximations to integrals, in particular those arising from Bayesian hierarchical modeling, are treated in detail by Tierney and Kadane (1986) and Tierney, Kass and Kadane (1989). Methods for obtaining quantities derived from marginal distributions abound, starting with the EM algorithm (Dempster, Laird and Rubin, 1977) and an accelerated strategy (Laird and Louis, 1987). Recent techniques include interesting work on applications of Gibbs sampling (e.g., Gelfand and Smith, 1990; Gelfand *et al.*, 1990), which can sometimes provide methods of obtaining estimates without doing the integrations that the formal derivations dictate.

The problem of efficient computation is being addressed by many researchers, and the solution to any particular problem is probably contained in some available strategy. Knowing where to look, however, may be a problem. The references in the previous paragraph should provide some guidelines. A good general introduction to statistical computing is Thisted (1988).

b. Hierarchical EM

The EM algorithm, in a particular form, can be readily applied to a hierarchy to yield a computational scheme that is conceptually straightforward. Recall a general hierarchy like (100),

$$\begin{aligned} \mathbf{X} \mid \boldsymbol{\theta}, \boldsymbol{\lambda} &\sim f(\mathbf{x} \mid \boldsymbol{\theta}, \boldsymbol{\lambda}), \\ \boldsymbol{\theta} \mid \boldsymbol{\lambda} &\sim \pi_{\boldsymbol{\theta}}(\boldsymbol{\theta} \mid \boldsymbol{\lambda}), \\ \boldsymbol{\lambda} &\sim \pi_{\boldsymbol{\lambda}}(\boldsymbol{\lambda}) . \end{aligned} \tag{143}$$

With the goal being estimation of $\boldsymbol{\theta}$ and $\boldsymbol{\lambda}$, we can apply the EM algorithm with the following definitions:

$$\begin{aligned} \text{incomplete data:} &\quad \mathbf{x}; \\ \text{complete data:} &\quad \mathbf{x}, \boldsymbol{\lambda} . \end{aligned}$$

The actual data are always the incomplete data, and the actual data and the parameter in the lowest level of the hierarchy is the complete data. The two steps of the EM algorithm are then given by

$$\begin{aligned} \text{E-step:} &\quad \text{calculate } \hat{\lambda} = E(\boldsymbol{\lambda} \mid \mathbf{x}, \hat{\boldsymbol{\theta}}); \\ \text{M-step:} &\quad \text{maximize } L(\boldsymbol{\theta} \mid \mathbf{x}, \hat{\boldsymbol{\lambda}}) \text{ to obtain } \hat{\boldsymbol{\theta}} . \end{aligned} \tag{144}$$

To implement (144), two distributions are required. The first is

$$f(\boldsymbol{\lambda} \mid \mathbf{x}, \boldsymbol{\theta}) = \frac{f(\mathbf{x} \mid \boldsymbol{\lambda}, \boldsymbol{\theta})\pi_{\boldsymbol{\theta}}(\boldsymbol{\theta} \mid \boldsymbol{\lambda})\pi_{\boldsymbol{\lambda}}(\boldsymbol{\lambda})}{\int f(\mathbf{x} \mid \boldsymbol{\lambda}, \boldsymbol{\theta})\pi_{\boldsymbol{\theta}}(\boldsymbol{\theta} \mid \boldsymbol{\lambda})\pi_{\boldsymbol{\lambda}}(\boldsymbol{\lambda})\, d\boldsymbol{\lambda}}, \tag{145}$$

which is used in the calculation of the conditional expected value. That is, in (144)

$$E(\boldsymbol{\lambda} \mid \mathbf{x}, \hat{\boldsymbol{\theta}}) = \int \boldsymbol{\lambda} f(\boldsymbol{\lambda} \mid \mathbf{x}, \hat{\boldsymbol{\theta}})\, d\boldsymbol{\lambda}\,. \tag{146}$$

The second distribution required is

$$f(\mathbf{x}, \boldsymbol{\lambda} \mid \boldsymbol{\theta}) = \frac{f(\mathbf{x} \mid \boldsymbol{\lambda}, \boldsymbol{\theta})\pi_{\boldsymbol{\theta}}(\boldsymbol{\theta} \mid \boldsymbol{\lambda})\pi_{\boldsymbol{\lambda}}(\boldsymbol{\lambda})}{\iint f(\mathbf{x} \mid \boldsymbol{\lambda}, \boldsymbol{\theta})\pi_{\boldsymbol{\theta}}(\boldsymbol{\theta} \mid \boldsymbol{\lambda})\pi_{\boldsymbol{\lambda}}(\boldsymbol{\lambda})\, d\mathbf{x}\, d\boldsymbol{\lambda}}, \tag{147}$$

which yields the likelihood function $L(\boldsymbol{\theta} \mid \mathbf{x}, \boldsymbol{\lambda})$, used in (144). [Formally, the M-step of (144) yields a posterior mode. It will give the ML estimate when $\pi_{\boldsymbol{\lambda}}(\boldsymbol{\lambda}) = 1$.] Of course, this application of the EM algorithm is reasonable only if either (145) or (147) is easy to derive. In particular, it should be expressible in closed form. Otherwise, it would probably be just as good to maximize the likelihood $L(\boldsymbol{\theta}, \boldsymbol{\lambda} \mid \mathbf{x}) = f(\mathbf{x} \mid \boldsymbol{\theta}, \boldsymbol{\lambda})$ in $\boldsymbol{\theta}$ and $\boldsymbol{\lambda}$ simultaneously.

For the mixed linear model, parallels between this construction and the EM construction given in Section 8.3b are straightforward. There, the complete data is $(\mathbf{y}, \mathbf{u})$, and the incomplete data is $\mathbf{y}$, which comes from the ML hierarchy [as in (22)]

$$\begin{aligned} \mathbf{y} \mid \boldsymbol{\beta}, \mathbf{u}, \mathbf{R} &\sim \mathcal{N}(\mathbf{X}\boldsymbol{\beta} + \mathbf{Z}\mathbf{u}, \mathbf{R}), \\ \mathbf{u} \mid \mathbf{D} &\sim \mathcal{N}(\mathbf{0}, \mathbf{D})\,. \end{aligned} \tag{148}$$

(Recall that when a parameter, or unobservable, doesn't have a distribution specified in the hierarchy, we take it to have a point-mass distribution.) The hierarchy (148) is actually simpler than (143), since there are only two levels. We thus have

$$\begin{aligned} &\text{E-step:} \quad \text{calculate } \hat{\mathbf{u}} = E(\mathbf{u} \mid \mathbf{y}, \hat{\boldsymbol{\beta}}, \hat{\mathbf{R}}, \hat{\mathbf{D}}); \\ &\text{M-step:} \quad \text{maximize } L(\boldsymbol{\beta}, \mathbf{R}, \mathbf{D} \mid \mathbf{y}, \hat{\mathbf{u}}) \text{ to obtain } \hat{\boldsymbol{\beta}}, \hat{\mathbf{R}}, \hat{\mathbf{D}}\,. \end{aligned} \tag{149}$$

As explained in Section 8.3, in the E-step we only need calculate the conditional expected value of the sufficient statistics, which often will provide a simplification.

As another example, consider the REML hierarchy of (24),

$$\begin{aligned} \mathbf{y} \mid \boldsymbol{\beta}, \mathbf{u}, \mathbf{R} &\sim \mathcal{N}(\mathbf{X}\boldsymbol{\beta} + \mathbf{Z}\mathbf{u}, \mathbf{R}) \\ \boldsymbol{\beta} \sim \text{uniform}(-\infty, \infty), &\quad \mathbf{u} \sim \mathcal{N}(\mathbf{0}, \mathbf{D}), \end{aligned} \tag{150}$$

which can also be written without $\boldsymbol{\beta}$ as

$$\begin{aligned} \mathbf{y} \mid \mathbf{u}, \mathbf{R} &\sim \int f(\mathbf{y} \mid \boldsymbol{\beta}, \mathbf{u}, \mathbf{R})\, d\boldsymbol{\beta}, \\ f_{\mathbf{u}}(\mathbf{u}) &\sim \mathcal{N}(\mathbf{0}, \mathbf{D})\,. \end{aligned} \tag{151}$$

We now have

$$\begin{aligned} \text{incomplete data:} &\quad \mathbf{y}; \\ \text{complete data:} &\quad \mathbf{y}, \mathbf{u}\,. \end{aligned}$$

Thus, the EM steps are

$$\begin{aligned} &\text{E-step:} \quad \text{calculate } \hat{\mathbf{u}} = E(\mathbf{u} \mid \mathbf{y}, \hat{\mathbf{R}}, \hat{\mathbf{D}}); \\ &\text{M-step:} \quad \text{maximize } L(\mathbf{R}, \mathbf{D} \mid \mathbf{y}, \hat{\mathbf{u}}) \text{ over } \mathbf{R} \text{ and } \mathbf{D} \text{ to obtain } \hat{\mathbf{R}} \text{ and } \hat{\mathbf{D}}; \end{aligned} \tag{152}$$

where, from (48),

$$E(\mathbf{u} \mid \mathbf{y}, \hat{\mathbf{R}}, \hat{\mathbf{D}}) = \hat{\mathbf{D}}\mathbf{Z}'\hat{\mathbf{V}}^{-1}(\mathbf{y} - \mathbf{X}\hat{\boldsymbol{\beta}}), \tag{153}$$

where $\hat{\mathbf{V}} = \mathbf{Z}\hat{\mathbf{D}}\mathbf{Z}' + \hat{\mathbf{R}}$ and $\hat{\boldsymbol{\beta}} = (\mathbf{X}'\hat{\mathbf{V}}^{-1}\mathbf{X})^{-1}\mathbf{X}'\hat{\mathbf{V}}^{-1}\mathbf{y}$, and

$$L(\mathbf{R}, \mathbf{D} \mid \mathbf{y}, \mathbf{u}) = \left[\int f(\mathbf{y} \mid \boldsymbol{\beta}, \mathbf{u}, \mathbf{R})\, d\boldsymbol{\beta} \right] f_{\mathbf{u}}(\mathbf{u}), \tag{154}$$

an easier expression than (25), the usual REML likelihood. The closed form (153) allows easy calculation of the E-step, making the EM algorithm reasonable in this situation. Thus, we have exchanged a single, difficult, likelihood problem [as in (24) and (25)] for an iterative sequence of easier problems.

9.6. PHILOSOPHICAL CONSIDERATIONS IN HIERARCHICAL MODELING

Specification of a hierarchical model results in conceptually straightforward estimation methods. All calculations result from applying the laws of probability to obtain some particular density (or likelihood). Once the density or likelihood is obtained, application of standard techniques yields estimates for all quantities of interest. A goal of this chapter is to illustrate many of these techniques, so once the hierarchy is specified (any hierarchy!) reasonable estimates can be obtained.

Of course, in order to gain all of these wonderful estimation principles, we had to *specify the hierarchy*. Furthermore, all our estimates are good only if the hierarchical specification is reasonable. Thus, we have gained so much only because we have assumed so much. If there is reason to believe that the hierarchy is wrong then it might be prudent to investigate other hierarchies. The subject of robust Bayes analysis (Berger, 1985) is concerned with such questions. In particular, a set of estimates would be regarded as robust if different hierarchies yielded similar values. [Angers (1987) investigates hierarchies that have some built-in robustness properties.]

The hierarchical model, along with some Bayesian interpretations, also brings along some ease of inference. (Although we say "Bayesian interpretations", this is really more than is needed. In fact, most of the inferences considered in this chapter do not need any Bayesian interpretation. A more precise description would be "conditional interpetation".) The key feature of a conditional inference is that it is made conditional on the observed value of the data. That is, the data are considered fixed, and the inference about the parameter is made in the face of uncertainty about the parameter, not uncertainty about the data. This is in direct contrast to classical statistics, where the inference is made in the face of uncertainty about the data, that is, over repeated trials of the experiment. It is possible to evaluate hierarchical estimates according to these criteria (for example, MLEs are hierarchical estimators, and they are often evaluated using classical criteria), but we have not done so.

Another advantage of the hierarchy is the ease of estimating both means and variances. Using the general structure of (143), we can estimate any quantity from the appropriate posterior distribution. For example, any inference about θ would come from the formal posterior distribution

$$\pi(\boldsymbol{\theta} \mid \mathbf{x}, \boldsymbol{\lambda}) = \frac{f(\mathbf{x} \mid \boldsymbol{\theta}, \boldsymbol{\lambda})\pi_{\boldsymbol{\theta}}(\boldsymbol{\theta} \mid \boldsymbol{\lambda})\pi_{\boldsymbol{\lambda}}(\boldsymbol{\lambda})}{\int f(\mathbf{x} \mid \boldsymbol{\theta}, \boldsymbol{\lambda})\pi_{\boldsymbol{\theta}}(\boldsymbol{\theta} \mid \boldsymbol{\lambda})\pi_{\boldsymbol{\lambda}}(\boldsymbol{\lambda})\, d\boldsymbol{\theta}}, \tag{155}$$

or its empirical Bayes counterpart $\pi(\boldsymbol{\theta} \mid \mathbf{x}, \hat{\boldsymbol{\lambda}})$, where $\boldsymbol{\lambda}$ is estimated from the marginal distribution (likelihood)

$$m(\mathbf{x} \mid \boldsymbol{\lambda}) = L(\boldsymbol{\lambda} \mid \mathbf{x}) = \int f(\mathbf{x} \mid \boldsymbol{\theta}, \boldsymbol{\lambda})\pi_{\boldsymbol{\theta}}(\boldsymbol{\theta} \mid \boldsymbol{\lambda})\pi_{\boldsymbol{\lambda}}(\boldsymbol{\lambda})\, d\boldsymbol{\theta}.$$

The hierarchy also cautions us about underestimation of variance, as long as we keep our notation straight. If we infer using $\pi(\boldsymbol{\theta} \mid \mathbf{x}, \hat{\boldsymbol{\lambda}})$ then our inference is conditional on $\boldsymbol{\lambda} = \hat{\boldsymbol{\lambda}}$, indicating the assumption we are making.

The ease of inference of the hierarchical model is also evident in its straightforward interpretations of its entities. For example, for inference from the hierarchical form of the classic mixed model

$$\mathbf{y} = \mathbf{X}\boldsymbol{\beta} + \mathbf{Z}\mathbf{u} + \mathbf{e}$$

we do not have to worry about what quantities are fixed or random, or whether we are trying to estimate or predict. We only have to worry about whether the quantity is observable (data) or unobservable (parameter), and worry about calculating the distribution of the unobservable given (conditional on) the observable. The strategies mentioned throughout the chapter having to do with variance estimation (in particular the caution about forgetting the "missing piece") are not formally a concern of hierarchical models, but rather a concern of statistical estimation in general. Perhaps it is an illustration of the strength of hierarchical models that this concern is brought to the forefront, and can be dealt with in a reasonably straightforward way.

Throughout this chapter we have continually shown the connection between hierarchical estimates and their classical counterparts, in particular noting that in many cases the "usual" estimates can be obtained by allowing a distribution in the hierarchy to have infinite variance. In particular, recall equation (48) and the resulting discussion. There it was shown in the normal hierarchy

$$\begin{gathered} \mathbf{y} \mid \boldsymbol{\beta}, \mathbf{u} \sim \mathcal{N}(\mathbf{X}\boldsymbol{\beta} + \mathbf{Z}\mathbf{u}, \mathbf{R}), \\ \boldsymbol{\beta} \sim \mathcal{N}(\boldsymbol{\beta}_0, \mathbf{B}), \quad \mathbf{u} \sim \mathcal{N}(\mathbf{0}, \mathbf{D}) \end{gathered} \tag{156}$$

that if we take $\mathbf{B} = \infty$ then the estimates of $\mathbf{u}$ (the random effects) and $\boldsymbol{\beta}$ (the fixed effects) are

$$E(\mathbf{u} \mid \mathbf{y}) = \mathbf{D}\mathbf{Z}'\mathbf{V}^{-1}(\mathbf{y} - \mathbf{X}\boldsymbol{\beta}) \qquad \text{(best linear unbiased predictor)}$$

and (157)

$$E(\boldsymbol{\beta} \mid \mathbf{y}) = \hat{\boldsymbol{\beta}} = (\mathbf{X}'\mathbf{V}^{-1}\mathbf{X})^{-1}\mathbf{X}'\mathbf{V}^{-1}\mathbf{y} \quad \text{(best linear unbiased estimator)}.$$

Thus, the estimator of random effects results from a prior specification with finite variance, while the estimator of fixed effects results from a prior specification with infinite variance. This can be interpreted as saying that such a specification shows that we know more about random effects than fixed effects! This is because we model more structure in a random effect than a fixed effect. For a random effect we usually assume knowledge of the probability distribution of the levels, an assumption not made for fixed effects.

The observations of the previous paragraph are similar to those of Robinson (1991), who gives a very readable account of BLUP in particular and the estimation of random effects in general. The ramifications of fixed versus random, and of finite versus infinite variance, are treated in detail by Robinson, so we will not repeat those arguments here. We will, however, give an example (adapted from Robinson's paper) that shows why estimation of random effects assuming a distribution with finite variance is a reasonable thing to do.

Example. The following small, fictitious, data set shows coded first lactation milk yields for 9 dairy cows in 3 herds, each sired by one of four sires.

Herd	Sire	Yield
1	A	110
1	D	100
2	B	110
2	D	100
2	D	100
3	C	110
3	C	110
3	D	100
3	D	100

(158)

We fit the usual mixed model

$$\mathbf{y} = \mathbf{X}\boldsymbol{\beta} + \mathbf{Z}\mathbf{u} + \mathbf{e},$$

where $\mathbf{e} \sim \mathcal{N}(\mathbf{0}, \mathbf{I}_9)$ and $\mathbf{u} \sim \mathcal{N}(\mathbf{0}, \frac{1}{10}\mathbf{I}_4)$, and

$$\mathbf{X} = \begin{bmatrix} 1 & 0 & 0 \\ 1 & 0 & 0 \\ 0 & 1 & 0 \\ 0 & 1 & 0 \\ 0 & 1 & 0 \\ 0 & 0 & 1 \\ 0 & 0 & 1 \\ 0 & 0 & 1 \\ 0 & 0 & 1 \end{bmatrix} \quad \text{and} \quad \mathbf{Z} = \begin{bmatrix} 1 & 0 & 0 & 0 \\ 0 & 0 & 0 & 1 \\ 0 & 1 & 0 & 0 \\ 0 & 0 & 0 & 1 \\ 0 & 0 & 0 & 1 \\ 0 & 0 & 1 & 0 \\ 0 & 0 & 1 & 0 \\ 0 & 0 & 0 & 1 \\ 0 & 0 & 0 & 1 \end{bmatrix},$$

with results

$$\hat{\boldsymbol{\beta}} = [106.64 \quad 104.29 \quad 105.46]' \quad \text{(BLUE of fixed herd effects)},$$
$$\tilde{\mathbf{u}} = [.40 \quad .52 \quad .76 \quad -1.67]' \quad \text{(BLUP of random sire effects)}. \tag{159}$$

The rankings in $\tilde{\mathbf{u}}$ are very sensible. Examination of the data shows that the cows sired by D all have the lowest yield, and in $\tilde{\mathbf{u}}$ sire D has the lowest value (-1.67). The other cows all have the same yield, and so their sires are somewhat equivalent. However, there are two daughters of C, and only one from each of A and B. Also, there are two daughters of D in herd 2, which contains the daughter of B. This gives slightly more information on B than on A. Thus, in terms of information (variance) we have the most information on C, second most on B, and least on A. This order is reflected in the ranking by $\hat{\mathbf{u}}$ of [A B C] according to the values [.40 .52 .76].

In contrast, if we had treated $\mathbf{u}$ as a fixed factor, and had performed least squares on the entire model, we would obtain the (non-full rank) solution

$$\hat{\boldsymbol{\beta}} = [100 \quad 100 \quad 100]' \quad \text{and} \quad \hat{\mathbf{u}} = [10 \quad 10 \quad 10 \quad 0]'.$$

Now the sires A, B and C receive equal ranking, even though there is a differing amount of information on them. This is because treating an effect as fixed is similar to assigning it infinite prior variance. The fact that we have slightly more information on C makes no difference to infinity. Each sire is now treated the same. Thus, allowing a factor to be random, and hence assigning it finite variance, allows the resulting estimator to be sensitive to small changes in the amounts of information in the data. An advantage of a hierarchical model is that it gives us a framework under which all of these models can be evaluated and compared.

9.7. SUMMARY

General hierarchy

$$\mathbf{X} \mid \boldsymbol{\theta} \sim f(\mathbf{x} \mid \boldsymbol{\theta}) \quad \text{and} \quad \boldsymbol{\theta} \sim \pi(\boldsymbol{\theta}). \tag{1}$$

Posterior distribution and mean

$$\pi(\boldsymbol{\theta} \mid \mathbf{x}) = \frac{f(\mathbf{x} \mid \boldsymbol{\theta})\pi(\boldsymbol{\theta})}{\int f(\mathbf{x} \mid \boldsymbol{\theta})\pi(\boldsymbol{\theta})\, d\boldsymbol{\theta}}, \quad E(\boldsymbol{\theta} \mid \mathbf{x}) = \frac{\int \boldsymbol{\theta} f(\mathbf{x} \mid \boldsymbol{\theta})\pi(\boldsymbol{\theta})\, d\boldsymbol{\theta}}{\int f(\mathbf{x} \mid \boldsymbol{\theta})\pi(\boldsymbol{\theta})\, d\boldsymbol{\theta}}. \quad (2), (3)$$

The mixed model hierarchy

1. Given $\mathbf{u} = \mathbf{u}_0$ and $\boldsymbol{\beta} = \boldsymbol{\beta}_0$, $\mathbf{y} = \mathbf{X}\boldsymbol{\beta}_0 + \mathbf{Z}\mathbf{u}_0 + \mathbf{e}$, $\mathbf{e} \sim f_{\mathbf{e}}(\cdot)$; (8)
2. $(\mathbf{u}, \boldsymbol{\beta}) \sim f_{\mathbf{u},\boldsymbol{\beta}}(\cdot, \cdot)$ (9)

Under normality

$$\mathbf{y} \mid \boldsymbol{\beta}, \mathbf{u} \sim \mathcal{N}(\mathbf{X}\boldsymbol{\beta} + \mathbf{Z}\mathbf{u}, \mathbf{R}),$$
$$\boldsymbol{\beta} \sim \mathcal{N}(\boldsymbol{\beta}_0, \mathbf{B}), \quad \mathbf{u} \sim \mathcal{N}(\mathbf{0}, \mathbf{D}), \tag{13}$$

with $\boldsymbol{\beta}$, $\mathbf{u}$ and $\mathbf{e}$ being independent.

Ordinary maximum likelihood (ML) hierarchy

$$\begin{aligned}\mathbf{y} \mid \boldsymbol{\beta}, \mathbf{u} &\sim \mathcal{N}(\mathbf{X}\boldsymbol{\beta} + \mathbf{Z}\mathbf{u}, \mathbf{R}), \\ \mathbf{u} &\sim \mathcal{N}(\mathbf{0}, \mathbf{D}) .\end{aligned} \tag{22}$$

Restricted maximum likelihood (REML) hierarchy:

$$\begin{gathered}\mathbf{y} \mid \boldsymbol{\beta}, \mathbf{u} \sim \mathcal{N}(\mathbf{X}\boldsymbol{\beta} + \mathbf{Z}\mathbf{u}, \mathbf{R}), \\ \boldsymbol{\beta} \sim \text{uniform}(-\infty, \infty), \quad \mathbf{u} \mid \mathbf{D} \sim \mathcal{N}(\mathbf{0}, \mathbf{D}) .\end{gathered} \tag{24}$$

Empirical Bayes estimation of $\boldsymbol{\tau}$

(i) Specify for $\boldsymbol{\tau}$ a distribution $\pi(\boldsymbol{\tau} \mid \boldsymbol{\eta})$, where $\boldsymbol{\eta}$ parameterizes the distribution.
(ii) Use the formal Bayes posterior of $\boldsymbol{\tau}$,

$$\pi(\boldsymbol{\tau} \mid \mathbf{y}, \boldsymbol{\eta}) = \frac{f(\mathbf{y} \mid \boldsymbol{\tau}, \boldsymbol{\eta})\pi(\boldsymbol{\tau} \mid \boldsymbol{\eta})}{\int f(\mathbf{y} \mid \boldsymbol{\tau}, \boldsymbol{\eta})\pi(\boldsymbol{\tau} \mid \boldsymbol{\eta})\, d\boldsymbol{\tau}}, \tag{26}$$

to estimate $\boldsymbol{\tau}$, for example by calculating $E(\boldsymbol{\tau} \mid \mathbf{y}, \boldsymbol{\eta})$.
(iii) Using ML, calculate estimates $\hat{\boldsymbol{\eta}}$ of any unknown (hyper)parameters from the marginal distribution

$$m(\mathbf{y} \mid \boldsymbol{\eta}) = \int f(\mathbf{y} \mid \boldsymbol{\tau}, \boldsymbol{\eta})\pi(\boldsymbol{\tau} \mid \boldsymbol{\eta})\, d\boldsymbol{\tau}. \tag{27}$$

(iv) The empirical Bayes estimate of $\boldsymbol{\tau}$ is $E(\boldsymbol{\tau} \mid \mathbf{y}, \hat{\boldsymbol{\eta}})$.

Means and variances in the normal mixed model hierarchy

Mean and variance of y

conditional on $\boldsymbol{\beta}$: $E(\mathbf{y} \mid \boldsymbol{\beta}) = \mathbf{X}\boldsymbol{\beta}$ and $\text{var}(\mathbf{y} \mid \boldsymbol{\beta}) = \mathbf{ZDZ}' + \mathbf{R} = \mathbf{V}$;
unconditional on $\boldsymbol{\beta}$: $E(\mathbf{y}) = \mathbf{X}\boldsymbol{\beta}_0$ and $\text{var}(\mathbf{y}) = \mathbf{XBX}' + \mathbf{ZDZ}' + \mathbf{R} = \mathbf{XBX}' + \mathbf{V}$.

Posterior mean and variance of $\boldsymbol{\beta}$

conditional on $\mathbf{u}$:

$$E(\boldsymbol{\beta} \mid \mathbf{u}, \mathbf{y}) = (\mathbf{X}'\mathbf{R}^{-1}\mathbf{X} + \mathbf{B}^{-1})^{-1}[\mathbf{X}'\mathbf{R}^{-1}(\mathbf{y} - \mathbf{Z}\mathbf{u}) + \mathbf{B}^{-1}\boldsymbol{\beta}_0], \tag{46}$$

$$\text{var}(\boldsymbol{\beta} \mid \mathbf{u}, \mathbf{y}) = (\mathbf{X}'\mathbf{R}^{-1}\mathbf{X} + \mathbf{B}^{-1})^{-1} \tag{60}$$

$$= \mathbf{B} - \mathbf{BX}'(\mathbf{XBX}' + \mathbf{R})^{-1}\mathbf{XB};$$

unconditional on $\mathbf{u}$:

$$E(\boldsymbol{\beta} \mid \mathbf{y}) = (\mathbf{X}'\mathbf{V}^{-1}\mathbf{X} + \mathbf{B}^{-1})^{-1}(\mathbf{X}'\mathbf{V}^{-1}\mathbf{y} + \mathbf{B}^{-1}\boldsymbol{\beta}_0), \tag{40}$$

$$\text{var}(\boldsymbol{\beta} \mid \mathbf{y}) = (\mathbf{X}'\mathbf{V}^{-1}\mathbf{X} + \mathbf{B}^{-1})^{-1} \tag{41}$$

$$= \mathbf{B} - \mathbf{BX}'(\mathbf{XBX}' + \mathbf{ZDZ}' + \mathbf{R})^{-1}\mathbf{XB}.$$

Posterior mean and variance of u

$$\text{conditional on } \boldsymbol{\beta}: \quad E(\mathbf{u} \mid \boldsymbol{\beta}, \mathbf{y}) = \mathbf{D}\mathbf{Z}'\mathbf{V}^{-1}(\mathbf{y} - \mathbf{X}\boldsymbol{\beta}), \tag{40}$$

$$\operatorname{var}(\mathbf{u} \mid \boldsymbol{\beta}, \mathbf{y}) = (\mathbf{D}^{-1} + \mathbf{Z}'\mathbf{R}^{-1}\mathbf{Z})^{-1} \tag{55}$$

$$= \mathbf{D} - \mathbf{D}\mathbf{Z}'\mathbf{V}^{-1}\mathbf{Z}\mathbf{D};$$

$$\text{unconditional on } \boldsymbol{\beta}: \quad E(\mathbf{u} \mid \mathbf{y}) = \mathbf{D}\mathbf{Z}'(\mathbf{V} + \mathbf{X}\mathbf{B}\mathbf{X}')^{-1}(\mathbf{y} - \mathbf{X}\boldsymbol{\beta}_0),$$

$$\operatorname{var}(\mathbf{u} \mid \mathbf{y}) = [\mathbf{Z}'(\mathbf{X}\mathbf{B}\mathbf{X}' + \mathbf{R})^{-1}\mathbf{Z} + \mathbf{D}^{-1}]^{-1} \tag{47}$$

$$= \mathbf{D} - \mathbf{D}\mathbf{Z}'(\mathbf{X}\mathbf{B}\mathbf{X}' + \mathbf{Z}\mathbf{D}\mathbf{Z}' + \mathbf{R})^{-1}\mathbf{Z}\mathbf{D}\,.$$

Note: The two expressions for the variances are obtained from one another using the identity

$$(\mathbf{P} + \mathbf{Q}'\mathbf{S}^{-1}\mathbf{Q})^{-1} = \mathbf{P}^{-1} - \mathbf{P}^{-1}\mathbf{Q}'(\mathbf{S} + \mathbf{Q}\mathbf{P}^{-1}\mathbf{Q}')^{-1}\mathbf{Q}\mathbf{P}^{-1}\,.$$

Special case $E(\mathbf{u}) = 0$, $\mathbf{R} = \sigma_e^2\mathbf{I}$, $\mathbf{B}^{-1} = 0$

Posterior mean and variance of $\boldsymbol{\beta}$

$$\text{conditional on } \mathbf{u}: \quad E(\boldsymbol{\beta} \mid \mathbf{u}, \mathbf{y}) = (\mathbf{X}'\mathbf{X})^{-1}\mathbf{X}'(\mathbf{y} - \mathbf{Z}\mathbf{u}),$$

$$\operatorname{var}(\boldsymbol{\beta} \mid \mathbf{u}, \mathbf{y}) = \sigma_e^2(\mathbf{X}'\mathbf{X})^{-1};$$

$$\text{unconditional on } \mathbf{u}: \quad E(\boldsymbol{\beta} \mid \mathbf{y}) = \hat{\boldsymbol{\beta}} = (\mathbf{X}'\mathbf{V}^{-1}\mathbf{X})^{-1}\mathbf{X}'\mathbf{V}^{-1}\mathbf{y}, \tag{43}$$

$$\operatorname{var}(\boldsymbol{\beta} \mid \mathbf{y}) = \operatorname{var}(\hat{\boldsymbol{\beta}}) = (\mathbf{X}'\mathbf{V}^{-1}\mathbf{X})^{-1}\,. \tag{43}$$

Posterior mean and variance of u

$$\text{conditional on } \boldsymbol{\beta}: \quad E(\mathbf{u} \mid \boldsymbol{\beta}, \mathbf{y}) = (\sigma_e^2\mathbf{D}^{-1} + \mathbf{Z}'\mathbf{Z})^{-1}\mathbf{Z}'(\mathbf{y} - \mathbf{X}\boldsymbol{\beta}),$$

$$\operatorname{var}(\mathbf{u} \mid \boldsymbol{\beta}, \mathbf{y}) = (\mathbf{D}^{-1} + \mathbf{Z}'\mathbf{Z}/\sigma_e^2)^{-1};$$

$$\text{unconditional on } \boldsymbol{\beta}: \quad E(\mathbf{u} \mid \mathbf{y}) = \mathrm{BLUP}(\mathbf{u}) = \mathbf{D}\mathbf{Z}'\mathbf{V}^{-1}(\mathbf{y} - \mathbf{X}\hat{\boldsymbol{\beta}}),$$

$$\operatorname{var}(\mathbf{u} \mid \mathbf{y}) = \mathbf{D} - \mathbf{D}\mathbf{Z}'[\mathbf{V}^{-1} - \mathbf{V}^{-1}\mathbf{X} \times (\mathbf{X}'\mathbf{V}^{-1}\mathbf{X})^{-1}\mathbf{X}'\mathbf{V}^{-1}]\mathbf{Z}\mathbf{D}\,.$$

Variance approximations (Kass–Steffey)

For the hierarchy

$$\begin{aligned} \mathbf{X} \mid \boldsymbol{\theta}, \boldsymbol{\lambda} &\sim f(\mathbf{x} \mid \boldsymbol{\theta}, \boldsymbol{\lambda}), \\ \boldsymbol{\theta} \mid \boldsymbol{\lambda} &\sim \pi_{\boldsymbol{\theta}}(\boldsymbol{\theta} \mid \boldsymbol{\lambda}), \\ \boldsymbol{\lambda} &\sim \pi_{\boldsymbol{\lambda}}(\boldsymbol{\lambda}) \end{aligned} \tag{100}$$

the mean and variance of any function $g(\theta_i)$ can be approximated by

$$\begin{aligned} E[g(\theta_i) \mid \mathbf{x}] &\simeq E[g(\theta_i) \mid \mathbf{x}, \tilde{\boldsymbol{\lambda}}], \\ \operatorname{var}[g(\theta_i) \mid \mathbf{x}] &\simeq \operatorname{var}[g(\theta_i) \mid \mathbf{x}, \tilde{\boldsymbol{\lambda}}] + \sum_{j,h} \tilde{\sigma}_{jh}\tilde{\delta}_j\tilde{\delta}_h, \end{aligned} \tag{104}$$

where $\tilde{\sigma}_{jh}$ is the (j,h) element of the inverse negative Hessian of $l(\boldsymbol{\lambda} \mid \mathbf{x}) = \log L(\boldsymbol{\lambda} \mid \mathbf{x})$

$$\tilde{\boldsymbol{\Sigma}} = \{\tilde{\sigma}_{jh}\} = \left\{ -\frac{\partial^2}{\partial \lambda_j \partial \lambda_h} l(\boldsymbol{\lambda} \mid \mathbf{x}) \bigg|_{\boldsymbol{\lambda} = \tilde{\boldsymbol{\lambda}}} \right\}^{-1}$$

and (105)

$$\tilde{\delta}_j = \frac{\partial}{\partial \lambda_j} E[g(\theta_i) \mid \mathbf{x}, \boldsymbol{\lambda}] \bigg|_{\boldsymbol{\lambda} = \tilde{\boldsymbol{\lambda}}},$$

where $L(\boldsymbol{\lambda} \mid \mathbf{x})$ is the marginal likelihood

$$L(\boldsymbol{\lambda} \mid \mathbf{x}) = \int f(\mathbf{x} \mid \boldsymbol{\theta}, \boldsymbol{\lambda}) \pi_\theta(\boldsymbol{\theta} \mid \boldsymbol{\lambda}) \, d\boldsymbol{\theta}. \tag{103}$$

The hierarchical EM algorithm

Estimation of $\boldsymbol{\theta}$ in the hierarchy

$$\begin{aligned} X \mid \boldsymbol{\theta}, \boldsymbol{\lambda} &\sim f(\mathbf{x} \mid \boldsymbol{\theta}, \boldsymbol{\lambda}), \\ \boldsymbol{\theta} \mid \boldsymbol{\lambda} &\sim \pi_{\boldsymbol{\theta}}(\boldsymbol{\theta} \mid \boldsymbol{\lambda}), \\ \boldsymbol{\lambda} &\sim \pi_{\boldsymbol{\lambda}}(\boldsymbol{\lambda}), \end{aligned} \tag{143}$$

with

incomplete data: $\mathbf{x}$;

complete data: $\mathbf{x}, \boldsymbol{\lambda}$;

implement

$$\begin{aligned} &\text{E-step:} \quad \text{calculate } \hat{\boldsymbol{\lambda}} = E(\boldsymbol{\lambda} \mid \mathbf{x}, \hat{\boldsymbol{\theta}}); \\ &\text{M-step:} \quad \text{maximize } L(\boldsymbol{\theta} \mid \mathbf{x}, \hat{\boldsymbol{\lambda}}) \text{ to get } \hat{\boldsymbol{\theta}}; \end{aligned} \tag{144}$$

where

$$E(\boldsymbol{\lambda} \mid \mathbf{x}, \hat{\boldsymbol{\theta}}) = \frac{\int \boldsymbol{\lambda} f(\mathbf{x} \mid \boldsymbol{\lambda}, \hat{\boldsymbol{\theta}}) \pi_{\boldsymbol{\theta}}(\hat{\boldsymbol{\theta}} \mid \boldsymbol{\lambda}) \pi_{\boldsymbol{\lambda}}(\boldsymbol{\lambda}) \, d\boldsymbol{\lambda}}{\int f(\mathbf{x} \mid \boldsymbol{\lambda}, \hat{\boldsymbol{\theta}}) \pi_{\boldsymbol{\theta}}(\hat{\boldsymbol{\theta}} \mid \boldsymbol{\lambda}) \pi_{\boldsymbol{\lambda}}(\boldsymbol{\lambda}) \, d\boldsymbol{\lambda}} \tag{146}$$

and

$$L(\boldsymbol{\theta} \mid \mathbf{x}, \hat{\boldsymbol{\lambda}}) = \frac{f(\mathbf{x} \mid \hat{\boldsymbol{\lambda}}, \boldsymbol{\theta}) \pi_{\boldsymbol{\theta}}(\boldsymbol{\theta} \mid \hat{\boldsymbol{\lambda}}) \pi_{\boldsymbol{\lambda}}(\hat{\boldsymbol{\lambda}})}{\iint f(\mathbf{x} \mid \boldsymbol{\lambda}, \boldsymbol{\theta}) \pi_{\boldsymbol{\theta}}(\boldsymbol{\theta} \mid \boldsymbol{\lambda}) \pi_{\boldsymbol{\lambda}}(\boldsymbol{\lambda}) \, d\mathbf{x} \, d\boldsymbol{\lambda}}. \tag{147}$$

Estimation of $\mathbf{u}$ using the EM algorithm in the REML hierarchy of (24):

incomplete data: $\mathbf{y}$;

complete data: $\mathbf{y}, \mathbf{u}$;

implement

$$\begin{aligned} &\text{E-step:} \quad \text{calculate } \hat{\mathbf{u}} = E(\mathbf{u} \mid \mathbf{y}, \hat{\mathbf{R}}, \hat{\mathbf{D}}) = \hat{\mathbf{D}}\mathbf{Z}'\hat{\mathbf{V}}^{-1}(\mathbf{y} - \mathbf{X}\hat{\boldsymbol{\beta}}); \\ &\text{M-step:} \quad \text{maximize } L(\mathbf{R}, \mathbf{D} \mid \mathbf{y}, \hat{\mathbf{u}}) \text{ over } \mathbf{R}, \mathbf{D} \text{ to get } \hat{\mathbf{R}} \text{ and } \hat{\mathbf{D}}; \end{aligned} \tag{152}$$

where

$$L(\mathbf{R}, \mathbf{D} \mid \mathbf{y}, \mathbf{u}) = \left[\int f(\mathbf{y} \mid \boldsymbol{\beta}, \mathbf{u}, \mathbf{R})\, d\boldsymbol{\beta} \right] f_{\mathbf{u}}(\mathbf{u})\,. \tag{154}$$

9.8. EXERCISES

E 9.1. For the hierarchy

$$X \mid \Lambda \sim \text{binomial}(\Lambda, p),$$
$$\Lambda \sim \text{Poisson}(\lambda)$$

(a) verify $E(X \mid \Lambda) = p\Lambda$ and $\text{var}(X \mid \Lambda) = p\Lambda$;

(b) show that the marginal distribution of X is Poisson with parameter $p\lambda$.

E 9.2. In a general hierarchical model

$$x_i \sim f(x_i \mid \theta_i),$$
$$\theta_i \sim \pi(\theta),$$

for $i = 1, \ldots, n$, show that the x_is have identical marginal distributions. (This is a basis of empirical Bayes estimation.)

E 9.3. For the hierarchy of (35), which leads to the classical mixed model, a formal Bayes estimator of $\boldsymbol{\beta}$ is given by $E(\boldsymbol{\beta} \mid \mathbf{y}) = \mathbf{C}^{-1}(\mathbf{X}'\mathbf{L}\mathbf{y} + \mathbf{B}^{-1}\boldsymbol{\beta}_0)$, where $\mathbf{C}$ is given in (39).

(a) Argue that the empirical Bayes principle of Section 9.2c dictates that estimates for $\mathbf{C}$, $\mathbf{B}$ and $\boldsymbol{\beta}_0$ in $E(\boldsymbol{\beta} \mid \mathbf{y})$ be obtained from

$$\int \left[\int f(\mathbf{y} \mid \boldsymbol{\beta}, \mathbf{u}, \mathbf{R}) f_{\boldsymbol{\beta}}(\boldsymbol{\beta} \mid \boldsymbol{\beta}_0, \mathbf{B}) f_{\mathbf{u}}(\mathbf{u} \mid \mathbf{D})\, d\mathbf{u} \right] d\boldsymbol{\beta}\,.$$

(b) Show that the marginal density in (a) is given by equation (83).

(c) Show that the result of (b) implies that the point estimates $\hat{\mathbf{B}}$, $\hat{\mathbf{D}}$ and $\hat{\mathbf{R}}$ of $\mathbf{B}$, $\mathbf{D}$ and $\mathbf{R}$ are the same whether they are obtained to estimate $E(\mathbf{u} \mid \mathbf{y})$ or $E(\boldsymbol{\beta} \mid \mathbf{y})$. Is this a good thing?

E 9.4. (a) Use the hierarchy of (22) together with Wishart distributions for $\mathbf{D}$ and $\mathbf{R}$ to obtain (non-closed form) expressions for empirical Bayes estimates of $\mathbf{D}$ and $\mathbf{R}$. Are there values of the Wishart hyperparameters for which the estimates obtained here are the same as ML estimates?

(b) For the hierarchy of (24) repeat (a) after replacing "ML estimates" by "REML estimates".

Note. The "ML–REML prior", namely $\pi(\mathbf{D}, \mathbf{R}) = 1$, can be thought of as a prior with infinite variance. Thus we would expect things to match if $\boldsymbol{\Sigma} = \infty$ in (32). However, remember that we must be careful about this case, as shown in E 9.8.

E 9.5. (a) Apply an empirical Bayes strategy, as outlined in (26) and (27), to the general hierarchy of (19), and show how to estimate **R** alone (the matrix **D** is to be integrated out).

(b) Apply your strategy to the hierarchy of (22). Is there any connection between your estimate of **R** and the ML estimate of **R**?

(c) Apply your strategy to the hierarchy of (24). Is there any connection between your estimate of **R** and the REML estimate of **R**?

E 9.6. Verify that the third exponential in (38), the basis for the marginal density of **y**, is a perfect square in that

$$\mathbf{y}'(\mathbf{L} - \mathbf{LXC}^{-1}\mathbf{X}'\mathbf{L})\mathbf{y} - 2\mathbf{y}'\mathbf{LXC}^{-1}\mathbf{B}^{-1}\boldsymbol{\beta}_0 + \boldsymbol{\beta}_0'(\mathbf{B}^{-1} - \mathbf{B}^{-1}\mathbf{C}^{-1}\mathbf{B}^{-1})\boldsymbol{\beta}_0$$
$$= [\mathbf{y} - E(\mathbf{y})]'(\mathbf{L} - \mathbf{LXC}^{-1}\mathbf{X}'\mathbf{L})[\mathbf{y} - E(\mathbf{y})],$$

where

$$E(\mathbf{y}) = (\mathbf{L} - \mathbf{LXC}^{-1}\mathbf{X}'\mathbf{L})^{-1}\mathbf{LXC}^{-1}\mathbf{B}^{-1}\boldsymbol{\beta}_0 = \mathbf{X}\boldsymbol{\beta}_0,$$

by establishing

$$\mathbf{B}^{-1}\mathbf{C}^{-1}\mathbf{X}'\mathbf{L}(\mathbf{L} - \mathbf{L}'\mathbf{XC}^{-1}\mathbf{X}'\mathbf{L})^{-1}\mathbf{L}'\mathbf{XC}^{-1}\mathbf{B}^{-1} = \mathbf{B}^{-1} - \mathbf{B}^{-1}\mathbf{C}^{-1}\mathbf{B}^{-1}.$$

(See Appendix M, or E 9.12.)

E 9.7. Verify (41). Results (1) and (2) of Appendix S.6 may be helpful.

E 9.8. Illustrate some of the problems alluded to following (43), when equating the statements "$\mathbf{B} = \infty$" and "$\mathbf{B}^{-1} = \mathbf{0}$".

(a) Show that the matrix

$$\mathbf{B}_n = \begin{bmatrix} n & n \\ n & 1 \end{bmatrix}$$

satisfies $\lim_{n\to\infty} \mathbf{B}_n^{-1} = \mathbf{0}$, a matrix of all zeros, but $\lim_{n\to\infty} \mathbf{B}_n$ is not a matrix with each element equal to infinity.

(b) What mathematical or statistical meaning would you attach to the statements "$\mathbf{B} = \infty$" and "$\mathbf{B}^{-1} = \mathbf{0}$" in order that they be equivalent?

E 9.9. Show that the BLUE (best linear unbiased estimator) of $\boldsymbol{\beta}$, $\hat{\boldsymbol{\beta}} = (\mathbf{X}'\mathbf{V}^{-1}\mathbf{X})^{-1}\mathbf{X}'\mathbf{V}^{-1}\mathbf{y}$, can be derived as a special case of $E(\boldsymbol{\beta} \mid \mathbf{y})$ of (39). That is, specify and interpret the values needed for the prior parameters in order to have

$$E(\boldsymbol{\beta} \mid \mathbf{y}) = \hat{\boldsymbol{\beta}} = (\mathbf{X}'\mathbf{V}^{-1}\mathbf{X})^{-1}\mathbf{X}'\mathbf{V}^{-1}\mathbf{y}.$$

Comment on the interpretation of these parameter values.

E 9.10. For the hierarchy in (35) establish that

$$E(\mathbf{u} \mid \boldsymbol{\beta}, \mathbf{y}) = \mathbf{u}_0 + \mathbf{A}^{-1}\mathbf{Z}'\mathbf{R}^{-1}(\mathbf{y} - \mathbf{X}\boldsymbol{\beta} - \mathbf{Z}\mathbf{u}_0) \quad \text{and}$$

$$\text{var}(\mathbf{u} \mid \boldsymbol{\beta}, \mathbf{y}) = \mathbf{A}^{-1} .$$

Methods similar to those used to establish (53) will work, along with (27) of Appendix M.5.

E 9.11. With (49) and $\mathbf{V} = \mathbf{ZDZ}' + \mathbf{R}$ verify

(a) that $\mathbf{V}_{12}\mathbf{V}_{22}^{-1}$ of (50) is $\mathbf{BX}'(\mathbf{XBX}' + \mathbf{V})^{-1}$;
(b) $E(\boldsymbol{\beta} \mid \mathbf{y})$ and $\text{var}(\boldsymbol{\beta} \mid \mathbf{y})$ of (52).

E 9.12. For the matrices $\mathbf{A} = \mathbf{D}^{-1} + \mathbf{Z}'\mathbf{R}^{-1}\mathbf{Z}$, $\mathbf{L} = \mathbf{R}^{-1} - \mathbf{R}^{-1}\mathbf{ZA}^{-1}\mathbf{Z}'\mathbf{R}^{-1}$, $\mathbf{C} = \mathbf{X}'\mathbf{LX} + \mathbf{B}^{-1}$ and $\mathbf{V} = \mathbf{ZDZ}' + \mathbf{R}$ establish the following identities [they can be used to establish the correspondence between the expressions in (39) and (40)]:

(a) $\mathbf{L} = \mathbf{V}^{-1}$;
(b) $\mathbf{L} - \mathbf{LXC}^{-1}\mathbf{X}'\mathbf{L} = (\mathbf{XBX}' + \mathbf{V})^{-1} = (\mathbf{XBX}' + \mathbf{ZDZ}' + \mathbf{R})^{-1}$;
(c) $\mathbf{A}^{-1}\mathbf{Z}'\mathbf{R}^{-1} = \mathbf{DZ}'\mathbf{V}^{-1}$;
(d) $(\mathbf{L} - \mathbf{LXC}^{-1}\mathbf{X}'\mathbf{L})^{-1}\mathbf{L}'\mathbf{XC}^{-1}\mathbf{B}^{-1} = \mathbf{X}$;
(e) $|\mathbf{V}| = |\mathbf{R}|\,|\mathbf{D}|\,|\mathbf{A}|$;
(f) $|\mathbf{V}| = |\mathbf{K}'\mathbf{VK}|\,|\mathbf{X}'\mathbf{V}^{-1}\mathbf{X}|$ for $\mathbf{K}$ satisfying $\mathbf{K}'\mathbf{X} = \mathbf{0}$ and $\mathbf{K}(\mathbf{K}'\mathbf{VK})^{-1}\mathbf{K}' = \mathbf{V}^{-1} - \mathbf{V}^{-1}\mathbf{X}(\mathbf{X}'\mathbf{V}^{-1}\mathbf{X})^{-1}\mathbf{X}'\mathbf{V}^{-1}$.

E 9.13. For the hierarchy (35), or from (49), use E 9.12 to establish for $\mathbf{u}_0 = \mathbf{0}$ and $\mathbf{V} = \mathbf{ZDZ}' + \mathbf{R}$ that

(a)
$$\begin{aligned} E(\mathbf{u} \mid \mathbf{y}) &= \mathbf{u}_0 + \mathbf{DZ}'(\mathbf{XBX}' + \mathbf{ZDZ}' + \mathbf{R})^{-1}(\mathbf{y} - \mathbf{X}\boldsymbol{\beta}_0 - \mathbf{Z}\mathbf{u}_0) \\ &= \mathbf{DZ}'(\mathbf{V} + \mathbf{XBX}')^{-1}(\mathbf{y} - \mathbf{X}\boldsymbol{\beta}_0); \end{aligned}$$

(b)
$$\mathbf{V}^{-1}\mathbf{X} - \mathbf{V}^{-1}\mathbf{X}(\mathbf{X}'\mathbf{V}^{-1}\mathbf{X})^{-1}\mathbf{X}'\mathbf{V}^{-1}\mathbf{X} = \mathbf{0};$$

(c)
$$\begin{aligned} E(\mathbf{u} \mid \mathbf{y}) &= \mathbf{DZ}'(\mathbf{V} + \mathbf{XBX}')^{-1}(\mathbf{y} - \mathbf{X}\boldsymbol{\beta}_0) \\ &= \mathbf{DZ}'\mathbf{V}^{-1}[\mathbf{y} - \mathbf{X}(\mathbf{X}'\mathbf{V}^{-1}\mathbf{X})^{-1}\mathbf{X}'\mathbf{V}^{-1}\mathbf{y}] \quad \text{for } \mathbf{B}^{-1} = \mathbf{0} \\ &= \mathbf{DZ}'\mathbf{V}^{-1}(\mathbf{y} - \mathbf{X}\hat{\boldsymbol{\beta}}) \quad \text{for } \hat{\boldsymbol{\beta}} \text{ of (43)} \\ &= \text{BLUP of } \mathbf{u}\,. \end{aligned}$$

Note that $\boldsymbol{\beta}_0$ is eliminated when $\mathbf{B}^{-1} = \mathbf{0}$.

E 9.14. For the hierarchy (67)

(a) show that

$$E(\boldsymbol{\alpha} \mid \mu, \mathbf{y}) = \frac{n\sigma_\alpha^2}{\sigma_e^2 + n\sigma_\alpha^2}(\bar{y} - \mu\mathbf{1}_a);$$

(b) derive $\text{var}(\boldsymbol{\alpha} \mid \mu, \mathbf{y})$.

E 9.15. This exercise establishes, in general, the identities used in Section 9.3d. For the hierarchical specification

$$X \sim f(x \mid \theta, \lambda),$$
$$\theta \sim \pi_\theta(\theta \mid \lambda),$$
$$\lambda \sim \pi_\lambda(\lambda)$$

show that for a function $g(\theta)$

(a) $E[g(\theta) \mid x] = E\{E[g(\theta) \mid x, \lambda]\}$;

(b) $\text{var}[g(\theta) \mid x] = E\{\text{var}[g(\theta) \mid x, \lambda]\} + \text{var}\{E[g(\theta) \mid x, \lambda]\}$.

In each case the outer expectation is over the distribution of λ. These results can be established by writing $\pi(\theta \mid x) = \int \pi(\theta \mid x, \lambda)\pi(\lambda \mid x)\, d\lambda$ and interchanging the order of integration.

E 9.16. Consider the normal hierarchy of (18), where $\boldsymbol{\beta}_0 \neq \mathbf{0}$, and assume that $\mathbf{R}$, $\mathbf{B}$ and $\mathbf{D}$ are all diagonal matrices; and use $\boldsymbol{\eta}$ as the vector of symbols $\boldsymbol{\beta}_0$, $\mathbf{R}$, $\mathbf{B}$ and $\mathbf{D}$.

(a) Derive an empirical Bayes estimate of $\mathbf{u}$, along with an approximation of its variance, starting from the proper Bayes estimates $E(\mathbf{u} \mid \boldsymbol{\beta}, \mathbf{y}, \boldsymbol{\eta})$ and $\text{var}(\mathbf{u} \mid \boldsymbol{\beta}, \mathbf{y}, \boldsymbol{\eta})$.

(b) Derive an empirical Bayes estimate of $\mathbf{u}$, along with an approximation of its variance, starting from the proper Bayes estimates $E(\mathbf{u} \mid \mathbf{y}, \boldsymbol{\eta})$ and $\text{var}(\mathbf{u} \mid \mathbf{y}, \boldsymbol{\eta})$. (Note that here we start with the posterior estimate of $\mathbf{u}$ after $\boldsymbol{\beta}$ has been integrated out.)

E 9.17. Consider the same normal hierarchy as in E 9.16, where $\boldsymbol{\beta}_0 \neq \mathbf{0}$, but no longer assume that $\mathbf{R}$, $\mathbf{B}$ and $\mathbf{D}$ are diagonal matrices. Repeat (a) and (b) of E 9.16 in this more general case.

E 9.18. The hierarchical specification that uses a point mass density for $\boldsymbol{\beta}$ is (22). If $\boldsymbol{\beta}$ is considered fixed but unknown, this hierarchy leads to ordinary maximum likelihood estimation.

(a) Derive an empirical Bayes estimate of $\mathbf{u}$, along with an approximation of its variance, based on the proper Bayes estimates $E(\mathbf{u} \mid \mathbf{y}, \boldsymbol{\beta}, \boldsymbol{\eta})$ and $\text{var}(\mathbf{u} \mid \mathbf{y}, \boldsymbol{\beta}, \boldsymbol{\eta})$, where $\boldsymbol{\eta}$ is the vector of symbols $\mathbf{R}$, $\mathbf{D}$, with $\mathbf{R}$ and $\mathbf{D}$ being diagonal matrices.

(b) Show how to implement your estimation strategy of (a) in the more general case of non-diagonal $\mathbf{R}$ and $\mathbf{D}$.

(c) Reconcile your answers in (a) and (b) with ordinary maximum likelihood estimation.

E 9.19. The hierarchical specification (24) leads to REML estimation.

(a) Derive an empirical Bayes estimate of $\mathbf{u}$, along with an approximation of its variance, based on the proper Bayes estimates $E(\mathbf{u} \mid \mathbf{y}, \boldsymbol{\eta})$ and $\text{var}(\mathbf{u} \mid \mathbf{y}, \boldsymbol{\eta})$, where $\boldsymbol{\eta}$ is the vector of symbols $\mathbf{R}$, $\mathbf{D}$, with $\mathbf{R}$ and $\mathbf{D}$ being diagonal matrices.

(b) Show how to implement your estimation strategy of (a) in the more general case of non-diagonal $\mathbf{R}$ and $\mathbf{D}$.

(c) Reconcile your answers in (a) and (b) with REML.

E 9.20. In this exercise derivations of empirical Bayes estimates in the 1-way random model, given in Section 9.3d, are to be completed.

(a) Derive (87); that is, obtain an explicit expression for the likelihood function $L(\mu_0, \sigma_e^2, \sigma_\mu^2, \sigma_\alpha^2 \mid \mathbf{y})$. (Note that this is a special case of derivations given in Section 9.3c.)

(b) Define $\tilde{\boldsymbol{\eta}}' = [\tilde{\eta}_1 \quad \tilde{\eta}_2 \quad \tilde{\eta}_3 \quad \tilde{\eta}_4] = [\tilde{\mu}_0 \quad \tilde{\sigma}_e^2 \quad \tilde{\sigma}_\mu^2 \quad \tilde{\sigma}_\alpha^2]$. Derive

$$-\frac{\partial^2}{\partial \eta_j\, \partial \eta_k} L(\boldsymbol{\eta} \mid \mathbf{y})\bigg|_{\boldsymbol{\eta}=\tilde{\boldsymbol{\eta}}} \quad \text{and} \quad \frac{\partial}{\partial \eta_j} E(\alpha_i \mid \mathbf{y}, \boldsymbol{\eta})\bigg|_{\boldsymbol{\eta}=\tilde{\boldsymbol{\eta}}}$$

for $j, k = 1, \ldots, 4$.

(c) For $\hat{\boldsymbol{\eta}}$ of (b), obtain an expression for the empirical Bayes estimate $E(\alpha_i \mid \mathbf{y}, \hat{\boldsymbol{\eta}})$.

(d) Use the Kass–Steffey approximation to show

$$\text{var}(\alpha_i \mid \mathbf{y}) \simeq \text{var}(\alpha_i \mid \mathbf{y}, \tilde{\boldsymbol{\eta}}) + \boldsymbol{\delta}'\boldsymbol{\Sigma}\boldsymbol{\delta}$$

for

$$\boldsymbol{\delta} = \frac{\partial}{\partial \boldsymbol{\eta}} E(\alpha_i \mid \mathbf{y}, \boldsymbol{\eta})\bigg|_{\boldsymbol{\eta}=\tilde{\boldsymbol{\eta}}} \quad \text{and} \quad \boldsymbol{\Sigma} = \left[-\frac{\partial^2}{\partial \boldsymbol{\eta}\, \partial \boldsymbol{\eta}'} l(\boldsymbol{\eta} \mid \mathbf{y})\bigg|_{\boldsymbol{\eta}=\tilde{\boldsymbol{\eta}}} \right]^{-1}.$$

E 9.21. Here we explore, in a simple case, the relationship between two variance expressions, using a special case of the beta–binomial hierarchy (116) with $n_{ij} = n$, $b_i = a = 1$. A Bayes estimator of p is $E(p \mid t)$, where $t = \Sigma_{k=1}^{n} y_k$, and a variance approximation is often based on $\text{var}(p \mid t)$, which is not $\text{var}[E(p \mid t)]$.

(a) Show that

$$\text{var}(t) = \frac{n\alpha\beta}{(\alpha+\beta)^2} \qquad \text{(classical)},$$

$$\text{var}[E(p \mid t)] = \frac{n\alpha\beta}{(\alpha+\beta)^2(n+\alpha+\beta)^2} \qquad \text{(classical)},$$

$$\text{var}(p \mid t) = \frac{(t+\alpha)(n-t+\beta)}{(n+\alpha+\beta)^2(n+\alpha+\beta+1)} \qquad \text{(Bayesian)}.$$

(b) As $n \to \infty$, show that

$$\frac{\text{var}(p \mid t)}{\text{var}[E(p \mid t)]} \to 1,$$

showing the asymptotic equivalence of the classical and Bayesian calculation. [Other examples of comparisons of classical and Bayesian variances are in (42) and (73).]

E 9.22. For the hierarchy of (116) estimates of α_i and β_i can be obtained by the method of moments. [This estimation method is less preferred than maximum likelihood, but sometimes has the advantage of yielding explicit answers. See Casella and Berger (1990, Chap. 7) for a complete discussion.] From (124) we can obtain the marginal mean and variance of $y_{ij\cdot}$, the mean and variance of the beta–binomial distribution, and equate these to the sample moments to obtain the method-of-moments estimates of $\boldsymbol{\alpha}$ and $\boldsymbol{\beta}$.

(a) Using (124), show that for $j = 1, \ldots, b_i$

$$E(y_{ij\cdot} \mid \alpha_i, \beta_i) = n_{ij} \frac{\alpha_i}{\alpha_i + \beta_i} \quad \text{and}$$

$$\operatorname{var}(y_{ij\cdot} \mid \alpha_i, \beta_i) = \frac{n_{ij}\alpha_i\beta_i}{(\alpha_i + \beta_i)^2} \frac{\alpha_i + \beta_i + n_{ij}}{\alpha_i + \beta_i + 1}.$$

(b) Let $n_{ij} = n_i$. Show that for each i the sample mean and variance of $y_{i1\cdot}, \ldots, y_{in_i\cdot}$ are given by (126).

(c) Equate these sample moments to the moments in (a) to show that the method-of-moments estimators of α_i and β_i are given by (127).

E 9.23. Hierarchical models and empirical Bayes methods are feasible only if there is enough replication to be able to estimate all parameters.

(a) For the hierarchy of (116) show that if $b_i = 1$ then the MLEs for p_i and $\alpha_i/(\alpha_i + \beta_i)$ are the same; and the method of moments (as in the previous exercise) fails.

(b) For the general mixed model hierarchy of (8) and (9) formulate some principles about how much replication (or data) is necessary to estimate all parameters of interest.

E 9.24. A useful variation of the beta–binomial hierarchy is the beta–geometric hierarchy

$$y_{ij} \sim \text{geometric}(p_{ij}),$$

$$p_{ij} \sim \text{beta}(\alpha_i, \beta_i) \quad \text{for } i = 1, \ldots, a, j = 1, \ldots, b.$$

Here there are ab combinations, and the ij combination has success probability p_{ij} and variance $\operatorname{var}(p_{ij})$. This model also arises in animal breeding experiments. If for each of ab cows to be artificially inseminated the process is a Bernoulli trial, with success probability p_{ij}, and the trials are repeated until a success occurs, then $y_{ij} \sim \text{geometric}(p_{ij})$, where $P(y_{ij} = k) = p_{ij}(1 - p_{ij})^k$, $k = 0, 1, 2, \ldots$. The interpretation of the rest of the model is similar to the beta–binomial hierarchy.

(a) Derive expressions for $\pi(p_{ij} \mid y_{ij}, \alpha_i, \beta_i)$, the posterior distribution, and $m(y_{ij} \mid \alpha_i, \beta_i)$, the marginal distribution.

(b) Derive the posterior mean and variance $E(p_{ij} \mid y_{ij}, \alpha_i, \beta_i)$ and $\text{var}(p_{ij} \mid y_{ij}, \alpha_i, \beta_i)$.

(c) Derive an expression for the full likelihood of the hierarchical model, and show how to obtain MLEs for α_i and β_i.

(d) Using either a prior distribution for α_i and β_i, or a Kass–Steffey approximation, obtain an estimate of $\text{var}(p_{ij} \mid y_{ij})$.

E 9.25. A model similar to the logit–normal hierarchy of (136) is the probit–normal hierarchy, also discussed in Chapter 10. This hierarchy uses normal distributions with a probit link function (see Section 10.5). The hierarchy is, for $i = 1, \ldots, n$,

$$\begin{aligned} y_i &\sim \text{Bernoulli}(p_i), \\ p_i &= E(y_i \mid \boldsymbol{\beta}, \mathbf{u}) = \Phi(\mathbf{x}_i'\boldsymbol{\beta} + \mathbf{z}_i'\mathbf{u}), \\ \boldsymbol{\beta} &\sim \mathcal{N}(\boldsymbol{\beta}_0, \mathbf{B}), \quad \mathbf{u} \sim \mathcal{N}(\mathbf{0}, \mathbf{D}), \end{aligned}$$

where Φ is the standard normal cumulative density function.

(a) Write an expression for $L(\boldsymbol{\beta}_0, \mathbf{B}, \mathbf{D} \mid \mathbf{y})$, the full likelihood for the hierarchy.

(b) Derive an expression for the REML likelihood $L(\mathbf{D} \mid \mathbf{y})$ and show how to obtain a REML estimate of $\mathbf{D}$.

(c) Derive a strategy for obtaining a point estimate of $\mathbf{u}$.

E 9.26. Another variation of the logit–normal model of (136) is based on the geometric distribution, similar to the use in E 9.24. The hierarchy can use either a logit or probit link function and is, for $i = 1, \ldots, n$,

$$\begin{aligned} y_i &\sim \text{geometric}(p_i), \\ p_i &= E(y_i \mid \boldsymbol{\beta}, \mathbf{u}), \end{aligned}$$

$$p_i = \Phi(\mathbf{x}_i'\boldsymbol{\beta} + \mathbf{z}_i'\mathbf{u}) \quad \text{or} \quad \text{logit}(p_i) = \log\left(\frac{p_i}{1 - p_i}\right) = \mathbf{x}_i'\boldsymbol{\beta} + \mathbf{z}_i'\mathbf{u},$$

$$\boldsymbol{\beta} \sim \mathcal{N}(\boldsymbol{\beta}_0, \mathbf{B}) \quad \text{and} \quad \mathbf{u} \sim \mathcal{N}(\mathbf{0}, \mathbf{D}).$$

Answer (a), (b) and (c) of E 9.25 using the logit hierarchy. The answers for the probit hierarchy are similar.

E 9.27. Based on the hierarchy (136), show how to obtain estimates of $\mathbf{B}$ and $\boldsymbol{\beta}$. Use a strategy similar to that used in (140)–(142).

E 9.28. To $(\mathbf{V} + \mathbf{XBX}')^{-1}$ apply the identity

$$(\mathbf{D} + \mathbf{CA}^{-1}\mathbf{B})^{-1} \equiv \mathbf{D}^{-1} - \mathbf{D}^{-1}\mathbf{C}(\mathbf{A} + \mathbf{BD}^{-1}\mathbf{C})^{-1}\mathbf{BD}^{-1}.$$

Then show that

(a) for $\mathbf{V}$, $\mathbf{X}$ and $\mathbf{B}$ being scalars v, x and b respectively, the limit as $b \to \infty$ is zero; but

(b) for $\mathbf{V}$, $\mathbf{X}$ and $\mathbf{B}$ as matrices the limit of the right-hand side is $\mathbf{V}^{-1} - \mathbf{V}^{-1}\mathbf{X}(\mathbf{X}'\mathbf{V}^{-1}\mathbf{X})^{-1}\mathbf{X}'\mathbf{V}^{-1}$, in contrast to scalar intuition which suggests that $(\mathbf{V} + \mathbf{XBX}')^{-1}$ tends to $\mathbf{0}$ as $\mathbf{B} \to \infty$.

CHAPTER 10

BINARY AND DISCRETE DATA

10.1. INTRODUCTION

Techniques for the estimation of variance components from binary (0/1) or discrete (categorical) data are much less widely developed than for continuous data. The lack of methods for such data is due in large part both to the difficulty of specifying realistic models and, once specified, to their computational intractability. In this chapter we explore the problems in identifying tractable models for binary and categorical data, and review some of the approaches that have been proposed to deal with them.

To see why models are more difficult for discrete data than for continuous data, we return to the construction of models for continuous data, and consider how we defined random effects and error terms. The latter were defined as $\mathbf{y} - E(\mathbf{y} \mid \mathbf{u})$ [see (60) of Section 4.6, for example] and to them we attributed a distribution, sometimes $\mathcal{N}(0, \sigma_e^2\mathbf{I})$ and, in all cases, a distribution having constant variance, independent of the value of the mean of $\mathbf{y}$. This is not a reasonable assumption for discrete data. Consider binary data where y_i takes on only the values zero and one. Then y_i is distributed as a Bernoulli random variable with probability of success $p_i = \Pr\{y_i = 1\} = E(y_i)$ and variance $\text{var}(y_i) = p_i(1 - p_i) = E(y_i)[1 - E(y_i)]$. As the mean of y_i approaches one or zero, the variance approaches zero and this dependence between mean and variance must be included in any reasonable model. Thus a model with an additive error component with fixed variance cannot capture the dependence between mean and variance and therefore is inadequate for categorical data.

Further problems arise when specifying the distribution of random effects. For simplicity, consider a model for a binary variable y_{ij} with a single fixed effect βx_{ij} and a single random effect α_i. Conditional on the random effects, the mean of y_{ij} will be taken as

$$E(y_{ij} \mid \alpha_i) = \beta x_{ij} + \alpha_i \,. \tag{1}$$

For the continuous data situation the α_i are usually assumed to be i.i.d. with variance σ_α^2 and are often assumed to have a normal distribution. For the binary data situation, since the mean or conditional mean of y_{ij} cannot be larger than one or less than zero, the α_i cannot have a normal distribution, and as the mean of y_{ij} approaches zero or one the variance of the α_i must approach zero. So the distribution of the α_i also cannot have a fixed variance. The usual way of accommodating these requirements is to consider nonlinear models which allow the random effects to enter into the conditional mean in a non-additive fashion.

A common model for binary data where y_{ij} has a Bernoulli distribution with probability of success of p_{ij} is the logistic regression model where $\text{logit}(p_{ij})$, defined as $\text{logit}(p_{ij}) = \log[p_{ij}/(1 - p_{ij})]$, is assumed to be linear in the fixed and random effects. Thus a mixed model analogous to (1) could be defined as

$$y_{ij} \mid \alpha_i \sim \text{independent Bernoulli}[E(y_{ij} \mid \alpha_i)],$$

with

$$\text{logit}[E(y_{ij} \mid \alpha_i)] = \beta x_{ij} + \alpha_i \quad \text{and} \quad \alpha_i \sim \text{i.i.d. } \mathcal{N}(0, \sigma_\alpha^2) \,. \tag{2}$$

Comparing this to the continuous data situation, we see that the distribution assumed for y_{ij}, conditional on the random effects, is a Bernoulli as opposed to a normal distribution, and $\text{logit}[E(y_{ij} \mid \alpha_i)]$ instead of $E(y_{ij} \mid \alpha_i)$ is modeled as linear in the fixed and random effects. In nonlinear models such as (2) the function (logit here) which connects the mean of y and the effects is called the *link function*. Otherwise the constructions are the same. The use of the Bernoulli distribution takes care of the connection between mean and variance. The logit transformation maps the interval $(0, 1)$ for p_{ij} on to the whole real line, where problems with the upper and lower limits of the p_{ij} disappear. It is then reasonable to assume a normal (or other unbounded) distribution for α_i.

This approach is not without its problems. As discussed in Section 10.3, the computations for ML or REML for model (2) are quite intensive; much more so than for continuous data. This approach also raises a conflict in interpretation of the parameters. In the continuous data model, (1), β is the amount of change in the mean of y_{ij} associated with a change of one unit in x_{ij}. This is true in the conditional distribution of y_{ij} given α_i, as well as in the marginal distribution, since

$$E(y_{ij}) = E[E(y_{ij} \mid \alpha_i)] = E(\beta x_{ij} + \alpha_i) = \beta x_{ij} \,.$$

This identical meaning in the marginal and conditional distributions holds because of the linear model. For (2) it no longer holds because the model is nonlinear (E 10.1). β represents the change on the logit scale of the conditional mean of y_{ij} for a change of one unit in x_{ij}. But the same is not true of the marginal mean of y_{ij} since, in general,

$$E(y_{ij}) = E[E(y_{ij} \mid \alpha_i)] = E\left[\frac{1}{1 + e^{-(\beta x_{ij} + \alpha_i)}}\right] \neq \frac{1}{1 + e^{-\beta x_{ij}}} \,. \tag{3}$$

In fact, no closed form expression exists for $E(y_{ij})$ under this model.

What does this mean in practical terms? Consider animal breeding data, where y_{ij} is one if a cow experiences difficulty in calving and zero otherwise. Further suppose that β represents the effect of birth order and the α_i are individual animal effects. Then β is interpreted as the effect of increasing birth order on the logit of the probability of calving difficulty *for an individual animal* (because this is conditional on the animal effects). However, because of (3), β does *not* represent the change in the logit of the probability of calving difficulty in the entire population. This would require averaging the conditional distribution over all animals to obtain the marginal distribution. Zeger, Liang and Albert (1988) give formulae for the marginal and conditional means of y, as functions of fixed effects, for a number of different models, including the logit–normal and probit–normal models described below in (9) and (11).

The nonlinear link (3) between the mean of y_{ij} and the fixed and random effects correctly models the fact that the variance of y_{ij} induced by the random effects is less as the mean of y_{ij} approaches zero or one. Yet this very fact increases the difficulty of interpretation since the variance in y_{ij} due to the random effects is dependent on the mean, i.e., the fixed effects. Thus separate interpretations of the influence of the fixed and random effects on y_{ij} are no longer possible.

10.2. ANOVA METHODS

Given these problems, what approaches have been proposed for analyzing discrete data? If the data consist of binomial proportions, all with a constant number of trials, n, then the usual recommendations are to analyze the proportions directly (or their arcsin transformation) using ANOVA methods, assuming they are approximately normally distributed and homoscedastic. However, analyzing proportions can only be recommended when the proportions are in the middle range (e.g., 0.2–0.8) and heteroscedasticity is unlikely to be a problem. With highly varying proportions the observations will have quite different variances and should be appropriately weighted in the analysis. Because of the presence of variance components, the weighting factor is no longer the binomial variance $p(1-p)/n$ and the proper weights depend on the relative size of the variance components and the binomial variance. Furthermore, the arcsin transformation is not necessarily appropriate when the binomial p is allowed to vary with the random effects as is the case in (2). See Cochran (1943) for a clear discussion of these points. Landis and Koch (1977) give the details of using MANOVA to estimate variance components for a one-way random effects model with categorical data. When the group sizes are unequal, or the proportions cover a wide range, then more sophisticated techniques are necessary.

10.3. BETA-BINOMIAL MODELS

a. Introduction

For a binary variable y a natural approach to capturing the variability in the mean of y is to model it directly rather than indirectly as in (2). That is, assume a parametric distribution for $p = E(y)$. A logical distribution is the beta distribution, since it is a flexible distribution on the interval (0, 1); it is also the conjugate prior density for the binomial distribution from Bayesian analysis and it leads to mathematically tractable results. If y is distributed as a binomial(n, p) variable, conditional on the value of p, and p has a beta distribution with parameters α and β, then the marginal distribution is beta–binomial, i.e.,

$$f(y) = \binom{n}{y} \frac{B(\alpha + y, n + \beta - y)}{B(\alpha, \beta)},$$

where $B(\alpha, \beta) = \int_0^1 x^{\alpha-1}(1-x)^{\beta-1}\,dx$ is the beta function.

b. Model specification

How do we allow the values of the parameters α and β to vary in order to form realistic models? Let us consider for continuous data the mixed model with a single fixed effect and nested random effects:

$$y_{ijk} = \mu + \eta_i + \gamma_{ij} + e_{ijk}, \quad \text{where the } \eta_i \text{ are fixed effects,}$$

$$\gamma_{ij} \sim \text{i.i.d. } \mathcal{N}(0, \sigma_\gamma^2)$$

and (4)

$$e_{ijk} \sim \text{i.i.d. } \mathcal{N}(0, \sigma_e^2), \quad \text{independently of the } \gamma_{ij}\,.$$

This model allows the mean of the y_{ijk} to vary with i and allows the y_{ijk} to be correlated within levels of i and j, i.e., $\rho(y_{ijk}, y_{ijk'}) = \sigma_\gamma^2/(\sigma_e^2 + \sigma_\gamma^2)$ for $k \neq k'$.

By following the hierarchical specification (see Chapter 9) of a model for the binary data, we can induce a correlation among all the ys that have the same p. Thus, to mimic the correlation structure in model (4), we would use the following specification:

$$y_{ijk} \mid p_{ij} \sim \text{independent Bernoulli}(p_{ij})$$

and (5)

$$p_{ij} \sim \text{independent beta}(\alpha_i, \beta_i)$$

for $i = 1, 2, \ldots, a$, $j = 1, 2, \ldots, b_i$ and $k = 1, 2, \ldots, n_{ij}$. This induces a correlation among all the ys within each (i, j) combination, i.e., among those with the same p_{ij} (E 10.3). Also, since the parameters of the beta distribution depend only on i, the mean of the conditional distribution of y_{ijk} given p_{ij} is allowed to vary with i. In this general form (5) also allows the variance of the conditional mean of y_{ijk} to vary from one level of i to the next, which (4) does not.

What is the variance component for this model? Since the variance of the conditional mean is allowed to have different variances depending on i, we need an estimate of the variance in each of the classes, given by the variances of the beta distribution,

$$\text{var}(p_{ij}) = \frac{\alpha_i \beta_i}{(\alpha_i + \beta_i)^2(\alpha_i + \beta_i + 1)}\,.$$

It does not really make sense to try to reparameterize the model to have a single variance since, as discussed above, the conditional mean and the variance must be related. Noting that (see E 10.3)

$$\operatorname{var}(p_{ij}) = \mu_i(1 - \mu_i)\sigma_i, \tag{6}$$

where $\mu_i = E(p_{ij}) = \alpha_i/(\alpha_i + \beta_i)$ as in (10) of Appendix S.6d and $\sigma_i = 1/(\alpha_i + \beta_i + 1)$, Crowder (1978) suggests restricting all the σ_i to have a common value σ. Note that (6) incorporates the need for the variance to decrease to zero as μ_i approaches zero or one. Also, σ_i is the intra-class correlation coefficient so that y_{ijk} and $y_{ijk'}$ are uncorrelated if and only if σ_i is zero. Thus σ is the analog of $\sigma_\gamma^2/(\sigma_\gamma^2 + \sigma_e^2)$, the intra-class correlation coefficient, for normal, linear models; i.e., equation (4). For some situations σ would therefore be a useful parameter of interest.

c. Likelihood

The likelihood for model (5) takes a relatively simple form. Denoting the number of successes within level (i,j) by $t_{ij} = y_{ij\cdot} = \Sigma_k\, y_{ijk}$, the log likelihood can be written as

$$l = \sum_i \sum_j \left[\sum_{r=0}^{t_{ij}-1} \log(\alpha_i + r) + \sum_{r=0}^{n_{ij}-t_{ij}-1} \log(\beta_i + r) - \sum_{r=0}^{n_{ij}-1} \log(\alpha_i + \beta_i + r) \right]. \tag{7}$$

For interpretational and numerical reasons Williams (1975) suggests reparameterizing l in terms of the mean of the beta distribution, $\mu_i = \alpha_i/(\alpha_i + \beta_i)$, and the parameter $\theta_i = 1/(\alpha_i + \beta_i)$. In this reparameterization

$$l = \sum_i \sum_j \left[\sum_{r=0}^{t_{ij}-1} \log(\mu_i + r\theta_i) + \sum_{r=0}^{n_{ij}-t_{ij}-1} \log(1 - \mu_i + r\theta_i) - \sum_{r=0}^{n_{ij}-1} \log(1 + r\theta_i) \right]. \tag{8}$$

Closed form maximum likelihood estimators for μ_i and θ_i do not exist for this model, so (7) or (8) needs to be maximized numerically.

d. Discussion

The beta–binomial approach is somewhat limited in its application to variance components estimation problems. Since we model the correlation by having the correlated Bernoulli variables all selected from a distribution with the same probability of success, we are limited to the type of model (5) where the random effects are nested within the fixed effects. This precludes any sort of regression model which has independent variables specific to each Bernoulli variable. Also, since we are capturing the variation in the conditional mean with a single distribution, the beta–binomial approach is not amenable to multiple random effects. Thus model (5) is about the most general model possible with this approach.

10.4. LOGIT–NORMAL MODELS

A more flexible approach to variance components for binary data is the approach outlined in the introduction. This approach uses a logit function to link the mean of $\mathbf{y}$ to the fixed and random effects and assumes the random effects are normally distributed. Conditional on the random effects $\mathbf{u}$,

$$y_i \mid \mathbf{u} \sim \text{independent Bernoulli}[E(y_i \mid \mathbf{u})], \quad i = 1, 2, \ldots, n,$$

$$\text{logit}[E(y_i \mid \mathbf{u})] = \mathbf{x}_i'\boldsymbol{\beta} + \mathbf{z}_i'\mathbf{u} \qquad (9)$$

and

$$\mathbf{u} \sim \mathcal{N}(\mathbf{0}, \mathbf{D}),$$

where, in the model for the vector of $\text{logit}[E(y_i \mid \mathbf{u})]$ for $i = 1, 2, \ldots, n$, $\mathbf{x}_i'$ and $\mathbf{z}_i'$ are the ith rows of $\mathbf{X}$ and $\mathbf{Z}$, the model matrices for the fixed and random effects, respectively. For certain specific situations this approach is developed in Pierce and Sands (1975), Stiratelli, Laird and Ware (1984), and Wong and Mason (1985).

A main drawback to this approach is computational. The likelihood based on (9) is proportional to

$$\int f_{\mathbf{y}\mid\mathbf{p}}(\mathbf{y} \mid \mathbf{p}) \exp(-\tfrac{1}{2}\mathbf{u}'\mathbf{D}^{-1}\mathbf{u}) \, |\mathbf{D}|^{-\frac{1}{2}} \, d\mathbf{u}, \qquad (10)$$

where

$$f_{\mathbf{y}\mid\mathbf{p}}(\mathbf{y} \mid \mathbf{p}) = \Pi_{i=1}^{n} \exp[y_i(\mathbf{x}_i'\boldsymbol{\beta} + \mathbf{z}_i'\mathbf{u})] \, [1 + \exp(\mathbf{x}_i'\boldsymbol{\beta} + \mathbf{z}_i'\mathbf{u})]^{-1} .$$

This cannot be simplified appreciably and all of the above authors suggest approximations in order to ease the computational burden of finding ML estimates. Stiratelli, Laird and Ware (1984) and Wong and Mason (1985) propose REML estimation for (9) by treating the fixed effects as random effects whose variances tend to infinity. Both of these papers advocate the use of the EM algorithm for estimation of the variance components.

Stiratelli *et al.* (1984) make the simplifying assumption that, for $\mathbf{u}' = [\mathbf{u}_1' \quad \mathbf{u}_2' \quad \ldots \quad \mathbf{u}_r']$, $\text{var}(\mathbf{u}) = \{_d \; \mathbf{D}\}_{i=1}^{r}$. Then, with $\mathbf{B}$ representing $\text{var}(\boldsymbol{\beta})$, the mth iterate of the EM algorithm for REML takes the form

$$\mathbf{D}^{(m)} = \hat{\mathbf{t}}^{(m)}/r \qquad \text{(M-step)},$$

$$\hat{\mathbf{t}}^{(m+1)} \equiv \sum_{i=1}^{r} E(\mathbf{u}_i\mathbf{u}_i' \mid \mathbf{y})\Big|_{\mathbf{D}=\mathbf{D}^{(m)},\, \mathbf{B}^{-1}=\mathbf{0}} \qquad \text{(E-step)} .$$

Iterations cease when the elements of $\mathbf{D}$ stabilize. While conceptually straightforward, the E-step requires hefty computation of the integral which is implicit in the expectation.

10.5. PROBIT–NORMAL MODELS

a. Introduction

Probit–normal models are a class of models very similar to logit–normal models that arise by replacing the logit function in (9) by the probit function $\Phi^{-1}(\cdot)$, where $\Phi(\cdot)$ is the standard normal c.d.f. This gives a model

$$y_i \mid \mathbf{u} \sim \text{independent Bernoulli}[E(y_i \mid \mathbf{u})], \quad i = 1, 2, \ldots, n,$$

$$E(y_i \mid \mathbf{u}) = \Phi(\mathbf{x}_i'\boldsymbol{\beta} + \mathbf{z}_i'\mathbf{u})$$

and (11)

$$\mathbf{u} \sim \mathcal{N}(\mathbf{0}, \mathbf{D}) .$$

This model retains the flexibility of the logit–normal models as well as most of the computational problems. The likelihood for (11) is proportional to

$$\int f_{\mathbf{y}|\mathbf{p}}(\mathbf{y} \mid \mathbf{p}) \exp(-\tfrac{1}{2}\mathbf{u}'\mathbf{D}^{-1}\mathbf{u})\, |\mathbf{D}|^{-\frac{1}{2}}\, d\mathbf{u},$$

where (12)

$$f_{\mathbf{y}|\mathbf{p}}(\mathbf{y} \mid \mathbf{p}) = \prod_{i=1}^{n} \Phi(\mathbf{x}_i'\boldsymbol{\beta} + \mathbf{z}_i'\mathbf{u})^{y_i}[1 - \Phi(\mathbf{x}_i'\boldsymbol{\beta} + \mathbf{z}_i'\mathbf{u})]^{1 - y_i} .$$

This model is used in Harville and Mee (1984), where it is extended for use with ordered categorical data, and in Gilmour, Anderson and Rae (1985) for a single random effect. It is also essentially that used in Ochi and Prentice (1984) for a model similar to (5). To overcome the computational problems, which were declared "insurmountable" for the general model by Harville and Mee, they resorted to *ad hoc* estimation methods, whereas Ochi and Prentice developed a complicated approximation scheme for finding the maximum likelihood estimators, and Gilmour *et al.* used quasi-likelihood methods. McCulloch (1990) shows how to adapt the EM algorithm to probit–normal models.

b. An example

We illustrate the use of the probit–normal model on a data set (courtesy of Professor S. Via at Cornell University) on reproductive success in aphids. Twenty-eight female aphids were collected in the field in both the early and late summer. Clonal lines were raised from each female in the laboratory in two separate chambers (sublines). For each clonal subline 0 to 2 females were raised on alfalfa and on clover. A total of 412 individuals were tested and each individual was recorded as surviving to reproduce or not. So there are two random effects (clone and subline nested within clone) and four fixed effects (constant term, crop, time and crop by time interaction).

Let y_{ijklm} represent the mth response on the ith clone, jth subline, crop k and time l, where $y_{ijklm} = 1$ if the aphid survived to reproduce and 0 otherwise. The model employed was

$$\Phi^{-1}[E(y_{ijklm} \mid \mathbf{u})] = \mu + \alpha_k + \beta_j + \gamma_{kl} + u_{1i} + u_{2j}$$

and

$$\mathbf{u}_1 \sim \mathcal{N}(\mathbf{0}, \mathbf{I}\sigma_c^2), \quad \mathbf{u}_2 \sim \mathcal{N}(\mathbf{0}, \mathbf{I}\sigma_s^2),$$

where $\mathbf{u}_1$ is the vector of clone random effects and $\mathbf{u}_2$ is the vector of subline random effects. The log-likelihood was numerically maximized, giving a maximum value of -181.667 and estimates $\hat{\sigma}_c^2 = .166$ and $\hat{\sigma}_s^2 = .035$. This would give an estimated within-clone, within-subline correlation of $(.166 + .035)/(1 + .166 + .035) = .17$ on the probit scale.

10.6. DISCUSSION

It should be clear from the preceding outline that methods for the analysis of binary or categorical data are only available for a limited variety of problems. For situations with binomial or categorical data with proportions in the mid-range and approximately equal n, ANOVA methods may be adequate. For simple situations the beta–binomial approach may be adequate or the logit–normal or probit–normal models may be computationally feasible. For more complicated situations the beta–binomial approach becomes inadequate and the logit–normal and probit–normal models become computationally limiting. Surely, as computers become more and more powerful, such models will come into greater use.

The logit–normal and probit–normal models are very similar. Zeger, Liang and Albert (1988) show how to approximate one from the other. However there are some slight differences. The probit–normal models reduce to the usual probit analysis when there is a single random effect and a single observation per level of the random effect. The logit–normal models do not reduce to a standard logistic regression analysis (E 10.5). Thus the logit–normal differs from the normal, linear model, which, with a single observation per level of the random effect, reduces to a fixed effects analysis. Also, the marginal mean of y_i in the probit–normal model is slightly simpler than the logit–normal. It can be shown (E 10.4) that

$$E(y_i) = \Phi\left[\frac{\mathbf{x}_i'\boldsymbol{\beta}}{(1 + \mathbf{z}_i'\mathbf{D}\mathbf{z}_i)^{\frac{1}{2}}}\right]$$

so that the marginal mean has a form similar to the conditional mean. On the other hand, for inferences about the fixed effects the logit–normal model may be simpler. It allows exact conditional inference for some balanced data situations (Conaway, 1989).

10.7. SUMMARY

Binary data variance components models

Beta–binomial

$$y_{ijk} \mid p_{ij} \sim \text{independent Bernoulli}(p_{ij}),$$
$$p_{ij} \sim \text{independent beta}(\alpha_i, \beta_i);$$

$$l = \sum_i \sum_j \left[\sum_{r=0}^{t_{ij}-1} \log(\alpha_i + r) + \sum_{r=0}^{n_{ij}-t_{ij}-1} \log(\beta_i + r) - \sum_{r=0}^{n_{ij}-1} \log(\alpha_i + \beta_i + r) \right];$$

$$\text{var}(p_{ij}) = \frac{\alpha_i \beta_i}{(\alpha_i + \beta_i)^2(\alpha_i + \beta_i + 1)};$$

$$\rho(y_{ijk}, y_{qrs}) = \begin{cases} 0 & (i \neq q \text{ or } j \neq r) \\ \dfrac{1}{\alpha_i + \beta_i + 1} & (i = q \text{ and } j = r,\ k \neq s), \\ 1 & (i = q, j = r, k = s)\,. \end{cases}$$

Logit–normal

$$y_i \mid \mathbf{u} \sim \text{independent Bernoulli}[E(y_i \mid \mathbf{u})],$$

where

$$\text{logit}[E(y_i \mid \mathbf{u})] = \mathbf{x}_i'\boldsymbol{\beta} + \mathbf{z}_i'\mathbf{u}$$

and

$$\mathbf{u} \sim \mathcal{N}(\mathbf{0}, \mathbf{D})\,.$$

The log likelihood is given in (10).

Probit–normal

$$y_i \mid \mathbf{u} \sim \text{independent Bernoulli}[E(y_i \mid \mathbf{u})],$$

where

$$E(y_i \mid \mathbf{u}) = \Phi(\mathbf{x}_i'\boldsymbol{\beta} + \mathbf{z}_i'\mathbf{u})$$

and

$$\mathbf{u} \sim \mathcal{N}(\mathbf{0}, \mathbf{D})\,.$$

The log likelihood is given in (12).

10.8. EXERCISES

E 10.1. Suppose $y = 1/(1 + e^{-\mu-\alpha})$, where $\alpha \sim (0, \sigma^2)$.

(a) Show that if $\mu = 0$ and the distribution of α is symmetric about zero then

$$E(y) = \frac{1}{1 + e^{-\mu - E(\alpha)}} = \frac{1}{1 + e^{-\mu}} = \tfrac{1}{2}\,.$$

(b) Argue by Taylor expansion when σ^2 is small that

$$E(y) > \frac{1}{1 + e^{-\mu - E(\alpha)}} = \frac{1}{1 + e^{-\mu}} \quad \text{when } \mu < 0$$

and

$$E(y) < \frac{1}{1 + e^{-\mu - E(\alpha)}} = \frac{1}{1 + e^{-\mu}} \quad \text{when } \mu > 0\,.$$

E 10.2. If $y \sim \text{binomial}(n, p)$ conditional on the value of p, and if $p \sim B(\alpha, \beta)$, show that the marginal distribution of y is beta–binomial, namely

$$f(y) = \frac{\binom{n}{y} B(\alpha + y, n + \beta - y)}{B(\alpha, \beta)}\,.$$

E 10.3. For model (5) show that

(a) the log likelihood is given by (7);

(b) the log likelihood can be rewritten as (8);

(c) the correlation of y_{ijk} with y_{qrs} is

$$\rho(y_{ijk}, y_{qrs}) = \begin{cases} 0 & (i \neq q \text{ or } j \neq r), \\ \dfrac{1}{\alpha_i + \beta_i + 1} & (i = q \text{ and } j = r,\ k \neq q), \\ 1 & (i = q,\ j = r \text{ and } k = s); \end{cases}$$

(d) $\text{var}\, E(y_{ij} \mid p_{ij}) = 0$ if and only if $\theta_i = 1/(\alpha_i + \beta_i) = 0$ for $E(y_{ij} \mid p_{ij}) \in (0, 1)$.

E 10.4. (a) For the model in equation (11) show that

$$E(y_i) = \Phi\left[\frac{\mathbf{x}_i'\boldsymbol{\beta}}{(1 + \mathbf{z}_i'\mathbf{D}\mathbf{z}_i)^{\frac{1}{2}}}\right].$$

(b) On the probit scale how do the coefficients of the fixed effects compare for the conditional and marginal means?

E 10.5. Consider a simple version of the probit–normal model:

$$y_i \mid \mathbf{u} \sim \text{independent Bernoulli}[\Phi(x_i\beta + u_i)],$$
$$\mathbf{u} \sim \mathcal{N}(\mathbf{0}, \mathbf{I}\sigma^2)\,.$$

In this model there is one level of the random effect for each observation. Show that y_i follows the usual probit model, i.e.,

$$y_i \sim \text{independent Bernoulli}[\Phi(x_i\beta^*)]$$

for a suitable definition of β^*. (*Hint*: See E 10.4.) The equivalent logit–normal model is

$$y_i \mid \mathbf{u} \sim \text{Bernoulli}\left[\frac{1}{1 + \exp(-x_i\beta + u_i)}\right],$$

Show that y_i does not follow the usual logit model, which is

$$y_i \sim \text{independent Bernoulli}\left[\frac{1}{1 + \exp(-x_i\beta^*)}\right].$$

CHAPTER 11

OTHER PROCEDURES

Estimation methods based on ANOVA, ML, REML and Bayes have been considered at length in preceding chapters. Nevertheless, there are other estimation topics that merit discussion, and from a wide array that is available we have chosen just three: (i) defining and estimating covariance components, which is important in applications as varied as animal breeding and educational testing; (ii) defining variance components in terms of a covariance structure, which models a variance component as a covariance so that a negative estimate has meaning as a negative covariance; (iii) criteria-based estimation (such as minimum norm and minimum variance estimation), which is somewhat more theoretical than other methods.

11.1. ESTIMATING COMPONENTS OF COVARIANCE

Suppose we measure weight and body length of piglets at two weeks of age. Let y_{1ij} be the weight of piglet j from sow i, and y_{2ij} its body length. The model equations for a 1-way classification random model for each of these observations can be taken as

$$y_{1ij} = \mu_1 + \alpha_{1i} + e_{1ij} \quad \text{and} \quad y_{2ij} = \mu_2 + \alpha_{2i} + e_{2ij}, \tag{1}$$

for $i = 1, 2, \ldots, a$ sows and $j = 1, 2, \ldots, n_i$ piglets from sow i, with a total of $n_. = N$ piglets. μ_1 and μ_2 represent overall means of weight and body length, α_{1i} and α_{2i} are the effects of sow i on the two variables, and e_{1ij} and e_{2ij} are the corresponding random error terms. Treating the α_1s and α_2s as random effects with zero means, the usual random model conditions are

$$\text{var}(\boldsymbol{\alpha}_1) = \sigma^2_{\alpha_1}\mathbf{I}_a, \quad \text{var}(\mathbf{e}_1) = \sigma^2_{e_1}\mathbf{I}_N \quad \text{and} \quad \text{cov}(\boldsymbol{\alpha}_1, \mathbf{e}'_1) = \mathbf{0}_{a \times N}$$

and (2)

$$\text{var}(\boldsymbol{\alpha}_2) = \sigma^2_{\alpha_2}\mathbf{I}_a, \quad \text{var}(\mathbf{e}_2) = \sigma^2_{e_2}\mathbf{I}_N \quad \text{and} \quad \text{cov}(\boldsymbol{\alpha}_2, \mathbf{e}'_2) = \mathbf{0}_{a \times N}\,.$$

The inclusion of components of covariance between sow effects and between error terms involves having the model also include

$$\operatorname{cov}(\boldsymbol{\alpha}_1, \boldsymbol{\alpha}_2') = \tau_\alpha \mathbf{I}_a \quad \text{and} \quad \operatorname{cov}(\mathbf{e}_1, \mathbf{e}_2') = \tau_e \mathbf{I}_N \,. \tag{3}$$

This is the assumption that $\operatorname{cov}(\alpha_{1i}, \alpha_{2i}) = \tau_\alpha$ for all i, but $\operatorname{cov}(\alpha_{1i}, \alpha_{2i'}) = 0$ for $i \neq i'$: similarly $\operatorname{cov}(e_{1ij}, e_{2ij}) = \tau_e$ but $\operatorname{cov}(e_{1ij}, e_{2i'j'}) = 0$ unless $i = i'$ and $j = j'$. Thus τ_α is the covariance between the two sow effects, one on piglet weight and the other on piglet body length. To the geneticist this is a multiple of the genetic covariance between the two traits, which, along with the variance components $\sigma^2_{\alpha_1}$ and $\sigma^2_{\alpha_2}$, leads to genetic correlation, a parameter of great interest.

A second example where this model might be suitable, with components of covariance being of interest, would be test scores on schoolchildren in different classes that had each taken an English test and a mathematics test. Another example would be fleece weight and staple length of the fleece obtained from shearing a thousand ewes, each of which was the daughter of one of, say, 30 different rams. Then the components of covariance between the sire effect on fleece weight and the sire effect on staple length could be of interest. These examples, which are similar and straightforward, represent only one of several ways that components of covariance between random effects can be included as part of a model. Along with considering estimation procedures, we therefore also indicate a variety of ways in which covariance components can be present.

a. Easy ANOVA estimation for certain models

The schoolchildren example just described is the simplest illustration of a class of components of variance and covariance models for which ANOVA estimation of the covariance components is based very easily upon whatever ANOVA estimation is used for the variance components. For that example y_{1jk} and y_{2jk} of (1) will be the English score and the mathematics score, respectively, of child j in class i.

Under conditions such as this, where every observational unit (a child, in this case) has observations on the same pair of variables, there is a very easy ANOVA method for estimating the covariance components of (3) when the same form of model is used for each variable as, for example, in (1). It is just a simple extension of whatever ANOVA method is chosen for estimating the variance components. If, for $\mathbf{B}$ being some symmetric matrix, $\mathbf{y}_1'\mathbf{B}\mathbf{y}_1$ is the ANOVA estimator of $\sigma^2_{\alpha_1}$ in the model for $\mathbf{y}_1$ then, of course, the same ANOVA estimator of $\sigma^2_{\alpha_2}$ in the model for $\mathbf{y}_2$ is $\mathbf{y}_2'\mathbf{B}\mathbf{y}_2$. As we now show, the corresponding ANOVA estimator of the covariance component σ_{12} is $\mathbf{y}_1'\mathbf{B}\mathbf{y}_2$. But one does not have to compute $\mathbf{y}_1'\mathbf{B}\mathbf{y}_2$ in that form. Because, in terms of our example, with $\sigma^2_{\alpha_{1+2}}$ and $\sigma^2_{e_{1+2}}$ representing the components of variance for the variable $y_1 + y_2$,

$$\tau_\alpha = \operatorname{cov}(\alpha_{1i}, \alpha_{2i}) = \tfrac{1}{2}(\sigma^2_{\alpha_{1+2}} - \sigma^2_{\alpha_1} - \sigma^2_{\alpha_2})$$

and

$$\mathbf{y}_1'\mathbf{B}\mathbf{y}_2 = \tfrac{1}{2}[(\mathbf{y}_1 + \mathbf{y}_2)'\mathbf{B}(\mathbf{y}_1 + \mathbf{y}_2) - \mathbf{y}_1'\mathbf{B}\mathbf{y}_1 - \mathbf{y}_2'\mathbf{B}\mathbf{y}_2],$$

it follows at once that

$$\hat{t}_\alpha = \tfrac{1}{2}(\hat{\sigma}^2_{\alpha_{1+2}} - \hat{\sigma}^2_{\alpha_1} - \hat{\sigma}^2_{\alpha_2}) \,. \tag{4}$$

The nature of this result is true in general for models of this form, as pointed out by Searle and Rounsaville (1974). All one has to do to estimate a component of covariance is to use the ANOVA estimates of components of variance of $\mathbf{y}_1$, of $\mathbf{y}_2$ and of $\mathbf{y}_1 + \mathbf{y}_2$. Thus for (4), using (82) and (83) of Chapter 3, we write

$$\hat{\sigma}^2_{e_1} = \text{MSE}_1, \quad \hat{\sigma}^2_{\alpha_1} = (\text{MSA}_1 - \text{MSE}_1)/f$$
$$\hat{\sigma}^2_{e_2} = \text{MSE}_2, \quad \hat{\sigma}^2_{\alpha_2} = (\text{MSA}_2 - \text{MSE}_2)/f \tag{5}$$

and

$$\hat{\sigma}^2_{e_{1+2}} = \text{MSE}_{1+2}, \quad \hat{\sigma}^2_{\alpha_{1+2}} = (\text{MSA}_{1+2} - \text{MSE}_{1+2})/f,$$

where $f = (N - \Sigma_i n_i^2/N)/(a-1)$, and, for example,

$$\text{MSE}_1 = \Sigma_i\Sigma_j(y_{1ij} - \bar{y}_{1i.})^2/(N-a)$$

and

$$\text{MSA}_{1+2} = \Sigma_i n_i[(\bar{y}_{1i.} + \bar{y}_{2i.}) - (\bar{y}_{1..} + \bar{y}_{2..})]^2/(a-1) \,.$$

Then using (5) in (4) and in

$$\hat{t}_e = \tfrac{1}{2}(\hat{\sigma}^2_{e_{1+2}} - \hat{\sigma}^2_{e_1} - \hat{\sigma}^2_{e_2}) \tag{6}$$

provides estimates of the covariance components based on estimated variance components. In this way, when a computing routine specifically calculates ANOVA variance components estimates, it can also be used [by means of (4) and (6), for example] for deriving ANOVA estimates of covariance components.

b. Examples of covariance components models

The model widely used throughout this book for data on a single variable is

$$\mathbf{y} = \mathbf{X}\boldsymbol{\beta} + \mathbf{Z}\mathbf{u} + \mathbf{e},$$

where $\boldsymbol{\beta}$ and $\mathbf{u}$ represent fixed effects and random effects, respectively; and $\mathbf{u}$ is partitioned as $\mathbf{u}' = [\mathbf{u}'_1 \quad \mathbf{u}'_2 \quad \ldots \quad \mathbf{u}'_r]$ into sub-vectors $\mathbf{u}_i$ of order q_i, with $E(\mathbf{u}_i) = \mathbf{0}$, $\text{var}(\mathbf{u}_i) = \sigma_i^2\mathbf{I}_{q_i}$, $\text{cov}(\mathbf{u}_i, \mathbf{u}'_{i'}) = \mathbf{0}_{q_i \times q_{i'}}$, and $\text{cov}(\mathbf{u}_i, \mathbf{e}') = \mathbf{0}_{q_i \times N}$. Thus

$$\text{var}(\mathbf{u}) = \mathbf{D} = \{_{\text{d}}\, \sigma_i^2\mathbf{I}_{q_i}\}_{i=1}^{r} \tag{7}$$

is a block diagonal matrix of diagonal matrices $\sigma_i^2\mathbf{I}_{q_i}$. This is because the covariance between every possible pair of (different) elements of $\mathbf{u}$ has been taken as zero. Generalizations of this model to allow for covariances between elements of $\mathbf{u}$ merely consist of having a form for $\mathbf{D}$ different from its block diagonal form in (7). At least two possibilities are available.

-i. Covariances between effects of the same random factor. Let u_{it} be an element of $\mathbf{u}_i$ for $t = 1, \ldots, q_i$. Suppose covariances between all pairs of elements of $\mathbf{u}_i$ are to be non-zero but covariances between elements of different **u**s are to be zero; then

$$\operatorname{cov}(u_{it}, u_{it'}) = d_{i,tt'} \quad \text{for} \quad t \neq t' \quad \text{and} \quad \operatorname{cov}(u_{it}, u_{i't'}) = 0 \quad \text{for } i \neq i' . \tag{8}$$

Hence

$$\operatorname{var}(\mathbf{u}_i) = \mathbf{D}_{ii} = \{_\text{m}\, d_{i,tt'}\}_{t,t'=1}^{q_i}$$

and (9)

$$\operatorname{var}(\mathbf{u}) = \mathbf{D} = \{_\text{d}\, \mathbf{D}_{ii}\}_{i=1}^{r} .$$

In this case the second equation of (9) shows the block-diagonal structure of the dispersion matrix of **u**, and the first equation in (9) defines the nature of those blocks. Situations in which $d_{i,tt'}$ is different for every t, t' pair seem unlikely, and certain patterns of values may be suitable on some occasions. For example, the intra-class correlation pattern of (28) in Chapter 3 might be appropriate:

$$\mathbf{D}_{ii} = \sigma_i^2[(1 - \rho_i)\mathbf{I}_{q_i} + \rho_i\mathbf{J}_{q_i}], \tag{10}$$

which has $d_{i,tt} = \sigma_i^2$ and $d_{i,tt'} = \rho_i\sigma_i^2$. Another possibility is

$$d_{i,tt'} = \sigma_i^2(\delta_{t,t'} + \rho_i\delta_{1,|t-t'|}),$$

where $\delta_{t,t'}$ is the Kronecker delta, $\delta_{t,t'} = \delta_{t',t} = 1$ for $t = t'$ and zero otherwise. This has $\operatorname{var}(u_{it}) = \sigma_i^2$ and $\operatorname{cov}(u_{it}, u_{it'}) = \rho_i\sigma_i^2$ for $|t - t'| = 1$ and zero otherwise, as illustrated following (28) of Chapter 3.

-ii. Covariances between effects of different random factors. The most general situation would be to have $\operatorname{cov}(u_{it}, u_{i't'}) = d_{ii',tt'}$ so that

$$\operatorname{cov}(\mathbf{u}_i, \mathbf{u}'_{i'}) = \mathbf{D}_{ii'} = \{_\text{m}\, d_{ii',tt'}\}_{t=1,t'=1}^{q_i \quad q_{i'}}$$

and (11)

$$\operatorname{var}(\mathbf{u}) = \mathbf{D} = \{_\text{m}\, \mathbf{D}_{ii'}\}_{i,i'=1}^{r} .$$

Again, it seems unlikely that every $d_{ii',tt'}$ would be different. One possibility is

$$\mathbf{D}_{ii} = \sigma_i^2[(1 - \rho_{ii})\mathbf{I}_{q_i} + \rho_{ii}\mathbf{J}_{q_i}]$$

and (12)

$$\mathbf{D}_{ii'} = \rho_{ii'}\sigma_i\sigma_{i'}\mathbf{J}_{q_i \times q_{i'}} \quad \text{for } i \neq i',$$

so that

$$\operatorname{var}(u_{it}) = \sigma_i^2 \quad \forall\, t = 1, \ldots, q_i,$$

$$\operatorname{cov}(u_{it}, u_{it'}) = \rho_{ii}\sigma_i^2 \quad \forall\, t \neq t' = 1, \ldots, q_i \tag{13}$$

and

$$\operatorname{cov}(u_{it}, u_{i't'}) = \rho_{ii'}\sigma_i\sigma_{i'} \quad \text{for } t = 1, \ldots, q_i,\ t' = 1, \ldots, q_{i'} \text{ and } i \neq i' .$$

-iii. Covariances between error terms. The usual variance–covariance structure taken for error terms is $\text{var}(\mathbf{e}) = \sigma_e^2\mathbf{I}_N$. This assumes that all error terms have the same variance, σ_e^2, and that covariances between all pairs of (different) error terms are zero. Clearly, though, one could posit any structure suited to the source of one's data, the most general being $\text{var}(\mathbf{e}) = \mathbf{R}$, a symmetric, positive definite matrix. Structures for var(**e**) other than $\sigma_e^2\mathbf{I}_N$ can be modeled in the same manner as for var(**u**) in (9) and (11), or in any manner suited to the situation at hand. Block diagonal **R**, or covariances arising in multi-trait models in genetics, are two such possibilities.

c. Combining variables into a single vector

The examples of Section 11.1a deal with two variables for which the model equations (1) can be written, with $\mathbf{Z}_1 = \{_d\, \mathbf{1}_{n_i}\}_{i=1}^{a}$, as

$$\mathbf{y}_1 = \mu_1\mathbf{1}_N + \mathbf{Z}_1\boldsymbol{\alpha}_1 + \mathbf{e}_1$$

and (14)

$$\mathbf{y}_2 = \mu_2\mathbf{1}_N + \mathbf{Z}_1\boldsymbol{\alpha}_2 + \mathbf{e}_2\ .$$

These can be combined into a single vector

$$\begin{bmatrix}\mathbf{y}_1\\ \mathbf{y}_2\end{bmatrix} = \begin{bmatrix}\mathbf{1}_N & \mathbf{0}\\ \mathbf{0} & \mathbf{1}_N\end{bmatrix}\begin{bmatrix}\mu_1\\ \mu_2\end{bmatrix} + \begin{bmatrix}\mathbf{Z}_1 & \mathbf{0}\\ \mathbf{0} & \mathbf{Z}_1\end{bmatrix}\begin{bmatrix}\boldsymbol{\alpha}_1\\ \boldsymbol{\alpha}_2\end{bmatrix} + \begin{bmatrix}\mathbf{e}_1\\ \mathbf{e}_2\end{bmatrix}$$

which can be written as

$$\mathbf{y} = \mathbf{X}\boldsymbol{\beta} + \mathbf{Z}\mathbf{u} + \mathbf{e}$$

with

$$\mathbf{y} = \begin{bmatrix}\mathbf{y}_1\\ \mathbf{y}_2\end{bmatrix},\quad \mathbf{X} = \begin{bmatrix}\mathbf{1}_N & \mathbf{0}\\ \mathbf{0} & \mathbf{1}_N\end{bmatrix},\quad \boldsymbol{\beta} = \begin{bmatrix}\mu_1\\ \mu_2\end{bmatrix},$$

$$\mathbf{Z} = \begin{bmatrix}\mathbf{Z}_1 & \mathbf{0}\\ \mathbf{0} & \mathbf{Z}_1\end{bmatrix},\quad \mathbf{u} = \begin{bmatrix}\boldsymbol{\alpha}_1\\ \boldsymbol{\alpha}_2\end{bmatrix}\quad \text{and}\quad \mathbf{e} = \begin{bmatrix}\mathbf{e}_1\\ \mathbf{e}_2\end{bmatrix}.$$

Then for the variances and covariances of (2) and (3)

$$\text{var}(\mathbf{u}) = \mathbf{D} = \begin{bmatrix}\sigma_{\alpha_1}^2\mathbf{I}_a & \tau_\alpha\mathbf{I}_a\\ \tau_\alpha\mathbf{I}_a & \sigma_{\alpha_2}^2\mathbf{I}_a\end{bmatrix} = \begin{bmatrix}\sigma_{\alpha_1}^2 & \tau_\alpha\\ \tau_\alpha & \sigma_{\alpha_2}^2\end{bmatrix}\otimes\mathbf{I}_a$$

and (15)

$$\text{var}(\mathbf{e}) = \mathbf{R} = \begin{bmatrix}\sigma_{e_1}^2\mathbf{I}_N & \tau_e\mathbf{I}_N\\ \tau_e\mathbf{I}_N & \sigma_{e_2}^2\mathbf{I}_N\end{bmatrix} = \begin{bmatrix}\sigma_{e_1}^2 & \tau_e\\ \tau_e & \sigma_{e_2}^2\end{bmatrix}\otimes\mathbf{I}_N\ .$$

The advantage of combining data vectors like this is that it puts data into a standard format to which ML and REML can be applied directly, as indicated in subsections e and f that follow. Moreover, the combining of just two vectors into a single vector can be easily extended into combining more than two vectors: and the standard format still applies.

d. Genetic covariances

In modeling biological data there is often interest in genetic relationships that arise from the biology of a situation. In the sheep example used earlier, for instance, we considered data from ewes that were daughters of a small group of sires. And in modeling the fleece weight data by the first equation in (13), the elements of $\boldsymbol{\alpha}_1$ are the sire effects on fleece weight. A matrix that quantifies whatever genetic relationships exist among such sires is called the relationship matrix, and is usually denoted by $\mathbf{A}$. [It can be calculated for any set of animals that are descended from some base population; and its inverse, $\mathbf{A}^{-1}$ (which can be calculated directly from genetic relationships without having to actually invert a matrix), is described in Henderson (1976).] Given that matrix $\mathbf{A}$, then, for the sires having daughters in our study, the variance of $\boldsymbol{\alpha}_1$, instead of being $\sigma^2_{\alpha_1}\mathbf{I}_a$, becomes $\text{var}(\boldsymbol{\alpha}_1) = \sigma^2_{\alpha_1}\mathbf{A}$. Here then, we are introducing covariances that arise from the genetics of the situation and which must be taken into account in estimating the variance component $\sigma^2_{\alpha_1}$.

The use of $\mathbf{A}$ also extends to where there are data $\mathbf{y}_1$ and $\mathbf{y}_2$ on two variables from the same animals, whereupon $\text{var}(\mathbf{u})$ of (15) becomes

$$\text{var}(\mathbf{u}) = \mathbf{D} = \begin{bmatrix} \sigma^2_{\alpha_1}\mathbf{A} & \tau_\alpha\mathbf{A} \\ \tau_\alpha\mathbf{A} & \sigma^2_{\alpha_2}\mathbf{A} \end{bmatrix} = \begin{bmatrix} \sigma^2_{\alpha_1} & \tau_\alpha \\ \tau_\alpha & \sigma^2_{\alpha_2} \end{bmatrix} \otimes \mathbf{A} . \tag{16}$$

And, of course, extension to more than two variables is clear. In this way not only are the covariances between sire effects for different traits taken into account, but so also are the genetic covariances due to relationships of the sires to one another.

e. Maximum likelihood (ML) estimation

-i. Estimation equations. We have seen in the preceding subsections how a variety of different occurrences of non-zero covariances incorporated in a model can all be represented very generally in the model

$$\mathbf{y} = \mathbf{X}\boldsymbol{\beta} + \mathbf{Z}\mathbf{u} + \mathbf{e}, \tag{17}$$

with $\text{cov}(\mathbf{u}, \mathbf{e}') = \mathbf{0}$ and

$$\mathbf{V} = \text{var}(\mathbf{y}) = \mathbf{Z}\mathbf{D}\mathbf{Z}' + \mathbf{R} . \tag{18}$$

We now assume normality. Then, as in (13) of Chapter 6, the log likelihood of $\mathbf{y}$ is

$$l = -\tfrac{1}{2}N \log 2\pi - \tfrac{1}{2}\log|\mathbf{V}| - \tfrac{1}{2}(\mathbf{y} - \mathbf{X}\boldsymbol{\beta})'\mathbf{V}^{-1}(\mathbf{y} - \mathbf{X}\boldsymbol{\beta}) . \tag{19}$$

Differentiating this with respect to $\boldsymbol{\beta}$ gives

$$2l_{\boldsymbol{\beta}} = \mathbf{X}'\mathbf{V}^{-1}\mathbf{y} - \mathbf{X}'\mathbf{V}^{-1}\mathbf{X}\boldsymbol{\beta}, \tag{20}$$

and equating this to $\mathbf{0}$ gives, with $\tilde{\mathbf{V}}$ and $\tilde{\boldsymbol{\beta}}$ representing the ML estimators of $\mathbf{V}$ and $\boldsymbol{\beta}$, respectively,

$$\mathbf{X}'\tilde{\mathbf{V}}^{-1}\mathbf{X}\tilde{\boldsymbol{\beta}} = \mathbf{X}'\tilde{\mathbf{V}}^{-1}\mathbf{y}\,. \tag{21}$$

Hence the ML estimator of the estimable vector $\mathbf{X}\boldsymbol{\beta}$ is

$$\mathbf{X}\tilde{\boldsymbol{\beta}} = \mathbf{X}(\mathbf{X}'\tilde{\mathbf{V}}^{-1}\mathbf{X})^{-}\mathbf{X}'\tilde{\mathbf{V}}^{-1}\mathbf{y}\,. \tag{22}$$

All this is similar to Section 6.2a.

But for estimating the variance and covariance components that make up the elements of $\mathbf{V}$ we must now be more general than in Chapter 6, wherein we took $\mathbf{V} = \Sigma_{i=1}^{r} \mathbf{Z}_i\mathbf{Z}_i'\sigma_i^2$ as in equation (10) of that chapter. That meant differentiating l of (19) with respect to just the σ_i^2s, which led in turn to equations (21) and (25) of Chapter 6. Now we need to cover a variety of forms for $\mathbf{V}$. Yet all of them when $\mathbf{u}$ and $\mathbf{e}$ are taken as having zero covariance are of the form $\mathbf{V} = \mathbf{ZDZ}' + \mathbf{R}$. And from Appendices M.7e and f we have the general results

$$\frac{\partial \mathbf{V}^{-1}}{\partial \theta} = -\mathbf{V}^{-1}\frac{\partial \mathbf{V}}{\partial \theta}\mathbf{V}^{-1} \quad \text{and} \quad \frac{\partial}{\partial \theta}\log|\mathbf{V}| = \mathrm{tr}\left(\mathbf{V}^{-1}\frac{\partial \mathbf{V}}{\partial \theta}\right) \tag{23}$$

where elements of $\mathbf{V}$ are considered as functions of θ.

Using those results, we array the variance and covariance components that occur in $\mathbf{V}$ as a vector $\boldsymbol{\theta} = \{_c\, \theta_h\}_{h=1}^{v}$, where v represents the total number of different components. For instance, in the sheep example $\boldsymbol{\theta}' = [\sigma_{\alpha_1}^2 \quad \sigma_{\alpha_2}^2 \quad \tau_\alpha \quad \sigma_{e_1}^2 \quad \sigma_{e_2}^2 \quad \tau_e]$ and $v = 6$. Then

$$l_{\theta_h} = \frac{\partial l}{\partial \theta_h} = -\tfrac{1}{2}\,\mathrm{tr}\left(\mathbf{V}^{-1}\frac{\partial \mathbf{V}}{\partial \theta_h}\right) + \tfrac{1}{2}(\mathbf{y} - \mathbf{X}\boldsymbol{\beta})'\mathbf{V}^{-1}\frac{\partial \mathbf{V}}{\partial \theta_h}\mathbf{V}^{-1}(\mathbf{y} - \mathbf{X}\boldsymbol{\beta}), \tag{24}$$

and equating this to zero gives

$$\mathrm{tr}\left[\tilde{\mathbf{V}}^{-1}\left(\left.\frac{\partial \mathbf{V}}{\partial \theta_h}\right|_{\boldsymbol{\theta}=\tilde{\boldsymbol{\theta}}}\right)\right] = (\mathbf{y} - \mathbf{X}\tilde{\boldsymbol{\beta}})'\tilde{\mathbf{V}}^{-1}\left(\left.\frac{\partial \mathbf{V}}{\partial \theta_h}\right|_{\boldsymbol{\theta}=\tilde{\boldsymbol{\theta}}}\right)\tilde{\mathbf{V}}^{-1}(\mathbf{y} - \mathbf{X}\tilde{\boldsymbol{\beta}}), \tag{25}$$

where

$$\left.\frac{\partial \mathbf{V}}{\partial \theta_h}\right|_{\boldsymbol{\theta}=\tilde{\boldsymbol{\theta}}} \quad \text{is} \quad \frac{\partial \mathbf{V}}{\partial \theta_h} \text{ written with } \tilde{\boldsymbol{\theta}} \text{ in place of } \boldsymbol{\theta}\,.$$

With $\mathbf{X}\tilde{\boldsymbol{\beta}}$ from (22) it is clear on defining

$$\mathbf{P} = \mathbf{V}^{-1} - \mathbf{V}^{-1}\mathbf{X}(\mathbf{X}'\mathbf{V}^{-1}\mathbf{X})^{-}\mathbf{X}'\mathbf{V}^{-1}, \tag{26}$$

as in preceding chapters, that $\tilde{\mathbf{V}}^{-1}(\mathbf{y} - \mathbf{X}\tilde{\boldsymbol{\beta}}) = \tilde{\mathbf{P}}\mathbf{y}$, and so we get the ML estimation equation as

$$\text{tr}\left[\tilde{\mathbf{V}}^{-1}\left(\frac{\partial \mathbf{V}}{\partial \theta_h}\bigg|_{\boldsymbol{\theta}=\tilde{\boldsymbol{\theta}}}\right)\right] = \mathbf{y}'\tilde{\mathbf{P}}\left(\frac{\partial \mathbf{V}}{\partial \theta_h}\bigg|_{\boldsymbol{\theta}=\tilde{\boldsymbol{\theta}}}\right)\tilde{\mathbf{P}}\mathbf{y}, \quad \text{for } h = 1, \ldots, v\,. \tag{27}$$

To further consider the derivative term, which occurs on both sides of this equation, let us now distinguish θ_d and θ_r as elements of θ that occur, respectively, in var$(\mathbf{u}) = \mathbf{D}$ and var$(\mathbf{e}) = \mathbf{R}$, whatever the forms of $\mathbf{D}$ and $\mathbf{R}$ may be. Then

$$\frac{\partial \mathbf{V}}{\partial \theta_d} = \mathbf{Z}\frac{\partial \mathbf{D}}{\partial \theta_d}\mathbf{Z}' \quad \text{and} \quad \frac{\partial \mathbf{V}}{\partial \theta_r} = \frac{\partial \mathbf{R}}{\partial \theta_r}\,. \tag{28}$$

Hence the ML equations (27) become

$$\text{tr}\left[\mathbf{Z}'\tilde{\mathbf{V}}^{-1}\mathbf{Z}\left(\frac{\partial \mathbf{D}}{\partial \theta_d}\bigg|_{\boldsymbol{\theta}=\tilde{\boldsymbol{\theta}}}\right)\right] = \mathbf{y}'\tilde{\mathbf{P}}\mathbf{Z}\left(\frac{\partial \mathbf{D}}{\partial \theta_d}\bigg|_{\boldsymbol{\theta}=\tilde{\boldsymbol{\theta}}}\right)\mathbf{Z}'\tilde{\mathbf{P}}\mathbf{y}, \tag{29}$$

for each parameter θ_d of $\mathbf{D}$, and

$$\text{tr}\left[\tilde{\mathbf{V}}^{-1}\left(\frac{\partial \mathbf{R}}{\partial \theta_r}\bigg|_{\boldsymbol{\theta}=\tilde{\boldsymbol{\theta}}}\right)\right] = \mathbf{y}'\tilde{\mathbf{P}}\left(\frac{\partial \mathbf{R}}{\partial \theta_r}\bigg|_{\boldsymbol{\theta}=\tilde{\boldsymbol{\theta}}}\right)\tilde{\mathbf{P}}\mathbf{y}, \tag{30}$$

for each parameter θ_r of $\mathbf{R}$.

At this point there appears to be no further tractable, algebraic, simplification of the general case. One now has to make use of the precise forms that $\mathbf{D}$ and $\mathbf{R}$ have for the task at hand, in order to know what the different elements θ_d and θ_r are, and where they occur in $\mathbf{D}$ and $\mathbf{R}$. Then the derivatives in (29) and (30) can be specified. For example, with $\mathbf{D}$ of (16)

$$\frac{\partial \mathbf{D}}{\partial \sigma^2_{\alpha_1}} = \frac{\partial}{\partial \sigma^2_{\alpha_1}}\begin{bmatrix} \sigma^2_{\alpha_1} & \tau_\alpha \\ \tau_\alpha & \sigma^2_{\alpha_2} \end{bmatrix} \otimes \mathbf{A} = \begin{bmatrix} \mathbf{A} & \mathbf{0} \\ \mathbf{0} & \mathbf{0} \end{bmatrix}.$$

Since there is a multitude of forms that $\mathbf{D}$ and $\mathbf{R}$ can have, this further simplification is left to the reader for whatever $\mathbf{D}$ and $\mathbf{R}$ are being used. Rather than algebraic simplification it may be possible to use computing packages that handle elementary differential calculus, in combination with those that carry out ML calculations.

-ii. Large-sample dispersion matrix. As in Chapter 6, the large sample dispersion matrix of the ML estimators is the inverse of the information matrix. Here, as there, in (39) of Chapter 6,

$$\text{var}(\tilde{\boldsymbol{\beta}}) \simeq (\mathbf{X}'\mathbf{V}^{-1}\mathbf{X})^{-1}\,.$$

And

$$\text{var}(\tilde{\boldsymbol{\sigma}}^2) \simeq \left[-E\left\{\frac{\partial^2 l}{\partial \theta_h\, \partial \theta_k}\right\}_{h,k=1}^{v}\right]^{-1}.$$

Denote $\partial \mathbf{V}/\partial\theta_h$ by $\mathbf{V}_h$ and $\partial^2\mathbf{V}/\partial\theta_h\partial\theta_k$ by $\mathbf{V}_{hk}$. Then from (24)

$$\frac{\partial l}{\partial\theta_h} = -\tfrac{1}{2}\,\mathrm{tr}(\mathbf{V}^{-1}\mathbf{V}_h) + \tfrac{1}{2}(\mathbf{y}-\mathbf{X}\boldsymbol{\beta})'\mathbf{V}^{-1}\mathbf{V}_h\mathbf{V}^{-1}(\mathbf{y}-\mathbf{X}\boldsymbol{\beta})$$

$$= \tfrac{1}{2}\,\mathrm{tr}[-\mathbf{V}^{-1}\mathbf{V}_h + \mathbf{V}^{-1}\mathbf{V}_h\mathbf{V}^{-1}(\mathbf{y}-\mathbf{X}\boldsymbol{\beta})(\mathbf{y}-\mathbf{X}\boldsymbol{\beta})']$$

and so, on using $E[(\mathbf{y}-\mathbf{X}\boldsymbol{\beta})(\mathbf{y}-\mathbf{X}\boldsymbol{\beta})'] = \mathbf{V}$,

$$-E\left(\frac{\partial^2 l}{\partial\theta_h\,\partial\theta_k}\right) = -\tfrac{1}{2}\,\mathrm{tr}[\mathbf{V}^{-1}\mathbf{V}_k\mathbf{V}^{-1}\mathbf{V}_h - \mathbf{V}^{-1}\mathbf{V}_{hk}$$

$$+(-\mathbf{V}^{-1}\mathbf{V}_k\mathbf{V}^{-1}\mathbf{V}_h\mathbf{V}^{-1} + \mathbf{V}^{-1}\mathbf{V}_{hk}\mathbf{V}^{-1} - \mathbf{V}^{-1}\mathbf{V}_h\mathbf{V}^{-1}\mathbf{V}_k\mathbf{V}^{-1})\mathbf{V}]$$

$$= \tfrac{1}{2}\,\mathrm{tr}(\mathbf{V}^{-1}\mathbf{V}_h\mathbf{V}^{-1}\mathbf{V}_k)\,.$$

Hence

$$\mathrm{var}(\tilde{\boldsymbol{\sigma}}^2) \simeq 2\left[\left\{_{\mathrm{m}}\;\mathrm{tr}\left(\mathbf{V}^{-1}\frac{\partial\mathbf{V}}{\partial\theta_h}\mathbf{V}^{-1}\frac{\partial\mathbf{V}}{\partial\theta_k}\right)\right\}_{h,k=1}^{v}\right]^{-1}. \tag{31}$$

Of the v variance and covariance parameters in $\mathbf{V}$, suppose v_d of them are in $\mathbf{D}$ and $v - v_d = v_r$ of them are in $\mathbf{R}$. Order elements of $\boldsymbol{\theta}$ so that $\boldsymbol{\theta}' = [\boldsymbol{\theta}_d' \;\; \boldsymbol{\theta}_r']$, with the v_d parameters pertaining to $\mathbf{D}$ being in $\boldsymbol{\theta}_d$ and the v_r of $\mathbf{R}$ being in $\boldsymbol{\theta}_r$. Let θ_{di} and θ_{rj} be the ith and jth elements of $\boldsymbol{\theta}_d$ and of $\boldsymbol{\theta}_r$, respectively. Then (31) is

$$\mathrm{var}(\tilde{\boldsymbol{\sigma}}^2) \simeq 2\begin{bmatrix} \left\{_{\mathrm{m}}\;\mathrm{tr}\left(\mathbf{V}^{-1}\mathbf{Z}\dfrac{\partial\mathbf{D}}{\partial\theta_{di}}\mathbf{Z}'\mathbf{V}^{-1}\mathbf{Z}\dfrac{\partial\mathbf{D}}{\partial\theta_{di'}}\mathbf{Z}'\right)\right\}_{i,i'=1}^{v_d} & \left\{_{\mathrm{m}}\;\mathrm{tr}\left(\mathbf{V}^{-1}\mathbf{Z}\dfrac{\partial\mathbf{D}}{\partial\theta_{di}}\mathbf{Z}'\mathbf{V}^{-1}\dfrac{\partial\mathbf{R}}{\partial\theta_{rj}}\right)\right\}_{i=1,j=1}^{v_d\;\;v_r} \\ \left\{_{\mathrm{m}}\;\mathrm{tr}\left(\mathbf{V}^{-1}\dfrac{\partial\mathbf{R}}{\partial\theta_{rj}}\mathbf{V}^{-1}\mathbf{Z}\dfrac{\partial\mathbf{D}}{\partial\theta_{di}}\mathbf{Z}'\right)\right\}_{j=1,i=1}^{v_r\;\;v_d} & \left\{_{\mathrm{m}}\;\mathrm{tr}\left(\mathbf{V}^{-1}\dfrac{\partial\mathbf{R}}{\partial\theta_{rj}}\mathbf{V}^{-1}\dfrac{\partial\mathbf{R}}{\partial\theta_{rj'}}\right)\right\}_{j,j'=1}^{v_r} \end{bmatrix}^{-1}$$

Depending on the form of $\mathbf{Z}$ and on whatever structure or pattern there is in $\mathbf{D}$ and $\mathbf{R}$, the trace terms in these matrices may simplify, and the arithmetic will also be aided by the standard results $\mathrm{tr}(\mathbf{X}'\mathbf{X}) = \Sigma_i\Sigma_j x_{ij}^2 = \mathrm{sesq}(\mathbf{X})$ and $\mathrm{tr}(\mathbf{X}'\mathbf{Y}) = \Sigma_i\Sigma_j x_{ij}y_{ij}$. Since specific details do depend so much on the exact form of $\mathbf{Z}$, $\mathbf{D}$ and $\mathbf{R}$, there is little or no merit in attempting any further simplification of these formulae for the general case.

f. Restricted maximum likelihood (REML) estimation

-i. Estimation equations. We make the transition from ML to REML by the same replacements as made in Section 6.7b, namely replace

$$\mathbf{y} \text{ by } \mathbf{K}'\mathbf{y}, \quad \mathbf{Z} \text{ by } \mathbf{K}'\mathbf{Z},$$

$$\mathbf{X} \text{ by } \mathbf{K}'\mathbf{X} = \mathbf{0}, \quad \mathbf{V} \text{ by } \mathbf{K}'\mathbf{V}\mathbf{K} = \mathbf{K}'\mathbf{Z}\mathbf{D}\mathbf{Z}'\mathbf{K} + \mathbf{K}'\mathbf{R}\mathbf{K}$$

and in doing so recall that $\mathbf{P}$ gets replaced by $(\mathbf{K}'\mathbf{V}\mathbf{K})^{-1}$ but that $\tilde{\mathbf{P}} = \mathbf{K}(\mathbf{K}'\tilde{\mathbf{V}}\mathbf{K})^{-1}\mathbf{K}'$. Then the ML equations (29) and (30) become

$$\mathrm{tr}\left[\mathbf{Z}'\mathbf{K}(\mathbf{K}'\tilde{\mathbf{V}}\mathbf{K})^{-1}\mathbf{K}'\mathbf{Z}\left(\frac{\partial\mathbf{D}}{\partial\theta_d}\bigg|_{\boldsymbol{\theta}=\tilde{\boldsymbol{\theta}}}\right)\right] = \mathbf{y}'\mathbf{K}(\mathbf{K}'\tilde{\mathbf{V}}\mathbf{K})^{-1}\mathbf{K}'\mathbf{Z}\left(\frac{\partial\mathbf{D}}{\partial\theta_d}\bigg|_{\boldsymbol{\theta}=\tilde{\boldsymbol{\theta}}}\right)\mathbf{Z}'\mathbf{K}(\mathbf{K}'\tilde{\mathbf{V}}\mathbf{K})^{-1}\mathbf{K}'\mathbf{y}$$

and

$$\operatorname{tr}\left[(\mathbf{K}'\tilde{\mathbf{V}}\mathbf{K})^{-1}\mathbf{K}'\left(\left.\frac{\partial \mathbf{R}}{\partial \theta_r}\right|_{\boldsymbol{\theta}=\tilde{\boldsymbol{\theta}}}\right)\mathbf{K}\right] = \mathbf{y}'\mathbf{K}(\mathbf{K}'\tilde{\mathbf{V}}\mathbf{K})^{-1}\mathbf{K}'\left(\left.\frac{\partial \mathbf{R}}{\partial \theta_r}\right|_{\boldsymbol{\theta}=\tilde{\boldsymbol{\theta}}}\right)\mathbf{K}(\mathbf{K}'\tilde{\mathbf{V}}\mathbf{K})^{-1}\mathbf{K}'\mathbf{y},$$

which reduce, for each parameter θ_d in $\mathbf{D}$, to

$$\operatorname{tr}\left[\mathbf{Z}'\tilde{\mathbf{P}}\mathbf{Z}\left(\left.\frac{\partial \mathbf{D}}{\partial \theta_d}\right|_{\boldsymbol{\theta}=\tilde{\boldsymbol{\theta}}}\right)\right] = \mathbf{y}'\tilde{\mathbf{P}}\mathbf{Z}\left(\left.\frac{\partial \mathbf{D}}{\partial \theta_d}\right|_{\boldsymbol{\theta}=\tilde{\boldsymbol{\theta}}}\right)\mathbf{Z}\tilde{\mathbf{P}}\mathbf{y}$$

and for each parameter θ_r in $\mathbf{R}$ to

$$\operatorname{tr}\left[\tilde{\mathbf{P}}\left(\left.\frac{\partial \mathbf{R}}{\partial \theta_r}\right|_{\boldsymbol{\theta}=\tilde{\boldsymbol{\theta}}}\right)\right] = \mathbf{y}'\tilde{\mathbf{P}}\left(\left.\frac{\partial \mathbf{R}}{\partial \theta_r}\right|_{\boldsymbol{\theta}=\tilde{\boldsymbol{\theta}}}\right)\tilde{\mathbf{P}}\mathbf{y}\,.$$

-ii. Large-sample dispersion matrix. In the sampling dispersion matrix for ML making the same replacements as in subsection i above gives

$$\operatorname{var}(\tilde{\boldsymbol{\sigma}}^2) \simeq 2\begin{bmatrix} \left\{_m \operatorname{tr}\left(\mathbf{PZ}\dfrac{\partial \mathbf{D}}{\partial \theta_{di}}\mathbf{Z}'\mathbf{PZ}\dfrac{\partial \mathbf{D}}{\partial \theta_{di'}}\mathbf{Z}'\right)\right\}_{i,i'=1}^{v_d} & \left\{_m \operatorname{tr}\left(\mathbf{PZ}\dfrac{\partial \mathbf{D}}{\partial \theta_{di}}\mathbf{Z}'\mathbf{P}\dfrac{\partial \mathbf{R}}{\partial \theta_{rj}}\right)\right\}_{i=1,j=1}^{v_d\ \ v_r} \\ \left\{_m \operatorname{tr}\left(\mathbf{P}\dfrac{\partial \mathbf{R}}{\partial \theta_{rj}}\mathbf{PZ}\dfrac{\partial \mathbf{D}}{\partial \theta_{di}}\mathbf{Z}'\right)\right\}_{r=1,d=1}^{v_r\ \ v_d} & \left\{_m \operatorname{tr}\left(\mathbf{P}\dfrac{\partial \mathbf{R}}{\partial \theta_{rj}}\mathbf{P}\dfrac{\partial \mathbf{R}}{\partial \theta_{rj'}}\right)\right\}_{j,j'=1}^{v_r} \end{bmatrix}^{-1}$$

Again, there is no merit in attempting further simplification of formulae such as these.

11.2. MODELING VARIANCE COMPONENTS AS COVARIANCES

The problem of sometimes getting negative values for estimated variance components has been seen to arise in the ANOVA methods of estimation. And even with ML and REML solutions this negativity can be a problem when it occurs in the midst of an iterative procedure, for then the calculated $\mathbf{D} = \{_d\ \sigma_i^2 \mathbf{I}_{q_i}\}$ will not be positive definite, and is singular if negative values for any σ^2s are replaced by zero. Nevertheless, because a variance component can also be interpreted as a covariance, negativity in that context is not necessarily out of place. For example, in the 1-way classification random model with model equation $y_{ij} = \mu + \alpha_i + e_{ij}$ the covariance between y_{ij} and $y_{ij'}$ for $j \neq j'$ is σ_α^2:

$$\operatorname{cov}(y_{ij}, y_{ij'}) = \operatorname{cov}(\mu + \alpha_i + e_{ij}, \mu + \alpha_i + e_{ij'}) = \operatorname{cov}(\alpha_i, \alpha_i) = \sigma_\alpha^2\,.$$

Since covariances can be negative, and because there are situations in which a covariance of the form $\operatorname{cov}(y_{ij}, y_{ij'})$ might truly be negative, such a situation would seem to throw doubt on the utility of a model that leads to a covariance, which can be negative, being identical to a variance, which cannot be negative. So maybe developing a model that circumvents this possibility is what is needed. This is what Green (1988) discusses in the framework of clinical trial data wherein the covariance between observations on the same person receiving different doses of the same drug could be negative if large doses of the drug were such as to produce adverse effects.

There have therefore been several papers recently (e.g., Smith and Murray, 1984; Green, 1988; Hocking, Green and Bremer, 1989) describing models for variance components and their estimation in terms of a covariance structure. The estimation method employed is essentially an ANOVA method and, indeed, for balanced data it is identical to ANOVA methodology. For balanced data and for all-cells-filled unbalanced data it provides excellent diagnostic opportunities for assessing the different covariance contributions to a variance component estimate, which is especially useful when that estimate is negative. These diagnostics are demonstrated by Hocking *et al.* (1989).

a. All-cells-filled data

We briefly illustrate the modeling and estimation method in terms of the 2-way crossed classification, random model, drawing heavily on Hocking (1985), Green (1988) and Hocking *et al.* (1989) to do so.

As usual (e.g., Chapters 4 and 5), the traditional random model equation for y_{ijk}, the kth observation in the ith row and jth column, is taken as

$$y_{ijk} = \mu + \alpha_i + \beta_j + \gamma_{ij} + e_{ijk}, \tag{32}$$

where $i = 1, \ldots, a, j = 1, \ldots, b$ and $k = 1, \ldots, n_{ij}$. All effects (except μ) are random, with zero means, zero covariances and variances $\sigma_\alpha^2, \sigma_\beta^2, \sigma_\gamma^2$ and σ_e^2, respectively. In contrast, the model used by Hocking *et al.* has no model equation but "is given simply by describing the mean and covariance structure implicit in" (32) in the following manner:

$$\begin{aligned} E(y_{ijk}) &= \mu, \\ \text{var}(y_{ijk}) &= \phi_0 + \phi_1 + \phi_2 + \phi_{12} \\ \text{cov}(y_{ijk}, y_{i'j'k'}) &= \begin{cases} \phi_1 & \text{for } i = i' \text{ and } j \neq j', \\ \phi_2 & \text{for } i \neq i' \text{ and } j = j', \\ \phi_1 + \phi_2 + \phi_{12} & \text{for } i = i', j = j' \text{ and } k \neq k' . \end{cases} \end{aligned} \tag{33}$$

As is easily seen "the two forms of model are mathematically equivalent" (Hocking *et al.* 1989, p. 228) in the sense of there being a one-to-one correspondence of the variance components of (32) to the covariances in (33), namely

$$\begin{aligned} \sigma_\alpha^2 &= \phi_1, \quad \sigma_\beta^2 = \phi_2, \\ \sigma_\gamma^2 &= \phi_{12}, \quad \sigma_e^2 = \phi_0 . \end{aligned} \tag{34}$$

But the two models are not statistically equivalent, because their parameter spaces are not the same. Variances cannot be negative, whereas covariances can, although the negativity of the covariances in (33) is restricted by needing the dispersion matrix of $\mathbf{y}$ to be positive definite, e.g., $\phi_0 + \phi_1 + \phi_2 + \phi_{12} > 0$.

Now observe that since

$$\operatorname{cov}(y_{ijk}, y_{ij'k'}) = \phi_1$$

for each i, and for every j, j' pair with $j \neq j'$, so also does $\operatorname{cov}(\bar{y}_{ij\cdot}, \bar{y}_{ij'\cdot}) = \phi_1$. And an estimate of this for the j, j' pair is $\Sigma_{i=1}^{a}(\bar{y}_{ij\cdot} - \tilde{y}_{\cdot j\cdot})(\bar{y}_{ij'\cdot} - \tilde{y}_{\cdot j'\cdot})/(a-1)$, where

$$\tilde{y}_{\cdot j\cdot} = \frac{1}{a}\sum_{i=1}^{a} \bar{y}_{ij\cdot} \; . \tag{35}$$

Therefore, on using all $\frac{1}{2}b(b-1)$ pairs j, j' for $j < j'$, the estimate of ϕ_1 is taken as

$$\hat{\phi}_1 = \frac{\sum\sum_{j<j'} \Sigma_i(\bar{y}_{ij\cdot} - \tilde{y}_{\cdot j\cdot})(\bar{y}_{ij'\cdot} - \tilde{y}_{\cdot j'\cdot})}{\frac{1}{2}(a-1)b(b-1)} \; . \tag{36}$$

It is to be emphasized that (36) applies only for all-cells-filled data. It is identical to $\hat{\sigma}_\alpha^2$ obtained from Yates' (1934) unweighted means analysis of (145) in Chapter 5. And it is unbiased for $\phi_1 = \sigma_\alpha^2$ (see E 11.3). Estimation of ϕ_2 is analogous to (36).

b. Balanced data

For balanced data (35) reduces to $\bar{y}_{\cdot j\cdot}$:

$$\tilde{y}_{\cdot j\cdot} = \bar{y}_{\cdot j\cdot} = \frac{1}{an}\sum_i\sum_k y_{ijk} \quad \text{for } n_{ij} = n \quad \forall\; i \text{ and } j \, . \tag{37}$$

Thus the mean of the cell means, $\tilde{y}_{\cdot j\cdot}$, of (35) reduces for balanced data to the regular column mean $\bar{y}_{\cdot j\cdot}$, as in (37). And then (36) reduces to the familiar ANOVA estimator

$$\hat{\phi}_1 = \hat{\sigma}_\alpha^2 = (\text{MSA} - \text{MSAB})/bn \tag{38}$$

of (26) in Chapter 4, derived from Table 4.5.

c. Diagnostic opportunities

The really interesting feature of $\hat{\phi}_1$ of (36) is that it demonstrates for all-cells-filled data that the ANOVA estimator of σ_α^2 for balanced data and the unweighted-means-analysis estimator for unbalanced data can each be expressed as a simple average of estimated covariances. And from this one can look at the individual estimated covariances that go into that average, and scrutinize them for any underlying patterns. Hocking *et al.* (1989) give an example of doing this, using (36) and noting that for having four levels of the column factor in their data, the values of the covariance estimate $\Sigma_i(\bar{y}_{ij\cdot} - \tilde{y}_{\cdot j\cdot})(\bar{y}_{ij'\cdot} - \tilde{y}_{\cdot j'\cdot})/(a-1)$ are as shown in Table 11.1. Although $\hat{\sigma}_\alpha^2 = \hat{\phi}_1 = (34.1 + 35.3 + 37.0 + 13.1 + 13.0 + 13.3)/6 = 24.3$, it is clear from Table 11.1 that the estimated covariances involving column 1 are all considerably larger than those among columns 2, 3 and 4. The diagnostic value of (36) is self-evident.

TABLE 11.1. VALUES OF THE ESTIMATED COVARIANCE $\frac{1}{3}\Sigma_i(\bar{y}_{ij\cdot} - \tilde{y}_{\cdot i\cdot})(\bar{y}_{ij'\cdot} - \tilde{y}_{\cdot j'\cdot})$

	$j' = 2$	$j' = 3$	$j' = 4$
$j = 1$	34.1	35.3	37.0
$j = 2$		13.1	13.0
$j = 3$			13.3

Source: Hocking *et al.* (1989, Table 3).

This, then, is the underlying idea of what Hocking *et al.* call their "Ave" method of estimation. That paper contains much more detail than is given here, including a weighted version of the Ave estimator, and presentation and discussion of efficiency values for the case of 115 observations in a 4-by-4 layout with 11 cells each having 10 observations and the other 5 cells each having one. This is considered for 14 different combinations of the values 0, 0.1, 0.5, 1 and 2 for σ_α^2 and σ_γ^2. They summarize by saying their "limited numerical evidence... suggests that in many cases these estimators are very efficient". Efficiencies are compared with those of a Henderson Method III estimator and of estimation from the weighted squares of means as in (147) of Chapter 5. No comparison is made with ML or REML estimation.

d. Some-cells-empty data

Green (1988) extends the estimation procedure typified by (36) to the case of some-cells-empty data. For each pair j, j' let $m_{jj'}$ be the number of rows in which the cells in columns j and j' have observations, i.e., in which both $n_{ij} > 0$ and $n_{ij'} > 0$, for $i = 1, \ldots, a$. Thus $m_{jj'} = a - \Sigma_{i=1}^{a} \delta_{n_{ij}n_{ij'},0}$ for δ being the Kronecker delta. Then define $\mathscr{M}_{jj'}$ as the set of indices i for those $m_{jj'}$ rows for which $n_{ij} > 0$ and $n_{ij'} > 0$, and also define

$$\tilde{y}_{\cdot j(j')\cdot} = \sum_{i \in \mathscr{M}_{jj'}} \bar{y}_{ij\cdot}/m_{jj'} \quad \text{and} \quad \tilde{y}_{\cdot j'(j)\cdot} = \sum_{i \in \mathscr{M}_{jj'}} \bar{y}_{ij'\cdot}/m_{jj'} .$$

Then ϕ_1 is estimated as

$$\tilde{\phi}_1 = \frac{1}{\frac{1}{2}b(b-1)} \sum_{j<j'}\sum \frac{\sum_{i \in \mathscr{M}_{jj'}} [\bar{y}_{ij\cdot} - \tilde{y}_{\cdot j(j')\cdot}][\bar{y}_{ij'\cdot} - \tilde{y}_{\cdot j'(j)\cdot}]}{m_{jj'} - 1} .$$

Example. Consider Table 11.2 as a set of observed cell means, with four empty cells. For Table 11.2 the values of $m_{jj'}$ and $\tilde{y}_{\cdot j(j')\cdot}$ are as follows:

$$m_{12} = 2, \quad \tilde{y}_{\cdot 1(2)\cdot} = (12 + 16)/2 = 14, \qquad \tilde{y}_{\cdot 2(1)\cdot} = (14 + 20)/2 = 17,$$

$$m_{13} = 3, \quad \tilde{y}_{\cdot 1(3)\cdot} = (12 + 21 + 24)/3 = 19, \quad \tilde{y}_{\cdot 3(1)\cdot} = (15 + 23 + 25)/3 = 21,$$

$$m_{23} = 2, \quad \tilde{y}_{\cdot 2(3)\cdot} = (14 + 16)/2 = 15, \qquad \tilde{y}_{\cdot 3(2)\cdot} = (15 + 21)/2 = 18 .$$

TABLE 11.2. OBSERVED CELL MEANS

	$j = 1$	$j = 2$	$j = 3$
$i = 1$	12	14	15
$i = 2$	—	16	21
$i = 3$	16	20	—
$i = 4$	21	—	23
$i = 5$	24	—	25

Using these and the means in the table in the equation for $\tilde{\phi}_1$ gives

$$\begin{aligned}\tfrac{1}{2}3(2)\tilde{\phi}_1 &= [(12-14)(14-17)+(16-14)(20-17)]/(2-1)\\ &\quad + [(12-19)(15-21)+(21-19)(23-21)+(24-19)(25-21)]/(3-1)\\ &\quad + [(14-15)(15-18)+(16-15)(21-18)]/(2-1) = 51,\end{aligned}$$

which leads to $\tilde{\phi}_1 = 17$.

A weakness of this method of estimation for data having many empty cells is that, depending on the pattern of empty cells throughout the data layout, it is possible for much of the data to be unused in the estimation process. For example, suppose the check marks in Table 11.3 represent cells that contain data. Then all values of $m_{jj'}$ are 0 or 1 and no estimation of ϕ_1 is possible; and the same is true for $m_{ii'}$ and the estimation of ϕ_2.

TABLE 11.3. CELLS CONTAINING DATA

	$j = 1$	$j = 2$	$j = 3$	$j = 4$
$i = 1$	√	√		
$i = 2$		√	√	
$i = 3$	√			√
$i = 4$	√		√	

11.3. CRITERIA-BASED PROCEDURES

ANOVA estimation originated from the empiricism of equating mean squares to their expected values. Although that implicitly yielded unbiasedness, there was no specification of desired criteria for estimating variance components with the object of developing estimators that satisfy those criteria. Certainly, minimum variance properties were established for ANOVA estimators from balanced data—but only long after such estimators were first suggested. And ML and REML estimators get their attractive properties such as consistency, efficiency and asymptotic normality through being the outcome of the maximum likelihood method, and not by specifying those properties as criteria at the outset and then developing estimators to satisfy those criteria. This is the methodology that is described now. It leads to a variety of procedures with acronymic names such as MINQUE, MIVQUE, MIMSQUE, MIVQUE(0) and I-MINQUE, the underlying philosophy being to derive estimators that have minimum norm (or variance, or mean square) and which are quadratic functions of the data and are unbiased. We begin with specifying these criteria.

a. Three criteria

A generalization of estimating a single variance component is estimating a linear combination of components, $\mathbf{p}'\boldsymbol{\sigma}^2$, where $\mathbf{p}'$ represents any known vector. Since a variance is a second moment, it seems natural to estimate it by a quadratic form (which is a homogeneous second-order function) of data. We therefore consider estimating $\mathbf{p}'\boldsymbol{\sigma}^2$ by $\mathbf{y}'\mathbf{A}\mathbf{y}$ for symmetric $\mathbf{A}$, but with $\mathbf{A}$ to be determined by whatever criteria we wish to impose on $\mathbf{y}'\mathbf{A}\mathbf{y}$ as an estimator of $\mathbf{p}'\boldsymbol{\sigma}^2$. With the model equation

$$\mathbf{y} = \mathbf{X}\boldsymbol{\beta} + \sum_{i=0}^{r} \mathbf{Z}_i\mathbf{u}_i \quad \text{and} \quad \mathbf{V} = \text{var}(\mathbf{y}) = \sum_{i=0}^{r} \mathbf{Z}_i\mathbf{Z}_i'\sigma_i^2, \tag{39}$$

three criteria come to mind.

-i. Unbiasedness. Since $E(\mathbf{y}'\mathbf{A}\mathbf{y}) = \text{tr}(\mathbf{AV}) + \boldsymbol{\beta}'\mathbf{X}'\mathbf{A}\mathbf{X}\boldsymbol{\beta}$ unbiasedness demands

$$\mathbf{p}'\boldsymbol{\sigma}^2 = \Sigma_i\, \text{tr}(\mathbf{A}\mathbf{Z}_i\mathbf{Z}_i')\sigma_i^2 + \boldsymbol{\beta}'\mathbf{X}'\mathbf{A}\mathbf{X}\boldsymbol{\beta}\,. \tag{40}$$

Requiring (40) to be true for all σ_i^2 and for all $\boldsymbol{\beta}$ leads to

$$p_i = \text{tr}(\mathbf{A}\mathbf{Z}_i\mathbf{Z}_i') \quad \text{and} \quad \mathbf{X}'\mathbf{A}\mathbf{X} = \mathbf{0}\,. \tag{41}$$

It is tempting to think that $\mathbf{X}'\mathbf{A}\mathbf{X} = \mathbf{0}$ of (41) leads to $\mathbf{AX} = \mathbf{0}$, but this is not necessarily so. It is true that $\mathbf{A}' = \mathbf{A}$ implies $\mathbf{A} = \mathbf{L}'\mathbf{L}$ for some $\mathbf{L}'$ of full column rank and then $\mathbf{X}'\mathbf{A}\mathbf{X} = \mathbf{X}'\mathbf{L}'\mathbf{L}\mathbf{X}$. Therefore $\mathbf{X}'\mathbf{A}\mathbf{X} = \mathbf{0}$ is $\mathbf{X}'\mathbf{L}'\mathbf{L}\mathbf{X} = \mathbf{0}$. But $\mathbf{X}'\mathbf{L}'\mathbf{L}\mathbf{X} = \mathbf{0}$ implies $\mathbf{L}\mathbf{X} = \mathbf{0}$ *only if* $\mathbf{L}$ is real; whereupon $\mathbf{L}\mathbf{X} = \mathbf{0}$ implies $\mathbf{AX} = \mathbf{0}$. But $\mathbf{L}$ is not always real. It *is* when $\mathbf{A}$ is non-negative definite (n.n.d.), so that for $\mathbf{A}$ n.n.d. and symmetric $\mathbf{X}'\mathbf{A}\mathbf{X} = \mathbf{0}$ does imply $\mathbf{AX} = \mathbf{0}$.

$\mathbf{A}$ being n.n.d. is only a sufficient condition for $\mathbf{X}'\mathbf{A}\mathbf{X} = \mathbf{0}$ to imply $\mathbf{AX} = \mathbf{0}$. As an example, when estimating the class variance component σ_α^2 from balanced data of a 1-way classification random model, the ANOVA estimator is $\hat{\sigma}_\alpha^2 = (\text{MSA} - \text{MSE})/n$, as in (55) of Chapter 3. In writing this as $\mathbf{y}'\mathbf{A}\mathbf{y}$ the matrix $\mathbf{A}$ is not n.n.d. Evidence for this is in the negative estimate $\hat{\sigma}_\alpha^2 = -10$ of (56) in Chapter 3. Of course, for ANOVA estimators from balanced data $\mathbf{AX} = \mathbf{0}$ is always true, even though $\mathbf{A}$ may not be n.n.d., as just noted.

-ii. Translation invariance. A quadratic form $\mathbf{y}'\mathbf{A}\mathbf{y}$ is said to be translation-invariant in the context of the model (39) if its value is unaltered by location changes in $\boldsymbol{\beta}$, i.e., if $\boldsymbol{\beta}$ becomes $\boldsymbol{\beta} + \boldsymbol{\delta}$ then $\mathbf{y}'\mathbf{A}\mathbf{y}$ is translation-invariant if $\mathbf{y}'\mathbf{A}\mathbf{y} = (\mathbf{y} - \mathbf{X}\boldsymbol{\delta})'\mathbf{A}(\mathbf{y} - \mathbf{X}\boldsymbol{\delta})$. This reduces to $(2\mathbf{y} + \mathbf{X}\boldsymbol{\delta})'\mathbf{A}\mathbf{X}\boldsymbol{\delta} = \mathbf{0}$ or, for $\mathbf{z} = 2\mathbf{y} + \mathbf{X}\boldsymbol{\delta}$ to $\mathbf{z}'\mathbf{A}\mathbf{X}\boldsymbol{\delta} = \mathbf{0}$. We want this to be true for all $\mathbf{z}'$ and $\boldsymbol{\delta}$. Among

those values will be the cases when $\mathbf{z}'$ and $\boldsymbol{\delta}$ are, respectively, a row and a column of an identity matrix. Thus each element of $\mathbf{AX}$ must be zero and so $\mathbf{AX} = \mathbf{0}$. Since $\mathbf{AX} = \mathbf{0}$ always implies $\mathbf{X}'\mathbf{AX} = \mathbf{0}$, we therefore also have unbiasedness of $\mathbf{y}'\mathbf{Ay}$ if $p_i = \text{tr}(\mathbf{AZ}_i\mathbf{Z}_i')\forall\, i$; but unbiasedness does not always imply translation invariance except if $\mathbf{A}$ is n.n.d.

-iii. Minimum variance. The variance of $\mathbf{y}'\mathbf{Ay}$ does, in general, involve third and fourth moments of elements of $\mathbf{y}$. But confining attention to $\mathbf{y}$ being normally distributed gives, from using (38) in Theorem S4 of Appendix S.5,

$$\text{var}(\mathbf{y}'\mathbf{Ay}) = 2\,\text{tr}[(\mathbf{AV})^2] + 4\boldsymbol{\beta}'\mathbf{X}'\mathbf{AVAX}\boldsymbol{\beta}\,. \tag{42}$$

And the mean squared error of $\mathbf{y}'\mathbf{Ay}$ as an estimator of $\mathbf{p}'\boldsymbol{\sigma}^2$ for given $\mathbf{p}$ is

$$\begin{aligned}\text{MSE}(\mathbf{y}'\mathbf{Ay}) &= \text{var}(\mathbf{y}'\mathbf{Ay}) + [E(\mathbf{y}'\mathbf{Ay}) - \mathbf{p}'\boldsymbol{\sigma}^2]^2\\ &= 2\,\text{tr}[(\mathbf{AV})^2] + 4\boldsymbol{\beta}'\mathbf{X}'\mathbf{AVAX}\boldsymbol{\beta} + [\text{tr}(\mathbf{AV}) + \boldsymbol{\beta}'\mathbf{X}'\mathbf{AX}\boldsymbol{\beta} - \mathbf{p}'\boldsymbol{\sigma}^2]^2\,.\end{aligned} \tag{43}$$

A criterion for deriving estimators can be to minimize (42) or (43).

b. LaMotte's minimum mean square procedures

LaMotte (1973b) considered five different classes of estimators, governed by different combinations of the criteria of the preceding subsection. In each he determined $\mathbf{A}$, subject to those criteria, by minimizing the mean squared error given in (43) assuming $\boldsymbol{\beta}$ and $\boldsymbol{\sigma}^2$ (and hence $\mathbf{V}$) known. His five classes and their estimation follow. In all cases the estimator involves pre-assigned values of $\boldsymbol{\beta}$ and $\boldsymbol{\sigma}^2$, which for that purpose are denoted by $\boldsymbol{\beta}_0$ and $\boldsymbol{\sigma}_0^2$. Replacing $\boldsymbol{\beta}$ and $\boldsymbol{\sigma}^2$ by $\boldsymbol{\beta}_0$ and $\boldsymbol{\sigma}_0^2$, we define

$$\mathbf{P}_0 = \mathbf{V}_0^{-1} - \mathbf{V}_0^{-1}\mathbf{X}(\mathbf{X}'\mathbf{V}_0^{-1}\mathbf{X})^-\mathbf{X}'\mathbf{V}_0^{-1}, \tag{44}$$

$$\mathbf{B}_0 = \mathbf{X}(\mathbf{X}'\mathbf{V}_0^{-1}\mathbf{X})^-\mathbf{X}' + \mathbf{S}_0 \quad \text{for } \mathbf{S}_0 = \mathbf{X}\boldsymbol{\beta}\mathbf{V}_0^{-1}\boldsymbol{\beta}_0'\mathbf{X}', \tag{45}$$

and

$$\mathbf{B}_0^- = \mathbf{V}_0^{-1} - \mathbf{P}_0 - \frac{\mathbf{V}_0^{-1}\mathbf{S}\mathbf{V}_0^{-1}}{1 + c_0}, \quad \text{with } c_0 = \boldsymbol{\beta}_0'\mathbf{X}'\mathbf{X}\boldsymbol{\beta}_0\,. \tag{46}$$

Using these expressions [for which LaMotte (1973b) has two useful lemmas; see E 11.4(b,c)], the five classes of estimation procedures given by LaMotte (1973b) are summarized as follows. In all cases the given form of $\mathbf{y}'\mathbf{Ay}$ is a best estimator of $\mathbf{p}'\boldsymbol{\sigma}^2$ at $\boldsymbol{\beta}_0, \boldsymbol{\sigma}_0^2$; i.e., when assuming $\boldsymbol{\beta} = \boldsymbol{\beta}_0$ and $\boldsymbol{\sigma}^2 = \boldsymbol{\sigma}_0^2$. The reader is referred to Appendix A0 of that LaMotte paper for his excellent discussion of describing an estimator as "best at $\boldsymbol{\theta}_0$".

-i. Class C_0: unrestricted

$$\mathbf{y}'\mathbf{Ay} = \frac{\mathbf{p}'\boldsymbol{\sigma}_0^2}{c_0^2 + (n+2)(2c_0+1)}\,\mathbf{y}'[(2c_0+1)\mathbf{V}^{-1} - \mathbf{V}_0^{-1}\mathbf{X}\boldsymbol{\beta}_0\boldsymbol{\beta}_0'\mathbf{X}'\mathbf{V}_0^{-1}]\mathbf{y}\,.$$

This estimator is neither unbiased nor translation-invariant.

-ii. ***Class C_1: expectation of* y′Ay *containing no* β.** A is confined to satisfying $\mathbf{X}'\mathbf{A}\mathbf{X} = \mathbf{0}$. This is not unbiasedness, because $\text{tr}(\mathbf{A}\mathbf{Z}_i\mathbf{Z}_i') = p_i$ is not also being demanded of $\mathbf{A}$.

$$\mathbf{y}'\mathbf{A}\mathbf{y} = \frac{\mathbf{p}'\boldsymbol{\sigma}_0^2\mathbf{y}'\mathbf{P}_0\mathbf{y}}{N - r_{\mathbf{X}} + 2} .$$

The only criterion imposed on this estimator is that its expected value does not contain $\boldsymbol{\beta}_0$.

-iii. Class C_2: translation-invariant. The estimator in this class is derived using $\mathbf{A}\mathbf{X} = \mathbf{0}$, and it turns out to be the same estimator as in class C_1. It is translation-invariant.

-iv. Class C_3: unbiased. Define

$$\mathbf{M}_{i,0} = \mathbf{P}_0\mathbf{Z}_i\mathbf{Z}_i'\mathbf{P}_0 + \mathbf{P}_0\mathbf{Z}_i\mathbf{Z}_i'\mathbf{B}_0^- + \mathbf{B}_0^-\mathbf{Z}_i\mathbf{Z}_i'\mathbf{P}_0$$

for $\mathbf{P}$ and $\mathbf{B}^-$ of (44) and (46). The estimator of $\mathbf{p}'\boldsymbol{\sigma}^2$ is $\mathbf{p}'\tilde{\boldsymbol{\sigma}}^2$ for $\tilde{\boldsymbol{\sigma}}^2$ satisfying

$$\{_m\ \text{tr}(\mathbf{M}_{i,0}\mathbf{Z}_j\mathbf{Z}_j')\}_{i,j=0}^{r}\tilde{\boldsymbol{\sigma}}^2 = \{_c\ \mathbf{y}'\mathbf{M}_{i,0}\mathbf{y}\}_{i=0}^{r} .$$

This is the best (at $\boldsymbol{\beta}_0, \boldsymbol{\sigma}_0^2$) estimator that is unbiased.

-v. Class C_4: translation-invariant and unbiased. The estimator of $\boldsymbol{\sigma}^2$ is given by

$$\{_m\ \text{tr}(\mathbf{P}_0\mathbf{Z}_i\mathbf{Z}_i'\mathbf{P}_0\mathbf{Z}_j\mathbf{Z}_j')\}_{i,j=0}^{r}\tilde{\boldsymbol{\sigma}}^2 = \{_c\ \mathbf{y}'\mathbf{P}_0\mathbf{Z}_i\mathbf{Z}_i'\mathbf{P}_0\mathbf{y}\}_{i=0}^{r} . \tag{47}$$

This estimator is unbiased, translation-invariant and best at $\boldsymbol{\beta}_0, \boldsymbol{\sigma}_0^2$.

LaMotte (1973b) gives extensive details for the derivation of these results, and also for mean squared errors and attainable lower bounds thereof.

Notice that the estimation equations (47) of Class C_4 are the same as those for REML in (104) of Chapter 6 except that $\dot{\mathbf{P}}$ there represents $\mathbf{P}$ with the solution $\dot{\boldsymbol{\sigma}}^2$ to those equations replacing $\boldsymbol{\sigma}^2$, whereas in (47) $\mathbf{P}_0$ is $\mathbf{P}$ with the pre-assigned $\boldsymbol{\sigma}_0^2$ replacing $\boldsymbol{\sigma}^2$. Thus the REML equations have to be solved iteratively, but equations (47) are just a simple set of linear equations in the elements of $\tilde{\boldsymbol{\sigma}}^2$, because $\mathbf{P}_0$ is a matrix with numerical elements. No iteration is required as with REML. Since, for unbiased estimators, mean squared error equals variance, equations (47) also represent minimum variance unbiased estimators, on assuming $\mathbf{P}$ is actually $\mathbf{P}_0$. We proceed to derive (47) *ab initio*.

c. Minimum variance estimation (MINVAR)

As an estimator of $\mathbf{p}'\boldsymbol{\sigma}^2$ for known $\mathbf{p}'$, we seek symmetric $\mathbf{A}$ such that $\mathbf{y}'\mathbf{A}\mathbf{y}$ has the following properties:

(i) translation invariance, which requires $\mathbf{A}\mathbf{X} = \mathbf{0}$;
(ii) unbiasedness, which additionally demands $\text{tr}(\mathbf{A}\mathbf{Z}_i\mathbf{Z}_i') = p_i$;
(iii) minimum variance, which means (under normality) that

$$\text{var}(\mathbf{y}'\mathbf{A}\mathbf{y}) = 2\ \text{tr}(\mathbf{A}\mathbf{V})^2 + 4\boldsymbol{\beta}'\mathbf{X}'\mathbf{A}\mathbf{V}\mathbf{A}\mathbf{X}\boldsymbol{\beta} = 2\ \text{tr}(\mathbf{A}\mathbf{V})^2$$

(using $\mathbf{A}\mathbf{X} = \mathbf{0}$) is to be a minimum.

Thus the problem is to choose $\mathbf{A}$ so that $\text{tr}(\mathbf{AV})^2$ is minimized, for $\mathbf{V}$ p.d., subject to $\mathbf{A} = \mathbf{A}'$, $\mathbf{AX} = \mathbf{0}$ and $\text{tr}(\mathbf{AZ}_i\mathbf{Z}_i') = p_i$ for $i = 0, 1, \ldots, r$. Since it is $\text{tr}(\mathbf{AV})^2$ that is to be minimized with respect to elements of $\mathbf{A}$, it is clear that the resulting $\mathbf{A}$ will have elements that are functions of elements of $\mathbf{V}$. But those elements of $\mathbf{V}$ are functions (usually various sums) of the variance components that we seek to estimate in the form $\mathbf{y}'\mathbf{Ay}$. Thus our anticipated estimators are to be functions of the parameters they are estimating. This is not acceptable. We circumvent this situation as follows.

Suppose the variance components were to be considered known, represented by $\boldsymbol{\sigma}^2$. Then ask the question "What value would $\boldsymbol{\sigma}^2$ have to be in order for the preceding minimization problem to be satisfied?" In other words, what equations would $\boldsymbol{\sigma}^2$ have to satisfy so that for any known $\mathbf{p}'$ we would have $\mathbf{p}'\boldsymbol{\sigma}^2 = \mathbf{y}'\mathbf{Ay}$ such that $\text{tr}(\mathbf{AV})^2$ is a minimum subject to $\mathbf{A} = \mathbf{A}'$, $\mathbf{AX} = \mathbf{0}$ and $\text{tr}(\mathbf{AZ}_i\mathbf{Z}_i') = p_i$ for $i = 0, 1, \ldots, r$? Solving this problem can be achieved in a variety of ways. We begin by redefining the problem to put it in a form for which the answer is well known.

Since $\mathbf{V}$ is p.d., non-singular $\mathbf{V}^{\frac{1}{2}}$ exists such that $\mathbf{V} = \mathbf{V}^{\frac{1}{2}}\mathbf{V}^{\frac{1}{2}}$. (See E 11.5.) Using $\mathbf{V}^{\frac{1}{2}}$, define

$$\breve{\mathbf{Z}}_i = \mathbf{V}^{-\frac{1}{2}}\mathbf{Z}_i \quad \text{and} \quad \breve{\mathbf{A}} = \mathbf{V}^{\frac{1}{2}}\mathbf{A}\mathbf{V}^{\frac{1}{2}}, \tag{48}$$

and

$$\breve{\mathbf{X}} = \mathbf{V}^{-\frac{1}{2}}\mathbf{X} \quad \text{and} \quad \breve{\mathbf{M}} = \mathbf{I} - \breve{\mathbf{X}}(\breve{\mathbf{X}}'\breve{\mathbf{X}})^{-}\breve{\mathbf{X}}' = \breve{\mathbf{M}}' = \breve{\mathbf{M}}^2, \tag{49}$$

noting that

$$\breve{\mathbf{M}}\breve{\mathbf{X}} = \mathbf{0} \quad \text{and} \quad \mathbf{V}^{-\frac{1}{2}}\breve{\mathbf{M}}\mathbf{V}^{-\frac{1}{2}} = \mathbf{P}\,. \tag{50}$$

Also observe that

$$\breve{\mathbf{A}} = \breve{\mathbf{A}}' = \breve{\mathbf{A}}\breve{\mathbf{M}} = \breve{\mathbf{M}}\breve{\mathbf{A}} = \breve{\mathbf{M}}\breve{\mathbf{A}}\breve{\mathbf{M}} \quad \text{if and only if} \quad \mathbf{AX} = \mathbf{0} \text{ and } \mathbf{A} = \mathbf{A}'\,. \tag{51}$$

The two trace terms of the minimization problem can now be written as

$$\text{tr}(\mathbf{AV})^2 = \text{tr}(\mathbf{AVAV}) = \text{tr}(\mathbf{V}^{\frac{1}{2}}\mathbf{AVAV}^{\frac{1}{2}}) = \text{tr}(\breve{\mathbf{A}}^2) = (\text{vec}\,\breve{\mathbf{A}})'\,\text{vec}\,\breve{\mathbf{A}},$$

and

$$\text{tr}(\mathbf{AZ}_i\mathbf{Z}_i') = (\text{vec}\,\breve{\mathbf{A}})'\,\text{vec}(\breve{\mathbf{M}}\breve{\mathbf{Z}}_i\breve{\mathbf{Z}}_i'\breve{\mathbf{M}})\,.$$

Then, because from (51), using $\breve{\mathbf{A}}$ and $\breve{\mathbf{M}}$ implicitly includes $\mathbf{AX} = \mathbf{0}$ and $\mathbf{A} = \mathbf{A}'$, we can rewrite the minimization problem as: find $\breve{\mathbf{A}}$ to minimize

$$(\text{vec}\,\breve{\mathbf{A}})'\,\text{vec}\,\breve{\mathbf{A}} \text{ subject to } (\text{vec}\,\breve{\mathbf{A}})'\{_r\ \text{vec}(\breve{\mathbf{M}}\breve{\mathbf{Z}}_i\breve{\mathbf{Z}}_i'\breve{\mathbf{M}})\} = \mathbf{p}'\,. \tag{52}$$

We solve this minimization problem with the following well-known lemma.

Lemma. $\mathbf{t}'\mathbf{t}$ is minimized, subject to $\mathbf{t}'\mathbf{W} = \boldsymbol{\lambda}'$, by $\mathbf{t} = \mathbf{W}\boldsymbol{\theta}$ for $\boldsymbol{\lambda} = \mathbf{W}'\mathbf{W}\boldsymbol{\theta}$, for some vector $\boldsymbol{\theta}$.

Proof. For $2\boldsymbol{\theta}$ being a vector of Lagrange multipliers, minimizing $\mathbf{t}'\mathbf{t} - 2\boldsymbol{\theta}'(\boldsymbol{\lambda} - \mathbf{W}'\mathbf{t})$ with respect to elements of $\mathbf{t}$ and of $\boldsymbol{\theta}$ leads, respectively, to $\mathbf{t} = \mathbf{W}\boldsymbol{\theta}$ and $\boldsymbol{\lambda} = \mathbf{W}'\mathbf{t}$, for which the latter is then $\boldsymbol{\lambda} = \mathbf{W}'\mathbf{W}\boldsymbol{\theta}$.

Comment. The lemma as stated is a simple result in mathematics. It does, of course, have an important application in statistics, in least squares estimation where for $\mathbf{z} \sim (\mathbf{W}\boldsymbol{\beta}, \mathbf{I})$ we seek $\mathbf{t}'\mathbf{z}$ as an unbiased estimator of $\boldsymbol{\lambda}'\boldsymbol{\beta}$ that has minimum variance.

We use the lemma to solve the minimization problem of (52) by writing

$$\mathbf{t} \equiv \operatorname{vec} \check{\mathbf{A}}, \quad \mathbf{W} \equiv \{_{\mathrm{r}} \operatorname{vec}(\check{\mathbf{M}}\check{\mathbf{Z}}_i\check{\mathbf{Z}}_i'\check{\mathbf{M}})\} \quad \text{and} \quad \boldsymbol{\lambda} = \mathbf{p} .$$

Then (52) is solved in the form

$$\operatorname{vec} \check{\mathbf{A}} = \{_{\mathrm{r}} \operatorname{vec}(\check{\mathbf{M}}\check{\mathbf{Z}}_i\check{\mathbf{Z}}_i'\check{\mathbf{M}})\}\boldsymbol{\theta} \tag{53}$$

and

$$\mathbf{p} = \{_{\mathrm{c}} [\operatorname{vec}(\check{\mathbf{M}}\check{\mathbf{Z}}_i\check{\mathbf{Z}}_i'\check{\mathbf{M}})]'\}\{_{\mathrm{r}} \operatorname{vec}(\check{\mathbf{M}}\check{\mathbf{Z}}_i\check{\mathbf{Z}}_i'\check{\mathbf{M}})\}\boldsymbol{\theta} . \tag{54}$$

These equations simplify. First, (53) is $\operatorname{vec} \check{\mathbf{A}} = \Sigma_i \theta_i \operatorname{vec}(\check{\mathbf{M}}\check{\mathbf{Z}}_i\check{\mathbf{Z}}_i'\check{\mathbf{M}})$, and because $\check{\mathbf{A}}$ and $\check{\mathbf{M}}$ have the same order, this is

$$\check{\mathbf{A}} = \Sigma_i \theta_i \check{\mathbf{M}}\check{\mathbf{Z}}_i\check{\mathbf{Z}}_i'\check{\mathbf{M}} .$$

Then, in substituting for $\check{\mathbf{A}}$, $\check{\mathbf{M}}$ and $\check{\mathbf{Z}}$ from (48) and (49) and using (50), this reduces to

$$\mathbf{A} = \Sigma_i \theta_i \mathbf{P}\mathbf{Z}_i\mathbf{Z}_i'\mathbf{P} . \tag{55}$$

Similarly, using $(\operatorname{vec} \mathbf{K})' \operatorname{vec} \mathbf{L} = \operatorname{tr}(\mathbf{KL})$ for any $\mathbf{K}$ and $\mathbf{L}$ of appropriate orders, (54) reduces to

$$\mathbf{p} = \{_{\mathrm{m}} \operatorname{tr}(\mathbf{P}\mathbf{Z}_i\mathbf{Z}_i'\mathbf{P}\mathbf{Z}_j\mathbf{Z}_j')\}_{i,j=0}^{r}\boldsymbol{\theta} . \tag{56}$$

Therefore, since we want $\boldsymbol{\sigma}^2$ to satisfy $\mathbf{p}'\boldsymbol{\sigma}^2 = \mathbf{y}'\mathbf{A}\mathbf{y}$, we have on using (55) and (56)

$$\boldsymbol{\theta}'\{_{\mathrm{m}} \operatorname{tr}(\mathbf{P}\mathbf{Z}_i\mathbf{Z}_i'\mathbf{P}\mathbf{Z}_j\mathbf{Z}_j')\}\boldsymbol{\sigma}^2 = \mathbf{y}'(\Sigma_i \theta_i \mathbf{P}\mathbf{Z}_i\mathbf{Z}_i'\mathbf{P})\mathbf{y} = \boldsymbol{\theta}'\{_{\mathrm{c}}\, \mathbf{y}'\mathbf{P}\mathbf{Z}_i\mathbf{Z}_i'\mathbf{P}\mathbf{y}\} . \tag{57}$$

Throughout this development $\mathbf{p}$ of $\mathbf{p}'\boldsymbol{\sigma}^2$ has been assumed known; and $\boldsymbol{\theta}$ depends on $\mathbf{p}$ through (54). Therefore, in wanting the development to apply to all $\mathbf{p}$, we also want it to apply for all $\boldsymbol{\theta}$. Hence in (57) we let $\boldsymbol{\theta}'$ be successive rows of $\mathbf{I}$ (of order $r + 1$) and so get the equations

$$\{_{\mathrm{m}} \operatorname{tr}(\mathbf{P}\mathbf{Z}_i\mathbf{Z}_i'\mathbf{P}\mathbf{Z}_j\mathbf{Z}_j')\}\boldsymbol{\sigma}^2 = \{_{\mathrm{c}}\, \mathbf{y}'\mathbf{P}\mathbf{Z}_i\mathbf{Z}_i'\mathbf{P}\mathbf{y}\} . \tag{58}$$

These equations have to be solved for $\boldsymbol{\sigma}^2$. Since $\mathbf{P}$ involves $\boldsymbol{\sigma}^2$, through $\mathbf{V}^{-1}$, a solution for $\boldsymbol{\sigma}^2$ has to be obtained by numerical techniques. Clearly not all of its elements will necessarily be non-negative.

The solution to (58) has been called the minimum variance, location-invariant, unbiased estimator of $\boldsymbol{\sigma}^2$. But, because of the iterative procedure involved, it will not be an unbiased estimator of $\boldsymbol{\sigma}^2$. Nor, even under normality, will it have minimum variance. It might better be called PSEUDO-MINVAR.

Notice that the form of (58) is exactly the same as the REML equations in (104) of Chapter 6. Thus REML solutions and solutions to (58) are identical. And, of course the form of (58) is also similar to (47) for LaMotte's Class C_4: replacing $\mathbf{P}$ in (58) with $\mathbf{P}_0$ gives (47), whereupon it is a minimum variance (under normality), location-invariant, unbiased estimator at $\boldsymbol{\sigma}_0^2$. But, of course, (58) has to be solved iteratively whereas (47) has just a single solution, one that depends on $\boldsymbol{\sigma}_0^2$.

d. Minimum norm estimation (MINQUE)

In a series of four papers, Rao (1970, 1971a,b, 1972) suggested a method of estimation that does not require the normality assumption that is the foundation of ML, REML and MINVAR (minimum variance). It has the same basis as LaMotte's approach, of estimating $\mathbf{p}'\boldsymbol{\sigma}^2$ by $\mathbf{y}'\mathbf{A}\mathbf{y}$ with $\mathbf{A} = \mathbf{A}'$, $\mathbf{A}\mathbf{X} = \mathbf{0}$ and $p_i = \text{tr}(\mathbf{A}\mathbf{Z}_i\mathbf{Z}_i')$, but instead of deriving $\mathbf{A}$ by minimizing an unknown variance (or mean square), it minimizes a known norm, a Euclidean norm, which is akin to a generalized variance. The derivation is as follows.

In the model $\mathbf{y} = \mathbf{X}\boldsymbol{\beta} + \Sigma_{i=0}^r \mathbf{Z}_i\mathbf{u}_i$ the random vectors $\mathbf{u}_i$ are unknown. They have mean zero. Therefore if $\mathbf{u}_i = \{_c\, \mathbf{u}_{ij}\}_j$ were known, a "natural" estimator (Rao's own word—1972, p. 113) of σ_i^2 would be $\mathbf{u}_i'\mathbf{u}_i/q_i$, where q_i is the order of $\mathbf{u}_i$. Thus a "natural" estimator of $\mathbf{p}'\boldsymbol{\sigma}^2$ would be

$$\mathbf{p}'\tilde{\boldsymbol{\sigma}}^2 = \Sigma_i p_i \frac{\mathbf{u}_i'\mathbf{u}_i}{q_i} = \mathbf{u}'\left\{_d\, \frac{p_i}{q_i}\mathbf{I}_{q_i}\right\}\mathbf{u} = \mathbf{u}'\boldsymbol{\Delta}\mathbf{u} \quad \text{for } \boldsymbol{\Delta} = \left\{_d\, \frac{p_i}{q_i}\mathbf{I}_{q_i}\right\}.$$

In contrast we are going to use as an estimator $\mathbf{p}'\hat{\boldsymbol{\sigma}}^2 = \mathbf{y}'\mathbf{A}\mathbf{y} = \mathbf{u}'\mathbf{Z}'\mathbf{A}\mathbf{Z}\mathbf{u}$. Hence the difference between the two estimators is $\mathbf{p}'\hat{\boldsymbol{\sigma}}^2 - \mathbf{p}'\tilde{\boldsymbol{\sigma}}^2 = \mathbf{u}'(\mathbf{Z}'\mathbf{A}\mathbf{Z} - \boldsymbol{\Delta})\mathbf{u}$.

Rao chose to minimize a weighted Euclidean norm based on this difference, using $\sigma_{0,i}^2$ as pre-assigned values of σ_i^2 in the form $\mathbf{D} = \{_d\, \sigma_{0,i}^2\mathbf{I}_{q_i}\}$. Thus $\mathbf{D}$ is simply var($\mathbf{u}$) with $\sigma_{0,i}^2$ in place of σ_i^2. Then the norm that gets minimized is $\text{tr}(\mathbf{F}\mathbf{F}')$ for $\mathbf{F} = \mathbf{D}^{\frac{1}{2}}(\mathbf{Z}'\mathbf{A}\mathbf{Z} - \boldsymbol{\Delta})\mathbf{D}^{\frac{1}{2}}$. Modest algebra, including the use of $p_i = \text{tr}(\mathbf{A}\mathbf{Z}_i\mathbf{Z}_i')$, reduces this (see E 11.7) to

$$\text{tr}(\mathbf{F}\mathbf{F}') = \text{tr}[(\mathbf{A}\mathbf{V}_0)^2] - \sum_i \frac{p_i^2\sigma_{0,i}^4}{q_i},$$

where $\mathbf{V}_0$ is $\mathbf{V}$ with σ_0^2 in place of $\boldsymbol{\sigma}^2$. It is $\text{tr}(\mathbf{F}\mathbf{F}')$ that is to be minimized with respect to elements of $\mathbf{A}$. But since those elements do not occur in p_i, $\sigma_{0,i}^2$ or q_i, we have only to minimize $\text{tr}[(\mathbf{A}\mathbf{V}_0)^2]$. Thus the minimization problem is to minimize $\text{tr}[(\mathbf{A}\mathbf{V}_0)^2]$ subject to $\mathbf{A} = \mathbf{A}'$, $\mathbf{A}\mathbf{X} = \mathbf{0}$ and $p_i = \text{tr}(\mathbf{A}\mathbf{Z}_i\mathbf{Z}_i')$. This is exactly the same minimization problem as in the preceding subsection, only with $\mathbf{V}_0$ replacing $\mathbf{V}$. Accordingly its solution is (58) with that same replacement, leading to the estimation equation

$$\{_m\, \text{tr}(\mathbf{P}_0\mathbf{Z}_i\mathbf{Z}_i'\mathbf{P}_0\mathbf{Z}_j\mathbf{Z}_j')\}\hat{\boldsymbol{\sigma}}^2 = \{_c\, \mathbf{y}'\mathbf{P}_0\mathbf{Z}_i\mathbf{Z}_i'\mathbf{P}_0\mathbf{y}\}\,. \tag{59}$$

Equations (59) yield what are known as MINQUE estimators: minimum norm, quadratic unbiased estimators. They are exactly the same as (47) for LaMotte's Class C_4. Moreover, they have the same *form* as the MINVAR equations of (58)—and the REML equations of (104) in Chapter 6. But there is that big difference: MINVAR and REML equations (they are the same) have to be solved by iteration: the LaMotte Class C_4 and the MINQUE equations (which are the same), do not. The reason for this is that in (59) the unknown variances occur only in $\hat{\boldsymbol{\sigma}}^2$. This is because $\mathbf{P}_0$ in (59) has elements that are all known numbers: it is $\mathbf{P}$ with $\boldsymbol{\sigma}^2$ replaced by $\boldsymbol{\sigma}_0^2$, where $\boldsymbol{\sigma}_0^2$ has been decided on as part of the estimation process. Thus (59) is simply a set of $r+1$ equations that are linear in the $r+1$ unknown variance components. They get solved, and the solutions are the MINQUE estimates; but they do, of course, depend on what has been used as $\boldsymbol{\sigma}_0^2$. And if n people had the same data and used the same model, but used n different vectors $\boldsymbol{\sigma}_0^2$, then (59) would yield n different MINQUE estimates. This is a distinctive feature of the MINQUE procedure. It is something we do not favor. Nevertheless, the estimators are locally minimum norm (minimum variance, under normality) in the neighborhood of $\boldsymbol{\sigma}_0^2$, and are locally unbiased.

e. REML, MINQUE and I-MINQUE

The REML equations are

$$\{_m \operatorname{tr}(\tilde{\mathbf{P}}\mathbf{Z}_i\mathbf{Z}_i'\tilde{\mathbf{P}}\mathbf{Z}_j\mathbf{Z}_j')\}\tilde{\boldsymbol{\sigma}}^2 = \{_c\ \mathbf{y}'\tilde{\mathbf{P}}\mathbf{Z}_i\mathbf{Z}_i'\tilde{\mathbf{P}}\mathbf{y}\}, \tag{60}$$

whereas the MINQUE equations are

$$\{_m \operatorname{tr}(\mathbf{P}_0\mathbf{Z}_i\mathbf{Z}_i'\mathbf{P}_0\mathbf{Z}_j\mathbf{Z}_j')\}\hat{\boldsymbol{\sigma}}^2 = \{_c\ \mathbf{y}'\mathbf{P}_0\mathbf{Z}_i\mathbf{Z}_i'\mathbf{P}_0\mathbf{y}\}\ . \tag{61}$$

Equations (61) get solved directly for $\hat{\boldsymbol{\sigma}}^2$. Equations (60) have to be solved iteratively. To start the iteration, an initial value has to be used for $\boldsymbol{\sigma}^2$; call it $\boldsymbol{\sigma}_0^2$. Then equations (60) yield the first iterate, $\tilde{\boldsymbol{\sigma}}_1^2$. But (60) using $\boldsymbol{\sigma}_0^2$ for $\boldsymbol{\sigma}^2$ in $\mathbf{P}$ is identical to (61). Therefore $\tilde{\boldsymbol{\sigma}}_1^2$ will be exactly the same as the solution of the MINQUE equations (61). Thus we have the relationship

$$\text{a MINQUE} = \text{a first iterate solution of REML}\ . \tag{62}$$

And notice that it is "a MINQUE", not "the MINQUE". For a given set of unbalanced data different values of $\boldsymbol{\sigma}_0^2$ used in $\mathbf{P}_0$ of (61) will not necessarily yield the same MINQUE $\hat{\boldsymbol{\sigma}}^2$. Indeed, one can expect the $\tilde{\boldsymbol{\sigma}}^2$-values to be different for each $\boldsymbol{\sigma}_0^2$. A genetic application of the connection between REML and MINQUE is discussed by Henderson (1985), and Rao (1979) also considers the connection more generally.

Consider the MINQUE equations (61) again. They yield an estimator, $\tilde{\boldsymbol{\sigma}}_{(1)}^2$, say. Bearing in mind that $\boldsymbol{\sigma}_0^2$ is a pre-assigned value of $\boldsymbol{\sigma}^2$, it would not be unnatural, having obtained $\tilde{\boldsymbol{\sigma}}_{(1)}^2$ based on $\boldsymbol{\sigma}_0^2$ from (61), to contemplate using $\tilde{\boldsymbol{\sigma}}_{(1)}^2$ in place of $\boldsymbol{\sigma}_0^2$ in (61) and solve, yielding what we may call $\tilde{\boldsymbol{\sigma}}_{(2)}^2$; and this process could be continued. It is called iterative MINQUE, or I-MINQUE. Clearly, if one uses I-MINQUE to convergence then, providing the starting values for iterating I-MINQUE are the same as for iterating REML,

$$\text{I-MINQUE estimates} = \text{REML (based on normality) solutions}\ . \tag{63}$$

True it is that neither MINQUE nor I-MINQUE require normality assumptions, but so far as estimates are concerned (63) is valid. Moreover, Brown (1976) has shown that I-MINQUE estimators, the basis of which require no normality assumptions, are asymptotically normally distributed; and Rich and Brown (1979) consider the effect of imposing non-negativity constraints on I-MINQUE estimators.

Equations (62) and (63) are both clearly statements about solutions of equations, so far as REML goes; i.e., non-negativity requirements of REML estimators have not been brought into play. This highlights the as-yet-unmentioned fact that MINQUE as a method contains no provision for precluding negative estimates. Choosing $\boldsymbol{\sigma}_0^2$ and solving (61) is no guarantee against getting one or more negative estimates.

f. REML for balanced data

The REML and MINVAR procedures are, under normality, the same—as discussed following (60). But we also know [e.g., Graybill (1956) and colleagues; see Section 4.4] for balanced data the ANOVA estimators are MINVAR. Hence

$$\text{REML (based on normality) solutions} = \text{ANOVA for balanced data}\,. \tag{64}$$

g. MINQUEO

A particular easy form of MINQUE is when $\boldsymbol{\sigma}_0^2$ is taken as a null vector except for $\sigma_{0,0}^2 = 1$. Then the MINQUE equations (61) reduce for $\mathbf{M} = \mathbf{I} - \mathbf{XX}^+$ to

$$\{_m\, \text{tr}(\mathbf{MZ}_i\mathbf{Z}_i'\mathbf{MZ}_j\mathbf{Z}_j')\}\hat{\boldsymbol{\sigma}}^2 = \{_c\, \mathbf{y}'\mathbf{MZ}_i\mathbf{Z}_i'\mathbf{My}\}\,. \tag{65}$$

The resulting estimators were suggested by Rao (1970) in the first of his four papers on MINQUE.

With minimum variance estimators being called MIVQUE, estimators obtained from (65) have been called MIVQUEO by Goodnight (1978)—but MINQUEO seems more general. Without using any name, Seely [1971, equation (6)] has MINQUEO as a method of estimation, Corbeil and Searle (1976a) have it as the starting point of the (iterative) REML procedure, and Hartley *et al.* [1978, equation (10)] espouse its use on grounds of relatively easy computability, a feature that is promoted by Goodnight (1978). Reconciliation with (62) of the Corbeil and Searle (1976a) description and of the Hartley *et al.* (1978) description are the topics of E 11.8 and E 11.9.

h. MINQUE for the 1-way classification

By way of example we outline derivation of the MINQUE equations (61) for the 1-way classification with unbalanced data. The full details are extensive and are left for the solutions manual that will be available from the authors; and the results are as in Swallow and Searle (1978). The task is to simplify (61) to be in the form

$$\begin{bmatrix} s_{00} & s_{01} \\ s_{10} & s_{11} \end{bmatrix} \begin{bmatrix} \sigma_e^2 \\ \sigma_\alpha^2 \end{bmatrix} = \begin{bmatrix} u_0 \\ u_1 \end{bmatrix}, \tag{66}$$

where $s_{00} = \text{sesq}(\mathbf{Z}_0'\mathbf{P}_0\mathbf{Z}_0)$, $s_{01} = s_{10} = \text{sesq}(\mathbf{Z}_0'\mathbf{P}_0\mathbf{Z}_1)$, $s_{11} = \text{sesq}(\mathbf{Z}_1'\mathbf{P}_0\mathbf{Z}_1)$, $u_0 = \text{sesq}(\mathbf{Z}_0'\mathbf{P}_0\mathbf{y})$ and $u_1 = \text{sesq}(\mathbf{Z}_1'\mathbf{P}_0\mathbf{y})$. In making these simplifications we use the model equation

$$\mathbf{y} = \mu\mathbf{1}_N + \mathbf{Z}_1\boldsymbol{\alpha} + \mathbf{e} \quad \text{with } \mathbf{X} = \mathbf{1}_N, \mathbf{Z}_0 = \mathbf{I}_N \text{ and } \mathbf{Z}_1 = \{_d\, \mathbf{1}_{n_i}\} .$$

This leads to

$$\mathbf{V}_0^{-1} = \left\{_d\, \frac{1}{\sigma_{e,0}^2}\mathbf{I}_{n_i} - \frac{\sigma_{\alpha,0}^2}{\sigma_{e,0}^2\lambda_i}\mathbf{J}_{n_i}\right\}_{i,i'=1}^{r} \quad \text{with } \lambda_i = \sigma_{e,0}^2 + n_i\sigma_{\alpha,0}^2,$$

where $\sigma_{e,0}^2$ and $\sigma_{\alpha,0}^2$ are the pre-assigned values of σ_e^2 and σ_α^2 that are to be the basis of the MINQUE procedure. Then

$$\mathbf{P}_0 = \mathbf{V}_0^{-1} - k\left\{_m\, \frac{\mathbf{J}_{n_i \times n_{i'}}}{\lambda_i\lambda_{i'}}\right\}_{i,i'=0}^{r} \quad \text{for } k = 1\Big/\sum_i \frac{n_i}{\lambda_i} .$$

After tedious algebra we find, for $k_i = n_i/\lambda_i$, that

$$s_{00} = \frac{N-a}{\sigma_e^2} + \sum_i \frac{k_i^2}{n_i^2} + k^2\left(\sum_i \frac{k_i^2}{n_i}\right)^2 - 2k\sum_i \frac{k_i^3}{n_i^2},$$

$$s_{01} = \sum_i \frac{k_i^2}{n_i} - 2k\sum_i \frac{k_i^3}{n_i} + k^2\sum_i k_i^2 \sum_i \frac{k_i^2}{n_i},$$

$$s_{11} = \Sigma_i k_i^2 - 2k\Sigma_i k_i^3 + k^2(\Sigma_i k_i^2)^2,$$

$$u_0 = \frac{\text{SSE}}{\sigma_e^2} + \sum_i \frac{k_i}{n_i}(\bar{y}_{i\cdot} - k\Sigma_i k_i \bar{y}_{i\cdot})^2,$$

$$u_1 = \Sigma_i k_i^2(\bar{y}_{i\cdot} - k\Sigma_i k_i \bar{y}_{i\cdot})^2 . \tag{67}$$

Various adaptations to the MINQUE equations are suggested by Chaubey (1984) for eliminating the possibility of MINQUE estimators being negative.

11.4. SUMMARY

Components of covariance: Section 11.1

Model: (1), (2), (3)

$$\hat{\sigma}_{\alpha_{12}} = \tfrac{1}{2}(\hat{\sigma}_{\alpha_{1+2}}^2 - \hat{\sigma}_{\alpha_1}^2 - \hat{\sigma}_{\alpha_2}^2) . \tag{4),(6}$$

Different forms for $\mathbf{D} = \text{var}(\mathbf{u})$:

$$\mathbf{D} = \{_d\, \mathbf{D}_{ii}\} = \{_d\, \sigma_i^2[(1-\rho_i)\mathbf{I}_{q_i} + \rho_i\mathbf{J}_{q_i}]\}; \tag{9),(10}$$

$$\mathbf{D} = \{_m\, \mathbf{D}_{ii'}\},$$

with (11), (12)

$$\mathbf{D}_{ii} = \sigma_i^2[(1-\rho_{ii})\mathbf{I}_{q_i} + \rho_{ii}\mathbf{J}_{q_i}], \quad \text{and} \quad \mathbf{D}_{ii'} = \rho_{ii'}\sigma_i\sigma_{i'}\mathbf{J}_{q_i \times q_{i'}} \quad \text{for } i \neq i' .$$

Also $\text{var}(\mathbf{e}) = \mathbf{R}$; left to the reader.

Genetic covariances: relationship matrix . (16)

Maximum likelihood:

$$\mathbf{V} = \mathbf{ZDZ}' + \mathbf{R}; \tag{18}$$

ML equations; (28), (29)

$$\text{var}(\tilde{\boldsymbol{\sigma}}^2) . \tag{31}$$

Restricted maximum likelihood: Section 11.1f

Modeling variance components as covariances: Section 11.2

Model: (32)

Estimation: (36)

Balanced data simplification: (38)

Diagnostics: Section 11.2c.

Some-cells empty data: Section 11.2d.

Criteria-based procedures: Section 11.3

Estimate $\mathbf{p}'\boldsymbol{\sigma}^2$ by $\mathbf{y}'\mathbf{Ay}$.

Unbiasedness:

$$p_i = \text{tr}(\mathbf{AZ}_i\mathbf{Z}_i') . \tag{41}$$

Translation invariance:

$$\mathbf{AX} = \mathbf{0} .$$

Minimum variance:

$$\text{var}(\mathbf{y}'\mathbf{Ay}); \tag{42}$$

$$\text{MSE}(\mathbf{y}'\mathbf{Ay}) . \tag{43}$$

LaMotte's procedures: Section 11.3b.

Minimum variance estimation (MINVAR):

$$\{_m \, \text{tr}(\mathbf{PZ}_i\mathbf{Z}_i'\mathbf{PZ}_j\mathbf{Z}_j')\}\boldsymbol{\sigma}^2 = \{_c \, \mathbf{y}'\mathbf{PZ}_i\mathbf{Z}_i'\mathbf{Py}\} . \tag{58}$$

Minimum norm estimation (MINQUE):

$$\{_m \operatorname{tr}(\mathbf{P}_0\mathbf{Z}_i\mathbf{Z}_i'\mathbf{P}_0\mathbf{Z}_j\mathbf{Z}_j')\}\hat{\boldsymbol{\sigma}}^2 = \{_c \mathbf{y}'\mathbf{P}_0\mathbf{Z}_i\mathbf{Z}_i'\mathbf{P}_0\mathbf{y}\} . \tag{59}$$

$$\text{a-MINQUE} = \text{a first iterate of REML} . \tag{62}$$

$$\text{I-MINQUE estimates} = \text{REML solutions} \tag{63}$$

$$\text{REML solutions} = \text{ANOVA for balanced data} \tag{64}$$

Special case (MINQUEO):

$$\{_m \operatorname{tr}(\mathbf{M}\mathbf{Z}_i\mathbf{Z}_i'\mathbf{M}\mathbf{Z}_j\mathbf{Z}_j')\}\hat{\boldsymbol{\sigma}}^2 = \{_c \mathbf{y}'\mathbf{M}\mathbf{Z}_i\mathbf{Z}_i'\mathbf{M}\mathbf{y}\} . \tag{65}$$

MINQUE for the 1-way classification: (66), (67)

11.5. EXERCISES

E 11.1. Apply the estimation method described in Section 11.1a to estimating components of covariance in the 2-way crossed classification, no interaction, random model, with one observation per cell.

E 11.2. Apply (29), (30) and (31) to the following variations of the 1-way classification.

(a) Model (1), just the y_{1ij} data.

(b) Model (1) and (2), without (3); i.e., $\tau_\alpha = \tau_e = 0$.

(c) Model (1), (2) and (3).

(d) Model (1), (2) and (10) with $\sigma_i^2 = \sigma_\alpha^2$ and $\rho_i = \rho \ \forall \ i$.

(e) The preceding case with $\rho = 0$.

(f) Model (1), (2) and (11) with $\sigma_i^2 = \sigma_\alpha^2$, $\rho_{ii} = \rho_1 \ \forall \ i$ and $\rho_{ii'} = \rho_2 \ \forall \ i \neq i'$.

(g) The preceding case, with $\rho_1 = 0$.

(h) Model (1) with $\mathbf{D}$ of (15), but $\mathbf{R} = \sigma_e^2\mathbf{I}$.

(i) Model (1) with $\mathbf{D}$ of (16).

Try each of the preceding cases (i) with balanced data and (ii) with unbalanced data. You may find some cases more difficult than others.

E 11.3. (a) Show that $\hat{\phi}_1$ of (36) is unbiased for σ_α^2.

(b) For balanced data reduce (36) to (38).

(c) Show that $\hat{\phi}_1$ of (36) is $\hat{\sigma}_\alpha^2$ of (145) in Chapter 5.

(d) Why are the data of Table 11.3 very unsatisfactory for estimating ϕ_2 in a manner analogous to (36)?

E 11.4. Matrices $\mathbf{P}, \mathbf{B}, \mathbf{S}$ and $\mathbf{B}^-$ and scalar c are defined in Section 11.3b.

(a) Show that $\mathbf{PX} = \mathbf{0}$, $\mathbf{PS} = \mathbf{0}$, $\mathbf{SV}^{-1}\mathbf{S} = c\mathbf{S}$, $\mathbf{PVP} = \mathbf{P}$ and $\mathbf{BB}^-\mathbf{B} = \mathbf{B}$.

(b) For symmetric $\mathbf{A}$ show that $\mathbf{X}'\mathbf{AX} = \mathbf{0}$ if and only if there exists a symmetric $\mathbf{C}$ such that $\mathbf{A} = \mathbf{PCP} + \mathbf{PCB}^- + \mathbf{B}^-\mathbf{CP}$.

(c) For symmetric $\mathbf{A}$ show that $\mathbf{AX} = \mathbf{0}$ if and only if there exists a symmetric $\mathbf{C}$ such that $\mathbf{A} = \mathbf{PCP}$.

E 11.5. For results in Section 11.3:

(a) The canonical form under orthogonal similarity of $\mathbf{V}$ is $\mathbf{U}'\mathbf{VU} = \{_d\ \lambda_i\}$ for orthogonal $\mathbf{U}$ and the λ_i being eignroots of $\mathbf{V}$ (e.g., Searle, 1987, p. 283). Show that this leads to the existence of nonsingular $\mathbf{V}^{\frac{1}{2}}$ such that $\mathbf{V} = \mathbf{V}^{\frac{1}{2}}\mathbf{V}^{\frac{1}{2}}$.

(b) Verify (51).

(c) Show that $\text{tr}(\mathbf{AZ}_i\mathbf{Z}_i') = (\text{vec}\,\breve{\mathbf{A}})'\,\text{vec}(\breve{\mathbf{M}}\mathbf{Z}_i\mathbf{Z}_i'\breve{\mathbf{M}})$.

(d) Confirm (55) and (56).

E 11.6. Verify the comment that follows the lemma in Section 11.3c.

E 11.7. For $\mathbf{F}$ and $\mathbf{D}_0$ defined in Section 11.3d show that

$$\text{tr}(\mathbf{FF}') = \text{tr}(\mathbf{AV}_0)^2 - \sum_i \frac{p_i^2\sigma_{0,i}^4}{q_i}\,.$$

E 11.8. [Corbeil and Searle (1976a).] Using a matrix $\mathbf{T}$ of full row rank $N - r_{\mathbf{X}}$ and such that $\mathbf{T}'(\mathbf{TT}')^{-1}\mathbf{T} = \mathbf{M}$ and $E(\mathbf{Ty}) = \mathbf{0}$, these authors define

$$\mathbf{L} = \mathbf{T}'(\mathbf{THT}')^{-1}\mathbf{T}$$

for $\mathbf{H} = \mathbf{V}/\sigma_e^2$ as in (28) of Chapter 6. Then their REML equations are

$$(N - r)\hat{\sigma}_0^2 = \mathbf{y}'\mathbf{Ly} \tag{68}$$

and

$$\sigma_0^2\,\text{tr}(\mathbf{Z}_i\mathbf{Z}_i'\mathbf{L}) = \mathbf{y}'\mathbf{LZ}_i\mathbf{Z}_i'\mathbf{Ly} \quad \text{for } i = 1, \ldots, r\,. \tag{69}$$

(a) Prove that $\mathbf{LHL} = \mathbf{L}$.

(b) Prove that $\mathbf{LH}$ is idempotent, of rank $\mathbf{N} - \mathbf{r_X}$.

(c) By multiplying the ith equation of (69) by σ_i^2 and summing over $i = 1, \ldots, r$, incorporate (68) into (69) to yield the equations

$$\{_m\ \text{tr}(\mathbf{Z}_i\mathbf{Z}_i'\mathbf{LZ}_j\mathbf{Z}_j'\mathbf{L}\}_{i,j=0}^{r}\hat{\boldsymbol{\sigma}}^2 = \{_c\ \mathbf{y}'\mathbf{LZ}_i\mathbf{Z}_i'\mathbf{Ly}\}_{i=0}^{r}\,. \tag{70}$$

(d) Show that $\boldsymbol{\sigma}_0^2 = [\theta \quad 0\mathbf{1}_r']$ used in place of $\boldsymbol{\sigma}^2$ reduces (67) to the MINQUEO equations (65).

(e) Explain why $\mathbf{L} = \mathbf{P}\sigma_e^2$.

E 11.9. [Hartley, Rao and LaMotte (1978).] These authors use a matrix that we call $\mathbf{W}$, which they describe as $\mathbf{X}$ orthogonalized "By a Gram–Schmidt orthogonalization process...", omitting any linearly dependent columns. Then for

$$\mathbf{V}_i = (\mathbf{I} - \mathbf{W}\mathbf{W}')\mathbf{Z}_i \tag{71}$$

their equations are

$$\{_{\mathrm{m}}\, \mathrm{tr}(\mathbf{V}_i\mathbf{V}_i'\mathbf{V}_j\mathbf{V}_j')\}_{i,j=0}^{4}\boldsymbol{\sigma}^2 = \{_{\mathrm{c}}\, \mathbf{y}'\mathbf{V}_i\mathbf{V}_i'\mathbf{y}\}_{i=0}^{4} . \tag{72}$$

Show that these are the MINQUE equations (65).

E 11.10. Derive (67) from (66), using (60).

E 11.11. (a) Write $\hat{\sigma}_\alpha^2 = (\mathrm{MSA} - \mathrm{MSE})/n$ of Section 3.5 as $\mathbf{y}'\mathbf{A}\mathbf{y}$ and derive $\mathbf{A}$ as a linear combination of the three matrices $\mathbf{I}_a \otimes \bar{\mathbf{J}}_n$, $\bar{\mathbf{J}}_{an}$ and $\mathbf{I}_{an}$.

(b) Show that $\mathbf{AX} = \mathbf{A1} = \mathbf{0}$.

(c) Use $\mathbf{A}$ derived in (a) to confirm $\hat{\sigma}_\alpha^2 = -10$ of (56) of Chapter 3.

(d) For data consisting of two observations in each of two classes, namely 4, 14 and 6, 16, calculate $\hat{\sigma}_\alpha^2 = (\mathrm{MSA} - \mathrm{MSE})/n$.

(e) For the data of (d) calculate the numerical value of $\mathbf{A}$ of (a) and use it to confirm $\hat{\sigma}_\alpha^2$ of (d); and show that $\mathbf{AX} = \mathbf{0}$.

CHAPTER 12

THE DISPERSION-MEAN MODEL

This chapter deals with the general mixed model restructured so as to be a linear model that has $\boldsymbol{\sigma}^2$ as the vector of parameters; i.e., for $\mathscr{X}$ and $\mathscr{Y}$ (which shall be defined), $E(\mathscr{Y}) = \mathscr{X}\boldsymbol{\sigma}^2$. It is called the dispersion-mean model and was first proposed by Pukelsheim (1974). It can also be viewed as an outcome of the seminal work of Seely (1971). A variation of it is used by Malley (1986).

12.1. THE MODEL

As in Chapter 11, we confine attention to estimating a linear function of variance components by a translation-invariant quadratic form $\mathbf{y}'\mathbf{A}\mathbf{y}$ with $\mathbf{A}$ symmetric. Hence we deal here only with $\mathbf{y}'\mathbf{A}\mathbf{y}$ where

$$\mathbf{y} = \mathbf{X}\boldsymbol{\beta} + \mathbf{Z}\mathbf{u} + \mathbf{e}$$

and

$$\mathbf{A} = \mathbf{A}' \quad \text{and} \quad \mathbf{A}\mathbf{X} = \mathbf{0} \tag{1}$$

as in Section 11.3a-ii. We also use $\mathbf{M}$ of (18) and (19) in Appendix M.4b, i.e.,

$$\mathbf{M} = \mathbf{I} - \mathbf{X}(\mathbf{X}'\mathbf{X})^{-}\mathbf{X}' = \mathbf{I} - \mathbf{X}\mathbf{X}^{+} = \mathbf{M}' = \mathbf{M}^2 \quad \text{with } \mathbf{M}\mathbf{X} = \mathbf{0}\,. \tag{2}$$

Then, along with the symmetry of both $\mathbf{A}$ and $\mathbf{M}$, it follows that

$$\mathbf{A}\mathbf{M} = \mathbf{A} = \mathbf{A}' = \mathbf{M}\mathbf{A}; \quad \text{and so} \quad \mathbf{M}\mathbf{A}\mathbf{M} = \mathbf{A}\mathbf{M} = \mathbf{A}\,. \tag{3}$$

Now consider $\mathbf{y}'\mathbf{A}\mathbf{y}$:

$$\begin{aligned}
\mathbf{y}'\mathbf{A}\mathbf{y} &= \mathbf{y}'\mathbf{M}\mathbf{A}\mathbf{M}\mathbf{y} && [\mathbf{A} = \mathbf{M}\mathbf{A}\mathbf{M}] \\
&= \text{vec}(\mathbf{y}'\mathbf{M}\mathbf{A}\mathbf{M}\mathbf{y}) && [\mathbf{y}'\mathbf{A}\mathbf{y} \text{ is scalar}] \\
&= [(\mathbf{M}\mathbf{y})' \otimes \mathbf{y}'\mathbf{M}]\,\text{vec}\,\mathbf{A} && [\text{vec}(\mathbf{A}\mathbf{B}\mathbf{C}) = (\mathbf{C}' \otimes \mathbf{A})\,\text{vec}\,\mathbf{B}] \\
&= (\text{vec}\,\mathbf{A})'(\mathbf{M}\mathbf{y} \otimes \mathbf{M}\mathbf{y}) && [\text{transposing a scalar}]\,. \qquad (4)
\end{aligned}$$

Therefore $E(\mathbf{y}'\mathbf{A}\mathbf{y})$ for any $\mathbf{A} = \mathbf{MAM}$ depends, apart from $\mathbf{A}$, only on

$$\begin{aligned}
E(\mathbf{My} \otimes \mathbf{My}) &= E[\mathbf{M}(\mathbf{y} - \mathbf{X}\boldsymbol{\beta}) \otimes \mathbf{M}(\mathbf{y} - \mathbf{X}\boldsymbol{\beta})] \qquad [\mathbf{MX} = \mathbf{0}] \\
&= (\mathbf{M} \otimes \mathbf{M})E[(\mathbf{y} - \mathbf{X}\boldsymbol{\beta}) \otimes (\mathbf{y} - \mathbf{X}\boldsymbol{\beta})] \quad [\mathbf{AB} \otimes \mathbf{RS} = (\mathbf{A} \otimes \mathbf{R})(\mathbf{B} \otimes \mathbf{S})] \\
&= (\mathbf{M} \otimes \mathbf{M})E\{\text{vec}[(\mathbf{y} - \mathbf{X}\boldsymbol{\beta})(\mathbf{y} - \mathbf{X}\boldsymbol{\beta})']\} \quad [\mathbf{t} \otimes \mathbf{t} = \text{vec}(\mathbf{tt}')] \\
&= (\mathbf{M} \otimes \mathbf{M}) \text{ vec } \mathbf{V} \\
&= (\mathbf{M} \otimes \mathbf{M}) \text{ vec}\left(\sum_{i=0}^{r} \mathbf{Z}_i\mathbf{Z}_i'\sigma_i^2\right) \\
&= (\mathbf{M} \otimes \mathbf{M})\Sigma_i \text{ vec}(\mathbf{Z}_i\mathbf{Z}_i')\sigma_i^2 \\
&= (\mathbf{M} \otimes \mathbf{M})\{_{\text{r}} \text{ vec}(\mathbf{Z}_i\mathbf{Z}_i')\}_{i=0}^{r}\{_{\text{c}}\, \sigma_i^2\}_{i=0}^{r} \\
&= \{_{\text{r}} \text{ vec}(\mathbf{MZ}_i\mathbf{Z}_i'\mathbf{M})\}_{i=0}^{r}\{_{\text{c}}\, \sigma_i^2\}_{i=0}^{r} \qquad [\text{vec}(\mathbf{ABC}) = (\mathbf{C}' \otimes \mathbf{A}) \text{ vec } \mathbf{B}] \,.
\end{aligned} \tag{5}$$

Now define

$$\mathscr{Y} = \mathbf{My} \otimes \mathbf{My} \quad \text{and} \quad \mathscr{X} = \{_{\text{r}} \text{ vec}(\mathbf{MZ}_i\mathbf{Z}_i'\mathbf{M})\} \,. \tag{6}$$

Then (5) is

$$E(\mathscr{Y}) = \mathscr{X}\boldsymbol{\sigma}^2, \tag{7}$$

which is a linear model for $\boldsymbol{\sigma}^2$, where, by virtue of (6) and the definition of $\mathbf{M}$ in (2), the elements of $\mathscr{Y}$ are squares and products of the residuals after fitting the fixed effects model $\mathbf{y} = \mathbf{X}\boldsymbol{\beta} + \mathbf{e}$ using ordinary least squares. Thus it is that, for estimating any linear function of the variance components by the translation-invariant quadratic form $\mathbf{y}'\mathbf{A}\mathbf{y}$, the model (7) is the underlying model for the variance components: and that quadratic form can always be expressed, using (6) and (7), as a linear function of elements of $\mathscr{Y}$; i.e., $\mathbf{y}'\mathbf{A}\mathbf{y} = (\text{vec } \mathbf{A})'\mathscr{Y}$ because

$$\begin{aligned}
\mathbf{y}'\mathbf{A}\mathbf{y} = \mathbf{y}'\mathbf{MAMy} &= \text{vec}(\mathbf{y}'\mathbf{MAMy}) = (\mathbf{y}'\mathbf{M} \otimes \mathbf{y}'\mathbf{M}) \text{ vec } \mathbf{A} \\
&= (\text{vec } \mathbf{A})'(\mathbf{My} \otimes \mathbf{My}) = (\text{vec } \mathbf{A})'\mathscr{Y} \,.
\end{aligned}$$

Hence the choice of any symmetric $\mathbf{A}$ (satisfying $\mathbf{AX} = \mathbf{0}$) to be used in $\mathbf{y}'\mathbf{A}\mathbf{y}$ is equivalent to the choice of the linear combination $(\text{vec } \mathbf{A})'\mathscr{Y}$ of elements of y. Therefore we can confine attention to (7). This is the key to much of the subsequent development of this chapter.

12.2. ORDINARY LEAST SQUARES (OLS) YIELDS MINQUEO

We show that the ordinary least squares (OLS) equations for estimating $\boldsymbol{\sigma}^2$ from (7), namely $\mathscr{X}'\mathscr{X}\boldsymbol{\sigma}^2 = \mathscr{X}'\mathrm{y}$, are in fact the MINQUEO equations of

Chapter 11. This is so because

$$\begin{aligned}\mathscr{X}'\mathscr{X} &= \{_c\, [\text{vec}(\mathbf{MZ}_i\mathbf{Z}_i'\mathbf{M})]'\}\{_r\, \text{vec}(\mathbf{MZ}_i\mathbf{Z}_i'\mathbf{M})\}\\ &= \{_m\, [\text{vec}(\mathbf{MZ}_i\mathbf{Z}_i'\mathbf{M})]'\, \text{vec}(\mathbf{MZ}_j\mathbf{Z}_j'\mathbf{M})\}\\ &= \{_m\, \text{tr}(\mathbf{MZ}_i\mathbf{Z}_i'\mathbf{MMZ}_j\mathbf{Z}_j'\mathbf{M})\} \qquad [(\text{vec}\,\mathbf{A}')'\,\text{vec}\,\mathbf{B} = \text{tr}(\mathbf{AB})]\\ &= \{_m\, \text{tr}(\mathbf{Z}_i\mathbf{Z}_i'\mathbf{MZ}_j\mathbf{Z}_j'\mathbf{M})\} \qquad [\mathbf{M} = \mathbf{M}^2]\,.\end{aligned}$$

Similar algebra reduces $\mathscr{X}'\mathbf{y}$, so that $\mathscr{X}'\mathscr{X}\boldsymbol{\sigma}^2 = \mathscr{X}'\mathbf{y}$ becomes (see E 12.1)

$$\{_m\, \text{tr}(\mathbf{Z}_i\mathbf{Z}_i'\mathbf{MZ}_j\mathbf{Z}_j'\mathbf{M})\}\boldsymbol{\sigma}^2 = \{_c\, \mathbf{y}'\mathbf{MZ}_i\mathbf{Z}_i'\mathbf{My}\},$$

which is the same as the MINQUEO equation in (59) of Chapter 11.

12.3. FOURTH MOMENTS IN THE MIXED MODEL

Having used OLS in the dispersion-mean model, we proceed to consider generalized least squares (GLS). This demands knowing var($\mathscr{Y}$). With $\mathscr{Y} = \mathbf{My} \otimes \mathbf{My} = (\mathbf{M} \otimes \mathbf{M})(\mathbf{y} \otimes \mathbf{y})$ from (6), it is clear that elements of $\mathscr{Y}$ involve squares of elements, and products of different elements of $\mathbf{y}$. Therefore var($\mathscr{Y}$) involves fourth moments of elements of $\mathbf{y}$. To derive var($\mathscr{Y}$), we begin with var($\mathbf{u} \otimes \mathbf{u}$) for $\mathbf{u}$ in the general mixed model

$$\mathbf{y} = \mathbf{X}\boldsymbol{\beta} + \mathbf{Zu} = \mathbf{X}\boldsymbol{\beta} + \sum_{i=0}^{r} \mathbf{Z}_i\mathbf{u}_i \quad \text{with } \text{var}(\mathbf{y}) = \mathbf{V} = \mathbf{ZDZ}' = \sum_{i=0}^{r} \mathbf{Z}_i\mathbf{Z}_i'\sigma_i^2, \tag{8}$$

having, as usual

$$\text{var}(\mathbf{u}) = E(\mathbf{uu}') = \mathbf{D} = \{_d\, \sigma_i^2\mathbf{I}_{q_i}\}_{i=0}^{r} \quad \text{and} \quad q = q. = \sum_{i=0}^{r} q_i\,. \tag{9}$$

a. Dispersion matrix of u ⊗ u

Noting that $\mathbf{u} \otimes \mathbf{u} = \text{vec}(\mathbf{uu}')$ gives

$$\begin{aligned}\text{var}(\mathbf{u} \otimes \mathbf{u}) &= E\;(\mathbf{u} \otimes \mathbf{u})(\mathbf{u} \otimes \mathbf{u})' - [E(\mathbf{u} \otimes \mathbf{u})][E(\mathbf{u} \otimes \mathbf{u})]'\\ &= E(\mathbf{uu}' \otimes \mathbf{uu}') - [E\;\text{vec}(\mathbf{uu}')][E\;\text{vec}(\mathbf{uu}')]'\\ &= E(\mathbf{uu}' \otimes \mathbf{uu}') - (\text{vec}\,\mathbf{D})(\text{vec}\,\mathbf{D})'\,.\end{aligned} \tag{10}$$

*-i. **A normalizing transformation.*** To simplify $E(\mathbf{uu}' \otimes \mathbf{uu}')$, define

$$\mathbf{w} = \mathbf{D}^{-\frac{1}{2}}\mathbf{u} = \{w_k\}_{k=1}^{q}, \quad \text{with } E(\mathbf{w}) = \mathbf{0} \text{ and } \text{var}(\mathbf{u}) = E(\mathbf{ww}') = \mathbf{I}_q;$$

and, for γ_k^* and γ_i being kurtosis parameters given by

$$E(w_k^4) = 3 + \gamma_k^* \quad \text{and} \quad E(u_{ij}^4) = \sigma_i^4(3 + \gamma_i) = E(u_{ij'}^4)\ \forall\ j',$$

we have

$$\{_d\, \gamma_k^*\}_{k=1}^{q} = \{_d\, \gamma_i\mathbf{I}_{q_i}\}_{i=0}^{r}\,. \tag{11}$$

This simply means that n_i of the γ_k^*s have the value γ_i; i.e., $\gamma_k^* = \gamma_i$ for $k = m_{i-1} + 1, \ldots, m_i$, where $m_i = \Sigma_{s=1}^{i} n_s$. Then

$$E(\mathbf{uu}' \otimes \mathbf{uu}') = (\mathbf{D}^{\frac{1}{2}} \otimes \mathbf{D}^{\frac{1}{2}})E(\mathbf{ww}' \otimes \mathbf{ww}')(\mathbf{D}^{\frac{1}{2}} \otimes \mathbf{D}^{\frac{1}{2}})\,. \tag{12}$$

The middle term in (12) is

$$E(\mathbf{ww}' \otimes \mathbf{ww}') = E\{_c\, w_k\{_r\, w_l \mathbf{ww}'\}_{l=1}^{q}\}_{k=1}^{q}$$

for which a typical term is $E(w_k w_l w_m w_n)$; and this, because the w_ks have zero mean and zero covariance, has only two non-zero values

$$E(w_k^4) = 3 + \gamma_k^* \quad \text{and} \quad E(w_k^2 w_l^2) = 1; \tag{13}$$

i.e., whenever $k = l = m = n$, $E(w_k w_l w_m w_n) = 3 + \gamma_k^*$, and if k, l, m and n are equal only in pairs $E(w_k w_l w_m w_n) = 1$. Otherwise, the value is zero. Using these values, we now illustrate $E(\mathbf{ww}' \otimes \mathbf{ww}')$ for a small example, and argue from that to the general case.

-ii. Example. Consider the model equation $y_{ij} = \mu + \alpha_i + e_{ij}$ for a 1-way classification with just one class, $i = 1$, having only two observations, $n_1 = 2$. This means $q = q_0 + q_1 = 2 + 1 = 3$, and

$$\mathbf{w}' = [e_{11}/\sigma_e \quad e_{12}/\sigma_e \quad \alpha_1/\sigma_\alpha] \,. \tag{14}$$

Then $\mathbf{ww}' \otimes \mathbf{ww}'$ has order 9, and the 9 pairs of subscripts on the $w_k w_l$ products in vec($\mathbf{ww}'$) are

$$[11 \quad 21 \quad 31 \quad 12 \quad 22 \quad 32 \quad 13 \quad 23 \quad 33]' \,.$$

The quartets of subscripts in $\mathbf{ww}' \otimes \mathbf{ww}'$, which is vec($\mathbf{ww}'$)[vec($\mathbf{ww}'$)]$'$, are therefore as shown in Table 12.1.

Each element in the matrix of Table 12.2 is the number of subscripts that are the same in the corresponding element of Table 12.1; e.g., in Table 12.2 the leading element is 4, corresponding to the element 1111 in Table 12.1. Thus the first five elements in the first row of Table 12.2 are 4, 3, 3, 3 and *tp*, the *tp* corresponding to the 1212 in Table 12.1, in which there are two pairs of equal subscripts. Then $E(\mathbf{ww}' \otimes \mathbf{ww}')$ is a 9×9 null matrix except that, in accord with (13), it has $3 + \gamma_k^*$ and unity corresponding to each element 4 and *tp*, respectively, in Table 12.2. Thus, on using (11) and (14) to get $\gamma_1^* = \gamma_2^* = \gamma_0$ and $\gamma_3^* = \gamma_1$,

$$E(\mathbf{ww}' \otimes \mathbf{ww}') = \left[\begin{array}{ccc|ccc|ccc} 3+\gamma_0 & \cdot & \cdot & \cdot & 1 & \cdot & \cdot & \cdot & 1 \\ \cdot & 1 & \cdot & 1 & \cdot & \cdot & \cdot & \cdot & \cdot \\ \cdot & \cdot & 1 & \cdot & \cdot & \cdot & 1 & \cdot & \cdot \\ \hline \cdot & 1 & \cdot & 1 & \cdot & \cdot & \cdot & \cdot & \cdot \\ 1 & \cdot & \cdot & \cdot & 3+\gamma_0 & \cdot & \cdot & \cdot & 1 \\ \cdot & \cdot & \cdot & \cdot & \cdot & 1 & \cdot & 1 & \cdot \\ \hline \cdot & \cdot & 1 & \cdot & \cdot & \cdot & 1 & \cdot & \cdot \\ \cdot & \cdot & \cdot & \cdot & \cdot & 1 & \cdot & 1 & \cdot \\ 1 & \cdot & \cdot & \cdot & 1 & \cdot & \cdot & \cdot & 3+\gamma_1 \end{array}\right] \tag{15}$$

TABLE 12.1. QUARTETS OF SUBSCRIPTS IN $\mathbf{ww}' \otimes \mathbf{ww}' = \text{vec}(\mathbf{ww}')[\text{vec}(\mathbf{ww}')]'$ for $\mathbf{w}' = [w_1 \quad w_2 \quad w_3]$

1111	1112	1113	1211	1212	1213	1311	1312	1313
1121	1122	1123	1221	1222	1223	1321	1322	1323
1131	1132	1133	1231	1232	1233	1331	1332	1333
2111	2112	2113	2211	2212	2213	2311	2312	2313
2121	2122	2123	2221	2222	2223	2321	2322	2323
2131	2132	2133	2231	2232	2233	2331	2332	2333
3111	3112	3113	3211	3212	3213	3311	3312	3313
3121	3122	3123	3221	3222	3223	3321	3322	3323
3131	3132	3133	3231	3232	3233	3331	3332	3333

TABLE 12.2. NUMBER OF EQUAL SUBSCRIPTS IN ELEMENTS OF TABLE 12.1, USING *tp* TO REPRESENT TWO PAIRS OF EQUALITIES

4	3	3	3	*tp*	2	3	2	*tp*
3	*tp*	2	*tp*	3	2	2	2	2
3	2	*tp*	2	2	2	*tp*	2	3
3	*tp*	2	*tp*	3	2	2	2	2
tp	3	2	3	4	3	2	3	*tp*
2	2	2	2	3	*tp*	2	*tp*	3
3	2	*tp*	2	2	2	*tp*	2	3
2	2	2	2	3	*tp*	2	*tp*	3
tp	2	3	2	*tp*	3	3	3	4

Scrutiny of (15) shows that

$$E(\mathbf{ww}' \otimes \mathbf{ww}') = \mathbf{T}_1 + \mathbf{T}_2 + \mathbf{T}_3 + \mathbf{T}_4, \tag{16}$$

where

$$\mathbf{T}_1 = \mathbf{I}_{3^2}, \tag{17}$$

$$\mathbf{T}_2 = \left[\begin{array}{ccc|ccc|ccc} 1 & \cdot & \cdot & \cdot & \cdot & \cdot & \cdot & \cdot & \cdot \\ \cdot & \cdot & \cdot & 1 & \cdot & \cdot & \cdot & \cdot & \cdot \\ \cdot & \cdot & \cdot & \cdot & \cdot & \cdot & 1 & \cdot & \cdot \\ \hline \cdot & 1 & \cdot & \cdot & \cdot & \cdot & \cdot & \cdot & \cdot \\ \cdot & \cdot & \cdot & \cdot & 1 & \cdot & \cdot & \cdot & \cdot \\ \cdot & \cdot & \cdot & \cdot & \cdot & \cdot & \cdot & 1 & \cdot \\ \hline \cdot & \cdot & 1 & \cdot & \cdot & \cdot & \cdot & \cdot & \cdot \\ \cdot & \cdot & \cdot & \cdot & \cdot & 1 & \cdot & \cdot & \cdot \\ \cdot & \cdot & \cdot & \cdot & \cdot & \cdot & \cdot & \cdot & 1 \end{array}\right] = \mathbf{I}_{(3,3)} \equiv \mathbf{S}_3 \quad \text{[Section M.9]}, \tag{18}$$

$$\mathbf{T}_3 = \left[\begin{array}{ccc|ccc|ccc} 1 & \cdot & \cdot & \cdot & 1 & \cdot & \cdot & \cdot & 1 \\ \cdot & \cdot & \cdot & \cdot & \cdot & \cdot & \cdot & \cdot & \cdot \\ \cdot & \cdot & \cdot & \cdot & \cdot & \cdot & \cdot & \cdot & \cdot \\ \hline \cdot & \cdot & \cdot & \cdot & \cdot & \cdot & \cdot & \cdot & \cdot \\ 1 & \cdot & \cdot & \cdot & 1 & \cdot & \cdot & \cdot & 1 \\ \cdot & \cdot & \cdot & \cdot & \cdot & \cdot & \cdot & \cdot & \cdot \\ \hline \cdot & \cdot & \cdot & \cdot & \cdot & \cdot & \cdot & \cdot & \cdot \\ \cdot & \cdot & \cdot & \cdot & \cdot & \cdot & \cdot & \cdot & \cdot \\ 1 & \cdot & \cdot & \cdot & 1 & \cdot & \cdot & \cdot & 1 \end{array}\right] = (\text{vec}\,\mathbf{I}_3)(\text{vec}\,\mathbf{I}_3)' \tag{19}$$

and

$$\mathbf{T}_4 = \left[\begin{array}{ccc|ccc|ccc} \gamma_0 & \cdot & \cdot & \cdot & \cdot & \cdot & \cdot & \cdot & \cdot \\ \cdot & \cdot & \cdot & \cdot & \cdot & \cdot & \cdot & \cdot & \cdot \\ \cdot & \cdot & \cdot & \cdot & \cdot & \cdot & \cdot & \cdot & \cdot \\ \hline \cdot & \cdot & \cdot & \cdot & \cdot & \cdot & \cdot & \cdot & \cdot \\ \cdot & \cdot & \cdot & \cdot & \gamma_0 & \cdot & \cdot & \cdot & \cdot \\ \cdot & \cdot & \cdot & \cdot & \cdot & \cdot & \cdot & \cdot & \cdot \\ \hline \cdot & \cdot & \cdot & \cdot & \cdot & \cdot & \cdot & \cdot & \cdot \\ \cdot & \cdot & \cdot & \cdot & \cdot & \cdot & \cdot & \cdot & \cdot \\ \cdot & \cdot & \cdot & \cdot & \cdot & \cdot & \cdot & \cdot & \gamma_1 \end{array}\right] = \{_{\text{d}}\ \gamma_0 \quad \mathbf{0}'_3 \quad \gamma_0 \quad \mathbf{0}'_3 \quad \gamma_1\}, \tag{20}$$

where $\mathbf{0}'_q$ is a row of q zeros. For the diagonal elements $3 + \gamma_0$ in (15) note that the 3 is accounted for in (16) because each of $\mathbf{T}_1$, $\mathbf{T}_2$ and $\mathbf{T}_3$ has 1 in the same element, and the γ_0 comes from $\mathbf{T}_4$. And for $\mathbf{T}_4$ of order $q^2 \times q^2$ rather than $3^2 \times 3^2$ of (20),

$$\begin{aligned}\mathbf{T}_4 = \{_{\text{d}}\ \gamma_0 \quad \mathbf{0}'_q \quad \gamma_0 \quad \mathbf{0}'_q \quad \cdots \quad \mathbf{0}'_q \quad \gamma_0 \quad \mathbf{0}'_q \quad \gamma_1 \quad \mathbf{0}'_q \quad \cdots \quad \gamma_1 \quad \mathbf{0}'_q \\ \cdots \quad \mathbf{0}'_q \quad \gamma_r \quad \mathbf{0}'_q \quad \cdots \quad \mathbf{0}'_q \quad \gamma_r\},\end{aligned} \tag{21}$$

with, from (11), γ_i occurring q_i times, for $i = 0, \ldots, r$. Inspection of (21) reveals that it is, using (11),

$$\mathbf{T}_4 = \{_{\text{d}}\ \text{vec}\{_{\text{d}}\ \gamma_i \mathbf{I}_{q_i}\}_{i=0}^{r}\} . \tag{22}$$

*-iii. **The general form of** $E(\mathbf{ww}' \otimes \mathbf{ww}')$.* The general form of (15) is now clear: using q^2-order forms of (17)–(19), together with (22), in (15) gives

$$E(\mathbf{ww}' \otimes \mathbf{ww}') = \mathbf{I} + \mathbf{S}_q + (\text{vec}\,\mathbf{I}_q)(\text{vec}\,\mathbf{I}_q)' + \mathbf{T}_{\text{A}} . \tag{23}$$

Therefore substituting (23) into (12) and then that into (10) gives

$$\begin{aligned}\text{var}(\mathbf{u} \otimes \mathbf{u}) &+ (\text{vec}\,\mathbf{D})(\text{vec}\,\mathbf{D})' \\ &= (\mathbf{D}^{\frac{1}{2}} \otimes \mathbf{D}^{\frac{1}{2}})[\mathbf{I} + \mathbf{S}_q + (\text{vec}\,\mathbf{I}_q)(\text{vec}\,\mathbf{I}_q)' + \mathbf{T}_{\text{A}}](\mathbf{D}^{\frac{1}{2}} \otimes \mathbf{D}^{\frac{1}{2}}) . \qquad (24)\end{aligned}$$

In multiplying out (24), the second term simplifies by using (58) of Appendix M.9, and the third term becomes $(\text{vec}\,\mathbf{D})(\text{vec}\,\mathbf{D})'$ based on (53) of Appendix M.9. Thus (24) becomes

$$\text{var}(\mathbf{u} \otimes \mathbf{u}) = (\mathbf{D} \otimes \mathbf{D})(\mathbf{I} + \mathbf{S}_q) + (\mathbf{D}^{\frac{1}{2}} \otimes \mathbf{D}^{\frac{1}{2}})\mathbf{T}_{\text{A}}(\mathbf{D}^{\frac{1}{2}} \otimes \mathbf{D}^{\frac{1}{2}}) . \qquad (25)$$

b. Fourth central moments of y

-i. General case. The fourth central moments of $\mathbf{y}$ are given by

$$\begin{aligned}\mathbf{F} &= \text{var}[(\mathbf{y} - \mathbf{X}\boldsymbol{\beta}) \otimes (\mathbf{y} - \mathbf{X}\boldsymbol{\beta})] \\ &= \text{var}(\mathbf{Zu} \otimes \mathbf{Zu}) \qquad (26) \\ &= (\mathbf{Z} \otimes \mathbf{Z})\,\text{var}(\mathbf{u} \otimes \mathbf{u})(\mathbf{Z}' \otimes \mathbf{Z}') \\ &= (\mathbf{Z} \otimes \mathbf{Z})[(\mathbf{D} \otimes \mathbf{D})(\mathbf{I} + \mathbf{S}_q) + (\mathbf{D}^{\frac{1}{2}} \otimes \mathbf{D}^{\frac{1}{2}})\mathbf{T}_{\text{A}}(\mathbf{D}^{\frac{1}{2}} \otimes \mathbf{D}^{\frac{1}{2}})](\mathbf{Z}' \otimes \mathbf{Z}'), \\ &\qquad \text{using (25)} \\ &= (\mathbf{V} \otimes \mathbf{V})(\mathbf{I} + \mathbf{S}_N) + (\mathbf{Z}\mathbf{D}^{\frac{1}{2}} \otimes \mathbf{Z}\mathbf{D}^{\frac{1}{2}})\{_{\text{d}}\, \text{vec}\{_{\text{d}}\, \gamma_i \mathbf{I}_{q_i}\}_{i=0}^{r}\}(\mathbf{D}^{\frac{1}{2}}\mathbf{Z}' \otimes \mathbf{D}^{\frac{1}{2}}\mathbf{Z}'), \qquad (27)\end{aligned}$$

using (8) and (22).

-ii. Under normality. Normality assumptions include $\gamma_i = 0$, which reduces (27) to

$$\mathbf{F}_{\mathcal{N}} = (\mathbf{V} \otimes \mathbf{V})(\mathbf{I} + \mathbf{S}_N) . \qquad (28)$$

c. Dispersion matrix of $\mathscr{Y}$

-i. General case

$$\begin{aligned}\text{var}(\mathscr{Y}) &= \text{var}(\mathbf{My} \otimes \mathbf{My}) \\ &= \text{var}[\mathbf{M}(\mathbf{y} - \mathbf{X}\boldsymbol{\beta}) \otimes \mathbf{M}(\mathbf{y} - \mathbf{X}\boldsymbol{\beta})] \quad [\mathbf{MX} = \mathbf{0}] \\ &= (\mathbf{M} \otimes \mathbf{M})\mathbf{F}(\mathbf{M} \otimes \mathbf{M}) \\ &= (\mathbf{MVM} \otimes \mathbf{MVM})(\mathbf{I} + \mathbf{S}_N) \qquad (29) \\ &\quad + (\mathbf{MZD}^{\frac{1}{2}} \otimes \mathbf{MZD}^{\frac{1}{2}})\{_{\text{d}}\, \text{vec}\{_{\text{d}}\, \gamma_i \mathbf{I}_{q_i}\}_{i=0}^{r}\}(\mathbf{D}^{\frac{1}{2}}\mathbf{Z}'\mathbf{M} \otimes \mathbf{D}^{\frac{1}{2}}\mathbf{Z}'\mathbf{M}),\end{aligned}$$

after using (58) of Section M.9. To facilitate subsequent discussion, especially when considering the case of zero kurtosis (e.g., normality), we label the two parts of the sum in (29) separately, and write

$$\text{var}(\mathscr{Y}) = \mathscr{V} + \mathscr{V}_k \qquad (30)$$

on writing $\mathbf{S}$ for $\mathbf{S}_N$ and defining $\mathscr{V}$ and $\mathscr{V}_k$ as

$$\mathscr{V} = (\mathbf{MVM} \otimes \mathbf{MVM})(\mathbf{I} + \mathbf{S}) \qquad (31)$$

and

$$\mathscr{V}_k = (\mathbf{M}\otimes\mathbf{M})(\mathbf{Z}\mathbf{D}^{\frac{1}{2}}\otimes\mathbf{Z}\mathbf{D}^{\frac{1}{2}})\{_d \operatorname{vec}\{_d\, \gamma_i \mathbf{I}_{q_i}\}_{i=0}^{r}\}(\mathbf{D}^{\frac{1}{2}}\mathbf{Z}'\otimes\mathbf{D}^{\frac{1}{2}}\mathbf{Z}')(\mathbf{M}\otimes\mathbf{M})\,. \tag{32}$$

-ii. ***Under normality.*** When $\mathbf{y}$ is normally distributed, i.e.,

$$\mathbf{y} \sim \mathscr{N}(\mathbf{X}\boldsymbol{\beta}, \mathbf{V}), \tag{33}$$

then $\gamma_i = 0 \;\forall\; i$ in (32) reduces (30) to $\mathscr{V}$ of (31) so that

$$\mathscr{Y} \sim (\mathscr{X}\sigma^2, \mathscr{V})\,. \tag{34}$$

d. Variance of a translation-invariant quadratic form

The quadratic form $\mathbf{y}'\mathbf{A}\mathbf{y}$ is translation-invariant when $\mathbf{A}$ is symmetric and $\mathbf{A}\mathbf{X} = \mathbf{0}$, whereupon its variance is

$$\begin{aligned}
\operatorname{var}(\mathbf{y}'\mathbf{A}\mathbf{y}) &= \operatorname{var}[(\mathbf{X}\boldsymbol{\beta}+\mathbf{Z}\mathbf{u})'\mathbf{A}(\mathbf{X}\boldsymbol{\beta}+\mathbf{Z}\mathbf{u})] \\
&= \operatorname{var}(\mathbf{u}'\mathbf{Z}'\mathbf{A}\mathbf{Z}\mathbf{u}) && [\mathbf{A}\mathbf{X}=\mathbf{0}] \\
&= \operatorname{var}[\operatorname{vec}(\mathbf{u}'\mathbf{Z}'\mathbf{A}\mathbf{Z}\mathbf{u})] && [\mathbf{u}'\mathbf{Z}'\mathbf{A}\mathbf{Z}\mathbf{u} \text{ is scalar}] \\
&= \operatorname{var}[(\mathbf{u}'\mathbf{Z}'\otimes\mathbf{u}'\mathbf{Z}')\operatorname{vec}\mathbf{A}] \\
&= \operatorname{var}[(\operatorname{vec}\mathbf{A})'(\mathbf{Z}\mathbf{u}\otimes\mathbf{Z}\mathbf{u})] \\
&= (\operatorname{vec}\mathbf{A})'\mathbf{F}(\operatorname{vec}\mathbf{A}), && \text{using (26)}\,.
\end{aligned} \tag{35}$$

Using (27) gives (35) as

$$\operatorname{var}(\mathbf{y}'\mathbf{A}\mathbf{y}) = v_1 + v_2$$

for

$$\begin{aligned}
v_1 &= (\operatorname{vec}\mathbf{A})'(\mathbf{V}\otimes\mathbf{V})(\mathbf{I}+\mathbf{S})\operatorname{vec}\mathbf{A} \\
&= 2(\operatorname{vec}\mathbf{A})'(\mathbf{V}\otimes\mathbf{V})\operatorname{vec}\mathbf{A} && [(57) \text{ of Section M.9}] \\
&= 2(\operatorname{vec}\mathbf{A})'\operatorname{vec}(\mathbf{V}\mathbf{A}\mathbf{V}) && [(53) \text{ of Section M.9}] \\
&= 2\operatorname{tr}[(\mathbf{A}\mathbf{V})^2] && [(54) \text{ of Section M.9}]
\end{aligned} \tag{36}$$

and for

$$v_2 = (\operatorname{vec}\mathbf{A})'(\mathbf{Z}\mathbf{D}^{\frac{1}{2}}\otimes\mathbf{Z}\mathbf{D}^{\frac{1}{2}})\{_d \operatorname{vec}\{_d\, \gamma_i\mathbf{I}_{q_i}\}_{i=0}^{r}\}(\mathbf{D}^{\frac{1}{2}}\mathbf{Z}'\otimes\mathbf{D}^{\frac{1}{2}}\mathbf{Z}')\operatorname{vec}\mathbf{A}\,. \tag{37}$$

Also, on using

$$\mathbf{H}_{q\times q} = \mathbf{D}^{\frac{1}{2}}\mathbf{Z}'\mathbf{A}\mathbf{Z}\mathbf{D}^{\frac{1}{2}} = \{h_{st}\}_{s,t=1}^{q} \quad\text{and}\quad \mathbf{L}_{q\times q} = \{_d\, \gamma_i\mathbf{I}_{q_i}\}_{i=0}^{r} = \{l_{st}\}_{s,t=1}^{q}, \tag{38}$$

(37) can, again, with the help of (53) of Section M.9, be written as

$$v_2 = (\operatorname{vec}\mathbf{H})'\{_d \operatorname{vec}\mathbf{L})\operatorname{vec}\mathbf{H} = \sum_{s=1}^{q}\sum_{t=1}^{q} h_{st}^2 l_{st}\,. \tag{39}$$

But, from (38), $\mathbf{L}$ is diagonal, with $l_{st} = 0$ except when $s = t$, for which l_{tt} is a γ_i. Therefore (39) is

$$v_2 = \sum_{i=0}^{r} \gamma_i \sum_{t=1}^{q_i} h_{tt}^2$$

$$= \sum_{i=0}^{r} \gamma_i \sum_{t=1}^{q_i} [\text{square of } (t,t)\text{th element of } (i,i)\text{th submatrix of } \mathbf{D}^{\frac{1}{2}}\mathbf{Z}'\mathbf{A}\mathbf{Z}\mathbf{D}^{\frac{1}{2}}]$$

$$= \sum_{i=0}^{r} \gamma_i \sigma_i^4 \sum_{t=1}^{q_i} [\text{square of } (t,t)\text{th element of } (i,i)\text{th submatrix of } \mathbf{Z}'\mathbf{A}\mathbf{Z}]$$

$$= \sum_{i=0}^{r} \gamma_i \sigma_i^4 (\text{sum of squares of diagonal elements of } \mathbf{Z}_i'\mathbf{A}\mathbf{Z}_i) \,. \tag{40}$$

Therefore from (36) and (40)

$$\text{var}(\mathbf{y}'\mathbf{A}\mathbf{y}) = 2\,\text{tr}(\mathbf{A}\mathbf{V})^2 + \sum_{i=0}^{r} \gamma_i \sigma_i^4 (\text{sum of squares of diagonal elements of } \mathbf{Z}_i'\mathbf{A}\mathbf{Z}_i) \,. \tag{41}$$

This is the variance, under non-normality, of a translation-invariant quadratic form $\mathbf{y}'\mathbf{A}\mathbf{y}$. Under normality, $\gamma_i = 0 \;\forall\; i$ and (41) reduces to the familiar form $\text{var}(\mathbf{y}'\mathbf{A}\mathbf{y}) = 2\,\text{tr}(\mathbf{A}\mathbf{V})^2$. Equation (41) is, of course, equivalent to the result given by Rao (1971b). See E 12.6.

12.4. GENERALIZED LEAST SQUARES (GLS)

a. GLS yields REML equations under normality

With normality, every γ_i is zero and so $\mathscr{V}_k$ of (32) is null. Then (30) gives $\text{var}(\mathscr{Y}) = \mathscr{V}$ for $\mathscr{V}$ of (31). For convenience we write

$$\mathbf{B} = \mathbf{M}\mathbf{V}\mathbf{M} \tag{42}$$

and then from (31)

$$\mathscr{V} = \text{var}(\mathscr{Y}) = (\mathbf{B} \otimes \mathbf{B})(\mathbf{I} + \mathbf{S}) = (\mathbf{I} + \mathbf{S})(\mathbf{B} \otimes \mathbf{B}), \tag{43a}$$

with

$$\mathscr{V}^- = \tfrac{1}{4}(\mathbf{I} + \mathbf{S})(\mathbf{B}^- \otimes \mathbf{B}^-) \,. \tag{43b}$$

These two expressions for $\mathscr{V}$ and $\mathscr{V}^-$ are based on (56) and (58) of Appendix M.9. GLSE on $E(\mathscr{Y}) = \mathscr{X}\boldsymbol{\sigma}^2$ then yields

$$\mathscr{X}'\mathscr{V}^-\mathscr{X}\hat{\boldsymbol{\sigma}}^2 = \mathscr{X}'\mathscr{V}^-\mathscr{Y} \,. \tag{44}$$

At first glance these equations seem innocent enough for calculating $\hat{\boldsymbol{\sigma}}^2$ as an estimator of $\boldsymbol{\sigma}^2$. But through $\mathscr{V}$ of (43) and $\mathbf{B}$ of (42) the $\mathscr{V}^-$ in (44) involves

$\boldsymbol{\sigma}^2$, which is unknown. Therefore (44) cannot be solved for $\hat{\boldsymbol{\sigma}}^2$. Nevertheless, by replacing $\hat{\boldsymbol{\sigma}}^2$ in (44) with $\boldsymbol{\sigma}^2$, we can think of the resulting equations, namely

$$\mathscr{X}'\mathscr{V}^-\mathscr{X}\boldsymbol{\sigma}^2 = \mathscr{X}'\mathscr{V}^-\mathscr{Y} \tag{45}$$

as being quasi-GLSE equations in $\boldsymbol{\sigma}^2$. Clearly their solution for $\boldsymbol{\sigma}^2$ will have to be obtained by numerical methods.

We proceed to reduce (45) to be in terms of $\mathbf{y}$, $\mathbf{X}$ and $\mathbf{Z}$ and functions thereof. First, from (6) and (43b)

$$\begin{aligned}\mathscr{X}'\mathscr{V}^- &= [(\mathbf{M}\otimes\mathbf{M})\{_r\,\text{vec}(\mathbf{Z}_i\mathbf{Z}_i')\}]'\tfrac{1}{4}(\mathbf{I}+\mathbf{S})(\mathbf{B}^-\otimes\mathbf{B}^-) &(46)\\ &= \tfrac{1}{4}[\{_r\,\text{vec}(\mathbf{M}\mathbf{Z}_i\mathbf{Z}_i'\mathbf{M})\}]'(\mathbf{I}+\mathbf{S})(\mathbf{B}^-\otimes\mathbf{B}^-), &\text{using (53) of Section M.8,}\\ &= \tfrac{1}{4}[(\mathbf{B}^-\otimes\mathbf{B}^-)(\mathbf{I}+\mathbf{S})\{_r\,\text{vec}(\mathbf{M}\mathbf{Z}_i\mathbf{Z}_i'\mathbf{M})\}_{i=0}^{r}]' &[\text{transpose}]\\ &= \tfrac{1}{2}[(\mathbf{B}^-\otimes\mathbf{B}^-)\{_r\,\text{vec}(\mathbf{M}\mathbf{Z}_i\mathbf{Z}_i'\mathbf{M})\}_{i=0}^{r}]' &[(57)\text{ of Section M.9 with }\theta=\tfrac{1}{2}]\\ &= \tfrac{1}{2}[\{_r\,\text{vec}(\mathbf{B}^-\mathbf{M}\mathbf{Z}_i\mathbf{Z}_i'\mathbf{M}\mathbf{B}^-)\}_{i=0}^{r}]', &\text{using (53) of Section M.8}\quad(47)\\ &= \tfrac{1}{2}\{_c\,[\text{vec}(\mathbf{B}^-\mathbf{M}\mathbf{Z}_i\mathbf{Z}_i'\mathbf{M}\mathbf{B}^-)]'\}_{i=0}^{r}\,.\end{aligned}$$

Therefore

$$\begin{aligned}\mathscr{X}'\mathscr{V}^-\mathscr{X} &= \tfrac{1}{2}\{_c\,[\text{vec}(\mathbf{B}^-\mathbf{M}\mathbf{Z}_i\mathbf{Z}_i'\mathbf{M}\mathbf{B}^-)]'\}\{_r\,\text{vec}(\mathbf{M}\mathbf{Z}_j\mathbf{Z}_j'\mathbf{M})\}\\ &= \tfrac{1}{2}\{_m\,\text{tr}[(\mathbf{B}^-\mathbf{M}\mathbf{Z}_i\mathbf{Z}_i'\mathbf{M}\mathbf{B}^-)\mathbf{M}\mathbf{Z}_j\mathbf{Z}_j'\mathbf{M}]\} \qquad &[(54)\text{ of Section M.8}]\\ &= \tfrac{1}{2}\{_m\,\text{tr}(\mathbf{Z}_i\mathbf{Z}_i'\mathbf{P}\mathbf{Z}_j\mathbf{Z}_j'\mathbf{P})\}_{i,j=0}^{r}\,. &[\text{Section M.4f}]\end{aligned}$$

Similarly, from (47) and (6)

$$\mathscr{X}'\mathscr{V}^-\mathscr{Y} = \tfrac{1}{2}[\{_c\,[\text{vec}(\mathbf{B}^-\mathbf{M}\mathbf{Z}_i\mathbf{Z}_i'\mathbf{M}\mathbf{B}^-]'\}_{i=0}^{r}](\mathbf{M}\mathbf{y}\otimes\mathbf{M}\mathbf{y}),$$

and this is a column of scalars. Therefore it is

$$\begin{aligned}\mathscr{X}'\mathscr{V}^-\mathscr{Y} &= \tfrac{1}{2}\{_c\,(\mathbf{y}'\mathbf{M}\otimes\mathbf{y}'\mathbf{M})\,\text{vec}(\mathbf{B}^-\mathbf{M}\mathbf{Z}_i\mathbf{Z}_i'\mathbf{M}\mathbf{B}^-)\}_{i=0}^{r}\\ &= \tfrac{1}{2}\{_c\,\mathbf{y}'\mathbf{M}(\mathbf{M}\mathbf{V}\mathbf{M})^-\mathbf{M}\mathbf{Z}_i\mathbf{Z}_i'\mathbf{M}(\mathbf{M}\mathbf{V}\mathbf{M})^-\mathbf{M}\mathbf{y}\}_{i=0}^{r}, &\text{using (53) of Section M.8,}\\ &= \tfrac{1}{2}\{_c\,\mathbf{y}'\mathbf{P}\mathbf{Z}_i\mathbf{Z}_i'\mathbf{P}\mathbf{y}\}_{i=0}^{r}\,. &[\text{Section M.4f}]\end{aligned}$$

Therefore the GLS equation $\mathscr{X}'\mathscr{V}^-\mathscr{X}\boldsymbol{\sigma}^2 = \mathscr{X}'\mathscr{V}^-\mathbf{y}$ of (44) is

$$\{_m\,\text{tr}(\mathbf{Z}_i\mathbf{Z}_i'\mathbf{P}\mathbf{Z}_j\mathbf{Z}_j'\mathbf{P})\}\boldsymbol{\sigma}^2 = \{_c\,\mathbf{y}'\mathbf{P}\mathbf{Z}_i\mathbf{Z}_i'\mathbf{P}\mathbf{y}\}, \tag{48}$$

which is the same as the REML estimation equation in (90) of Chapter 6.

What we have dealt with here is the case of normality—of assuming $\mathbf{y}\sim\mathcal{N}(\mathbf{X}\boldsymbol{\beta},\mathbf{V})$. That leads to $\text{var}(\mathscr{Y})=\mathscr{V}$ as in (34), which, via (45), produced (48). But nothing prevents us conceptually from considering exactly the same approach with non-normal data. The sole change in (45) will be that $\mathscr{V}$ will be replaced by $\mathscr{V} + \mathscr{V}_k$ of (30)–(32). This offers an extension of the ideas of REML to situations other than normality, and is an alternative to the marginal likelihood interpretation discussed in Section 9.2.

We next show that REML equations are the same as BLUE equations in the dispersion-mean model $E(\mathscr{Y}) = \mathscr{X}\boldsymbol{\sigma}^2$. To do so, we need some results from estimating the fixed effects when $\mathbf{V}$ is assumed known.

b. Excursus on estimating fixed effects

In the general linear model $E(\mathbf{y}) = \mathbf{X}\boldsymbol{\beta}$ with $\text{var}(\mathbf{y}) = \mathbf{V}$ and $\mathbf{V}$ non-singular it is well known that the GLS estimator of $\mathbf{X}\boldsymbol{\beta}$, denoted GLSE($\mathbf{X}\boldsymbol{\beta}$), is the same as the best linear unbiased estimator BLUE($\mathbf{X}\boldsymbol{\beta}$):

$$\text{GLSE}(\mathbf{X}\boldsymbol{\beta}) = \mathbf{X}(\mathbf{X}'\mathbf{V}^{-1}\mathbf{X})^{-}\mathbf{X}'\mathbf{V}^{-1}\mathbf{y} = \text{BLUE}(\mathbf{X}\boldsymbol{\beta})\,.$$

When $\mathbf{V}$ is singular, it can be shown that using $\mathbf{V}^{-}$ in place of $\mathbf{V}^{-1}$ yields $\text{GLSE}(\mathbf{X}\boldsymbol{\beta}) = \mathbf{X}(\mathbf{X}'\mathbf{V}^{-}\mathbf{X})^{-}\mathbf{X}'\mathbf{V}^{-}\mathbf{y}$. For this case we have two important theorems concerning the equality of GLSE($\mathbf{X}\boldsymbol{\beta}$), OLSE($\mathbf{X}\boldsymbol{\beta}$) and BLUE($\mathbf{X}\boldsymbol{\beta}$). The first is from Zyskind and Martin (1969), that

$$\text{GLSE}(\mathbf{X}\boldsymbol{\beta}) = \text{BLUE}(\mathbf{X}\boldsymbol{\beta}) \quad \text{if and only if} \quad \mathbf{V}\mathbf{V}^{-}\mathbf{X} = \mathbf{X}\,. \tag{49}$$

The proof of this is omitted, because of its length—primarily because it requires the derivation of BLUE($\mathbf{X}\boldsymbol{\beta}$) for singular $\mathbf{V}$ as

$$\text{BLUE}(\mathbf{X}\boldsymbol{\beta}) = (\mathbf{I} - \mathbf{M})[\mathbf{I} - \mathbf{V}\mathbf{M}(\mathbf{M}\mathbf{V}\mathbf{M})^{-}\mathbf{M}]\mathbf{y}$$

(see Pukelsheim, 1974; Albert, 1976). The second theorem we use is that

$$\text{BLUE}(\mathbf{X}\boldsymbol{\beta}) = \text{OLSE}(\mathbf{X}\boldsymbol{\beta}) \quad \text{iff} \quad \mathbf{V}\mathbf{X} = \mathbf{X}\mathbf{Q} \text{ for some } \mathbf{Q}\,. \tag{50}$$

This comes from Zyskind (1967), wherein $\mathbf{V}\mathbf{X} = \mathbf{X}\mathbf{Q}$ is only one form of the theorem's necessary and sufficient condition. The theorem applies for both nonsingular and singular $\mathbf{V}$. The proof for nonsingular $\mathbf{V}$ is left to the reader as E 12.4; that for singular $\mathbf{V}$ is lengthy (see Searle and Pukelsheim, 1989).

c. REML is BLUE

Having shown in (48) that quasi-GLS estimation in the model $E(\mathscr{Y}) = \mathscr{X}\boldsymbol{\sigma}^2$ leads to REML equations, we now show that REML equations are BLUE equations in that model by showing under normality that (49) is satisfied; i.e., that $\mathscr{V}\mathscr{V}^{-}\mathscr{X} = \mathscr{X}$. We see this as follows.

$$\begin{aligned}\mathscr{V}\mathscr{V}^{-}\mathscr{X} &= \mathscr{V}(\mathscr{X}'\mathscr{V}^{-})' \\ &= (\mathbf{MVM} \otimes \mathbf{MVM})(\mathbf{I} + \mathbf{S})\tfrac{1}{2}[\{_r\, \text{vec}(\mathbf{MVM})^{-}\mathbf{MZ}_i\mathbf{Z}_i'\mathbf{M}(\mathbf{MVM})^{-}\}]\end{aligned}$$

from $\mathscr{V}$ of (43) and $\mathscr{X}'\mathscr{V}^{-}$ of (47). Hence

$$\mathscr{V}\mathscr{V}^{-}\mathscr{X} = \tfrac{1}{2}(\mathbf{I} + \mathbf{S})\{_r\, \text{vec}[\mathbf{MVM}(\mathbf{MVM})^{-}\mathbf{MZ}_i\mathbf{Z}_i'\mathbf{M}(\mathbf{MVM})^{-}\mathbf{MVM}]\},$$

and because $\mathbf{V}$ is n.n.d., $\mathbf{MVM}(\mathbf{MVM})^{-}\mathbf{MZ}_i = \mathbf{MZ}_i$ and so

$$\mathscr{V}\mathscr{V}^{-}\mathscr{X} = \tfrac{1}{2}(\mathbf{I} + \mathbf{S})\{_r\, \text{vec}[\mathbf{MZ}_i\mathbf{Z}_i'\mathbf{M})\} = \{_r\, \text{vec}(\mathbf{MZ}_i\mathbf{Z}_i'\mathbf{M})\} = \mathscr{X}\,. \tag{51}$$

Q.E.D.

Therefore, by the Zyskind and Martin (1969) result in (49), the REML equations are BLUE equations for the estimation of $\boldsymbol{\sigma}^2$ from the dispersion-mean model.

12.5. MODIFIED GLS YIELDS ML

Suppose we knew $\boldsymbol{\beta}$. Then consider

$$\mathscr{Y}_0 = (\mathbf{y} - \mathbf{X}\boldsymbol{\beta}) \otimes (\mathbf{y} - \mathbf{X}\boldsymbol{\beta}) \tag{52}$$

with

$$\begin{aligned} E(\mathscr{Y}_0) &= E\{\text{vec}[(\mathbf{y} - \mathbf{X}\boldsymbol{\beta})(\mathbf{y} - \mathbf{X}\boldsymbol{\beta})']\} = \text{vec}\,\mathbf{V} = \text{vec}(\Sigma_i \mathbf{Z}_i\mathbf{Z}_i'\sigma_i^2) \\ &= \{_\text{r}\, \text{vec}(\mathbf{Z}_i\mathbf{Z}_i')\}\boldsymbol{\sigma}^2 = \mathbf{C}\boldsymbol{\sigma}^2, \quad \text{for } \mathbf{C} = \{_\text{r}\, \text{vec}(\mathbf{Z}_i\mathbf{Z}_i')\}\,. \end{aligned} \tag{53}$$

Hence, on assuming normality of $\mathbf{y}$ and using (26) and (28), we have

$$\mathscr{Y}_0 \sim (\mathbf{C}\boldsymbol{\sigma}^2, \mathbf{F}_{\mathscr{N}}) \quad \text{for } \mathbf{F}_{\mathscr{N}} = (\mathbf{V} \otimes \mathbf{V})(\mathbf{I} + \mathbf{S})\,. \tag{54}$$

In comparing $\mathscr{Y} \sim (\mathscr{X}\boldsymbol{\sigma}^2, \mathscr{V})$ with $\mathscr{Y}_0 \sim (\mathbf{C}\boldsymbol{\sigma}^2, \mathbf{F}_{\mathscr{N}})$ we see from (6) and (53) that

$$\mathscr{X} = \{_\text{r}\, \text{vec}(\mathbf{M}\mathbf{Z}_i\mathbf{Z}_i'\mathbf{M})\} \quad \text{and} \quad \mathbf{C} = \{_\text{r}\, \text{vec}(\mathbf{Z}_i\mathbf{Z}_i')\},$$

and from (43a) and (54) that

$$\mathscr{V} = (\mathbf{B} \otimes \mathbf{B})(\mathbf{I} + \mathbf{S}), \quad \text{with } \mathbf{B} = \mathbf{MVM}, \quad \text{and} \quad \mathbf{F}_{\mathscr{N}} = (\mathbf{V} \otimes \mathbf{V})(\mathbf{I} + \mathbf{S})\,.$$

Therefore $\mathbf{C}$ is $\mathscr{X}$ (and $\mathbf{F}_{\mathscr{N}}$ is $\mathscr{V}$) with $\mathbf{M}$ replaced by $\mathbf{I}$. Thus, because $\mathscr{V}\mathscr{V}^-\mathscr{X} = \mathscr{X}$, as in (51) it is clear that

$$\mathbf{F}_{\mathscr{N}}\mathbf{F}_{\mathscr{N}}^-\mathbf{C} = \mathbf{C}, \tag{55}$$

and so GLS applied to $\mathscr{Y}_0$ of (54) yields the BLUE equations for $\boldsymbol{\sigma}^2$.

Furthermore, because (48) is the quasi-GLSE equation for $\boldsymbol{\sigma}^2$ obtained from $\mathscr{Y}$, replacing $\mathbf{M}$ by $\mathbf{I}$ in (48) gives the quasi-GLSE equation for $\boldsymbol{\sigma}^2$ obtainable from $\mathscr{Y}_0$. One must also replace $\mathscr{Y} = \mathbf{My} \otimes \mathbf{My}$ by $\mathscr{Y}_0 = (\mathbf{y} - \mathbf{X}\boldsymbol{\beta}) \otimes (\mathbf{y} - \mathbf{X}\boldsymbol{\beta})$, which (after replacing $\mathbf{M}$ by $\mathbf{I}$), is equivalent to replacing $\mathbf{y}$ by $\mathbf{y} - \mathbf{X}\boldsymbol{\beta}$; and note that replacing $\mathbf{M}$ by $\mathbf{I}$ in $\mathbf{P} = \mathbf{M}(\mathbf{MVM})^-\mathbf{M}$ (see E 12.3) means replacing $\mathbf{P}$ by $\mathbf{V}^{-1}$ (assuming $\mathbf{V}$ to be non-singular). Thus (48) becomes

$$\{_\text{m}\, \text{tr}(\mathbf{Z}_i\mathbf{Z}_i'\mathbf{V}^{-1}\mathbf{Z}_j\mathbf{Z}_j'\mathbf{V}^{-1})\}_{i,j=0}^{r}\boldsymbol{\sigma}^2 = \{_\text{c}\, (\mathbf{y} - \mathbf{X}\boldsymbol{\beta})'\mathbf{V}^{-1}\mathbf{Z}_i\mathbf{Z}_i'\mathbf{V}^{-1}(\mathbf{y} - \mathbf{X}\boldsymbol{\beta})\}_{i=0}^{r}, \tag{56}$$

as the GLSE equations based on $\mathscr{Y}_0$. And because of (55) they are the BLUE equations for estimating $\boldsymbol{\sigma}^2$ from $\mathscr{Y}_0$.

An impracticality of (56) is that $\boldsymbol{\beta}$ is unknown. Replacing $\mathbf{X}\boldsymbol{\beta}$ by $\mathbf{X}\boldsymbol{\beta}^0 = \mathbf{X}(\mathbf{X}'\mathbf{V}^{-1}\mathbf{X})^{-1}\mathbf{X}'\mathbf{V}^{-1}\mathbf{y}$ leads to replacing $\mathbf{V}^{-1}(\mathbf{y} - \mathbf{X}\boldsymbol{\beta})$ by $\mathbf{V}^{-1}(\mathbf{y} - \mathbf{X}\boldsymbol{\beta}^0) = \mathbf{Py}$, and then the equations are

$$\{_\text{m}\, \text{tr}(\mathbf{V}^{-1}\mathbf{Z}_i\mathbf{Z}_i'\mathbf{V}^{-1}\mathbf{Z}_j\mathbf{Z}_j')\}_{i,j=0}^{r} = \{_\text{c}\, \mathbf{y}'\mathbf{P}\mathbf{Z}_i\mathbf{Z}_i'\mathbf{P}\mathbf{y}\}_{i=0}^{r}, \tag{57}$$

which are the ML equations in Chapter 6. This result was first obtained by Anderson (1978).

12.6. BALANCED DATA

Estimation of variance components from balanced data involves numerous relationships emanating from the incidence matrices **X** and **Z** being partitionable into submatrices that are Kronecker products (KPs) of identity matrices and summing vectors; and with $\mathbf{V} = \Sigma_i \sigma_i^2 \mathbf{Z}_i \mathbf{Z}_i'$ having each $\mathbf{Z}_i \mathbf{Z}_i'$ as a KP of **I**- and **J**-matrices (details are shown in Section 4.6). These relationships lead to estimators from balanced data that have several attractive properties.

Seely (1971) has very general results that are salient to establishing some of these properties. They lead, for example, to the result under normality (i.e., when every kurtosis parameter $\gamma_i = 0$) that

$$\text{ANOVA} = \text{UMVUTIQ}, \tag{58}$$

meaning ANOVA estimators have the property of being uniformly minimum variance, unbiased, translation-invariant, quadratic. We adopt freely from Anderson *et al.* (1984) in discussing this topic further.

a. Estimation under zero kurtosis

-i. History. The variances of the minimum variance estimators of (58) do, of course, depend on $\text{var}(\mathbf{y}) = \mathbf{V} = \Sigma_i \sigma_i^2 \mathbf{V}_i$, where $\mathbf{V}_i = \mathbf{Z}_i \mathbf{Z}_i'$. More than that, existence of UMVUTIQ estimators of the σ_i^2s comes from **V** and the $\mathbf{V}_i$s having a certain structure. To be precise, let

$$\mathscr{B} = \{\Sigma_i t_i \mathbf{V}_i \mid t_r \in \mathbb{R}\}, \tag{59}$$

be the set of all matrices that are linear combinations $\Sigma_i t_i \mathbf{V}_i$ of the $\mathbf{V}_i$s for the t_is being any real scalars (represented by $\mathbb{R}$). Then $\mathscr{B}$ is defined by Seely (1971) as a quadratic subspace of symmetric matrices when every member **B** of $\mathscr{B}$ has $\mathbf{B}^2$ also in $\mathscr{B}$.

Seely's (1971, p. 715) results on uniform minimum variance unbiased estimation are established on the basis of two assumptions:

(a) that $\mathscr{B}$ is a quadratic subspace of symmetric matrices; and
(b) that matrices $\mathbf{H}_i$ exist such that $\mathbf{V}_i \mathbf{X} = \mathbf{X} \mathbf{H}_i$ for each i.

These assumptions certainly hold in most fixed effects model, as in Atiqullah (1962), wherein $\mathbf{V} = \sigma_0^2 \mathbf{V}_0$ and $\mathbf{V}_0 = \mathbf{I}_N$. They also hold for the random effects model in Theorem 7 of Graybill and Hultquist (1961), since their requirement that an analysis of variance exists leads to Seely's assumption (a), while their assumption (iv) is Seely's assumption (b). Since Seely (1971, p. 717) shows that his assumptions (a) and (b) necessitate invariance of the resulting estimator, neither Atiqullah (1962) nor Graybill and Hultquist (1961) need a restriction to invariant quadratic estimators.

In general, however, an ANOVA model with balanced data does not necessarily satisfy Seely's assumption (a) for the same kind of reasons that Seely's (1971, p. 719) example of the balanced incomplete block design does not, and as further evidenced in Example 1 of Kleffe and Pincus (1974, p. 53). Another demonstration that $\mathscr{B}$ is not always a quadratic subspace is given by Searle and Henderson (1979) for the 2-way crossed classification where both $\mathbf{V}^{-1}$ and $\mathbf{V}^2$ include a term in $\mathbf{J}_N$ whereas $\mathbf{V}$ itself does not. But the $\mathbf{V}_i$s of $\mathbf{V}$, together with $\mathbf{J}_N$ do form a quadratic subspace and $\mathbf{V}$ is a member of it. Indeed, there are typically two distinct situations.

(1) For some models (e.g., crossed classification models having no nested factors) the $\mathbf{V}_i$s do not define a quadratic subspace. This is because, by the crossed nature of the factors, there is a product of two $\mathbf{V}_i$s that yields $\mathbf{J}_N$, and $\mathbf{J}_N$ has to be included in $\mathscr{B}$.
(2) For other models (e.g., completely nested models, and mixed models having random factors that are, within themselves, effectively nested) the $\mathbf{V}_i$s define a quadratic subspace and no product $\mathbf{V}_i\mathbf{V}_j$ yields $\mathbf{J}_N$, and so there is no need to include $\mathbf{J}_N$.

In contrast to $\mathbf{V} = \text{var}(\mathbf{y})$, consider the variance of $\mathbf{My}$ from which $\mathscr{Y} = \mathbf{My} \otimes \mathbf{My}$ is formed:

$$\text{var}(\mathbf{My}) = \mathbf{MVM} = \sum_{i=0}^{r} \sigma_i^2 \mathbf{MV}_i\mathbf{M} . \tag{60}$$

The analogous form of $\mathscr{B}$ for matrices $\mathbf{MV}_i\mathbf{M}$ is then

$$\mathscr{B}_{\mathrm{M}} = \{\Sigma_i t_i \mathbf{MV}_i\mathbf{M} \mid t_i \in \mathbb{R}\} . \tag{61}$$

Concerning $\mathscr{B}_{\mathrm{M}}$, Theorem 6 of Kleffe and Pincus (1974, p. 52) shows that in any linear model the quadratic subspace property that is not always evident in $\mathbf{V}$ is needed only of $\mathscr{B}_{\mathrm{M}}$. For balanced data this is always the case, i.e., $\mathscr{B}_{\mathrm{M}}$ defines a quadratic subspace, resulting from the fact that $\mathbf{M}$ and the $\mathbf{V}_i$s are all linear combinations of KPs of **I**s and **J**s. No matrix such as $\mathbf{J}_N$ ever has to be included with the the $\mathbf{MV}_i\mathbf{M}$s. This is so because $\mathbf{MJ}_N$ is null. Note, too, that the analogue of Seely's assumption (b) is trivially satisfied, since $\mathbf{My}$ has expectation zero.

Theorems 1 and 3 of Seely (1971) assert that for balanced data with zero kurtosis there exists an unbiased invariant quadratic estimator of the variance components that has uniformly minimum variance in its class (UMVUIQ). Under normality this estimator retains the UMV property among all unbiased invariant estimators, whether they are quadratic or not (UMVUI). We now show that this estimator also coincides with the ANOVA estimator, thus justifying (60) and (61).

*-ii. **The model.*** Under zero kurtosis (every $\gamma_i = 0$) we have, as in (34),

$$\mathscr{Y} = (\mathbf{My} \otimes \mathbf{My}) \sim (\mathscr{X}\boldsymbol{\sigma}^2, \mathscr{V}) \qquad (62)$$

for $\mathscr{X}$ of (6) and $\mathscr{V}$ of (31). We now show two important properties of this model.

Property A of the model: $\mathbf{Z}_i\mathbf{Z}_i'\mathbf{X} = \mathbf{XQ}_i$ for some $\mathbf{Q}_i$. As described in Chapter 4, when there are m main effect factors in a model, $\mathbf{X}$ and $\mathbf{Z}$ can each be partitioned, $\mathbf{X}$ into f submatrices (for f fixed effects factors) and $\mathbf{Z}$ into $r + 1$ submatrices (for r random effects factors plus error). Each submatrix is a KP of $m + 1$ matrices, each of which is an $\mathbf{I}$ or a $\mathbf{1}$. Therefore there are 2^{m+1} possible matrices that can be submatrices of $\mathbf{X}$ or $\mathbf{Z}$. Furthermore, each of them (typified for convenience as $\mathbf{Z}_h$, be it a submatrix of $\mathbf{X}$ or of $\mathbf{Z}$) is such that $\mathbf{Z}_h\mathbf{Z}_k'$ is a KP of $m + 1$ matrices that are each $\mathbf{I}$ or $\mathbf{J}$. Hence, because $\mathbf{I}$ and $\mathbf{J}$ matrices commute in multiplication, so do $\mathbf{Z}_h\mathbf{Z}_h'$ and $\mathbf{Z}_k\mathbf{Z}_k'$. Moreover, since $\mathbf{I}^2 = \mathbf{I}$, $\mathbf{IJ} = \mathbf{JI} = \mathbf{J}$ and $\mathbf{J}^2 = n\mathbf{J}$,

$$\mathbf{Z}_h\mathbf{Z}_h'\mathbf{Z}_k\mathbf{Z}_k' = \mathbf{Z}_k\mathbf{Z}_k'\mathbf{Z}_h\mathbf{Z}_h' = \phi_l\mathbf{Z}_l\mathbf{Z}_l' \qquad (63)$$

for every pair, $h, k = 1, 2, \ldots, 2^{m+1}$, and for $\phi_l > 0$ being a scalar and l being some integer in the range $1, 2, \ldots, 2^{m+1}$. Also, (63) means that, for example

$$\mathbf{Z}_i\mathbf{Z}_i'\mathbf{V} = \Sigma_j\sigma_j^2\mathbf{Z}_i\mathbf{Z}_i'\mathbf{Z}_j\mathbf{Z}_j' = \Sigma_j\sigma_j^2\phi_{ij}\mathbf{Z}_j\mathbf{Z}_j' \qquad (64)$$

for ϕ_{ij} being some scalar. Hence

$$\text{for } \mathbf{V} = \sum_{i=0}^{r} \sigma_i^2\mathbf{Z}_i\mathbf{Z}_i' \quad \text{we have} \quad \mathbf{V}^2 = \sum_{l=1}^{2^{m+1}} \theta_l\mathbf{Z}_l\mathbf{Z}_l', \qquad (65)$$

where θ_l may be zero for some values of l in the range $1, 2, \ldots, 2^{m+1}$.

Now define $\mathbf{V}_i$, $\mathbf{V}_{it}$, $\mathbf{X}_p$ and $\mathbf{X}_{pt}$ by the following equations, where $\mathbf{V}_{i,m+1}$ and $\mathbf{X}_{p,m+1}$ are, respectively, $\mathbf{V}_{it}$ and $\mathbf{V}_{pt}$ for $t = m + 1$:

$$\mathbf{X} = [\mathbf{X}_1 \quad \ldots \quad \mathbf{X}_p \quad \ldots \quad \mathbf{X}_f],$$

$$\mathbf{V}_i = \mathbf{Z}_i\mathbf{Z}_i' = (\mathbf{V}_{i1} \otimes \cdots \otimes \mathbf{V}_{it} \otimes \cdots \otimes \mathbf{V}_{i,m+1}), \qquad (66)$$

$$\mathbf{X}_p = (\mathbf{X}_p \otimes \cdots \otimes \mathbf{X}_{pt} \otimes \cdots \otimes \mathbf{X}_{p,m+1})\,. \qquad (67)$$

Then

$$\mathbf{Z}_i\mathbf{Z}_i'\mathbf{X} = \mathbf{V}_i\{_r\,\mathbf{X}_p\}_{p=1}^{f} = \{_r\,\mathbf{V}_i\mathbf{X}_p\}_{p=1}^{f}\,.$$

Hence using (66) and (67) and applying $(\mathbf{A} \otimes \mathbf{B})(\mathbf{R} \otimes \mathbf{S}) = \mathbf{AR} \otimes \mathbf{BS}$ gives

$$\begin{aligned}\mathbf{Z}_i\mathbf{Z}_i'\mathbf{X} &= \{_r\,\mathbf{V}_{i1}\mathbf{X}_{p1} \otimes \mathbf{V}_{i2}\mathbf{X}_{p2} \otimes \cdots \otimes \mathbf{V}_{i,m+1}\mathbf{V}_{p,m+1}\}_{p=1}^{f} \\ &= \{_r\,\bigotimes_{t=1}^{m+1} \mathbf{V}_{it}\mathbf{X}_{pt}\}_{p=1}^{f}\,. \end{aligned} \qquad (68)$$

TABLE 12.3. PRODUCTS $\mathbf{V}_{it}\mathbf{X}_{pt} = \mathbf{X}_{pt}\mathbf{G}_{ipt}$

$\mathbf{V}_{it}$	$\mathbf{X}_{pt}$	$\mathbf{V}_{it}\mathbf{X}_{pt} = \mathbf{X}_{pt}\mathbf{G}_{ipt}$	$\mathbf{G}_{ipt}$
$\mathbf{I}$	$\mathbf{I}$	$\mathbf{I} = \mathbf{I}\mathbf{I}$	$\mathbf{I}$
$\mathbf{I}$	$\mathbf{1}$	$\mathbf{1} = \mathbf{1}1$	1
$\mathbf{J}$	$\mathbf{I}$	$\mathbf{J} = \mathbf{I}\mathbf{J}$	$\mathbf{J}$
$\mathbf{J}$	$\mathbf{1}$	$N_t\mathbf{1} = \mathbf{1}N_t$	N_t

Now in (66), each $\mathbf{V}_{it}$ is either $\mathbf{I}$ or $\mathbf{J}$, and in (67) each $\mathbf{X}_{pt}$ is either $\mathbf{I}$ or $\mathbf{1}$, all of order N_t, the number of levels of the tth main effect. Therefore the four possible values of the product $\mathbf{V}_{it}\mathbf{X}_{pt}$ in (68), together with a matrix $\mathbf{G}_{ipt}$ defined such that $\mathbf{V}_{it}\mathbf{V}_{pt} = \mathbf{X}_{pt}\mathbf{G}_{ipt}$ in each case, are as shown in Table 12.3. Therefore from (68)

$$\mathbf{Z}_i\mathbf{Z}_i'\mathbf{X} = \{_{\mathrm{r}} \bigotimes_{t=1}^{m+1} \mathbf{X}_{it}\mathbf{G}_{ipt}\}_{p=1}^{f} = \{_{\mathrm{r}}\, \mathbf{X}_i\mathbf{G}_{ip}\}_{p=1}^{f} \quad \text{for } \mathbf{G}_{ip} = \bigotimes_{t=1}^{m+1} \mathbf{G}_{ipt} \tag{69}$$

$$= [\mathbf{X}_1 \quad \mathbf{X}_2 \quad \cdots \quad \mathbf{X}_f]\begin{bmatrix} \mathbf{G}_{i1} & & & \\ & \mathbf{G}_{i2} & & \\ & & \ddots & \\ & & & \mathbf{G}_{if} \end{bmatrix} = \mathbf{X}\mathbf{Q}_i \tag{70}$$

for $\mathbf{Q}_i = \{_{\mathrm{d}}\, \mathbf{G}_{ip}\}_{p=1}^{f}$. Thus $\mathbf{Z}_i\mathbf{Z}_i'\mathbf{X} = \mathbf{X}\mathbf{Q}_i$.

Conformability for the product $\mathbf{X}_{it}\mathbf{G}_{ipt}$ in (69) might seem to be lacking in some cases because, in Table 12.3, two values of $\mathbf{G}_{ipt}$ are scalar. However, matrix products do exist even when a scalar is involved; e.g., for scalar θ, both $\mathbf{A}\theta$ and $(\mathbf{A} \otimes \mathbf{B})(\theta \otimes \mathbf{L}) = \mathbf{A}\theta \otimes \mathbf{BL}$ exist. Therefore (69) does exist.

Property B of the model: $\mathscr{V}\mathscr{X} = \mathscr{X}\mathbf{Q}$ for some $\mathbf{Q}$. Equation (47) shows $\mathscr{X}'\mathscr{V}^-$. And from (43) and (45) we see that $\mathscr{V}$ is simply $\mathscr{V}^-$ with $\mathbf{B} \otimes \mathbf{B}$ replacing $\frac{1}{4}(\mathbf{B}^- \otimes \mathbf{B}^-)$. Making this replacement in $\mathscr{X}'\mathscr{V}^-$ of (47) with $\mathbf{B} = \mathbf{MVM}$ of (42) and transposing therefore immediately gives

$$2\mathscr{V}\mathscr{X} = \{_{\mathrm{r}}\, \mathrm{vec}(\mathbf{MVMZ}_i\mathbf{Z}_i'\mathbf{MVM})\}_{i=0}^{r}\,. \tag{71}$$

In doing this note that $\mathbf{B} = \mathbf{MVM}$ and hence $\mathbf{BM} = \mathbf{B}$. But with $\mathbf{Z}_i\mathbf{Z}_i'\mathbf{X} = \mathbf{X}\mathbf{Q}_i$ from (70)

$$\mathbf{Z}_i\mathbf{Z}_i'\mathbf{M} = \mathbf{Z}_i\mathbf{Z}_i'(\mathbf{I} - \mathbf{X}\mathbf{X}^+) = \mathbf{Z}_i\mathbf{Z}_i' - \mathbf{X}\mathbf{Q}_i\mathbf{X}^+\,.$$

Therefore with $\mathbf{MX} = \mathbf{0}$

$$\mathbf{MZ}_i\mathbf{Z}_i'\mathbf{M} = \mathbf{MZ}_i\mathbf{Z}_i' - \mathbf{MXQ}_i\mathbf{X}^+ = \mathbf{MZ}_i\mathbf{Z}_i' = \mathbf{Z}_i\mathbf{Z}_i'\mathbf{M}; \tag{72}$$

this last equality coming from the symmetry of $\mathbf{MZ}_i\mathbf{Z}_i'\mathbf{M}$. Hence in (71)

$$\begin{aligned}\mathbf{MVMZ}_i\mathbf{Z}_i'\mathbf{MVM} &= \mathbf{MVMZ}_i\mathbf{Z}_i'\mathbf{VM}, \quad \text{using } (72),\\ &= \mathbf{MVM}\Sigma_j\sigma_j^2\phi_{ij}\mathbf{Z}_j\mathbf{Z}_j'\mathbf{M}, \quad \text{using } (64),\\ &= \mathbf{M}\Sigma_j\sigma_j^2\phi_{ij}\mathbf{VMZ}_j\mathbf{Z}_j'\mathbf{M}\\ &= \mathbf{M}\Sigma_j\sigma_j^2\phi_{ij}\mathbf{VZ}_j\mathbf{Z}_j'\mathbf{M}, \quad \text{using } (72),\\ &= \mathbf{M}\Sigma_j\sigma_j^2\phi_{ij}\Sigma_t\sigma_t^2\phi_{jt}\mathbf{Z}_t\mathbf{Z}_t'\mathbf{M}, \quad \text{using } (64),\\ &= \Sigma_t(\Sigma_j\sigma_j^2\sigma_t^2\phi_{ij}\phi_{jt})\mathbf{MZ}_t\mathbf{Z}_t'\mathbf{M}\\ &= \Sigma_t\theta_{it}\mathbf{MZ}_t\mathbf{Z}_t'\mathbf{M} \quad \text{for } \theta_{it} = \Sigma_j\sigma_j^2\sigma_t^2\phi_{ij}\phi_{jt}\,.\end{aligned}$$

Therefore in (72)

$$\begin{aligned}\mathscr{V}\mathscr{X} &= 2\{_{\mathrm{r}}\ \mathrm{vec}(\Sigma_t\theta_{it}\mathbf{MZ}_t\mathbf{Z}_t'\mathbf{M})\}_{i=0}^{r}\\ &= 2\{_{\mathrm{r}}\ \sum_{t=0}^{r}\theta_{it}\ \mathrm{vec}(\mathbf{MZ}_t\mathbf{Z}_t'\mathbf{M})\}_{i=0}^{r}\\ &= 2\{_{\mathrm{r}}\ \mathrm{vec}(\mathbf{MZ}_i\mathbf{Z}_i'\mathbf{M})\}_{i=0}^{r}\mathbf{Q} = \mathbf{XQ} \quad \text{for } \mathbf{Q} = \{_{\mathrm{m}}\ \theta_{ii'}\}_{i,i'=0}^{r}\,. \qquad (73)\end{aligned}$$

-iii. ***Conclusion.*** With zero kurtosis and $\mathscr{Y} \sim (\mathscr{X}\boldsymbol{\sigma}^2, \mathscr{V})$ we have $\mathscr{V}\mathscr{X} = \mathscr{X}\mathbf{Q}$ for some $\mathbf{Q}$. Therefore, by (50), the BLUE of $\boldsymbol{\sigma}^2$ in this model is the OLSE in the same model. But OLSE, as we have seen in Section 12.2, has equation

$$\{_{\mathrm{m}}\ \mathrm{tr}(\mathbf{MZ}_i\mathbf{Z}_i'\mathbf{MZ}_j\mathbf{Z}_j')\}_{i,j=0}^{r}\hat{\boldsymbol{\sigma}}^2 = \{_{\mathrm{r}}\ \mathbf{y}'\mathbf{MZ}_i\mathbf{Z}_i'\mathbf{My}\}\,. \qquad (74)$$

Then, since $\mathscr{B}_{\mathrm{M}}$ of (61) is a quadratic subspace of symmetric matrices, the result of Seely (1971) discussed following (59) shows that the estimators $\hat{\boldsymbol{\sigma}}^2$ of (74) are UMVUQ—and because they are also translation-invariant they are thus UMVUIQ; and, under normality, they are UMVUI. Furthermore, because in ANOVA models with balanced data, ANOVA estimators have these same properties, as discussed in Section 12.1, the estimators in (74) are the ANOVA estimators.

We now turn to the case of non-zero kurtosis, which uses $\mathscr{V} + \mathscr{V}_k$ of (30) in place of $\mathscr{V}$. Otherwise we follow the same line of reasoning.

b. Estimation under non-zero kurtosis

-i. ***The model.*** As was done by Seely (1970, 1971), Pukelsheim (1976, 1977, 1979), Brown (1976, 1978) and Anderson (1978, 1979a,b), we have, for non-zero kurtosis, from (7) and (30)

$$\mathscr{Y} = \mathbf{My} \otimes \mathbf{My} \sim (\mathscr{X}\boldsymbol{\sigma}^2, \mathscr{V} + \mathscr{V}_k), \qquad (75)$$

with $\mathscr{X}$, $\mathscr{V}$ and $\mathscr{V}_k$ given by (6), (31) and (32), respectively, the latter being

$$\mathscr{V}_k = (\mathbf{MZ} \otimes \mathbf{MZ})(\mathbf{D}^{\frac{1}{2}} \otimes \mathbf{D}^{\frac{1}{2}})\{_{\mathrm{d}}\ \mathrm{vec}(\{_{\mathrm{d}}\ \gamma_i\mathbf{I}_{q_i}\})\}_{i=0}^{r})\}(\mathbf{D}^{\frac{1}{2}} \otimes \mathbf{D}^{\frac{1}{2}})(\mathbf{Z}'\mathbf{M} \otimes \mathbf{Z}'\mathbf{M})\,. \qquad (76)$$

This can be written as

$$\mathscr{V}_k = (\mathbf{MZ} \otimes \mathbf{MZ})\mathbf{G}\boldsymbol{\Delta}\mathbf{G}'(\mathbf{Z}'\mathbf{M} \otimes \mathbf{Z}'\mathbf{M}), \tag{77}$$

using

$$\mathbf{G} = \{_r\, \mathbf{e}_t \otimes \mathbf{e}_t\}_{t=0}^{q} \tag{78}$$

for $\mathbf{e}_t$ being the tth column of $\mathbf{I}_q$ (for $q = \Sigma_{i=0}^r q_i$) and for

$$\boldsymbol{\Delta} = \{_d\, \sigma_i^4 \gamma_i \mathbf{I}_{q_i}\}_{i=1}^{r} = \{_d\, \delta_t\}_{t=1}^{q},$$

where for $i = 0, 1, \ldots, r$

$$\delta_t = \sigma_i^4 \gamma_i \quad \text{for } t = \left(\sum_{s=0}^{i-1} q_s\right) + 1, \left(\sum_{s=0}^{i-1} q_s\right) + 2, \ldots, \sum_{s=0}^{i} q_s\,.$$

Example. For $q_0 = 2$, $q_1 = 1$ so $q = 2 + 1 = 3$,

$$(\mathbf{D}^{\frac{1}{2}} \otimes \mathbf{D}^{\frac{1}{2}})\{_d \operatorname{vec}(\{_d\, \gamma_i \mathbf{I}_{q_i}\})\}(\mathbf{D}^{\frac{1}{2}} \otimes \mathbf{D}^{\frac{1}{2}})$$

$$= \begin{bmatrix} \sigma_0 & \cdot & \cdot \\ \cdot & \sigma_0 & \cdot \\ \cdot & \cdot & \sigma_1 \end{bmatrix} \otimes \begin{bmatrix} \sigma_0 & \cdot & \cdot \\ \cdot & \sigma_0 & \cdot \\ \cdot & \cdot & \sigma_1 \end{bmatrix} \left\{_d \operatorname{vec} \begin{bmatrix} \gamma_0 & \cdot & \cdot \\ \cdot & \gamma_0 & \cdot \\ \cdot & \cdot & \gamma_1 \end{bmatrix} \right\}$$

$$\times \begin{bmatrix} \sigma_0 & \cdot & \cdot \\ \cdot & \sigma_0 & \cdot \\ \cdot & \cdot & \sigma_1 \end{bmatrix} \otimes \begin{bmatrix} \sigma_0 & \cdot & \cdot \\ \cdot & \sigma_0 & \cdot \\ \cdot & \cdot & \sigma_1 \end{bmatrix}$$

$$= \mathbf{U}\{_d\, \gamma_0 \quad 0 \quad 0 \quad 0 \quad \gamma_0 \quad 0 \quad 0 \quad 0 \quad \gamma_1\}\mathbf{U}$$

$$\text{for } \mathbf{U} = \{_d\, \sigma_0^2 \quad \sigma_0^2 \quad \sigma_0\sigma_1 \quad \sigma_0^2 \quad \sigma_0^2 \quad \sigma_0\sigma_1 \quad \sigma_0\sigma_1 \quad \sigma_0\sigma_1 \quad \sigma_1^2\},$$

$$= \{_d\, \sigma_0^4\gamma_0 \quad 0 \quad 0 \quad 0 \quad \sigma_0^4\gamma_0 \quad 0 \quad 0 \quad 0 \quad \sigma_1^4\gamma_1\}$$

$$= \begin{bmatrix} 1 & \cdot & \cdot \\ \cdot & \cdot & \cdot \\ \cdot & \cdot & \cdot \\ \cdot & \cdot & \cdot \\ \cdot & 1 & \cdot \\ \cdot & \cdot & \cdot \\ \cdot & \cdot & \cdot \\ \cdot & \cdot & \cdot \\ \cdot & \cdot & 1 \end{bmatrix} \begin{bmatrix} \sigma_0^4\gamma_0 & \cdot & \cdot \\ \cdot & \sigma_0^4\gamma_0 & \cdot \\ \cdot & \cdot & \sigma_1^4\gamma_1 \end{bmatrix} \begin{bmatrix} 1 & \cdot & \cdot & \cdot & \cdot & \cdot & \cdot & \cdot & \cdot \\ \cdot & \cdot & \cdot & \cdot & 1 & \cdot & \cdot & \cdot & \cdot \\ \cdot & \cdot & \cdot & \cdot & \cdot & \cdot & \cdot & \cdot & 1 \end{bmatrix} = \mathbf{G}\boldsymbol{\Delta}\mathbf{G}'\,.$$

-ii. ANOVA estimation. For non-zero kurtosis ANOVA models with balanced data we now verify (58) by exhibiting a matrix $\mathbf{H}$ that satisfies

$\mathscr{V}_k\mathscr{X} = \mathscr{X}\mathbf{H}$, where $\mathscr{X}$ and $\mathscr{V}_k$ are defined in (6) and (32), respectively. Then since $\mathscr{V}\mathscr{X} = \mathscr{X}\mathbf{Q}$, we will have $(\mathscr{V} + \mathscr{V}_k)\mathscr{X} = \mathscr{X}\mathbf{Q}^*$ for $\mathbf{Q}^* = \mathbf{Q} + \mathbf{H}$, and so the condition for ordinary least squares estimation being the same as best linear unbiased estimation will be satisfied for the non-zero kurtosis case. Theorem 4.5 of Pukelsheim (1977), Theorem 6 of Kleffe (1977) and Theorem 1.4 of Drygas (1980) point out the need for a matrix $\mathbf{Q}$; we substantiate this, as in Anderson *et al.* (1984), by showing its existence for the non-zero kurtosis case in ANOVA estimation from balanced data.

With $\mathscr{X}$ of (6) and $\mathscr{V}_k$ of (76)

$$\mathscr{V}_k\mathscr{X} = \mathscr{V}_k\{_r \operatorname{vec}(\mathbf{MZ}_s\mathbf{Z}_s'\mathbf{M})\}_{s=0}^{r} = \{_r\, \mathscr{V}_k \operatorname{vec}(\mathbf{MZ}_s\mathbf{Z}_s'\mathbf{M})\}_{s=0}^{r} .$$

In order to show that this is $\mathscr{X}\mathbf{Q}_k$ for some $\mathbf{Q}_k$, we only need show that $\mathscr{V}_k\mathscr{X} = \{_r\, \mathscr{X}\mathbf{h}_s\}_{s=0}^{r}$ for $\mathbf{h}_s$ being some vector. We therefore use (77) for $\mathscr{V}_k$ and consider $\mathbf{u}_s$ defined by $\mathscr{V}\mathscr{X} = \{\mathbf{u}_s\}_{s=0}^{r}$. Then

$$\begin{aligned}
\mathbf{u}_s &= \mathscr{V}_k \operatorname{vec}(\mathbf{MZ}_s\mathbf{Z}_s'\mathbf{M}) \\
&= (\mathbf{MZ} \otimes \mathbf{MZ})\mathbf{G}\boldsymbol{\Delta}\mathbf{G}'(\mathbf{Z}'\mathbf{M} \otimes \mathbf{Z}'\mathbf{M}) \operatorname{vec}(\mathbf{MZ}_s\mathbf{Z}_s'\mathbf{M}) \\
&= (\mathbf{MZ} \otimes \mathbf{MZ})\mathbf{G}\boldsymbol{\Delta}\mathbf{G}' \operatorname{vec}(\mathbf{Z}'\mathbf{MZ}_s\mathbf{Z}_s'\mathbf{MZ}) \\
&\qquad\qquad [(53) \text{ of Section M.8, and } \mathbf{M}^2 = \mathbf{M}] \\
&= (\mathbf{MZ} \otimes \mathbf{MZ}) \operatorname{vec}[\boldsymbol{\Delta} \operatorname{diag}(\mathbf{Z}'\mathbf{MZ}_s\mathbf{Z}_s'\mathbf{MZ})], \qquad (79)
\end{aligned}$$

on using (55) of Section M.8, and where the notation diag($\mathbf{A}$) represents a diagonal matrix of the diagonal elements of $\mathbf{A}$.

Now suppose there are scalars λ_{is} for $s = 0, 1, \ldots, r$ such that $\operatorname{diag}(\mathbf{Z}'\mathbf{MZ}_s\mathbf{Z}_s'\mathbf{MZ})$ in (79) has the form

$$\operatorname{diag}(\mathbf{Z}'\mathbf{MZ}_s\mathbf{Z}_s'\mathbf{MZ}) = \{_d\, \lambda_{is}\mathbf{I}_{q_i}\}_{i=0}^{r} = \boldsymbol{\Gamma}_s, \quad \text{say} . \qquad (80)$$

Then from (79)

$$\begin{aligned}
\mathbf{u}_s = (\mathbf{MZ} \otimes \mathbf{MZ}) \operatorname{vec}(\boldsymbol{\Delta}\boldsymbol{\Gamma}_s) &= \operatorname{vec}(\mathbf{MZ}\boldsymbol{\Delta}\boldsymbol{\Gamma}_s\mathbf{Z}'\mathbf{M}) \\
&= \operatorname{vec}\Big(\mathbf{M} \sum_{i=0}^{r} \sigma_i^4\gamma_i\lambda_{is}\mathbf{Z}_i\mathbf{Z}_i'\mathbf{M}\Big) \\
&= \sum_{i=0}^{r} \operatorname{vec}(\mathbf{MZ}_i\mathbf{Z}_i'\mathbf{M})\, \sigma_i^4\gamma_i\lambda_{is} \\
&= \mathscr{X}\mathbf{h}_s \quad \text{for } \mathbf{h}_s = \{_c\, \sigma_i^4\gamma_i\lambda_{is}\}_{i=0}^{r} . \qquad (81)
\end{aligned}$$

Therefore it remains to show that there are scalars λ_{is} satisfying (81). To this end, partition $\mathbf{Z}_i$ into its columns $\mathbf{z}_{ij}$ for $j = 1, \ldots, q_i$ and define

$$\begin{aligned}
\lambda_{ijs} &= \mathbf{z}_{ij}'\mathbf{MZ}_s\mathbf{Z}_s'\mathbf{Mz}_{ij} = \mathbf{z}_{ij}'\mathbf{MZ}_s\mathbf{Z}_s'\mathbf{z}_{ij}, \quad \text{from (72)} \\
&= \mathbf{z}_{ij}'\mathbf{Z}_s\mathbf{Z}_s'\mathbf{z}_{ij} - \mathbf{z}_{ij}'\mathbf{XX}^+\mathbf{Z}_s\mathbf{Z}_s'\mathbf{z}_{ij} \quad [\mathbf{M} = \mathbf{I} - \mathbf{XX}^+] . \qquad (82)
\end{aligned}$$

Every $\mathbf{Z}_i$ is a KP of **1**s and **I**s, so that partitioning the **I**s into their columns, denoted by **e**-vectors, gives each

$$\mathbf{z}_{ij} \quad \text{as a KP of } \mathbf{e}\text{s and } \mathbf{1}\text{s}; \tag{83}$$

and every

$$\mathbf{Z}_k\mathbf{Z}_k' \quad \text{is a KP of } \mathbf{I}\text{s and } \mathbf{J}\text{s}; \tag{84}$$

and

$$\mathbf{X}\mathbf{X}^+ \quad \text{is a sum of KPs of } \mathbf{I}\text{s}, \bar{\mathbf{J}}\text{s and } \mathbf{C}\text{s} \quad [\mathbf{C} = \mathbf{I} - \bar{\mathbf{J}}]\,. \tag{85}$$

All these KPs are conformable, whereupon each of the two terms of (82) is also a KP. Therefore, on applying (83), (84) and (85) to (82), with each term in (82) being a scalar, it is clear that that scalar is a KP of scalars (and hence a product of scalars), with the scalar that is in position t of the KP having, for some matrix $\mathbf{Q}_t$, one of the forms

$$\mathbf{e}_j'\mathbf{Q}_t\mathbf{e}_j = j\text{th diagonal element of } \mathbf{Q}_t$$

or

$$\mathbf{e}_j'\mathbf{Q}_t\mathbf{1} = j\text{th row sum of } \mathbf{Q}_t \tag{86}$$

or

$$\mathbf{1}'\mathbf{Q}_t\mathbf{1} = \text{sum of all elements of } \mathbf{Q}_t\,.$$

And, from applying (86) to (82) we see from (84) and (85) that $\mathbf{Q}_t$ is either the matrix in position t of $\mathbf{Z}_k\mathbf{Z}_k'$ (and so is either an **I** or a **J**), or else it is a product of matrices in position t of $\mathbf{X}\mathbf{X}^+$ and $\mathbf{Z}_k\mathbf{Z}_k'$ (and so is either an **I**, a $\bar{\mathbf{J}}$, a **C** or a **0**). Hence $\mathbf{Q}_t$ does not depend on j. Therefore neither do the scalars in (86), and hence λ_{ijk} of (82) does not depend on j. Therefore (81) holds and so

$$\mathscr{V}_k\mathscr{X} = \mathscr{X}\{\mathbf{h}_s\}_{s=0}^{r} = \mathscr{X}\mathbf{H} \quad \text{for } \mathbf{H} = \{\mathbf{h}_s\}_{s=0}^{r}\,.$$

-iii. Conclusion. We now have

$$\mathscr{V}\mathscr{X} + \mathscr{V}_k\mathscr{X} = \mathscr{X}\mathbf{Q} + \mathscr{X}\mathbf{H} = \mathscr{X}(\mathbf{Q} + \mathbf{H}),$$

and so in the dispersion mean model again, by (50), the BLUE of $\boldsymbol{\sigma}^2$ is the same as the OLSE with balanced data. But this OLSE is ANOVA: and since BLUE in the dispersion mean model is UMVUI, we have ANOVA = UMVUI. And since we know this is true under normality, when kurtosis is zero, and we have now just shown that it is also true for non-zero kurtosis, we can therefore say it is true always.

12.7. NON-NEGATIVE ESTIMATION

In the general form of both random and mixed models there has been a long-time interest in the conditions under which non-negative quadratic unbiased estimators of variance components are available. [See, e.g., Pukelsheim (1978) and Styan and Pukelsheim (1981).] Almost all reports on this topic and subsets thereof [e.g., dropping unbiasedness; see Hartung (1981)] involve balanced data and quadratic subspaces of real symmetric matrices based on **X** and **Z** and their submatrices. An early discussion of these is Seely (1970), and their use in estimating variance components is to be found in such papers as Pukelsheim (1981a,b). Mathew (1984), Baksalary and Molinska (1984), Gnot *et al.* (1985) and Mathew *et al.* (1991a,b). In particular, Pukelsheim (1981a) shows that, in the presence of a quadratic subspace condition, there is the dichotomy that either the ANOVA estimator (derived without paying attention to non-negativity) is automatically non-negative, or else the two properties of unbiasedness and non-negativity cannot be achieved simultaneously. In the latter case, unbiasedness needs to be replaced by some other meaningful statistical properties, and a variety of ways of comparing estimators emerges, as in, for example, Mathew *et al.* (1991a,b).

12.8. SUMMARY

Model

$$\mathbf{y} = \mathbf{X}\boldsymbol{\beta} + \mathbf{Z}\mathbf{u} + \mathbf{e} = \mathbf{X}\boldsymbol{\beta} + \sum_{i=0}^{r} \mathbf{Z}_i\mathbf{u}_i;$$

$$\text{var}(\mathbf{u}) = \mathbf{D} = \{_{\text{d}}\, \sigma_i^2\mathbf{I}_{q_i}\}; \quad \text{var}(\mathbf{y}) = \mathbf{V} + \mathbf{ZDZ}' .$$

Dispersion-mean model

$$\mathbf{M} = \mathbf{I} - \mathbf{X}(\mathbf{X}'\mathbf{X})^{-}\mathbf{X} = \mathbf{I} - \mathbf{X}\mathbf{X}^{+} = \mathbf{M}' = \mathbf{M}^2, \quad \text{with } \mathbf{MX} = \mathbf{0}; \tag{2}$$

$$\mathscr{Y} = \mathbf{My} \otimes \mathbf{My} \quad \text{and} \quad \mathscr{X} = \{_{\text{r}}\, \text{vec}(\mathbf{MZ}_i\mathbf{Z}_i'\mathbf{M})\}; \tag{6}$$

$$E(\mathscr{Y}) = \mathscr{X}\boldsymbol{\sigma}^2 . \tag{7}$$

OLSE → MINQUEO: Section 12.2

$$\mathscr{X}'\mathscr{X} = \mathscr{X}'\mathscr{Y} \rightarrow \{_{\text{m}}\, \text{tr}(\mathbf{Z}_i\mathbf{Z}_i'\mathbf{MZ}_j\mathbf{Z}_j'\mathbf{M})\}\boldsymbol{\sigma}^2 = \{_{\text{c}}\, \mathbf{y}'\mathbf{MZ}_i\mathbf{Z}_i'\mathbf{My}\} .$$

Fourth moments

$$\mathbf{w} \sim \mathscr{N}(\mathbf{0}, \mathbf{I}_q), \quad q = \sum_{i=0}^{r} q_i;$$

$$E(w_k^4) = 3 + \gamma_k^* \quad \text{and} \quad E(u_{ij}^4) = \sigma_i^4(3 + \gamma_i) \quad \text{with } \{_{\text{d}}\, \gamma_k^*\}_{r=1}^{q} = \{_{\text{d}}\, \gamma_i I_{q_i}\}_{i=0}^{r}; \tag{11}$$

$$\mathbf{T}_{\text{A}} = \{_{\text{d}}\, \text{vec}\{_{\text{d}}\, \gamma_i\mathbf{I}_{q_i}\}_{i=0}^{r}\}; \tag{22}$$

$$\text{var}(\mathbf{u} \otimes \mathbf{u}) = (\mathbf{D} \otimes \mathbf{D})(\mathbf{I} + \mathbf{S}_q) + (\mathbf{D}^{\frac{1}{2}} \otimes \mathbf{D}^{\frac{1}{2}})\mathbf{T}_{\text{A}}(\mathbf{D}^{\frac{1}{2}} \otimes \mathbf{D}^{\frac{1}{2}}); \tag{25}$$

$$\begin{aligned} &\text{var}[(\mathbf{y} - \mathbf{X}\boldsymbol{\beta}) \otimes (\mathbf{y} - \mathbf{X}\boldsymbol{\beta})] \\ &\quad = (\mathbf{V} \otimes \mathbf{V})(\mathbf{I} + \mathbf{S}_N) + (\mathbf{ZD}^{\frac{1}{2}} \otimes \mathbf{ZD}^{\frac{1}{2}})\mathbf{T}_{\text{A}}(\mathbf{D}^{\frac{1}{2}}\mathbf{Z}' \otimes \mathbf{D}^{\frac{1}{2}}\mathbf{Z}') \qquad (27) \\ &\quad = (\mathbf{V} \otimes \mathbf{V})(\mathbf{I} + \mathbf{S}_N), \quad \text{under normality;} \qquad (28) \end{aligned}$$

$$\begin{aligned}\text{var}(\mathscr{Y}) &= (\mathbf{MVM} \otimes \mathbf{MVM})(\mathbf{I} + \mathbf{S}_N) \\ &\quad + (\mathbf{MZD}^{\frac{1}{2}} \otimes \mathbf{MZD}^{\frac{1}{2}})\mathbf{T}_{\mathrm{A}}(\mathbf{D}^{\frac{1}{2}}\mathbf{Z}'\mathbf{M} \otimes \mathbf{D}^{\frac{1}{2}}\mathbf{Z}'\mathbf{M}) \qquad (29) \\ &= (\mathbf{MVM} \otimes \mathbf{MVM})(\mathbf{I} + \mathbf{S}_N), \quad \text{under normality;} \quad (31), (34)\end{aligned}$$

$$\begin{aligned}\text{var}(\mathbf{y}'\mathbf{A}\mathbf{y}) &= 2\,\text{tr}(\mathbf{AV})^2 \\ &\quad + \sum_{i=0}^{r} \gamma_i \sigma_i^4 \text{ (sum of squares of diagonal elements of } \mathbf{Z}_i'\mathbf{A}\mathbf{Z}_i) \quad (41) \\ &= 2\,\text{tr}(\mathbf{AV})^2], \quad \text{under normality}.\end{aligned}$$

Estimating $\boldsymbol{\sigma}^2$ from the dispersion-mean model, with non-singular V

GLSE($\boldsymbol{\sigma}^2$) yields REML:

$$\mathscr{X}'\mathscr{V}^-\mathscr{X}\boldsymbol{\sigma}^2 = \mathscr{X}'\mathscr{V}^-\mathbf{y} \rightarrow \{_{\text{m}}\, \text{tr}(\mathbf{Z}_i\mathbf{Z}_i'\mathbf{P}\mathbf{Z}_j\mathbf{Z}_j'\mathbf{P})\}\boldsymbol{\sigma}^2 = \{_{\text{c}}\, \mathbf{y}'\mathbf{P}\mathbf{Z}_i\mathbf{Z}_i'\mathbf{P}\mathbf{y}\}. \quad (48)$$

BLUE($\boldsymbol{\sigma}^2$) yields REML: (51)

GLSE on $\mathscr{Y}_0 = (\mathbf{y} - \mathbf{X}\boldsymbol{\beta}^0) \otimes (\mathbf{y} - \mathbf{X}\boldsymbol{\beta}^0)$ yields ML. (57)

Balanced data

ANOVA estimators are UMVUI. (74)

12.9. EXERCISES

E 12.1. For Section 8.2 reduce $\mathscr{X}'\mathbf{y}$ to a column of scalars $\mathbf{y}'\mathbf{M}\mathbf{Z}_i\mathbf{Z}_i'\mathbf{M}\mathbf{y}$.

E 12.2. Verify $\mathscr{V}^-$ of (45).

E 12.3. By partitioning $\mathbf{M}$ of (2) as $\mathbf{M} = [\mathbf{K} \quad \mathbf{K}\mathbf{T}']'$ for $\mathbf{K}'$ of Section M4e, prove the theorem of Section M.4f through showing that the only non-null part of $(\mathbf{MVM})^-$ is $(\mathbf{K}'\mathbf{V}\mathbf{K})^{-1}$.

E 12.4. Prove the theorem at the end of Section 12.4b, assuming that $\mathbf{V}$ is non-singular.

E 12.5. Give direct proofs, without recourse to results in Section 12.4c, of the following results for $\mathscr{Y}_0$, $\mathbf{C}$ and $\mathbf{F}$ of Section 12.5:

(a) $\mathbf{F}^- = \frac{1}{4}(\mathbf{I} + \mathbf{S})(\mathbf{V}^- \otimes \mathbf{V}^-)$;
(b) $\mathbf{F}\mathbf{F}^-\mathbf{C} = \mathbf{C}$;
(c) $\mathbf{C}'\mathbf{F}^- = \frac{1}{2}\{_{\text{c}}\, [\text{vec}(\mathbf{V}^{-1}\mathbf{Z}_i\mathbf{Z}_i'\mathbf{V}^{-1})]'\}$;
(d) the GLS equations (48).

E 12.6. Show that (41) is

$$\text{var}(\mathbf{y}'\mathbf{A}\mathbf{y}) = 2\,\text{tr}[(\mathbf{B}\boldsymbol{\Delta}_1)^2] + 2\,\text{tr}(\tilde{\mathbf{B}}\boldsymbol{\Delta}_2\tilde{\mathbf{B}}),$$

where $\mathbf{B} = \mathbf{Z}'\mathbf{A}\mathbf{Z}$, $\boldsymbol{\Delta}_1 = \mathbf{D}$, $\tilde{\mathbf{B}}$ is $\mathbf{B}$ with all off-diagonal elements changed to zero, and $\boldsymbol{\Delta}_2 = \{_{\text{d}}\, \gamma_i\sigma_i^4\mathbf{I}_{q_i}\}_{i=0}^{r}$. This is Rao's (1971b) form.

APPENDIX F

ESTIMATION FORMULAE FOR UNBALANCED DATA

Catalogued here are detailed formulae for estimating variance components from unbalanced data for three nested random models and four forms of model for the 2-way crossed classification. The formulae are given without comment, and thus are to be viewed simply as a reference source. Most of them are from Searle (1971, Chap. 11) with some improved layout, although a number given there have not been reproduced here. Only those considered to be the most useful are shown. For example, the 23 pages of the 3-way crossed classification are not included.

Subscript ranges are shown as part of each model, but are not included as limits in summations, e.g., $\Sigma_{i=1}^{a}$ occurs as Σ_i.

The first three models are nested random models: the 1-, 2- and 3-way cases. (The 1-way classification is usually not thought of as a nested classification but it can be: error, nested within classes.) In all nested models the three Henderson methods are all the same and are usually called the ANOVA method. This is true not only for nested models that are random models but also those that are mixed models.

PART I. THREE NESTED MODELS

F.1. THE 1-WAY CLASSIFICATION

a. Model

$$y_{ij} = \mu + \alpha_i + e_{ij};$$

$$i = 1, 2, \ldots, a \text{ and } j = 1, 2, \ldots, n_i, \quad \text{with } n = \Sigma_i n_i\,.$$

b. Analysis of variance estimators

Calculate

$$T_0 = \sum_{i=1}^{a} \sum_{j=1}^{n_i} y_{ij}^2, \quad T_A = \sum_{i=1}^{a} \frac{y_{i.}^2}{n_i}, \quad T_\mu = \frac{y_{..}^2}{N},$$

$$S_2 = \Sigma_i n_i^2 \quad \text{and} \quad S_3 = \Sigma_i n_i^3$$

Then

$$\hat{\sigma}_e^2 = (T_0 - T_A)/(N - a)$$

and

$$\hat{\sigma}_\alpha^2 = [T_A - T_\mu - (a-1)\hat{\sigma}_e^2]/(N - S_2/N)\,.$$

c. Variances of analysis of variance estimators (under normality)

$$\text{var}(\hat{\sigma}_e^2) = 2\sigma_e^4/(N - a),$$

$$\text{var}(\hat{\sigma}_\alpha^2) = \frac{2\sigma_e^4 N^2(N-1)(a-1)}{(N-a)(N^2 - S_2)^2} + \frac{4\sigma_e^2\sigma_\alpha^2 N}{N^2 - S_2} + \frac{2\sigma_\alpha^4(N^2 S_2 + S_2^2 - 2NS_3)}{(N^2 - S_2)^2},$$

$$\text{cov}(\hat{\sigma}_\alpha^2, \hat{\sigma}_e^2) = -N(a-1)\,\text{var}(\hat{\sigma}_e^2)/(N^2 - S_2)$$

(Searle, 1956).

d. Maximum likelihood estimation (under normality)

Solve iteratively, as in (133), (134) and (135) of Chapter 3 (and see Chapter 8 also),

$$\tilde{\mu} = \sum_i \frac{n_i \bar{y}_{i.}}{\tilde{\sigma}_e^2 + n_i\tilde{\sigma}_\alpha^2} \Bigg/ \sum_i \frac{n_i}{\tilde{\sigma}_e^2 + n_i\tilde{\sigma}_\alpha^2},$$

$$\frac{N-a}{\tilde{\sigma}_e^2} + \sum_i \frac{1}{\tilde{\sigma}_e^2 + n_i\tilde{\sigma}_\alpha^2} - \frac{\text{SSE}}{\tilde{\sigma}_e^4} = \sum_i \frac{n_i(\bar{y}_{i.} - \tilde{\mu})^2}{(\tilde{\sigma}_e^2 + n_i\tilde{\sigma}_\alpha^2)^2},$$

$$\sum_i \frac{n_i}{\tilde{\sigma}_e^2 + n_i\tilde{\sigma}_\alpha^2} = \sum_i \frac{n_i^2(\bar{y}_{i.} - \tilde{\mu})^2}{(\tilde{\sigma}_e^2 + n_i\tilde{\sigma}_\alpha^2)^2}$$

e. Large-sample variances of maximum likelihood estimators (under normality)

$$w_i = n_i/(1 + n_i\sigma_\alpha^2/\sigma_e^2)$$

and

$$D = N\Sigma_i w_i^2 - (\Sigma_i w_i)^2\,.$$

Then

$$\operatorname{var}(\tilde{\sigma}_e^2) = 2\sigma^4(\Sigma_i w_i^2)/D,$$
$$\operatorname{var}(\tilde{\sigma}_\alpha^2) = 2\sigma_e^4(N - a + \Sigma_i w_i^2/n_i^2)/D$$

and

$$\operatorname{cov}(\tilde{\sigma}_\alpha^2, \tilde{\sigma}_e^2) = -2\sigma_e^4(\Sigma_i w_i^2/n_i)/D$$

(Crump, 1951; Searle, 1956).

F.2. THE 2-WAY NESTED CLASSIFICATION

a. Model

$$y_{ijk} = \mu + \alpha_i + \beta_{ij} + e_{ijk};$$
$$i = 1, 2, \ldots, a, \quad j = 1, 2, \ldots, b_i \quad \text{and} \quad k = 1, 2, \ldots, n_{ij},$$

with

$$b_{.} = \Sigma_i b_i \quad \text{and} \quad N = \Sigma_i \Sigma_j n_{ij}\,.$$

b. Analysis of variance estimators

Calculate

$$k_1 = \Sigma_i n_{i.}^2/N, \quad k_3 = \Sigma_i \Sigma_j n_{ij}^2/N, \quad k_{12} = \Sigma_i(\Sigma_j n_{ij}^2/n_{i.}),$$
$$T_A = \Sigma_i y_{i..}^2/n_{i.}, \quad T_{AB} = \Sigma_i \Sigma_j y_{ij.}^2/n_{ij},$$
$$T_0 = \Sigma_i \Sigma_j \Sigma_k y_{ijk}^2 \quad \text{and} \quad T_\mu = y_{...}^2/N\,.$$

Then

$$\hat{\sigma}_e^2 = (T_0 - T_{AB})/(N - b_{.}),$$
$$\hat{\sigma}_\beta^2 = [T_{AB} - T_A - (b_{.} - a)\hat{\sigma}_e^2]/(N - k_{12}),$$
$$\hat{\sigma}_\alpha^2 = [T_A - T_\mu - (k_{12} - k_3)\hat{\sigma}_\beta^2 - (a - 1)\hat{\sigma}_e^2]/(N - k_1)$$

(Searle, 1961).

c. Variances of analysis of variance estimators (under normality)

$$\operatorname{var}(\hat{\sigma}_e^2) = 2\sigma_e^4/(N - b_{.})\,.$$

Calculate

$$k_4 = \Sigma_i \Sigma_j n_{ij}^3, \quad k_5 = \Sigma_i(\Sigma_j n_{ij}^3/n_{i.}),$$
$$k_6 = \Sigma_i(\Sigma_j n_{ij}^2)^2/n_{i.}, \quad k_7 = \Sigma_i(\Sigma_j n_{ij}^2)^2/n_{i.}^2,$$
$$k_8 = \Sigma_i n_{i.}(\Sigma_j n_{ij}^2), \quad k_9 = \Sigma_i n_{i.}^3$$

and

$$\lambda_1 = (N - k_{12})^2[k_1(N + k_1) - 2k_9/N],$$

$$\lambda_2 = k_3[N(k_{12} - k_3)^2 + k_3(N - k_{12})^2] + (N - k_3)^2 k_7$$
$$- 2(N - k_3)[(k_{12} - k_3)k_5 + (N - k_{12})k_6/N]$$
$$+ 2(N - k_{12})(k_{12} - k_3)k_4/N,$$

$$\lambda_3 = [(N - k_{12})^2(N - 1)(a - 1) - (N - k_3)^2(a - 1)(b. - a)$$
$$+ (k_{12} - k_3)^2(N - 1)(b. - a)]/(N - b.),$$

$$\lambda_4 = (N - k_{12})^2[k_3(N + k_1) - 2k_8/N],$$

$$\lambda_5 = (N - k_{12})^2(N - k_1) \quad \text{and} \quad \lambda_6 = (N - k_{12})(N - k_3)(k_{12} - k_3)\,.$$

Then

$$\operatorname{var}(\hat{\sigma}_\alpha^2) = \frac{2(\lambda_1\sigma_\alpha^4 + \lambda_2\sigma_\beta^4 + \lambda_3\sigma_e^4 + 2\lambda_4\sigma_\alpha^2\sigma_\beta^2 + 2\lambda_5\sigma_\alpha^2\sigma_e^2 + 2\lambda_6\sigma_\beta^2\sigma_e^2}{(N - k_1)^2(N - k_{12})^2},$$

$$\operatorname{var}(\hat{\sigma}_\beta^2) = \frac{2(k_7 + Nk_3 - 2k_5)\sigma_\beta^4 + 4(N - k_{12})\sigma_\beta^2\sigma_e^2 + 2(b. - a)(N - a)\sigma_e^4/(N - b.)}{(N - k_{12})^2},$$

$$\operatorname{cov}(\hat{\sigma}_\alpha^2, \hat{\sigma}_e^2) = [(k_{12} - k_3)(b. - a)/(N - k_{12}) - (a - 1)] \operatorname{var}(\hat{\sigma}_e^2)/(N - k_1),$$

$$\operatorname{cov}(\hat{\sigma}_\beta^2, \hat{\sigma}_e^2) = -(b. - a) \operatorname{var}(\hat{\sigma}_e^2)/(N - k_{12}),$$

$$\operatorname{cov}(\hat{\sigma}_\alpha^2, \hat{\sigma}_\beta^2) = \{2[k_5 - k_7 + (k_6 - k_4)/N]\sigma_\beta^4 + 2(a - 1)(b. - a)\sigma_e^4/(N - b)$$
$$- (N - k_{12})(k_{12} - k_3) \operatorname{var}(\hat{\sigma}_\beta^2)\}/(N - k_1)(N - k_{12})$$

(Searle, 1961).

d. Large-sample variances of maximum likelihood estimators (under normality)

$$\begin{bmatrix} \operatorname{var}(\tilde{\sigma}_\alpha^2) & \operatorname{cov}(\tilde{\sigma}_\alpha^2, \tilde{\sigma}_\beta^2) & \operatorname{cov}(\tilde{\sigma}_\alpha^2, \tilde{\sigma}_e^2) \\ \operatorname{cov}(\tilde{\sigma}_\alpha^2, \tilde{\sigma}_\beta^2) & \operatorname{var}(\tilde{\sigma}_\beta^2) & \operatorname{cov}(\tilde{\sigma}_\beta^2, \tilde{\sigma}_e^2) \\ \operatorname{cov}(\tilde{\sigma}_\alpha^2, \tilde{\sigma}_e^2) & \operatorname{cov}(\tilde{\sigma}_\beta^2, \tilde{\sigma}_e^2) & \operatorname{var}(\tilde{\sigma}_e^2) \end{bmatrix} = 2 \begin{bmatrix} t_{\alpha\alpha} & t_{\alpha\beta} & t_{\alpha e} \\ t_{\alpha\beta} & t_{\beta\beta} & t_{\beta e} \\ t_{\alpha e} & t_{\beta e} & t_{ee} \end{bmatrix}^{-1}$$

with

$$m_{ij} = n_{ij}\sigma_\beta^2 + \sigma_e^2,$$

$$A_{ipq} = \Sigma_j(n_{ij}^p/m_{ij}^q), \quad \text{for integers } p \text{ and } q,$$

$$q_i = 1 + \sigma_\alpha^2 A_{i11},$$

and

$$t_{\alpha\alpha} = \Sigma_i A_{i11}^2/q_i^2, \quad t_{\alpha\beta} = \Sigma_i A_{i22}^2/q_i^2, \quad t_{\alpha e} = \Sigma_i A_{i12}^2/q_i^2,$$
$$t_{\beta\beta} = \Sigma_i (A_{i22} - 2\sigma_\alpha^2 A_{i33}/q_i + \sigma_\alpha^4 A_{i22}^2/q_i^2),$$
$$t_{\beta e} = \Sigma_i (A_{i12} - 2\sigma_\alpha^2 A_{i23}/q_i + \sigma_\alpha^4 A_{i12} A_{i22}/q_i^2)$$

and

$$t_{ee} = \Sigma_i (A_{i02} - 2\sigma_\alpha^2 A_{i13}/q_i + \sigma_\alpha^4 A_{i12}^2/q_i^2) + (N - b.)/\sigma_e^4$$

(Searle, 1970).

F.3. THE 3-WAY NESTED CLASSIFICATION

a. Model

$$y_{ijkm} = \mu + \alpha_i + \beta_{ij} + \gamma_{ijk} + e_{ijkm};$$
$$i = 1, 2, \ldots, a, \quad \text{and} \quad j = 1, 2, \ldots, b_i, \quad k = 1, 2, \ldots, c_{ij}$$

and

$$m = 1, 2, \ldots, n_{ijk},$$

with

$$b. = \Sigma_i b_i, \quad c_{i.} = \Sigma_j c_{ij}, \quad c_{.j} = \Sigma_i c_{ij} \quad \text{and} \quad N = \Sigma_i \Sigma_j \Sigma_k n_{ijk}\,.$$

b. Analysis of variance estimators

Calculate

$$k_1 = \Sigma_i n_{i..}^2/N, \qquad k_2 = \Sigma_i \Sigma_j n_{ij.}^2/N,$$
$$k_3 = \Sigma_i \Sigma_j \Sigma_k n_{ijk}^2/N, \quad k_4 = \Sigma_i \Sigma_j n_{ij.}^2/n_{i..},$$
$$k_5 = \Sigma_i \Sigma_j \Sigma_k n_{ijk}^2/n_{i..}, \quad k_6 = \Sigma_i \Sigma_j \Sigma_k n_{ijk}^2/n_{ij.}$$

and

$$v_1 = N - k_1, \quad v_2 = k_4 - k_2, \quad v_3 = k_5 - k_2, \quad v_4 = a - 1,$$
$$v_5 = N - k_4, \quad v_6 = k_6 - k_5, \quad v_7 = b. - a, \quad v_8 = N - k_6,$$
$$v_9 = c_{..} - b., \quad v_{10} = N - c_{..}\,.$$

Then with

$$T_0 = \Sigma_i \Sigma_j \Sigma_k \Sigma_m y_{ijkm}^2, \quad T_A = \Sigma_i y_{i...}^2/n_{i..},$$
$$T_{AB} = \Sigma_i \Sigma_j y_{ij..}^2/n_{ij.}, \quad T_{ABC} = \Sigma_i \Sigma_j \Sigma_k y_{ijk.}^2/n_{ijk} \quad \text{and} \quad T_\mu = y_{....}^2/N,$$
$$\hat{\sigma}_e^2 = (T_0 - T_{ABC})/v_{10}, \quad \hat{\sigma}_\gamma^2 = (T_{ABC} - T_{AB} - v_9 \hat{\sigma}_e^2)/v_8,$$
$$\hat{\sigma}_\beta^2 = (T_{AB} - T_A - v_7 \hat{\sigma}_e^2 - v_6 \hat{\sigma}_\gamma^2)/v_5$$

and

$$\hat{\sigma}_\alpha^2 = (T_A - T_\mu - v_4 \hat{\sigma}_e^2 - v_3 \hat{\sigma}_\gamma^2 - v_2 \hat{\sigma}_\beta^2)/v_1\,.$$

(Mahamunulu, 1963).

c. Variances of analysis of variance estimators (under normality)

$$\mathrm{var}(\hat{\sigma}_e^2) = 2\sigma_e^4/v_{10},$$

$$\mathrm{cov}(\hat{\sigma}_\alpha^2, \hat{\sigma}_e^2) = [v_2(v_7v_8 - v_6v_9) + v_5(v_3v_9 - v_4v_8)]\ \mathrm{var}(\hat{\sigma}_e^2)/v_1v_5v_8,$$

$$\mathrm{cov}(\hat{\sigma}_\beta^2, \hat{\sigma}_e^2) = -(v_7v_8 - v_6v_9)\ \mathrm{var}(\hat{\sigma}_e^2)/v_5v_8,$$

$$\mathrm{cov}(\hat{\sigma}_\gamma^2, \hat{\sigma}_e^2) = -v_9\ \mathrm{var}(\hat{\sigma}_e^2)/v_8\ .$$

Calculate

$$k_7 = \Sigma_i n_{i..}^3, \qquad k_8 = \Sigma_i\Sigma_j n_{ij.}^3,$$

$$k_9 = \Sigma_i\Sigma_j\Sigma_k n_{ijk}^3, \qquad k_{10} = \Sigma_i(\Sigma_j\Sigma_k n_{ijk}^3)/n_{i..},$$

$$k_{11} = \Sigma_i\Sigma_j(\Sigma_k n_{ijk}^3)/n_{ij.}, \qquad k_{12} = \Sigma_i(\Sigma_j n_{ij.}^3)/n_{i..},$$

$$k_{13} = \Sigma_i(\Sigma_j n_{ij.}^2)^2/n_{i..}, \qquad k_{14} = \Sigma_i(\Sigma_j\Sigma_k n_{ijk}^2)^2/n_{i..},$$

$$k_{15} = \Sigma_i\Sigma_j(\Sigma_k n_{ijk}^2)^2/n_{ij.}, \qquad k_{16} = \Sigma_i[\Sigma_j n_{ij.}(\Sigma_k n_{ijk}^2)]/n_{i..},$$

$$k_{17} = \Sigma_i(\Sigma_j n_{ij.}^2)(\Sigma_j\Sigma_k n_{ijk}^2)/n_{i..}, \qquad k_{18} = \Sigma_i[\Sigma_j(\Sigma_k n_{ijk}^2)^2 n_{ij.}]/n_{i..},$$

$$k_{19} = \Sigma_i\Sigma_j(\Sigma_k n_{ijk}^2)^2/n_{ij.}^2, \qquad k_{20} = \Sigma_i(\Sigma_j n_{ij.}^2)(\Sigma_j\Sigma_k n_{ijk}^2)/n_{i..}^2,$$

$$k_{21} = \Sigma_i(\Sigma_j\Sigma_k n_{ijk}^2)^2/n_{i..}^2, \qquad k_{22} = \Sigma_i(\Sigma_j n_{ij.}^2)^2/n_{i..}^2,$$

$$k_{23} = \Sigma_i n_{i..}(\Sigma_j n_{ij.}^2), \qquad k_{24} = \Sigma_i n_{i..}(\Sigma_j\Sigma_k n_{ijk}^2),$$

$$k_{25} = \Sigma_i\Sigma_j n_{ij.}(\Sigma_k n_{ijk}^2)$$

and

$$\Delta_1 = k_{19} + k_{21} - 2k_{18}, \quad \Delta_2 = Nk_3 + k_{19} - 2k_{11},$$

$$\Delta_3 = k_{10} - k_{18}, \quad \Delta_4 = k_{11} - k_{19} \quad \text{and} \quad \Delta_5 = (k_9 - k_{15})/N$$

$$d_1 = v_8^2(Nk_2 + k_{22} - 2k_{12}),$$

$$d_2 = v_8^2\Delta_1 + v_6^2\Delta_2 + 2v_6v_8(\Delta_3 - \Delta_4),$$

$$d_3 = (v_7v_8 - v_6v_9)^2/v_{10} + v_7v_8^2 + v_6^2v_9,$$

$$d_4 = v_8^2(Nk_3 + k_{20} - 2k_{16}), \quad d_5 = v_5v_8^2 \quad \text{and} \quad d_6 = v_6v_8(v_6 + v_8)$$

and

$$g_1 = v_5 d_5(Nk_1 + k_1^2 - 2k_7/N),$$
$$g_2 = v_5 d_5(k_{22} + k_2^2 - 2k_{13}/N) + v_2^2 d_1 - 2v_2 d_5[k_{12} - k_{22} - (k_8 - k_{13})/N],$$
$$g_3 = v_5 d_5(k_{21} + k_3^2 - 2k_{14}/N) + v_2^2 v_8^2 \Delta_1 + (v_2 v_6 - v_3 v_5)^2 \Delta_2$$
$$- 2v_2 d_5[k_{18} + k_{21} - (k_{15} - k_{14})]/N)$$
$$+ 2v_8(v_2 v_6 - v_3 v_5)[v_5(\Delta_3 - \Delta_5) - v_2(\Delta_4 - \Delta_3)],$$
$$g_4 = v_5 d_5(a - 1) + v_2^2 v_7 v_8^2 + v_9(v_2 v_6 - v_3 v_5)^2$$
$$+ [v_4 v_5 v_8 - v_2 v_7 v_8 + v_9(v_2 v_6 - v_3 v_5)]^2/v_{10}$$
$$g_5 = v_5 d_5(Nk_2 + k_1 k_2 - 2k_{23}/N), \quad g_6 = v_5 d_5(Nk_3 + k_1 k_3 - 2k_{24}/N),$$
$$g_7 = v_1 v_5 d_5,$$
$$g_8 = v_5 d_5(k_{20} + k_2 k_3 - 2k_{17}/N) + v_2^2 v_8^2(Nk_3 + k_{20} - 2k_{16})$$
$$- 2v_2 d_5[k_{16} - k_{20} - (k_{25} - k_{17})/N]$$
$$g_9 = v_2 d_5(v_2 + v_5) \quad \text{and} \quad g_{10} = v_8[v_8(v_3 v_5^2 + v_2^2 v_6) + (v_2 v_6 - v_3 v_5)^2].$$

Then

$$\text{var}(\hat\sigma_\alpha^2) = 2(g_1\sigma_\alpha^4 + g_2\sigma_\beta^4 + g_3\sigma_\gamma^4 + g_4\sigma_e^4 + 2g_5\sigma_\alpha^2\sigma_\beta^2 + 2g_6\sigma_\alpha^2\sigma_\gamma^2 + 2g_7\sigma_\alpha^2\sigma_e^2$$
$$+ 2g_8\sigma_\beta^2\sigma_\gamma^2 + 2g_9\sigma_\beta^2\sigma_e^2 + 2g_{10}\sigma_\gamma^2\sigma_e^2)/v_1^2 v_5^2 v_8^2,$$

$$\text{var}(\hat\sigma_\beta^2) = 2(d_1\sigma_\beta^4 + d_2\sigma_\gamma^4 + d_3\sigma_e^4 + 2d_4\sigma_\beta^2\sigma_\gamma^2 + 2d_5\sigma_\beta^2\sigma_e^2 + 2d_6\sigma_\gamma^2\sigma_e^2)/v_5^2 v_8^2,$$

$$\text{var}(\hat\sigma_\gamma^2) = 2[\Delta_2\sigma_\gamma^4 + v_9(v_9 + v_{10})\sigma_e^4/v_{10} + 2v_8\sigma_\gamma^2\sigma_e^2]/v_8^2,$$

$$\text{cov}(\hat\sigma_\beta^2, \hat\sigma_\gamma^2) = [2(\Delta_4 - \Delta_3)\sigma_\gamma^4 + 2v_7 v_9\sigma_e^4/v_{10} - v_6 v_8 \,\text{var}(\hat\sigma_\gamma^2)]/v_5 v_8,$$

$$\text{cov}(\hat\sigma_\alpha^2, \hat\sigma_\gamma^2) = \{2[v_5(\Delta_3 - \Delta_5) - v_2(\Delta_4 - \Delta_3)]\sigma_\gamma^4 + 2v_9(v_4 v_5 - v_2 v_7)\sigma_e^4/v_{10}$$
$$- v_8(v_3 v_5 - v_2 v_6)\,\text{var}(\hat\sigma_\gamma^2)\}/v_1 v_5 v_8$$

and

$$v_1 v_5 v_8 \,\text{cov}(\hat\sigma_\alpha^2, \hat\sigma_\beta^2) = 2[k_{12} - k_{22} - (k_8 - k_{13})/N]\sigma_\beta^4$$
$$+ 2[k_{18} - k_{21} - (k_{15} - k_{14})/N - v_6(\Delta_3 - \Delta_5) - v_3(\Delta_4 - \Delta_3)]\sigma_\gamma^4$$
$$+ 2[k_{16} - k_{20} - (k_{25} - k_{17})/N]\sigma_\beta^2\sigma_\gamma^2 + 2[v_4 v_7 v_8 - v_9(v_4 v_6 + v_3 v_7)]\sigma_e^4/v_{10}$$
$$- v_2 v_5 v_8 \,\text{var}(\hat\sigma_\beta^2) + v_3 v_6 v_8 \,\text{var}(\hat\sigma_\gamma^2)\,.$$

(Mahamunulu, 1963).

PART II. THE 2-WAY CROSSED CLASSIFICATION

F.4. WITH INTERACTION, RANDOM MODEL

a. Model

$$y_{ijk} = \mu + \alpha_i + \beta_j + \gamma_{ij} + e_{ijk};$$
$$i = 1, 2, \ldots, a, \quad j = 1, 2, \ldots, b \quad \text{and} \quad k = 1, 2, \ldots, n_{ij},$$

with

$$n_{ij} > 0 \quad \text{for} \quad s\,(i, j)\text{-cells} \quad \text{and} \quad \Sigma_i\Sigma_j n_{ij} = N\,.$$

b. Henderson Method I estimators

Calculate Table F.1 and, for $n_{ij} > 0$,

$$T_0 = \Sigma_i\Sigma_j\Sigma_k y_{ijk}^2, \quad T_A = \Sigma_i y_{i..}^2/n_{i.}, \quad T_B = \Sigma_j y_{.j.}^2/n_{.j},$$
$$T_{AB} = \Sigma_i\Sigma_j y_{ij.}^2/n_{ij} \quad \text{and} \quad T_\mu = y_{...}^2/N;$$
$$\text{SSA} = T_A - T_\mu, \quad \text{SSB} = T_B - T_\mu,$$
$$\text{SSAB}^* = T_{AB} - T_A - T_B + T_\mu,$$
$$\text{SSE} = T_0 - T_{AB}\,.$$

TABLE F.1. ANALYSIS OF VARIANCE ESTIMATION OF VARIANCE COMPONENTS IN THE 2-WAY CROSSED CLASSIFICATION, INTERACTION, RANDOM MODEL

Terms needed for calculating estimators and their variances. For estimators only, calculate k_1, k_2, k_3, k_4 and k_{23}.

$k_1 = \Sigma_i n_{i.}^2$	$k_2 = \Sigma_j n_{.j}^2$
$k_3 = \Sigma_i(\Sigma_j n_{ij}^2)/n_{i.}$	$k_4 = \Sigma_j(\Sigma_i n_{ij}^2)/n_{.j}$
$k_5 = \Sigma_i n_{i.}^3$	$k_6 = \Sigma_j n_{.j}^3$
$k_7 = \Sigma_i(\Sigma_j n_{ij}^2)^2/n_{i.}$	$k_8 = \Sigma_j(\Sigma_i n_{ij}^2)^2/n_{.j}$
$k_9 = \Sigma_i(\Sigma_j n_{ij}^2)^2/n_{i.}^2$	$k_{10} = \Sigma_j(\Sigma_i n_{ij}^2)^2/n_{.j}^2$
$k_{11} = \Sigma_i(\Sigma_j n_{ij}^3)/n_{i.}$	$k_{12} = \Sigma_j(\Sigma_i n_{ij}^3)/n_{.j}$
$k_{13} = \Sigma_i(\Sigma_j n_{ij}^2)(\Sigma_j n_{ij}n_{.j})/n_{i.}$	$k_{14} = \Sigma_j(\Sigma_i n_{ij}^2)(\Sigma_i n_{ij}n_{i.})/n_{.j}$
$k_{15} = \Sigma_i(\Sigma_j n_{ij}n_{.j})^2 n_{i.}$	$k_{16} = \Sigma_j(\Sigma_i n_{ij}n_{i.})^2/n_{.j}$
$k_{17} = \Sigma_i(\Sigma_j n_{ij}^2 n_{.j})/n_{i.}$	$k_{18} = \Sigma_j(\Sigma_i n_{ij}^2 n_{i.})/n_{.j}$
$k_{19} = \Sigma_i(\Sigma_j n_{ij}^2)n_{i.}$	$k_{20} = \Sigma_j(\Sigma_i n_{ij}^2)n_{.j}$
$k_{21} = \Sigma_i\Sigma_{i' \neq i}(\Sigma_j n_{ij}n_{i'j})^2/n_{i.}n_{i'.}$	$k_{22} = \Sigma_j\Sigma_{j' \neq j}(\Sigma_i n_{ij}n_{ij'})^2/n_{.j}n_{.j'}$
$k_{23} = \Sigma_i\Sigma_j n_{ij}^2$	$k_{24} = \Sigma_i\Sigma_j n_{ij}^3$
$k_{25} = \Sigma_i\Sigma_j n_{ij}n_{i.}n_{.j}$	$k_{26} = \Sigma_i\Sigma_j n_{ij}^2/n_{i.}n_{.j}$
$k_{27} = \Sigma_i\Sigma_j n_{ij}^3/n_{i.}n_{.j}$	$k_{28} = \Sigma_i\Sigma_j n_{ij}^4/n_{i.}n_{.j}$

$k_r' = k_r/N$ for all r.

Then

$$\hat{\sigma}_e^2 = \text{SSE}/(N-s) = \text{MSE}$$

and with

$$\mathbf{P} = \begin{bmatrix} N-k_1' & k_3-k_2' & k_3-k_{23}' \\ k_4-k_1' & N-k_2' & k_4-k_{23}' \\ k_1'-k_4 & k_2'-k_3 & N-k_3-k_4+k_{23}' \end{bmatrix}$$

$$\hat{\boldsymbol{\sigma}}^2 = \begin{bmatrix} \hat{\sigma}_\alpha^2 \\ \hat{\sigma}_\beta^2 \\ \hat{\sigma}_\gamma^2 \end{bmatrix} = \mathbf{P}^{-1} \begin{bmatrix} \text{SSA} - (a-1)\text{MSE} \\ \text{SSB} - (b-1)\text{MSE} \\ \text{SSAB*} - (s-a-b+1)\text{MSE} \end{bmatrix}$$

as in (32) of Section 5.3b. This is equivalent to calculating

$$\delta_A = [\text{SSB} + \text{SSAB*} - (s-a)\text{MSE}]/(N-k_3)$$

and

$$\delta_B = [\text{SSA} + \text{SSAB*} - (s-b)\text{MSE}]/(N-k_4)$$

with which

$$\hat{\sigma}_\gamma^2 = [(N-k_1')\delta_B + (k_3-k_2')\delta_A - \{\text{SSA} - (a-1)\text{MSE}\}]/(N-k_1'-k_2'+k_{23}'),$$

$$\hat{\sigma}_\beta^2 = \delta_A - \hat{\sigma}_\gamma^2 \quad \text{and} \quad \hat{\sigma}_\alpha^2 = \delta_B - \hat{\sigma}_\gamma^2 \,.$$

(Searle, 1958).

c. Variances of Henderson Method I estimators (under normality)

$$\text{var}(\hat{\sigma}_e^2) = 2\sigma_e^4/(N-s)\,.$$

For $\mathbf{P}$ given above and for $\mathbf{H}$ and $\mathbf{f}$ being

$$\mathbf{H} = \begin{bmatrix} 1 & 0 & 0 & -1 \\ 0 & 1 & 0 & -1 \\ -1 & -1 & 1 & 1 \end{bmatrix} \quad \text{and} \quad \mathbf{f} = \begin{bmatrix} a-1 \\ b-1 \\ s-a-b+1 \end{bmatrix}$$

$$\text{var}(\hat{\boldsymbol{\sigma}}^2) = \mathbf{P}^{-1}[\mathbf{H}\,\text{var}(\mathbf{t})\,\mathbf{H}' + \text{var}(\hat{\sigma}_e^2)\,\mathbf{f}\mathbf{f}']\mathbf{P}^{-1\prime},$$

and

$$\text{cov}(\hat{\boldsymbol{\sigma}}^2, \hat{\sigma}_e^2) = -\mathbf{P}^{-1}\mathbf{f}\,\text{var}(\hat{\sigma}_e^2),$$

where

$$\text{var}(\mathbf{t}) = \text{var}[T_A \quad T_B \quad T_{AB} \quad T_\mu]'\,.$$

Var($\mathbf{t}$) has 10 different elements; each element is a function of the 10 squares and products of σ_α^2, σ_β^2, σ_γ^2 and σ_e^2. The 10×10 matrix of these coefficients is shown in Table F.2. Apart from N, a, b, s and unity, Table F.2 involves only

TABLE F.2. ANALYSIS OF VARIANCE ESTIMATION OF VARIANCE COMPONENTS IN THE 2-WAY CROSSED CLASSIFICATION, INTERACTION, RANDOM MODEL

Coefficients of squares and products of variance components in var(**t**), the variance–covariance matrix of the Ts, the uncorrected sums of squares

	Variance of				Covariance of					
	T_A	T_B	T_{AB}	T_μ	T_A, T_B	T_A, T_{AB}	T_A, T_μ	T_B, T_{AB}	T_B, T_μ	T_{AB}, T_μ
$2\sigma_\alpha^4$	k_1	$k_{22} + k_{10}$	k_1	$(k'_1)^2$	k_{18}	k_1	k'_5	k_{18}	k'_{16}	k'_5
$2\sigma_\beta^4$	$k_{21} + k_9$	k_2	k_2	$(k'_2)^2$	k_{17}	k_{17}	k'_{15}	k_2	k'_6	k'_6
$2\sigma_\gamma^4$	k_9	k_{10}	k_{23}	$(k'_{23})^2$	k_{28}	k_{11}	k'_7	k_{12}	k'_8	k'_{24}
$2\sigma_e^4$	a	b	s	1	k_{26}	a	1	b	1	1
$4\sigma_\alpha^2\sigma_\beta^2$	k_{23}	k_{23}	k_{23}	$k'_1 k'_2$	k_{23}	k_{23}	k'_{25}	k_{23}	k'_{25}	k'_{25}
$4\sigma_\alpha^2\sigma_\gamma^2$	k_{23}	k_{10}	k_{23}	$k'_1 k'_{23}$	k_{12}	k_{23}	k'_{19}	k_{12}	k'_{14}	k'_{19}
$4\sigma_\alpha^2\sigma_e^2$	N	k_4	N	k'_1	k_4	N	k'_1	k_4	k'_1	k'_1
$4\sigma_\beta^2\sigma_\gamma^2$	k_9	k_{23}	k_{23}	$k'_2 k'_{23}$	k_{11}	k_{11}	k'_{13}	k_{23}	k'_{20}	k'_{20}
$4\sigma_\beta^2\sigma_e^2$	k_3	N	N	k'_2	k_3	k_3	k'_2	N	k'_2	k'_2
$4\sigma_\gamma^2\sigma_e^2$	k_3	k_4	N	k'_{23}	k_{27}	k_3	k'_{23}	k_4	k'_{23}	k'_{23}

k_r for $r = 1, 2, \ldots, 28$ is given in Table F.1.
$k'_r = k_r/N$ for all r.

28 different terms. These are shown in Table F.1. An example of using Table F.2 is

$$\begin{aligned}\operatorname{var}(T_{AB}) = 2[&k_1\sigma_\alpha^4 + k_2\sigma_\beta^4 + k_{23}\sigma_\gamma^4 + s\sigma_e^4 \\ &+ 2(k_{23}\sigma_\alpha^2\sigma_\beta^2 + k_{23}\sigma_\alpha^2\sigma_\gamma^2 + N\sigma_\alpha^2\sigma_e^2 + k_{23}\sigma_\beta^2\sigma_\gamma^2 + N\sigma_\beta^2\sigma_e^2 + N\sigma_\gamma^2\sigma_e^2)].\end{aligned}$$

d. Henderson Method III estimators

Label the factor having the smaller number of levels in the data as the β-factor, with b levels.

Calculate $R(\mu, \boldsymbol{\alpha}, \boldsymbol{\beta})$ and h_6 as in Table F.3. Also, using Table F.1, calculate

$$h_1 = N - k_1', \quad h_2 = N - k_2', \quad h_3 = N - k_{23}',$$
$$h_4 = N - k_3 = h_5$$

and

$$h_7 = N - k_4 = h_8 .$$

TABLE F.3. COMPUTING FORMULAE FOR THE TERMS NEEDED IN USING HENDERSON'S METHOD III FOR ESTIMATING VARIANCE COMPONENTS ADDITIONAL TO THOSE NEEDED IN HENDERSON'S METHOD I: FOR THE 2-WAY CLASSIFICATION, MIXED OR RANDOM MODELS

To calculate $R(\mu, \boldsymbol{\alpha}, \boldsymbol{\beta})$ compute	To calculate h_6 compute
For $j = 1, \ldots, b$	For $i = 1, \ldots, a$
$c_{jj} = n_{.j} - \sum_{i=1}^{a} n_{ij}^2/n_{i.}$,	$\lambda_i = \sum_{j=1}^{b} \frac{n_{ij}^2}{n_{i.}}$.
$c_{jj'} = -\sum_{i=1}^{a} \frac{n_{ij}n_{ij'}}{n_{i.}}, \quad j \neq j'$	For $i = 1, \ldots, a$ and $j, j' = 1, \ldots, b$
$\left(\text{Check: } \sum_{j'=1}^{b} c_{jj'} = 0\right)$;	$f_{i,jj} = (n_{ij}^2/n_{i.})(\lambda_i + n_{i.} - 2n_{ij})$,
$r_j = y_{.j.} - \sum_{i=1}^{a} n_{ij}\bar{y}_{i..}$	$f_{i,jj'} = (n_{ij}n_{ij'}/n_{i.})(\lambda_i - n_{ij} - n_{ij'}) \quad \text{for } j \neq j'$
$\left(\text{Check: } \sum_{j=1}^{b} r_j = 0\right)$.	$\left(\text{Check: } \sum_{j=1}^{b} f_{i,jj'} = 0\right)$.
For $j, j' = 1, 2, \ldots, (b-1)$	For $i = 1, \ldots, a$ and $j, j' = 1, \ldots, (b-1)$
$\mathbf{C} = \{c_{jj'}\}$ and $\mathbf{r} = \{r_j\}$.	$\mathbf{F}_i = \{f_{i,jj'}\}$.
Then	Then
$t_B = \mathbf{r}'\mathbf{C}^{-1}\mathbf{r} = R(\boldsymbol{\beta} \mid \mu, \boldsymbol{\alpha})$	$k^* = \sum_{i=1}^{a} \lambda_i + \operatorname{tr}\left(\mathbf{C}^{-1} \sum_{i=1}^{a} \mathbf{F}_i\right)$
and $R(\mu, \boldsymbol{\alpha}, \boldsymbol{\beta}) = T_A + t_B$.	and $h_6 = N - k^*$.

The available estimation equations, from (142) of Chapter 5, are

$$\hat{\sigma}_e^2 = \frac{\Sigma_i\Sigma_j\Sigma_k(y_{ijk} - \bar{y}_{ij\cdot})^2}{N - s}, \tag{142a}$$

$$\hat{\sigma}_\gamma^2 = \frac{1}{h_6}[T_{AB} - R(\mu, \boldsymbol{\alpha}, \boldsymbol{\beta}) - (s - a - b + 1)\hat{\sigma}_e^2] \tag{142b}$$

and any two of

$$\hat{\sigma}_\alpha^2 = \frac{1}{h_7}[T_{AB} - T_B - (s - b)\hat{\sigma}_e^2] - \hat{\sigma}_\gamma^2, \tag{142c}$$

$$\hat{\sigma}_\beta^2 = \frac{1}{h_4}[T_{AB} - T_A - (s - a)\hat{\sigma}_e^2] - \hat{\sigma}_\gamma^2 \tag{142d}$$

and

$$h_1\hat{\sigma}_\alpha^2 + h_2\hat{\sigma}_\beta^2 = T_{AB} - T_\mu - h_3\hat{\sigma}_\gamma^2 - (s - 1)\hat{\sigma}_e^2\,. \tag{142e}$$

Calculation of h_6 and $R(\mu, \boldsymbol{\alpha}, \boldsymbol{\beta})$ needed for (124b) is shown in Table F.3.

Because any two of equations (142c, d and e) can be used, there are three different ways of using these equations, as shown in Table 5.4.

F.5. WITH INTERACTION, MIXED MODEL

a. Model

$$y_{ijk} = \mu + \alpha_i + \beta_j + \gamma_{ij} + e_{ijk}, \quad \beta_j\text{s taken as fixed effects;}$$

$$i = 1, 2, \ldots, a, \quad j = 1, 2, \ldots, b \quad \text{and} \quad k = 1, 2, \ldots, n_{ij},$$

with

$$n_{ij} > 0 \quad \text{for } s \quad (i, j)\text{-cells} \quad \text{and} \quad \Sigma_i\Sigma_j n_{ij} = N\,.$$

The model is exactly the same as the random model case of the preceding section, except that the βs are taken as fixed effects. They are assumed to be fewer in number than the random effects in the data.

b. Henderson Method III

Method I cannot be used because it is a mixed model; and Method II cannot be used because the model contains interactions between the fixed and random main effects. And in Method III, equations (142d and e) come from sums of squares whose expectations contain β. Therefore, Method III for this model has the prescription

$$\text{use (142a, b and c)} \quad \text{and Table F.3 for } h_6 \text{ and } R(\mu, \boldsymbol{\alpha}, \boldsymbol{\beta})\,.$$

F.6. NO INTERACTION, RANDOM MODEL

a. Model

$$y_{ijk} = \mu + \alpha_i + \beta_j + e_{ijk};$$

$$i = 1, 2, \ldots, a, \quad j = 1, 2, \ldots, b \quad \text{and} \quad k = 1, 2, \ldots, n_{ij},$$

with

$$n_{ij} > 0 \quad \text{for } s \quad (i,j)\text{-cells} \quad \text{and} \quad \Sigma_i \Sigma_j n_{ij} = N .$$

b. Henderson Method I

Calculate

$$T_0 = \Sigma_i \Sigma_j \Sigma_k y_{ijk}^2, \quad T_\mu = y_{...}^2/N,$$

$$T_A = \Sigma_i y_{i..}^2/n_{i.} \quad \text{and} \quad T_B = \Sigma_j y_{.j.}^2/n_{.j} .$$

Using Table F.1, calculate

$$\lambda_1 = (N - k'_1)/(N - k_4) \quad \text{and} \quad \lambda_2 = (N - k'_2)/(N - k_3) .$$

Then

$$\hat{\sigma}_e^2 = \frac{\lambda_2(T_0 - T_A) + \lambda_1(T_0 - T_B) - (T_0 - T_\mu)}{\lambda_2(N - a) + \lambda_1(N - b) - (N - 1)},$$

$$\hat{\sigma}_\alpha^2 = [T_0 - T_B - (N - b)\hat{\sigma}_e^2]/(N - k_4)$$

and

$$\hat{\sigma}_\beta^2 = [T_0 - T_A - (N - a)\hat{\sigma}_e^2]/(N - k_3) .$$

c. Variances of Henderson Method I estimators (under normality)

Writing

$$\mathbf{Q} = \begin{bmatrix} N - k'_1 & k_3 - k'_2 & a - 1 \\ k_4 - k'_1 & N - k'_2 & b - 1 \\ k'_1 - k_4 & k'_2 - k_3 & N - a - b + 1 \end{bmatrix} \quad \text{and} \quad \hat{\boldsymbol{\sigma}}^2 = \begin{bmatrix} \hat{\sigma}_\alpha^2 \\ \hat{\sigma}_\beta^2 \\ \hat{\sigma}_e^2 \end{bmatrix},$$

It can be shown that the estimators are solutions to

$$\mathbf{Q}\boldsymbol{\sigma}^2 = \begin{bmatrix} T_A - T_\mu \\ T_B - T_\mu \\ T_0 - T_A - T_B + T_\mu \end{bmatrix} = \mathbf{Ht} + \begin{bmatrix} 0 \\ 0 \\ T_0 - T_{AB} \end{bmatrix}$$

for **Ht** of Section F.4c.

When every $n_{ij} = 0$ or 1, $T_{AB} = T_0$ and $\mathbf{Q}\hat{\boldsymbol{\sigma}}^2 = \mathbf{Ht}$, so that

$$\text{var}(\hat{\boldsymbol{\sigma}}^2) = \mathbf{Q}^{-1}\mathbf{H}\,\text{var}(\mathbf{t})\,\mathbf{H}'\mathbf{Q}^{-1\prime} .$$

var(**t**) will be calculated exactly as in Tables F.1 and F.2 except with $\sigma_\gamma^2 = 0$.

When some $n_{ij} \geq 1$, T_{AB} exists even though it is not used in the estimation procedure. Nevertheless,

$$\hat{\boldsymbol{\sigma}}^2 = \mathbf{Q}^{-1}\left(\mathbf{Ht} + \begin{bmatrix} 0 \\ 0 \\ T_0 - T_{AB} \end{bmatrix}\right).$$

Furthermore, $T_0 - T_{AB}$ has variance $2\sigma_e^4(N - s)$ and is independent of every element in $\mathbf{Ht}$, whether $\sigma_\gamma^2 = 0$ or not. Therefore

$$\text{var}(\hat{\boldsymbol{\sigma}}^2) = \mathbf{Q}^{-1}\mathbf{H}\,\text{var}(\mathbf{t})\,\mathbf{H}'\mathbf{Q}^{-1\prime} + 2\mathbf{q}_3\mathbf{q}_3'\sigma_e^4(N - s)$$

where $\mathbf{q}_3$ is column 3 of $\mathbf{Q}^{-1}$. As with the $n_{ij} = 0$ or 1 case, var($\mathbf{t}$) is calculated from Tables F.1 and F.2 using $\sigma_\gamma^2 = 0$.

d. Henderson Method III

Calculate $R(\mu, \boldsymbol{\alpha}, \boldsymbol{\beta})$ of Table F.3, and from Table F.1 calculate

$$h_1 = N - k_1', \quad h_2 = N - k_2',$$
$$h_4 = N - k_3, \quad h_7 = N - k_4 .$$

The available estimation equations, from (124), are

$$\hat{\sigma}_e^2 = \frac{T_0 - R(\mu, \boldsymbol{\alpha}, \boldsymbol{\beta})}{N - a - b + 1} \tag{124a}$$

and any two of

$$\hat{\sigma}_\alpha^2 = \frac{1}{h_7}[R(\mu, \boldsymbol{\alpha}, \boldsymbol{\beta}) - T_B - (a - 1)\hat{\sigma}_e^2], \tag{124b}$$

$$\hat{\sigma}_\beta^2 = \frac{1}{h_4}[R(\mu, \boldsymbol{\alpha}, \boldsymbol{\beta}) - T_A - (b - 1)\hat{\sigma}_e^2] \tag{124c}$$

and

$$h_1\hat{\sigma}_\alpha^2 + h_2\hat{\sigma}_\beta^2 = R(\mu, \boldsymbol{\alpha}, \boldsymbol{\beta}) - T_\mu - (a + b - 2)\hat{\sigma}_e^2 . \tag{124d}$$

Because any two of equations (124b, c and d) can be used, there are three different ways of using these equations, as shown in Table 5.3.

e. Variances of Henderson Method III estimators (under normality)

For estimators obtained using equations (124a, b and c) of the preceding section, Low (1964) derives the following variances and covariances. Calculate

$$N' = N - a - b + 1$$

and, with the aid of Table F.1,

$$f_1 = k_1 - 2k_{18} + \Sigma_i\Sigma_{i'}(\Sigma_j n_{ij}n_{i'j}/n_{.j})^2$$

and

$$f_2 = k_2 - 2k_{17} + \Sigma_j \Sigma_{j'} (\Sigma_i n_{ij} n_{ij'} / n_{i.})^2 .$$

Then

$$\begin{aligned}
\text{var}(\hat{\sigma}_e^2) &= 2\sigma_e^4 / N', \\
\text{cov}(\hat{\sigma}_\alpha^2, \hat{\sigma}_e^2) &= -(a-1)\,\text{var}(\hat{\sigma}_e^2)/h_7, \quad \text{cov}(\hat{\sigma}_\beta^2, \hat{\sigma}_e^2) = -(b-1)\,\text{var}(\hat{\sigma}_e^2)/h_4, \\
\text{var}(\hat{\sigma}_\alpha^2) &= 2[\sigma_e^4 (N-b)(a-1)/N' + 2h_7 \sigma_e^2 \sigma_\alpha^2 + f_1 \sigma_\alpha^4]/h_7^2, \\
\text{var}(\hat{\sigma}_\beta^2) &= 2[\sigma_e^4 (N-a)(b-1)/N' + 2h_4 \sigma_e^2 \sigma_\beta^2 + f_2 \sigma_\beta^4]/h_4^2
\end{aligned}$$

and

$$\text{cov}(\hat{\sigma}_\alpha^2, \hat{\sigma}_\beta^2) = 2\sigma_e^4 [k_{26} - 1 + (a-1)(b-1)/N']/h_4 h_7 .$$

(Low, 1964).

F.7. NO INTERACTION, MIXED MODEL

a. Model

$$y_{ijk} = \mu + \alpha_i + \beta_j + e_{ijk}, \quad \beta_j\text{s taken as fixed effects;}$$
$$i = 1, 2 \ldots, a, \quad j = 1, 2, \ldots, b \quad \text{and} \quad k = 1, 2, \ldots, n_{ij},$$

with

$$n_{ij} > 0 \text{ for } s \quad (i,j)\text{-cells} \quad \text{and} \quad \Sigma_i \Sigma_j n_{ij} = N .$$

b. Henderson Method III

Method I cannot be used because this is a mixed model. Method II could be used because there are no interactions between fixed and random effects. But Method III is much easier because it simply involves using just two of the equations in the preceding model:

use (124a and b); and Table F.3 for $R(\mu, \boldsymbol{\alpha}, \boldsymbol{\beta})$.

APPENDIX M

SOME RESULTS IN MATRIX ALGEBRA

Readers of this book are assumed to have a working knowledge of matrix algebra. Nevertheless, a few reminders are provided in this appendix.

M.1. SUMMING VECTORS, AND J-MATRICES

Vectors having every element equal to unity are called *summing vectors* and are denoted by $\mathbf{1}$, using a subscript to represent order when necessary; e.g., $\mathbf{1}'_3 = [1 \quad 1 \quad 1]$. They are called summing vectors because, with $\mathbf{x}' = [x_1 \quad x_2 \quad x_3]$, for example, $\mathbf{1}'\mathbf{x} = \Sigma_{i=1}^{3} x_i$. In particular, the inner product of $\mathbf{1}_n$ with itself is n: $\mathbf{1}'_n\mathbf{1}_n = n$. A product of a summing vector with a matrix yields a vector of either column totals or row totals, of the matrix involved: for $\mathbf{B}$ having elements b_{ij}, the product $\mathbf{1}'\mathbf{B}$ is a row vector of column totals $b_{.j}$, and $\mathbf{B1}$ is a column vector of row totals $\mathbf{b}_{i.}$.

Outer products of summing vectors with each other are matrices having every element unity. They are denoted by $\mathbf{J}$. For example,

$$\mathbf{1}_2\mathbf{1}'_3 = \begin{bmatrix} 1 \\ 1 \end{bmatrix} [1 \quad 1 \quad 1] = \begin{bmatrix} 1 & 1 & 1 \\ 1 & 1 & 1 \end{bmatrix} = \mathbf{J}_{2\times 3} .$$

$\mathbf{J}$-matrices that are square are the most common form:

$$\mathbf{1}_n\mathbf{1}'_n = \mathbf{J}_n .$$

Product of $\mathbf{J}$s with each other and with $\mathbf{1}$s are, respectively, $\mathbf{J}$s and $\mathbf{1}$s multiplied by scalars. For square $\mathbf{J}$s

$$\mathbf{J}_n^2 = n\mathbf{J}_n \quad \text{and} \quad \mathbf{J}_n\mathbf{1}_n = n\mathbf{1}_n; \quad \text{and } \operatorname{tr}(\mathbf{J}_n) = n .$$

Two useful variants of $\mathbf{J}_n$ are

$$\bar{\mathbf{J}}_n = \tfrac{1}{n}\mathbf{J}_n \quad \text{and} \quad \mathbf{C}_n = \mathbf{I}_n - \bar{\mathbf{J}}_n,$$

with

$$\operatorname{tr}(\bar{\mathbf{J}}_n) = 1 \quad \text{and} \quad \operatorname{tr}(\mathbf{C}_n) = n - 1$$

and products (omitting the subscript n)

$$\mathbf{J1} = \mathbf{1}, \quad \mathbf{C1} = \mathbf{0}, \quad \bar{\mathbf{J}}^2 = \bar{\mathbf{J}} \quad \text{and } \mathbf{C}^2 = \mathbf{C} .$$

Thus $\bar{\mathbf{J}}$ and $\mathbf{C}$ are idempotent. $\mathbf{C}$ is called the *centering matrix* because $\mathbf{Cx}$ is a vector of elements $x_i - \bar{x}$ for $\bar{x} = \Sigma_{i=1}^n x_i/n$.

Illustration. The mean and sum of squares of data $x_1, x_2, \ldots, x_n$ are easily expressed in terms of the preceding matrices. Thus

$$\bar{x} = \sum_{i=1}^{n} \frac{x_i}{n} = \frac{\mathbf{1}_n'\mathbf{x}}{n} = \frac{\mathbf{x}'\mathbf{1}_n}{n}, \quad s^2 = \sum_{i=1}^{n} (x_i - \bar{x})^2 = \mathbf{x}'\mathbf{C}\mathbf{x} \quad \text{and} \quad n\bar{x}^2 = \mathbf{x}'\bar{\mathbf{J}}\mathbf{x} .$$

Linear combinations of $\mathbf{I}$ (an identity matrix) and $\mathbf{J}$ arise in a variety of circumstances, for which the following results are often found useful.

(i) $(a\mathbf{I}_n + b\mathbf{J}_n)(\alpha\mathbf{I}_n + \beta\mathbf{J}_n) = a\alpha\mathbf{I}_n + (a\beta + b\alpha + b\beta n)\mathbf{J}_n$.

(ii) $(a\mathbf{I}_n + b\mathbf{J}_n)^{-1} = \dfrac{1}{a}\left(\mathbf{I}_n - \dfrac{b}{a + nb}\mathbf{J}_n\right)$, for $a \neq 0$ and $a \neq -nb$.

(iii) $|a\mathbf{I}_n + b\mathbf{J}_n| = a^{n-1}(a + nb)$.

(iv) Eigenroots of $a\mathbf{I}_n + b\mathbf{J}_n$ are a, with multiplicity $n - 1$, and $a + nb$.

M.2. DIRECT SUMS AND PRODUCTS

The matrix

$$\mathbf{B}_1 \oplus \mathbf{B}_2 = \begin{bmatrix} \mathbf{B}_1 & \mathbf{0} \\ \mathbf{0} & \mathbf{B}_2 \end{bmatrix}$$

is the *direct sum* of $\mathbf{B}_1$ and $\mathbf{B}_2$, where those matrices can be of any order. This operation extends immediately to any number of matrices:

$$\bigoplus_{i=1}^{k} \mathbf{B}_i = \mathbf{B}_1 \oplus \mathbf{B}_2 \oplus \mathbf{B}_3 \oplus \cdots \oplus \mathbf{B}_k = \begin{bmatrix} \mathbf{B}_1 & \mathbf{0} & \cdots & \mathbf{0} \\ \mathbf{0} & \mathbf{B}_2 & & \vdots \\ \vdots & & \ddots & \mathbf{0} \\ \mathbf{0} & & & \mathbf{B}_k \end{bmatrix} .$$

For $\mathbf{A}$ of order $r \times c$ with elements a_{ij} for $i = 1, \ldots, r$ and $j = 1, \ldots, c$, the matrix

$$\mathbf{A} \otimes \mathbf{B} = \begin{bmatrix} a_{11}\mathbf{B} & a_{12}\mathbf{B} & \cdots & a_{1c}\mathbf{B} \\ a_{21}\mathbf{B} & a_{22}\mathbf{B} & \cdots & a_{2c}\mathbf{B} \\ \vdots & & & \\ a_{r1}\mathbf{B} & a_{r2}\mathbf{B} & \cdots & a_{rc}\mathbf{B} \end{bmatrix}$$

is the *direct product* of $\mathbf{A}$ and $\mathbf{B}$. For $\mathbf{A}$ being $r \times c$ and $\mathbf{B}$ being $s \times d$, the order of $\mathbf{A} \otimes \mathbf{B}$ is $rs \times cd$. (Whereas the preceding formulation is in terms of $a_{ij}\mathbf{B}$, there is also an alternative in terms of $b_{ij}\mathbf{A}$, but it is very rarely used today, and when it is it is denoted $\mathbf{B} \otimes \mathbf{A}$ in keeping with the above.) The matrix $\mathbf{A} \otimes \mathbf{B}$ often goes by the name Kronecker product (KP) because of Kronecker's association with the determinant of $\mathbf{A} \otimes \mathbf{B}$, although in this regard Henderson *et al.* (1983) suggest that "Zehfuss product" would be more appropriate historically.

The definition of $\mathbf{A} \otimes \mathbf{B}$ extends very naturally to more than two matrices; e.g.,

$$\mathbf{A} \otimes \mathbf{B} \otimes \mathbf{C} = \mathbf{A} \otimes (\mathbf{B} \otimes \mathbf{C}),$$

and

$$\bigotimes_{i=1}^{k} \mathbf{A}_i = \mathbf{A}_1 \otimes \mathbf{A}_2 \otimes \mathbf{A}_3 \otimes \cdots \otimes \mathbf{A}_k .$$

One particularly useful application is that $\mathbf{I}$ can always be expressed as Kronecker products of $\mathbf{I}$s of lesser order:

$$\mathbf{I}_{an} = \mathbf{I}_a \otimes \mathbf{I}_n \quad \text{and} \quad \mathbf{I}_{abn} = \mathbf{I}_a \otimes \mathbf{I}_b \otimes \mathbf{I}_n .$$

Some useful properties of direct products follow.

(i) In transposing products the reversal rule does not apply: i.e.,

$$(\mathbf{A} \otimes \mathbf{B})' = \mathbf{A}' \otimes \mathbf{B}' .$$

(ii) For $\mathbf{x}$ and $\mathbf{y}$ being vectors: $\mathbf{x}' \otimes \mathbf{y} = \mathbf{y}\mathbf{x}' = \mathbf{y} \otimes \mathbf{x}'$.

(iii) For λ being a scalar: $\lambda \otimes \mathbf{A} = \lambda\mathbf{A} = \mathbf{A} \otimes \lambda = \mathbf{A}\lambda$.

(iv) For partitioned matrices, although

$$[\mathbf{A}_1 \quad \mathbf{A}_2] \otimes \mathbf{B} = [\mathbf{A}_1 \otimes \mathbf{B} \quad \mathbf{A}_2 \otimes \mathbf{B}],$$

$$\mathbf{A} \otimes [\mathbf{B}_1 \quad \mathbf{B}_2] \neq [\mathbf{A} \otimes \mathbf{B}_1 \quad \mathbf{A} \otimes \mathbf{B}_2] .$$

(v) Provided conformability requirements for regular matrix multiplication are satisfied, $(\mathbf{A} \otimes \mathbf{B})(\mathbf{X} \otimes \mathbf{Y}) = \mathbf{AX} \otimes \mathbf{BY}$.

(vi) For $\mathbf{A}$ and $\mathbf{B}$ square and nonsingular, $(\mathbf{A} \otimes \mathbf{B})^{-1} = \mathbf{A}^{-1} \otimes \mathbf{B}^{-1}$.

(vii) Rank and trace obey product rules. For $r_{\mathbf{A}}$ and $\text{tr}(\mathbf{A})$ being the rank and trace, respectively, of $\mathbf{A}$,

$$r_{\mathbf{A}\otimes\mathbf{B}} = r_{\mathbf{A}} r_{\mathbf{B}} \quad \text{and} \quad \text{tr}(\mathbf{A} \otimes \mathbf{B}) = \text{tr}(\mathbf{A})\,\text{tr}(\mathbf{B}) .$$

(viii) Provided $\mathbf{A}$ and $\mathbf{B}$ are square, $|\mathbf{A}_{p\times p} \otimes \mathbf{B}_{m\times m}| = |\mathbf{A}|^m |\mathbf{B}|^p$.

(ix) Eigenroots of $\mathbf{A} \otimes \mathbf{B}$ are all possible products of an eigenroot of $\mathbf{A}$ and an eigenroot of $\mathbf{B}$.

M.3. A MATRIX NOTATION IN TERMS OF ELEMENTS

Familiar notation for a matrix **A** or order $p \times q$ is

$$\mathbf{A} = \{a_{ij}\} \quad \text{for } i = 1, \ldots, p \text{ and } j = 1, \ldots, q,$$

where a_{ij} is the element that is in the ith row and jth column of **A**. We abbreviate this to

$$\mathbf{A} = \{_\mathrm{m}\, a_{ij}\}_{i=1,j=1}^{p\quad q} = \{_\mathrm{m}\, a_{ij}\}_{i,j} = \{_\mathrm{m}\, a_{ij}\},$$

using m to indicate that the elements inside the braces are being arrayed as a matrix; and sufficient detail of subscripts follows the braces as is necessary, depending on context.

This notation is extended to row and column vectors and to diagonal matrices with the use of r, c and d as follows. First, a column vector is

$$\mathbf{u} = \begin{bmatrix} u_1 \\ u_2 \\ \vdots \\ u_t \end{bmatrix} = \{_\mathrm{c}\, u_i\}_{i=1}^{t} = \{_\mathrm{c}\, u_i\},$$

the c being used to show that it is a column vector. Similarly

$$\mathbf{u}' = \{_\mathrm{r}\, u_i\}_{i=1}^{t} = \{_\mathrm{r}\, u_i\}$$

is a row vector, and a diagonal matrix is

$$\begin{bmatrix} a_1 & 0 & 0 & \cdots & 0 \\ 0 & a_2 & 0 & \cdots & 0 \\ \vdots & & \ddots & \vdots & 0 \\ 0 & 0 & \cdots & & a_k \end{bmatrix} = \{_\mathrm{d}\, a_i\}_{i=1}^{k} = \{_\mathrm{d}\, a_i\}$$

where each of the last two symbols are used interchangeably. Extension to partitioned matrices is straightforward. For example, a direct sum is

$$\bigoplus_{i=1}^{3} \mathbf{A}_i = \begin{bmatrix} \mathbf{A}_1 & \mathbf{0} & \mathbf{0} \\ \mathbf{0} & \mathbf{A}_2 & \mathbf{0} \\ \mathbf{0} & \mathbf{0} & \mathbf{A}_3 \end{bmatrix} = \{_\mathrm{d}\, \mathbf{A}_i\}_{i=1}^{3}.$$

This notation has a variety of uses: e.g.,

$$\bigoplus_{i=1}^{k} \mathbf{A}_i = \{_\mathrm{d}\, \mathbf{A}_i\} \quad \text{and} \quad \mathbf{A} \otimes \mathbf{B} = \{_\mathrm{m}\, a_{ij}\mathbf{B}\}_{i,j}.$$

It can also be used in a nested manner. For example, with

$$\mathbf{y}_i = \{_\mathrm{c}\, y_{ij}\}_{j=1}^{n_i},$$

$$\mathbf{y} = \{_\mathrm{c}\, \mathbf{y}_i\}_{i=1}^{a} = \{_\mathrm{c}\, \{_\mathrm{c}\, y_{ij}\}_{j=1}^{n_i}\}_{i=1}^{a} .$$

And it is especially helpful in algebraic simplifications when typical elements of matrices are easily specified, but giving each matrix its own symbol is not needed. For example,

$$\{_\mathrm{d}\, \mathbf{1}_{n_i}\}\{_\mathrm{d}\, \mathbf{1}'_{n_i}\} = \{_\mathrm{d}\, \mathbf{J}_{n_i}\} . \tag{1}$$

In this manner, it is an especially economic notation when successively introducing or developing new matrices in terms of already-defined symbols, but where one does not wish, or need, to have individual symbols for the matrices themselves.

An adaptation of the block diagonal notation of (1) is useful for accommodating a situation that occurs with some-cells-empty data in the 2-way crossed classification. Consider the following two sets of n_{ij}-values for a 2×3 layout:

Grid 1

n_{ij}		
2	3	7
4	5	6

Grid 2

n_{ij}		
2	3	7
4	0	6

For grid 1 we have a matrix

$$\mathbf{A}_1 = \{_\mathrm{c}\, \{_\mathrm{d}\, \mathbf{1}_{n_{ij}}\}_j\}_i = \begin{bmatrix} \mathbf{1}_2 & \cdot & \cdot \\ \cdot & \mathbf{1}_3 & \cdot \\ \cdot & \cdot & \mathbf{1}_7 \\ \mathbf{1}_4 & \cdot & \cdot \\ \cdot & \mathbf{1}_5 & \cdot \\ \cdot & \cdot & \mathbf{1}_6 \end{bmatrix},$$

where dots represent null matrices (in this case vectors). For grid 2 the corresponding matrix that we want is

$$\mathbf{A}_2 = \begin{bmatrix} \mathbf{1}_2 & \cdot & \cdot \\ \cdot & \mathbf{1}_3 & \cdot \\ \cdot & \cdot & \mathbf{1}_7 \\ \mathbf{1}_4 & \cdot & \cdot \\ \cdot & \cdot & \mathbf{1}_6 \end{bmatrix}.$$

In trying to use the block diagonal notation of $\mathbf{A}_1$ for $\mathbf{A}_2$ we would have $\mathbf{1}_0$ in place of $\mathbf{1}_5$:

$$\{_{\mathrm{c}}\{_{\mathrm{d}}\,\mathbf{1}_{n_{ij}}\}_j\}_i = \begin{bmatrix} \mathbf{1}_2 & \cdot & \cdot \\ \cdot & \mathbf{1}_3 & \cdot \\ \cdot & \cdot & \mathbf{1}_7 \\ \mathbf{1}_4 & \cdot & \cdot \\ \cdot & \mathbf{1}_0 & \cdot \\ \cdot & \cdot & \mathbf{1}_6 \end{bmatrix}.$$

We rewrite this as

$$\{_{\mathrm{c}}\{_{\mathrm{d}*}\,\mathbf{1}_{n_{ij}}\}_{\mathrm{j}}\}_{\mathrm{i}} = \begin{bmatrix} \mathbf{1}_2 & \cdot & \cdot \\ \cdot & \mathbf{1}_3 & \cdot \\ \cdot & \cdot & \mathbf{1}_7 \\ \mathbf{1}_4 & \cdot & \cdot \\ \cdot & \cdot & \mathbf{1}_6 \end{bmatrix} = \mathbf{A}_2 .$$

The d* means that when $n_{ij} = 0$ the symbol $\mathbf{1}_0$ is used but then the row that it occurs in is deleted. $\mathbf{1}_0$ is like having a column vector that has no rows: it has position but no dimension.

M.4. GENERALIZED INVERSES

a. Definitions

Readers will be familiar with a nonsingular matrix $\mathbf{T}$ being a square matrix that has an inverse $\mathbf{T}^{-1}$ such that $\mathbf{TT}^{-1} = \mathbf{T}^{-1}\mathbf{T} = \mathbf{I}$. More generally, for any non-null matrix $\mathbf{A}$, be it rectangular, or square and singular, there are always matrices $\mathbf{A}^-$ satisfying

$$\mathbf{AA}^-\mathbf{A} = \mathbf{A} . \tag{2}$$

When $\mathbf{A}$ is non-singular, (2) leads to $\mathbf{A}^- = \mathbf{A}^{-1}$, but otherwise there is an infinite number of matrices $\mathbf{A}^-$ that for each $\mathbf{A}$ satisfy (2). Each such $\mathbf{A}^-$ is called a *generalized inverse of* $\mathbf{A}$.

Example. For

$$\mathbf{A} = \begin{bmatrix} 1 & 2 & 3 & 2 \\ 3 & 7 & 11 & 4 \\ 4 & 9 & 14 & 6 \end{bmatrix}, \quad \mathbf{A}^- = \begin{bmatrix} 7-t & -2-t & t \\ -3+2t & 1+2t & -2t \\ -t & -t & t \\ 0 & 0 & 0 \end{bmatrix} \tag{3}$$

Calculation of $\mathbf{AA}^-\mathbf{A}$ yields $\mathbf{A}$ no matter what value is used for t, thus illustrating the existence of infinitely many matrices $\mathbf{A}^-$ satisfying (2).

Two useful matrices involving products of $\mathbf{A}$ and $\mathbf{A}^-$ are

$$\mathbf{A}^-\mathbf{A}, \quad \text{idempotent, of rank } r_{\mathbf{A}}, \tag{4}$$

and

$$\mathbf{A}^{\sim} = \mathbf{A}^-\mathbf{AA}^-, \quad \text{for which} \quad \mathbf{AA}^{\sim}\mathbf{A} = \mathbf{A} \text{ and } \mathbf{A}^{\sim}\mathbf{AA}^{\sim} = \mathbf{A}^{\sim}\,. \tag{5}$$

Any matrix $\mathbf{A}^*$ satisfying $\mathbf{AA}^*\mathbf{A} = \mathbf{A}$ and $\mathbf{A}^*\mathbf{AA}^* = \mathbf{A}^*$ is called a *reflexive generalized inverse* of $\mathbf{A}$. A simple example is $\mathbf{A}^{\sim} = \mathbf{A}^-\mathbf{AA}^-$ of (5), which provides a simple way of deriving a generalized inverse of $\mathbf{A}$ that is reflexive from one that is not.

An important special case of both $\mathbf{A}^-$ and $\mathbf{A}^{\sim}$ is the unique (for given $\mathbf{A}$) matrix $\mathbf{A}^+$, which satisfies what are known as the four Penrose conditions:

$$\begin{array}{llll} \text{(i)} & \mathbf{AA}^+\mathbf{A} = \mathbf{A}, & \text{(iii)} & \mathbf{A}^+\mathbf{AA}^+ = \mathbf{A}^+, \\ \text{(ii)} & \mathbf{AA}^+ \text{ symmetric}, & \text{(iv)} & \mathbf{A}^+\mathbf{A} \text{ symmetric}\,. \end{array} \tag{6}$$

Named after its originators, Moore (1920) and Penrose (1955), the matrix $\mathbf{A}^+$ is called the Moore–Penrose inverse. Matrices $\mathbf{A}^-$ satisfying (2) are matrices that satisfy just Penrose condition (i), in (6), and reflexive generalized inverses $\mathbf{A}^{\sim}$ of (5) satisfy (i) and (iii). The satisfying of all four conditions in (6) produces the matrix $\mathbf{A}^+$ that is not only unique for given $\mathbf{A}$ but which also plays a role for rectangular and for square singular matrices that is similar to that played by the regular inverse of nonsingular (square) matrices. A convenient derivation of $\mathbf{A}^+$ is

$$\mathbf{A}^+ = \mathbf{A}'(\mathbf{AA}')^-\mathbf{A}(\mathbf{A}'\mathbf{A})^-\mathbf{A}', \tag{7}$$

where $\mathbf{A}'$ represents the transpose of $\mathbf{A}$. Notice also that

$$\mathbf{G} = \mathbf{A}^-\mathbf{AA}^- + (\mathbf{I} - \mathbf{A}^-\mathbf{A})\mathbf{T} + \mathbf{S}(\mathbf{I} - \mathbf{AA}^-)$$

is a generalized inverse of $\mathbf{A}$ for any (conformable) matrices $\mathbf{T}$ and $\mathbf{S}$.

b. Generalized inverses of X′X

Matrices of the form $\mathbf{X}'\mathbf{X}$ play an important role in linear models. Clearly, $\mathbf{X}'\mathbf{X}$ is square and symmetric and, for $\mathbf{X}$ having elements that are real numbers (i.e., do not involve $\sqrt{-1}$), $\mathbf{X}'\mathbf{X}$ is positive semi-definite (p.s.d.). Solutions for $\boldsymbol{\beta}$ to equations $\mathbf{X}'\mathbf{X}\boldsymbol{\beta} = \mathbf{X}'\mathbf{y}$ occur frequently in linear model work, and are often in terms of generalized inverses of $\mathbf{X}'\mathbf{X}$, which we denote as $(\mathbf{X}'\mathbf{X})^-$ and $\mathbf{G}$ interchangeably. Then $\mathbf{G}$ is defined by

$$\mathbf{X}'\mathbf{XGX}'\mathbf{X} = \mathbf{X}'\mathbf{X}\,. \tag{8}$$

Sometimes we also use $\mathbf{H}$ defined as

$$\mathbf{H} = \mathbf{GX}'\mathbf{X}, \quad \text{idempotent, of rank } r_{\mathbf{X}}\,. \tag{9}$$

Note that although $\mathbf{X}'\mathbf{X}$ is symmetric, $\mathbf{G}$ need not be symmetric. For example,

$$\mathbf{X}'\mathbf{X} = \begin{bmatrix} 7 & 3 & 2 & 2 \\ 3 & 3 & \cdot & \cdot \\ 2 & \cdot & 2 & \cdot \\ 2 & \cdot & \cdot & 2 \end{bmatrix} \quad \text{has} \quad \mathbf{G} = \begin{bmatrix} 9 & 0 & 0 & 3 \\ 5 & -13\frac{2}{3} & -14 & -17 \\ 1 & -10 & -9\frac{1}{2} & -13 \\ 0 & -9 & -9 & -11\frac{1}{2} \end{bmatrix} \tag{10}$$

as a generalized inverse, and $\mathbf{G}$ is certainly not symmetric. Despite this, transposing (8) shows that when $\mathbf{G}$ is a generalized inverse of $\mathbf{X}'\mathbf{X}$, then so also is $\mathbf{G}'$. As a consequence, as may be easily verified,

$$(\mathbf{X}'\mathbf{X})^{\sim} = \mathbf{G}\mathbf{X}'\mathbf{X}\mathbf{G}' \tag{11}$$

is a symmetric, reflexive generalized inverse of $\mathbf{X}'\mathbf{X}$ as defined in (5).

The following theorem is a cornerstone for many results in linear model theory.

Theorem M.1. When $\mathbf{G}$ is a generalized inverse of $\mathbf{X}'\mathbf{X}$:

$$\mathbf{G}' \text{ is also a generalized inverse of } \mathbf{X}'\mathbf{X}, \tag{12}$$

$$\mathbf{X}\mathbf{G}\mathbf{X}'\mathbf{X} = \mathbf{X}, \tag{13}$$

$$\mathbf{X}\mathbf{G}\mathbf{X}' \text{ is invariant to } \mathbf{G}; \quad \text{i.e., } \mathbf{X}\mathbf{G}\mathbf{X}' \text{ has the same value for every } \mathbf{G}, \tag{14}$$

$$\mathbf{X}\mathbf{G}\mathbf{X}' \text{ is symmetric, whether } \mathbf{G} \text{ is or not}, \tag{15}$$

$$\mathbf{X}\mathbf{G}\mathbf{X}'\mathbf{1} = \mathbf{1} \text{ when } \mathbf{1} \text{ is a column of } \mathbf{X}, \tag{16}$$

$$\mathbf{X}\mathbf{G}\mathbf{X}' = \mathbf{X}\mathbf{X}^{+}, \quad \text{where } \mathbf{X}^{+} \text{ is the Moore–Penrose inverse of } \mathbf{X}. \tag{17}$$

Proof. Condition (12) comes from transposing (8). Result (13) is true because for real matrices there is a theorem [e.g., Searle (1982), p. 63] indicating that if $\mathbf{P}\mathbf{X}'\mathbf{X} = \mathbf{Q}\mathbf{X}'\mathbf{X}$ then $\mathbf{P}\mathbf{X} = \mathbf{Q}\mathbf{X}$; applying this to the transpose of (8) and then transposing yields (13); and applying it to $\mathbf{X}\mathbf{G}\mathbf{X}'\mathbf{X} = \mathbf{X} = \mathbf{X}\mathbf{F}\mathbf{X}'\mathbf{X}$ for $\mathbf{F}$ being any other generalized inverse of $\mathbf{X}'\mathbf{X}$ yields (14). Using $(\mathbf{X}'\mathbf{X})^{\sim}$ of (11) in place of $\mathbf{G}$ in $\mathbf{X}\mathbf{G}\mathbf{X}'$ demonstrates the symmetry of (15) which, by (14), therefore holds for any $\mathbf{G}$. Finally, (16) follows from considering an individual column of $\mathbf{X}$ in (13), and (17) is established by using (7) for $\mathbf{X}^{+}$. Q.E.D.

Notice that (12) and (13) spawn three other results similar to (13): $\mathbf{X}\mathbf{G}'\mathbf{X}'\mathbf{X} = \mathbf{X}$, $\mathbf{X}'\mathbf{X}\mathbf{G}\mathbf{X}' = \mathbf{X}'$ and $\mathbf{X}'\mathbf{X}\mathbf{G}'\mathbf{X}' = \mathbf{X}'$. These and (12)–(17) are used frequently in some of the chapters. They have the effect of making $\mathbf{G}$ behave very like (but not exactly the same as) a regular inverse.

A particularly useful matrix is $\mathbf{M} = \mathbf{I} - \mathbf{X}\mathbf{G}\mathbf{X}'$. Theorem M.1 provides the means for verifying that $\mathbf{M}$ has the following properties: $\mathbf{M}$ is symmetric, indempotent, invariant to $\mathbf{G}$, of rank $N - r_{\mathbf{X}}$ when $\mathbf{X}$ has N rows, and its products with $\mathbf{X}$ and $\mathbf{X}'$ are null. Thus, with $\mathbf{M}$ having three equivalent forms,

$$\mathbf{M} = \mathbf{I} - \mathbf{X}\mathbf{G}\mathbf{X}' = \mathbf{I} - \mathbf{X}(\mathbf{X}'\mathbf{X})^{-}\mathbf{X}' = \mathbf{I} - \mathbf{X}\mathbf{X}^{+}, \tag{18}$$

we have

$$\mathbf{M} = \mathbf{M}' = \mathbf{M}^2, \quad r_{\mathbf{M}} = N - r_{\mathbf{X}}, \quad \mathbf{MX} = \mathbf{0} \quad \text{and} \quad \mathbf{X}'\mathbf{M} = \mathbf{0}\,. \tag{19}$$

c. Partitioning $\mathbf{X}'\mathbf{X}$

With $\mathbf{X}$ partitioned as $\mathbf{X} = [\mathbf{X}_1 \quad \mathbf{X}_2]$,

$$\mathbf{X}'\mathbf{X} = \begin{bmatrix} \mathbf{X}_1'\mathbf{X}_1 & \mathbf{X}_1'\mathbf{X}_2 \\ \mathbf{X}_2'\mathbf{X}_1 & \mathbf{X}_2'\mathbf{X}_2 \end{bmatrix}. \tag{20}$$

Then one form of generalized inverse of $\mathbf{X}'\mathbf{X}$ is (using Searle, 1982, p. 263)

$$\mathbf{G} = \begin{bmatrix} (\mathbf{X}_1'\mathbf{X}_1)^- & \mathbf{0} \\ \mathbf{0} & \mathbf{0} \end{bmatrix} + \begin{bmatrix} -(\mathbf{X}_1'\mathbf{X}_1)^-\mathbf{X}_1'\mathbf{X}_2 \\ \mathbf{I} \end{bmatrix} (\mathbf{X}_2'\mathbf{M}_1\mathbf{X}_2)^-[-\mathbf{X}_2'\mathbf{X}_1(\mathbf{X}_1'\mathbf{X}_1)^- \quad \mathbf{I}] \tag{21}$$

for

$$\mathbf{M}_1 = \mathbf{I} - \mathbf{X}_1(\mathbf{X}_1'\mathbf{X}_1)^-\mathbf{X}_1'$$

being the same function of $\mathbf{X}_1$ as $\mathbf{M}$ of (18) is of $\mathbf{X}$, and hence

$$\mathbf{M}_1 = \mathbf{M}_1' = \mathbf{M}_1^2 \quad \text{and} \quad \mathbf{M}_1\mathbf{X}_1 = \mathbf{0}\,.$$

[Note that it is the symmetry of $\mathbf{X}'\mathbf{X}$ that contributes to $\mathbf{G}$ of (21) being one form of $(\mathbf{X}'\mathbf{X})^-$. Partitioning a nonsymmetric matrix into four submatrices does not, in general, lead to the resulting form of (21) being valid—see Searle (1982, Sec. 10.5).]

Another form of $(\mathbf{X}'\mathbf{X})^-$ for partitioned $\mathbf{X}'\mathbf{X}$, and different from $\mathbf{G}$, is

$$\mathbf{F} = \begin{bmatrix} \mathbf{0} & \mathbf{0} \\ \mathbf{0} & (\mathbf{X}_2'\mathbf{X}_2)^- \end{bmatrix} + \begin{bmatrix} \mathbf{I} \\ -(\mathbf{X}_2'\mathbf{X}_2)^-\mathbf{X}_2'\mathbf{X}_1 \end{bmatrix} (\mathbf{X}_1'\mathbf{M}_2\mathbf{X}_1)^-[\mathbf{I} \quad -\mathbf{X}_1'\mathbf{X}_2(\mathbf{X}_2'\mathbf{X}_2)^-], \tag{22}$$

where

$$\mathbf{M}_2 = \mathbf{I} - \mathbf{X}_2(\mathbf{X}_2'\mathbf{X}_2)^-\mathbf{X}_2'\,.$$

Verification that (21) and (22) are each generalized inverses of $\mathbf{X}'\mathbf{X}$ of (20) demands using (13); and although we find that

$$\mathbf{XGX}' = \mathbf{X}_1(\mathbf{X}_1'\mathbf{X}_1)^-\mathbf{X}_1' + \mathbf{M}_1\mathbf{X}_2(\mathbf{X}_2'\mathbf{M}_1\mathbf{X}_2)^-\mathbf{X}_2'\mathbf{M}_1 \tag{23}$$

and

$$\mathbf{XFX}' = \mathbf{X}_2(\mathbf{X}_2'\mathbf{X}_2)^-\mathbf{X}_2' + \mathbf{M}_2\mathbf{X}_1(\mathbf{X}_1'\mathbf{M}_2\mathbf{X}_1)^-\mathbf{X}_1'\mathbf{M}_2, \tag{24}$$

which look different, we know from (14) that they are the same. That each is invariant to the generalized inverses it involves is nevertheless clear. In (23) the first term is invariant to the choice of $(\mathbf{X}_1'\mathbf{X}_1)^-$—by (14); and by the symmetry and idempotency of $\mathbf{M}_1$ the second term is $\mathbf{M}_1\mathbf{X}_2[(\mathbf{M}_1\mathbf{X}_2)'\mathbf{M}_1\mathbf{X}_2]^-(\mathbf{M}_1\mathbf{X}_2)'$, and so it too, by (14), has the invariance property. Nevertheless, a direct development of the equality of (23) to (24) without appealing to (14) seems difficult.

The preceding results of this section are all in terms of generalized inverses. When $\mathbf{X}'\mathbf{X}$ of (20) is non-singular, all of those results still apply, with the generalized inverses being regular inverses.

d. Rank results

The standard result for the rank of a product matrix is $r_{\mathbf{AB}} \leqslant r_{\mathbf{B}}$. Thus using $r(\mathbf{X})$ and $r_{\mathbf{X}}$ interchangeably to represent the rank of $\mathbf{X}$, we have $r(\mathbf{AA}^-) \leqslant r_{\mathbf{A}}$; and from $\mathbf{A} = \mathbf{AA}^-\mathbf{A}$ we have $r_{\mathbf{A}} \leqslant r(\mathbf{AA}^-)$. Therefore $r(\mathbf{AA}^-) = r_{\mathbf{A}}$. Also, because $\mathbf{AA}^-$ is idempotent its trace and rank are equal. In particular, $\text{tr}(\mathbf{AA}^+) = r_{\mathbf{A}}$. Therefore from (17)

$$\text{tr}[\mathbf{A}(\mathbf{A}'\mathbf{A})^-\mathbf{A}'] = \text{tr}(\mathbf{AA}^+) = r_{\mathbf{A}}\,. \tag{25}$$

Applying (25) to each term in (23), using the indempotency and symmetry of $\mathbf{M}_1$ in doing so, gives

$$r_{\mathbf{X}} = r_{\mathbf{X}_1} + r_{\mathbf{X}_2'\mathbf{M}_1\mathbf{X}_2},$$

which, on using $r_{\mathbf{AA}'} = r_{\mathbf{A}}$ for $\mathbf{A}$ being real, leads to

$$r_{\mathbf{X}_2'\mathbf{M}_1\mathbf{X}_2} = r_{\mathbf{M}_1\mathbf{X}_2} = r_{[\mathbf{X}_1\ \ \mathbf{X}_2]} - r_{\mathbf{X}_1}\,. \tag{26}$$

This result is useful in the context of degrees of freedom for sums of squares based on (23), as in (98) of Chapter 5. A particular case of (26) is when $\mathbf{X}$ has full column rank: then so does $\mathbf{M}_1\mathbf{X}_2$; and, of course, $\mathbf{M}_2\mathbf{X}_1$ also.

e. Vectors orthogonal to columns of X

Suppose $\mathbf{k}'$ is such that $\mathbf{k}'\mathbf{X} = \mathbf{0}$. Then $\mathbf{X}'\mathbf{k} = \mathbf{0}$ and, from the theory of solving linear equations (e.g., Searle, 1982, Sec. 9.4b), $\mathbf{k} = [\mathbf{I} - (\mathbf{X}')^-\mathbf{X}']\mathbf{c}$ for any vector $\mathbf{c}$, of appropriate order. Therefore, since $(\mathbf{X}^-)'$ is a generalized inverse of $\mathbf{X}'$ we can write $\mathbf{k}' = \mathbf{c}'(\mathbf{I} - \mathbf{XX}^-)$. Moreover, because $(\mathbf{X}'\mathbf{X})^-\mathbf{X}'$ is a generalized inverse of $\mathbf{X}$ another form for $\mathbf{k}'$ is $\mathbf{k}' = \mathbf{c}'[\mathbf{I} - \mathbf{X}(\mathbf{X}'\mathbf{X})^-\mathbf{X}']$; as is $\mathbf{c}'(\mathbf{I} - \mathbf{XX}^+)$ since $\mathbf{X}(\mathbf{X}'\mathbf{X})^-\mathbf{X}' = \mathbf{XX}^+$. Thus two forms of $\mathbf{k}'$ are

$$\mathbf{k}' = \mathbf{c}'(\mathbf{I} - \mathbf{XX}^-), \quad \text{or} \quad \mathbf{k}' = \mathbf{c}'[\mathbf{I} - \mathbf{X}(\mathbf{X}'\mathbf{X})^-\mathbf{X}'] = \mathbf{c}'(\mathbf{I} - \mathbf{XX}^+)\,.$$

With $\mathbf{M}$ defined in (18), as $\mathbf{M} = \mathbf{I} - \mathbf{XX}^+ = \mathbf{I} - \mathbf{X}(\mathbf{X}'\mathbf{X})^-\mathbf{X}'$, we therefore have $\mathbf{k}' = \mathbf{c}'\mathbf{M}$.

With $\mathbf{X}$ of order $N \times p$ of rank r, there are only $N - r$ linearly independent vectors $\mathbf{k}'$ satisfying $\mathbf{k}'\mathbf{X} = \mathbf{0}$ (e.g., Searle, 1982, Sec. 9.7a). Using a set of such $N - r$ linearly independent vectors $\mathbf{k}'$ as rows of $\mathbf{K}'$, we then have the following theorem, for $\mathbf{K}'\mathbf{X} = \mathbf{0}$ with $\mathbf{K}'$ having maximum row rank $N - r$ and $\mathbf{K}' = \mathbf{C}'\mathbf{M}$ for some $\mathbf{C}$.

f. A theorem involving K′ of maximum row rank, for K′X being null

Theorem. If $\mathbf{K}'\mathbf{X} = \mathbf{0}$, where $\mathbf{K}'$ has maximum row rank, and $\mathbf{V}$ is positive definite then

$$\mathbf{K}(\mathbf{K}'\mathbf{VK})^{-1}\mathbf{K}' = \mathbf{P} \quad \text{for } \mathbf{P} \equiv \mathbf{V}^{-1} - \mathbf{V}^{-1}\mathbf{X}(\mathbf{X}'\mathbf{V}^{-1}\mathbf{X})^-\mathbf{X}'\mathbf{V}^{-1}\,.$$

Khatri's (1966) proof of this is for $\mathbf{X}$ having full column rank. For the more general case considered here, of $\mathbf{X}$ not of full column rank, we offer a shorter proof (due to Pukelsheim, personal communication, 1986) than that given by Khatri.

Proof. Both $\mathbf{KK}^+ = \mathbf{K}(\mathbf{K}'\mathbf{K})^{-1}\mathbf{K}'$ and $\mathbf{XX}^+ = \mathbf{X}(\mathbf{X}'\mathbf{X})^-\mathbf{X}'$ are symmetric and idempotent, and $\mathbf{K}'\mathbf{X} = \mathbf{0}$. Therefore $\mathbf{KK}^+\mathbf{X} = \mathbf{0}$ and $\mathbf{XX}^+\mathbf{K} = \mathbf{0}$. Hence $\mathbf{T} = \mathbf{I} - \mathbf{XX}^+ - \mathbf{KK}^+$ is symmetric, and idempotent. Therefore

$$\begin{aligned}\operatorname{tr}(\mathbf{TT}') = \operatorname{tr}(\mathbf{T}^2) = \operatorname{tr}(\mathbf{T}) &= \operatorname{tr}(\mathbf{I}) - \operatorname{tr}(\mathbf{XX}^+) - \operatorname{tr}(\mathbf{KK}^+)\\ &= N - r_{\mathbf{X}} - r_{\mathbf{K}}\\ &= N - r_{\mathbf{X}} - (N - r_{\mathbf{X}})\\ &= 0\,.\end{aligned}$$

But $\mathbf{T}$ is real, so that $\operatorname{tr}(\mathbf{TT}') = 0$ implies $\mathbf{T} = \mathbf{0}$. Therefore $\mathbf{I} - \mathbf{XX}^+ = \mathbf{KK}^+$.

Because $\mathbf{V}$ is positive definite, a symmetric matrix $\mathbf{V}^{\frac{1}{2}}$ always exists such that $\mathbf{V} = (\mathbf{V}^{\frac{1}{2}})^2$. Then, since $(\mathbf{V}^{\frac{1}{2}}\mathbf{K})'\mathbf{V}^{-\frac{1}{2}}\mathbf{X} = \mathbf{0}$, because $\mathbf{K}'\mathbf{X} = \mathbf{0}$, the preceding result applies for $\mathbf{K}$ and $\mathbf{X}$ replaced by $\mathbf{V}^{\frac{1}{2}}\mathbf{K}$ and $\mathbf{V}^{-\frac{1}{2}}\mathbf{X}$, respectively. Making these replacements after writing $\mathbf{I} - \mathbf{XX}^+ = \mathbf{KK}^+$ as

$$\mathbf{I} - \mathbf{X}(\mathbf{X}'\mathbf{X})^-\mathbf{X} = \mathbf{K}(\mathbf{K}'\mathbf{K})^{-1}\mathbf{K}'$$

gives

$$\mathbf{I} - \mathbf{V}^{-\frac{1}{2}}\mathbf{X}(\mathbf{X}'\mathbf{V}^{-1}\mathbf{X})^-\mathbf{X}'\mathbf{V}^{-\frac{1}{2}} = \mathbf{V}^{\frac{1}{2}}\mathbf{K}(\mathbf{K}'\mathbf{V}\mathbf{K})^{-1}\mathbf{K}'\mathbf{V}^{\frac{1}{2}};$$

i.e.,

$$\mathbf{P} = \mathbf{V}^{-1} - \mathbf{V}^{-1}\mathbf{X}(\mathbf{X}'\mathbf{V}^{-1}\mathbf{X})^-\mathbf{X}'\mathbf{V}^{-1} = \mathbf{K}(\mathbf{K}'\mathbf{V}\mathbf{K})^{-1}\mathbf{K}'\,. \qquad \text{Q.E.D.}$$

An extension of this result is that $\mathbf{P} = \mathbf{M}(\mathbf{MVM})^-\mathbf{M}$. This is established by first noting that with $\mathbf{M} = \mathbf{I} - \mathbf{XX}^+ = \mathbf{KK}^+$, as in the preceding proof,

$$\begin{aligned}\mathbf{K}'\mathbf{MVMK}[\mathbf{K}^+(\mathbf{MVM})^-\mathbf{K}^{+\prime}]\mathbf{K}'\mathbf{MVMK} &= \mathbf{K}'\mathbf{MVM}(\mathbf{MVM})^-\mathbf{MVMK}\\ &= \mathbf{K}'\mathbf{MVMK}\,.\end{aligned}$$

Therefore

$$(\mathbf{K}'\mathbf{MVMK})^- = \mathbf{K}^+(\mathbf{MVM})^-\mathbf{K}^{+\prime}\,.$$

Hence, starting from $\mathbf{P} = \mathbf{K}(\mathbf{K}'\mathbf{VK})^{-1}\mathbf{K}'$ and using $\mathbf{MK} = \mathbf{K}$ gives

$$\begin{aligned}\mathbf{P} &= \mathbf{MK}(\mathbf{K}'\mathbf{MVMK})^{-1}\mathbf{K}'\mathbf{M}\\ &= \mathbf{MK}[\mathbf{K}^+(\mathbf{MVM})^-\mathbf{K}^{+\prime}]\mathbf{K}'\mathbf{M}\\ &= \mathbf{M}(\mathbf{MVM})^-\mathbf{M}\,.\end{aligned}$$

M.5. THE SCHUR COMPLEMENT

In the inverse of a nonsingular partitioned matrix

$$\begin{bmatrix} \mathbf{A} & \mathbf{B} \\ \mathbf{C} & \mathbf{D} \end{bmatrix}^{-} = \begin{bmatrix} \mathbf{A}^{-1} & \mathbf{0} \\ \mathbf{0} & \mathbf{0} \end{bmatrix} + \begin{bmatrix} -\mathbf{A}^{-1}\mathbf{B} \\ \mathbf{I} \end{bmatrix} (\mathbf{D} - \mathbf{C}\mathbf{A}^{-1}\mathbf{B})^{-1} [-\mathbf{C}\mathbf{A}^{-1} \quad \mathbf{I}] \tag{27}$$

the matrix $\mathbf{D} - \mathbf{C}\mathbf{A}^{-1}\mathbf{B}$ is known as the Schur complement of $\mathbf{A}$. Marsaglia and Styan (1974a,b) give numerous results concerning Schur complements, of which we use primarily two. The first is

$$(\mathbf{D} - \mathbf{C}\mathbf{A}^{-1}\mathbf{B})^{-1} = \mathbf{D}^{-1} + \mathbf{D}^{-1}\mathbf{C}(\mathbf{A} - \mathbf{B}\mathbf{D}^{-1}\mathbf{C})^{-1}\mathbf{B}\mathbf{D}^{-1}, \tag{28a}$$

as may be verified by multiplying the right-hand side by $\mathbf{D} - \mathbf{C}\mathbf{A}^{-1}\mathbf{B}$. Similarly

$$(\mathbf{D} + \mathbf{C}\mathbf{A}^{-1}\mathbf{B})^{-1} = \mathbf{D}^{-1} - \mathbf{D}^{-1}\mathbf{C}(\mathbf{A} + \mathbf{B}\mathbf{D}^{-1}\mathbf{C})^{-1}\mathbf{B}\mathbf{D}^{-1}, \tag{28b}$$

on replacing $\mathbf{A}$ by $-\mathbf{A}$ in (28a); and a useful special case of (28b) is

$$(\mathbf{D} + \lambda\mathbf{t}\mathbf{t}')^{-1} = \mathbf{D}^{-1} - \frac{\mathbf{D}^{-1}\mathbf{t}\mathbf{t}'\mathbf{D}^{-1}}{1/\lambda + \mathbf{t}'\mathbf{D}^{-1}\mathbf{t}} . \tag{29}$$

The determinant of $\mathbf{D} - \mathbf{C}\mathbf{A}^{-1}\mathbf{B}$ is derived as follows. To begin, observe that

$$\begin{vmatrix} \mathbf{R} & \mathbf{0} \\ \mathbf{X} & \mathbf{T} \end{vmatrix} = |\mathbf{R}|\,|\mathbf{T}| = \begin{vmatrix} \mathbf{R}' & \mathbf{X}' \\ \mathbf{0} & \mathbf{T}' \end{vmatrix}, \tag{30}$$

wherein the first equality comes from performing row operations on the rows through $\mathbf{R}$ to triangularize $\mathbf{R}$. The second equality comes simply from transposing the matrix. Next, it is clear that

$$\begin{bmatrix} \mathbf{A} & \mathbf{B} \\ \mathbf{C} & \mathbf{D} \end{bmatrix} = \begin{bmatrix} \mathbf{A} & \mathbf{0} \\ \mathbf{C} & \mathbf{D} - \mathbf{C}\mathbf{A}^{-1}\mathbf{B} \end{bmatrix} \begin{bmatrix} \mathbf{I} & \mathbf{A}^{-1}\mathbf{B} \\ \mathbf{0} & \mathbf{I} \end{bmatrix} .$$

Taking determinants and using (30) gives

$$\begin{vmatrix} \mathbf{A} & \mathbf{B} \\ \mathbf{C} & \mathbf{D} \end{vmatrix} = |\mathbf{A}|\,|\mathbf{D} - \mathbf{C}\mathbf{A}^{-1}\mathbf{B}| .$$

In similar manner

$$\begin{vmatrix} \mathbf{A} & \mathbf{B} \\ \mathbf{C} & \mathbf{D} \end{vmatrix} = |\mathbf{D}|\,|\mathbf{A} - \mathbf{B}\mathbf{D}^{-1}\mathbf{C}|$$

and so

$$|\mathbf{D} - \mathbf{C}\mathbf{A}^{-1}\mathbf{B}| = (|\mathbf{D}|/|\mathbf{A}|)|\mathbf{A} - \mathbf{B}\mathbf{D}^{-1}\mathbf{C}| . \tag{31}$$

This is particularly useful when $\mathbf{A}$ is a scalar:

$$|\mathbf{D} - \mathbf{x}\mathbf{y}'/a| = |\mathbf{D}|(a - \mathbf{y}'\mathbf{D}^{-1}\mathbf{x})/a . \tag{32}$$

M.6. THE TRACE OF A MATRIX

The trace of a matrix is the sum of its diagonal elements:

$$\mathrm{tr}(\mathbf{A}) = \Sigma_i a_{ii} \,.$$

Thus tr(**A**) is defined only for **A** being square. The trace of a product has a useful property:

$$\mathrm{tr}(\mathbf{AB}) = \mathrm{tr}(\mathbf{BA})$$

because

$$\mathrm{tr}(\mathbf{AB}) = \Sigma_i(\Sigma_j a_{ij} b_{ji}) = \Sigma_j(\Sigma_i b_{ji} a_{ij}) = \mathrm{tr}(\mathbf{BA}) \,.$$

And for computing purposes a useful result for any matrix **M** is

$$\mathrm{tr}(\mathbf{MM}') = \mathrm{sesq}(\mathbf{M}), \tag{33}$$

where sesq(**M**) represents the sum of squares of elements of **M**. We use the abbreviation sesq rather than ssqe to avoid any possible confusion of the latter with a sum of squares of data. Verification of (33) is

$$\mathrm{tr}(\mathbf{MM}') = \Sigma_i[\Sigma_j m_{ij}(m')_{ji}] = \Sigma_i\Sigma_j m_{ij}^2 = \mathrm{sesq}(\mathbf{M}) \,.$$

A useful special case is when **M** is symmetric:

$$\mathrm{tr}(\mathbf{M}^2) = \mathrm{sesq}(\mathbf{M}) \quad \text{when } \mathbf{M} = \mathbf{M}' \,. \tag{34}$$

Another useful result is

$$\mathrm{tr}(\mathbf{JA}) = \mathrm{tr}(\mathbf{11}'\mathbf{A}) = \mathrm{tr}(\mathbf{1}'\mathbf{A1}) = \mathbf{1}'\mathbf{A1} = \Sigma_i\Sigma_j a_{ij} \,.$$

M.7. DIFFERENTIATION OF MATRIX EXPRESSIONS

a. Scalars

Beginning with an example

$$\lambda = 3x_1 + 5x_2,$$

$\partial\lambda/\partial\mathbf{x}$ is defined as

$$\frac{\partial\lambda}{\partial\mathbf{x}} = \begin{bmatrix} \dfrac{\partial\lambda}{\partial x_1} \\ \dfrac{\partial\lambda}{\partial x_2} \end{bmatrix} = \begin{bmatrix} 3 \\ 5 \end{bmatrix}.$$

In general this extends to

$$\frac{\partial}{\partial\mathbf{x}}(\mathbf{a}'\mathbf{x}) = \mathbf{a} = \frac{\partial}{\partial\mathbf{x}}(\mathbf{x}'\mathbf{a}), \tag{35}$$

the second equality arising from $\mathbf{a}'\mathbf{x} = \mathbf{x}'\mathbf{a}$. Also

$$\frac{\partial}{\partial \mathbf{x}'}(\mathbf{a}'\mathbf{x}) = \left[\frac{\partial}{\partial \mathbf{x}}(\mathbf{a}'\mathbf{x})\right]' = \mathbf{a}' = \frac{\partial}{\partial \mathbf{x}'}(\mathbf{x}'\mathbf{a}) . \tag{36}$$

b. Vectors

Beginning generally with $\mathbf{y}_{r \times 1}$ and $\mathbf{x}_{p \times 1}$, where elements of $\mathbf{y}$ are differentiable functions of elements of $\mathbf{x}$, we define

$$\frac{\partial \mathbf{y}'}{\partial \mathbf{x}} = \left\{_{\mathrm{m}} \frac{\partial y_j}{\partial x_i}\right\}_{i=1,j=1}^{p\quad r}, \quad \text{a matrix of order } p \times r . \tag{37}$$

Similarly, the transpose of this is

$$\left(\frac{\partial \mathbf{y}'}{\partial \mathbf{x}}\right)' = \left\{_{\mathrm{m}} \frac{\partial y_j}{\partial x_i}\right\}_{j=1,i=1}^{r\quad p} = \frac{\partial \mathbf{y}}{\partial \mathbf{x}'}, \quad \text{a matrix of order } r \times p . \tag{38}$$

In particular

$$\frac{\partial \mathbf{x}}{\partial \mathbf{x}'} = \frac{\partial \mathbf{x}'}{\partial \mathbf{x}} = \mathbf{I} .$$

Therefore, for $\mathbf{A}$ and $\mathbf{B}$ not functions of $\mathbf{x}$

$$\frac{\partial}{\partial \mathbf{x}'}(\mathbf{A}\mathbf{x}) = \mathbf{A}\frac{\partial \mathbf{x}}{\partial \mathbf{x}'} = \mathbf{A} \tag{39}$$

and

$$\frac{\partial}{\partial \mathbf{x}}\mathbf{x}'\mathbf{B} = \frac{\partial \mathbf{x}'}{\partial \mathbf{x}}\mathbf{B} = \mathbf{B} . \tag{40}$$

c. Inner products

Consider $\mathbf{u}'\mathbf{v}$, where each element of $\mathbf{u}'$ and $\mathbf{v}$ is a function of elements of $\mathbf{x}$. Then

$$\frac{\partial(\mathbf{u}'\mathbf{v})}{\partial \mathbf{x}} = \frac{\partial}{\partial \mathbf{x}}\Sigma_i u_i v_i = \Sigma_i \frac{\partial u_i}{\partial \mathbf{x}} v_i + \Sigma_i u_i \frac{\partial v_i}{\partial \mathbf{x}} .$$

Each term in each sum is a column vector. Consideration of conformability therefore leads to having

$$\frac{\partial \mathbf{u}'\mathbf{v}}{\partial \mathbf{x}} = \frac{\partial \mathbf{u}'}{\partial \mathbf{x}}\mathbf{v} + \frac{\partial \mathbf{v}'}{\partial \mathbf{x}}\mathbf{u} . \tag{41}$$

d. Quadratic forms

Utilizing the preceding results yields

$$\begin{aligned} \frac{\partial}{\partial \mathbf{x}}\mathbf{x}'\mathbf{A}\mathbf{x} &= \frac{\partial \mathbf{x}'}{\partial \mathbf{x}}(\mathbf{A}\mathbf{x}) + \frac{\partial}{\partial \mathbf{x}}(\mathbf{A}\mathbf{x})'\mathbf{x} \\ &= \mathbf{A}\mathbf{x} + \mathbf{A}'\mathbf{x} . \end{aligned} \tag{42}$$

When $\mathbf{A}$ is symmetric, which it usually is in this context of a quadratic form,

$$\frac{\partial}{\partial \mathbf{x}}(\mathbf{x}'\mathbf{A}\mathbf{x}) = 2\mathbf{A}\mathbf{x} \quad \text{for symmetric } \mathbf{A}\,. \tag{43}$$

e. Inverses

With scalar t, we define

$$\frac{\partial \mathbf{A}}{\partial t} = \left\{ _{\text{m}} \frac{\partial a_{ij}}{\partial t} \right\}.$$

With $\mathbf{A}$ nonsingular, $\mathbf{A}\mathbf{A}^{-1} = \mathbf{I}$ gives

$$\frac{\partial \mathbf{A}}{\partial t}\mathbf{A}^{-1} + \mathbf{A}\frac{\partial \mathbf{A}^{-1}}{\partial t} = \mathbf{0}$$

and so

$$\frac{\partial \mathbf{A}^{-1}}{\partial t} = -\mathbf{A}^{-1}\frac{\partial \mathbf{A}}{\partial t}\mathbf{A}^{-1}\,. \tag{44}$$

f. Determinants

Suppose $\mathbf{A}$ is a square matrix having elements that are not functionally related. Then denoting the cofactor of a_{ij} in $|\mathbf{A}|$ by $|\mathbf{A}_{ij}|$, we have

$$\frac{\partial |\mathbf{A}|}{\partial a_{ij}} = |\mathbf{A}_{ij}|, \tag{45}$$

one particular case of which is

$$\frac{\partial |\mathbf{A}|}{\partial a_{ii}} = |\mathbf{A}_{ii}|\,. \tag{46}$$

Whereas (46) applies when $\mathbf{A}$ is symmetric, (45) does not, for $i \neq j$, because then elements of $\mathbf{A}$ are functionally related; e.g., for some i and j write

$$a_{ij} = a_{ji} = \theta, \quad \text{say}\,.$$

Then in place of (45) we have

$$\begin{aligned} \frac{\partial |\mathbf{A}|}{\partial \theta} &= \frac{\partial |\mathbf{A}|}{\partial a_{ij}}\frac{\partial a_{ij}}{\partial \theta} + \frac{\partial |\mathbf{A}|}{\partial a_{ji}}\frac{\partial a_{ji}}{\partial \theta} \\ &= |\mathbf{A}_{ij}| + |\mathbf{A}_{ji}| \\ &= 2|\mathbf{A}_{ij}| \quad \text{because } \mathbf{A} \text{ is symmetric}\,. \end{aligned} \tag{47}$$

Hence, in general

$$\frac{\partial |\mathbf{A}|}{\partial a_{ij}} = (2 - \delta_{ij})|\mathbf{A}_{ij}| \quad \text{for symmetric } \mathbf{A}, \tag{48}$$

where δ_{ij} is the Kronecker delta, $\delta_{ij} = 0$ for $i \neq j$ and $\delta_{ij} = 1$ for $i = j$.

Suppose that elements of **A** are functions of the scalar t. Then

$$\begin{aligned}\frac{\partial}{\partial t}\log|\mathbf{A}| &= \frac{1}{|\mathbf{A}|}\frac{\partial|\mathbf{A}|}{\partial t} = \frac{1}{|\mathbf{A}|}\sum_{i\leqslant j}\sum\frac{\partial|\mathbf{A}|}{\partial a_{ij}}\frac{\partial a_{ij}}{\partial t} \\ &= \frac{1}{|\mathbf{A}|}\sum_{i\leqslant j}\sum(2-\delta_{ij})|\mathbf{A}_{ij}|\frac{\partial a_{ij}}{\partial t} \\ &= \frac{1}{|\mathbf{A}|}\Sigma_i\Sigma_j|\mathbf{A}_{ij}|\frac{\partial a_{ij}}{\partial t} = \Sigma_i\Sigma_j\frac{|\mathbf{A}_{ij}|}{|\mathbf{A}|}\frac{\partial a_{ij}}{\partial t} \\ &= \Sigma_i\Sigma_j a^{ij}\frac{\partial a_{ij}}{\partial t} = \mathrm{tr}\left[(\mathbf{A}^{-1})'\frac{\partial\mathbf{A}}{\partial t}\right] \\ &= \mathrm{tr}\left(\mathbf{A}^{-1}\frac{\partial\mathbf{A}}{\partial t}\right) \text{ for } \mathbf{A}^{-1} = \{_{\mathrm{m}}\, a^{ij}\}\,. \end{aligned} \tag{49}$$

This result is used in deriving maximum likelihood equations for estimating variance components, in Section 6.2a.

g. Traces

When $\mathrm{tr}(\mathbf{XP})$ exists, its value is $\Sigma_i\Sigma_j x_{ij}p_{ji}$. Hence $\dfrac{\partial}{\partial x_{ij}}\mathrm{tr}(\mathbf{XP}) = p_{ji}$ and so

$$\frac{\partial}{\partial\mathbf{X}}\mathrm{tr}(\mathbf{PX}) = \frac{\partial}{\partial\mathbf{X}}\mathrm{tr}(\mathbf{XP}) \equiv \left\{\frac{\partial}{\partial x_{ij}}\mathrm{tr}(\mathbf{XP})\right\} = \{_{\mathrm{m}}\, p_{ji}\}_{j,i} = \mathbf{P}' \tag{50}$$

because $\mathrm{tr}(\mathbf{XP}) = \mathrm{tr}(\mathbf{PX})$. And

$$\frac{\partial}{\partial\mathbf{X}}\mathrm{tr}(\mathbf{PX}') = \frac{\partial}{\partial\mathbf{X}}\mathrm{tr}(\mathbf{X}'\mathbf{P}) = \left[\frac{\partial}{\partial\mathbf{X}'}\mathrm{tr}(\mathbf{X}'\mathbf{P})\right]' = \mathbf{P}\,. \tag{51}$$

Hence, using $\mathrm{tr}(\mathbf{TS}) = \mathrm{tr}(\mathbf{ST})$ and (50) and (51),

$$\frac{\partial}{\partial\mathbf{X}}\mathrm{tr}(\mathbf{XPX}') = \mathbf{XP}' + \mathbf{XP}\,. \tag{52}$$

An alternative derivation of (50) based on (37) is as follows. First, when $\mathbf{y}'$ in (37) is a scalar, $\mathbf{a}'\mathbf{x}$ say, (37) reduces to (35). Second, for the scalar λ and with $\mathbf{x}_k$ being the kth column of **X**, we define $\partial\lambda/\partial\mathbf{X}$ as

$$\frac{\partial\lambda}{\partial\mathbf{X}} = \left\{_{\mathrm{r}}\,\frac{\partial\lambda}{\partial\mathbf{x}_k}\right\}_k = \left\{_{\mathrm{m}}\,\frac{\partial\lambda}{\partial x_{ik}}\right\}_{i,k}.$$

Then, for $\boldsymbol{\pi}'_j$ being the jth row of **P**

$$\frac{\partial\,\mathrm{tr}(\mathbf{PX})}{\partial\mathbf{X}} = \left\{_{\mathrm{r}}\,\frac{\partial\Sigma_j\boldsymbol{\pi}'_j\mathbf{x}_j}{\partial\mathbf{x}_k}\right\}_k = \{_{\mathrm{r}}\,\boldsymbol{\pi}_j\} = \mathbf{P}',$$

the penultimate equality being based on (35).

M.8. THE OPERATORS vec AND vech

The matrix operation vec **X** creates a column vector from the columns of **X** by locating them one under the other:

$$\text{vec}\begin{bmatrix} 1 & 11 & 21 \\ 5 & 15 & 25 \end{bmatrix} = \begin{bmatrix} 1 \\ 5 \\ 11 \\ 15 \\ 21 \\ 25 \end{bmatrix}.$$

Thus for **X** of order $p \times q$

$$\mathbf{X} = \{_{\text{r}}\, \mathbf{x}_j\}_{j=1}^{q}, \quad \text{vec}\, \mathbf{X} = \{_{\text{c}}\, \mathbf{x}_j\}_{j=1}^{q},$$

whereupon vec **X** is $pq \times 1$.

Similarly vech **X** for symmetric **X** creates a column vector from the columns of **X**, starting at the diagonal elements.

$$\text{vech}\begin{bmatrix} 0 & 1 & 2 \\ 1 & 22 & 47 \\ 2 & 47 & 50 \end{bmatrix} = \begin{bmatrix} 0 \\ 1 \\ 2 \\ 22 \\ 47 \\ 50 \end{bmatrix}.$$

Searle (1982, Sec. 12.10) indicates some of the many results pertaining to these operators, with more details being available in Henderson and Searle (1979, 1981).

Three results involving the vec operator that get repeated use in Chapter 12 are

$$\text{vec}(\mathbf{ABC}) = (\mathbf{C}' \otimes \mathbf{A})\, \text{vec}\, \mathbf{B}, \tag{53}$$

$$\text{tr}(\mathbf{AB}) = (\text{vec}\, \mathbf{A}')'\, \text{vec}\, \mathbf{B} \quad \text{and} \quad (\mathbf{t} \otimes \mathbf{t}) = \text{vec}(\mathbf{tt}')\,. \tag{54}$$

Proof of (53) is to be found in Searle (1982, p. 333); and, after a moment's reflection, (54) is self-evident.

A final result involving the vec operator and diagonal matrices is as follows. Define

$$\mathbf{e}_t = t\text{th column of } \mathbf{I}_q, \quad \mathbf{G} = \{_{\text{r}}\, \mathbf{e}_t \otimes \mathbf{e}_t\}_{t=1}^{q} \quad \text{and} \quad \mathbf{\Delta} = \{_{\text{d}}\, \delta_t\}_{t=1}^{q}\,.$$

Then for a square matrix **A** of order q define

$$\text{diag}(\mathbf{A}) = \{_{\text{d}}\, a_{tt}\}_{t=1}^{q},$$

and we have

$$
\begin{aligned}
\mathbf{G\Delta G}' \text{ vec } \mathbf{A} &= \{_{\mathrm{r}}\, \mathbf{e}_t \otimes \mathbf{e}_t\}\{_{\mathrm{d}}\, \delta_t\}\{_{\mathrm{c}}\, \mathbf{e}'_t \otimes \mathbf{e}'_t\} \text{ vec } \mathbf{A} \\
&= \Sigma_t \delta_t(\mathbf{e}_t \otimes \mathbf{e}_t)(\mathbf{e}'_t \otimes \mathbf{e}'_t) \text{ vec } \mathbf{A} \\
&= \Sigma_t \delta_t(\mathbf{e}_t\mathbf{e}'_t \otimes \mathbf{e}_t\mathbf{e}'_t) \text{ vec } \mathbf{A}, \quad \text{by (v) of Section M.2}, \\
&= \Sigma_t \delta_t \text{ vec}(\mathbf{e}_t\mathbf{e}'_t\mathbf{A}\mathbf{e}_t\mathbf{e}'_t), \quad \text{by (53)}, \\
&= \Sigma_t \delta_t \text{ vec}(\mathbf{e}_t a_{tt} \mathbf{e}'_t) \quad \text{because } \mathbf{e}'_t\mathbf{A}\mathbf{e}_t = a_{tt}, \\
&= \text{vec}(\Sigma_t \delta_t \mathbf{e}_t a_{tt} \mathbf{e}'_t) \\
&= \text{vec}[\mathbf{\Delta} \text{ diag}(\mathbf{A})], \qquad (55)
\end{aligned}
$$

because $\mathbf{e}_t a_{tt} \mathbf{e}'_t$ is a null matrix except for a_{tt} as its tth diagonal element; and diag($\mathbf{A}$) represents a diagonal matrix of the diagonal elements of $\mathbf{A}$.

M.9. vec PERMUTATION MATRICES

A particular form of permutation matrix ($\mathbf{I}$ with its rows permuted in any fashion) is that known as the *vec permutation matrix*, or *commutation matrix*, to be denoted equivalently as $\mathbf{S}_n$ or $\mathbf{I}_{(n,n)}$. It can be described in a variety of ways, one being that it is an identity matrix of order n^2 with its rows (columns) permuted in such a way that $\mathbf{I}_{(n,n)}$ can be partitioned as an $n \times n$ matrix of submatrices of order $n \times n$, the (s, t)th of which is null except that its (t, s)th element is unity. Other descriptions and names can be found in MacRae (1974), Henderson and Searle (1979) and Magnus and Neudecker (1979). An example, for $n = 3$, is $\mathbf{T}_2$ in Section 12.3.

A number of useful results are the following:

$$
\mathbf{S}_n \equiv \mathbf{I}_{(n,n)};
$$

$$
\mathbf{S}_n = \mathbf{S}'_n, \quad \mathbf{S}_n^2 = \mathbf{I}_{n^2} \quad (\mathbf{I} + \mathbf{S}_n)^2 = 2(\mathbf{I} + \mathbf{S}_n); \qquad (56)
$$

for $\mathbf{A}_{n \times n}$

$$
\text{vec } \mathbf{A}' = \mathbf{S}_n \text{ vec } \mathbf{A};
$$

for symmetric $\mathbf{A}_{n \times n}$

$$
\text{vec } \mathbf{A} = \text{vec } \mathbf{A}' = \mathbf{S}_n \text{ vec } \mathbf{A} = [\theta \mathbf{I} + (1 - \theta)\mathbf{S}_n] \text{ vec } \mathbf{A}; \qquad (57)
$$

for $\mathbf{A}$ and $\mathbf{B}$ of the same order

$$
\left.
\begin{aligned}
\mathbf{S}_n(\mathbf{A}_{n \times k} \otimes \mathbf{B}_{n \times k})\mathbf{S}_k &= \mathbf{B}_{n \times k} \otimes \mathbf{A}_{n \times k}, \\
\mathbf{S}_n(\mathbf{A}_{n \times k} \otimes \mathbf{B}_{n \times k}) &= (\mathbf{B}_{n \times k} \otimes \mathbf{A}_{n \times k})\mathbf{S}_k \,.
\end{aligned}
\right\} \qquad (58)
$$

Details of these and other results can be found in the references at the end of the preceding paragraph.

M.10. THE EQUALITY $\mathbf{VV^-X} = \mathbf{X}$

Theorem. If $\mathbf{VV^-X} = \mathbf{X}$ then for $\mathbf{y} \sim (\mathbf{X\beta}, \mathbf{V})$

(i) $\mathbf{VV^{\sim}X} = \mathbf{X}$ for $\mathbf{V^{\sim}}$ being any generalized inverse of $\mathbf{V}$;
(ii) $\mathbf{VV^-y} = \mathbf{y}$ almost everywhere, for $E(\mathbf{y}) = \mathbf{X\beta}$;
(iii) $\mathbf{X'V^-X}$ and $\mathbf{X'V^-y}$ are invariant to $\mathbf{V^-}$.

Proof.

(i) $$\mathbf{X} = \mathbf{VV^-X} = (\mathbf{VV^{\sim}V})\mathbf{V^-X} = \mathbf{VV^{\sim}}(\mathbf{VV^-X}) = \mathbf{VV^{\sim}X}\,.$$

(ii)
$$\begin{aligned}\mathbf{0} &= (\mathbf{I} - \mathbf{VV^-})\mathbf{V}(\mathbf{I} - \mathbf{VV^-})' \\ &= (\mathbf{I} - \mathbf{VV^-})[E(\mathbf{y} - \mathbf{X\beta})(\mathbf{y} - \mathbf{X\beta})'](\mathbf{I} - \mathbf{VV^-})' \\ &= \mathrm{E}(\mathbf{zz}') \quad \text{for } \mathbf{z} = (\mathbf{I} - \mathbf{VV^-})(\mathbf{y} - \mathbf{X\beta})\,.\end{aligned}$$

But $E(\mathbf{zz}') = \mathbf{0}$ implies $\mathbf{z} = \mathbf{0}$ almost everywhere. Therefore, almost everywhere,

$$\mathbf{0} = \mathbf{z} = (\mathbf{I} - \mathbf{VV^-})(\mathbf{y} - \mathbf{X\beta}) = (\mathbf{I} - \mathbf{VV^-})\mathbf{y} \text{ when } \mathbf{VV^-X} = \mathbf{X}.$$

Hence $\mathbf{VV^-y} = \mathbf{y}$.

(iii) $\mathbf{X'V^{\sim}X} = \mathbf{X'V^{\sim}}(\mathbf{VV^-X}) = (\mathbf{X'V^{\sim}V})\mathbf{V^-X} = \mathbf{X'V^-X}$, and the same for $\mathbf{X'V^-y}$ with the $\mathbf{y}$ in place of the final $\mathbf{X}$ of $\mathbf{X'V^{\sim}X}$. Q.E.D.

As a result of (i), note that $\mathbf{VV^-X} = \mathbf{X}$ is a condition on $\mathbf{V}$, not on $\mathbf{V^-}$.

APPENDIX S

SOME RESULTS IN STATISTICS

The assumption is that a reader's background knowledge includes familiarity with matrix algebra and basic mathematical statistics. Nevertheless, just as with Appendix M, so here, a few reminders are provided.

S.1. CONDITIONAL FIRST AND SECOND MOMENTS

The joint density function of two random variables G and Y, say $f_{G,Y}(g, y)$, will be abbreviated notationally to $f(g, y)$; and the conditional density $f_{G|Y=y}(g, y)$ will be denoted $f(g \mid y)$. With $E(g)$ denoting the expected value of g, and using E_y to represent expectation over y, we then have the two well-known results

$$E(g) = E_y[E(g \mid y)]$$

and

$$\text{var}(g) = E_y[\text{var}(g \mid y)] + \text{var}_y[E(g \mid y)] \,.$$

For h being some other random variable, var(g) is the special case of

$$\text{cov}(g, h) = E_y[\text{cov}(g \mid y, h \mid y)] + \text{cov}_y[E(g \mid y), E(h \mid y)]$$

when g and h are the same.

Verification of the $E(g)$ result is straightforward:

$$\begin{aligned} E(g) &= \iint gf(g, y)\, dg\, dy = \iint gf(g \mid y) f(y)\, dg\, dy \\ &= \int\left[\int gf(g \mid y)\, dg\right] f(y) dy = \int E(g \mid y) f(y) dy \\ &= E_y E(g \mid y) \,. \end{aligned}$$

Replacing g by $[g - E(g)][h - E(h)]$ in this result gives $\text{cov}(g, h)$ as

$$\begin{aligned}
\text{cov}(g, h) &= E\{[g - E(g)][h - E(h)]\} \\
&= E_y E(\{[g - E(g)][h - E(h)]\} \mid y) \\
&= E_y E(\{[g - E(g \mid y) + E(g \mid y) - E(g)][h - E(h \mid y) + E(h \mid y) - E(h)]\} \mid y) \\
&= E_y E(\{[g - E(g \mid y)][h - E(h \mid y)] + [E(g \mid y) - E(g)][E(h \mid y) - E(h)] \\
&\quad + [g - E(g \mid y)][E(h \mid y) - E(h)] + [E(g \mid y) - E(g)][h - E(h \mid y) \mid y] .
\end{aligned}$$

The last term in this expression is

$$\begin{aligned}
&E_y E(\{[E(g \mid y) - E(g)][h - E(h \mid y)]\} \mid y) \\
&\quad = E_y([E(g \mid y) - E(g)]E\{[h - E(h \mid y)] \mid y\}),
\end{aligned}$$

because $E(g \mid y)$ and $E(g)$ are constant w.r.t. the E outside the curly braces,

$$\begin{aligned}
&\quad = E_y([E(g \mid y) - E(g)][E(h \mid y) - E(h \mid y)]) \\
&\quad = 0 .
\end{aligned}$$

In similar manner the third term of $\text{cov}(g, h)$ is zero. Therefore

$$\begin{aligned}
\text{cov}(g, h) &= E_y E(\{[g - E(g \mid y)][h - E(h \mid y)]\} \mid y) \\
&\quad + E_y E(\{[E(g \mid y) - E(g)][E(h \mid y) - E(h)]\} \mid y) \\
&= E_y E\{[g \mid y - E(g \mid y)][h \mid y - E(h \mid y)] \\
&\quad + E_y E\{[E(g \mid y) - E(g)][E(h \mid y) - E(h)]\} \\
&= E_y[\text{cov}(g \mid y, h \mid y)] + E_y\{[E(g \mid y) - E(g)][E(h \mid y) - E(h)]\} \\
&= E_y[\text{cov}(g \mid y, h \mid y)] + \text{cov}_y[E(g \mid y), E(h \mid y)] .
\end{aligned}$$

And when $h = g$ this covariance becomes the variance result for $\text{var}(g)$.

S.2. LEAST SQUARES ESTIMATION

Estimation by the method of least squares is an ancient topic. In exceedingly brief form, we develop just two aspects of the method here. Both are designed for estimating $\boldsymbol{\beta}$ in the linear model having model equation $\mathbf{y} = \mathbf{X}\boldsymbol{\beta} + \mathbf{e}$ and $E(\mathbf{y}) = \mathbf{X}\boldsymbol{\beta}$. The first is the method of ordinary least squares (OLS), which fleetingly treats $S = (\mathbf{y} - \mathbf{X}\boldsymbol{\beta})'(\mathbf{y} - \mathbf{X}\boldsymbol{\beta})$ as a function of $\boldsymbol{\beta}$ and takes as the estimator (call it $\hat{\boldsymbol{\beta}}$) the value of $\boldsymbol{\beta}$ that minimizes S. This leads to equations $\mathbf{X}'\mathbf{X}\hat{\boldsymbol{\beta}} = \mathbf{X}'\mathbf{y}$, with a solution $\hat{\boldsymbol{\beta}} = (\mathbf{X}'\mathbf{X})^{-}\mathbf{X}'\mathbf{y}$. Since $\hat{\boldsymbol{\beta}}$ is not invariant to the choice of $(\mathbf{X}'\mathbf{X})^{-}$, whereas $\mathbf{X}\hat{\boldsymbol{\beta}} = \mathbf{X}(\mathbf{X}'\mathbf{X})^{-}\mathbf{X}'\mathbf{y}$ is [see (14) of Appendix M.4], attention is confined to $\mathbf{X}\hat{\boldsymbol{\beta}}$ and linear combinations of its elements. Thus $\mathbf{X}\hat{\boldsymbol{\beta}}$

is the OLS estimator of $\mathbf{X\beta}$, which is summarized as

$$\text{OLSE}(\mathbf{X\beta}) = \mathbf{X}(\mathbf{X}'\mathbf{X})^{-}\mathbf{X}'\mathbf{y} .$$

On denoting var($\mathbf{y}$) by $\mathbf{V}$, an adaptation of OLS when $\mathbf{V}$ is known and non-singular is to use $(\mathbf{y} - \mathbf{X\beta})'\mathbf{V}^{-1}(\mathbf{y} - \mathbf{X\beta})$ as S. By exactly the same procedure as is used in deriving OLSE($\mathbf{X\beta}$), this yields what is called the generalized least squares (GLS) estimator of $\mathbf{X\beta}$:

$$\text{GLSE}(\mathbf{X\beta}) = \mathbf{X}(\mathbf{X}'\mathbf{V}^{-1}\mathbf{X})^{-}\mathbf{X}'\mathbf{V}^{-1}\mathbf{y} .$$

This is also known as BLUE($\mathbf{X\beta}$), the best, linear, unbiased estimator of $\mathbf{X\beta}$. The meaning of this is that it is a linear function of the elements of the data vector $\mathbf{y}$, it is unbiased for $\mathbf{X\beta}$, and that it is best in the sense that of all linear functions of $\mathbf{y}$ that are unbiased for $\mathbf{X\beta}$ this one has minimum variance.

The numerous details that can attend these estimators are discussed in varying degrees of generality in a multitude of books, e.g., Rao (1973) and Searle (1987), and there is a vast array of research papers on these topics. Clearly, for $\mathbf{V}$ nonsingular, GLSE($\mathbf{X\beta}$) equals OLSE($\mathbf{X\beta}$) when $\mathbf{V} = \boldsymbol{\sigma}^2\mathbf{I}$. However, when $\mathbf{V}$ is singular (symmetric and positive semi-definite) derivation of GLSE($\mathbf{X\beta}$) is more difficult: under certain conditions it consists of the preceding expression with $\mathbf{V}^{-1}$ replaced by $\mathbf{V}^{-}$; otherwise it has an entirely different form. Puntanen and Styan (1989) have an excellent review of this topic, and Searle and Pukelsheim (1989) have many of the details.

These estimators are defined in terms of estimating $\mathbf{X\beta}$ because only linear combinations of elements of $\mathbf{X\beta}$ are estimable; i.e., for $\boldsymbol{\lambda}'$ being any vector, $\boldsymbol{\lambda}'\mathbf{X\beta}$ is estimable. Then $\boldsymbol{\lambda}'\mathbf{X\beta}$ is said to be an *estimable function*, meaning that there exists a linear function of the observations that is unbiased for $\boldsymbol{\lambda}'\mathbf{X\beta}$. This implies, for $\boldsymbol{\beta}^0$ being a solution of the normal equations $\mathbf{X}'\mathbf{X}\boldsymbol{\beta}^0 = \mathbf{X}'\mathbf{y}$ and for a given $\boldsymbol{\lambda}$, that $\boldsymbol{\lambda}'\mathbf{X}\boldsymbol{\beta}^0$ has the same value for every $\boldsymbol{\beta}^0$ and is the OLSE of $\boldsymbol{\lambda}'\mathbf{X\beta}$. The same is true for $\boldsymbol{\beta}^*$ being a solution of the GLSE equations $\mathbf{X}'\mathbf{V}^{-1}\mathbf{X}\boldsymbol{\beta}^* = \mathbf{X}'\mathbf{V}^{-1}\mathbf{y}$; for given λ the expression $\boldsymbol{\lambda}'\mathbf{X}\boldsymbol{\beta}^*$ has the same value for every $\boldsymbol{\beta}^*$ and is the GLSE of $\boldsymbol{\lambda}'\mathbf{X\beta}$.

S.3. NORMAL AND χ^2-DISTRIBUTIONS

The scalar random variable x is said to be normally distributed with mean μ and variance σ^2 when it has probability density function

$$\frac{e^{-\frac{1}{2}(x-\mu)^2/\sigma^2}}{\sqrt{2\pi\sigma^2}} .$$

We often represent this by the notation $x \sim \mathcal{N}(\mu, \sigma^2)$.

The vector of n random variables $\mathbf{x}' = [x_1 \quad x_2 \quad \dots \quad x_n]$ is said to have a multivariate normal distribution with mean vector $\boldsymbol{\mu}$ and non-singular dispersion matrix $\mathbf{V}$ when it has probability density function

$$\frac{e^{-\frac{1}{2}(\mathbf{x}-\boldsymbol{\mu})'\mathbf{V}^{-1}(\mathbf{x}-\boldsymbol{\mu})}}{(2\pi)^{\frac{1}{2}N}|\mathbf{V}|^{\frac{1}{2}}} .$$

This is represented as $\mathbf{x} \sim \mathcal{N}_n(\boldsymbol{\mu}, \mathbf{V})$, often with the subscript n omitted when it is evident from the context. Searle (1971, 1987) and many other texts have numerous details about these distributions. Certain properties useful to the purposes of this book are as follows.

For $\mathbf{x} \sim \mathcal{N}(\boldsymbol{\mu}, \mathbf{V})$

(i) $E(\mathbf{x}) = \boldsymbol{\mu}$ and $\text{var}(\mathbf{x}) = \mathbf{V}$;

(ii) $\mathbf{Kx} \sim \mathcal{N}(\mathbf{K}\boldsymbol{\mu}, \mathbf{KVK}')$.

On writing

$$\mathbf{x} = \begin{bmatrix} \mathbf{x}_1 \\ \mathbf{x}_2 \end{bmatrix} \sim \mathcal{N}\left(\begin{bmatrix} \boldsymbol{\mu}_1 \\ \boldsymbol{\mu}_2 \end{bmatrix}, \begin{bmatrix} \mathbf{V}_{11} & \mathbf{V}_{12} \\ \mathbf{V}_{21} & \mathbf{V}_{22} \end{bmatrix}\right),$$

(iii) the marginal distribution of $\mathbf{x}_1$ is

$$\mathbf{x}_1 \sim \mathcal{N}(\boldsymbol{\mu}_1, \mathbf{V}_{11});$$

(iv) the conditional distribution of $\mathbf{x}_1$ given $\mathbf{x}_2$ is

$$\mathbf{x}_1 \,|\, \mathbf{x}_2 \sim \mathcal{N}[\boldsymbol{\mu}_1 + \mathbf{V}_{12}\mathbf{V}_{22}^{-1}(\mathbf{x}_2 - \boldsymbol{\mu}_2), \mathbf{W}_{11}^{-1}]$$
$$\text{for } \mathbf{W}_{11} = (\mathbf{V}_{11} - \mathbf{V}_{12}\mathbf{V}_{22}^{-1}\mathbf{V}_{21})^{-1} .$$

Properties of quadratic forms $\mathbf{x}'\mathbf{Ax}$ when $\mathbf{x} \sim \mathcal{N}(\boldsymbol{\mu}, \mathbf{V})$ are given in Appendix S.5.

a. Central χ^2

The simplest variable having a χ^2-distribution is the sum of squares of n independently normally distributed variables having zero mean and unit variance: when

$$\mathbf{x} \sim \mathcal{N}(\mathbf{0}, \mathbf{I}_n), \quad u = \mathbf{x}'\mathbf{x} = \sum_{i=1}^{n} x_i^2 \sim \chi_n^2, \quad \text{with } E(u) = n \text{ and } \text{var}(u) = 2n .$$

This is the central χ^2-distribution; n is known as the degrees of freedom of the distribution

A well-known result of special interest is that

$$\mathbf{x} \sim \mathcal{N}(\mu\mathbf{1}, \sigma^2\mathbf{I}_n) \quad \text{implies} \quad \sum_{i=1}^{n} (x_i - \bar{x})^2/\sigma^2 \sim \chi_{n-1}^2 .$$

b. Mean squares

Suppose SS is a sum of squares on f degrees of freedom, and MS is the corresponding mean square. Then MS = SS/f. Therefore for expected values

$E(\text{SS}) = f E(\text{MS})$. There are many situations where

$$\frac{\text{SS}}{E(\text{MS})} \sim \chi_f^2, \quad \text{whereupon} \quad \text{var}\left(\frac{\text{SS}}{E(\text{MS})}\right) = 2f\,.$$

Hence

$$\text{var}(\text{SS}) = 2f[E(\text{MS})]^2, \quad \text{and so} \quad \text{var}(\text{MS}) = \frac{\text{var}(\text{SS})}{f^2} = \frac{2[E(\text{MS})]^2}{f}\,.$$

Furthermore, by the definition of variance,

$$\text{var}(\text{MS}) = E[\text{MS} - E(\text{MS})]^2 = E[(\text{MS})^2] - [E(\text{MS})]^2\,.$$

Equating these two expressions for var(MS) gives

$$E[(\text{MS})^2] = [E(\text{MS})]^2\left(1 + \frac{2}{f}\right),$$

so that

$$\frac{E[(\text{MS})^2]}{f+2} = \frac{[E(\text{MS})]^2}{f}\,.$$

Therefore

$$\frac{(\text{MS})^2}{f+2} \quad \text{is an unbiased estimator of} \quad \frac{[E(\text{MS})]^2}{f}\,.$$

This is used in deriving an unbiased estimator of a variance of an estimated variance (e.g., Sections 4.5f and 5.2e).

c. Non-central χ^2

More general is the non-central χ^2 distribution, definable through the sum of squares of independently distributed normal variables having a non-zero mean:

$$\mathbf{x} \sim \mathcal{N}(\boldsymbol{\mu}, \mathbf{I}_n) \quad \text{defines} \quad u = \sum_{i=1}^{n} x_i^2 \sim \chi^{2\prime}(n, \lambda),$$

with, for $\lambda = \frac{1}{2}\boldsymbol{\mu}'\boldsymbol{\mu}$,

$$E(u) = n + 2\lambda \quad \text{and} \quad \text{var}(u) = 2n + 8\lambda\,.$$

n is the degrees of freedom and λ is called the non-centrality parameter. Having $\lambda = 0$ causes $\chi^{2\prime}(n, \lambda)$ to simplify to χ_n^2.

S.4. F-DISTRIBUTIONS

Ratios of two independent χ^2-variables, each divided by its degrees of freedom, have F-distributions. They come in three forms, of which we consider but two.

First is the central F-distribution for

$$u \sim \chi^2_n, \quad \text{and, independently,} \quad v \sim \chi^2_m$$

$$F = \frac{u}{n} \Big/ \frac{v}{m} \sim \mathscr{F}^n_m,$$

where n and m are called the numerator and denominator degrees of freedom of the F-distribution. Similarly, for

$$u' \sim \chi^{2\prime}(n, \lambda), \quad \text{and, independently,} \quad v \sim \chi^2_m,$$

$$F' = \frac{u'}{n} \Big/ \frac{v}{m} \sim \mathscr{F}'(n, m, \lambda)$$

characterizes the non-central F-distribution.

Means and variances for the central $\mathscr{F}$-distribution are

$$E(F) = \frac{m}{m-2} \quad \text{and} \quad \text{var}(F) = \frac{2m^2(n+m-2)}{n(m-2)^2(m-4)}$$

and those for the non-central F are

$$E(F') = \frac{m}{m-2}\left(1 + \frac{2\lambda}{n}\right)$$

and

$$\text{var}(F') = \frac{2m^2}{n^2(m-2)}\left[\frac{(n+2\lambda)^2}{(m-2)(m-4)} + \frac{n+4\lambda}{m-4}\right].$$

Having $\lambda = 0$ reduces the non-central $\mathscr{F}$ to the central $\mathscr{F}$. This is the basis of using an F-statistic to test a hypothesis. If F' that has an $\mathscr{F}'$-distribution is such that when some hypothesis is true the λ of that distribution is zero, then F' has an $\mathscr{F}$-distribution under that hypothesis, and comparing the computed F' with tabulated values of the central $\mathscr{F}$-distribution provides a test of the hypothesis.

S.5. QUADRATIC FORMS

A quadratic form is $\mathbf{y}'\mathbf{A}\mathbf{y}$, where $\mathbf{A}$ can always be taken as symmetric. It is useful in statistics because every sum of squares of data represented as $\mathbf{y}$ can be written as $\mathbf{y}'\mathbf{A}\mathbf{y}$ for some $\mathbf{A}$; and because there are theorems about quadratic forms that provide useful statistical properties of sums of squares. We quote four such theorems, confining ourselves to $\text{var}(\mathbf{y}) = \mathbf{V}$ being nonsingular. (Singular $\mathbf{V}$ can be handled, but it is considerably more complicated.)

Theorem S1.

For $\mathbf{y} \sim (\boldsymbol{\mu}, \mathbf{V})$, meaning that $E(\mathbf{y}) = \boldsymbol{\mu}$ and $\text{var}(\mathbf{y}) = \mathbf{V}$,

$$E(\mathbf{y}'\mathbf{A}\mathbf{y}) = \text{tr}(\mathbf{A}\mathbf{V}) + \boldsymbol{\mu}'\mathbf{A}\boldsymbol{\mu}\ .$$

Theorem S2.

If $\mathbf{y} \sim \mathcal{N}(\boldsymbol{\mu}, \mathbf{V})$ then $\mathbf{y}'\mathbf{A}\mathbf{y} \sim \chi^{2'}(r_{\mathbf{A}}, \frac{1}{2}\boldsymbol{\mu}'\mathbf{A}\boldsymbol{\mu})$ if and only if $\mathbf{AV}$ is idempotent .

Theorem S3.

If $\mathbf{y} \sim \mathcal{N}(\boldsymbol{\mu}, \mathbf{V})$ then $\mathbf{y}'\mathbf{A}\mathbf{y}$ and $\mathbf{y}'\mathbf{B}\mathbf{y}$ are independent if and only if $\mathbf{AVB} = \mathbf{0}$.

Theorem S4.

$$\text{If } \mathbf{y} \sim \mathcal{N}(\boldsymbol{\mu}, \mathbf{V}) \text{ then } \operatorname{var}(\mathbf{y}'\mathbf{A}\mathbf{y}) = 2\operatorname{tr}[(\mathbf{AV})^2] + 4\boldsymbol{\mu}'\mathbf{AVA}\boldsymbol{\mu} .$$

Theorem S4 is a special case of the more general result that the kth cumulant of $\mathbf{y}'\mathbf{A}\mathbf{y}$ is $2^{k-1}(k-1)![\operatorname{tr}(\mathbf{AV})^k + k\boldsymbol{\mu}'\mathbf{A}(\mathbf{VA})^{k-1}\boldsymbol{\mu}]$. Through consideration of $\operatorname{var}[\mathbf{y}'(\mathbf{A}+\mathbf{B})\mathbf{y}]$, this theorem readily yields the covariance result

$$\operatorname{cov}(\mathbf{y}'\mathbf{A}\mathbf{y}, \mathbf{y}'\mathbf{B}\mathbf{y}) = 2\operatorname{tr}(\mathbf{AVBV}) .$$

Details and proofs of these widely known theorems can be found in Searle (1971, Chap. 2). The sufficient condition in each of Theorems S2 and S3 is easily proven, whereas the necessity conditions are not so easy to prove. Driscoll and Gundberg (1986) have an interesting history of these necessity conditions, and the first straightforward proof of that for Theorem S3 is given by Reid and Driscoll (1989).

Calculating the trace terms of Theorems S1 and S4 is often simplified by the results derived in Appendix M.6:

$$\operatorname{tr}(\mathbf{MM}') = \operatorname{sesq}(\mathbf{M}), \quad \text{which is } \operatorname{tr}(\mathbf{M}^2) = \operatorname{sesq}(\mathbf{M}) \text{ for } \mathbf{M} = \mathbf{M}' .$$

S.6. BAYES ESTIMATION

A brief introduction to the ideas of Bayes estimation is given here, including an elementary example. We begin with some definitions of density functions.

a. Density functions

The *cumulative density function* of a random variable X is

$$F_X(x) = \Pr(X \leqslant x),$$

and the *joint cumulative density* function of two random variables X and Y is

$$F_{X,Y}(x, y) = \Pr(X \leqslant x, Y \leqslant y) .$$

From this comes the *joint density function* of X and Y:

$$f_{X,Y}(x, y) = \frac{\partial^2}{\partial x\, \partial y} F_{X,Y}(x, y) .$$

The *marginal density function* of X is the joint density function of X and Y after integrating out y (or summing over y for discrete densities):

$$f_X(x) = \int_{R_y} f_{X,Y}(x, y)\, dy, \tag{1}$$

where R_y represents the range of y-values that Y can take. The *conditional density function* of X, given y, is

$$f_{X|Y}(x \mid y) = \frac{f_{X,Y}(x, y)}{f_Y(y)} . \tag{2}$$

For notational simplicity the subscripts X and Y can be dropped from the preceding representations; their presence emphasizes that, for example, $f_X(x)$ is not necessarily the same function of x as $f_Y(y)$ is of y. However, if in dropping the X and Y subscripts we adopt a convention that the f of $f(\cdot)$ always represents a density function then we accept the fact that $f(x)$ and $f(y)$ are not necessarily the same functions of their respective arguments and we have a less cumbersome notation:

$$f(x) = \int_{R_y} f(x, y)\, dy \quad \text{and} \quad f(x \mid y) = \frac{f(x, y)}{f(y)} . \tag{3}$$

This is a particularly simpler notation when, for example, the random variable X is to be an estimated variance component such as $\hat{\sigma}^2_\alpha$.

b. Bayes Theorem

Just as $f(x \mid y)$ is as defined above, so is

$$f(y \mid x) = \frac{f(x, y)}{f(x)} . \tag{4}$$

Therefore

$$f(x, y) = f(x \mid y) f(y) = f(y \mid x) f(x) . \tag{5}$$

Hence

$$f(x \mid y) = \frac{f(x, y)}{f(y)} = \frac{f(x, y)}{\int_{R_x} f(x, y)\, dx} = \frac{f(y \mid x) f(x)}{\int_{R_x} f(y \mid x) f(x)\, dx} . \tag{6}$$

This is Bayes Theorem. It is used in estimation in the context of a density function $f(y)$ being a function of a parameter θ, so that the density function can be represented as $f(y \mid \theta)$. Bayes estimation is based on assuming that we can specify a range of values within which θ lies, and over the range we have some feeling for the probabilities of θ taking those possible values. Thus we treat θ as a random variable, with a density to be denoted $\pi(\theta)$, which is called the *prior density* of θ. Then Bayes estimation is based on y representing data and on using Bayes Theorem to derive what is the conditional density $\pi(\theta \mid y)$, which in this context is called the *posterior density* of θ: from (6) it is

$$\pi(\theta \mid y) = \frac{f(y, \theta)}{f(y)} = \frac{f(y, \theta)}{\int f(y, \theta)\, d\theta}$$

$$= \frac{f(y \mid \theta)\pi(\theta)}{\int f(y \mid \theta)\pi(\theta)\, d\theta} . \tag{7}$$

This can be thought of as $\pi(\theta)$ updated by the data through the use of $f(y \mid \theta)$.

c. Bayes estimation

Once the posterior density $\pi(\theta \mid y)$ has been derived, it can be used for estimating θ in any way one wishes; e.g., the mean $E(\theta \mid y) = \int_{R_\theta} \theta \pi(\theta \mid y)\, d\theta$ will be a function only of y and can be used as an estimator of θ; so can the median or mode of $\pi(\theta \mid y)$.

d. Example

Our example involves the beta distribution, for which the density function is

$$f(x) = \frac{x^{a-1}(1-x)^{b-1}}{B(a, b)} \quad \text{for } 0 \leqslant x \leqslant 1 . \tag{8}$$

where

$$B(a, b) = \int_0^1 x^{a-1}(1-x)^{b-1}\, dx = \frac{\Gamma(a)\Gamma(b)}{\Gamma(a+b)} . \tag{9}$$

$\Gamma(a)$ is the gamma function of a, namely

$$\Gamma(a) = \int_0^\infty x^{a-1}e^{-x}\, dx,$$

with $\Gamma(a) = (a-1)(a-2)\dots 2(1) = (a-1)!$ when a is a positive integer greater than unity. The mean and variance of this density are

$$E(x) = \frac{a}{a+b} \quad \text{and} \quad \text{var}(x) = \frac{ab}{(a+b)^2(a+b+1)} . \tag{10}$$

Our example of Bayes estimation does not explicitly concern variance components, but it is related to the variance components model of Section 10.3. However, the distribution functions involved provide easy illustration of Bayes methodology. The example is that of estimating p from n independent Bernoulli trials yielding realized values of the random variables $X_1, X_2, \dots, X_n$ where $\Pr(X_i = 1) = p$ and $\Pr(X_i = 0) = 1 - p$ for $i = 1, \dots, n$. Define the random variable

$$Y = \sum_{i=1}^{n} X_i . \tag{11}$$

Then

$$f(y \mid p) = \Pr(Y = y) = \binom{n}{y} p^y (1 - p)^{n-y} . \tag{12}$$

Maximum likelihood estimation (see Section S.7) would use (12) as the likelihood, relabeling it $L(p \mid y)$, and would take as the estimator of p the value of p that maximizes

$$\log L(p \mid y) = \log \binom{n}{y} + y \log p + (n - y) \log (1 - p) .$$

Equating

$$\frac{\partial \log L(p \mid y)}{\partial p} = \frac{y}{p} - \frac{n - y}{1 - p},$$

to zero and denoting the solution for p by $\tilde{p}$ gives the estimator as

$$\tilde{p} = y/n = \bar{x} .$$

To illustrate Bayes estimation, we use for $\pi(p)$ the beta density (4) with $a = 2$ and $b = 2$. Then (9) is $B(a, b) = 1/3! = \frac{1}{6}$, and so from (8)

$$\pi(p) = 6p(1 - p) . \tag{13}$$

Then from (7)

$$\pi(p \mid y) = \frac{f(y \mid p)\pi(p)}{\int_0^1 f(y \mid p)\pi(p)\, dp},$$

and on using (12) and (13) this is

$$\pi(p \mid y) = \frac{\binom{n}{y} p^y (1 - p)^{n-y} 6p(1 - p)}{\int_0^1 \binom{n}{y} p^y (1 - p)^{n-y} 6p(1 - p)\, dp} = \frac{p^{y+1}(1 - p)^{n-y+1}}{\int_0^1 p^{y+1}(1 - p)^{n-y+1}\, dp} .$$

Applying (9) to the denominator gives

$$\pi(p \mid y) = \frac{p^{y+1}(1 - p)^{n-y+1}}{B(y + 2, n - y + 2)} = \frac{\Gamma(n + 4)}{\Gamma(y + 2)\Gamma(n - y + 2)} p^{y+1}(1 - p)^{n-y+1} . \tag{14}$$

This is the posterior density of p. It is, by comparison with (8), a beta density with $a = y + 2$ and $b = n - y + 2$. Hence from (10) its mean is

$$E(p \mid y) = \frac{y + 2}{y + 2 + n - y + 2} = \frac{y + 2}{n + 4} . \tag{15}$$

This is now available as a possible estimator of p. It is a Bayes estimator:

$$\hat{p} = \frac{y+2}{n+4}. \tag{16}$$

Note in passing that $\lim_{n\to\infty} \hat{p} \to y/n = \bar{x} = \tilde{p}$; i.e., as n becomes large, the Bayes estimator tends to the ML (maximum likelihood) estimator. Note too that

$$\begin{aligned}\hat{p} &= \frac{1}{1+4/n}\,\bar{x} + \frac{4}{n+4}\left(\tfrac{1}{2}\right)\\ &= \frac{1}{1+4/n}\,\tilde{p} + \frac{4/n}{1+4/n}\,E(p),\end{aligned} \tag{17}$$

where, from (13) [or from (10), for $a = 2 = b$] the mean of the prior density $\pi(p)$ of p in (13) is $E(p) = \frac{1}{2}$. We see that (17) shows $\hat{p}$, the Bayes estimator, as being a weighted mean of $\tilde{p}$, the ML estimator, and of $E(p)$, the mean of the prior density $\pi(p)$ in (13).

The preceding results are special cases of the more general result when in place of (13) we take $\pi(p)$ as the general beta density with parameters a and b, similar to (8):

$$\pi(p) = \frac{p^{a-1}(1-p)^{b-1}}{B(a,b)}. \tag{18}$$

Then $\pi(p \mid y)$ becomes the beta density with parameters $y + a$ and $n - y + b$:

$$\pi(p \mid y) = \frac{p^{y+a-1}(1-p)^{n-y+b-1}}{B(y+a, n-y+b)}. \tag{19}$$

Thus on taking $\hat{p} = E(p \mid y)$ as the Bayes estimator, it is, from (10)

$$\hat{p} = \frac{y+a}{n+a+b} = \frac{1}{1+(a+b)/n}\,\tilde{p} + \frac{(a+b)/n}{1+(a+b)/n}\,E(p) \tag{20}$$

where, from (10) and (18), $E(p) = a/(a+b)$.

e. Empirical Bayes estimation

Suppose in (18) and (19) that a and b are unknown. Then, because

$$f(y,p) = \pi(p \mid y)f(y) = f(y \mid p)\pi(p), \tag{21}$$

$$f(y) = \frac{f(y \mid p)\pi(p)}{\pi(p \mid y)} = \frac{\binom{n}{y}B(a+y, n-y+b)}{B(a,b)}, \tag{22}$$

after substitution from (12), (18) and (19). If (22) is used to provide estimates $\dot{a}$ and $\dot{b}$ of a and b, which are then used in (20) in place of a and b, the resulting expression $\dot{p} = (y + \dot{a})/(n + \dot{a} + \dot{b})$ is an empirical Bayes estimator of p.

An analogy of the preceding example with estimating variance components is as follows. In the example we have

$$X_i \mid p_i \sim \text{Binomial}(1, p_i) \quad \text{and} \quad p_i \sim \text{Beta}(\alpha, \beta) .$$

In the 1-way classification, variance components model we have $y_{ij} \mid (\mu + \alpha_i) \sim \mathcal{N}(\mu + \alpha_i, \sigma_e^2)$ and $\mu + \alpha_i \sim \mathcal{N}(\mu, \sigma_\alpha^2)$.

S.7. MAXIMUM LIKELIHOOD

a. The likelihood function

Suppose a vector of random variables, $\mathbf{x}$, has density function $f(\mathbf{x})$. Let $\boldsymbol{\theta}$ be the vector of parameters involved in $f(\mathbf{x})$. Then $f(\mathbf{x})$ is a function of both $\mathbf{x}$ and $\boldsymbol{\theta}$. As a result, it can be thought of in at least two different contexts. The first is as above, as a density function, in which case $\boldsymbol{\theta}$ is usually assumed to be known. With this in mind we use the symbol $f(\mathbf{x} \mid \boldsymbol{\theta})$ in place of $f(\mathbf{x})$ to explicitly emphasize that $\boldsymbol{\theta}$ is being taken as known.

A second context is where $\mathbf{x}$ represents a known vector of data and where $\boldsymbol{\theta}$ is unknown. Then $f(\mathbf{x})$ will be a function of just $\boldsymbol{\theta}$. It is called the *likelihood function* for the data $\mathbf{x}$; and because in this context $\boldsymbol{\theta}$ is unknown and $\mathbf{x}$ is known, we use the symbol $L(\boldsymbol{\theta} \mid \mathbf{x})$. Thus although $f(\mathbf{x} \mid \boldsymbol{\theta})$ and $L(\boldsymbol{\theta} \mid \mathbf{x})$ represent the same thing mathematically, i.e.,

$$f(\mathbf{x} \mid \boldsymbol{\theta}) \equiv L(\boldsymbol{\theta} \mid \mathbf{x}),$$

it is convenient to use each in its appropriate context.

b. Maximum likelihood estimation

The likelihood function $L(\boldsymbol{\theta} \mid \mathbf{x})$ is the foundation of the widely used method of estimation known as maximum likelihood estimation. It yields estimators that have many good properties. ML is used as abbreviation for maximum likelihood and MLE for maximum likelihood estimate—with whatever suffix is appropriate to the context: estimate, estimator (and their plurals) or estimation.

The essence of the ML method is to view $L(\boldsymbol{\theta} \mid \mathbf{x})$ as a function of the mathematical variable $\boldsymbol{\theta}$ and to derive $\tilde{\boldsymbol{\theta}}$ as that value of $\boldsymbol{\theta}$ which maximizes $L(\boldsymbol{\theta} \mid \mathbf{x})$. The only proviso is that this maximization must be carried out within the range of permissible values for $\boldsymbol{\theta}$. For example, if one element of $\boldsymbol{\theta}$ is a variance then permissible values for that variance are non-negative values. This aspect of ML estimation is very important in estimating variance components.

Under widely existing regularity conditions on $f(\mathbf{x} \mid \boldsymbol{\theta})$, a general method of establishing equations that yield MLEs is to differentiate L with respect to $\boldsymbol{\theta}$ and equate the derivative to $\mathbf{0}$. But maximizing L is equivalent to maximizing the natural logarithm of L, which we denote by l, and it is often easier to use l rather than L. Thus for

$$l = \log L(\boldsymbol{\theta} \mid \mathbf{x}) \text{ the equations } \left.\frac{\partial l}{\partial \boldsymbol{\theta}}\right|_{\boldsymbol{\theta} = \dot{\boldsymbol{\theta}}} = \mathbf{0}$$

are known as the ML equations, with $\dot{\boldsymbol{\theta}}$ their solution being called an ML solution. When there is only one value of $\dot{\boldsymbol{\theta}}$ satisfying equations (23) then, provided it is within the permissible range of $\boldsymbol{\theta}$, the ML estimator of $\boldsymbol{\theta}$, to be denoted $\tilde{\boldsymbol{\theta}}$, is $\dot{\boldsymbol{\theta}}$; i.e., $\tilde{\boldsymbol{\theta}} = \dot{\boldsymbol{\theta}}$. When $\dot{\boldsymbol{\theta}}$ is not within the permissible range of $\boldsymbol{\theta}$ then adjustments have to be made to $\boldsymbol{\theta}$ and the nature of these adjustments depends upon the context and form of $f(\mathbf{x} \mid \boldsymbol{\theta})$.

c. Asymptotic dispersion matrix

A useful property of the ML estimator $\tilde{\boldsymbol{\theta}}$ is that its large-sample, or asymptotic (as $N \to \infty$), dispersion matrix is known. For $\mathbf{I}(\boldsymbol{\theta})$, known as the information matrix, and defined as

$$\mathbf{I}(\boldsymbol{\theta}) = E\left(\frac{\partial l}{\partial \boldsymbol{\theta}} \frac{\partial l}{\partial \boldsymbol{\theta}'}\right) = E\left\{{}_{\mathrm{m}} \frac{\partial l}{\partial \theta_i} \frac{\partial l}{\partial \theta_j}\right\}_{i,j},$$

the asymptotic dispersion matrix is

$$\operatorname{var}(\tilde{\boldsymbol{\theta}}) \simeq [\mathbf{I}(\boldsymbol{\theta})]^{-1} .$$

Note that this is always available without even needing the ML estimator $\tilde{\boldsymbol{\theta}}$ itself, or its density function. An alternative form of the information matrix that is valid in many situations is

$$\mathbf{I}(\boldsymbol{\theta}) = -E\left(\frac{\partial^2 l}{\partial \boldsymbol{\theta}\, \partial \boldsymbol{\theta}'}\right) = -E \quad \left\{{}_{\mathrm{m}} \frac{\partial^2 l}{\partial \theta_i\, \partial \theta_j}\right\}_{i,j} .$$

Proof of this is as follows. The (i,j)th element of $\mathbf{I}(\boldsymbol{\theta})$ is

$$\begin{aligned}
E\left(\frac{\partial^2 l}{\partial \theta_i\, \partial \theta_j}\right) &= E \frac{\partial}{\partial \theta_i}\left(\frac{\partial \log L}{\partial \theta_j}\right) \\
&= E \frac{\partial}{\partial \theta_i}\left(\frac{1}{L} \frac{\partial L}{\partial \theta_j}\right) \\
&= E\left(\frac{-1}{L^2} \frac{\partial L}{\partial \theta_i} \frac{\partial L}{\partial \theta_j} + \frac{1}{L} \frac{\partial^2 L}{\partial \theta_i\, \partial \theta_j}\right),
\end{aligned}$$

by the product rule of differentiation,

$$= -\int \frac{\partial L}{\partial \theta_i} \frac{\partial L}{\partial \theta_j} \frac{1}{L^2} f(x, \theta)\, dx + \int \frac{\partial^2 L}{\partial \theta_i\, \partial \theta_j} \frac{1}{L} f(x, \theta)\, dx,$$

by the definition of expectation

$$= -\int \frac{1}{L} \frac{\partial L}{\partial \theta_i} \frac{1}{L} \frac{\partial L}{\partial \theta_j} f(x, \theta)\, dx + \frac{\partial^2}{\partial \theta_i\, \partial \theta_j} \int f(x, \theta)\, dx,$$

by interchanging, in the second term, the derivative and integral operations, which is permissible under regularity conditions.

On recognizing that $\int f(x,\theta)\,dx = 1$, this gives

$$E\frac{\partial^2 l}{\partial\theta_i\,\partial\theta_j} = -\int \frac{\partial \log L}{\partial\theta_i}\frac{\partial \log L}{\partial\theta_j} f(x,\theta)\,dx + \frac{\partial^2(1)}{\partial\theta_i\,\partial\theta_j} = -E\frac{\partial l}{\partial\theta_i}\frac{\partial l}{\partial\theta_j}\,.$$

Thus the two forms of $\mathbf{I}(\boldsymbol{\theta})$ are equivalent.

d. Transforming parameters

Suppose parameters represented by $\boldsymbol{\theta}$ are transformed in a one-to-one manner to the vector $\boldsymbol{\Delta}$. Then the matrix

$$\mathbf{J}_{\boldsymbol{\theta}\to\boldsymbol{\Delta}} = \left\{_{\mathrm{m}}\ \frac{\partial\theta_i}{\partial\Delta_j}\right\}_{i,j}$$

is the Jacobian matrix of the transformation $\boldsymbol{\theta}\to\boldsymbol{\Delta}$.

Theorem. After the one-to-one transformation $\boldsymbol{\theta}\to\boldsymbol{\Delta}$,

$$\mathbf{I}(\boldsymbol{\Delta}) = (\mathbf{J}_{\boldsymbol{\theta}\to\boldsymbol{\Delta}})'\mathbf{I}(\boldsymbol{\theta})\mathbf{J}_{\boldsymbol{\theta}\to\boldsymbol{\Delta}}\,.$$

Proof. For notational convenience denote $\mathbf{J}_{\boldsymbol{\theta}\to\boldsymbol{\Delta}}$ by $\mathbf{H}$. From the preceding section

$$\mathbf{I}(\boldsymbol{\Delta}) = E\left(\frac{\partial l}{\partial\boldsymbol{\Delta}}\frac{\partial l}{\partial\boldsymbol{\Delta}'}\right).$$

But

$$\frac{\partial l}{\partial\boldsymbol{\Delta}} = \left\{_{\mathrm{c}}\ \frac{\partial l}{\partial\Delta_j}\right\} = \left\{_{\mathrm{c}}\ \sum_i \frac{\partial l}{\partial\theta_i}\frac{\partial\theta_i}{\partial\Delta_j}\right\} = \left\{_{\mathrm{c}}\ \sum_i \frac{\partial\theta_i}{\partial\Delta_j}\frac{\partial l}{\partial\theta_i}\right\} = \mathbf{H}\frac{\partial l}{\partial\boldsymbol{\theta}}\,.$$

Therefore

$$\mathbf{I}(\boldsymbol{\Delta}) = E\left[\mathbf{H}\frac{\partial l}{\partial\boldsymbol{\theta}}\left(\mathbf{H}\frac{\partial l}{\partial\boldsymbol{\theta}}\right)'\right] = \mathbf{H}E\left(\frac{\partial l}{\partial\boldsymbol{\theta}}\frac{\partial l}{\partial\boldsymbol{\theta}'}\right)\mathbf{H}' = \mathbf{H}\mathbf{I}(\boldsymbol{\theta})\mathbf{H}'\,.$$

Q.E.D.

REFERENCES

Ahrens, H. (1965). Standardfehler geschatzter Varianzkomponenten eines unbalanzie Versuchplanes in r-stufiger hierarchischer Klassifikation. *Monatsc. Deut. Akad. Wiss. Berlin* **7**, 89–94.

Airy, G.B. (1861). *On the Algebraical and Numerical Theory of Errors of Observations and the Combinations of Observations.* MacMillan, London.

Albert, A. (1976). When is a sum of squares an analysis of variance? *Ann. Stat.* **4**, 775–778.

Albert, J.H. (1988). Computation methods using a Bayesian hierarchical generalized linear model. *J. Amer. Stat. Assoc.* **83**, 1037–1044.

Anderson, R.D. (1978). Studies on the estimation of variance components. Ph.D. Thesis, Cornell University, Ithaca, New York.

Anderson, R.D. (1979a). On the history of variance component estimation. In *Variance Components and Animal Breeding* (L.D. VanVleck and S.R. Searle, eds.), 19–42. Animal Science Department, Cornell University, Ithaca, New York.

Anderson, R.D. (1979b). Estimating variance components from balanced data: Optimum properties of REML solutions and MIVQUE estimators. In *Variance Components and Animal Breeding* (L.D. VanVleck and S.R. Searle, eds.), 205–216. Animal Science Department, Cornell University, Ithaca, New York.

Anderson, R.D., Henderson, H.V., Pukelsheim, F. and Searle, S.R. (1984). Best estimation of variance components from balanced data, with arbitrary kurtosis. *Math. Operationsforsch. Stat., Ser. Stat.* **15**, 163–176.

Anderson, R.L. (1975). Designs and estimators for variance components. In *Statistical Design and Linear Models* (J.N. Srivastava, ed.), 1–30. North-Holland, Amsterdam.

Anderson, R.L. and Bancroft, T.A. (1952). *Statistical Theory in Research.* McGraw-Hill, New York.

Anderson, R.L. and Crump, P.P. (1967). Comparisons of designs and estimation procedures for estimating parameters in a two-stage nested process. *Technometrics* **9**, 499–516.

Anderson, T.W. (1973). Asymptotically efficient estimation of covariance matrices with linear structure. *Ann. Stat.* **1**, 135–141.

Anderson, T.W. (1984). *An Introduction to Multivariate Analysis*, 2nd edn. John Wiley & Sons, New York.

Andrade, D.F. and Helms, R.W. (1984). Maximum likelihood estimates in the multivariate normal with patterned mean and covariance via the EM algorithm. *Commun. Stat. A: Theory & Methods* **13**, 2239–2251.

Angers, J.-F. (1987). Development of robust Bayes estimators for a multivariate normal mean. Ph.D. Thesis, Department of Statistics, Purdue University, West Lafayette, Indiana.

Arnold, S.F. (1981). *The Theory of Linear Models and Multivariate Statistics.* John Wiley & Sons, New York.

Atiqullah, M. (1962). The estimation of residual variance in quadratically balanced least-squares problems and the robustness of the F-test. *Biometrika* **49**, 83–91.

Babb, J.S. (1986). Pooling maximum likelihood estimates of variance components obtained from subsets of unbalanced data. M.S. Thesis, Biometrics Unit, Cornell University, Ithaca, New York.

Baksalary, J.K. and Molinska, A. (1984). Non-negative unbiased estimability of linear combinations of two variance components. *J. Stat. Planning & Inference* **10**, 1–8.

Balestra, P. (1973). Best quadratic unbiased estimators of the variance–covariance matrix in normal regression. *J. Econometrics* **2**, 67–78.

Bard, Y. (1974). *Nonlinear Parameter Estimation.* Academic Press, New York.

Bates, D.M. and Watts, D.G. (1981). A relative offset orthogonality convergence criterion for nonlinear least squares. *Technometrics* **23**, 179–183.

Bennett, C.A. and Franklin, N.L. (1954). *Statistical Analysis in Chemistry and the Chemical Industry.* John Wiley & Sons, New York.

Berger, J.O. (1985). *Statistical Decision Theory and Bayesian Analysis*, 2nd edn. Springer-Verlag, New York.

Bessel, I. (1820). Beschreibung des auf der Königsberger Sternwart aufgestellten Reichenbachschen Meridiankrieses, dessen Anwendung und Geranigkeit imgleichen der Repoldschen Uhr. *Astronomisches Jahrbuch für das Jahr 1823.* Berlin.

Blischke, W.R. (1966). Variances of estimates of variance components in a three-way classification. *Biometrics* **22**, 553–565.

Blischke, W.R. (1968). Variances of moment estimators of variance components in the unbalanced r-way classification. *Biometrics* **24**, 527–540.

BMDP (1988). BMDP Statistical Software, Inc., 1440 Sepulveda Blvd, Los Angeles, California.

Boardman, T.J. (1974). Confidence intervals for variance components—a comparative Monte Carlo study. *Biometrics* **30**, 251–262.

Broemeling, L.D. (1969). Confidence regions for variance ratios of random models. *J. Amer. Stat. Assoc.* **64**, 660–664.

Brown, K.G. (1976). Asymptotic behavior of MINQUE-type estimators of variance components. *Ann. Stat.* **4**, 746–754.

Brown, K.G. (1978). Estimation of variance components using residuals. *J. Amer. Stat. Assoc.* **73**, 141–146.

Buck, R.C. (1978). *Advanced Calculus.* McGraw-Hill, New York.

Bulmer, M.G. (1980). *The Mathematical Theory of Quantitative Genetics*, Oxford University Press.

Burdick, R.K., Birch, N.J. and Graybill, F.A. (1986). Confidence intervals on measures of variability in an unbalanced two-fold nested design, and with equal subsampling. *J. Stat. Comp. & Simul.* **25**, 259–272.

Burdick, R.K. and Eickman, J.E. (1986). Confidence intervals on the among-group variance component in the unbalanced one-fold nested design. *J. Stat. Comp. & Simul.* **26**, 205–219.

Burdick, R.K. and Graybill, F.A. (1984). Confidence intervals on linear combinations of variance components in the unbalanced one-way classification. *Technometrics* **26**, 131–136.

Burdick, R.K. and Graybill, F.A. (1985). Confidence intervals on the total variance in an unbalanced two-fold nested classification with equal subsampling. *Commun. Stat. A: Theory & Methods* **14**, 761–774.

Burdick, R.K. and Graybill, F.A. (1988). The present status of confidence interval estimation on variance components in balanced and unbalanced random models. *Commun. Stat.: Theory & Methods* (Special Issue on Analysis of the Unbalanced Mixed Model) **17**, 1165–1195.

Burdick, R.K., Maqsood, F. and Graybill, F.A. (1986). Confidence intervals on the intraclass correlation in the unbalanced 1-way classification. *Commun. Stat. A: Theory & Methods* **15**, 3353–3378.

Bush, N. and Anderson, R.L. (1963). A comparison of three different procedures for estimating variance components. *Technometrics* **5**, 421–440.

Callanan, T.P. and Harville, D.A. (1989). Some new algorithms for computing maximum likelihood estimates of variance components. In *Computer Science and Statistics: Proceedings of the 21st Symposium on the Interface* (K. Berk and L. Malone, eds.), 435–444. Amer. Stat. Assoc., 1429 Duke Street, Alexandria, Virginia.

Casella, G. and Berger, R.L. (1990). *Statistical Inference.* Wadsworth and Brooks/Cole, Pacific Grove, California.

Chakravorti, S.R. and Grizzle, J.E. (1975). Analysis of data from multiclinic experiments. *Biometrics* **31**, 325–338.

Chatterjee, S.K. and Das, K. (1983). Estimation of variance components in an unbalanced 1-way classification. *J. Stat. Planning & Inference* **8**, 27–41.

Chaubey, Y.B. (1984). On the comparison of some non-negative estimators of variance components for two models. *Commun. Stat. B: Simul. & Comp.* **13**, 619–633.

Chauvenet, W. (1863). *A Manual of Spherical and Practical Astronomy, 2: Theory and Use of Astronomical Instruments.* Lippincott, Philadelphia.

Cochran, W.G. (1939). The use of analysis of variance in enumeration by sampling. *J. Amer. Stat. Assoc.* **34**, 492–510.

Cochran, W.G. (1943). Analysis of variance for percentages based on unequal numbers. *J. Amer. Stat. Assoc.* **38**, 287–301.

Cochran, W.G. (1951). Improvement by means of selection. *Proceedings of the 2nd Berkeley Symposium* (J. Neyman, ed.), 449–470. University of California Press, Los Angeles, California.

Conaway, M.R. (1989). Analysis of repeated categorical measurements with conditional likelihood methods. *J. Amer. Stat. Assoc.* **84**, 53–62.

Corbeil, R.R. and Searle, S.R. (1976a). Restricted maximum likelihood (REML) estimation of variance components in the mixed model. *Technometrics* **18**, 31–38.

Corbeil, R.R. and Searle, S.R. (1976b). A comparison of variance component estimators. *Biometrics* **32**, 779–791.

Cornfield, J. and Tukey, J.W. (1956). Average values of mean squares in factorials. *Ann. Math. Stat.* **27**, 907–949.

Crowder, M.J. (1978). Beta-binomial ANOVA for proportions. *Appl. Stat.* **27**, 34–37.

Crump, S.L. (1946). The estimation of variance components in analysis of variance. *Biometrics Bull.* **2**, 7–11.

Crump, S.L. (1947). The estimation of variance in multiple classification. Ph.D. Thesis, Iowa State University, Ames, Iowa.

Crump, S.L. (1951). The present status of variance components analysis. *Biometrics* **7**, 1–16.

Daniels, H.E. (1939). The estimation of components of variance. *J. R. Stat. Soc. Suppl.* **6**, 186–197.

Das, K. (1987). Estimation of variance components in an unbalanced 2-way mixed model under heteroscedasticity. *J. Stat. Planning & Inference* **21**, 285–291.

Das, R. and Sinha, B.K. (1987). Robust optimum invariant unbiased tests for variance components. In *Proceedings, Second International Tampere Conference on Statistics* (T. Pukkila and S. Puntanen, eds.), 317–342. University of Tampere, Tampere, Finland.

Dempfle, L. (1977). Relation entre BLUP (best linear unbiased prediction) et estimateurs Bayesiens. *Annales de Génétique et de Sélection Animale* **9**, 27–32.

Dempster, A.P., Laird, N.M. and Rubin, D.B. (1977). Maximum likelihood from incomplete data via the EM algorithm. *J. R. Stat. Soc., Ser. B* **39**, 1–38.

Dempster, A.P., Rubin, D.B. and Tsutakawa, R.K. (1981). Estimation in covariance component models. *J. Amer. Stat. Assoc.* **76**, 341–353.

Dempster, A.P., Selwyn, M.R., Patel, C.M. and Roth, A.J. (1984). Statistics and computational aspects of mixed model analysis. *Appl. Stat.* **33**, 203–214.

Donner, A. (1986). A review of inference procedures for the intra-class correlation coefficient in the one-way random effects model. *Int. Stat. Rev.* **54**, 67–82.

Driscoll, M.F. and Gundberg, W.R. (1986). The history of the development of Craig's theorem. *The Amer. Stat.* **40**, 65–71.

Drygas, H. (1980). Hsu's theorem in variance component models. In *Mathematical Statistics, Banach Center Publication* **6**, 95–107. PWN – Polish Scientific Publishers, Warsaw.

DuCrocq, V., Quaas, R.L., Pollak, E.J. and Casella, G. (1988a). Length of productive life of dairy cows, 1. Justification of a Weibull model. *J. Dairy Sci.* **71**, 3061–3070.

DuCrocq, V., Quaas, R.L., Pollak, E.J. and Casella, G. (1988b). Length of productive life of dairy cows. 2. Variance component estimation and sire evaluation. *J. Dairy Sci.* **71**, 3071–3079.

Durbin, J. (1953). A note on regression when there is extraneous information about one of the coefficients. *J. Amer. Stat. Assoc.* **48**, 799–808.

Efron, B. (1982). *The Jackknife, Bootstrap, and Other Resampling Plans.* Society for Industrial and Applied Mathematics, Philadelphia, Pennsylvania.

Eisenhart, C. (1947). The assumptions underlying the analysis of variance. *Biometrics* **3**, 1–21.

Fisher, R.A. (1918). The correlation between relatives on the supposition of Medelian inheritance. *Trans. R. Soc. Edinburgh* **52**, 399–433.

Fisher, R.A. (1922). On the mathematical foundations of theoretical statistics. *Phil. Trans. R. Soc. Lond., Ser. A* **222**, 309–368.

Fisher, R.A. (1925). *Statistical Methods for Research Workers*, 1st edn. Oliver & Boyd, Edinburgh and London.

Fisher, R.A. (1935). Discussion of Neyman *et al.*, 1935. *J. R. Stat. Soc., Ser. B* **2**, 154–155.

Fleiss, J.L. (1971). On the distribution of a linear combination of independent chi squares. *J. Amer. Stat. Assoc.* **66**, 142–144.

Ganguli, M. (1941). A note on nested sampling. *Sankhyā* **5**, 449–452.

Gauss, K.F. (1809). *Theoria Motus Corporum Celestrium in Sectionibus Conics Solem Ambientium.* Perthes and Besser, Hamburg.

Gaylor, D.W. and Hartwell, T.D. (1969). Expected mean squares for nested classifications. *Biometrics* **25**, 427–430.

Gaylor, D.W. and Hopper, F.N. (1969). Estimating the degrees of freedom for linear combinations of mean squares by Satterthwaite's formula. *Technometrics* **11**, 691–706.

Gelfand, A.E., Hills, S.E., Racine-Poon, A. and Smith, A.F.M. (1990). Illustration of Bayesian inference in normal data models using Gibbs sampling. *J. Amer. Stat. Assoc.* **85**, 972–985.

Gelfand, A.E. and Smith, A.F.M. (1990). Sampling-based approaches to calculating marginal densities. *J. Amer. Stat. Assoc.* **85**, 398–409.

Gianola, D. and Fernando, R.L. (1986). Bayesian methods in animal breeding theory. *J. Animal Sci.* **63**, 217–244.

Gianola, D. and Goffinet, B. (1982). Sire evaluation with best linear predictors. *Biometrics* **38**, 1085–1088.

Gianola, D. and Hammond, K. (1990). *Advances in Statistical Methods for Genetic Improvement of Livestock.* Springer-Verlag, New York.

Giesbrecht, F.G. (1983). An efficient procedure for computing MINQUE of variance components and generalized least squares estimates of fixed effects. *Commun. Stat. A: Theory & Methods* **12**, 2169–2177.

Giesbrecht, F.G. (1985). MIXMOD, a SAS procedure for analysing mixed models. Technical report 1659, Institute of Statistics, North Carolina State University, Raleigh, North Carolina.

Giesbrecht, F.G. and Burns, J.C. (1985). Two-stage analysis based on a mixed model: large-sample asymptotic theory and small-sample simulation results. *Biometrics* **41**, 477–486.

Giesbrecht, F.G. and Burrows, P.M. (1978). Estimating variance components in hierarchical structures using MINQUE and restricted maximum likelihood. *Commun. Stat. A: Theory & Methods* **7**, 891–904.

Gill, P.E., Murray, W. and Wright, M.H. (1981). *Practical Optimization*. Academic Press, New York.

Gilmour, A.R., Anderson, R.D. and Rae, A.L. (1985). The analysis of binomial data by a generalized linear mixed model. *Biometrika* **72**, 593–599.

Gnot, S. and Kleffe, J. (1983). Quadratic estimation in mixed linear models with two variance components. *J. Stat. Planning & Inference* **8**, 267–279.

Gnot, S., Kleffe, J. and Zmyślony, R. (1985). Nonnegativity of admissible invariant quadratic estimates in mixed linear models with two variance components. *J. Stat. Planning & Inference* **12**, 249–258.

Goodnight, J.H. (1978). Computing MIVQUEO estimates of variance components. Technical Report R-105, SAS Institute Inc., Carey, North Carolina.

Gosslee, D.G. and Lucas, H.L. (1965). Analysis of variance of disproportionate data when interaction is present. *Biometrics* **21**, 115–133.

Graser, H.-U., Smith, S.P. and Tier, B. (1987). A derivative-free approach for estimating variance components in animal models by restricted maximum likelihood. *J. Animal Sci.* **64**, 1362–1370.

Graybill, F.A. (1954). On quadratic estimates of variance components. *Ann. Math. Stat.* **25**, 367–372.

Graybill, F.A. (1961). *An Introduction to Linear Statistical Models*. McGraw-Hill, New York.

Graybill, F.A. (1976). *Theory and Application of the Linear Model*. Duxbury, North Scituate, Massachusetts.

Graybill, F.A. and Hultquist, R.A. (1961). Theorems concerning Eisenhart's Model II. *Ann. Math. Stat.* **32**, 261–269.

Graybill, F.A. and Wang, C.-M. (1980). Confidence intervals on non-negative linear combinations of variances. *J. Amer. Stat. Assoc.* **75**, 869–873.

Graybill, F.A. and Wortham, A.W. (1956). A note on uniformly best unbiased estimators for variance components. *J. Amer. Stat. Assoc.* **51**, 266–268.

Green, J.W. (1988). Diagnostic methods for repeated measures experiments with missing cells. Technical Report, Department of Mathematical Sciences, University of Delaware, Newark, Delaware.

Guttman, I. (1982). *Linear Models, an Introduction*. John Wiley & Sons, New York.

Hammersley, J.M. (1949). The unbiased estimate and standard error of the interclass variance. *Metron* **15**, 189–205.

Hartley, H.O. (1967). Expectations, variances and covariances of ANOVA mean squares by 'synthesis'. *Biometrics* **23**, 105–114. Corrigenda **23**, 853.

Hartley, H.O. and Rao, J.N.K. (1967). Maximum likelihood estimation for the mixed analysis of variance model. *Biometrika* **54**, 93–108.

Hartley, H.O., Rao, J.N.K. and LaMotte, L.R. (1978). A simple synthesis-based method of estimating variance components. *Biometrics* **34**, 233–243.

Hartley, H.O. and Searle, S.R. (1969). A discontinuity in mixed model analysis. *Biometrics* **25**, 573–576.

Hartung, J. (1981). Non-negative minimum biased invariant estimation in variance components models. *Ann. Stat.* **9**, 278–292.

Hartung, J. and Voet, B. (1986). Best invariant unbiased estimators for the mean squared error of variance components estimators. *J. Amer. Stat. Assoc.* **81**, 689–691.

Harville, D.A. (1967). Statistical dependence between random effects and the numbers of observations on the effects for the balanced one-way classification. *J. Amer. Stat. Assoc.* **62**, 1375–1386.

Harville, D.A. (1968). Statistical dependence between subclass means and the numbers of observations in the subclass for the two-way completely random classification. *J. Amer. Stat. Assoc.* **63**, 1484–1494.

Harville, D.A. (1969a). Quadratic unbiased estimation of variance components for the one-way classification. *Biometrika* **56**, 313–326. Corrigenda **57**, 226.

Harville, D.A. (1969b). Variances of variance component estimators for the unbalanced two-way cross classification with application to balanced incomplete block designs. *Ann. Math. Stat.* **40**, 408–416.

Harville, D.A. (1976). Extension of the Gauss–Markov theorem to include the estimation of random effects. *Ann. Stat.* **2**, 384–395.

Harville, D.A. (1977). Maximum-likelihood approaches to variance component estimation and to related problems. *J. Amer. Stat. Assoc.* **72**, 320–340.

Harville, D.A. (1985). Decomposition of prediction error. *J. Amer. Stat. Assoc.* **80**, 132–138.

Harville, D.A. (1986). Using ordinary least squares software to compute combined intrablock estimates of treatment contrasts. *The Amer. Stat.* **40**, 153–157.

Harville, D.A. and Mee, R.W. (1984). A mixed model procedure for analyzing ordered categorical data. *Biometrics* **40**, 393–408.

Hemmerle, W.J. and Hartley, H.O. (1973). Computing maximum likelihood estimates for the mixed A.O.V. model using the W-transformation. *Technometrics* **15**, 819–831.

Henderson, C.R. (1948). Estimation of general, specific and maternal combining abilities in crosses among inbred lines of swine. Ph.D. Thesis, Iowa State University, Ames, Iowa.

Henderson, C.R. (1950). Estimation of genetic parameters (Abstract). *Ann. Math. Stat.* **21**, 309–310.

Henderson, C.R. (1953). Estimation of variance and covariance components. *Biometrics* **9**, 226–252.

Henderson, C.R. (1963). Selection index and expected genetic advance. In *Statistical Genetics and Plant Breeding* (W.D. Hanson and H.F. Robinson, eds.), 141–163. National Academy of Sciences and National Research Council Publication No. 982, Washington, D.C.

Henderson, C.R. (1969). Design and analysis of animal husbandry experiments. In *Techniques and Procedures in Animal Science Research*, 2nd edn, Chapter 1. American Society of Animal Science Monograph, Quality Corporation, Albany, New York.

Henderson, C.R. (1973a). Maximum likelihood estimation of variance components. Unpublished manuscript, Department of Animal Science, Cornell University, Ithaca, New York.

Henderson, C.R. (1973b). Sire evaluation and genetic trends. In *Proceedings Animal Breeding and Genetics Symposium in Honor of Dr Jay L. Lush*, 10–41. American Society of Animal Science, and American Dairy Science Association, Champaign, Illinois.

Henderson, C.R. (1975). Best linear unbiased estimation and prediction under a selection model. *Biometrics* **31**, 423–447.

Henderson, C.R. (1976). A simple method for computing the inverse of a numerator matrix used in prediction of breeding values. *Biometrics* **32**, 69–84.

Henderson, C.R. (1977). Best linear unbiased prediction of breeding values not in the model for records. *J. Dairy Sci.* **60**, 783–787.

Henderson, C.R. (1984). *Applications of Linear Models in Animal Breeding.* University of Guelph, Guelph, Ontario.

Henderson, C.R. (1985). MIVQUE and REML estimation of additive and non-additive genetic variances. *J. Animal Sci.* **61**, 113–121.

Henderson, C.R., Kempthorne, O., Searle, S.R. and von Krosigk, C.N. (1959). Estimation of environmental and genetic trends from records subject to culling. *Biometrics* **15**, 192–218.

Henderson, C.R., Searle, S.R. and Schaeffer, L.R. (1974). The invariance and calculation of method 2 for estimating variance components. *Biometrics* **30**, 583–588.

Henderson, H.V., Pukelsheim, F. and Searle, S.R. (1983). On the history of the Kronecker product. *Linear & Multilinear Algebra* **14**, 113–120.

Henderson, H.V. and Searle, S.R. (1979). Vec and vech operators for matrices, with some uses for Jacobians and multivariate statistics. *Can. J. Stat.* **7**, 65–81.

Henderson, H.V. and Searle, S.R. (1981). The vec-permutation matrix, the vec operator and Kronecker products: a review. *Linear & Multilinear Algebra* **9**, 271–288.

Herbach, L.H. (1959). Properties of Model II type analysis of variance tests, A: optimum nature of the *F*-test for Model II in the balanced case. *Ann. Math. Stat.* **30**, 939–959.

Hill, B.M. (1965). Inference about variance components in the one-way model. *J. Amer. Stat. Assoc.* **60**, 806–825.

Hill, B.M. (1967). Correlated errors in the random model. *J. Amer. Stat. Assoc.* **62**, 1387–1400.

Hocking, R.R. (1973). A discussion of the two-way mixed model. *The Amer. Stat.* **27**, 148–152.

Hocking, R.R. (1985). *The Analysis of Linear Models.* Brooks/Cole, Monterey, California.

Hocking, R.R., Green, J.W. and Bremer, R.H. (1989). Variance component estimation with model-based diagnostics. *Technometrics* **31**, 227–240.

Hocking, R.R. and Kutner, M.H. (1975). Some analytical and numerical comparisons of estimators for the mixed A.O.V. model. *Biometrics* **31**, 19–28.

Hoerl, R.W. (1985). Ridge analysis 25 years later. *The Amer. Stat.* **39**, 186–192.

Hoerl, A.E. and Kennard, R.W. (1970). Ridge regression: biased estimation for non-orthogonal problems. *Technometrics* **12**, 55–67.

Hultquist, R.A. and Graybill, F.A. (1965). Minimal sufficient statistics for the two-way classification mixed model design. *J. Amer. Stat. Assoc.* **60**, 182–192.

Imhof, J.P. (1960). A mixed model for the 3-way layout with two random effect factors. *Ann. Math. Stat.* **31**, 906–928.

Jackson, R.W.B. (1939), Reliability of mental tests. *Brit. J. Psychol.* **29**, 267–287.

Jennrich, R.J. and Sampson, P.F. (1976). Newton–Raphson and related algorithms for maximum likelihood variance component estimation. *Technometrics* **18**, 11–17.

Jennrich, R.J. and Schluchter, M.D. (1986). Unbalanced repeated measures models with structured covariance matrices. *Biometrics* **42**, 805–820.

Kackar, R.N. and Harville, D.A. (1981). Unbiasedness of two-stage estimation and prediction procedures for mixed linear models. *Commun. Stat. A: Theory & Methods* **10**, 1249–1261.

Kackar, R.N. and Harville, D.A. (1984). Approximations for standard errors of estimators of fixed and random effects in mixed linear models. *J. Amer. Stat. Assoc.* **79**, 853–861.

Kass, R.E. and Steffey, D. (1986). Approximate Bayesian inference in conditionally independent hierarchical models (parametric empirical Bayes models). Technical Report No. 386, Department of Statistics, Carnegie–Mellon University, Pittsburgh, Pennsylvania.

Kass, R.E. and Steffey, D. (1989). Approximate Bayesian inference in conditionally independent hierarchical models (parametric empirical Bayes models). *J. Amer. Stat. Assoc.* **84**, 717–726.

Kempthorne, O. (1968). Discussion of Searle (1968). *Biometrics* **24**, 782–784.

Kempthorne, O. (1975). Fixed and mixed models in the analysis of variance. *Biometrics* **31**, 473–486.

Kempthorne, O. (1977). Status of quantitative genetic theory. In *Proceedings International Conference Quantitative Genetics* (E. Pollak *et al.*, eds.), 719–760. Iowa State University Press, Ames, Iowa.

Kempthorne, O. (1980). The term design matrix. *The Amer. Stat.* **34**, 249.

Kennedy, W.J. and Gentle, J.E. (1980). *Statistical Computing.* Marcel Dekker, New York.

Khatri, G.C. (1966). A note on a MANOVA model applied to problems in growth curves. *Ann. Inst. Stat. Math.* **18**, 75–78.

Khattree, R. and Gill, D.S. (1988). Comparison of some estimates of variance components using Pitman's nearness criterion. *Proceedings, American Statistical Association, Statistical Computing Section*, 133–136.

Khattree, R. and Naik, D.N. (1990). Optimum tests for random effects in unbalanced nested designs. *Statistics (Math. Operationsforsch. Stat)* **21**, 163–168.

Khuri, A.I. (1981). Simultaneous confidence intervals for functions of variance components in random models. *J. Amer. Stat. Assoc.* **76**, 878–885.

Khuri, A.I. (1984). Interval estimation of fixed effects and of functions of variance components in balanced mixed models. *Sankhyā B*, **46**, 10–28.

Khuri, A.I. and Littell, R.C. (1987). Exacts tests for the main effects variance components in an unbalanced random two-way model. *Biometrics* **43**, 545–560.

Khuri, A.I. and Sahai, H. (1985). Variance components analysis: a selective literature survey. *Int. Stat. Rev.* **53**, 279–300.

Kirk, R.E. (1968). *Experimental Design Procedures for the Behavioral Sciences.* Brooks/Cole, Belmont, California.

Kleffe, J. (1977). Invariant methods for estimating variance components in mixed linear models. *Math. Operationsforsch. Stat., Ser. Stat.* **8**, 233–250.

Kleffe, J. and Pincus, R. (1974). Bayes and best quadratic unbiased estimators for parameters of the covariance matrix in a normal linear model. *Math. Operationsforsch. Stat.* **5**, 43–67.

Kleffe, J. and Rao, J.N.K. (1986). The existence of asymptotically unbiased non-negative quadratic estimates of variance components in ANOVA models. *J. Amer. Stat. Assoc.* **81**, 692–698.

Koch, G.G. (1967a). A general approach to the estimation of variance components. *Technometrics* **9**, 93–118.

Koch, G.G. (1967b). A procedure to estimate the population mean in random effects models. *Technometrics* **9**, 577–586.

Koch, G.G. (1968). Some further remarks concerning "A general approach to the estimation of variance components". *Technometrics* **10**, 551–558.

Kussmaul, K. and Anderson, R.L. (1967). Estimation of variance components in two-stage nested designs with composite samples. *Technometrics* **9**, 373–389.

Laird, N.M. (1978). Empirical Bayes methods for two-way contingency tables. *Biometrika* **65**, 581–590.

Laird, N.M. (1982). Computation of variance components using the EM algorithm. *J. Stat. Comp. & Simul.* **14**, 295–303.

Laird, N.M., Lange, N. and Stram, D. (1987). Maximum likelihood computations with repeated measures: application of the EM algorithm. *J. Amer. Stat. Assoc.* **82**, 97–105.

Laird, N.M. and Louis, T.A. (1987). Empirical Bayes confidence intervals based on bootstrap samples. *J. Amer. Stat. Assoc.* **82**, 739–757.

Laird, N.M. and Ware, J.H. (1982). Random-effects models for longitudinal data. *Biometrics* **38**, 963–974.

LaMotte, L.R. (1970). A class of estimators of variance components. Technical Report 10, Department of Statistics, University of Kentucky, Lexington, Kentucky.

LaMotte, L.R. (1971). Locally best quadratic estimators of variance components. Technical Report 22, Department of Statistics, University of Kentucky, Lexington, Kentucky.

LaMotte, L.R. (1972). Notes on the covariance matrix of a random, nested ANOVA model. *Ann. Math. Stat.* **43**, 659–662.

LaMotte, L.R. (1973a). On non-negative quadratic unbiased estimation of variance components. *J. Amer. Stat. Assoc.* **68**, 728–730.

LaMotte, L.R. (1973b). Quadratic estimation of variance components. *Biometrics* **29**, 311–330.

LaMotte, L.R. (1976). Invariant quadratic estimators in the random, one-way ANOVA model. *Biometrics* **32**, 793–804.

LaMotte, L.R., McWhorter, A., Jr and Prasad, R.A. (1988). Confidence intervals and tests on the variance ratio in random models with two variance components. *Commun. Stat. A; Theory & Methods* **17**, 1135–1164.

Landis, J.R. and Koch, G.G. (1977). A one-way component of variance model for categorical data. *Biometrics* **33**, 671–679.

Legendre, A.M. (1806). *Nouvelles Méthodes pour la Determination des Orbites des Cométes; avec un Supplément Contenant Divers Perfectionnements de ces Méthodes et leur Application aux deux Cométes de 1805.* Courcier, Paris.

Lehman, E.L. (1961). Some Model I problems of selection. *Ann. Math. Stat.* **32**, 990–1012.

Leonard, T. (1975). Bayesian estimation methods for two-way contingency tables. *J. R. Stat. Soc., Ser. B* **37**, 23–37.

Leone, F.C., Nelson, L.S., Johnson, N.L. and Eisenstat, S. (1968). Sampling distributions of variance components. II. Empirical studies of unbalanced nested designs. *Technometrics* **10**, 719–738.

Levenberg, K. (1944). A method for the solution of certain non-linear problems in least squares. *Q. Appl. Math.* **2**, 164–168.

Li, S.H. and Klotz, J.H. (1978). Components of variance estimation for the split-plot design. *J. Amer. Stat. Assoc.* **73**, 147–152.

Lindley, D.V. and Smith, A.F.M. (1972). Bayes estimates for the linear model (with discussion). *J. R. Stat. Soc. Ser. B* **34**, 1–41.

Lindstrom, M.J. and Bates, D.M. (1988). Newton–Raphson and EM algorithms for linear mixed-effects models for repeated measures data. *J. Amer. Stat. Assoc.* **83**, 1014–1022.

Little, R.J.A. and Rubin, D.B. (1987). *Statistical Analyses with Missing Data*, John Wiley & Sons, New York.

Loh, W.Y. (1986). Improved estimators for ratios of variance components. *J. Amer. Stat. Assoc.* **81**, 699–702.

Louis, T.A. (1982). Finding the observed information matrix when using the EM algorithm. *J. R. Stat. Soc., Ser. B* **44**, 226–233.

Low, L.Y. (1964). Sampling variances of estimates of components of variance from a non-orthogonal two-way classification. *Biometrika* **51**, 491–494.

Lum, M.D. (1954). Rules for determining error terms in hierarchical and partially hierarchical models. Technical Report, Wright Air Development Center, Dayton, Ohio.

MacRae, E.C. (1974). Matrix derivatives with an application to an adaptive linear decision problem. *Ann. Stat.* **2**, 337–346.

Magnus, J.R. and Neudecker, H. (1979). The commutation matrix: some properties and applications. *Ann. Stat.* **7**, 381–394.

Mahamunulu, D.M. (1963). Sampling variances of the estimates of variance components in the unbalanced three-way nested classification. *Ann. Math. Stat.* **34**, 521–527.

Malley, J.D. (1986). *Optimal Unbiased Estimation of Variance Components.* Lecture Notes in Statistics, Vol. 39, Springer-Verlag, Berlin.

Marquardt, D.W. (1963). An algorithm for least squares estimation of nonlinear parameters. *SIAM J.* **11**, 431–441.

Marsaglia, J. and Styan, G.P.H. (1974a). Equalities and inequalities for ranks of matrices. *Linear & Multilinear Algebra* **2**, 269–292.

Marsaglia, J. and Styan, G.P.H. (1974b). Rank conditions for generalized inverses of partitioned matrices. *Sankhyā* **36**, 437–442.

Mathew, T. (1984). On non-negative quadratic unbiased estimability of variance components. *Ann. Stat.* **12**, 1566–1569.

Mathew, T., Sinha, B.K. and Sutradhar, B.C. (1991a). Nonnegative estimation of variance components in balanced mixed models with two variance components. *J. Amer. Stat. Assoc.* (in press).

Mathew, T., Sinha, B.K. and Sutradhar, B.C. (1991b). Nonnegative estimation of variance components in unbalanced mixed models with two variance components. *J. Multivariate Analysis* (in press).

McCullagh, P. and Nelder, J.A. (1983). *Generalized Linear Models*. Chapman and Hall, London.

McCulloch, C.E. (1990). Maximum likelihood variance components estimation for binary data. Biometrics Unit Technical Report BU-1037-M, Cornell University, Ithaca, New York.

Miller, J.J. (1973). Asymptotic properties and computation of maximum likelihood estimates in the mixed model of the analysis of variance. Technical Report No. 12, Department of Statistics, Stanford University, Stanford, California.

Miller, J.J. (1977). Asymptotic properties of maximum likelihood estimates in the mixed model of the analysis of variance. *Ann. Stat.* **5**, 746–762.

Millman, J. and Glass, G.V. (1967). Rules of thumb for writing the ANOVA table. *J. Educational Measurement* **4**, 41–57.

Mood, A.M. (1950). *Introduction to the Theory of Statistics*. McGraw-Hill, New York.

Mood, A.M. and Graybill, F.A. (1963). *Introduction to the Theory of Statistics*, 2nd edn. McGraw-Hill, New York.

Mood, A.M., Graybill, F.A. and Boes, D.C. (1974). *Introduction to the Theory of Statistics*, 3rd edn. McGraw-Hill, New York.

Moore, E.H. (1920). On the reciprocal of the general algebraic matrix. *Bull. Amer. Math. Soc.* **26**, 394–395.

Moré, J.J. (1977). The Levenberg–Marquardt algorithm: implementation and theory. In *Numerical Analysis* (G.A. Watson, ed.), 105–116. Springer-Verlag, Berlin and New York.

Morris, C. (1983). Parametric empirical Bayes inference: theory and applications (with discussion). *J. Amer. Stat. Assoc.* **78**, 47–65.

Muse, H.D., Anderson, R.L. and Thitakamol, B. (1982). Additional comparisons of designs to estimate variance components in a two-way classification model. *Commun. Stat. A: Theory & Methods* **11**, 1403–1425.

Neter, J. and Wasserman, W. (1974). *Applied Linear Statistical Models*. Richard D. Irwin, Homewood, Illinois.

Nelder, J.A. (1965a). The analysis of randomized experiments with orthogonal block structure, I. Block structure and the null analysis of variance. *Proc. R. Soc. Lond., Ser. A* **283**, 147–162.

Nelder, J.A. (1965b). The analysis of randomized experiments with orthogonal block structure, II. Treatment structure and the general analysis of variance. *Proc. R. Soc. Lond., Ser. A* **283**, 163–178.

Nerlove, M. (1971). A note on error component models. *Econometrics* **39**, 383–396.

Neyman, J., Iwaszkiewicz, K. and Kolodziejczyk, S.T. (1935). Statistical problems in agricultural experimentation. *J. R. Stat. Soc. Suppl.* **2**, 107–154.

Ochi, Y. and Prentice, R.L. (1984). Likelihood inference in a correlated probit regression model. *Biometrika* **71**, 531–543.

Patterson, H.D. and Thompson, R. (1971). Recovery of inter-block information when block sizes are unequal. *Biometrika* **58**, 545–554.

Patterson, H.D. and Thompson, R. (1974) Maximum likelihood estimation of components of variance. *Proc. Eighth Internat. Biom. Conf.*, 197–209.

Peixoto, J.L. and Harville, D.A. (1986). Comparisons of alternative predictors under the balanced one-way random model. *J. Amer. Stat. Assoc.* **81**, 431–436.

Penrose, R.A. (1955). A generalized inverse for matrices. *Proc. Camb. Phil. Soc.* **51**, 406–413.

Piegorsch, W.W. and Casella, G. (1989). The early use of matrix diagonal increments in statistical problems. *SIAM Rev.* **31**, 428–434.

Piegorsch, W.W. and Casella, G. (1990). Empirical Bayes estimation for generalized linear models. Technical Report BU-1067-M, Biometrics Unit, Cornell University, Ithaca, New York.

Pierce, B.J. (1852). Criterion for the rejection of doubtful observations. *Astron. J.* **2**, 161–163.

Pierce, D.A. and Sands, B.R. (1975). Extra-Bernoulli variation in binary data. Technical Report 46, Department of Statistics, Oregon State University.

Plackett, R.L. (1972). Studies in the history of probability and statistics. XXIX: The discovery of the method of least squares. *Biometrika* **59**, 239–251.

Portnoy, S. (1982). Maximizing the probability of correctly ordering random variables using linear predictors. *J. Multivariate Analysis* **12**, 256–269.

Preitschopf, F. (1987). Personal communication.

Pukelsheim, F. (1974). Schätzen von Mittelwert und Streuungsmatrix in Gauss–Markov Modellen. Diplomarbeit, Freiburg im Breisgau, Germany.

Pukelsheim, F. (1976). Estimating variance components in linear models. *J. Multivariate Analysis* **6**, 626–629.

Pukelsheim, F. (1977). On Hsu's model in regression analysis. *Math. Operationsforsch. Stat., Ser. Stat.* **8**, 323–331.

Pukelsheim, F. (1978). Examples for unbiased non-negative estimation in variance component models. Technical Report 113, Department of Statistics, Stanford University, Stanford, California.

Pukelsheim, F. (1979). Classes of linear models. *Proceedings of Variance Components and Animal Breeding: A Conference in Honor of C.R. Henderson*, (L.D. Van Vleck and S.R. Searle, eds.), 69–83. Animal Science Department, Cornell University, Ithaca, New York.

Pukelsheim, F. (1981a). On the existence of unbiased nonnegative estimates of variance covariance components. *Ann. Stat.* **9**, 293–299.

Pukelsheim, F. (1981b). Linear models and convex geometry: aspects of nonnegative variance estimation. *Math. Operationsforsch. Stat. Ser. Stat.* **12**, 271–286.

Pukelsheim, F. and Styan, G.P.H. (1979). Nonnegative definiteness of the estimated dispersion matrix in a multivariate linear model. *Bull. Acad. Polon. Sci., Sér. Sci. Math.* **XXVII**, 327–330.

Puntanen, S. and Styan, G.P.H. (1989). On the equality of the ordinary least squares estimator and the best linear unbiased estimator. *The Amer. Stat.* **43**, 153–164.

Rao, C.R. (1965). *Linear Statistical Inference and its Applications*, 1st edn. John Wiley & Sons, New York.

Rao, C.R. (1970). Estimation of heteroscedastic variances in linear models. *J. Amer. Stat. Assoc.* **65**, 161–172.

Rao, C.R. (1971a). Estimation of variance and covariance components—MINQUE theory. *J. Multivariate Analysis* **1**, 257–275.

Rao, C.R. (1971b). Minimum variance quadratic unbiased estimation of variance components. *J. Multivariate Analysis* **1**, 445–456.

Rao, C.R. (1972). Estimation of variance and covariance components in linear models. *J. Amer. Stat. Assoc.* **67**, 112–115.

Rao, C.R. (1973). *Linear Statistical Inference and its Applications*, 2nd edn. John Wiley & Sons, New York.

Rao, C.R. (1979). MINQUE theory and its relation to ML and MML estimation of variance components. *Sankhyā B* **41**, 138–153.

Rao, C.R. and Kleffe, J. (1988). *Estimation of Variance Components and Applications*. North-Holland, Amsterdam.

Reid, J.G. and Driscoll, M.F. (1988). An accessible proof of Craig's Theorem in the noncentral case. *The American Statistician* **42**, 139–142.

Rich, D.K. and Brown, K.G. (1979). Estimation of variance components using residuals: Some empirical evidence. *Proceedings of Variance Components and Animal Breeding: A Conference in Honor of C.R. Henderson*, (L.D. Van Vleck and S.R. Searle, eds.), 139–154. Animal Science Department, Cornell University, Ithaca, New York.

Robinson, G.K. (1991). That BLUP is a good thing—the estimation of random effects. *Stat. Sci.* **6**, 15–51.

Robinson, J. (1965). The distribution of a general quadratic form in normal variables. *Austral. J. Stat.* **7**, 110–114.

Rohde, C.A. and Tallis, G.M. (1969). Exact first- and second-order moments of estimates of components of covariance. *Biometrika* **56**, 517–525.

Russell, T.S. and Bradley, R.A. (1958). One-way variances in the two-way classification. *Biometrika* **45**, 111–129.

Sahai, H. (1979). A bibliography on variance components. *Int. Stat. Rev.* **47**, 177–222.

Sahai, H., Khuri, A.I. and Kapadia, C.H. (1985). A second bibliography on variance components. *Commun. Stat. A: Theory & Methods* **14**, 63–115.

Samuels, M.L., Casella, G. and McCabe, G.P. (1991). Interpreting blocks and random factors. *J. Amer. Stat. Assoc.* **86**, 798–821.

Satterthwaite, F.E. (1941). Synthesis of variance. *Psychometrika* **6**, 309–316.

Satterthwaite, F.E. (1946). An approximate distribution of estimates of variance components. *Biometrics Bull.* **2**, 110–114.

Scheffé, H. (1956). Alternative models for the analysis of variance. *Ann. Math. Stat.* **27**, 251–271.

Scheffé, H. (1959). *The Analysis of Variance.* John Wiley & Sons, New York.

Schultz, E.F., Jr (1955). Rules of thumb for determining expectations of mean squares in analysis of variance. *Biometrics* **11**, 123–135.

Searle, S.R. (1956). Matrix methods in components of variance and covariance analysis. *Ann. Math. Stat.* **27**, 737–748.

Searle, S.R. (1958). Sampling variances of estimates of components of variance. *Ann. Math. Stat.* **29**, 167–178.

Searle, S.R. (1961). Variance components in the unbalanced two-way nested classification. *Ann. Math. Stat.* **32**, 1161–1166.

Searle, S.R. (1968). Another look at Henderson's methods of estimating variance components. *Biometrics* **24**, 749–778.

Searle, S.R. (1970). Large sample variances of maximum likelihood estimators of variance components using unbalanced data. *Biometrics* **26**, 505–524.

Searle, S.R. (1971). *Linear Models.* John Wiley & Sons, New York.

Searle, S.R. (1974). Prediction, mixed models and variance components. In *Reliability and Biometry* (F. Proschan and R.J. Serfling, eds.), 229–266. Society for Industrial and Applied Mathematics, Philadelphia.

Searle, S.R. (1979). Notes on variance components estimation. A detailed account of maximum likelihood and kindred methodology. Technical Report BU-673-M, Biometrics Unit, Cornell University, Ithaca, New York.

Searle, S.R. (1982). *Matrix Algebra Useful for Statistics.* John Wiley & Sons, New York.

Searle, S.R. (1987). *Linear Models for Unbalanced Data.* John Wiley & Sons, New York.

Searle, S.R. (1988a). Mixed models and unbalanced data: wherefrom, whereat and whereto. *Commun. Stat. A: Theory & Methods (Special Issue on Analysis of the Unbalanced Mixed Model)* **17**, 935–968.

Searle, S.R. (1988b). Best linear unbiased estimation in mixed linear models of the analysis of variance. In *Probability and Statistics: Essays in Honor of Franklin A. Graybill* (J. Srivastava, ed.), 233–241. North-Holland, Amsterdam.

Searle, S.R. (1989). Variance components—some history and a summary account of estimation methods. *J. Animal Breeding & Genetics* **106**, 1–29.

Searle, S.R. and Fawcett, R.F. (1970). Expected mean squares in variance components models having finite populations. *Biometrics* **26**, 243–254.

Searle, S.R. and Henderson, C.R. (1961). Computing procedures for estimating components of variance in the two-way classification, mixed model. *Biometrics* **17**, 607–616. Corrigenda, **23**, 852 (1967).

Searle, S.R. and Henderson, H.V. (1979). Dispersion matrices for variance components models. *J. Amer. Stat. Assoc.* **74**, 465–470.

Searle, S.R. and Pukelsheim, F. (1986). Establishing χ^2 properties of sums of squares using induction. *The Amer. Stat.* **39**, 301–303.

Searle, S.R. and Pukelsheim, F. (1989). On least squares and best linear unbiased estimation. Technical Report BU-997-M, Biometrics Unit, Cornell University, Ithaca, New York.

Searle, S.R. and Rounsaville, T.R. (1974). A note on estimating covariance components. *The Amer. Stat.* **28**, 67–68.

Seber, G.A.F. (1977). *Linear Regression Analysis.* John Wiley & Sons, New York.

Seely, J. (1970). Linear spaces and unbiased estimation—application to the mixed model. *Ann. Math. Stat.* **41**, 1735–1748.

Seely, J. (1971). Quadratic subspaces and completeness. *Ann. Math. Stat.* **42**, 710–721.

Seifert, B. (1979). Optimal testing for fixed effects in general balanced mixed classifications models. *Math. Operationsforsch. Stat., Ser. Stat.* **10**, 237–256.

Seifert, B. (1981). Explicit formulae of exact tests in mixed balanced ANOVA models. *Biometrical J.* **23**, 535–550.

Shoukri, M.M. and Ward, R.H. (1984). On the estimation of the intra-class correlation. *Commun. Stat. A: Theory & Methods* **13**, 1239–1258.

Singh, B. (1989). Moments of group variance estimator in 1-way unbalanced classification. *J. Indian Soc. Agric. Stat.* **41**, No. 1, 77–84.

Singha, R.A. (1984). Effect of non-normality on the estimation of functions of variance components. *J. Indian Soc. Agric. Stat.* **35**, No. 1, 89–98.

Smith, A.F.M. (1983). Discussion of an article by DuMouchel and Harris. *J. Amer. Stat. Assoc.* **78**, 310–311.

Smith, D.W. and Hocking, R.R. (1978). Maximum likelihood analyses of the mixed model: the balanced case. *Commun. Stat. A: Theory & Methods* **7**, 1253–1266.

Smith, D.W. and Murray, L.W. (1984). An alternative to Eisenhart's Model II and mixed model in the case of negative variance estimates. *J. Amer. Stat. Assoc.* **79**, 145–151.

Smith, S.P. and Graser, H.-U. (1986). Estimating variance components in a class of mixed models by restricted maximum likelihood. *J. Dairy Sci.* **69**, 1156–1165.

Snedecor, G.W. (1934). *Calculation and Interpretation of Analysis of Variance and Covariance.* Collegiate Press, Ames, Iowa.

Snedecor, G.W. (1937, 1940, 1946). *Statistical Methods*, 1st, 3rd and 4th edns. Iowa State College Press, Ames, Iowa.

Snedecor, G.W. and Cochran, W.G. (1989). *Statistical Methods*, 8th edn. Iowa State College Press, Ames, Iowa.

Speed, T.P. (1983). General balance. In *Encyclopedia of Statistical Sciences*, Vol. 3 (S. Kotz, N.L. Johnson and C.B. Read, eds.), 320–326. John Wiley & Sons, New York.

Spjøtvoll, E. (1967). Optimum invariant tests in unbalanced variance components models. *Ann. Math. Stat.* **38**, 422–429.

Sprott, D.A. (1975). Marginal and conditional sufficiency. *Biometrika*, **62**, 599–605.

Steel, R.G.D. and Torrie, J.H. (1960). *Principles and Procedures of Statistics.* McGraw-Hill, New York.

Steel, R.G.D. and Torrie, J.H. (1980). *Principles and Procedures of Statistics: A Biometrical Approach*, 2nd edn. McGraw-Hill, New York.

Steffey, D. and Kass, R.E. (1991). Comment on Robinson's paper. *Statist. Sci.* **6**, 45–47.

Stiratelli, R., Laird, N.M. and Ware, J.H. (1984). Random-effects models for serial observations with binary response. *Biometrics* **40**, 961–971.

Styan, G.P.H. and Pukelsheim, F. (1981). Nonnegative definiteness of the estimated dispersion matrix in a multivariate linear model. *Bull. Acad. Polon. Sci., Sér. Sci. Math., Astron. Phys.* **27**, 327–330.

Swallow, W.H. and Monahan, J.F. (1984). Monte-Carlo comparison of ANOVA, MINQUE, REML, and ML estimators of variance components. *Technometrics* **26**, 47–57.

Swallow, W.H. and Searle, S.R. (1978). Minimum variance quadratic unbiased estimation (MIVQUE) of variance components. *Technometrics* **20**, 265–272.

Szatrowski, T.H. and Miller, J.J. (1980). Explicit maximum likelihood estimates from balanced data in the mixed model of the analysis of variance. *Ann. Stat.* **8**, 811–819.

Tan, W.Y. and Cheng, S.S. (1984). On testing variance components in three-stage unbalanced nested random effects. *Sankhyā* **46**, 188–200.

Thisted, R.A. (1988). *Elements of Statistical Computing.* Chapman and Hall, New York.

Thompson, R. (1980). Maximum likelihood estimation of variance components. *Math. Operationsforsch. Stat. Ser. Stat.* **11**, 545–561.

Thompson, R. and Meyer, K. (1986). Estimation of variance components: what is missing in the EM algorithm? *J. Stat. Computation and Simulation* **24**, 215–230.

Thompson, W.A., Jr (1961). Negative estimates of variance components: an introduction. *Bull. Internat. Statist. Inst.* **34**, 1–4.

Thompson, W.A., Jr (1962). The problem of negative estimates of variance components. *Ann. Math. Stat.* **33**, 273–289.

Thompson, W.A., Jr and Moore, J.R. (1963). Non-negative estimates of variance components. *Technometrics* **5**, 441–450.

Thomsen, I. (1975). Testing hypotheses in unbalanced variance components models for two-way layouts. *Ann. Stat.* **3**, 257–265.

Tiao, G.C. and Tan, W.Y. (1965). Bayesian analysis of random effects models in the analysis of variance I: posterior distribution of the variance components. *Biometrika* **52**, 37–53.

Tiao, G.C. and Tan, W.Y. (1966). Bayesian analysis of random effects models in the analysis of variance II: effect of autocorrelated errors. *Biometrika* **53**, 477–495.

Tierney, L. and Kadane, J.B. (1986). Accurate approximations for posterior moments and marginal densities. *J. Amer. Stat. Assoc.* **81**, 82–86.

Tierney, L., Kass, R.E. and Kadane, J.B. (1989). Fully exponential Laplace approximations to expectations and variances of nonpositive functions. *J. Amer. Stat. Assoc.* **84**, 710–716.

Tippett, L.H.C. (1931). *The Methods of Statistics*, 1st edn. Williams and Norgate, London.

Townsend, E.C. (1968). Unbiased estimators of variance components in simple unbalanced designs. Ph.D. Thesis, Cornell University, Ithaca, New York.

Townsend, E.C. and Searle, S.R. (1971). Best quadratic unbiased estimation of variance components from unbalanced data in the one-way classification. *Biometrics* **27**, 643–657.

Urquhart, N.S., Weeks, D.L. and Henderson, C.R. (1973). Estimation associated with linear models: a revisitation. *Commun. Stat. A: Theory & Methods* **1**, 303–330.

Verdooren, L.R. (1982). How large is the probability for the estimate of a variance component to be negative? *Biometrical J.* **24**, 339–360.

Wald, A. (1940). A note on the analysis of variance with unequal class frequencies. *Ann. Math. Stat.* **11**, 96–100.

Wald, A. (1941). On the analysis of variance in case of multiple classifications with unequal class frequencies. *Ann. Math. Stat.* **12**, 346–350.

Wang, Y.Y. (1967). A comparison of several variance component estimators. *Biometrika* **54**, 301–305.

Welch, B.L. (1936). The specification of rules for rejecting too-variable a product, with particular reference to an electric lamp problem. *J. R. Stat. Soc. Suppl.* **3**, 29–48.

Welch, B.L. (1956). On linear combinations of several variances. *J. Amer. Stat. Assoc.* **51**, 132–148.

Wesolowska-Janczarek, M.T. (1984). Estimation of covariance matrices in unbalanced random and mixed multivariate models. *Biometrical J.* **26**, 665–674.

Westfall, P.H. (1986). Asymptotic normality of the ANOVA estimates of components of variance in the non-normal hierarchical mixed model. *Ann. Stat.* **14**, 1572–1582.

Westfall, P.H. (1987). A comparison of variance component estimates for arbitrary underlying distributions. *J. Amer. Stat. Assoc.* **82**, 866–874.

Westfall, P.H. (1988). Robustness and power of tests for a null variance ratio. *Biometrika* **75**, 107–214.

Westfall, P.H. (1989). Power comparisons for invariant variance ratio tests in mixed ANOVA models. *Ann. Stat.* **17**, 318–326.

Wilk, M.B. and Kempthorne, O. (1955). Fixed, mixed and random models. *J. Amer. Stat. Assoc.* **50**, 1144–1167.

Wilk, M.B. and Kempthorne, O. (1956). Some aspects of the analysis of factorial experiments in a completely randomized design. *Ann. Math. Stat.* **27**, 950–985.

Williams, D.A. (1975). The analysis of binary responses from toxicological experiments involving reproduction and teratogenicity. *Biometrics* **31**, 949–952.

Williams, J.S. (1962). A confidence interval for variance components. *Biometrika* **49**, 278–281.

Winsor, C.P. and Clarke, G.L. (1940). Statistical study of variation in the catch of plankton nets. *Sears Foundation J. Marine Res.* **3**, 1–34.

Wong, G.Y. and Mason, W.M. (1985). The hierarchical logistic regression model for multilevel analysis. *J. Amer. Stat. Assoc.* **80**, 513–524.

Wu, C.F. (1983). On the convergence properties of the EM algorithm. *Ann. Stat.* **11**, 95–103.

Yates, F. (1934). The analysis of multiple classifications with unequal numbers in the different classes. *J. Amer. Stat. Assoc.* **29**, 51–66.

Yates, F. (1967). A fresh look at the basic principles of the design and analysis of experiments. *Proceedings Fifth Berkeley Symposium*, Part IV (L. LeCam and J. Neyman, eds.), 777–790. University of California Press, Berkeley, California.

Yates, F. and Zacopanay, I. (1935). The estimation of the efficiency of sampling with special reference to sampling for yield in cereal experiments. *J. Agric. Sci.* **25**, 545–577.

Yu, H., Searle, S.R. and McCulloch, C.E. (1991). Properties of maximum likelihood estimators of variance components in the one-way-classification model, balanced data. Technical Report BU-1134-M, Biometrics Unit, Cornell University, Ithaca, New York.

Yule, G.U. (1911). *An Introduction to the Theory of Statistics.* Charles Griffin, London.

Zeger, S.L., Liang, K.-Y. and Albert, D.S. (1988). Models for longitudinal data: A generalized estimating equation approach. *Biometrics* **44**, 1049–1060.

Zyskind, G. (1967). On canonical forms, non-negative covariance matrices and best and simple least squares linear estimators in linear models. *Ann. Math. Stat.* **38**, 1092–1109.

Zyskind, G. and Martin, F.B. (1969). On best linear estimation and a general Gauss–Markoff theorem in linear models with arbitrary non-negative structure. *SIAM J. Appl. Math.* **17**, 1190–1202.

LIST OF TABLES AND FIGURES

CHAPTER 4:

CHAPTER 5:

CHAPTER 6–NONE

CHAPTER 7–NONE

CHAPTER 8:

CHAPTER 9–NONE

CHAPTER 10–NONE

CHAPTER 11:

CHAPTER 12:

APPENDIX F:

APPENDIX M–NONE

APPENDIX S–NONE

AUTHOR INDEX

SUBJECT INDEX

Notes: (1) This index should be used in conjunction with the Table of Contents, which contains the titles of all sections and sub-sections. For index entries that refer to chapters, these titles are relevant sub-entries for the index.

(2) For brevity's sake, "1-way classification," is often abbreviated to "1-way" and "2-way crossed classification," to "2-way."

WILEY SERIES IN PROBABILITY AND STATISTICS

ESTABLISHED BY WALTER A. SHEWHART AND SAMUEL S. WILKS

Editors: *David J. Balding, Noel A. C. Cressie, Nicholas I. Fisher, Iain M. Johnstone, J. B. Kadane, Geert Molenberghs. Louise M. Ryan, David W. Scott, Adrian F. M. Smith, Jozef L. Teugels*
Editors Emeriti: *Vic Barnett, J. Stuart Hunter, David G. Kendall*

The ***Wiley Series in Probability and Statistics*** is well established and authoritative. It covers many topics of current research interest in both pure and applied statistics and probability theory. Written by leading statisticians and institutions, the titles span both state-of-the-art developments in the field and classical methods.

Reflecting the wide range of current research in statistics, the series encompasses applied, methodological and theoretical statistics, ranging from applications and new techniques made possible by advances in computerized practice to rigorous treatment of theoretical approaches.

This series provides essential and invaluable reading for all statisticians, whether in academia, industry, government, or research.

† ABRAHAM and LEDOLTER · Statistical Methods for Forecasting
AGRESTI · Analysis of Ordinal Categorical Data
AGRESTI · An Introduction to Categorical Data Analysis
AGRESTI · Categorical Data Analysis, *Second Edition*
ALTMAN, GILL, and McDONALD · Numerical Issues in Statistical Computing for the Social Scientist
AMARATUNGA and CABRERA · Exploration and Analysis of DNA Microarray and Protein Array Data
ANDĚL · Mathematics of Chance
ANDERSON · An Introduction to Multivariate Statistical Analysis, *Third Edition*
* ANDERSON · The Statistical Analysis of Time Series
ANDERSON, AUQUIER, HAUCK, OAKES, VANDAELE, and WEISBERG · Statistical Methods for Comparative Studies
ANDERSON and LOYNES · The Teaching of Practical Statistics
ARMITAGE and DAVID (editors) · Advances in Biometry
ARNOLD, BALAKRISHNAN, and NAGARAJA · Records
* ARTHANARI and DODGE · Mathematical Programming in Statistics
* BAILEY · The Elements of Stochastic Processes with Applications to the Natural Sciences
BALAKRISHNAN and KOUTRAS · Runs and Scans with Applications
BARNETT · Comparative Statistical Inference, *Third Edition*
BARNETT and LEWIS · Outliers in Statistical Data, *Third Edition*
BARTOSZYNSKI and NIEWIADOMSKA-BUGAJ · Probability and Statistical Inference
BASILEVSKY · Statistical Factor Analysis and Related Methods: Theory and Applications
BASU and RIGDON · Statistical Methods for the Reliability of Repairable Systems
BATES and WATTS · Nonlinear Regression Analysis and Its Applications
BECHHOFER, SANTNER, and GOLDSMAN · Design and Analysis of Experiments for Statistical Selection, Screening, and Multiple Comparisons
BELSLEY · Conditioning Diagnostics: Collinearity and Weak Data in Regression

*Now available in a lower priced paperback edition in the Wiley Classics Library.
†Now available in a lower priced paperback edition in the Wiley–Interscience Paperback Series.

† BELSLEY, KUH, and WELSCH · Regression Diagnostics: Identifying Influential Data and Sources of Collinearity
BENDAT and PIERSOL · Random Data: Analysis and Measurement Procedures, *Third Edition*
BERRY, CHALONER, and GEWEKE · Bayesian Analysis in Statistics and Econometrics: Essays in Honor of Arnold Zellner
BERNARDO and SMITH · Bayesian Theory
BHAT and MILLER · Elements of Applied Stochastic Processes, *Third Edition*
BHATTACHARYA and WAYMIRE · Stochastic Processes with Applications
† BIEMER, GROVES, LYBERG, MATHIOWETZ, and SUDMAN · Measurement Errors in Surveys
BILLINGSLEY · Convergence of Probability Measures, *Second Edition*
BILLINGSLEY · Probability and Measure, *Third Edition*
BIRKES and DODGE · Alternative Methods of Regression
BLISCHKE AND MURTHY (editors) · Case Studies in Reliability and Maintenance
BLISCHKE AND MURTHY · Reliability: Modeling, Prediction, and Optimization
BLOOMFIELD · Fourier Analysis of Time Series: An Introduction, *Second Edition*
BOLLEN · Structural Equations with Latent Variables
BOLLEN and CURRAN · Latent Curve Models: A Structural Equation Perspective
BOROVKOV · Ergodicity and Stability of Stochastic Processes
BOULEAU · Numerical Methods for Stochastic Processes
BOX · Bayesian Inference in Statistical Analysis
BOX · R. A. Fisher, the Life of a Scientist
BOX and DRAPER · Empirical Model-Building and Response Surfaces
* BOX and DRAPER · Evolutionary Operation: A Statistical Method for Process Improvement
BOX, HUNTER, and HUNTER · Statistics for Experimenters: Design, Innovation, and Discovery, *Second Editon*
BOX and LUCEÑO · Statistical Control by Monitoring and Feedback Adjustment
BRANDIMARTE · Numerical Methods in Finance: A MATLAB-Based Introduction
BROWN and HOLLANDER · Statistics: A Biomedical Introduction
BRUNNER, DOMHOF, and LANGER · Nonparametric Analysis of Longitudinal Data in Factorial Experiments
BUCKLEW · Large Deviation Techniques in Decision, Simulation, and Estimation
CAIROLI and DALANG · Sequential Stochastic Optimization
CASTILLO, HADI, BALAKRISHNAN, and SARABIA · Extreme Value and Related Models with Applications in Engineering and Science
CHAN · Time Series: Applications to Finance
CHARALAMBIDES · Combinatorial Methods in Discrete Distributions
CHATTERJEE and HADI · Sensitivity Analysis in Linear Regression
CHATTERJEE and PRICE · Regression Analysis by Example, *Third Edition*
CHERNICK · Bootstrap Methods: A Practitioner's Guide
CHERNICK and FRIIS · Introductory Biostatistics for the Health Sciences
CHILÈS and DELFINER · Geostatistics: Modeling Spatial Uncertainty
CHOW and LIU · Design and Analysis of Clinical Trials: Concepts and Methodologies, *Second Edition*
CLARKE and DISNEY · Probability and Random Processes: A First Course with Applications, *Second Edition*
* COCHRAN and COX · Experimental Designs, *Second Edition*
CONGDON · Applied Bayesian Modelling
CONGDON · Bayesian Statistical Modelling
CONOVER · Practical Nonparametric Statistics, *Third Edition*
COOK · Regression Graphics

*Now available in a lower priced paperback edition in the Wiley Classics Library.
†Now available in a lower priced paperback edition in the Wiley–Interscience Paperback Series.

COOK and WEISBERG · Applied Regression Including Computing and Graphics
COOK and WEISBERG · An Introduction to Regression Graphics
CORNELL · Experiments with Mixtures, Designs, Models, and the Analysis of Mixture Data, *Third Edition*
COVER and THOMAS · Elements of Information Theory
COX · A Handbook of Introductory Statistical Methods
* COX · Planning of Experiments
CRESSIE · Statistics for Spatial Data, *Revised Edition*
CSÖRGŐ and HORVÁTH · Limit Theorems in Change Point Analysis
DANIEL · Applications of Statistics to Industrial Experimentation
DANIEL · Biostatistics: A Foundation for Analysis in the Health Sciences, *Eighth Edition*
* DANIEL · Fitting Equations to Data: Computer Analysis of Multifactor Data, *Second Edition*
DASU and JOHNSON · Exploratory Data Mining and Data Cleaning
DAVID and NAGARAJA · Order Statistics, *Third Edition*
* DEGROOT, FIENBERG, and KADANE · Statistics and the Law
DEL CASTILLO · Statistical Process Adjustment for Quality Control
DEMARIS · Regression with Social Data: Modeling Continuous and Limited Response Variables
DEMIDENKO · Mixed Models: Theory and Applications
DENISON, HOLMES, MALLICK and SMITH · Bayesian Methods for Nonlinear Classification and Regression
DETTE and STUDDEN · The Theory of Canonical Moments with Applications in Statistics, Probability, and Analysis
DEY and MUKERJEE · Fractional Factorial Plans
DILLON and GOLDSTEIN · Multivariate Analysis: Methods and Applications
DODGE · Alternative Methods of Regression
* DODGE and ROMIG · Sampling Inspection Tables, *Second Edition*
* DOOB · Stochastic Processes
DOWDY, WEARDEN, and CHILKO · Statistics for Research, *Third Edition*
DRAPER and SMITH · Applied Regression Analysis, *Third Edition*
DRYDEN and MARDIA · Statistical Shape Analysis
DUDEWICZ and MISHRA · Modern Mathematical Statistics
DUNN and CLARK · Basic Statistics: A Primer for the Biomedical Sciences, *Third Edition*
DUPUIS and ELLIS · A Weak Convergence Approach to the Theory of Large Deviations
* ELANDT-JOHNSON and JOHNSON · Survival Models and Data Analysis
ENDERS · Applied Econometric Time Series
† ETHIER and KURTZ · Markov Processes: Characterization and Convergence
EVANS, HASTINGS, and PEACOCK · Statistical Distributions, *Third Edition*
FELLER · An Introduction to Probability Theory and Its Applications, Volume I, *Third Edition,* Revised; Volume II, *Second Edition*
FISHER and VAN BELLE · Biostatistics: A Methodology for the Health Sciences
FITZMAURICE, LAIRD, and WARE · Applied Longitudinal Analysis
* FLEISS · The Design and Analysis of Clinical Experiments
FLEISS · Statistical Methods for Rates and Proportions, *Third Edition*
† FLEMING and HARRINGTON · Counting Processes and Survival Analysis
FULLER · Introduction to Statistical Time Series, *Second Edition*
FULLER · Measurement Error Models
GALLANT · Nonlinear Statistical Models
GEISSER · Modes of Parametric Statistical Inference
GEWEKE · Contemporary Bayesian Econometrics and Statistics

*Now available in a lower priced paperback edition in the Wiley Classics Library.
†Now available in a lower priced paperback edition in the Wiley–Interscience Paperback Series.

GHOSH, MUKHOPADHYAY, and SEN · Sequential Estimation
GIESBRECHT and GUMPERTZ · Planning, Construction, and Statistical Analysis of Comparative Experiments
GIFI · Nonlinear Multivariate Analysis
GIVENS and HOETING · Computational Statistics
GLASSERMAN and YAO · Monotone Structure in Discrete-Event Systems
GNANADESIKAN · Methods for Statistical Data Analysis of Multivariate Observations, *Second Edition*
GOLDSTEIN and LEWIS · Assessment: Problems, Development, and Statistical Issues
GREENWOOD and NIKULIN · A Guide to Chi-Squared Testing
GROSS and HARRIS · Fundamentals of Queueing Theory, *Third Edition*
* HAHN and SHAPIRO · Statistical Models in Engineering
HAHN and MEEKER · Statistical Intervals: A Guide for Practitioners
HALD · A History of Probability and Statistics and their Applications Before 1750
HALD · A History of Mathematical Statistics from 1750 to 1930
† HAMPEL · Robust Statistics: The Approach Based on Influence Functions
HANNAN and DEISTLER · The Statistical Theory of Linear Systems
HEIBERGER · Computation for the Analysis of Designed Experiments
HEDAYAT and SINHA · Design and Inference in Finite Population Sampling
HELLER · MACSYMA for Statisticians
HINKELMANN and KEMPTHORNE · Design and Analysis of Experiments, Volume 1: Introduction to Experimental Design
HINKELMANN and KEMPTHORNE · Design and Analysis of Experiments, Volume 2: Advanced Experimental Design
HOAGLIN, MOSTELLER, and TUKEY · Exploratory Approach to Analysis of Variance
* HOAGLIN, MOSTELLER, and TUKEY · Exploring Data Tables, Trends and Shapes
* HOAGLIN, MOSTELLER, and TUKEY · Understanding Robust and Exploratory Data Analysis
HOCHBERG and TAMHANE · Multiple Comparison Procedures
HOCKING · Methods and Applications of Linear Models: Regression and the Analysis of Variance, *Second Edition*
HOEL · Introduction to Mathematical Statistics, *Fifth Edition*
HOGG and KLUGMAN · Loss Distributions
HOLLANDER and WOLFE · Nonparametric Statistical Methods, *Second Edition*
HOSMER and LEMESHOW · Applied Logistic Regression, *Second Edition*
HOSMER and LEMESHOW · Applied Survival Analysis: Regression Modeling of Time to Event Data
† HUBER · Robust Statistics
HUBERTY · Applied Discriminant Analysis
HUNT and KENNEDY · Financial Derivatives in Theory and Practice
HUSKOVA, BERAN, and DUPAC · Collected Works of Jaroslav Hajek—with Commentary
HUZURBAZAR · Flowgraph Models for Multistate Time-to-Event Data
IMAN and CONOVER · A Modern Approach to Statistics
† JACKSON · A User's Guide to Principle Components
JOHN · Statistical Methods in Engineering and Quality Assurance
JOHNSON · Multivariate Statistical Simulation
JOHNSON and BALAKRISHNAN · Advances in the Theory and Practice of Statistics: A Volume in Honor of Samuel Kotz
JOHNSON and BHATTACHARYYA · Statistics: Principles and Methods, *Fifth Edition*
JOHNSON and KOTZ · Distributions in Statistics

*Now available in a lower priced paperback edition in the Wiley Classics Library.
†Now available in a lower priced paperback edition in the Wiley–Interscience Paperback Series.

JOHNSON and KOTZ (editors) · Leading Personalities in Statistical Sciences: From the Seventeenth Century to the Present
JOHNSON, KOTZ, and BALAKRISHNAN · Continuous Univariate Distributions, Volume 1, *Second Edition*
JOHNSON, KOTZ, and BALAKRISHNAN · Continuous Univariate Distributions, Volume 2, *Second Edition*
JOHNSON, KOTZ, and BALAKRISHNAN · Discrete Multivariate Distributions
JOHNSON, KEMP, and KOTZ · Univariate Discrete Distributions, *Third Edition*
JUDGE, GRIFFITHS, HILL, LÜTKEPOHL, and LEE · The Theory and Practice of Econometrics, *Second Edition*
JUREČKOVÁ and SEN · Robust Statistical Procedures: Aymptotics and Interrelations
JUREK and MASON · Operator-Limit Distributions in Probability Theory
KADANE · Bayesian Methods and Ethics in a Clinical Trial Design
KADANE AND SCHUM · A Probabilistic Analysis of the Sacco and Vanzetti Evidence
KALBFLEISCH and PRENTICE · The Statistical Analysis of Failure Time Data, *Second Edition*
KASS and VOS · Geometrical Foundations of Asymptotic Inference
† KAUFMAN and ROUSSEEUW · Finding Groups in Data: An Introduction to Cluster Analysis
KEDEM and FOKIANOS · Regression Models for Time Series Analysis
KENDALL, BARDEN, CARNE, and LE · Shape and Shape Theory
KHURI · Advanced Calculus with Applications in Statistics, *Second Edition*
KHURI, MATHEW, and SINHA · Statistical Tests for Mixed Linear Models
* KISH · Statistical Design for Research
KLEIBER and KOTZ · Statistical Size Distributions in Economics and Actuarial Sciences
KLUGMAN, PANJER, and WILLMOT · Loss Models: From Data to Decisions, *Second Edition*
KLUGMAN, PANJER, and WILLMOT · Solutions Manual to Accompany Loss Models: From Data to Decisions, *Second Edition*
KOTZ, BALAKRISHNAN, and JOHNSON · Continuous Multivariate Distributions, Volume 1, *Second Edition*
KOTZ and JOHNSON (editors) · Encyclopedia of Statistical Sciences: Volumes 1 to 9 with Index
KOTZ and JOHNSON (editors) · Encyclopedia of Statistical Sciences: Supplement Volume
KOTZ, READ, and BANKS (editors) · Encyclopedia of Statistical Sciences: Update Volume 1
KOTZ, READ, and BANKS (editors) · Encyclopedia of Statistical Sciences: Update Volume 2
KOVALENKO, KUZNETZOV, and PEGG · Mathematical Theory of Reliability of Time-Dependent Systems with Practical Applications
LACHIN · Biostatistical Methods: The Assessment of Relative Risks
LAD · Operational Subjective Statistical Methods: A Mathematical, Philosophical, and Historical Introduction
LAMPERTI · Probability: A Survey of the Mathematical Theory, *Second Edition*
LANGE, RYAN, BILLARD, BRILLINGER, CONQUEST, and GREENHOUSE · Case Studies in Biometry
LARSON · Introduction to Probability Theory and Statistical Inference, *Third Edition*
LAWLESS · Statistical Models and Methods for Lifetime Data, *Second Edition*
LAWSON · Statistical Methods in Spatial Epidemiology
LE · Applied Categorical Data Analysis
LE · Applied Survival Analysis
LEE and WANG · Statistical Methods for Survival Data Analysis, *Third Edition*
LEPAGE and BILLARD · Exploring the Limits of Bootstrap

*Now available in a lower priced paperback edition in the Wiley Classics Library.
†Now available in a lower priced paperback edition in the Wiley–Interscience Paperback Series.

LEYLAND and GOLDSTEIN (editors) · Multilevel Modelling of Health Statistics
LIAO · Statistical Group Comparison
LINDVALL · Lectures on the Coupling Method
LIN · Introductory Stochastic Analysis for Finance and Insurance
LINHART and ZUCCHINI · Model Selection
LITTLE and RUBIN · Statistical Analysis with Missing Data, *Second Edition*
LLOYD · The Statistical Analysis of Categorical Data
LOWEN and TEICH · Fractal-Based Point Processes
MAGNUS and NEUDECKER · Matrix Differential Calculus with Applications in Statistics and Econometrics, *Revised Edition*
MALLER and ZHOU · Survival Analysis with Long Term Survivors
MALLOWS · Design, Data, and Analysis by Some Friends of Cuthbert Daniel
MANN, SCHAFER, and SINGPURWALLA · Methods for Statistical Analysis of Reliability and Life Data
MANTON, WOODBURY, and TOLLEY · Statistical Applications Using Fuzzy Sets
MARCHETTE · Random Graphs for Statistical Pattern Recognition
MARDIA and JUPP · Directional Statistics
MASON, GUNST, and HESS · Statistical Design and Analysis of Experiments with Applications to Engineering and Science, *Second Edition*
McCULLOCH and SEARLE · Generalized, Linear, and Mixed Models
McFADDEN · Management of Data in Clinical Trials
* McLACHLAN · Discriminant Analysis and Statistical Pattern Recognition
McLACHLAN, DO, and AMBROISE · Analyzing Microarray Gene Expression Data
McLACHLAN and KRISHNAN · The EM Algorithm and Extensions
McLACHLAN and PEEL · Finite Mixture Models
McNEIL · Epidemiological Research Methods
MEEKER and ESCOBAR · Statistical Methods for Reliability Data
MEERSCHAERT and SCHEFFLER · Limit Distributions for Sums of Independent Random Vectors: Heavy Tails in Theory and Practice
MICKEY, DUNN, and CLARK · Applied Statistics: Analysis of Variance and Regression, *Third Edition*
* MILLER · Survival Analysis, *Second Edition*
MONTGOMERY, PECK, and VINING · Introduction to Linear Regression Analysis, *Third Edition*
MORGENTHALER and TUKEY · Configural Polysampling: A Route to Practical Robustness
MUIRHEAD · Aspects of Multivariate Statistical Theory
MULLER and STOYAN · Comparison Methods for Stochastic Models and Risks
MURRAY · X-STAT 2.0 Statistical Experimentation, Design Data Analysis, and Nonlinear Optimization
MURTHY, XIE, and JIANG · Weibull Models
MYERS and MONTGOMERY · Response Surface Methodology: Process and Product Optimization Using Designed Experiments, *Second Edition*
MYERS, MONTGOMERY, and VINING · Generalized Linear Models. With Applications in Engineering and the Sciences
† NELSON · Accelerated Testing, Statistical Models, Test Plans, and Data Analyses
† NELSON · Applied Life Data Analysis
NEWMAN · Biostatistical Methods in Epidemiology
OCHI · Applied Probability and Stochastic Processes in Engineering and Physical Sciences
OKABE, BOOTS, SUGIHARA, and CHIU · Spatial Tesselations: Concepts and Applications of Voronoi Diagrams, *Second Edition*
OLIVER and SMITH · Influence Diagrams, Belief Nets and Decision Analysis
PALTA · Quantitative Methods in Population Health: Extensions of Ordinary Regressions

*Now available in a lower priced paperback edition in the Wiley Classics Library.
†Now available in a lower priced paperback edition in the Wiley–Interscience Paperback Series.

PANKRATZ · Forecasting with Dynamic Regression Models
PANKRATZ · Forecasting with Univariate Box-Jenkins Models: Concepts and Cases
* PARZEN · Modern Probability Theory and Its Applications
PEÑA, TIAO, and TSAY · A Course in Time Series Analysis
PIANTADOSI · Clinical Trials: A Methodologic Perspective
PORT · Theoretical Probability for Applications
POURAHMADI · Foundations of Time Series Analysis and Prediction Theory
PRESS · Bayesian Statistics: Principles, Models, and Applications
PRESS · Subjective and Objective Bayesian Statistics, *Second Edition*
PRESS and TANUR · The Subjectivity of Scientists and the Bayesian Approach
PUKELSHEIM · Optimal Experimental Design
PURI, VILAPLANA, and WERTZ · New Perspectives in Theoretical and Applied Statistics
† PUTERMAN · Markov Decision Processes: Discrete Stochastic Dynamic Programming
QIU · Image Processing and Jump Regression Analysis
* RAO · Linear Statistical Inference and Its Applications, *Second Edition*
RAUSAND and HØYLAND · System Reliability Theory: Models, Statistical Methods, and Applications, *Second Edition*
RENCHER · Linear Models in Statistics
RENCHER · Methods of Multivariate Analysis, *Second Edition*
RENCHER · Multivariate Statistical Inference with Applications
* RIPLEY · Spatial Statistics
* RIPLEY · Stochastic Simulation
ROBINSON · Practical Strategies for Experimenting
ROHATGI and SALEH · An Introduction to Probability and Statistics, *Second Edition*
ROLSKI, SCHMIDLI, SCHMIDT, and TEUGELS · Stochastic Processes for Insurance and Finance
ROSENBERGER and LACHIN · Randomization in Clinical Trials: Theory and Practice
ROSS · Introduction to Probability and Statistics for Engineers and Scientists
† ROUSSEEUW and LEROY · Robust Regression and Outlier Detection
* RUBIN · Multiple Imputation for Nonresponse in Surveys
RUBINSTEIN · Simulation and the Monte Carlo Method
RUBINSTEIN and MELAMED · Modern Simulation and Modeling
RYAN · Modern Regression Methods
RYAN · Statistical Methods for Quality Improvement, *Second Edition*
SALEH · Theory of Preliminary Test and Stein-Type Estimation with Applications
* SCHEFFE · The Analysis of Variance
SCHIMEK · Smoothing and Regression: Approaches, Computation, and Application
SCHOTT · Matrix Analysis for Statistics, *Second Edition*
SCHOUTENS · Levy Processes in Finance: Pricing Financial Derivatives
SCHUSS · Theory and Applications of Stochastic Differential Equations
SCOTT · Multivariate Density Estimation: Theory, Practice, and Visualization
SEARLE · Linear Models for Unbalanced Data
SEARLE · Matrix Algebra Useful for Statistics
† SEARLE, CASELLA, and McCULLOCH · Variance Components
SEARLE and WILLETT · Matrix Algebra for Applied Economics
SEBER and LEE · Linear Regression Analysis, *Second Edition*
† SEBER · Multivariate Observations
† SEBER and WILD · Nonlinear Regression
SENNOTT · Stochastic Dynamic Programming and the Control of Queueing Systems
* SERFLING · Approximation Theorems of Mathematical Statistics
SHAFER and VOVK · Probability and Finance: It's Only a Game!
SILVAPULLE and SEN · Constrained Statistical Inference: Inequality, Order, and Shape Restrictions

*Now available in a lower priced paperback edition in the Wiley Classics Library.
†Now available in a lower priced paperback edition in the Wiley–Interscience Paperback Series.

SMALL and McLEISH · Hilbert Space Methods in Probability and Statistical Inference
SRIVASTAVA · Methods of Multivariate Statistics
STAPLETON · Linear Statistical Models
STAUDTE and SHEATHER · Robust Estimation and Testing
STOYAN, KENDALL, and MECKE · Stochastic Geometry and Its Applications, *Second Edition*
STOYAN and STOYAN · Fractals, Random Shapes and Point Fields: Methods of Geometrical Statistics
STYAN · The Collected Papers of T. W. Anderson: 1943–1985
SUTTON, ABRAMS, JONES, SHELDON, and SONG · Methods for Meta-Analysis in Medical Research
TAKEZAWA · Introduction to Nonparametric Regression
TANAKA · Time Series Analysis: Nonstationary and Noninvertible Distribution Theory
THOMPSON · Empirical Model Building
THOMPSON · Sampling, *Second Edition*
THOMPSON · Simulation: A Modeler's Approach
THOMPSON and SEBER · Adaptive Sampling
THOMPSON, WILLIAMS, and FINDLAY · Models for Investors in Real World Markets
TIAO, BISGAARD, HILL, PEÑA, and STIGLER (editors) · Box on Quality and Discovery: with Design, Control, and Robustness
TIERNEY · LISP-STAT: An Object-Oriented Environment for Statistical Computing and Dynamic Graphics
TSAY · Analysis of Financial Time Series, *Second Edition*
UPTON and FINGLETON · Spatial Data Analysis by Example, Volume II: Categorical and Directional Data
VAN BELLE · Statistical Rules of Thumb
VAN BELLE, FISHER, HEAGERTY, and LUMLEY · Biostatistics: A Methodology for the Health Sciences, *Second Edition*
VESTRUP · The Theory of Measures and Integration
VIDAKOVIC · Statistical Modeling by Wavelets
VINOD and REAGLE · Preparing for the Worst: Incorporating Downside Risk in Stock Market Investments
WALLER and GOTWAY · Applied Spatial Statistics for Public Health Data
WEERAHANDI · Generalized Inference in Repeated Measures: Exact Methods in MANOVA and Mixed Models
WEISBERG · Applied Linear Regression, *Third Edition*
WELSH · Aspects of Statistical Inference
WESTFALL and YOUNG · Resampling-Based Multiple Testing: Examples and Methods for p-Value Adjustment
WHITTAKER · Graphical Models in Applied Multivariate Statistics
WINKER · Optimization Heuristics in Economics: Applications of Threshold Accepting
WONNACOTT and WONNACOTT · Econometrics, *Second Edition*
WOODING · Planning Pharmaceutical Clinical Trials: Basic Statistical Principles
WOODWORTH · Biostatistics: A Bayesian Introduction
WOOLSON and CLARKE · Statistical Methods for the Analysis of Biomedical Data, *Second Edition*
WU and HAMADA · Experiments: Planning, Analysis, and Parameter Design Optimization
WU and ZHANG · Nonparametric Regression Methods for Longitudinal Data Analysis
YANG · The Construction Theory of Denumerable Markov Processes
* ZELLNER · An Introduction to Bayesian Inference in Econometrics
ZHOU, OBUCHOWSKI, and McCLISH · Statistical Methods in Diagnostic Medicine

*Now available in a lower priced paperback edition in the Wiley Classics Library.
†Now available in a lower priced paperback edition in the Wiley–Interscience Paperback Series.